Fabrication, Implementation, and Applications

MICROFLUIDICS AND NANOFLUIDICS HANDBOOK

FABRICATION, IMPLEMENTATION, AND APPLICATIONS

EDITED BY

SUSHANTA K. MITRA
SUMAN CHAKRABORTY

CRC Press
Taylor & Francis Group
Boca Raton London New York

CRC Press is an imprint of the
Taylor & Francis Group, an **informa** business

CRC Press
Taylor & Francis Group
6000 Broken Sound Parkway NW, Suite 300
Boca Raton, FL 33487-2742

© 2012 by Taylor and Francis Group, LLC
CRC Press is an imprint of Taylor & Francis Group, an Informa business

International Standard Book Number: 978-1-4398-1672-1 (Hardback)
International Standard Book Number: 978-1-138-07238-1 (Paperback)

Library of Congress Cataloging-in-Publication Data

Microfluidics and nanofluidics handbook : fabrication, implementation, and applications / edited by
   Sushanta K. Mitra and Suman Chakraborty.
      p. cm.
   Includes bibliographical references and index.
   ISBN 978-1-4398-1672-1 (hardback)
   1. Microfluidic devices--Design and construction--Handbooks, manuals, etc. 2.
Microfluidics--Handbooks, manuals, etc. 3. Nanofluids--Handbooks, manuals, etc. I. Mitra, Sushanta
K. II. Chakraborty, Suman, 1973- III. Title.

TJ853.4.M53M5436 2011
620.1'06--dc22   2011005405

Contents

Part I   Experimental and Numerical Methods

Part II   Fabrication and Other Applications

Preface

Microfluidics has been an important research area for the last twenty years. A number of textbooks have been printed in the last couple of years because of demand in a large variety of practical engineering and science applications. When the publisher approached us with the task to generate a handbook for this multidisciplinary area of microfluidics and nanofluidics, we tried our best to ensure that it really captures the cross-disciplinary breadth of this subject encompassing biological sciences, chemistry, physics, and engineering applications. It is often a challenge, particularly for those with engineering backgrounds, to work in this area that requires fundamental knowledge of the basic sciences. We tried our best to fill in the knowledge gap that exists with available publications by pulling together key individuals, well known in their respective areas, to author chapters that will help graduate students, scientists, and practicing engineers to understand the overall area of microfluidics and nanofluidics.

This handbook is published in two volumes. Volume I consists of a section on Physics and Transport Phenomena and a section on Life Sciences and Related Applications. Hopefully, Volume I will provide readers with the fundamental science background that is required for the study of microfluidics and nanofluidics. This volume, Volume II, begins by focusing on topics related to Experimental and Numerical Methods before moving to another section on Fabrications and Other Applications, the chapters of which vary from aerospace to biological systems. The efforts have been to include as much interdisciplinary knowledge as possible, reflecting the inherent nature of this area.

Editing a handbook of this wide breadth is not possible without active help, mentorships, and support from a large number of individuals, which include all authors and reviewers who spent long hours going through the different chapters of this handbook. Dr. Mitra specifically acknowledges the mentorship he received from Dr. M. Yovanovich during this editorial process. He also thanks the support he received from his wife Jayeeta and son Neil, who never complained about him spending long hours on this handbook. Dr. Chakraborty acknowledges with gratitude the continuous moral support that he received from his parents and his wife, without which this project could not have been realized. He also dedicates this book to his son, who saw the light of the earth for the first time very recently. Both the editors are very thankful to Michael Slaughter, Jill Jurgensen, and the rest of the publishing team at CRC Press for their cooperation and support.

Editors

Dr. Sushanta K. Mitra is an associate professor in the Department of Mechanical Engineering at the University of Alberta, and he is the director of "Micro and Nano-scale Transport Laboratory" located at the National Institute for Nanotechnology. He received his bachelor's degree in mechanical engineering from Jadavpur University, India; master's degree from the University of Victoria, British Columbia, Canada; and PhD in mechanical engineering from the University of Waterloo, Ontario, Canada. His research areas include micro- or nanoscale transport processes, flow in porous media, and fuel cells. He has authored and coauthored more than 85 papers in peer-reviewed journals and conference proceedings. He is a registered Professional Engineer of Ontario and APEGGA.

E-mail: sushanta.mitra@ualberta.ca
Web page: www.mece.ualberta.ca/mntl

Dr. Suman Chakraborty is currently a professor in the Mechanical Engineering Department of the Indian Institute of Technology, Kharagpur, India. He has research interests in the area of microfluidics and micro- or nanoscale transport processes, including their theoretical, computational, and experimental modeling, encompassing the underlying fundamentals as well as the biomedical, biotechnological, chip cooling, and energy-related applications. He has been elected as a Fellow of the Indian National Academy of Science (FNASc) and Fellow of the Indian National Academy of Engineering (FNAE). He is recipient of the Indo–US Research Fellowship, the Scopus Young Scientist Award for high citation of his research in scientific/technical journals, and the Young Scientist/Young Engineer Awards from various National Academies of Science and Engineering. He has also been an Alexander von Humboldt Fellow and a visit-

ing professor at Stanford University. He has 160+ international journal publications. More details on his research can be obtained from the following URLs:
http://www.iitkgp.ac.in/fac-profiles/showprofile. php?empcode=bTmVW&depts_name=ME
http://sites.google.com/site/sumanchakraborty microfluidics/home

Contributors

Mona Abdolrazaghi
Department of Mechanical Engineering
The University of Alberta
Edmonton, Alberta, Canada

Morteza Ahmadi
University of Waterloo
Waterloo, Ontario, Canada

I. Yucel Akkutlu
University of Oklahoma
Norman, Oklahoma

S. AlShakhshir
University of Waterloo
Waterloo, Ontario, Canada

Rong Bai
University of Waterloo
Waterloo, Ontario, Canada

Ali Beskok
Department of Aerospace Engineering
Old Dominion University
Norfolk, Virginia

Kenneth S. Breuer
Division of Engineering
Brown University
Providence, Rhode Island

Mery Diaz Campos
Schlumberger
Houston, Texas

V.F. Cardoso
Department of Industrial Electronics
University of Minho
Braga, Portugal

Bayram Celik
Department of Aerospace Engineering
Old Dominion University
Norfolk, Virginia

J.C. Chai
The Petroleum Institute
Abu Dhabi, United Arab Emirates

Suman Chakraborty
Department of Mechanical Engineering
Indian Institute of Technology
  Kharagpur
Kharagpur, India

P. Chen
University of Waterloo
Waterloo, Ontario, Canada

Chan Hee Chon
Department of Mechanical and
  Mechatronics Engineering
University of Waterloo
Waterloo, Ontario, Canada

J.H. Correia
Department of Industrial Electronics
University of Minho
Braga, Portugal

Pradip Dutta
Department of Mechanical Engineering
Indian Institute of Science
Bangalore, India

Ranjan Ganguly
Department of Power Engineering
Jadavpur University
Kolkata, India

Animangsu Ghatak
Department of Chemical Engineering
Indian Institute of Technology
Kanpur, India

Bonnie L. Gray
School of Engineering Science
Simon Fraser University
Burnaby, British Columbia, Canada

Nicolas G. Green
School of Electronics and Computer
    Science
University of Southampton
Southampton, United Kingdom

Jeffrey S. Guasto
Department of Physics
Haverford College
Haverford, Pennsylvania

Peter Huang
Department of Mechanical Engineering
Binghamton University
Binghamton, New York

Jaesung Jang
School of Mechanical and Advanced
    Materials Engineering
Ulsan National Institute of Science and
    Technology
Ulsan, Korea

Yong-Hwan Kim
Growth and Investment Division
POSCO
Seoul, Korea

Aloke Kumar
Birck Nanotechnology Center and School
    of Mechanical Engineering
Purdue University
West Lafayette, Indiana
and
Biosciences Division
Oak Ridge National Laboratory
Oak Ridge, Tennessee

Sang-Youp Lee
Biomedical Research Institute
Korea Instituet of Science and Technology
Seoul, Korea

Dongqing Li
Department of Mechanical and
    Mechatronics Engineering
University of Waterloo
Waterloo, Ontario, Canada

X. Li
University of Waterloo
Waterloo, Ontario, Canada

Marc J. Madou
Mechanical and Aerospace Engineering
    Department
University of California, Irvine
Irvine, California

Rodrigo Martinez-Duarte
Mechanical and Aerospace Engineering
    Department
University of California, Irvine
Irvine, California

G. Minas
Department of Industrial
    Electronics
University of Minho
Braga, Portugal

Sushanta K. Mitra
Department of Mechanical
    Engineering
The University of Alberta
Edmonton, Alberta, Canada

Achintya Mukhopadhyay
Department of Mechanical
    Engineering
Jadavpur University
Kolkata, India

N.T. Nguyen
The Petroleum Institute
Abu Dhabi, United Arab Emirates

David S. Nobes
Department of Mechanical
    Engineering
The University of Alberta
Edmonton, Alberta, Canada

Xinxiang Pan
Department of Marine Engineering
Dalian Maritime University
Dalian, China

Ishwar K. Puri
Department of Engineering Science and
  Mechanics
Virginia Polytechnic Institute and State
  University
Blacksburg, Virginia

Shizhi Qian
Department of Aerospace
  Engineering
Old Dominion University
Norfolk, Virginia

Surya Raghu
Advanced Fluidics LLC
Columbia, Maryland

Mehdi Shahini
University of Waterloo
Waterloo, Ontario, Canada

Richard F. Sigal
University of Oklahoma
Norman, Oklahoma

Ashok Sinha
Center of Smart Interfaces
Darmstadt University of Technology
Darmstadt, Germany

Steven T. Wereley
Birck Nanotechnology Center and School
  of Mechanical Engineering
Purdue University
West Lafayette, Indiana

Stuart J. Williams
School of Mechanical Engineering
University of Louisville
Louisville, Kentucky

T.N. Wong
The Petroleum Institute
Abu Dhabi, United Arab Emirates

Y.F. Yap
The Petroleum Institute
Abu Dhabi, United Arab Emirates

John T.W. Yeow
University of Waterloo
Waterloo, Ontario, Canada

Hongpeng Zhang
Department of Marine Engineering
Dalian Maritime University
Dalian, China

Junfeng Zhang
School of Engineering
Laurentian University
Sudbury, Ontario, Canada

Part I

Experimental and
Numerical Methods

1

Image-Based Photonic Techniques for Microfluidics

David S. Nobes, Mona Abdolrazaghi, and Sushanta K. Mitra

CONTENTS

1.1 Introduction

Microfluidics is the science of studying the behavior of fluid flow through and around structures at the microscale. At this scale, flows behave differently, as inertial forces become smaller in comparison with surface effects (Masliyah and Bhattacharjee, 2006). Effects of friction, electrostatic forces, and viscous forces increase with decrease in the size of a microfluidic device. In these cases, properties that are a function of area increase faster than volumetric properties. This can be expressed by the "square-cube" law (Galileo, 1638/1954; Karniadakis et al., 2005) as

$$\frac{\text{Area}}{\text{Volume}} \propto \frac{L^2}{L^3} = \frac{1}{L},$$

(1.1)

where L is a characteristic dimension of the microdevice with a common order of $\sim 10^{-6}$ m (1 µm). Typically, channels and devices in the range of 1 mm down to 1 µm are termed microfluidic, and those less than 1 µm are termed nanofluidic.

At the microscale, flow dynamics and interactions within the microdevice at the surface differ from that in macroscale systems, and this is identified by terming liquid flows as *granular* and gas flows as *rarefied* (Karniadakis et al., 2005). This distinction is a result of the importance of processes and phenomena at the microscale compared with the macroscale. Flow movement at the wall, surface roughness of the wall, thermal creep, electrokinetic interaction, viscous heating, anomalous diffusion, and even quantum and chemical effects can be significant (Masliyah and Bhattacharjee, 2006). As a result, the continuum hypothesis, an essential tool in modeling macroscale flows, cannot be applied in modeling microscale flows. Therefore, constitutive laws such as conservation of mass and energy are generally modified when applied at the microscale (Karniadakis et al., 2005).

Flow fields at the microscale are investigated in different areas of health science and engineering to develop devices for aerospace, computer, mechanical, and biomedical applications (Raffel et al., 2007). As an example, in colloidal hydrodynamics (Masliyah and Bhattacharjee, 2006), investigation of the effect of the dielectrophoretic phenomenon on microflow can provide information about the levitation of colloidal particles (Molla and Bhattacharjee, 2007) transported in the flow. As a result of the presence of different forces effecting particle motion and the general motion of the particles themselves, a unique measurement technique was developed specifically to investigate near-wall phenomenon (Homeniuk et al., 2008).

In the following chapters of this handbook, the authors present the use of numerical and experimental techniques to investigate microfluidic phenomenon. Under the experimental techniques, general optical diagnostic methods in fluid mechanics are described, and an in-depth overview of their application is given. These discussions also include experimental procedures on instrumentation, image interrogation, and data reduction processes. There are also extensive discussions on the theories, measurement designs, and experimental procedures that are essential for successfully performing near-surface particle-tracking velocimetry (PTV) for nanofluidics.

The experimental procedure to investigate the balance of forces on fluids at the microscale is performed with all forces present in microscale flows. With the advent of modern electronics, computers, and other devices, experimental measurement techniques, which in the past would have been termed flow visualization, have become quantitative. Although there are a number of single-point measurement systems, the majority of modern techniques are now on imaging the microscale flow. The discussion in the remainder of this chapter is, therefore, restricted to image-based measurement techniques that use light (photons) to measure the velocity field in the region of interest. The next section discusses the system-level design of an image-based measurement system using photonics and relevant specifications of hardware elements. This is followed by a review of several applications of these types of hardware to develop microscale measurement techniques that have now become, in many cases, the standard experimental investigation tool. The intent of this chapter is to create the foundation for the design of experimental techniques related to microscale flows, which will help readers to understand more advanced flow-characterization techniques that are discussed in subsequent chapters.

1.2 System-Level Design Considerations

At the system level, subelements of image-based microscale measurement systems generally have similar characteristics. An outline of these subelements is shown in the concept map depicted in Figure 1.1. The map shows a microflow system coupled to an image-based photonic measurement system shown in the central highlighted box. The end result of any experimental investigation is the experimental result itself, which is highlighted in the last box. To achieve this experimental result, processing of data from the measurement system is required. The principle of processing is generally based on the approach used to develop the measurement system. As a result, the hardware present in the highlighted box can be reconfigured to provide suitable data that can be processed into a result depending on the measurement principle.

The microflow system comprises all devices used to supply fluid, alternating electric field, magnetic field, and so forth to the microfluidic device. The probe tool discussed here is a stream of photons. This necessitates that optical access is available to the region of interest in the microflow system. The microflow system also needs to be held in such a manner that optical access to the measurement system is available. Typically, the microchannel containing the flow is significantly smaller than the measurement system. In many cases, the flow channel is moved relative to the measurement system during experiments, and in these cases, a suitable device is required to traverse and translate the channel.

A number of elements often comprising many individual hardware components can be built into a measurement device that is used to collect data. These are outlined in the highlighted box in Figure 1.1. A light source is used to illuminate the region of interest within the microchannel through a suitable set of optics. The same region of interest is imaged with imaging optics, and an image of data is projected onto and collected by an image capture device, which is typically a camera. Suitable control software is used to control this process to manage the timing between different devices and, given the nature of the data, to manage computer memory. Although online processing is potentially possible, processing of the data is typically done as a separate postprocess operation, which is why this is shown in a separate box in Figure 1.1.

The actual arrangement of the hardware is a function of the measurement principle and how the data will be processed. The design specification of individual elements needs to be taken into consideration to achieve this end. To highlight this, a discussion of the

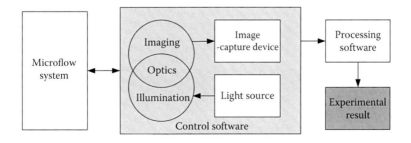

FIGURE 1.1
A concept map of the different components used in developing an image-based measurement device for investigating fluid flow in microfluidic devices.

important design specifications for these individual elements is presented in the next subsections.

1.2.1 Illumination, Light Scattering, and Fluorescence

1.2.1.1 Illumination Sources

The light source used for illumination is usually characterized by the wavelength of light used, the time for which it is available, and the overall power used in the measurement. In classical microscopy, a broadband light source is used, which has a continuous range of wavelengths over the visible spectrum. The light used for illumination can be introduced such that scattered light is reflected off objects in the field of view to the image-capture device, which might be the eye of a camera. Therefore, these objects appear bright in the image and the background appears dark, which is often referred to as *dark-field imaging* (Hecht, 2002). The light used for illumination can also be introduced along the optical axis of the imaging system and directed toward the image-capture device. This is termed *bright-field imaging*, which results in objects in the region of interest casting a shadow on the image-capture device.

Lasers have become an important light source for microimaging applications. Lasers are unique as a light source because they emit light over a narrow wavelength range and are typically considered to be of single wavelength (Saleh and Teich, 2007). Photons travel along the same vector, are spatially coherent, and have the same polarization. Lasers can emit light continuously or a continuous wavelength (CW) where a steady stream of photons is available for the measurement. In other cases, the laser can be pulsed so that photons interrogate the region of interest in a short duration, typically with an aim to freeze motion. Lasers produce significantly more power than broadband light sources, and pulse lasers typically produce high-power density that is an order of magnitude greater than that produced by CW lasers. Lasers are characterized by the material used to generate the laser beam within the optical cavity, the wavelength of emission, and the power of the beam.

As an example, argon ion CW lasers (Bridges, 1964) have many experimental applications. They can emit 13 individual wavelengths at the ultraviolet, visible, and near-infrared spectra. These lasers can operated in a multimode, that is, with all lines together or in a single mode so that only a single wavelength is available. Useful wavelengths in the visible spectrum are 488.0 and 514.5 nm because they produce maximum power. These laser units are commercially available in a range up to 20 W. With the advent of solid-state technology, diode-pumped solid-state lasers are available in various wavelengths and have the advantage of being compact. Depending on the laser configuration, they can be operated either as CW lasers or as pulsed lasers. For example, pulsed laser technology often used in measurement systems uses neodymium-doped yttrium aluminum garnet (Nd:YAG) crystal as the lazing medium (Geusic et al., 1964). These lasers have a fundamental wavelength of 1064 nm in the infrared spectrum and are often frequency doubled to 532 nm in the green part of the visible spectrum. These lasers are characterized by the amount of energy per pulse (joules/pulse) and typically have a low repetition rate in the range of 0–30 Hz, although high rates are available. The main characteristic of this style of laser is the short pulse duration, ~10 ns, and the high energy density per pulse (up to 400 mJ/pulse). This allows for bright illumination over a significantly shorter time scale than the majority of time scales found in typical microscale flows.

1.2.1.2 Light Scattering from Particles

Light can be scattered from particles seeded in the flow to characterize flow motion. The particle size must be sufficiently small to follow the flow and sufficiently large enough to scatter enough light for detection. At the macroscale, particle size d is in the order of 1–40 μm, depending on flow velocity and on whether the fluid is either a gas or a liquid, and light scattering can be described by the Mie theory (Mie, 1908; van de Hulst, 1981), which varies with d^{-2}. At the microscale where channel sizes are in the order of 10–100 μm, particles of this size constitute a significant proportion of the flow area. Smaller particles are therefore needed to characterize motion. As the size of the particles approach and become smaller than the wavelength λ of the light used for illumination, the amount of light scattered reduces significantly. In this size range, light scattering can be described by the Rayleigh theory (Rayleigh, 1881/1899, 1899/1903), which is most suitable for $d \ll \lambda$. In this regime, the amount of light scattered varies with d^{-6} (Born and Wolf, 1997). As flow channels become smaller, particles in the 50- to 100-nm size range are typically used, which is 1/5 to 1/10 the size of the illumination wavelength λ in the case of illumination with a frequency-doubled Nd:YAG laser, where $\lambda = 532$ nm. This highlights that as particles become smaller compared with the characteristic flow dimensions, the amount of light scattered quickly diminishes. Operating in this range requires significant care to minimize background noise and maximize signal-to-noise ratio.

1.2.1.3 Fluorescence of Molecules

Rather than scattering light from seeded particles, the specific wavelength characteristics of laser allow the spectral properties of individual molecules to be used to investigate the region of interest within a microscale flow. The fundamental wavelength of the laser can be used to excite or pump electrons in a molecule to a higher energy state. This process is shown in Figure 1.2a that depicts the energy levels within a molecule. The energy of a laser photon (E) is given as

$$E = h\nu, \tag{1.2}$$

where ν is the frequency of the light and h is Planck's constant. The frequency of the laser can be related to its wavelength (λ) with the classical equation of a wave:

$$c = \nu\lambda, \tag{1.3}$$

where c is the speed of light in a vacuum. The energy of the photon can therefore be expressed as

$$E = \frac{hc}{\lambda}, \tag{1.4}$$

which highlights that photons of a shorter wavelength have higher energy. The electron within the molecule is pumped to the higher energy state by the laser photon. The electron can return to the ground state through many paths, including collisional quenching; however, an important process is the de-excitation of the molecule through the emission of a photon. This process is known as fluorescence (Stokes, 1852), and it differs from the phosphorescence process in that the time scale for the emitting photon is in the order of nanoseconds compared with minutes to hours for phosphorescence (Guilbault, 1990).

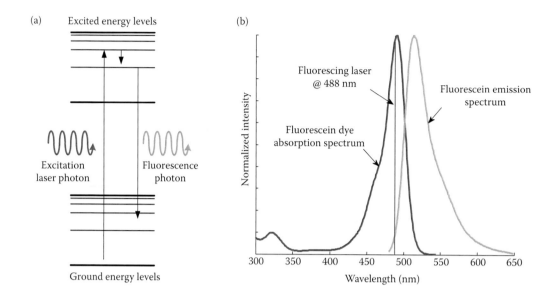

FIGURE 1.2
(See color insert.) Schematic of the fluorescence process: (a) the excitation process transfer level and (b) the adsorption and emission characteristics of fluorescein.

An important by-product of this process is that the wavelength of the emitted photon is either at the same energy or lower energy or is at the same wavelength or longer wavelength than the excitation photon. This is highlighted in Equation 1.4 for energies of the emitted photon that are less than that of the excitation photon.

For microflow investigations, a number of specific dyes that have suitable characteristics are used to take advantage of the fluorescence process (Horobin and Kiernan, 2002). These dyes have absorption bands that overlap the emission wavelength of available lasers and emit within the wavelength-detection range of detectors. The spectral characteristics of an example dye are shown in Figure 1.2b (Mota et al., 1991). Fluorescein sodium salt, a common dye used in experiments, has an absorption band that overlaps the emission wavelength of 488 nm of an argon ion laser and 473 nm of an Nd:YAG diode-pumped solid-state laser. The emission spectrum is in the range of 500–600 nm, well within the visible spectrum and the detection characteristics of most cameras.

The number of fluorescence photons or the intensity of the fluorescence signal (I_f) can be described by (Walker, 1987)

$$I_f = I_l\, f_{\text{optic}}\, \varepsilon\, Q_\lambda V C, \tag{1.5}$$

where I_l is the intensity of the laser beam, f_{optic} is the factors and losses related to the imaging optics, ε is the excitation coefficient of the dye, Q_λ is the quantum efficiency of the dye, V is the volume of the dye fluid mix under interrogation, and C is the concentration of the dye. During the design of an experimental system, a number of these factors can be optimized. However, during an experiment, there is usually one controlling factor, which is the laser intensity I_l that can be dynamically adjusted to control the fluorescence intensity. There is an upper limit to the number of fluorescence photons that are available corresponding to the concentration of the dye in the region of interest, at which point increasing illumination power will result in no further increase in fluorescence signal. A number of other factors that also influence the fluorescence signal include photobleaching, pH, temperature, local heating of the fluid, and characteristics of the microfluidic device (Walker, 1987).

1.2.2 Optics for Illumination and Imaging

Published measurement techniques for imaging of flows fields at the microscale have often been adapted from microscope technology used in biological and medical fields. The components of a classical microscope (Bradbury and Bracegirdle, 1998; Hecht, 2002; Abramowitz and Davidson, 2004) are shown in Figure 1.3a. Light from a broadband source is collimated with an optical device and reflected through a mirror into the region of interest for bright-field imaging. An objective lens gathers light and projects an image through a tube or a condenser lens onto an image-capture device, which in the case shown in Figure 1.3 a is a charge-coupled device (CCD) camera array. The schematic shows that the image is inverted as it passes between the objective lens and the tube lens that must be placed at a specific distance (L_1) from the objective lens (typically 160 mm) depending on the system's focal length and individual optics.

The classical approach to microscopy has a limitation in that all components must be at exactly specified locations within the system. There is flexibility with the introduction of infinity-corrected objectives (Goulette et al., 2010). An example of an optical system using these objectives is shown in Figure 1.3b. Although this has essentially the same setup of components as that in the classical system (Figure 1.3a), a notable difference is that light exiting the back of the objective is parallel rather than converging. Conceptually, this allows the tube lens to be placed at any location on the optical axis from the objective. Practically, however, the image rays exiting the back of the objective converge to a focal point at some distance far from the objective (>10 m), which results in a limited but flexible range (~50 to ~250 mm) over which the parallel ray assumption is valid within which to locate the tube lens.

The use of infinity-corrected optics allows multiple optical elements to be introduced in the system, an example of which is shown in Figure 1.3c. Light from the illumination

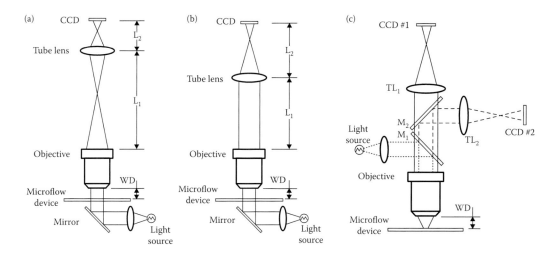

FIGURE 1.3
Schematic of optical systems used for imaging microflow devices on the basis of (a) a traditional microscope, (b) an infinity-corrected microscope, and (c) an example of multiple imaging with an infinity-corrected microscope. L_1, objective to tube lens distance; L_2, tube lens to imaging device distance; TL, tube lens; M, mirror; CCD, charge-coupled device; WD, microscope objective working distance.

source is projected onto the optical axis of the objective via a mirror (M_1) to generate epi-illumination. Light gathered by the same objective is then imaged onto two separate image collection devices using their tube lenses and a mirror (M_2). The mirrors perform an important task in the redirection of light rays within the system. If 50/50 beam splitter mirrors are used, then only 50% of the light from the light source will be available for illumination. Of the light collected using scattering processes, only 50% will pass through M_1 and the remaining 50% will pass through and be reflected by M_2, resulting in a maximum 25% of the scattered light reaching either of the two detectors. An alternative approach is to use mirrors that reflect and transmit light as a function of wavelength. Using a single wavelength light source such as a laser (λ_{laser}) and a dichroic mirror at M_1 would result in full reflection and transmission through the objective into the region of interest. If a fluorescence process is then used to identify objects within the field of view, the emitted light would be at a different wavelength (λ_{emit}). In the typical situation of $\lambda_{emit} > \lambda_{laser}$ for a fluorescence process, a correctly specified dichroic mirror at M_1 would allow the total fluorescence signal to pass through and would then allow maximum transmission of light to the detectors, thus maximizing the signal.

The image that is captured from any of the systems shown in Figure 1.3 is collected from a thin focal plane at a working distance from the end of the objective. The amount of light that is collected by the objective is a function of the numerical aperture (NA) of the objective and is defined as

$$NA = n\sin(\alpha), \tag{1.6}$$

where n is the refractive index of the medium between the objective front lens and the region of interest, and α is the half light-collection angle of the objective.

The depth of field (DOF) is the dimension parallel to the imaging axis within a region of interest over which a projected image is brought into focus on the image-capture device. This defines the out-of-plane resolution of the image and can be determined from

$$DOF = \frac{\lambda_0 n}{NA^2} + \frac{n}{M \times NA}e, \tag{1.7}$$

where λ_0 is the wavelength of the illumination light, n is the refractive index of the medium between the imaging objective and the microflow device, and e is the smallest distance that can be measured by a detector that has been placed in the image plane of the objective that has some lateral magnification of M. As the magnification of the objective increases, so does its NA, which results in a smaller DOF. This trend is also seen to reduce objective working distances with increase in magnification.

The resolution of an optical system can be defined as the shortest distance between two points that can be distinguished as individual objects by an image-capture device. At the limit of resolution, light from a point source will appear as an airy diffraction pattern. An example of this pattern is shown in Figure 1.4. When the point sources are well apart, the pattern can be resolved; however, as they approach each other, the pattern becomes unresolved. The resolution limit of an optical system can be determined from

$$Resolution = \frac{1.22\lambda_0}{NA_{obj} + NA_{tube}}, \tag{1.8}$$

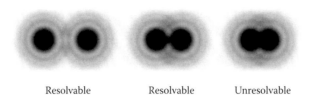

Resolvable   Resolvable   Unresolvable

FIGURE 1.4
Images of light scattered from objects, showing the generated airy diffraction pattern and the resolvable limit.

where λ_0 is the wavelength of light source used, NA_{obj} is the numerical aperture of the objective, and NA_{tube} is the numerical aperture of the tube lens. The effect of the generated airy disk pattern on resolution in the resolvable limit is shown in Figure 1.4. As the point scatter sources come closer together, individual airy disks overlap. Once the central point function of the disks overlaps, the location of the individual sources become unresolvable.

A number of other optics can be included within the optical path to perform different operations to facilitate the measurement. Although conceptually the optical signal is only influenced in a fundamental way by different optics, practically each new optic in the train will attenuate the signal by a small amount. Filters can be introduced to control the wavelength and the amount of light passing through the system. Polarization of the light can also be controlled using polarizers, polarization-sensitive beam splitters, and opto mechanical devices. The illumination can be ported to the system using fiber optics, and imaging fiber bundles can be used to transport the image from the objective to a remote image detector (Nobes et al., 2004). The overall arrangement of optical components is dependent on the measurement principle and how data will be processed to determine the parameter under investigation in the experiment.

1.2.3 Image Collection Devices

The development of new digital technology for cameras has replaced the use of film in image collection devices for experiments. This technology uses charge-coupled devices (CCD) (Boyle and Smith, 1970) and complementary metal-oxide-semiconductor (CMOS) (Baker, 2010) devices. Although the active material and electronics are different in each case, the main components that need to be considered for the design of the measurement system are similar. The optics of the measurement system image a field of view onto a two-dimensional array of pixels. Photons that interact with each pixel are converted into electrons, which are then converted into a digital count using an analog-to-digital converter (A/D). This array of digital number is then stored as an image file.

The pixel array spatially samples the image and converts the light intensity (I) in W/m² into a digital number (count), the value of which can be determined from

$$\text{count} = \frac{IA_{pix}Q}{N},\tag{1.9}$$

where A_{pix} is the area of the pixel (m²), Q is the quantum efficiency of the camera at a specific wavelength, and N is the number of intervals that the voltage of the signal is divided into by the A/D. This can be given by $N = 2^M$, where M is the A/D's resolution in bits. For example, an A/D with a resolution of 12 bits results in $2^{12} = 4096$ levels that the voltage is divided into or a digital gray scale range of 0 to 4095.

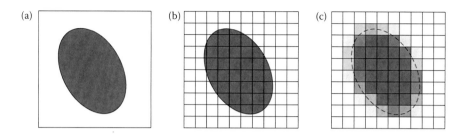

FIGURE 1.5
Images showing the effect of pixelation on the imaging of an object: (a) the object in the field of view, (b) the grid images by individual pixels, and (c) the relative gray scale assigned to each pixel.

The effect of spatial sampling on image resolution of a continuous image is shown in Figure 1.5. A continuous image shown in Figure 1.5a is subsampled by the pixel array grid, which is overlapped in Figure 1.5b. In this example, light scattered from the object is assumed to be of the same intensity across the object. The intensity measured by each pixel overlaying the object is averaged over the region of the image by each pixel. This results in a digital image shown in Figure 1.5c, where a gray scale level has been assigned to the different digital counts. Pixels that are fully within the object have the maximum number of counts and hence the darkest gray scale. Pixels that overlap the border region of the object that capture less photons of light have a consequently low digital count and are represented with a lighter gray scale. This discretization of the continuous image results in a pixelation of the boundary of the image, thereby blurring the shape of the image.

Increasing the magnification of the system will result in more pixels for imaging the object, thus allowing a finer determination of the boundary of the object. Magnification (M_{sys}) in a digital imaging system can be described by

$$M_{sys} = \frac{W_{camera}}{W_{FOV}},\tag{1.10}$$

where W_{camera} is the width of the camera sensor in pixels and W_{FOV} is the physical (m) field of view imaged by the optical system. The magnification is also used to determine the scaling for physical measurement of objects in the field of view. Magnification can also be determined directly from the pixel spacing of the array and the optical magnification of the imaging optics. Caution is needed when using this approach because the size of pixels quoted by CCD manufacturers is not the distance between pixel sensors. A pixel array has a fill factor that accounts for the spacing between pixels, which can be as low as 40%. To determine the magnification and scaling factor for a digital imaging system, it is the best practice to use a target of higher resolution than required. This can also be used to determine and document the optical distortion of the imaging system as it varies across the array. All optical systems have a distortion and are most commonly radially symmetric about the optical axis of the system. These distortions are typically classified as barrel distortion, where the magnification decreases with increasing distance from the optical axis, or pincushion distortion, which has the opposite effect.

When selecting a camera for a particular measurement application, a number of factors need to be taken into account. These are usually the number of pixels in the array that discretizes the image, the frame rate of the camera, the bit depth of the A/D, the noise characteristics of the system, and the signal response across the array. Although increasing the number of pixels will give a better representation of the image, it will slow the frame rate of the camera. This is because all pixels are read out through typically the same A/D, which will operate at a constant frequency. The frame rate and other framing options such as the use of double framing are intrinsically linked to the capture of transient features within the flow. Therefore, information on the maximum speed expected in the flow is needed to help specify the camera frame. Typically, for microfluidic experiments, flow velocities within microchannels are approximately in the order of micrometers per second, which is usually limited by the pressure drop for large flow rates. Increasing the bit depth of the A/D generally slows the camera frame rate but has the benefit of increased image contrast and generally lower noise characteristics. Noise in the system can be optical (stray photons) or electronic. Electronic noise comprises thermal, dark current, and readout noise. Reducing the temperature of the CCD will improve thermal and dark current noise characteristics, and reducing the frequency of the A/D will improve readout noise. These will impact the cost of the system and the maximum frame rate of the camera. In low-light-level situations, a compromise is sought sort between the amount of noise intrinsic to the camera and the signal-to-noise ratio needed to perform the measurement. In measurement techniques where signal intensity is a fundamental parameter of measurement, such as measuring concentration, the signal response across the array is also important. This is a measure of how individual pixels respond to the same number of interacting photons. Better signal response across the array is usually associated with higher bit depth, low frame-rate cameras that generally have lower noise. High frame-rate cameras have typically lower bit depth, high A/D speeds, and high intrinsic noise. Selection of a camera for a specific measurement technique is therefore a design trade-off of the different specifications.

1.2.4 Summation of the System Design

The discussion so far has highlighted some of the design aspects that need to be taken into account when selecting hardware for an image-based photonic measurement system for microscale flows. Ultimately, the measurement principle determines the arrangement of hardware at the system level. Properties of the microscale flow such as minimum expected time scale and the length scale over which the measurement is to be performed would then be used to determine individual hardware specifications. The principle of the measurement will also determine how data will be postprocessed to generate the derived parameter. The following discussion outlines a number of different principles that are aimed at deriving the fundamental properties of the microscale flow.

1.3 Examples of Applications of Experimental-System Design

The different components that have been discussed can be arranged in many different configurations to measure, resolve, and investigate different properties of microscale flows. A measurement principle is needed so that data that are collected can be processed to determine the parameter of interest. Since the momentum field is one of the main parameters used in the microscale flow theory (Karniadakis et al., 2005) and is often investigated, the following discussion focuses on how velocity measurement systems can be designed. A brief discussion also follows on how other parameters can be determined using a similar set of equipment but by changing the measurement principle.

1.3.1 Measurement of Flow Velocity (Particle-Based Techniques)

Fluid velocity is a fundamental parameter that is used to derive and define fluid motion and transport. It is difficult to determine the velocity of individual fluid molecules; however, a number of measurement techniques have been developed to measure fluid velocity by monitoring the motion of tracer particles that are seeded into the flow (Sinton, 2004; Lindken et al., 2009; Lee and Seok, 2009). The assumption here is that the seeded particles faithfully follow the flow (Melling, 1997). Laser Doppler velocimetry (Yeh and Cummins, 1964; Durst et al., 1976) is a technique that determines velocity by measuring the Doppler frequency shift of a scattered laser beam at a single point. It was originally developed for macroscale flows (Yeh and Cummins, 1964) and has been used at the microscale (Minor et al., 1997). However, as it is not image based, it is not discussed further here. There are a number of different image-based methods that have been developed in terms of both image acquisition and analysis of data to measure fluid flow velocity. These include particle image velocimetry (PIV), particle tracking velocimetry (PTV), and molecular tagging velocimetry (MTV). The next section is a discussion on the measurement and data processing principle used in each method, with some examples from the literature, and on the parameters that impact the design of the measurement system.

1.3.1.1 The Principle of PIV

This technique has been developed extensively at the macroscale to measure both the two-dimensional (Adrian, 1991; Willert and Gharib, 1991) and three-dimensional velocity vectors (Prasad and Adrian, 1993; Willert, 1997) of the flow field. The premise of PIV (Raffel et al., 2007) is that the average displacement in a region of the flow can be determined from the maximum of a cross-correlation function of two projected images of light scattered off seeded particles and captured by a single camera. To demonstrate this approach, an example of an image of a seeded flow field is shown in Figure 1.6a, with a highlighted subregion for interrogation. The light scattered from two pulses from a laser that freezes particle motion is captured on separate image frames at a known Δt apart (Figure 1.6b). The image frames are cross-correlated (Figure 1.6c) to determine the average displacement within subregions of the imaged area. The peak in the cross-correlation function defines the magnitude (Δs) and direction of the displacement vector shown in Figure 1.6d, which highlights the size of the region of interrogation and surrounding calculated vectors. Note that this displacement vector is based at the center of the interrogation region. Local

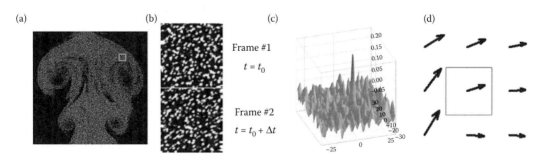

FIGURE 1.6
Schematic of velocity determination using PIV: (a) an example raw image, (b) the two subregions extracted from the main field captured at a known Δt, (c) the corresponding correlation map of the two subregions used to determines the average displacement, and (d) the displacement averaged over the subregions.

velocity (v) can then simply be determined from $v = \Delta s/\Delta t$. Vectors from these subregions then define a discretized map of the continuous velocity field across the imaged region. A discretized cross-correlation function ψ_{fg} for a discretized image field can be defined as

$$\psi_{fg} = \frac{\sum_{k=-\infty}^{\infty}\sum_{l=-\infty}^{\infty} f(k,l)g(k+m,l+n)}{\sum_{k=-\infty}^{\infty}\sum_{i=-\infty}^{\infty} f(k,l)\sum_{k=-\infty}^{\infty}\sum_{l=-\infty}^{\infty} g(k+m,l+n)}, \tag{1.11}$$

where $f(m, n)$ and $g(m, n)$ are the pixel intensities in sample regions in the image of size (m,n) and k and l are indices within the image subregion array. This correlation will approach a maximum when particle images match on an average with their spatially shifted partners in the second image. Particle image intensity and contrast of the particle from the background noise are therefore important to process a high-quality correlation peak for determining the average displacement and hence velocity of particles in the subregion.

In the conventional macro-PIV, the tracer particles are seeded in the flow and illuminated by a thin laser sheet as shown in Figure 1.7a. This laser sheet overlaps the region of interest that is imaged by a digital array camera, typically orthogonal to the light sheet. At the macroscale, in-plane resolution, Δx and Δy, is defined by the size of pixels in the array and the magnification of the optical system. Out-of-plane resolution, Δz, is defined by the thickness of the laser sheet used. If a single camera images the region of interest, the two in-plane components of velocity can be determined. Part of the assumption in the calculations is that the particles remain within the laser sheet and are imaged in both frames. This is therefore best suited for a two-dimensional flow field, with the region of interest and laser sheet aligned so that the main components of the flow are within the light sheet. If a third flow velocity component is present, more information is required. This can be provided by a second perspective view of the flow. With two perspectives, stereo-PIV (Prasad and Adrian, 1993; Willert, 1997) can determine all three components of velocity in a single plane.

A number of other approaches to PIV have been developed with specific aims to measure different parameters in the flow. Transient phenomenon can be investigated by using a high-speed camera to collect time-resolved images of the flow. High-speed PIV (Juhany et al., 2007) processes image data in a similar manner to conventional dual-frame PIV. However, the correlation is between successive frames in the data set. Velocity gradients can be easily measured in-plane as velocity is spatially resolved across the imaged region.

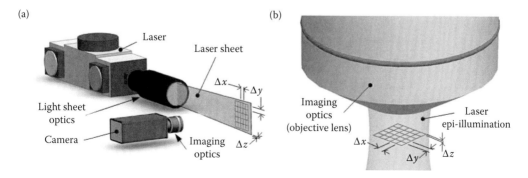

FIGURE 1.7
(See color insert.) Schematic comparing the arrangement of hardware and defined region of interest and resolution for (a) macroscale PIV and (b) microscale PIV.

To measure out-of-plane velocity gradients a dual-plane PIV (Ganapathisubramani et al., 2005) has been developed where in data are collected at two separate planes that are at a known separation distance within a time scale faster than the flow. The two stereo-PIV camera pairs image the planes, and the three-component three-dimensional data fields are postprocessed to determine gradients. Spatially resolving the three components of velocity over a volume requires the development of a different measurement principle. In tomographic PIV (Elsinga et al., 2006), multiple images of an illuminated volume are used to reconstruct the three-dimensional locations of seed particles. Two reconstructed volumes collected at a known Δt are correlated using a three-dimensional cross-correlation algorithm. In these examples, essentially the same basic hardware is used that has specific specifications to match the measurement principle.

1.3.1.2 Applications of PIV at the Microscale

The same basic principle can be used at the microscale to determine flow velocity. However, a new concept for the illumination of the flow is needed. There is a minimum thickness that a light sheet can be focused to, which is often larger than the size of the microflow channel. In microscale devices, typically no optical axis is available to introduce the light sheet orthogonally to the viewing direction. To overcome these drawbacks, a typical configuration, as shown in Figures 1.3b and c, that has single-sided optical access for both illumination and imaging of the flow is used. An infinity-corrected microscope images the region of interest, and an epi-illumination configuration illuminates the flow. Interchangeable objective lenses allow control of optical magnification. In-plane resolution, Δx and Δy as shown in Figure 1.7b, is again defined by the size of pixels in the array and the magnification of the optical system, whereas out-of-plane resolution, Δz, is defined by the focal plane thickness of the objective used. It is only within this range that seeded particles are brought into sharp focus. The focal plane thickness can be determined from (Meinhart et al., 2000a)

$$\Delta z = \frac{3n\lambda_{emit}}{NA^2} + \frac{2.16d}{\tan\theta} + d, \tag{1.12}$$

where λ_{emit} is the reflected or emitted wavelength of light from the particle, n is the refractive index of the medium between the objective lens and the microchannel flow, NA is the numerical aperture of the optical system, θ is the angle at which light is collected in the optical system, and d is the size of the particle being imaged. This method accounts for diffraction and use of geometric optics and assumes that the pixel size of the CCD is sufficient to resolve the particle. Focal plane thickness, along with other characteristics such as magnification and NA, is also specified for objective lenses by manufactures. However, these are valid only in situations where matching tube lenses are used in conjunction with the objectives.

This type of illumination can result in significant reflection from surfaces. To avoid this, fluorescent particles are typically used as seed particles. This allows high levels of bulk illumination of the region of interest to be used to generate the particle fluorescence signal that is sourced only from individual particle locations. A dichroic mirror and/or a laser line filter are used to allow the laser to illuminate the region of interest while allowing only the fluorescence signal to pass through to the camera. A pulsed or CW laser light source that has a laser frequency that is matched to excite the fluorescent particles can be used. A typical laser-particle pair is a frequency-doubled Nd:YAG laser at $\lambda_{laser} = 532$ nm and a rhodamine-coated particle. These particles will fluoresce at a longer wavelength in the range of 600 nm, allowing for good spectral separation of illumination laser and fluorescence signal.

1.3.1.3 Approaches to Micro-PIV

The first micro-PIV system was introduced by Santiago et al. (1998) to investigate the velocity field of a liquid Hele-Shaw flow. A bulk flow velocity of ~50 μm/s around a 30-μm diameter cylinder was approximated using a cross-correlation PIV algorithm. Fluorescent tagged polyester particles with $d = 300$ nm were visualized by an intensified CCD camera with a spatial resolution of $6.9 \times 6.9 \times 1.5$ μm. Illumination of the experiment was provided by a mercury lamp coupled to an epi-fluorescent microscope. The Hg-arc lamp of the microscope was sufficient for illuminating the field of view due to the low velocity of the flow. With this low velocity and small particle size, Brownian motion of the seed particles was significant. The obtained velocity vector field was improved by averaging several captured data sets. This approach to micro-PIV was limited to low-velocity flows because of long exposure time (~2 μs) and the time delay between captured images. Both these were hardware limitations that were quickly overcome by improved technology. Within a short period of time, Meinhart et al. (1999) was able to measure velocities three orders of magnitude faster by the use of a double-pulsed Nd:YAG laser for illumination. This significantly increased the signal available for measurement and allowed the use of a cooled CCD camera, with lower noise characteristics to be used to image the field.

This approach to determining the velocity field was extended to measure all the three components of velocity in a microchannel by using stereo-PIV. Lindken et al. (2006) used a standard stereomicroscope to investigate the flow in a T-shaped mixer using a double-pulsed Nd:YLF laser to excite fluorescent microbeads. Two double-frame cameras collected the two perspective views of the flow, which have a view angle, α, between them. Introducing two perspective views, however, also introduces other sources of errors in the measurement that are not present in macroscale stereo-PIV. These are related to the errors in determining the out-of-plane component because of the low angle between the cameras, the volume over which the measurement is resolved, and the number of particles present for PIV evaluation.

Lawson and Wu (1997) have shown that the error in determining the out-of-plane component of velocity is a function of the angle between the views (α) of the two cameras of the region of interest. This is at a minimum when the angle $\alpha = 45°$. Typically, however, the viewing angle in commercial stereomicroscopes is lower than this and is in the order of $\alpha \approx 10°–20°$. This can result in a significant increase in the error of the out-of-plane component. Lawson and Wu (1997) estimated that for $\alpha = 13.8°$, the out-of-plane error was approximately four times higher than the in-plane error. Lindken et al. (2006) indicate that the error in their out-of-plane measurement was less than 10% of the main flow velocity. They conclude, however, that higher accuracies similar to those found in macro-stereo-PIV will be difficult to achieve because of the achievable limit of the view angle.

Stereomicroscopes also have limits in the maximum total magnification of the system and the larger depths of focus of their objectives compared with the objectives of similar magnification in conventional single-view microscopes. This leads to a reduction in the spatial resolution of the out-of-plane, Δz, component. The measured velocity is also affected by the variation in the velocity gradients in the out-of-plane direction (Lindken et al., 2006). Stereomicro-PIV is therefore suited for larger microchannel flows because of the low optical-system magnification requirement. The larger depth of focus and the low magnification also impact the number of particles that can be used to seed the flow. Larger particles are typically used, and at nominal seeding levels required for PIV analysis, these result in a large background signal. Alternative PIV processing methods can be used to account for seeding densities. An ensemble correlation approach (Meinhart et al., 2000b) rather than an average of individual vector fields results in an average velocity field that is based on strong correlation peaks. This removes biasing from individual particle motion or lack of data present for the correlation algorithm to process.

The limitations of conventional PIV are that it requires good optical access and a transparent fluid medium. Therefore, the study of commercial microfluidic devices required alternative approaches. Using an X-ray source, Lee and Kim (2003) collected images of the shadows of particles flowing within an opaque Teflon tube that was suitable for processing with a PIV algorithm. Alumina microspheres with a diameter of 2 μm were used as seed particles. The change in the distribution of the X-ray beam was converted to visible light by passing the beam through a $CdWO_4$ scintillator crystal. Imaging optics focused on the scintillator reimaged light onto a double-frame CCD camera. A mechanical shutter was used to produce double X-ray pulses in a short period of time, ~10–40 ms. A cross-correlation PIV algorithm was then applied between the two images to obtain instantaneous velocity vector field. Since this initial work, several applications of X-ray PIV at the microscale have been achieved. Lee and Kim (2005) measured the velocity field of human blood flow by improving the diffraction-based and stochastic characteristics of blood cells without any contrast agents. Fouras et al. (2007) investigated the three-dimensional velocity field within a cylindrical tube by combining phase-contrast X-ray imaging and PIV. Im et al. (2007) extended this approach to three dimensions to measure the particle-laden flow in an opaque conduit by applying a tomographic reconstruction method. Further discussions related to the recent trends in micro-PIV are found in the following chapters.

1.3.1.4 The Principle of PTV

Low particle density and a need to measure all the three components of the flow caused the development of several techniques on the basis of PTV. The concept was developed at the macroscale, and most approaches were multicamera methods (Maas et al., 1993; Virant and Dracos, 1997; Ortiz-Villafuerte et al., 2000) that used three or more cameras to define the flow field. These approaches required multiple scientific-grade cameras and good optical access from multiple views. Volume illumination of a sparsely seeded flow field was needed to allow individual particles to track through a three-dimensional space.

A general PTV algorithm consists of a number of steps, including camera calibration, particle identification, three-dimensional particle reconstruction, and tracking of particles in the three-dimensional space (Maas et al., 1993). Camera calibration determines each camera's unique projection matrix, which specifies the viewing direction and camera position. Images from the multicamera system are then collected and processed in a general scheme that is outlined in Figure 1.8. The scheme consists of anywhere from one to n_c cameras that collect synchronized images of the field of view at a known Δt apart. Figure 1.8 shows an arbitrary series of acquired images, with particles visible in each image. Particles are identified within the two-dimensional images, and a list of their centroid locations is developed. Particles are then reconstructed from these positions and each camera's unique projection matrix. This process takes the particle locations from images acquired from all cameras at a single time step and determines the three-dimensional positions of particles using epipolar lines to match the particles and their positions from different views

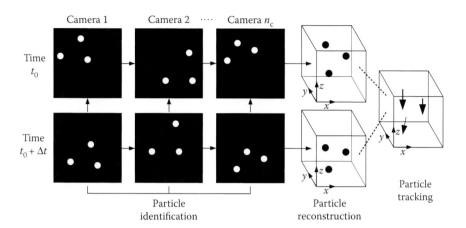

FIGURE 1.8
A flowchart of the steps used to determine three-dimensional particle motion in macro-PTV.

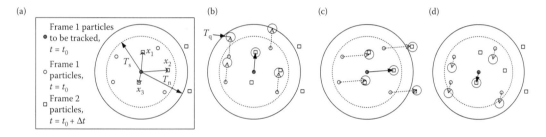

FIGURE 1.9
A schematic of the implementation of a probability-based PTV scheme: (a) the particle to be tracked, the region to be investigated, and identified possible matches, (b) search for x_1 particles, (c) search for x_2 particles, and (d) search for x_3 particles.

(Kurada et al., 1997; Ouellette et al., 2006). Particle reconstruction is completed for each individual time step.

Once particle locations have been determined, a particle-tracking algorithm can then be used to determine the most probable link of the particles between the time steps. An outline of a two-dimensional probability-based particle-tracking scheme (Baek and Lee, 1996) is shown in Figure 1.9. As outlined in Figure 1.9a, a single particle in Frame 1 is selected for tracking, around which a search radius, T_s, that is related to the maximum expected velocity in the flow is defined. Other particles within this radius for Frame 1 or $t = t_0$ are defined as nearest neighbors of the core particle. Potential trajectories can then be defined from Frame 1 to Frame 2 from the core particle to the particles that appear in Frame 2 within the search radius. In the example given here, these are shown as vectors x_1, x_2, and x_3 in Figure 1.9a. The algorithm then loops through each of these potential trajectories to determine the number of particles from Frame 1 that appear in Frame 2 at the end of this vector and that also appear within a neighborhood radius T_n. As there may be a variation in particle motion, a quasirigid radius, T_q, within which to search and capture particles is defined. Probability is then used to determine the most likely displacement vector. As shown in the example, there are two matches in Figure 1.9b, three in Figure 1.9c, and one in Figure 1.9d. The most likely vector for the particle to be tracked is therefore shown in Figure 1.9c.

The algorithm then loops through and considers this approach for each particle found within the region of interest and results in an irregular matrix of velocity vectors. This particle-tracking approach can easily be extended to three dimensions. Other approaches have been developed to search for and track particles, which include the use of multiple frames to track particles through time (Malik et al., 1993), the use of neural networks to find particle trajectories (Labonté, 1999), and the use of a method based on velocity gradient tensor in the case of strong deformation in the fluid (Ishikawa et al., 2000).

A two-dimensional configuration of micro-PTV is easily achieved with the use of hardware similar to that used in two-dimensional micro-PIV. The choice of using a PIV or a PTV algorithm for processing the data is based on the seeding density of imaged particles. PTV requires particles to be uniquely determined from one image frame to another, whereas PIV averages the velocity over a group of particles. Seeding densities are therefore significantly lower for PTV than for PIV.

At the microscale, multiple camera approaches to three-dimensional micro-PTV are difficult to implement because each camera requires a unique optical access to the volume, and each must be able to view the same microvolume in space. Therefore, a method to find the three-dimensional location of particles through the use of a single camera before a particle-tracking algorithm can be implemented is needed. A few methods reported in the literature achieve this on the basis of particle image defocusing, particle off-focusing, holographic reconstruction, and scanning of the focal plane through the volume of interest.

1.3.1.5 Reconstruction of Particle Locations for Three-Dimensional Micro-PTV

Particle defocusing methods (macroscale: Willert and Gharib, 1992; microscale: Yoon and Kim, 2006a) find the three-dimensional locations of particles in a microsized volume by spatially filtering the light that is projected by the imaging optics onto the imaging camera. A plate with a specified pattern of holes is placed in the optical path. Without this plate within the optical path, particles that are in focus would appear as single bright spots and out-of-focus particles would appear blurred depending on their distance from the focal plane. With this plate in the optical system, light from a particle will appear as a pattern of holes in the same geometry that appears on the plate. The out-of-plane distance of a particle, relative to the focal plane of the system, is then measured by calculating the distance between the spots created by a single particle. As the spots from one particle become further away from one another, the particle is found further away from the focal plane. The disadvantages of this approach are that optical alignment of the system can introduce errors (Lee and Kim, 2009), and the use of a spatial filter limits the light collected and results in a lower signal for image detection. This approach has been used to investigate a microflow with a sudden expansion (Yoon and Kim, 2006b). A 20× imaging objective was used to track 3-μm fluorescent particles over a region 768 μm wide × 388 μm high × 50 μm deep, with a spatial resolution of 5 μm in plane and 1 μm out of plane.

Off-focus methods use fluorescent particles, but and the diffraction rings that occur when they are imaged are not within the focal plane (Speidel et al., 2003; Wu et al., 2005; Luo et al., 2006). As a particle's distance from the focal plane increases, the diffraction rings around the particle will also increase in size. A calibration is used to find a linear relation between the diffraction ring size and the distance from the focal plane, which provides a measure of the out-of-plane position of the particle. Speidel et al. (2003) tracked 216-nm fluorescent latex beads through a volume of dimensions $40 \times 60 \times 3$ μm to investigate three-dimensional flows through a porous polymer network. Park and Kihm (2005) tracked 500-nm fluorescent particles as they passed around a 95-μm sphere snugly fit into a 100×100-μm channel. The volumetric measurement resolution was evaluated to be 5.16 μm in each component with an overall measurement uncertainty for a single vector of ±7.5%. This approach was also being used by Wu et al. (2005) to track the trajectories of bacteria in the range of 1–3 μm in a free solution.

The three-dimensional volumetric field of an object's, location was collected with a digital hologram (Schnars and Juptner, 1994). The use of this method to investigate flow phenomenon was first developed at the macroscale (Meng et al., 2004). A hologram image is formed by the interference of the superposition of a forward scattering beam that interferes with seed particles and a reference beam on the imaging plane. Image processing steps are used to eliminate backscatter noise from the hologram, by applying a box filter, a Gaussian filter, and a high-pass filter. A numerical reconstruction of the hologram is performed on the basis of the Huygen–Fresnel principle. This results in a three-dimensional field of particle locations. Successive holograms at a time interval, Δt, can then be used to track particles in the flow. Holographic reconstruction of particle locations was used at progressively smaller volumes of interest (Sheng et al., 2003, 2006, 2008), with high-speed cameras to collect time-resolved data over a volume as small as 1400 μm³.

An alternative approach is to image in-focus sharp particles at different planes within the microchannel and then reconstruct the three-dimensional location of particles from these multiple slices (Homeniuk et al., 2008; Nobes et al., 2010). An objective lens of suitable magnification is coupled to a piezostage that translates the objective normal to the focal plane. This allows the focal plane to be traversed rapidly through the volume of interest, forming a single scan of image collected at predetermined z-locations with a high-speed CCD camera. A single scan resolves a three-dimensional data set of particle locations. Successive scans can then be used to either correlate groups of particles or track individual particles. The volume over which the measurement can be made is dependent on the magnification of the objective lens used and the scan distance of the piezostage. This approach was used to investigate the levitation of 5-μm particles by dielectrophoresis over a region of $400 \times 400 \times 100$ μm (Homeniuk et al., 2008). The flow within a mixing microchannel was also investigated; however, a three-dimensional cross-correlation PIV approach was used to process data (Nobes et al., 2010). The maximum velocities that can be determined are limited by the frame rate of the CCD camera and the scan rate of the optics.

1.3.2 Velocity Measurement: Molecular Based

An appreciable limitation of particle-based velocity measurement techniques for microfluidics is that the particles have a significant size compared with the channel and flow dimensions. This concern is also valid at the macroscale and has led to the development of several approaches that use the spectral properties of different dyes used as flow markers. First developed at the macroscale, these approaches are now being used at the microscale.

1.3.2.1 The Principle of MTV

Molecular tagging velocimetry (MTV) (Lempert et al., 1995; Gendrich et al., 1997) uses caged fluorescent molecules premixed in the flow instead of seed particles to trace the motion of the fluid. The use of fluorescent molecules eliminates particle-induced disturbances from the measurement. Fluorescence of the dye, known as caged fluorescein or photoactivated fluorophore (PAF), is prevented because of the chemical attachment of a caging group. An ultraviolet laser known as the write laser is used to irreversibly remove the caging group from the dye and tag a group of molecules. This laser is used to write a pattern into the flow that has the dye homogeneously premixed in it. Exposure of the uncaged dye to an appropriate wavelength laser causes these uncaged molecules to fluoresce. The uncaged molecule is invisible until exposure to the fluorescing laser and is termed the read laser. The temporal evolution of this fluorescing dye with the flow is captured onto a camera for further image processing to determine flow velocity. A correlation-based approach is used to determine the motion of the fluid pattern to derive velocity.

This technique that was also developed under the acronym PHANTOMM (photoactivated nonintrusive tracking of molecular motion) by Lempert et al. (1995) was used to study flows both at the macroscale (Lempert et al., 1995; Gendrich et al., 1997) and at the microscale (Paul et al., 1998). Lempert et al. (1995) tagged a small portion of the flow seeded with the caged fluorescent dye, and the uncaging was performed by a frequency-tripled Nd:YAG laser (10 mJ at 355 nm). Excitation for fluorescence was carried out with either a continuous argon ion laser (0.5–1.5 W at 488 nm) or a pulsed dye laser with long pulse duration (50–100 mJ, ~2-μs pulse at 490 nm for fluorescein). The concept of measurement was proved by carrying out flow measurements for both a fully developed flow in a 1.4-cm inner diameter quartz tubing and wing tip vortices in a large-scale water channel (Lempert et al., 1995). Paul et al. (1998) first applied MTV to microscale flows. Images of pressure and electrokinetically driven flows, with homogeneously distributed caged fluorescent dye, in 75 and 100-μm inner-diameter circular cross-section capillaries were presented. Sinton and Li (2003) also used MTV for electro-osmotic velocity profile measurements in circular and square cross-section microchannels of various diameters and side lengths, respectively.

A common tagging method used for MTV is the laser line tagging approach (Paul et al., 1998; Sinton and Li, 2003). In this technique, a sheet of ultraviolet laser is used to write a line perpendicular to the capillary axis in a periodic fashion. This causes the uncaging of the caged dye in a line perpendicular to the flow direction. The transport of this uncaged dye with the flow results in a measurement of one component of velocity parallel to the flow direction by measuring the displacement of the tagged molecule's line centers. The line tagging method does not provide any information about the tangential component of velocity. Other tagging methods include the use of a grid and a structured mask for macroscale and microscale flows, respectively. The two components of velocity are calculated by locating the grid line centers using these different methods. Gendrich et al. (1997) applied an aluminum grid tagging method, with two detector systems to study the two-dimensional flow behavior in a macroscale vortex core. Garbe et al. (2008) used a structured mask to obtain two-dimensional velocity vector fields for an inhomogeneous flow in a micromixing chamber. The grid-based approach is difficult to apply to microfluidic applications; however, the structured mask approach has been successfully implemented to microscale flows. One of the limitations of this approach is that velocity information is available for a predefined pattern only, and a sparse matrix of velocity vectors is used to describe the flow field.

1.3.3 Measurement of Mixing

Mixing in microfluidic systems is an important and challenging problem. At the microscale, flow mixing is typically dominated by diffusion. It is only with the introduction of tangential velocities that the time scales of the mixing process can be reduced. Mixing can be improved by introducing three-dimensional flow geometry or with the use of segmented flow. Investigation of the mixing process at the microscale has focused on the evolution of concentration for bulk mixing and distribution of temperature to characterize heat transfer.

Information on mixing can be derived from the fluorescence intensity of dye molecules that are seeded in a flow that is mixing with another flow. As highlighted at the beginning of this chapter, the fluorescence intensity is primarily a function of the intensity of the probe laser used and the concentration of the dye in the fluid. Images of the spatial distribution of this intensity were calibrated (Walker, 1987) to determine the concentration of a two-dimensional region of the microscale flow.

Temperature can also be measured by using a dye that is sensitive to temperature. The fluorescence intensity is therefore a function of laser power, concentration, and temperature. A number of approaches were used to measure temperature on the basis of a single dye (Ross et al., 2001). However, most approaches now use different dyes to measure temperature, one that is insensitive to temperature and one that is sensitive to temperature (Sakakibara and Adrian, 1999). This allows the ratios of the fluorescence intensities to be determined, and this is only a function of temperature. The two-dye approach was used at the microscale to measure temperature (Natrajan and Christensen, 2009) distribution and heat transfer in microchannels.

In both of the above approaches, the number of photons or the intensity of the signal reaching the CCD is used to determine the desired parameter. The signal measured is assumed to be proportional to either the concentration or the temperature within the microscale flow. The response of the CCD is therefore an important hardware specification requirement for this type of measurement. In this measurement, both the linearity of the response of a single pixel and how this response changes across the array have an impact on errors. For this type of measurement, high-quality scientific-grade CCD cameras that have high bit depth and low noise characteristics are used.

1.4 Conclusions

Before peforming an experiment, it is useful to do a system-level design analysis to identify weaknesses. An analysis of image-based photonic techniques used for investigation of microfluidic flow shows that the concept of these techniques and the main components used have many similarities. The hardware of the measurement system consists of a light source to illuminate the flow, which is typically a laser, an optical system to image the region of interest, which is typically an infinity-corrected microscope, and an image-capture device that can use different technologies, which is often referred to in the literature as a CCD camera. A measurement principle to process the data in a defined manner is applied to this hardware to determine the parameter of interest.

Although only a small number of the different types of flow velocity measurement techniques that use images have been discussed here, the main characteristic of each is that contrast within the image is used to determine motion in the flow. Algorithms on the basis of PIV or PTV use contrast in the images to track groups or individual particles from which local velocities are determined. To process the data, clear images of all the individual particles are needed. To achieve a high quality of image contrast, the particles are frozen in the flow using pulsed lasers at a time scale significantly faster than the time scale of the flow. Large-format array detectors coupled to an optical system with a magnification to match the region of interest allow several pixels to characterize individual particles.

This allows the determination of a large vector field to characterize flow velocity within the region of interest.

Measurement of scalar quantities such as concentration or temperature at the system level requires a similar set of hardware equipment. Since the number of photons or the intensity of light is used to derive the parameters of interest, a different set of system hardware specifications is needed. This is mostly related to the performance of the detector and its response to light both for an individual pixel and across an array of pixels that make up the CCD.

When designing an experimental system to investigate microscale flows, a clear objective of which parameter to investigate is needed. The scale of the flow in terms of both the physical dimensions of the microscale system and the expected flow velocities is an important information for the design process. This is used with a given measurement principle for the parameter under investigation to determine system hardware and specifications.

Acknowledgments

The authors acknowledge the help of Sina Ghaemi, Darren Homenuik, Adam Madej, Xudong (Terry) Song, and Farhan Ahmad in compiling this chapter. The authors also acknowledge funding support for their research from the Alberta Ingenuity Fund, the Natural Sciences and Research Council (NSERC) of Canada, and the Canadian Foundation for Innovation.

References

Abramowitz M, Davidson MW. (2004). Introduction to microscopy. *Molecular Expressions*. Available from: http://micro.magnet.fsu.edu/primer/anatomy/introduction.html.

Adrian RJ. (1991). Particle-imaging techniques for experimental fluid-mechanics. *Ann Rev Fluid Mech*. 23:261–304.

Baker RJ. (2010). CMOS: *Circuit Design, Layout, and Simulation*. 3rd ed. Piscataway, NJ: Wiley-IEEE.

Baek SJ, Lee SJ. (1996). A new two-frame particle tracking algorithm using match probability. *Exp Fluids*. 22(1):23–32.

Born M, Wolf E. (1997). *Principles of Optics*. Oxford: Pergamon; p. 661.

Boyle WS, Smith GE. (1970). Charge coupled semiconductor devices. *Bell Syst Tech J*. 49(4):587–93.

Bradbury S, Bracegirdle B. (1998). *Introduction to Light Microscopy*. Oxford: BIOS Scientific Publishers.

Bridges WB. (1964). Laser oscillation in singly ionized argon in the visible spectrum. *Appl Phys Lett*. 4:128–30.

Durst F, Melling A, Whitelaw JH. (1976). *Principles and Practice of Laser Doppler Anemometry*. London: Academic Press.

Elsinga GE, Wieneke B, Scarano F, van Oudheusden BW. (2006). Tomographic particle image velocimetry. *Exp Fluids*. 41:933–47.

Fouras A, Dusting J, Lewis R, Hourigan K. (2007). Three-dimensional synchrotron x-ray particle image velocimetry. *J Appl Physiol*. 102(6):064916.

Galileo G. (1954). *Dialogues Concerning Two New Sciences* [Crew H, de Salvio A, Trans.]. New York: Dover Publications Inc. [Original work published 1638].

Ganapathisubramani B, Longmire EK, Marusic I, Pothos S. (2005). Dual-plane PIV technique to determine the complete velocity gradient tensor in a turbulent boundary layer. *Exp Fluids*. 39(2):222–31.

Garbe CS, Roetmann K, Beushausen V, Jähne B. (2008). An optical flow MTV based technique for measuring microfluidic flow in the presence of diffusion and Taylor dispersion. *Exp Fluids*. 44(3):439–50.

Gendrich CP, Koochesfahani MM, Nocera DG. (1997). Molecular tagging velocimetry and other novel applications of a new phosphorescent supramolecule. *Exp Fluids*. 23(5):361–72.

Geusic JE, Marcos HM, Van Uitert LG. (1964). Laser oscillations in Nd-doped yttrium aluminum, yttrium gallium and gadolinium garnets. *Appl Phys Lett*. 4(10):182–184.

Goulette T, Howard CD, Davidson MW. (2010). Infinity Optical Systems. Nikon MicroscopyU. Available from: http://www.microscopyu.com/articles/optics/cfintro.html.

Guilbault GG. (1990). *Practical Fluorescence*. 2nd ed. New York: Marcel Dekker, Inc.

Hecht E. (2002). *Optics*. 4th ed. San Francisco: Addison Wesley.

Homeniuk DLN, Nobes DS, Molla S, Bhattacharjee S. (2008). Volume particle tracking in three-dimensional micro-channel flows. 6th Joint ASME/JSME Fluids Engineering Conference (FEDSM2008), Jacksonville, Florida, USA, August 10–14.

Horobin R, John Kiernan J, editors. (2002). *Conn's Biological Stains: A Handbook of Dyes, Stains and Fluorochromes for Use in Biology and Medicine*. Oxford: BIOS Scientific Publishers LTD.

Im K-S, Fezzaa K, Wang YJ, Liu X, Wang J, Lai M-C. (2007). Particle tracking velocimetry using fast x-ray phase-contrast imaging. *Appl Phys Lett*. 90(9):091919.

Ishikawa M, Murai Y, Wada A, Iguchi M, Okamoto K, Yamamoto F. (2000). A novel algorithm for particle tracking velocimetry using the velocity gradient tensor. *Exp Fluids*. 29(6):519–31.

Juhany KA, Darji A, Zahrani MG, Husieni H, Behairi M, Dierksheide U. (2007). Preliminary Tests of High Speed PIV Measurements in a Ludwieg-Tube Supersonic Wind Tunnel. AIAA 2007–54, 45th AIAA Aerospace Sciences Meeting and Exhibit, Reno, Nevada.

Karniadakis G, Beskok A, Aluru N. (2005). *Microflows and Nanoflows: Fundamentals and Simulation*. New York: Springer Verlag.

Kurada S, Rankin GW, Sridhar K. (1997). A new particle image velocimetry technique for three-dimensional flows. *Opt Lasers Eng*. 28:343–76.

Labonté G. (1999). A new neural network for particle-tracking velocimetry. *Exp Fluids*. 26(4):340–6.

Lempert WR, Magee K, Ronney P, Gee KR, Haugland RP. (1995). Flow tagging velocimetry in incompressible flow using photo-activated non intrusive tracking of molecular motion (PHANTOMM). *Exp Fluids*. 18:249–57.

Lawson NJ, Wu J. (1997). Three-dimensional particle image velocimetry: error analysis of stereoscopic techniques. *Meas Sci Technol*. 8:894–900.

Lee SJ, Kim G-B. (2003). X-ray particle image velocimetry for measuring quantitative flow information inside opaque objects. *J Appl Physiol*. 94(5):3620.

Lee SJ, Kim G-B. (2005). Synchrotron microimaging technique for measuring the velocity fields of real blood flows. *J Appl Physiol*. 97(6):064701.

Lee SJ, Seok K. (2009). Advanced particle-based velocimetry techniques for microscale flows. *Microfluid Nanofluidics*. 6(5):577–88.

Lindken R, Westerweel J, Wieneke B. (2006). Stereoscopic micro particle image velocimetry. *Exp Fluids*. 41(2):161–71.

Lindken R, Rossi M, Grosse S, Westerweel J. (2009). Micro-particle image velocimetry (microPIV): recent developments, applications, and guidelines. *Lab Chip*. 9(17):2551–67.

Luo R, Yang XY, Peng XF, Sun YF. (2006). Three-dimensional tracking of fluorescent particles applied to micro-fluidic measurements. *J Micromech Microeng*. 16(8):1689–99.

Maas HG, Gruen A, Papantoniou D. (1993). Particle tracking velocimetry in three-dimensional flows: part I. Photogrammetric determination of particle coordinates. *Exp Fluids*. 15:133–46.

Malik NA, Dracos Th, Papantoniou DA. (1993). Particle tracking velocimetry in three-dimensional flows: part II. Particle tracking. *Exp Fluids*. 15:279–94.

Masliyah JH, Bhattacharjee S. (2006). *Electrokinetic and Colloid Transport Phenomena.* Hoboken (NJ): John Wiley and Sons; 736 p.

Meinhart C, Wereley ST, Santiago JG. (1999). PIV measurements of a microchannel flow. *Exp Fluids.* 27(5):414–9.

Meinhart CD, Wereley ST, Gray MHB. (2000a). Volume illumination for two-dimensional particle image velocimetry. *Meas Sci Technol.* 11:809–14.

Meinhart CD, Wereley ST, Santiago JG. (2000b). A PIV algorithm for estimating time-averaged velocity fields. *J Fluid Eng.* 122(2):285.

Meng H, Pan G, Pu Y,Woodward SH. (2004). Holographic particle image velocimetry: from film to digital recording. *Meas Sci Technol.* 15:673–85.

Mie G. (1908). BeiträgezurOptiktrüberMedien, speziellkolloidalerMetallösungen, Leipzig. *Ann Phys.* 330:377–445.

Minor M, van der Linde AJ, van Leeuwen HP, Lyklema J. (1997). Dynamic aspects of electrophoresis and electroosmosis: a new fast method for measuring particle mobilities. *J Colloid Interface Sci.* 189:370–5.

Melling A. (1997). Tracer particles and seeding for particle image velocimetry. *Meas Sci Technol.* 8:1406–16.

Molla S, Bhattacharjee S. (2007). Dielectrophoretic levitation in the presence of shear flow: implications for colloidal fouling of filtration membranes. *Langmuir.* 23(21):10618–27.

Mota MC, Carvalho P, Ramalho J, Leite E. (1991). Spectrophotometric analysis of sodium fluorescein aqueous solutions. Determination of molar absorption coefficient. *Int Ophthalmol.* 15(5):321–6.

Natrajan VK, Christensen KT. (2009). Two-color laser-induced fluorescent thermometry for microfluidic systems. *Meas Sci Technol.* 20(1):015401.

Nobes DS, Ford HD, Tatam RP. (2004). Instantaneous three component planar Doppler velocimetry using imaging fibre bundles. *Exp Fluids.* 36:3–10.

Nobes DS, Abdolrazaghi M, Homeniuk DLN. (2010). Three-component, three-dimensional velocity measurement technique for micro-channel applications using a scanning μPIV. 15th International Symposium on Applications of Laser Techniques to Fluid Mechanics; July 05–08, 2010; Lisbon, Portugal.

Ouellette NT, Xu H, Bodenschatz E. (2006). A quantitative study of three-dimensional Lagrangian particle tracking algorithms. *Exp Fluids.* 40:301–13.

Ortiz-Villafuerte J, Schmidl WD, Hassan YA. (2000). Three-dimensional PTV study of the surrounding flow and wake of a bubble rising in a stagnant liquid. *Exp Fluids.* Suppl:S202–10.

Park JS, Kihm KD. (2005). Three-dimensional micro-PTV using deconvolution microscopy. *Exp Fluids.* 40(3):491–9.

Paul PH, Garguilo MG, Rakestraw DJ. (1998). Imaging of pressure- and electrokinetically driven flows through open capillaries. *Anal Chem.* 70(13): 2459–67.

Prasad AK, Adrian RJ. (1993). Stereoscopic particle image velocimetry applied to liquid flows. *Exp Fluids.* 15(1):49–60.

Raffel M, Willert CE, Wereley ST, Kompenhans J. (2007). *Particle Image Velocimetry.* 2nd ed. Berlin: Springer Verlag.

Rayleigh (J. W. Strutt). (1881). Scientific Papers, Vol. I; 1899. p. 518. Reprinted from: *Philos Mag.* 12(5):81.

Rayleigh (J. W. Strutt). (1899). Scientific Papers, Vol. IV; 1903. p. 397. Reprinted from: *Philos Mag.* 47(5):375.

Ross D, Gaitan M, Locascio LE. (2001). Temperature measurement in microfluidic systems using a temperature-dependent fluorescent dye. *Anal Chem.* 17:4117–23.

Sakakibara J, Adrian RJ. (1999). Whole field measurement of temperature in water using two-color laser induced fluorescence. *Exp Fluids.* 26:7–15.

Saleh B, Teich MC. (2007). *Fundamentals of Photonics.* New Jersey: Wiley-Interscience.

Santiago JG, Wereley ST, Meinhart CD, Beebe DJ, Adrian RJ. (1998). A particle image velocimetry system for microfluidics. *Exp Fluids.* 25(4):316–9.

Schnars U, Jüptner W. (1994). Direct recording of holograms by a CCD target and numerical reconstruction. *Appl Opt*. 33(2):179.

Sheng J, Malkiel E, Katz J. (2003). Single beam two-views holographic particle image velocimetry. *Appl Opt*. 42(2):235–50.

Sheng J, Malkiel E, Katz J. (2006). Digital holographic microscope for measuring three-dimensional particle distributions and motions. *Appl Opt*. 45(16):3893–901.

Sheng J, Malkiel E, Katz J. (2008). Using digital holographic microscopy for simultaneous measurements of 3D near wall velocity and wall shear stress in a turbulent boundary layer. *Exp Fluids*. 45(6):1023–35.

Sinton D. (2004). Microscale flow visualization. *Microfluid Nanofluidics*. 1(1):2–21.

Sinton D, Li D. (2003). Electroosmotic velocity profiles in microchannels. *Colloids Surf A Physicochem Eng Asp*. 222(1–3):273–83.

Speidel M, Jonás A, Florin EL. (2003). Three-dimensional tracking of fluorescent nanoparticles with subnanometer precision by use of off-focus imaging. *Opt Lett*. 28(2):69–71.

Stokes GG. (1852). On the change of refrangibility of light. *Philos Trans R Soc Lond B Biol Sci*. 142:463–562.

van de Hulst HC. (1981). *Light Scattering by Small Particles*. New York: Dover.

Virant M, Dracos T. (1997). 3D PTV and its application on Lagrangian motion. *Meas Sci Technol*. 8:1539–52.

Walker DA. (1987). A fluorescence technique for measurement of concentration in mixing liquids. *J Phys E*. 20:217–24.

Willert CE. (1997). Stereoscopic digital particle image velocimetry for application in wind tunnel flows. *Meas Sci Technol*. 8:353.

Willert CE, Gharib M. (1991). Digital particle image velocimetry. *Exp Fluids*. 10:181.

Willert CE, Gharib M. (1992). Three-dimensional particle imaging with a single camera. *Exp Fluids*. 12:353–8.

Wu M, Roberts JW, Buckley M. (2005). Three-dimensional fluorescent particle tracking at microscale using a single camera. *Exp Fluids*. 38:461–5.

Yeh Y, Cummins H. (1964). Localised fluid flows measurements with a He-Ne laser spectrometer. *Appl Phys Lett*. 4:176–8.

Yoon SY, Kim KC. (2006a). 3D particle position and 3D velocity field measurement in a micro-volume via the defocusing concept. *Meas Sci Technol*. 17(11):2897–905.

Yoon SY, Kim KC. (2006b). Signal intensity enhancement of μ-LIF by using ultra-thin laser sheet illumination and aqueous mixture with ethanol/methanol for micro-channel applications. *Opt Lasers Eng*. 44:224–39.

2

Recent Developments in Microparticle Image Velocimetry

Sang-Youp Lee, Jaesung Jang, Yong-Hwan Kim, and Steven T. Wereley

CONTENTS

2.1 Introduction

Microparticle image velocimetry (μPIV) was developed by Santiago et al. (1999) to investigate microscale flow field, and it soon became one of the most versatile experimental tools in the field of microfluidic research. μPIV is a full-field flow measurement system in small-scale devices with high resolving power. PIV experiments basically consist of two major steps: image acquisition and postprocess (image interrogation to obtain velocity vectors). The major differences of μPIV from macro-PIV are that the flow fields of interest are much smaller and are in two-dimensional microfluidic devices (in most cases).

Small flow fields of interest cause instrumentation problems both in selective illumination of the measurement volume and in multiple optical accesses to measure three-dimensional flow fields. Such limitations result in high background noise from acquired images and therefore accuracy problem in the measured velocity fields. μPIV techniques have been developed to overcome such shortcomings in instrumentation and flow-field interrogation. For example, various μPIV techniques to acquire three-dimensional flow fields (Bown et al., 2006; Lindken et al., 2006; Yoon and Kim, 2006; Kim and Lee, 2007; Pereira et al., 2007; Arroyo and Hinsch, 2008; Hagsäter et al., 2008; Peterson et al., 2008; Lee and Kim, 2009) and evanescent wave PIV (Jin et al., 2003; Zettner and Yoda, 2003) that avoid high background noise inevitably associated with volume illumination were developed. Ensemble correlation method (Meinhart et al., 1999b) and μPIV image filters (Gui et al., 2002) were developed to enhance the acquired image and increase the accuracy of velocity fields by strengthening the correlation function peaks, respectively. Also, the spatial resolutions of velocity vector fields were greatly increased to single-pixel level (Westerweel et al., 2004).

Other than the basic function of μPIV as a full-field measurement system, it has also been widely applied for various purposes such as direct temperature measurement on the basis of the property of correlation function (Hohreiter et al., 2002), biosensing to detect virus on the basis of particle diffusivity measurement (Kumar et al., 2008), and particle or single-cell tracking (Chung et al., 2009).

In this chapter, we discuss about the general optical diagnostic method in fluid mechanics and give an in-depth overview of μPIV in the first two sections. Then, two examples of μPIV experiments for flow fields of fixed and moving boundaries are provided in the third section to discuss how the actual experimental procedures such as instrumentations, image interrogations, and data reduction processes are accomplished. Finally, the extension and recent achievements of μPIV are introduced in the last section.

2.2 Optical Diagnostics Metrology in Microscale Fluid Mechanics

Optical diagnostics has played a significant role in experimental fluid mechanics, beginning with the development of schlieren and shadowgraphy imaging of compressible flows in the early twentieth century. The use of optical diagnostics was greatly accelerated in the second half of the twentieth century, with the development of lasers and high-speed, high-resolution electronic cameras. Begun in the 1990s and continuing till date, many of these optical diagnostic techniques have been extended to microscale fluid mechanics. These techniques can be divided into pointwise and full-plane techniques. The pointwise techniques measure one spatial point at a time (although often a dense temporal stream can be acquired at each spatial point), whereas the full-plane techniques measure many spatial points simultaneously.

2.2.1 Pointwise Methods

Laser Doppler velocimetry (LDV) has been a standard optical measurement technique in fluid mechanics since the 1970s. In the case of a dual-beam LDV system, two coherent laser beams are aligned so that they intersect at some region. The volume of intersection of the two laser beams defines the measurement volume. In the measurement volume, the two coherent laser beams interfere with each other, producing a pattern of light and dark fringes. When a seed particle passes through these fringes, a pulsing reflection is created that is collected by a photomultiplier, processed, and turned into a velocity measurement. Traditionally, the measurement volumes of standard LDV systems have characteristic dimensions in the order of a few millimeters. Compton and Eaton (1996) used short focal length optics to obtain a measurement volume of 35×66 µm. Using lenses of very short focal length, Tieu et al. (1995) built a dual-beam solid-state LDA system with a measurement volume of approximately 5×10 µm. Their micro-LDV system was used to measure the flow through a 175-µm-thick channel, producing time-averaged measurements that compare well with the expected parabolic velocity profile, except within 18 µm of the wall. Advancements in microfabrication technology are expected to facilitate the development of new generations of self-contained solid-state LDV systems with micron-scale probe volumes. These systems will likely serve an important role in the diagnosis and monitoring of microfluidic systems (Gharib et al., 2002). However, the size of the probe volume significantly limits the number of fringes that it can contain, which subsequently limits the accuracy of the velocity measurements.

Optical Doppler tomography (ODT) has been developed to measure micron-scale flows embedded in a highly scattering medium. In the medical community, the ability to measure *in vivo* blood flow under the skin allows clinicians to determine the location and depth of burns (Chen et al., 1997). ODT combines single-beam Doppler velocimetry with heterodyne mixing from a low-coherence Michelson interferometer. The lateral spatial resolution of the probe volume is determined by the diffraction spot's size. The Michelson interferometer is used to limit the effective longitudinal length of the measurement volume to that of the coherence length of the laser. The ODT system developed by Chen et al. (1997) had lateral and longitudinal spatial resolutions of 5 and 15 µm, respectively. The system was applied to measure flow through a 580-µm-diameter conduit.

2.2.2 Full-Field Methods

Full-field experimental velocity measurement techniques are those that generate velocities that are minimally two-component velocity measurements distributed within a two-
dimensional plane. These types of velocity measurements are essential in microfluidics
for several reasons. First, global measurements, such as the pressure drop along a length
of channel, can reveal the dependence of flow physics on length scale by showing that the
pressure drop for flow through a small channel is smaller or larger than the flow through
a large channel. However, global measurements are not very useful for pointing to the
precise cause of why the physics might change, such as losing the no-slip boundary for
high-Knudsen-number gas flows. A detailed view of the flow, such as that provided by
full-field measurement techniques, is indispensable for establishing the reasons why flow
behavior changes at small scales. Full-field velocity measurement techniques are also useful for optimizing complicated processes such as mixing, pumping, or filtering—typical
microfluidic processes. Several of the common macroscopic full-field measurement techniques have been extended to microscopic length scales. These are scalar image velocimetry, molecular tagging velocimetry (MTV), and PIV. These techniques are introduced
briefly in this section and then discussed in detail in the following sections.

Scalar image velocimetry (SIV) refers to the determination of velocity vector fields by
recording images of a passive-scalar quantity and by inverting the transport equation for
a passive scalar. Dahm et al. (1992) originally developed SIV for measuring turbulent jets
at macroscopic length scales. Successful velocity measurements depend on having sufficient spatial variations in the passive-scalar field and relatively high Schmidt numbers.
Because SIV uses molecular tracers to follow the flow, it has several advantages at the
microscale over measurement techniques such as PIV or LDV, which use discrete flow-
tracing particles. For instance, the molecular tracers will not become trapped in even the
smallest passages within a microelectromechanical system (MEMS) device. In addition,
the discrete flow-tracing particles used in PIV can acquire a charge and move in response
to not only hydrodynamic forces but also electrical forces in a process called electrophoresis. However, molecular tracers typically have much higher diffusion coefficients than
discrete particles, which can significantly lower the spatial resolution and velocity resolution of the measurements.

Paul et al. (1998) analyzed fluid motion using a novel dye that, although not normally
fluorescent, can be made fluorescent by exposure to the appropriate wavelength of light.
Dyes of this class are typically called *caged dyes* because the fluorescent nature of the dye
is made ineffective by a photoreactive bond that can be easily broken. This caged dye was
used in a microscopic SIV procedure to estimate velocity fields for pressure- and electrokinetically driven flows in 75-μm-diameter capillary tubes. A 20×500-μm sheet of light
from a $\lambda = 355$-nm frequency-tripled Nd:YAG laser was used to uncage a 20-μm-thick cross-
sectional plane of the dye in the capillary tube. In this technique, only the uncaged dye is
excited when the test section is illuminated with a shuttered beam from a continuous wave
Nd:YVO$_4$ laser. The excited fluorescent dye is imaged using a 10×, NA = 0.3, objective lens
onto a CCD camera at two known time exposures. The velocity field is then inferred from
the motion of the passive scalar. We approximate the spatial resolution of this experiment
to be in the order of $100 \times 20 \times 20$ μm, on the basis of the displacement of the fluorescent dye
between exposures and the thickness of the light sheet used to uncage the fluorescent dye.

MTV is another technique that has shown promise in microfluidics research. In this
technique, flow-tracing molecules fluoresce or phosphoresce after being excited by a light
source. The excitement is typically in the form of a pattern such as a line or grid written

into the flow. The glowing grid lines are imaged twice with a short time delay between the two images. Local velocity vectors are estimated by correlating the grid lines between the two images (Koochesfahani et al., 1997). MTV has the same advantages and disadvantages, at least with respect to the flow-tracing molecules, as SIV. In contrast to SIV, MTV infers velocity in much the same way as PIV—a pattern is written into the flow and the evolution of that pattern allows for inferring the velocity field. MTV was demonstrated at microscopic length scales by Maynes and Webb (2002) in their investigation of liquid flow through capillary tubes, as well as by Lempert et al. (2001) in an investigation of supersonic micronozzles. Maynes and Webb (2002) investigated aqueous glycerin solutions flowing at Reynolds numbers ranging from 600 to 5000 through a 705-μm-diameter fused silica tube of circular cross section. They state that the spatial resolution of their technique is 10 μm across the diameter of the tube and 40 μm along the axis of the tube. The main conclusion of this work was that the velocity measured in their submillimeter tube agreed quite well with laminar flow theory and that the flow showed a transition to turbulence beginning at a Reynolds number of 2100. Lempert et al. (2001) flowed a mixture of gaseous nitrogen and acetone through a 1-mm straight-walled "nozzle" at pressure ratios ranging from highly underexpanded to perfectly matched. Because the nozzle was not transparent, the measurement area was limited to positions outside the nozzle. A single line was written in the gas, normal to the axis of the nozzle, by a frequency-quadrupled Nd:YAG (266-nm) laser. The time evolution of the line was observed by an intensified CCD camera. Lempert et al. report measurements at greater than Mach 1 with an accuracy of ±8 m/s and a spatial resolution of 10 μm perpendicular to the nozzle axis.

The machine vision community developed a class of velocimetry algorithms, called *optical-flow* algorithms, to determine the motion of rigid objects. The technique can be extended to fluid flows by assuming that the effect of molecular diffusion is negligible and requiring that the velocity field is sufficiently smooth. Because the velocity field is computed from temporal and spatial derivatives of the image field, the accuracy and the reliability of the velocity measurements are strongly influenced by noise in the image field. This technique imposes a smoothness criterion on the velocity field, which effectively low-pass filters the data and can lower the spatial resolution of the velocity measurements (Wildes et al., 1997). Amabile et al. (1996) applied the optical-flow algorithms to infer velocity fields from 500- to 1000-μm-diameter microtubes by indirectly imaging 1- to 20-μm-diameter x-ray-scattering emulsion droplets in a liquid flow. High-speed x-ray microimaging techniques were presented by Leu et al. (1997). A synchrotron is used to generate high-intensity x-rays that scatter the emulsion droplets onto a phosphorous screen. A CCD camera imaging the phosphorous screen detects variations in the scattered x-ray field. The primary advantage of the x-ray imaging technique is that one can obtain structural information about the flow field, without having optical access. Hitt et al. (1996) applied the optical flow algorithm to *in vivo* blood flow in microvascular networks, with diameters ~100 μm. The algorithm spectrally decomposes subimages into discrete spatial frequencies by correlating the different spatial frequencies to obtain flow-field information. The advantage of this technique is that it does not require discrete particle images to obtain reliable velocity information. Hitt et al. (1995) obtained *in vivo* images of blood cells flowing through a microvascular network using a 20× water-immersion lens with a spatial resolution in the order of 20 μm in all directions.

PIV has been used since the mid-1980s to obtain high spatial resolution two-dimensional velocity fields in macroscopic flows. The experimental procedure is, at its core, conceptually simple to understand. A flow is made visible by seeding it with particles. The particles are photographed at two different times. The images are sectioned into many smaller regions called *interrogation regions*. The motion of the group of particles within each interrogation region is determined using a statistical technique called a *cross-correlation*. If the array of gray values comprising the first image is called $f(i,j)$ and of the second image is called $g(i,j)$, the cross-correlation is given by

$$\Phi(m,n) = \sum_{j=1}^{q}\sum_{i=1}^{p} f(i,j) \times g(i+m, j+n). \tag{2.1}$$

The cross-correlation for a high-quality set of PIV measurements should look similar to that shown in Figure 2.1. The location of the peak indicates how far the particles have moved between the two images. Curve fitting with an appropriate model is used to obtain displacement results accurate to 0.1 pixels.

A PIV bibliography by Adrian (1996) lists more than 1200 references describing various PIV methods and the problems to which PIV methods have been applied for investigation. For a good reference describing many of the technical issues pertinent at macroscopic length scales, see the text (Raffel et al., 1998). This section provides a brief explanation of how PIV works in principle and then concentrates on how PIV is different at small length scales.

Santiago et al. (1998) demonstrated the first µPIV system—a PIV system with a spatial resolution sufficiently small enough to be able to make measurements in microscopic systems. Their system was capable of measuring slow flows—velocities on the order of hundreds of microns per second—with a spatial resolution of 6.9 × 6.9 × 1.5 µm. The system used an epifluorescent microscope and an intensified CCD camera to record 300-nm-diameter polystyrene flow-tracing particles. The particles were illuminated using a continuous Hg-arc lamp. The continuous Hg-arc lamp was chosen for situations that required low levels of illumination light (e.g., flows containing living biological specimens) and where the velocity was sufficiently small enough so that the particle motion can be frozen by the CCD camera's electronic shutter.

Koutsiaris et al. (1999) demonstrated a system suitable for slow flows that used 10-µm glass spheres for tracer particles and a low spatial resolution, high-speed video system to record the particle images, yielding a spatial resolution of 26.2 µm. They measured the flow of water inside 236-µm round glass capillaries and found agreement between the measurements and the analytical solution within the measurement uncertainty.

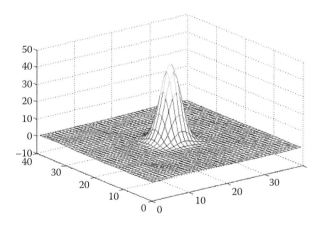

FIGURE 2.1
PIV cross-correlation peak.

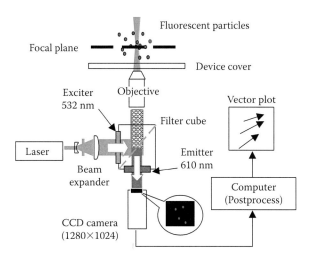

FIGURE 2.2
Schematic of a µPIV system. A pulsed Nd:YAG laser is used to illuminate fluorescent flow-tracing particles, and a cooled CCD camera is used to record the particle images.

Later applications of the µPIV technique moved steadily toward faster flows. The Hg-arc lamp was replaced with a New Wave two-headed Nd:YAG laser that allowed cross-correlation analysis of singly exposed image pairs acquired with submicrosecond time steps between images. At macroscopic length scales, this short time step would allow analysis of supersonic flows. However, because of the high magnification, the maximum velocity measurable with this time step is in the order of meters per second.

Meinhart et al. (1999a) applied µPIV to measure the flow field in a 30-µm-high × 300-µm-wide rectangular channel, with a flow rate of 50 µL/h, which is equivalent to a centerline velocity of 10 mm/s or three orders of magnitude greater than the initial effort a year before. The experimental apparatus, shown in Figure 2.2, images the flow with a 60×, NA = 1.4, oil-immersion lens. The 200-nm-diameter polystyrene flow-tracing particles that were chosen were small enough so that they faithfully followed the flow and were 150 times smaller than the smallest channel dimension. A subsequent investigation of the flow inside a microfabricated inkjet printer head by Meinhart and Zhang (2000) yielded very high-speed µPIV measurements. Using a slightly lower magnification (40×) and consequently lower spatial resolution, measurements of velocities as high as 8 m/s were made.

2.3 Overview of µPIV

2.3.1 Fundamental Physics Considerations of µPIV

Three fundamental problems differentiate µPIV from conventional macroscopic PIV: the particles are small compared with the wavelength of the illuminating light, the particles are small enough that the effects of the Brownian motion must be addressed, and the illumination source is typically not a light sheet but rather an illuminated volume of the flow.

2.3.1.1 Particles Small (λ)

Flow-tracing particles must be large enough to scatter sufficient light so that their images can be recorded. In the Rayleigh scattering regime, where the particle diameter d is much smaller than the wavelength of light, $d \ll \lambda$, the amount of light scattered by a particle varies as d^{-6} (Born and Wolf, 1997). Because the diameter of the flow-tracing particles must be small enough so that the particles do not disturb the flow being measured, the diameter can frequently be in the order of 50–100 nm. The diameters are then 1/10 to 1/5 the wavelength of green light, $\lambda = 532$ nm, and are therefore approaching the Rayleigh scattering criteria. This places significant constraints on the image-recording optics, making it extremely difficult to record particle images.

One solution to the imaging problem is to use epifluorescence imaging to record light emitted from fluorescently labeled particles, using an optical filter to remove the background light. This technique was used successfully in liquid flows to record images of 200–300-nm-diameter fluorescent particles (Adrian, 1996; Meinhart et al., 1999a). Although fluorescently labeled particles are well suited for μPIV studies in liquid flows, they are not readily applicable to high-speed gas flows for several reasons. First, commercially available fluorescently labeled submicron particles are available only in aqueous solutions. Furthermore, the emission decay time of many fluorescent molecules is in the order of several nanoseconds, which may cause streaking of the particle images for high-speed flows. Presently, seeding gas flows remain a significant problem in μPIV.

2.3.1.2 Effects of the Brownian Motion

When the seed particle size becomes small, the collective effect of collisions between the particles and a moderate number of fluid molecules is unbalanced, preventing the particle from following the flow to some degree (Probstein, 1994). This phenomenon, commonly called the *Brownian motion*, has two potential implications for μPIV: one is to cause an error in the measurement of the flow *velocity*, and the other is to cause an uncertainty in the *location* of the flow-tracing particles—although this problem is eliminated by using pulse lasers for illumination. To assess the effects of the Brownian motion, it is first necessary to establish how particles suspended in flows behave.

2.3.1.2.1 Flow/Particle Dynamics

In stark contrast to many macroscale fluid mechanics experiments, the hydrodynamic size of a particle (a measure of its ability to follow the flow based on the ratio of inertial to drag forces) is usually not a concern in microfluidic applications because of the large surface to volume ratios at small length scales. A simple model for the response time of a particle subjected to a step change in local fluid velocity can be used to gauge particle behavior. On the basis of a simple first-order inertial response to a constant flow acceleration (assuming Stokes flow for the particle drag), the response time τ_p of a particle is as follows:

$$\tau_p = \frac{d_p^{\,2} \rho_p}{18\eta}, \tag{2.2}$$

where d_p and ρ_p are the diameter and density of the particle, respectively, and η is the dynamic viscosity of the fluid. For typical μPIV experimental parameters of 300-nm-diameter polystyrene latex spheres immersed in water, the particle response time would be 10^{-9} s. This response time is much smaller than the time scales of any realistic liquid or low-speed gas flow field.

In the case of high-speed gas flows, the particle response time may be an important consideration when designing a system for microflow measurements. For example, a 400-nm particle seeded into an air micronozzle that expands from the sonic at the throat to Mach 2 over a 1-mm distance may experience a particle-to-gas relative flow velocity of more than 5% (assuming a constant acceleration and a stagnation temperature of 300 K). Particle response to flow through a normal shock would be significantly worse. Another consideration in gas microchannels is the breakdown of the no-slip and continuum assumptions as the particle Knudsen number Kn_p, defined as the ratio of the mean free path of the gas to the particle diameter, approaches (and exceeds) 1. For the case of the slip flow regime (10^{-3} < Kn_p < 0.1), it is possible to use corrections to the Stokes drag relation to quantify particle dynamics (Beskok et al., 1996). For example, a correction offered by Melling (1986) suggests the following relation for the particle response time:

$$\tau_p = (1 + 2.76Kn_p)\frac{d_p^2 \rho_p}{18\eta}. \tag{2.3}$$

2.3.1.2.2 Velocity Errors

Santiago et al. (1998) briefly considered the effect of the Brownian motion on the accuracy of µPIV measurements. Devasenathipathy et al. (2003) had in-depth consideration of the phenomenon of the Brownian motion necessary to completely explain the effects of Brownian motion in µPIV measurements. For time intervals Δt much larger than the particle's inertial response time, the dynamics of Brownian displacement is independent of inertial parameters such as particle and fluid density, and the mean square distance of diffusion is proportional to $D\Delta t$, where D is the diffusion coefficient of the particle. For a spherical particle in an unbounded medium subject to Stokes drag law, the diffusion coefficient D was first given by Einstein (1905) as

$$D = \frac{\kappa T}{3\pi\eta d_p}, \tag{2.4}$$

where d_p is the particle diameter, κ is the Boltzmann's constant, T is the absolute temperature of the fluid, and η is the dynamic viscosity of the fluid.

The random Brownian displacements cause particle trajectories to fluctuate about the deterministic path lines of the fluid flow field. Assuming that the flow field is steady over the time of measurement and the local velocity gradient is small, the imaged Brownian particle motion can be considered a fluctuation about a streamline that passes through the particle's initial location. An ideal, non-Brownian (i.e., deterministic) particle following a particular streamline for a time period Δt has x- and y-displacements of

$$\Delta x = u\Delta t, \tag{2.5a}$$

$$\Delta y = v\Delta t, \tag{2.5b}$$

where u and v are the x- and y-components of the time-averaged local fluid velocity, respectively. The relative errors, ε_x and ε_y, incurred as a result of imaging the Brownian particle displacements in a two-dimensional measurement of the x- and y-components of particle velocity, are given as

$$\varepsilon_x = \frac{\sigma_x}{\Delta x} = \frac{1}{u}\sqrt{\frac{2D}{\Delta t}}, \tag{2.6a}$$

$$\varepsilon_y = \frac{\sigma_y}{\Delta y} = \frac{1}{v}\sqrt{\frac{2D}{\Delta t}}. \tag{2.6b}$$

This Brownian error establishes a lower limit on the measurement time interval Δt because, for shorter times, the measurements are dominated by the uncorrelated Brownian motion. These quantities describe the relative magnitudes of the Brownian motion and are referred to here as Brownian intensities. The errors estimated by Equations 2.6a and 2.6b show that the relative Brownian intensity error decreases as the time of measurement increases. Larger time intervals produce flow displacements proportional to Δt while the root mean square of the Brownian particle displacements grows as $\Delta t^{1/2}$. In practice, the Brownian motion is an important consideration when tracing 50- to 500-nm particles in flow-field experiments with flow velocities of less than approximately 1 mm/s. For a velocity on the order of 0.5 mm/s and a 500-nm seed particle, the lower limit for the time spacing is approximately 100 μs for a 20% error due to the Brownian motion. This error can be reduced by both averaging over several particles in a single interrogation spot and ensemble averaging over several realizations. The diffusive uncertainty decreases as $1/\sqrt{N}$, where N is the total number of particles in the average (Bendat and Piersol, 1986).

Equation 2.6 demonstrates that the effect of the Brownian motion is relatively less important for faster flows. However, for a given measurement, when u increases, Δt will generally decrease. Equations 2.6a and 2.6b also demonstrate that when all conditions but Δt are fixed, increasing Δt will decrease the relative error introduced by the Brownian motion. Unfortunately, a longer Δt will decrease the accuracy of the results because the PIV measurements are based on a first-order accurate approximation of the velocity. Using a second-order accurate technique allows for a longer Δt to be used without increasing this error (Wereley and Meinhart, 2001).

2.3.1.3 Volume Illumination of the Flow

Another significant difference between μPIV and macroscopic PIV is that because of the lack of optical access and a significant diffraction in light-sheet-forming optics, light sheets are typically not a practical source of illumination for microflows. Consequently, the flow must be volume illuminated, leaving two choices for how to visualize the seed particles— with an optical system whose depth of focus exceeds the depth of the flow being measured or with an optical system whose depth of focus is small compared with that of the flow. Both of these techniques have been used in various applications of μPIV. Cummings (2001) used a large-depth-focus imaging system to explore electrokinetic and pressure-driven flows. The advantage of the large-depth-focus optical system is that all particles in the field of view of the optical system are well focused. The disadvantage of this scheme is that all depth information is lost and the resulting velocity fields are completely depth-averaged. Cummings (2001) addressed this problem with advanced processing techniques that are not covered here.

The second choice of imaging systems is one whose depth of focus is smaller than that of the flow domain. The optical system will then focus those particles that are within the depth of focus of the imaging system while the remaining particles will be unfocused and contribute to the background noise level. Because the optical system is used to define thickness of the measurement domain, it is important to characterize exactly how thick the depth of focus, or more appropriately, the *depth of correlation*, δz_m, is. Meinhart et al. (2000b) have considered this question in detail by starting from the basic principles of how small particles are imaged.

2.3.1.3.1 Depth of Field

The depth of field of a standard microscope objective lens is given by Inoué and Spring (1997) as

$$\delta z = \frac{n\lambda_0}{\mathrm{NA}^2} + \frac{ne}{\mathrm{NA} \times M},$$

(2.7)

where n is the refractive index of the fluid between the microfluidic device and the objective lens, λ_0 is the wavelength of light in a vacuum being imaged by the optical system, NA is the numerical aperture of the objective lens, M is the total magnification of the system, and e is the smallest distance that can be resolved by a detector located in the image plane of the microscope (for the case of a CCD sensor, e is the spacing between pixels). Equation 2.7 is the summation of the depths of field resulting from diffraction (first term on the right-hand side) and geometric effects (second term on the right-hand side).

The cutoff for the depth of field due to diffraction, that is, first term on the right-hand side of Equation 2.7, is chosen by convention to be one-quarter of the out-of-plane distance between the first two minima in the three-dimensional point-spread function, $u = \pm\pi$, in diffraction variables. Substituting $\mathrm{NA} = n \sin \theta = na/f$, and $\lambda_0 = n\lambda$ yields the first term on the right-hand side of Equation 2.7.

If a CCD sensor is used to record particle images, the geometric term in Equation 2.7 can be derived by projecting the CCD array into the flow field, and then considering the out-of-plane distance, the CCD sensor can be moved before the geometric shadow of the point source occupies more than a single pixel. This derivation is valid for small light collection angles, where $\tan \theta \sim \sin \theta = \mathrm{NA}/n$.

2.3.1.3.2 In-Plane Spatial Resolution Limits

The overall goal of μPIV is to obtain reliable two-dimensional velocity fields in microfluidic devices with high accuracy and high spatial resolution. In this section, we discuss the theoretical requirements for achieving both these goals and address the relative trade-offs between velocity accuracy and spatial resolution.

The most common mode of PIV is to record two successive images of flow-tracing particles that are introduced into the working fluid and that accurately follow the local motion of the fluid. The two particle images are separated by a known time delay, Δt. Typically, the two particle image fields are divided into uniformly spaced interrogation regions, which are cross-correlated to determine the most probable local displacement of the particles.

High spatial resolution is achieved by recording the images of flow-tracing particles with sufficiently small diameters, d_p, so that they faithfully follow the flow in microfluidic devices. The particle should be imaged with high-resolution optics and with sufficiently high magnification so that the particles are resolved with at least 3–4 pixels per particle diameter. Following Adrian (1991), the diffraction-limited spot size of a point source of light, d_s, imaged through a circular aperture is given by

$$d_s = 2.44(M+1)f^\# \lambda,$$ (2.8)

where M is the magnification, $f^\#$ is the f number of the lens, and λ is the wavelength of light. For infinity-corrected microscope objective lenses, $f^\# \approx 1/2[(n/NA)^2 - 1]^{1/2}$. The numerical aperture, NA, is defined as $NA = n \sin \theta$, where n is the index of refraction of the record-ing medium and θ is the half-angle subtended by the aperture of the recording lens. The actual recorded image can be estimated as the convolution of point-spread function with the geometric image. Approximating both these images as Gaussian functions, the effec-tive image diameter, d_e, can be written as (Adrian and Yao, 1985)

$$d_e = [d_s^2 + M^2 d_p^2]^{1/2}.$$ (2.9)

The effective particle image diameter places a bound on the spatial resolution that can be obtained by µPIV. Assuming that the particle images are sufficiently resolved by the CCD array, the location of the correlation peak can be sufficiently resolved to be within 1/10th the particle image diameter (Prasad et al., 1993). Therefore, the uncertainty of the correla-tion peak location for a particle with a diameter $d_p = 0.2$ µm recorded with an $NA = 1.4$ lens is $\delta x \sim d_e/10M = 35$ nm. Table 2.1 gives effective particle diameters recorded through a circular aperture and then projected back into the flow, d_e/M.

2.3.1.3.3 Out-of-Plane Spatial Resolution

It is common practice in PIV to use a sheet of light to illuminate the flow-tracing particles. In principle, the light sheet illuminates only particles contained within the depth of focus of the recording lens. This provides reasonably high-quality in-focus particle images to be recorded with low levels of background noise being emitted from the out-of-focus par-ticles. The out-of-plane spatial resolution of the velocity measurements is defined clearly by the thickness of the illuminating light sheet.

Because of the small length scales associated with µPIV, it is difficult if not impossible to form a light sheet that is only a few microns thick and even more difficult to align a light

TABLE 2.1

Effective Particle Image Diameters When Projected Back into the Flow, d_e/M (µm)

Particle Size, d_p (µm)	Microscope Objective Lens Characteristics				
	$M = 60$, NA = 1.4	$M = 40$, NA = 0.75	$M = 40$, NA = 0.6	$M = 20$, NA = 0.5	$M = 10$, NA = 0.25
0.01	0.29	0.62	0.93	1.24	2.91
0.10	0.30	0.63	0.94	1.25	2.91
0.20	0.35	0.65	0.95	1.26	2.92
0.30	0.42	0.69	0.98	1.28	2.93
0.50	0.58	0.79	1.06	1.34	2.95
0.70	0.76	0.93	1.17	1.43	2.99
1.00	1.04	1.18	1.37	1.59	3.08
3.00	3.01	3.06	3.14	3.25	4.18

sheet with the object plane of a microscope objective lens. Consequently, it is common practice in µPIV to illuminate the test section with a volume of light and rely on the depth of field of the lens to define the out-of-plane thickness of the measurement plane.

The effective particle image diameter in volume illumination can be derived as follows. Following the analysis of Olsen and Adrian (2000a), using $f^\# \approx 1/2[(n/NA)^2 - 1]^{1/2}$, the effective image diameter of a particle displaced a distance z from the objective plane can be approximated by combining Equations 2.8 and 2.9 and by adding a third term to account for the geometric spreading of a slightly out-of-focus particle

$$d_e = \left[M^2 d_p^2 + 1.49(M+1)^2 \lambda^2 \left[\left(\frac{n}{NA} \right)^2 - 1 \right] + \frac{M^2 D_a^2 z^2}{(s_o + z)^2} \right]^{1/2}, \qquad (2.10)$$

where s_o is the object distance and D_a is the diameter of the recording lens aperture.

The out-of-plane spatial resolution can be determined in terms of the *depth of correlation*. The depth of correlation is defined as the axial distance, z_{corr}, from the object plane in which a particle becomes sufficiently out of focus so that it no longer contributes significantly to the signal peak in the particle image correlation function. Following the analysis of Olsen and Adrian (2000a), the expression of the depth of correlation is derived as follows:

$$z_{corr} = \left[\left(\frac{1 - \sqrt{\varepsilon}}{\sqrt{\varepsilon}} \right) \left[\frac{d_p^2 [(n/NA)^2 - 1]}{4} + \frac{1.49(M+1)^2 \lambda^2 [(n/NA)^2 - 1]^2}{4M^2} \right] \right]^{1/2}. \qquad (2.11)$$

The depth of correlation, z_{corr}, is strongly dependent on numerical aperture, NA, and particle size, d_p, and is weakly dependent on magnification, M. ε represents the relative contribution of a particle displaced a distance z from the object plane compared with a particle located at the object plane. Table 2.2 gives the thickness of the measurement plane, $2z_{corr}$, for various microscope objective lenses and particle sizes. The highest out-of-plane resolution for these parameters is $2z_{corr} = 0.36$ µm for an NA = 1.4, $M = 60$ lens, and particle sizes $d_p < 0.1$ µm. For these calculations, it is important to note that the effective numerical aperture of an oil-immersion lens is reduced when imaging particles suspended in fluids such as water, where the refractive index is less than that of the immersion oil.

TABLE 2.2

Thickness of the Measurement Plane for Typical Experimental Parameters, $2z_{corr}$ (µm)

Particle Size, d_p (µm)	Microscope Objective Lens Characteristics				
	$M = 60$, NA = 1.4	$M = 40$, NA = 0.75	$M = 40$, NA = 0.6	$M = 20$, NA = 0.5	$M = 10$, NA = 0.25
0.01	0.36	1.6	3.7	6.5	34
0.10	0.38	1.6	3.8	6.5	34
0.20	0.43	1.7	3.8	6.5	34
0.30	0.52	1.8	3.9	6.6	34
0.50	0.72	2.1	4.2	7.0	34
0.70	0.94	2.5	4.7	7.4	35
1.00	1.3	3.1	5.5	8.3	36
3.00	3.7	8.1	13	17	49

2.3.1.3.4 Particle Visibility

The ability to obtain highly reliable velocity data depends significantly on the quality of the recorded particle images. In macroscopic PIV experiments, it is customary to use a sheet of light to illuminate only those particles that are within the depth of field of the recording lens. This minimizes background noise resulting from light emitted by out-of-focus particles. However, in μPIV, the small-length scales and poor optical access necessitate the use of volume illumination.

Experiments using the μPIV technique must be designed so that in-focus particle images can be observed despite the background light produced by out-of-focus particles and the test section surfaces. The background light from test section surfaces can be removed by using fluorescent techniques to filter out elastically scattered light (Santiago et al., 1998).

Background light from unfocused particles is not so easily removed but can be lowered to acceptable levels by choosing judiciously proper experimental parameters. Olsen and Adrian (2000a) present a theory to estimate peak particle visibility, defined as the ratio of the intensity of an in-focus particle image to the average intensity of the background light produced by the out-of-focus particles.

Assuming light is emitted uniformly from the particle, the light of a single particle reaching the image plane can be written as

$$J(z) = \frac{J_p D_a^2}{16(s_o + z)^2},$$ (2.12)

where J_p is total light flux emitted by a single particle. Approximating the intensity of an in-focus particle image as Gaussian,

$$I(r) = I_0 \exp\left(\frac{-4\beta^2 r^2}{d_e^2}\right),$$ (2.13)

where the unspecified parameter, β, is chosen to determine the cutoff level that defines the edge of the particle image. Approximating the airy distribution by a Gaussian distribution, with the area of the two axisymmetric functions being equal, the first zero in the airy distribution corresponds to $I/I_0 = \exp(-\beta^2 = -3.67)$ (Adrian and Yao, 1985). Because the total light flux reaching the image plane is $J = \int I(r)\,dA$, Equations 2.12 and 2.13 can be combined yielding (Olsen and Adrian, 2000a)

$$I(r,z) = \frac{J_p D_a^2 \beta^2}{4\pi d_e^2 (s_o + z)^2} \exp\left(\frac{-4\beta^2 r^2}{d_e^2}\right).$$ (2.14)

Idealizing particles that are located at a distance $|z| > \delta/2$ from the object plane as being out of focus and contributing uniformly to background intensity, while particles located within a distance $|z| < \delta/2$ as being completely in focus, the total flux of background light, J_B, can be approximated by

$$J_B = A_v C \left\{ \int_{-a}^{-\delta/2} J(z)\,dz + \int_{\delta/2}^{L-a} J(z)\,dz \right\},$$ (2.15)

where C is the number of particles per unit volume of fluid, L is the depth of the device, and A_v is the average cross-sectional area contained within the field of view.

Combining Equations 2.12 and 2.15, correcting for the effect of magnification, and assuming $s_o \gg \delta/2$, the intensity of the background glow can be expressed as (Olsen and Adrian, 2000a)

$$I_B = \frac{CJ_P LD_a^2}{16M^2(s_o - a)(s_o - a + L)}.$$ (2.16)

Following Olsen and Adrian (2000a), the visibility of an in-focus particle, V, can be obtained by combining Equations 2.10 and 2.14, dividing by Equation 2.16, and setting $r = 0$ and $z = 0$,

$$V = \frac{I(0,0)}{I_B} = \frac{4M^2\beta^2(s_o - a)(s_o - a + L)}{\pi C L s_o^2 (M^2 d_p^2 + 1.49(M+1)^2 \lambda^2 [(n/\mathrm{NA})^2 - 1])}.$$ (2.17)

For a given set of recording optics, particle visibility can be increased by decreasing particle concentration, C, or by decreasing test section thickness, L. For a fixed particle concentration, the visibility can be increased by decreasing the particle diameter or by increasing the numerical aperture of the recording lens. Visibility depends only weakly on magnification and object distance, s_o.

An expression for the volume fraction, V_{fr}, of particles in solution that produce a specific particle visibility can be obtained by rearranging Equation 2.17 and multiplying by the volume occupied by a spherical particle

$$V_{fr} = \frac{2d_p^3 M^2 \beta^2 (s_o - a)(s_o - a + L)}{3 V L s_o^2 (M^2 d_p^2 + 1.49(M+1)^2 \lambda^2 [(n/\mathrm{NA})^2 - 1])} \times 100\%.$$ (2.18)

Reasonably high-quality particle image fields require visibilities of, say, $V \sim 1.5$. For the purpose of illustration, assume that we are interested in measuring the flow at the centerline, $a = L/2$, of a microfluidic device with a characteristic depth, $L = 100$ μm. It is also important to seed the flow so that the particle volume fraction of seed particles is kept below a suitable level. This ensures that the particle loading on the fluid is not too large. Table 2.3 shows the maximum volume fraction of particles that can be in the fluid while maintaining an in-focus particle visibility, $V = 1.5$, for various experimental parameters. Here, the object distance, s_o, is estimated by adding the working distance of the lens to the designed coverslip thickness.

Meinhart et al. (2000b) demonstrated these competing signal-to-noise ratio and spatial resolution issues with a series of experiments using known particle concentrations and flow depths. The measured signal-to-noise ratio (equivalent to particle visibility above) was estimated from particle image fields taken of four different particle concentrations and four different device depths. A particle solution was prepared by diluting $d_p = 200$-nm-diameter polystyrene particles in deionized water. Test sections were formed using two feeler gauges of known thickness sandwiched between a glass microscope slide and a coverslip. The particle images were recorded near one of the feeler gauges to minimize errors associated with variations in the test section thickness, which could result from surface tension or deflection of the coverslip. The images were recorded with an oil-immersion $M = 60\times$, NA = 1.4, objective lens. The remainder of the μPIV system was as described above.

Meinhart et al. (2000b) defined the signal-to-noise ratio of the image to be the ratio of the peak image intensity of an average in-focus particle divided by the average background intensity. The maximum possible spatial resolution is defined as the smallest square

TABLE 2.3

Maximum Volume Fraction of Particles, V_{fr}, (%) While Maintaining an In-Focus Visibility, $V = 1.5$, for Imaging the Center of an $L = 100$ μm Deep Device

Particle Size, d_p (μm)	Microscope Objective Lens Characteristics				
	$M = 60$, NA = 1.4, $s_o = 0.38$ mm	$M = 40$, NA = 0.75, $s_o = 0.89$ mm	$M = 40$, NA = 0.6, $s_o = 3$ mm	$M = 20$, NA = 0.5, $s_o = 7$ mm	$M = 10$, NA = 0.25, $s_o = 10.5$ mm
0.01	2.0E − 5	4.3E − 6	1.9E − 6	1.1E − 6	1.9E − 7
0.10	1.7E − 2	4.2E − 3	1.9E − 3	1.1E − 3	1.9E − 4
0.20	1.1E − 1	3.1E − 2	1.4E − 2	8.2E − 3	1.5E − 3
0.30	2.5E − 1	9.3E − 2	4.6E − 2	2.7E − 2	5.1E − 3
0.50	6.0E − 1	3.2E − 1	1.8E − 1	1.1E − 1	2.3E − 2
0.70	9.6E − 1	6.4E − 1	4.1E − 1	2.8E − 1	6.2E − 2
1.00	1.5E + 0	1.2E + 0	8.7E − 1	6.4E − 1	1.7E − 1
3.00	4.8E + 0	4.7E + 0	4.5E + 0	4.2E + 0	2.5E + 0

TABLE 2.4

The Effect of Background Noise on Image Quality (i.e., Signal-to-Noise Ratio)

Depth (μm)	Particle Concentration (by Volume)			
	0.01%	0.02%	0.04%	0.08%
25	2.2	2.1	2.0	1.9
50	1.9	1.7	1.4	1.2
125	1.5	1.4	1.2	1.1
170	1.3	1.2	1.1	1.0

Source: Meinhart, C.D., Wereley, S.T., and Gray, M.H.B. Volume illumination for two-dimensional particle image velocimetry, in *Measurement, Science and Technology*, 11, 809–814 (2000).

interrogation region that would on average contain three particle images and is directly related to the number concentration of the particle images in each PIV image.

The measured signal-to-noise ratio is shown in Table 2.4. As expected, the results indicate that, for a given particle concentration, a higher signal-to-noise ratio is obtained by imaging a flow in a thinner device. This occurs because decreasing the thickness of the test section decreases the number of out-of-focus particles while the number of in-focus particles remains constant. In general, thinner test sections allow higher particle concentrations to be used, which can be analyzed using smaller interrogation regions. Consequently, the seed particle concentration must be chosen judiciously so that the desired spatial resolution can be obtained while maintaining adequate image quality (i.e., signal-to-noise ratio).

2.3.2 Special Processing Methods for μPIV Recordings

When evaluating digital PIV recordings with conventional correlation-based algorithms or image-pattern-tracking algorithms, a sufficient number of particle images is required in the interrogation window or the tracked image pattern to ensure reliable and accurate measurement results. However, in many cases, especially in μPIV measurements, the particle image density in the PIV recordings is usually not high enough. These PIV recordings are called *low image density* (LID) recordings and are usually evaluated with particle-tracking algorithms. When using particle-tracking algorithms, the velocity vector is determined with only one particle and hence the reliability and the accuracy of the technique are limited. In addition, interpolation procedures are usually necessary to obtain velocity vectors on the desired regular grid points from the random distributed particle-tracking results and therefore additional uncertainties are added to the final results. Fortunately, special processing methods can be used to evaluate the μPIV recordings, so that the errors resulting from the LID can be avoided (Wereley et al., 2002b). In this section, two methods are introduced to improve measurement accuracy of μPIV: by improving the evaluation algorithm and by using a digital image processing technique.

2.3.2.1 Ensemble Correlation Method

For correlation-based PIV evaluation algorithms, the correlation function at a certain interrogation spot is usually represented as:

$$\Phi_k(m,n) = \sum_{j=1}^{q}\sum_{i=1}^{p} f_k(i,j)\cdot g_k(i+m,j+n) \qquad (2.19)$$

where $f_k(i,j)$ and $g_k(i,j)$ are the gray value distributions of the first and second exposures, respectively, in the kth PIV recording pair at a certain interrogation spot of a size $p \times q$ pixels. The correlation function for a singly exposed PIV image pair has a peak at the position of the particle image displacement in the interrogation spot (or window), which should be the highest among all the peaks of Φ_k. The subpeaks, which result from noise or mismatch of particle images, are usually obviously lower than the main peak (i.e., the peak of the particle image displacement). However, when the interrogation window does not contain enough particle images or the noise level is too high, the main peak will become weak and may be lower than some of the subpeaks and, consequently, an erroneous velocity vector is generated. In the laminar and steady flows often measured by μPIV systems, the velocity field is independent of the measurement time. That means the main peak of $\Phi_k(m, n)$ is always at the same position for PIV recording pairs taken at different times while the subpeaks appear with random intensities and positions in different recording pairs. Therefore, when averaging Φ_k over a large number of PIV recording pairs (N), the main peak will remain at the same position in each correlation function, but the noise peaks, which occur randomly, will average to zero. The averaged (or ensemble) correlation function is given as

$$\Phi_{ens}(m,n) = \frac{1}{N}\sum_{k=1}^{N}\Phi_k(m,n). \qquad (2.20)$$

The ensemble correlation requires a steady flow. The concept of averaging correlation functions can also be applied to other evaluation algorithms such as correlation tracking and the MQD method. This method was first proposed and demonstrated by Meinhart et al. (1999b).

The ensemble correlation function technique is demonstrated for 101 LID-PIV recording pairs (Φ_{ens}) in Figure 2.3 in comparison with the correlation function for one of the single-recording pair (Φ_k). These PIV recording pairs are chosen from the flow measurement in a microfluidic biochip for impedance spectroscopy of biological species (Gomez et al., 2001). With the conventional evaluation function in Figure 2.3a, the main peak cannot easily be

(a) $\Phi_k(m, n)$ (b) $\Phi_{\text{ens}}(m, n)$

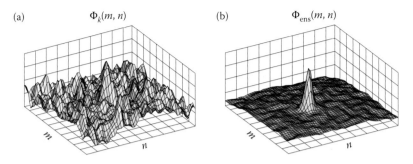

FIGURE 2.3
Effect of ensemble correlation: (a) results with conventional correlation for one of the PIV recording pairs and (b) results with ensemble correlation for 101 PIV recording pairs. (From Wereley, S.T., Gui, L., and Meinhart, C.D., *AIAA J.*, 40, 1047–1055, 2002b. With permission.)

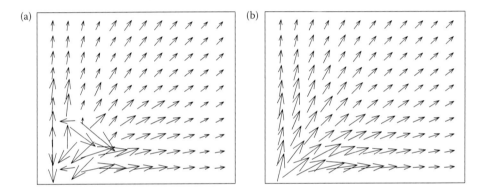

FIGURE 2.4
Comparison of the evaluation function of (a) a single PIV recording pair with (b) the average of 101 evaluation functions. (From Wereley, S.T., Gui, L., and Meinhart, C.D., *AIAA J.*, 40, 1047–1055, 2002b. With permission.)

identified among the subpeaks, so the evaluation result is neither reliable nor accurate. However, the ensemble correlation function in Figure 2.3b shows a very clear peak at the particle image displacement, and the subpeaks can hardly be recognized.

The effect of the ensemble correlation technique on the resulting velocity field is demonstrated in Figure 2.4 with the PIV measurement of flow in the microfluidic biochip. All the obvious evaluation errors resulting from the LID and strong background noise (Figure 2.4a) are avoided by using the ensemble correlation method on the basis of 101 PIV recording pairs (Figure 2.4b). One important note here is that because the bad vectors in Figure 2.4a all occur at the lower left corner of the flow domain, the removal of these bad vectors and subsequent replacement by interpolated vectors will only coincidentally generate results that bear any resemblance to the true velocity field in the device. In addition, if the problem leading to low signal levels in the lower left-hand corner of the images is systematic (i.e., larger background noise), even a large collection of images will not generate better results because they will all have bad vectors at the same location.

2.3.2.2 Removing Background Noise

When using the ensemble correlation technique, a large number of μPIV recording pairs are usually obtained, enabling the removal of the background noise from the μPIV recording

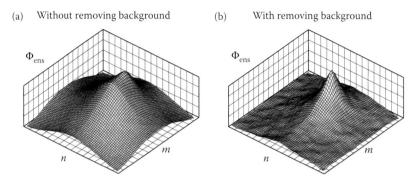

FIGURE 2.5
Ensemble correlation function for 100 image sample pairs (a) without and (b) with background removal. (From Wereley, S.T., Gui, L., and Meinhart, C.D., *AIAA J.*, 40, 1047–1055, 2002b. With permission.)

pairs. One of the possibilities for obtaining an image of the background from plenty of PIV recordings is averaging these recordings (Gui et al., 1997). Because the particles are randomly distributed and quickly move through the camera view area, their images will disappear in the averaged recording. However, the image of the background (including the boundary, the contaminants on the glass cover, and the particles adhered to the wall) maintains the same brightness distribution in the averaged recording because it does not move or change. Another method is building a minimum of the ensemble of PIV recordings at each pixel location because the minimal gray value at each pixel may reflect the background brightness in the successively recorded images (Cowen and Monismith, 1997). The background noise may be successfully removed by subtracting the background image from the PIV recordings.

A data set from a flow in a microchannel is used to demonstrate this point. The size of the interrogation regions is 64×64 pixels, and the total sample number is 100 pairs. The mean particle image displacement is approximately 12.5 pixels from left to right. In one particular interrogation region in the images, the particle images in a region at the left side of the interrogation region look darker than those outside this region. This may result from an asperity on the glass cover of the microchannel.

The ensemble correlation function for the 100 image sample pairs without background removal is given in Figure 2.5a, which shows a dominant peak near zero displacement because the fleck does not move. When the background image is built with the minimum gray value method and subtracted from the image sample pairs, the influence of the asperity is reduced so that the peak of the particle image displacement appears clearly in the evaluation function in Figure 2.5b.

2.4 Examples of Macroscopic and Microscopic PIV Measurements

2.4.1 Phase-Resolved Flow Field Produced by a Vibrating Cantilever Plate

Everyday experience tells us that flow can be generated simply by swinging a plate back and forth. Fanning with a plate is surely the simplest and most popular wind-generation method developed by our predecessors. With the development of the electric motor, these manual plate fans have largely been replaced by electric rotary fans. Electric rotary fans are very convenient—they do not require human power and continuously generate flow as long as electricity is provided. Electric rotary fans are widely used as convection-cooling devices on electrical systems such as CPUs, power supplies, electric circuit boards, and so forth. With the trend in the electronics industry continually toward smaller electronics components, ever smaller rotary fans are needed. However, electric rotary fans are difficult to make small because they are composed of many complex parts. Furthermore, rotary fans are based on permanent magnets, the power of which scales as the cube of the length scale of the system. Hence, MEMS-scale rotary fans, even if they could be produced, would compare very poorly with other types of flow actuation.

To produce fans for smaller applications, a different type of fan should be used and it should be composed of only a few simple parts. One alternative to electric rotary fans for small size ranges is a plate fan driven by a piezoelectric element. As shown in Figure 2.6, a simple piezoelectric fan is composed of a thin metal plate, clamped at one end and free at the other, onto which a piezoelectric material is deposited. The plate is deflected as an electrical potential difference is applied across the piezoelectric material causing a mechanical strain. When an AC potential is applied to the piezoelectric material, especially at the fundamental natural frequency of the cantilever, a flow is generated.

For the efficient design of piezoelectric fan, the mechanism by which a vibrating cantilever generates a flow must be thoroughly understood. It is difficult to understand unsteady flows generated by a vibrating cantilever using time-averaged velocity data alone. Phase-resolved velocity information is required to understand the flow generation mechanism. This study presents both qualitative and quantitative vortex flow fields acquired using smoke flow visualization and PIV measurements.

2.4.1.1 Experimental Setup

Although this study is designed to understand flow generation mechanism of micro- or miniscale cantilever plate, the size of the fan is not too critical (at least for small deflections)

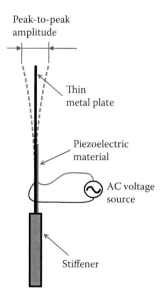

FIGURE 2.6
Schematic of a piezoelectric fan.

because the flow is determined by the vortices after they have separated from the tip of the fan. Figure 2.7 shows a schematic of the experimental setup. The piezoelectric fan is composed of a thin metal plate, the bottom portion of which is coated with a piezoelectric material. The plate is fixed to the apparatus in a cantilevered manner with its bottom end clamped and top end free. The width of the piezoelectric fan, defined as b, is 38.1 mm. The overall length of the flexible cantilever extending beyond its rigidly clamped base is 31 mm and its thickness is 0.13 mm. It is fixed between two endplates: a transparent front wall and a rear wall. Sidewalls are also placed on either side of the cantilever. H represents the distance from the flat plate to the sidewall. Half of the peak-to-peak amplitude of the cantilever tip vibration is represented as h_0, and the maximum tip speed is represented as V_T. The vibrating frequency f_0 studied here is fixed to 180 Hz, the fundamental natural frequency of the plate.

To make sure that no unusual dynamics attributable only to this particular cantilever cause the flow behaviors investigated here, the cantilever deflection is measured using a laser vibrometer (Polytec OFV511/OFV3001). The cantilever deflections are measured in situ. Because the cantilever is a composite of a piezoelectric patch and a thin metal plate, an effective cantilever length is found by minimizing the difference between the measured deflections and the analytical mode shape for the fundamental natural frequency of a uniform cantilever. The effective length, which is defined as c, is 25.4 mm.

Figure 2.8 shows the cantilever deflection as measured along the center of the plate for several phase angles. The solid curves in Figure 2.9 representing the analytical mode shape compare quite well (approximately 3% difference) with the experimentally measured vibration shape. To ascertain whether there are any lateral modes of vibration at the fundamental frequency, the motion at the center of the tip ($z/b = 0$) is compared with the motion at the edge ($z/b = 0.46$) and plotted in Figure 2.9. Clearly, the motion of the tip is very nearly

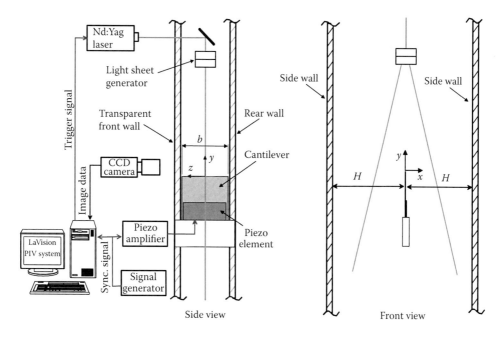

FIGURE 2.7
Schematic of experimental setup. (From Kim, Y.H., Wereley, S.T., and Chun, C.H., *Phys. Fluids.*, 26, 145–162, 2004. With permission.)

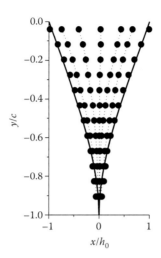

FIGURE 2.8
Shapes of deflected cantilever plate vibrating at 180 Hz for $h_0/c = 0.054$. Solid curves represent the analytical deflection shape at the fundamental natural frequency that most closely agrees with measured deflection shapes. The root mean square difference between the measured x/h_0 and the analytical shape at the maximum deflection is 0.033. (From Kim, Y.H., Wereley, S.T., and Chun, C.H., *Phys. Fluids.*, 26, 145–162, 2004. With permission.)

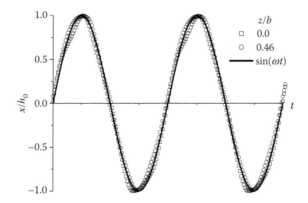

FIGURE 2.9
Deflection trajectories at the center and at the edge of the cantilever tip for $h_0/c = 0.054$. (From Kim, Y.H., Wereley, S.T., and Chun, C.H., *Phys. Fluids.*, 26, 145–162, 2004. With permission.)

sinusoidal and shows no lateral vibration modes at this frequency and amplitude. The tip deflection is measured at all five amplitudes used in this work to determine whether the above behavior holds for all the measurements to be presented here. Although slightly asymmetric deflections of the cantilever plate are shown in Figure 2.10a, the tip moves almost harmonically for all amplitudes according to the time series data (Figure 2.10a), as well as its Fourier transform (Figure 2.10b). The Fourier transform exhibits only one main peak and very little in the way of harmonics.

The cantilever flow fields are studied quantitatively using PIV and qualitatively using a smoke flow visualization method. A studio fog machine (Rosco, Inc.) is used for PIV particle seeding. A large transparent chamber is built to contain the seed particles. The

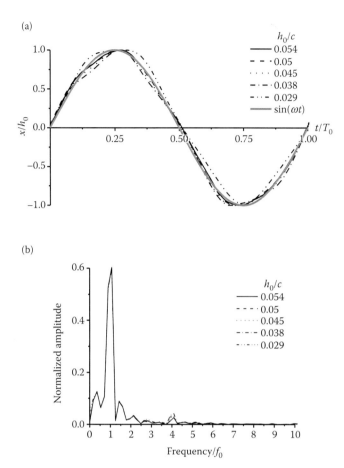

FIGURE 2.10
Deflections at the center of the cantilever tip for different amplitudes. (a) Trajectories of the tip; (b) FFT result of the trajectories. (From Kim, Y.H., Wereley, S.T., and Chun, C.H., *Phys. Fluids.*, 26, 145–162, 2004. With permission.)

cantilever plate and the two sidewalls are set up in this chamber. In the smoke visualizations, an array of smoke streams is continually supplied upstream of the cantilever plate throughout the visualization. In the PIV measurements, after the chamber is filled with fog, mixed well, and allowed to settle for about a minute, images are recorded using a 12-bit gray-level 1280- × 1024-pixel interline transfer CCD camera (LaVision Flowmaster 3S) and 60-mm Macro Lens (Nikon). A dual-cavity 120 mJ Nd:YAG Laser (NewWave Gemini) is used to illuminate the flow.

Davis image acquisition and analysis software (LaVision) is used for capturing images, and Edpiv is used for image interrogation and velocity evaluation. Edpiv is a custom-written PIV interrogation package written by Dr.-Ing. Gui and partially developed at the Microfuidics Laboratory at Purdue University. A multipass adaptive window shifting (32 × 32 → 16 × 16 pixels) and deformation technique known as the *central difference image correction* (CDIC) were used to evaluate the experimental images. The CDIC technique is a cross-correlation-based method by which Wereley et al. (2001) obtained very accurate results in the evaluation of flow fields having high-velocity gradients or vortices.

An average background removal technique is used to remove background noise images caused by blooming of CCD pixels and dirt particles occasionally attached to the transparent front and rear walls.

Phase-resolved velocity fields are acquired for eight different phases (steps of 45°) through the cycle of vibration. The electrical signal supplied to the piezoelectric amplifier to drive the cantilever is also used as an external trigger signal by the PIV timing electronics that control the acquisition timing of the camera and illumination timing of the laser. By setting a particular time delay between the external trigger and the acquisition time of the camera, any desired phase of the cantilever can be synchronized. Figure 2.11 shows the relationship of the phase angles to the deflection of the cantilever tip.

The two main parameters that determine the flow are the tip amplitude, h_0, and the maximum tip velocity, V_T. The Reynolds number Re_h, defined as $V_T h_0 / \nu$, is the dimensionless parameter that determines the flow field. The vibrating frequency f_0 studied here is fixed to 180 Hz, the fundamental natural frequency of the plate. Flow-field measurements are done for tip deflections of $h_0/c = 0.054, 0.05, 0.045, 0.038$, and 0.029 to observe the effect of the tip vibration amplitude. The corresponding Reynolds numbers, Re_h, for these amplitudes are 146, 126, 101, 72, and 43, respectively.

2.4.1.2 Sequence of Vortex Generation at the Tip

Smoke visualization of the time sequence of vortex generation at the tip of the vibrating cantilever is shown in Figure 2.12. A pair of counterrotating vortices develops at the cantilever tip during a single cycle of vibration. Although these smoke visualizations show the

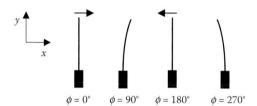

$\phi = 0°$ $\phi = 90°$ $\phi = 180°$ $\phi = 270°$

FIGURE 2.11
Definition of the phase angles of the deflected cantilever tip.

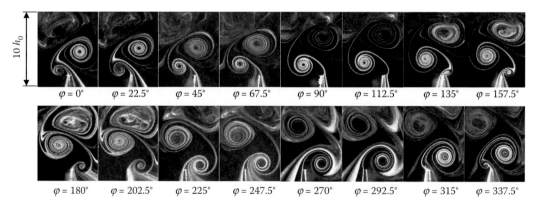

$\varphi = 0°$ $\varphi = 22.5°$ $\varphi = 45°$ $\varphi = 67.5°$ $\varphi = 90°$ $\varphi = 112.5°$ $\varphi = 135°$ $\varphi = 157.5°$

$\varphi = 180°$ $\varphi = 202.5°$ $\varphi = 225°$ $\varphi = 247.5°$ $\varphi = 270°$ $\varphi = 292.5°$ $\varphi = 315°$ $\varphi = 337.5°$

FIGURE 2.12
Smoke flow visualization around the cantilever tip for $Re_h = 146$. (From Kim, Y.H., Wereley, S.T., and Chun, C.H., *Phys. Fluids.*, 26, 145–162, 2004. With permission.)

sequence of vortex generation clearly, they do not provide quantitative velocity information. The locations and strengths of vortices cannot be identified exactly because streak lines in unsteady flow fields are different from streamlines. Moreover, the flow could be affected by the small steady mass flux injected into the upstream reservoir by the smoke seeder during the visualization.

Flow fields measured using PIV can be used to more precisely locate vortex centers and provide quantitative velocity information. Phase-resolved velocity fields of eight different phases are shown in Figure 2.13. The time sequence of counterrotating vortex pair development revealed by the PIV measurements corresponds very well to the smoke visualization results. High-velocity regions inclined from the vertical are observed between neighboring counterrotating vortices. The maximum velocity in the high-velocity region is almost four times the maximum cantilever tip speed. The net effect of these unsteady vortices is explained in the following paragraph.

2.4.1.3 Effect of the Tip Vibration Amplitude

Cantilever tip vibration amplitude is one parameter that can be controlled easily by varying the voltage supplied to the piezoelectric element. Because the cantilever vibration frequency is held constant, reducing the tip vibration amplitude reduces the maximum tip speed, which is given by $V_T = \omega h_0 = 2\pi f_0 h_0$.

Velocity fields averaged over one cycle of the tip deflection for five different amplitudes of the tip are shown in Figure 2.14. A high-velocity region is formed between these two time-averaged vortices, and the flow resembles a jet ejecting from the tip of the cantilever. Slightly asymmetric flow fields are found for all the Reynolds numbers. The flow fields at $Re_h = 72$ and 43 are more asymmetric than those at the other Reynolds numbers.

Farther downstream of the tip, the velocity fields become less similar to each other. For each of the Reynolds numbers, the flow acquires a preferred inclination, drifting away from the centerline of $y = 0$ as the downstream distance increases. The inclined flows are formed by small disturbances such as slight experimental imperfections, misalignments, or slightly distorted cantilever deflections.

The average flow over one period of oscillation of pitching airfoil, which is similar to the vibrating cantilever, is known to be unstable and to act as a narrow-band amplifier of perturbations (Anderson et al., 1998). The asymmetric flow patterns observed in Figure 2.14 seem to be caused by this unstable flow, which could amplify even slight disturbances in the wake of a vibrating cantilever. Jones at al. (1998) showed that an inclined flow field is formed in an airfoil wake when the airfoil oscillation speed is much larger than the free stream velocity (quite comparable with the conditions in the present investigation) using flow visualizations and numerical computations. An inclined flow was found to have dual-mode patterns in which the inclination direction randomly alternates between the modes, suggesting that small disturbances might be sufficient to switch the mode. In their numerical result, the mode (i.e., flow inclination direction) was determined by the initial conditions of the numerical method and remained fixed throughout the simulation, suggesting that the direction of flow inclination does not switch in the absence of random disturbances.

Analogous to this result, in the present work, the inclined flow fields for $Re_h = 72$ and 43 could be formed by small disturbances, such as a slight misalignment in the experimental apparatus or slightly asymmetric deflection of the cantilever plate because of the slightly distorted harmonic motion as shown in Figure 2.10a. The trajectory curve of the tip is smoother near the maximum value of x/h_0 than near the minimum of x/h_0 in case of

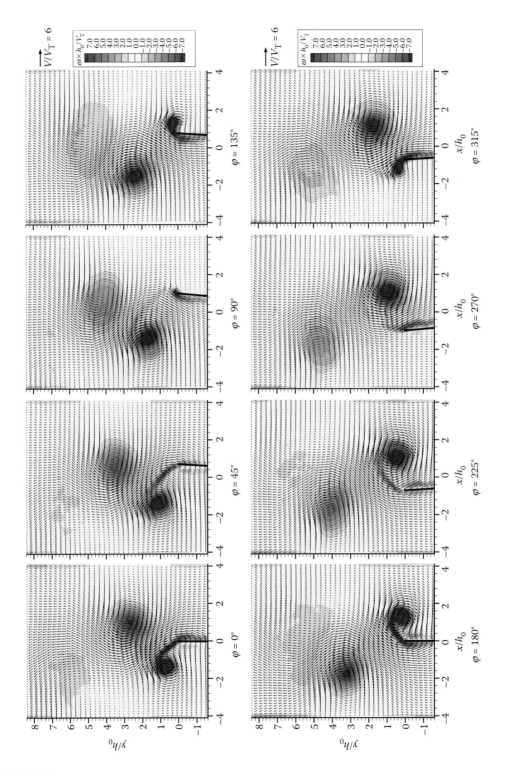

FIGURE 2.13
Phase-resolved velocity fields for $Re_h = 146$. (From Kim, Y.H., Wereley, S.T., and Chun, C.H., *Phys. Fluids.*, 26, 145–162, 2004. With permission.)

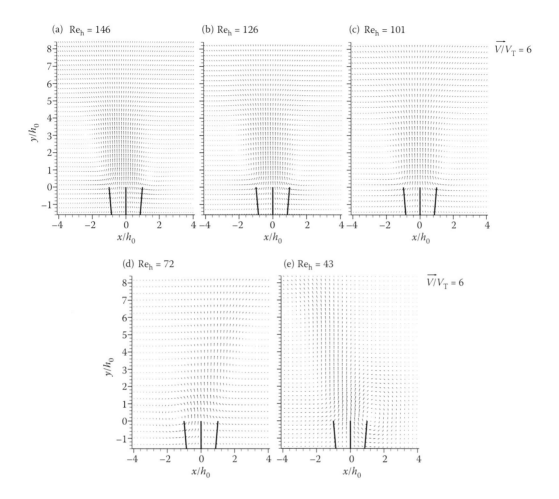

FIGURE 2.14
Velocity fields averaged over one cycle of cantilever deflection for five different amplitudes. (a) $Re_h = 146$; (b) $Re_h = 126$; (c) $Re_h = 101$; (d) $Re_h = 72$; (e) $Re_h = 43$. (From Kim, Y.H., Wereley, S.T., and Chun, C.H., *Phys. Fluids.*, 26, 145–162, 2004. With permission.)

$h_0/c = 0.038$, whereas the opposite occurs in case of $h_0/c = 0.029$. This asymmetric motion of the tip seems to form the different inclined direction between the two cases of $h_0/c = 0.038$ and 0.029. These disturbances remain fixed throughout the measurement of a single h_0 (or Reynolds number) because the experimental apparatus and the slightly distorted harmonic motion of the cantilever are not changed except when the value of h_0 is changed. These slight imperfections or misalignments are unavoidable in any real experimental apparatus and can be altered while assembling the apparatus after cleaning between experimental runs or even just on changing the driving voltage to the piezoelectric element. More flow-field measurements and analysis are presented by Kim et al. (2004).

2.4.1.4 Conclusion

Flow generation mechanism induced by vibrating cantilever plates is investigated by phase-resolved velocity measurements using smoke visualization and PIV techniques. A pair of counterrotating vortices develops at the cantilever tip during a single cycle of vibration. A high-velocity region is formed between these two counterrotating vortices.

A high-velocity region is formed in time-averaged flow field, and the flow resembles a jet ejecting from the tip of the cantilever The nondimensionalized velocity fields for five different tip vibration amplitudes (equivalent to five different Reynolds numbers) are nearly identical and symmetric near the cantilever tip, but asymmetric flows are formed and the nondimensionalized velocity fields are no longer identical further downstream of the tip. The inclined flows in the present work are formed by small disturbances such as slight experimental imperfections, misalignments, or slightly distorted cantilever deflections.

2.4.2 Entrance Length Measurement in Microchannel Flow

The entrance region in a channel flow is very important because the transport properties of the flow, such as the pressure gradient and the heat transfer coefficient, show drastic change in this region. Entrance length is most frequently defined as the axial distance required for the centerline velocity to reach 99% of the fully developed centerline velocity, usually assuming a uniform flow at the channel entrance (Shah and London, 1978). A uniform entrance velocity profile is, however, seldom achieved in microchannels because of flow development in the inlet region (Beavers et al., 1970) and the velocity overshoot caused by the abrupt velocity gradients because of the sharp turn of the flow at the entrance (Sparrow and Anderson, 1977) and nonslip condition at wall (Shah and London, 1978). For moderate Reynolds number flows, because the dimensionless entrance length, $L^+(=L_e/D_h)$, is approximately proportional to Reynolds number, the entrance length correlation is calculated from a boundary-layer-type governing equation and the entrance length is independent of inlet velocity profile (Atkinson et al., 1969; Shah and London, 1978). For a circular duct, the dimensionless entrance length L^+ is given as ~6% of the Reynolds number (Fox and McDonald, 1998) when $Re_D > 400$ (Shah and London, 1978). For low Reynolds number flows, however, the axial diffusion of vorticity determines the velocity development in the upstream reservoir and in the entrance (Vrentas et al., 1966) and therefore the entrance length depends on inlet velocity profiles.

Entrance length correlations accounting for the nonlinear relationship between L^+ and Reynolds number were proposed by Atkinson et al. (1969) and Chen (1973) for a tube and parallel plates to be

$$L^+ = C_1 + C_2 Re_D, \tag{2.21}$$

$$L^+ = \frac{C_1}{C_2 Re_D + 1} + C_3 Re_D, \tag{2.22}$$

where C_1, C_2, and C_3 are appropriate constants as listed in Table 2.5. Both correlations show that when the Reynolds number approaches zero, there exists a constant portion of entrance length. Chen's type of correlation indicates that the entrance length has a rational relation with the Reynolds number when the Reynolds number is small. When the

TABLE 2.5

Constants in Equations 2.21 and 2.22

Equations	Channel Type	C_1	C_2	C_3
Atkinson et al. (1969)	Circular	0.59	0.056	–
	Parallel	0.625	0.044	–
Chen (1973)	Circular	0.6	0.035	0.056
	Parallel	0.63	0.035	0.044

Reynolds number becomes large, both types of correlations are dominated by a proportional relationship between the dimensionless entrance length and the Reynolds number.

In a rectangular channel, an exact solution does not exist because of nonlinear terms in the momentum equations, and further the flow field in rectangular channels is more complicated than in circular channels because the flow field depends on two cross-sectional coordinates, that is, a function of aspect ratio. The correlations for a circular channel as shown in Equations 2.21 and 2.22 are often used with the hydraulic diameter approach. This approximation gives acceptably accurate results for rectangular channels with the aspect ratios $0.25 < \alpha < 4$ (Fox and McDonald, 1998).

In a small-scale channel, the importance of entrance length study can be emphasized further because the entrance length can be a significant portion of the total length of the channel. Planar microfabrication techniques used for a small-scale channel fabrication generally constrain the reservoir upstream of the microchannel to be of the same height as the channel—the marked departures from the geometries used at macroscopic length scales to study entrance length. Therefore, the flow conditions upstream of the channel entrance cannot be simply assumed as a uniform velocity profile. Further, such upstream flow condition makes two channels with aspect ratios of H/W and W/H, respectively, classified into totally different two inlet geometries.

In this section, the entrance lengths of two different aspect ratio microchannels are investigated for low Reynolds numbers using μPIV. The two channels yield the reciprocal aspect ratios (H/W and W/H) of ~2.75 (MC-I) and ~0.37 (MC-II), respectively. These aspect ratios must be treated separately because of the planar upstream reservoir geometry. The entrance lengths are found by comparing the widthwise velocity profile as a function of the axial distance with the fully developed velocity profile and the local centerline velocity with the fully developed centerline velocity.

2.4.2.1 Microchannel Fabrication

To ensure that the microchannels have sufficient length to fully develop, the microchannels should have large L/D ratios. One microchannel (MC-I) is fabricated from transparent acrylic using conventional machining, a precision sawing technique, and measures 120.0 mm (L) × 0.252 mm (W) × 0.694 mm (H). The other microchannel (MC-II) is fabricated with silicon using DRIE and measures 5020 μm (L) × 104.6 μm (W) × 38.6 μm (H). MC-I was fabricated from transparent acrylic. Figure 2.15a shows the fabrication procedure. The microchannel groove and the cover plate are made using conventional machining—a precision sawing technique. The 1.6-mm-thick cover plate is laminated with a 25.4-μm-thick pressure-sensitive adhesive layer (Adhesive Research Inc.) having an optical transmission rate of more than 95%. This plate is bonded on top of the microchannel groove. Further, this bonded piece is compressed by an external clamp to ensure that the channel is leak proof. Figure 2.15b shows the completed microchannel structure. Two deep reservoirs or plenums, leading to the shallow plenums, are built at both ends of the microchannel to ensure even distribution of the flow. The microchannel dimension is optically measured and found as 252 ± 6 and 694 ± 7 μm, respectively.

The fabrication procedure of MC-II is shown in Figure 2.16. First, 525-μm-thick P-type (100) 4-in. silicon wafers were thermally oxidized by 0.6 μm on both sides. The microchannels and pressure measurement reservoirs on the front side were defined by photolithography. Then the DRIE process (Cornell University) was used to form the microchannels and pressure measurement reservoirs. The wafers were oxidized again, and lithography using AZ 4620 was done on the backside. AZ 4620 is a thick photoresist used as a barrier layer

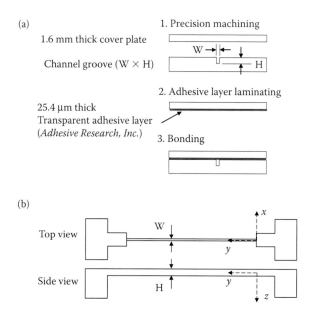

FIGURE 2.15
Fabrication procedure and geometry of MC-I: (a) fabrication process; (b) microchannel geometry. (From Lee, S.-Y., Jang, J., and Wereley, S.T., *Microfluid. Nanofluidics.*, 5, 1–12, 2008. With permission.)

for the backside DRIE process, which etched through the wafers to the bottom surfaces of the reservoirs, connecting the reservoirs to the backside of the chip where tubes were epoxy bonded to supply flow, remove flow, and measure pressure. After the remaining photoresist and oxide were removed, anodic bonding was done with 1-mm-thick Pyrex 7740 glass wafers at 350°C and 1000 V. Finally, the bonded wafers were diced (Figure 2.16a). MC-II consists of the main microchannel, inlet and outlet reservoirs, and five reservoirs for measuring pressure along the main microchannel. It has short narrow channels connecting the main microchannel with the three pressure reservoirs (Figure 2.16b). The five pressure reservoirs are located at 0 μm (inlet) and at 260, 1670, 3350, and 5020 μm (outlet) from the inlet. In the current study, the 5020-μm channel is tested, and the cross section is found to be 104.6 μm (W) × 38.6 μm (H).

2.4.2.2 The μPIV System

Figure 2.2 shows a schematic of the μPIV system (Santiago et al., 1998; Wereley et al., 1998) used to make the measurements. A two-cavity frequency-doubled Nd:YAG laser (New Wave Inc.) is used for the illumination. The wavelength and the pulse width are 532 nm and approximately 3–5 ns, respectively. The laser beam is delivered into the inverted epi-fluorescent microscope (Nikon, TE200) through the beam expander assembly, which is located between the laser aperture and the microscope back aperture. This beam expander assembly is carefully designed for laser beam to have characteristics similar to that of the original mercury lamp. A negative and a positive lens in a Galilean telescope arrangement are used to expand the beam, and a 5° diffuser is located between two lenses to disorder the collimation of the laser beam so that laser focal points do not damage internal optics. The laser beam is guided to the flow field of the microchannel device by passing through an epifluorescence filter cube and an objective lens. The filter cube, located below

(a)

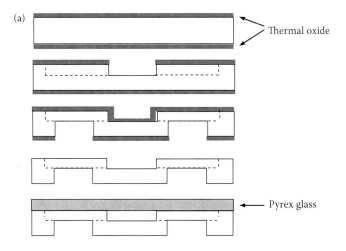

Thermal oxide

Pyrex glass

(b)   The holes made by DRIE for pressure measurements along the microchannel

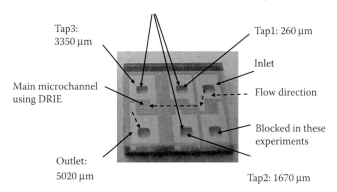

Tap3: 3350 µm

Tap1: 260 µm

Inlet

Main microchannel using DRIE

Flow direction

Blocked in these experiments

Outlet: 5020 µm

Tap2: 1670 µm

FIGURE 2.16
The schematics of (a) fabrication procedure and (b) photo of MC-II. (From Jang, J. and Wereley, S.T., *J. Micromech. Microeng.*, 16, 493–504, 2006. With permission.)

the objective lens, is an assembly of the exciter, emitter, and dichroic mirror. Because the dichroic mirror only transmits light in the range >585 nm, the beam is redirected to the objective lens. The beam coming out the objective lens illuminates a large volume of fluid within the microchannel in which the seeding particles are suspended.

Fluorescent particles (Duke Scientific Co., Palo Alto, CA) are used for flow seeding. For MC-I and MC-II, 1- and 0.69-µm-diameter polystyrenes were used, respectively. The fluorescent particles absorb the illuminating laser beam ($\lambda \sim 532$ nm) and emit a longer wavelength ($\lambda \sim 610$ nm). The signal from the measurement region includes the emitted light from both in-focus and out-of-focus particles and the reflection from the background. The reflection from the background is eliminated by the emitter filter and the dichroic mirror, while both the focused and unfocused particle images are imaged on the interline transfer CCD camera. After a specified time delay Δt, the same process as described above is performed to acquire the second image frame for the cross-correlation-based interrogation. Two different CCD cameras were used for MC-I and MC-II. For MC-I, Flow Master 3s, LaVision, Inc., was used. The camera resolution is 1280×1024 pixel². The field of view

$428.8 \times 343.0 \, \mu m^2$ is produced using a 20× objective lens (NA = 0.45; Nikon, Plan Fluor). For MC-II, Imager Intense, LaVision, Inc., is used. The camera resolution is 1376×1040 pixel^2, producing a field of view of $221.9 \times 167.7 \, \mu m^2$ with a 40× objective lens (NA = 0.60; Nikon, Plan Fluor).

The particle concentration must be considered carefully in μPIV because the background noise keeps increasing as the particle concentration increases, whereas the particle image intensity remains about same value (Meinhart et al., 1999b). Because MC-I has a large depth (694 μm), the ratio of the out-of-focus volume to focused volume is also large. The particle concentration is optimized to be 0.038% (by volume) by balancing the valid detection rate and the background noises from out-of-plane particles. For MC-II, the fluorescent particle concentration is 0.057% (by volume).

2.4.2.3 *Experimental Procedure*

The flow was pumped into the inlet of the microchannels using a syringe pump (Harvard Apparatus, 22). Because the syringe pump works with a worm gear and stepper motor, its actuation is not continuous at low speeds. Also, the friction between the piston and syringe wall is large when flow rate is small. The velocity fluctuations for each Reynolds number show are less than 3% and 4% in transverse and axial components, respectively, whereas the axial velocity fluctuation for $Re_D = 1$ is slightly lower than 9%. Consequently, the pump stability is better than 3%–4%, except for the slowest speed measured.

For MC-I, experiments were performed at eight Reynolds numbers, $Re_D = 1, 10, 20, 30, 40, 50, 70,$ and 100. Seventy image pairs, each showing the entire entrance region of the channel, were acquired. In the case of MC-II, eight Reynolds numbers were also chosen: 5, 10, 18, 24, 36, 48, 62, and 76. At each Reynolds number, 125 image pairs were acquired. Optical access was available from the narrow side (i.e., x–y plane) and wide side for MC-I and MC-II, respectively, at the middle planes of the microchannel depths. The evaluation accuracy was improved by using advanced interrogation algorithms, that is, ensemble average correlation method (Meinhart et al., 1999b) and CDIC (Wereley and Gui, 2003). Also a specially developed image processing technique, the μPIV image filter (Gui et al., 2002), is used. The 32- × 32-pixel interrogation windows were used.

2.4.2.4 *Results and Discussion*

2.4.2.4.1 *Velocity Profiles*

Figure 2.17 shows velocity profile development for Reynolds numbers (a) 1, (b) 20, (c) 40, and (d) 70 in MC-I. Each velocity profile represents the average of seven neighboring axial measurement locations. Normalization was performed with the fully developed centerline velocity. For all Reynolds numbers, the velocity fields develop quickly and the last four profiles (filled symbols) do not show significant differences.

When the Reynolds number is small, the flow develops very quickly because the vorticity diffusion is greater than the momentum. This can be seen with the profiles at $y = 17.4 \, \mu m$. As the Reynolds number increases, the inlet velocity profiles become more blunt, and for $Re_D = 70$, the inlet profiles are rather close to the inviscid core flat profiles. In spite of this, the overall development lengths in this Reynolds number range are shorter than those predicted by existing correlations, except for $Re_D = 1$. For example, the entrance lengths for Reynolds numbers from 10 to 100 are expected to vary from 0.372 to 2.0 mm with Chen's correlation, which means the entrance lengths are longer than one field of view for

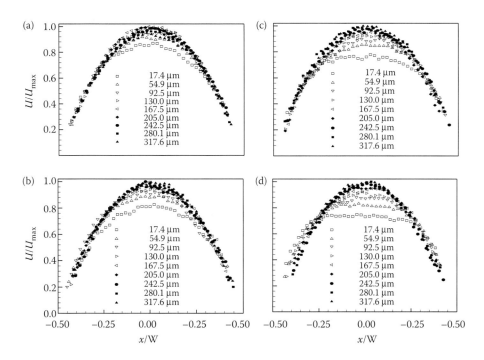

FIGURE 2.17
Normalized velocity profiles for Reynolds numbers (a) 1, (b) 20, (c) 40, and (d) 70 in MC-I. The last four profiles are shown with filled symbols. (From Lee, S.-Y., Jang, J., and Wereley, S.T., *Microfluid. Nanofluidics.*, 5, 1–12, 2008. With permission.)

$Re_D > 20$. When the Reynolds number is small, $Re_D = 1$, the constant portion of the entrance length correlation dominates the total entrance length.

Figure 2.18 shows the normalized velocity profiles in MC-II from the entrance to the fully developed region at Reynolds numbers (a) 18, (b) 36, and (c) 62. The normalization was performed in same manner as for Figure 2.17. The solid lines indicate the fully developed theoretical velocity profiles. The maximum velocities at the inlet are to the left of the center in both cases. This skewed inlet flow is due to the asymmetric flow path from the reservoir to the channel entrance. These asymmetric results at the inlet were also shown by Zhao (2003) using ANSYS simulations. The skewness decreases as the Reynolds number increases. In MC-II, a slight velocity overshoot is seen in the Reynolds number 18 and becomes more significant as Reynolds number increases. Fully developed profiles are observed within one field of view as in MC-I.

2.4.2.4.2 Centerline Velocity

The entrance length was investigated quantitatively using a curve fit of the normalized centerline velocity as a function of the dimensionless axial distance from the microchannel entrance. An exponential decaying function is used for curve fitting the centerline velocity development as shown,

$$\frac{U_{CL}}{U_{CL,fd}} = N_0 - N_1 e^{[-N_2 y/(Re_D D_h)]},$$

(2.23)

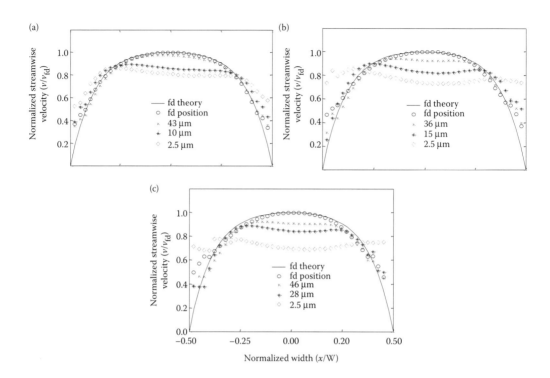

FIGURE 2.18
Normalized velocity profiles for Reynolds numbers (a) 18, (b) 36, and (c) 62. The solid lines indicate the theoretical fully developed velocity profiles. (From Lee, S.-Y., Jang, J., and Wereley, S.T., *Microfluid. Nanofluidics*, 5, 1–12, 2008. With permission.)

where N_0, N_1, and N_2 are positive constants, and the dimensionless entrance lengths were found with the condition $U_{CL}/U_{CL,fd} = 0.99 N_0$ (Lee et al., 2002). Figures 2.19 and 2.20 show the centerline developments in MC-I and MC-II, respectively. The data sets for the various Reynolds numbers do not collapse onto a universal curve using the data reduction variable $y/(Re_D D_h)$—for either microchannel. This is because the flow fields in the low Reynolds number range do not have a strong dependence on Reynolds number. In MC-I, the Reynolds numbers 1 and 10 show the strongest deviations, and in MC-II, Reynolds numbers 5 and 10 show the strongest deviations.

As Reynolds number increases, the initial slopes of data sets increase and the dimensionless entrance length, $L_e/(Re_D D_h)$, decreases. For MC-I, for example, $L_e/(Re_D D_h)$ is ~0.0705 for $Re_D = 10$ and 0.0285 for $Re_D = 20$. Using 99% of the fully developed centerline velocity, the entrance lengths were found to be $y/D_h = 0.6194, 0.7054, 0.5695, 0.7191, 0.8604, 1.0080,$ 1.0314, and 1.358. For $Re_D = 1$, there are large fluctuations after the dimensionless entrance length at approximately 0.6. For $Re_D = 100$, the entrance length is found at approximately 500 μm, which is slightly out of one field of view.

Figure 2.20 shows the curve fit results for MC-II. The normalized entrance lengths, y/D_h, are found to be 1.12, 1.24, 1.26, 1.28, 1.49, 1.87, 2.1, and 2.43 for each Reynolds number. Although Reynolds numbers do not exactly match with MC-I, the y/D_h values for MC-II are two or three times as large as the dimensionless entrance length in MC-I.

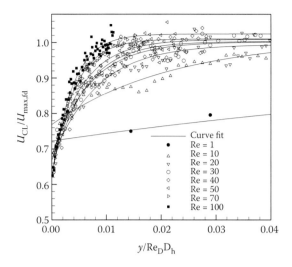

FIGURE 2.19
Normalized centerline velocities for various Re_Ds in MC-I (aspect ratio, H/W, 2.75). (From Lee, S.-Y., Jang, J., and Wereley, S.T., *Microfluid. Nanofluidics.*, 5, 1–12, 2008. With permission.)

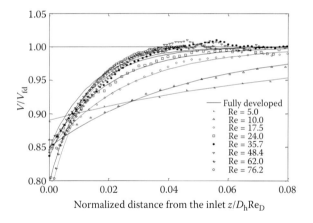

FIGURE 2.20
Normalized centerline velocity for various Re_Ds in MC-II (aspect ratio, H/W, 0.37). (From Lee, S.-Y., Jang, J., and Wereley, S.T., *Microfluid. Nanofluidics.*, 5, 1–12, 2008. With permission.)

2.4.2.4.3 Entrance Length Correlation

New entrance length correlations are attempted for the current experiments in the form of Atkinson's (see Equation 2.21 and Figure 2.21a) and Chen's (see Equation 2.22 and Figure 2.21b) correlations. Also, the existing entrance length correlations, [Atkinson et al. (1969) and Chen (1973)] are compared with the current experimental data.

Overall, Figure 2.21 shows that the existing correlations do not show good agreement with the experimental results. For Figure 2.21a, the fitting functions have the slopes of ~0.008 and ~0.018 for MC-I and MC-II, respectively, which indicates the weak Reynolds number dependence, and the constant portions are ~0.55 and ~0.96, respectively. Both slopes are much smaller than those of Atkinson's correlation (i.e., 0.056 for circular

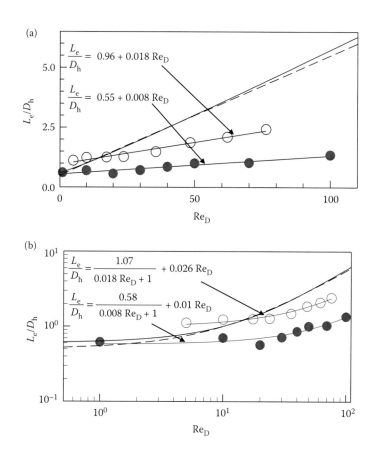

FIGURE 2.21
Comparisons of the entrance length correlations. ● and ○ indicate the experimental results of MC-I and MC-II, respectively. (a) Curve fits in the form of Atkinson's correlation. (b) Curve fits in the form of Chen's correlation. - - - - and —— indicate Atkinson's and Chen's correlations, respectively. (From Lee, S.-Y., Jang, J., and Wereley, S.T., *Microfluid. Nanofluidics.*, 5, 1–12, 2008. With permission.)

channel or 0.044 for parallel plates). The constant portion of MC-I is close to that of Atkinson's correlation (i.e., 0.59 for circular channel or 0.625 for parallel plates). For MC-II, the constant portion is found to be 0.96, which is greater than any of the established correlations. Constant entrance length in MC-II is greater than that in MC-I due to the mixed effect of the aspect ratio and the planar reservoir geometry. Because the aspect ratio (H/W) of MC-II is 0.37, the widthwise vorticity propagation should take longer than in the MC-I, whereas the vorticity should be fully developed in depthwise direction because of the planar reservoir geometry. For Figure 2.21b, the Reynolds number coefficients in the denominator of the first term, that is, 0.008 and 0.018, are from the low Reynolds number linear fit of Figure 2.21a. The estimates of the constant portion entrance lengths are close to the values of Figure 2.21a. As the Reynolds number increases, the second terms on the right-hand side, that is, slopes, dominate the entrance length. The slopes are found to be 0.010 and 0.026 for MC-I and MC-II, respectively, compared with Chen's correlations, which give the slope of 0.056 for circular channel and 0.044 for parallel plates.

2.4.2.5 Conclusion

The entrance length of microchannels with planar upstream reservoir geometries has been studied using µPIV experiments for a broad Reynolds number range, $1 < Re_D < 100$. Two microchannels with reciprocal aspect ratios of 2.75 and 0.37 are used for the experiments. The dimensionless entrance length for each Reynolds number was found using a decaying exponential function curve fit on the basis of the entrance length definition. From both the velocity profiles and the centerline velocities, the entrance lengths were found to have significant differences compared with those predicted by the existing entrance length correlations.

Entrance length correlations have been found in the form of Atkinson's and Chen's correlations. For both microchannels, the correlations showed weaker dependence on Reynolds number, that is, slope, than the existing correlations in the linear portion of the entrance length. The slopes were only 18% and 46% of Chen's correlation for MC-I and MC-II, respectively. The constant portion of entrance length agrees approximately in MC-I with the existing correlations, whereas MC-II shows a greater value than the estimates from Atkinson's and Chen's correlations. Thus, for MC-II, the experimental results are greater than the estimates when the Reynolds number is less than 15. The difference in correlations of MC-I and MC-II is no doubt due to the aspect ratio. Even if two microchannels have reciprocal aspect ratios, they must be considered completely different aspect ratios because of the planar inlet geometries. The results indicate that the planar reservoir geometry and the aspect ratio have mixed effects on the entrance lengths and must be considered together. Further, the depthwise velocity profile at the inlet needs to be measured to investigate the effect of the planar plenum geometry more thoroughly.

2.4.3 Wall Shear Stress Measurements

2.4.3.1 Wall Shear Stress in the Pulsating Blood Flows

In blood flow studies, the wall shear stress is an important parameter in fundamental areas of angiogenesis and cardiology. The endothelial cells show various responses to the shear stress, and such responses are crucial to the development of the cardiovascular system and for the proper functioning of an organism (Hierck et al., 2008). Medical and biological fluid dynamic researches require a nonintrusive technique that has both spatial and temporal resolving power. Blood flows are, however, highly unsteady because of pulsations associated with blood pumping. Although most blood flow fields are at lager scales compared with the common microfluidic applications, µPIV has been successfully used for velocity profile measurements in pulsating blood flows (Tsukada et al., 2000; Sugii et al., 2002; Hove et al., 2003; Vennemann et al., 2006; Hierck et al., 2008).

Because the wall shear stresses are extracted from wall-normal velocity gradients, very detailed near-wall flow velocity measurements are required to take spatial derivative. Vennemann et al. (2006) investigated the possibilities of such accurate measurements *in vivo* in the so-called vitelline network of a chick embryo using µPIV as shown in Figure 2.22. An ultrasound Doppler velocimeter and an additional computer system for trigger were used to trigger the timing unit with the heartbeat of the embryo. Part of the egg shell was removed to establish optical access to the embryo. Bio-inert, fluorescent liposomes with a nominal diameter of 400 nm were seeded as a tracer. Because of small dimension and neutral buoyancy of fluorescent liposomes, velocity gradient close to the wall was able to be resolved. To provide time-resolved ensemble-averaged velocity distribution in the developing ventricle and atrium of the embryo at different stages within the cardiac cycle, an acoustic pulsed Doppler velocimeter probe was used. The overview

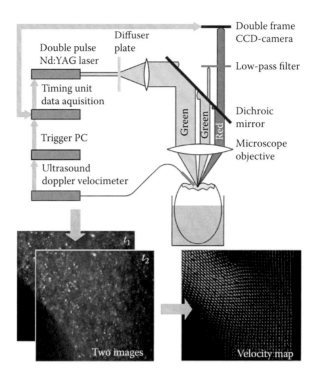

FIGURE 2.22
The µPIV setup, using a fluorescence microscope. A dichroic mirror and a low-pass filter allow only the light emitted by the fluorescent tracer particles to reach the camera. (From Vennemann, P., Kiger, K.T., Lindken, R., Groenendijk, B.C.W., Stekelenburgde Vos, S., Ten Hagen, T.L.M., et al. *J. Biomech.*, 39, 1191–1200, 2006. With permission.)

of the measured velocity distributions at different points in the embryonic cardiac cycle is shown in Figure 2.23. The maximum wall shear stress at a section where the flow is perfectly aligned with the flow at t = 300 ms is found to be 5 Pa. An effective dynamic blood viscosity is assumed to be 5 mPa in the shear stress calculation.

Hierck et al. (2008) performed an experiment similar to that performed by to Vennemann et al. (2006). However, Hierck et al. (2008) used a relatively long pulse duration (40 ns) laser, which increases the yield of the fluorescent dye in the tracer particles by a factor of 2 to 3 compared with a conventional Nd:YAG laser. The temporal evolution of the pulsatile flow is reconstructed by sorting the image pairs on the basis of a phase estimate instead of using external trigger device for phase-locked velocity measurement. The reconstructed method decreases the measurement time by an order of magnitude compared with using external trigger device, so changes in the embryo are minimized during recording data sets of sufficient length.

2.4.3.2 Second-Order Accuracy Wall Shear Stress Measurement with Single-Pixel Resolution

Conventional single-pixel evaluation (SPE) significantly improves the spatial resolution of PIV measurements to the physical limit of a CCD camera on the basis of the forward difference interrogation. This computational algorithm can be enhanced to second-order accuracy by simply modifying the numerical scheme with the central difference interrogation (CDI). Because the accuracy of wall shear stress measurements depends on the resolution and accuracy of near-wall velocity field, increasing both resolution and accuracy is crucial for wall shear stress measurement. Second-order accuracy, single-pixel resolution evaluation algorithm can

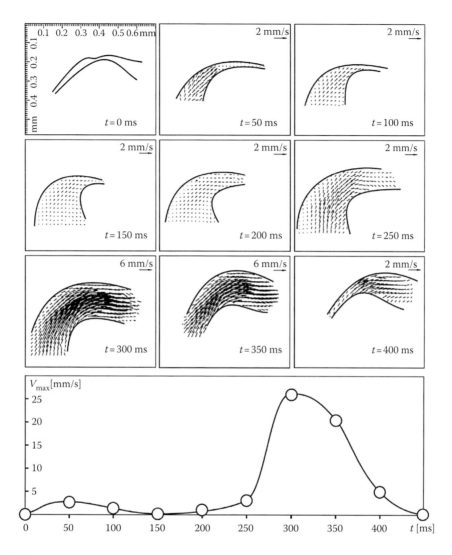

FIGURE 2.23
Velocity distribution in the developing ventricle at nine succeeding points of the cardiac cycle. Note that the vector scale for $t = 300$ and 350 ms is reduced. The maximal velocity of each measurement is plotted over the time in the bottom part of the graph. The connecting curve serves to guide the eye. (From Vennemann, P., Kiger, K.T., Lindken, R., Groenendijk, B.C.W., Stekelenburgde Vos, S., Ten Hagen, T.L.M., et al. *J. Biomech.*, 39, 1191–1200, 2006. With permission.)

increase both the resolution and the accuracy by at least one order of magnitude compared with the conventional fast Fourier transform (FFT)-based spatial cross-correlation.

Chuang and Wereley (2007) proposed central difference scheme at a single-pixel resolution. This method basically superposes the forward-time and the backward-time correlation domains, resulting in reduced bias error and rapid background noise elimination. The resulting correlation function can be given as

$$\Phi^*_{spe}(m,n) = \sum_{k=1}^{N} \left[f_k(x,y) \cdot g_k(x+m, y+n) + g_k(x,y) \cdot f_k(x-m, y-n) \right], \qquad (2.24)$$

where N is the number of image pairs (Gui et al., 2002), f and g are the intensity functions of the first and the second interrogation windows, respectively, and m and n are the searching displacements in two perpendicular directions.

The algorithm was demonstrated in a serpentine microchannel made of polydimethyl siloxane for both CDI SPE and spatial cross-correlation. The accuracy was compared by calculating R2 with the simulation result as a reference, and the CDI SPE algorithm showed two to six times higher accuracy than the FFT-based spatial cross-correlation algorithm.

2.5 μPIV Techniques Applied

The μPIV technique is very versatile and can be extended in several meaningful ways to make different but related measurements of use in characterizing microflows. One extension, called *microparticle image thermometry*, involves using μPIV to measure fluid temperature in the same high spatial resolution sense as the velocity is measured. Another extension involves using a similar measurement technique but having wavelengths longer than visual, called the *near infrared* (NIR), to measure flows completely encased in silicon—a real benefit to the MEMS field. The third extension is to avoid the low signal-to-noise ratio caused by the volume illumination by using evanescent waves to illuminate only a very thin region near the flow boundary.

2.5.1 Infrared μPIV

Another extension of μPIV that has potential to benefit microfluidics research in general and silicon MEMS research in particular is that of infrared (IR)-PIV (Han and Breuer, 2001). The main difference between the established technique of μPIV and IR-PIV is the wavelength of the illumination, which is increased from visible wavelengths to IR wavelengths to take advantage of silicon's relative transparency at IR wavelengths. Although this difference may seem trivial, it requires several important changes to the technique while enabling important new types of measurements to be made.

2.5.1.1 Differences between μPIV and IR-PIV

The fluorescent particles that allow the use of epifluorescent microscopes for μPIV are not available with both absorption and emission bands at IR wavelengths (Han and Breuer, 2001). Consequently, elastic scattering must be used in which the illuminating light is scattered directly by the seed particles with no change in wavelength. Using this mode of imaging, it is not possible to separate the images of the particles from that of the background using colored barrier filters as in the μPIV. The elastic scattering intensity I of a small particle of diameter d varies according to

$$I \propto \frac{d^6}{\lambda^4},$$ (2.25)

where λ is the wavelength of the illuminating light (Born and Wolf, 1997). Thus, a great price is exacted for imaging small particles with long wavelengths. The main implication of Equation 2.25 is that there is a trade-off between using longer wavelengths where silicon is more transparent and using shorter wavelengths where the elastic scattering is more efficient. Typically, IR cameras are also more efficient at longer wavelengths. Han and Breuer (2001) found a good compromise among these competing factors by using 1-μm polystyrene particles and an illumination wavelength of $\lambda = 1200$ nm.

An experimental apparatus suitable for making IR-PIV measurements is described by Han and Breuer (2001) and is shown in Figure 2.24. As with μPIV, a dual-headed Nd:YAG laser is used to illuminate the particles. However, in this case, the 532-nm laser light is used to drive an optoparametric oscillator (OPO)—a nonlinear crystal system that transforms the 532-nm light into any wavelength between 300 and 2000 nm. The laser light retains its short pulse duration when passing through the OPO. The output of the OPO is delivered via fiber optics to the microfluidic system being investigated. Han and Breuer (2001) used an off-axis beam delivery, as shown in Figure 2.24, with an angle of 65° between the normal surface of the device and the axis of the beam. Alternatively, dark-field illumination could be used. The light scattered by the particles is collected by a Mititoyo NIR microscope objective (50×, NA = 0.42) mounted on a 200-mm microscope tube and delivered to an Indigo Systems indium gallium arsenide (InGaAs) NIR camera.

The camera has a 320- × 256-pixel array with 30-μm pixels—a small number of large pixels compared with the high-resolution cameras typically used for μPIV applications. The NIR is a video rate camera that cannot be triggered, meaning that the PIV technique needs to be modified slightly. Instead of using the computer as the master for the PIV system, the camera is the master, running at its fixed frequency of 60 Hz. The laser pulses are synchronized to the video sync pulses generated by the camera and can be programmed to occur at any point within a video frame. For high-speed measurements, a process called frame straddling is used in which the first laser pulse is timed to occur at the very end of one frame and the second laser pulse is timed to occur at the very beginning of the next video frame. Using the frame straddling technique, the time between images can be reduced to as little as 0.12 ms-suitable for measuring velocities on the order of centimeters per second. Higher speed flows can be measured by recording the images from both laser pulses on a single video frame. The signal-to-noise ratio of the double-exposed images is decreased somewhat when compared with the single-exposed images, but flows on the order of hundreds of meters per second can be measured with the double-exposed technique.

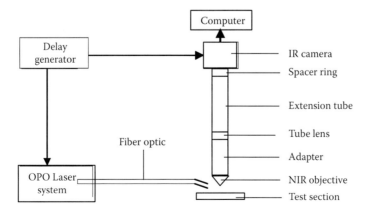

FIGURE 2.24
Schematic of the experimental apparatus for IR-PIV. (After Han, G. and Breuer, K.S., Proceedings of 4th International Symposium on Particle Image Velocimetry, Göttingen, Germany, Sept., paper number 1146, 2001.)

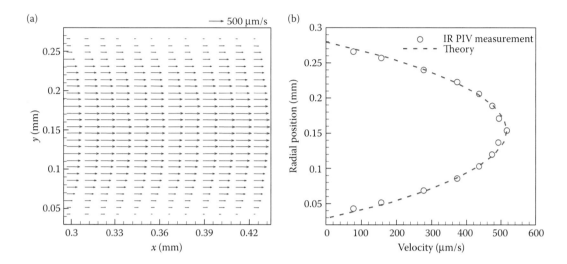

FIGURE 2.25
IR-PIV results for very low-speed flow: (a) velocity vectors and (b) comparison of the measurements with the theoretical profile. (From Liu, D., Garimella, S.V., and Wereley, S.T., *Exp. Fluids* 38, 385–392, 2005. With permission.)

Liu et al. (2005) successfully made IR-PIV measurement for laminar flow of water in a circular microcapillary tube of hydraulic diameter 255 μm. Figure 2.25 shows (a) the velocity field and (b) the velocity profile in the tube and Figure 2.25b shows that the experimental measurements agree very well with velocity profiles predicted from laminar theory.

2.5.2 Particle-Tracking Velocimetry

A typical method for increasing the resolution of PIV measurements at macroscopic length scales is called super-resolution PIV (Keane et al., 1995). In this technique, a typical PIV analysis is first used to determine the average velocity of the particles within each interrogation region. This spatially averaged velocity information is then used by a particle-tracking velocimetry (PTV) algorithm as an initial estimate of the velocity of each particle in the interrogation region. This estimate is progressively refined by one of several methods (Guezennec et al., 1994; Ohmi and Li, 2000; Cowen et al., 2001; Stitou and Riethmuller, 2001) to eventually determine the displacement of as many individual seed particles as possible. Recently, this super-resolution technique has been applied at microscopic length scales by Devasenathipathy et al. (2003) to produce very high spatial resolution velocity measurements of flow through microfabricated devices. The super-resolution PIV algorithm presented in their work is based on Kalman filtering for prediction of the particle trajectory and χ^2 testing for validation of particle identification (Takehara et al., 2000). An estimate of the velocity field is obtained from μPIV, and individual particle images are identified and located in both images by a particle mask correlation method. The approximate velocity of each particle in the first image is then found by interpolating the μPIV data. This velocity is used as a predictor for the particle pairing search in the second image by Kalman filtering. An iterative procedure is implemented until the number of particle pairs matched is maximized.

Devasenathipathy et al. (2003) measured the flow in prototypical microfluidic devices made of both silicon and acrylic in several shapes. The acrylic microfluidic devices used

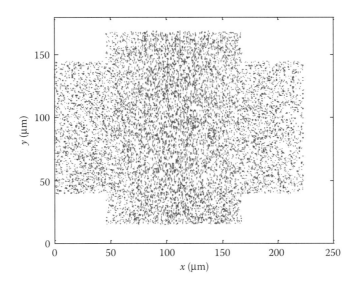

FIGURE 2.26
Velocity field at the intersection of a cross-channel system obtained by applying a combined PIV and PTV analysis. More than 5200 velocity measurements were recorded using 10 image pairs, 10 times the number produced by µPIV analysis. (From Devasenathipathy, S., Santiago, J.G., Wereley, S.T., Meinhart, C.D., and Takehara, K., *Exp. Fluids* 34, 504–514, 2003. With permission.)

were fabricated by ACLARA BioSciences. These microchannels were embossed on an acrylic substrate and have the D-shaped cross section typical of an isotropic etch (depth at centerline = 50 µm, top width = 120 µm). A 115-µm-thick acrylic sheet thermally laminated to the top of the channels acted as an optical coverslip and as the top wall of the channels. The flow in these microchannels was driven using the hydrostatic pressure head of liquid columns established in reservoirs mounted at the ends of the microchannels.

The velocity field is unstructured with the vectors nonuniformly spaced because of the random distribution of seed particles in the flow. The main advantages with this combined analysis are the higher sampling density and the higher spatial resolution achieved. The number of vectors obtained by µPIV is 480, whereas in the same area, the number of vectors produced by the super-resolution Kalman filtering χ^2 method is 5200, more than a tenfold increase. Note that the vectors from the PTV analysis are individual particle displacements, randomly positioned in the flow field, and include the full Brownian motion component. The spatial resolution of these PTV measurements along the flow direction is equal to the local particle displacement, which ranges from negligible values in the side channels to 10 µm at the centerline. Normal to the local flow direction the spatial resolution is well approximated by the seed particle size, which is 500 nm. The number density of particle displacements in the test section is 20/100 µm² (for the combined data set consisting of all 10 image pairs) so that we can expect a mean vector spacing on the order of 2 µm.

Both the µPIV and the PTV measurements compare very well with the predicted velocity distributions for their geometry. This agreement shows that a volume illumination may be used in conjunction with the depth of field of the objective lens to resolve two-dimensional velocity measurements in at least weakly three-dimensional velocity fields using µPIV.

2.5.3 Microparticle Image Thermometry

Microparticle image thermometry is minimally invasive and is a high-resolution temperature measurement technique based on the particle diffusion due to the Brownian motion. The square of the expected distance traveled by a particle with diffusivity D (recall Equation 2.4) in some time window Δt is given by

$$\langle s^2 \rangle = 2D\Delta t. \tag{2.26}$$

Combining Equations 2.6 and 2.26, it can be observed that an increase in fluid temperature, with all other factors held constant, will result in a greater expected particle displacement, $\sqrt{\langle s^2 \rangle}$.

Olsen and Adrian (2000b) derived analytical equations describing the shape and height of the cross-correlation function in the presence of the Brownian motion for both light-sheet illumination and volume illumination (as is used in μPIV). In both cases, the signal peak in the cross-correlation has a Gaussian shape with the peak located at the mean particle displacement. One of the key differences between light-sheet PIV and μPIV (volume illumination PIV) lies in the images formed by the seed particles. In light-sheet PIV, if the depth of focus of the camera is set to be greater than the thickness of the laser sheet, then all of the particle images will (theoretically) have the same diameter and intensity.

From their analysis of cross-correlation PIV, Olsen and Adrian (2000b) found that one effect of the Brownian motion on cross-correlation PIV is to increase the correlation peak width, Δs_o—taken as the $1/e$ diameter of the Gaussian peak (see Figure 2.27). For light-sheet PIV, they found that

$$\Delta s_{o,a} = \sqrt{2}\,\frac{d_e}{\beta} \tag{2.27}$$

when the Brownian motion is negligible, and

$$\Delta s_{o,c} = \sqrt{2}\,\frac{(d_e^2 + 8M^2\beta^2 D\Delta t)^{1/2}}{\beta} \tag{2.28}$$

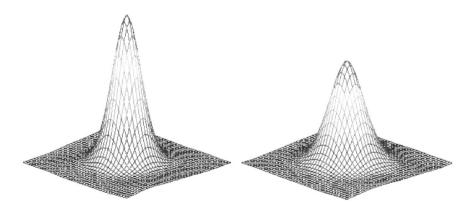

FIGURE 2.27
A pair of correlation functions demonstrates possible variations in peak width due to the Brownian motion. The autocorrelation (no Brownian motion) is on the left, and the cross-correlation (contains the Brownian motion) is shown on the right. Both figures are shown in the same scale.

when the Brownian motion is significant (note that in any experiment, even one with a significant Brownian motion, $\Delta s_{o,a}$ can be determined by computing the autocorrelation of one of the PIV image pairs). Here d_e is the particle image diameter, which is given in Equation 2.9 for light-sheet illumination and in Equation 2.10 for volume illumination; M is the magnification; D is the diffusivity; and Δt is the time delay between two frames of images.

The constant β is a parameter arising from the approximation of the airy point-response function as a Gaussian function (see Equation 2.13). Adrian and Yao (1985) found a best fit to occur for $\beta^2 = 3.67$. It can be seen that Equation 2.28 reduces to Equation 2.27 in cases where the Brownian motion is a negligible contributor to the measurement (i.e., when $D\Delta t \rightarrow 0$).

Hohreiter et al. (2002) derived the temperature term explicitly from diffusivity by manipulating Equations 2.27 and 2.28. To avoid the complicated calculations in Equation 2.9, squaring Equations 2.27 and 2.28, taking their difference, and multiplying by the quantity $\pi/4$ will convert the individual peak width (peak diameter for a three-dimensional peak) expressions to the difference of two correlation peak areas—namely, the difference in area between the auto- and cross-correlation peaks. Performing this operation and substituting Equation 2.4 in for D yields

$$\Delta A = \frac{\pi}{4}\left(\Delta s_{o,c}^2 - \Delta s_{o,a}^2\right) = C_0 \frac{T}{\mu}\Delta t, \tag{2.29}$$

where C_0 is the parameter $2M^2\kappa/3d_p$. κ and d_p are Boltzmann's constant and the physical particle diameter, respectively.

Hohreiter et al. (2002) successfully demonstrated that local temperature can be measured experimentally on the basis of the particle diffusion theory as in Equation 2.29. An Olympus BX50 system microscope with BX-FLA fluorescence light attachment (housing the dichroic mirror/prism and optical filter cube) was used to image the particle-laden solution. All experiments were carried out with a 50× objective (NA = 0.8). A Cohu (model 4915–3000) 8-bit CCD video camera was used—the CCD array consisted of 768 (horizontal) × 494 (vertical) pixels and a total image area of 6.4 × 4.8 mm. The particles—700-nm-diameter polystyrene latex microspheres (Duke Scientific Co.)—had a peak excitation wavelength at 542 nm and peak emission at 612 nm. A variable-intensity halogen lamp was used for illumination—optical filters were used to isolate the wavelength bands most applicable to the particles—520–550 nm for the incident illumination and greater than 580 nm for the particle fluorescence.

In the experiments, the power supplied to a patch heater (Figure 2.28) was adjusted incrementally and several successive images of the random particle motion were captured at each of several temperature steps. At each temperature, the system was allowed 5 min to reach a steady temperature as recorded by a thermocouple. Temperatures measured vary from 20°C to 50°C. Figure 2.29 shows the experimental results. Temperatures measured with correlation peak width change show a good agreement with the data from the thermocouple measurement. The average error over all the test cases is approximately ±3°C.

Chamarthy et al. (2009) successfully measured temperature field both in stationary fluid and in parabolic flow field using three different nonintrusive temperature measurements: PIV-based thermometry, LID-PIV, and particle-tracking thermometry. The LID-PIV tracking method was found to perform better than cross-correlation PIV and single

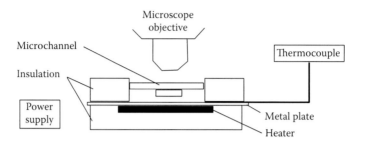

FIGURE 2.28
Microchannel, heater, and thermocouple arrangement used for experimentation. (After Hohreiter, V., Wereley, S.T., Olsen, M.,and Chung, J., *Meas. Sci. Technol.*, 13, 1072–1078, 2002. With permission.)

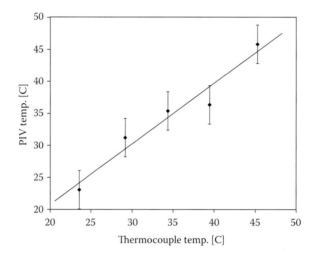

FIGURE 2.29
Temperature inferred from PIV measurements plotted versus thermocouple-measured temperature. Error bars indicate the range of average experimental uncertainty, ±3°C. (After Hohreiter, V., Wereley, S.T., Olsen, M., and Chung, J., *Meas. Sci. Technol.*, 13, 1072–1078, 2002. With permission.)

particle-tracking thermometry methods for measuring temperatures on the basis of the Brownian motion of particles in a stationary well of liquid. In the presence of a flow, the correlation function peak broadening associated with the Brownian motion in flowing fluid was calculated as

$$\Delta s_{\text{BM}}^2 = \Delta s_{\text{Total}}^2 - \Delta s_{\text{Vel}}^2. \tag{2.30}$$

The velocity vectors measured using μPIV were used to calculate the peak broadening induced due to the velocity gradients present in each interrogation window. This was used to obtain the peak width increase due to the Brownian motion alone, from which temperature was measured. The average difference between the measured and predicted temperatures was found to be ±2.6°C.

Chung et al. (2009) measured two-dimensional field temperature on the basis of Brownian diffusion around live cells. They also showed that z-directional temperature gradient can be measured by z-stacking method. They measured mean square displacements of 500-nm nanoparticles from 1500 frames of images using particle-tracking algorithm to trace the particles instead of using correlation function broadening effect. The results showed excellent agreements with temperature measured with thermocouple as shown in Figure 2.30. In the range of 1°C to 50°C, the accuracy of this technique was less than 1°C. Two- and three-dimensional temperature gradients were formed in the center microchannel by parallel cooling and/or heating microchannels next to the center microchannel (see Figure 2.31). Two-dimensional temperature field measurement and z-directional temperature measurement were successfully measured as shown in Figure 2.31. The spatial resolution of two-dimensional temperature field measurement was approximately 12 μm.

2.5.4 Evanescent Wave PIV

As explained in Section 3.1.3, one of the limitations of μPIV is that a volume of light is used to illuminate the flow, only a small fraction of which coincides with the depth of field of the imaging optics. This leads to out-of-focus particle images being present in the recorded images. These out-of-focus particle images reduce the signal-to-noise ratio of the measurements and limit the concentration of seed particles that can be used without completely obscuring the focused particle images. One solution to this problem is to use evanescent wave illumination to restrict the illumination to a region within a few hundred nanometers of the wall.

Although the light undergoes total internal reflection (TIR), some fraction of the incident light penetrates into the less dense medium and propagates parallel to the surface. This parallel light wave is called *evanescent wave* (Jin et al., 2003). In this illumination, the field amplitude decays exponentially over a distance comparable with the wavelength in the sample medium, allowing the information depth of optical signals to be limited to

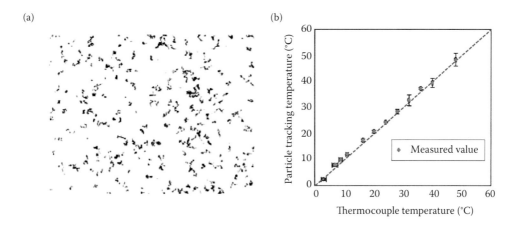

FIGURE 2.30
Brownian motion of fluorescent polystyrene nanoparticles and measured temperature. (a) Trajectories of the tracked 500-nm particles for a duration of 49.5 s (1500 frames). (b) Comparison of temperature data from particle tracking with simultaneously measured thermocouple temperature. (From Chung, K., Cho, J.K., Park, E., Breedveld, V., and Lu, H., *Anal. Chem.*, 81, 991–999, 2009. With permission.)

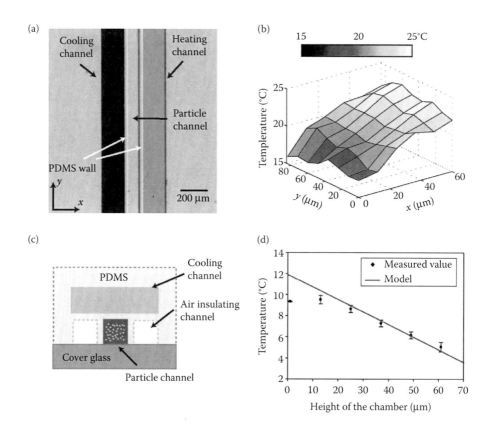

FIGURE 2.31
(See color insert.) (a) Optical micrograph of the microdevice fabricated using soft lithography. (b) Temperature distribution measurement. Hot and cold streams were at ~35°C and ~5°C, respectively. (c) Cross-sectional view of microdevice to generate z-depth temperature distribution. (d) Measured temperature distribution in z-direction. (From Chung, K., Cho, J.K., Park, E., Breedveld, V., and Lu, H., *Anal. Chem.*, 81, 991–999, 2009. With permission.)

scales of tens or hundreds of nanometers, a far higher resolution that can be achieved with conventional imaging optics. The penetration depth, z_p, is given as:

$$z_p = \frac{\lambda_0}{2\pi}(n_2^2 \sin^2 \theta - n_1^2)^{-1/2},\qquad(2.31)$$

where n_1 and n_2 are the refractive indices of two media of $n_1 < n_2$, λ_0 is the wavelength of the light in vacuum, and θ is the incident angle greater than the critical angle, $\theta_c = \sin^{-1}(n_2/n_1)$, to generate TIR.

Zettner and Yoda (2003) made a PIV measurement in a rotating Couette flow using evanescent wave. The fluorescent particles few hundreds of nanometers in diameter were used, and the spatial resolution in depth direction was approximately 380 nm and the field of view was 200 × 150 µm. Their results showed a reasonable agreement with the exact solution for single-phase Newtonian fluid flow.

Jin et al. (2003) used a TIR fluorescent microscopy to measure the slip velocity on hydrophilic and hydrophobic surfaces. Figure 2.32 shows the schematic of the objective-based TIR

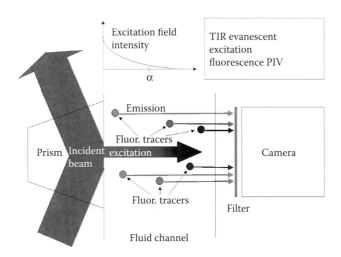

FIGURE 2.32
Schematic diagram of evanescent wave PIV system.

fluorescent microscopy. The penetration depth ranges from 98 to 232 nm. Fluorescent particles 200–300 nm in diameter were used for PTV evaluation. The minimal slip difference was found between a hydrophilic surface and a hydrophobic surface. Although evanescent wave illumination is a promising tool to measure near-surface flow, it is still difficult to measure the slip velocity for several reasons such as the interaction between seed particles and channel surface, relatively large particle size, and surface roughness.

2.5.5 Single-Pixel Resolution PIV

As mentioned in Section 2.3, the Brownian motions of tracer particles and the low particle density of µPIV images are often observed in µPIV, and these limitations can be addressed using ensemble correlation method. The ensemble averaging increases the effective image density for the correlation and reduces the measurement errors because of the Brownian motion of the tracer particles (Westerweel et al., 2004). One of the drawbacks in the correlation is that the spatial resolution of the velocity field depends strongly on the window size used for the ensemble correlation. As a result, velocity evaluation using ensemble correlation suffers from inaccuracies in the regions of higher velocity gradient than the spatial correlation and velocity biasing effects near the wall.

Westerweel et al. (2004) developed single-pixel ensemble correlation in which the smallest window size can be reduced to a single pixel to greatly improve the spatial resolution of velocity field to be measured—one vector per pixel. The basic principle is that the spatial averaging can be reduced down to the size of single pixel if the number of images is large enough, generally 1000 images or more. That is, the same signal-to-noise ratio can be obtained even with smaller window size of the ensemble correlation if we increase the number of images (see Figure 2.33). In fact, the results from synthetic PIV images demonstrated that the single-pixel ensemble correlation provided improved spatial resolution and reliability in high-velocity gradient flows and boundary layer flows (Figure 2.34). As shown in the Figure 2.34, single-pixel ensemble correlation can resolve minute flows near the wall. However, the single-pixel correlation works only when the flow is stationary during the full period. They also applied the algorithm to microchannel flows and showed

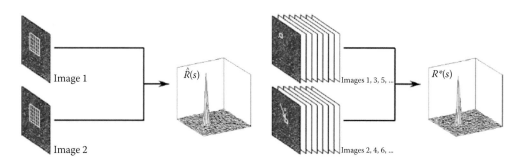

FIGURE 2.33
(Left) In conventional PIV, a spatial correlation over an $N \times N$-pixel interrogation domain is computed for a single image pair. (Right) The single-pixel ensemble correlation is computed over 1-pixel domains averaged over multiple image pairs to achieve the same signal-to-noise ratios. (From Westerweel, J., Geelhoed, P.F., and Lindken, R., *Exp. Fluids*, 37, 375–384, 2004. With permission.)

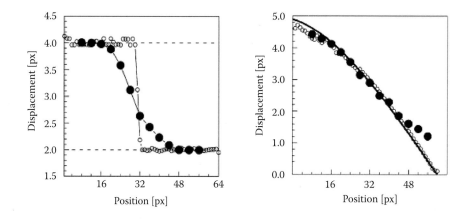

FIGURE 2.34
(Left) The result for the profile of the displacement for an infinitely thin shear layer (solid circles: 32×32 spatial correlation; empty circles: single-pixel ensemble correlation). (Right) The result for the profile of the displacement for a boundary layer (solid circles: 32×32 spatial correlation; empty circles: single-pixel ensemble correlation). The wall is located at 60 pixels. (From Westerweel, J., Geelhoed, P.F., and Lindken, R., *Exp. Fluids*, 37, 375–384, 2004. With permission.)

that the spatial resolution of the single-pixel ensemble correlation was as small as 0.3 µm. This was significantly enhanced compared with the spatial resolution of 2.4 µm in conventional ensemble correlation.

Billy et al. (2004) applied the single-pixel ensemble correlation algorithm to find the velocity field in the steady average Poiseuille flow in a small grooved channel. They showed that the single-pixel resolution algorithm was not a good one for nonlaminar flows that were steady flows on average and it needed more images or more pixels in the first frame to improve the correlation peaks.

2.5.6 Multidimensional µPIV

The µPIV has been successful in measuring two-dimensional velocity components (2C) on a measurement plane (two dimensional). Recently, several velocimetry techniques such as stereoscopic μPIV, defocusing μPIV/μPTV, and holographic μPIV/μPTV have been introduced to measure three-dimensional velocity components of microscale flows (Lee and Kim, 2009; Arroyo and Hinsch, 2008).

2.5.6.1 Stereoscopic μPIV

The stereoscopic μPIV system consists of a stereomicroscope equipped with a filter set for fluorescence imaging and a large-diameter stereo-objective lens with two off-axis beam paths, two digital cameras, and a double-pulse laser (Figure 2.35). Basically, two pairs of images are recorded on two cameras (e.g., camera 1 and 2 in Figure 2.35) at different angles just like human eyes, and the respective flow mappings obtained by two-dimensional μPIV method are processed to find the three velocity components of the flow fields in the microdevices. The stereoscopic μPIV is based on two-dimensional plane measurements in nature and hence requires some specialized hardware and/or calibration device or methods to yield three velocity components on the two-dimensional plane (Peterson et al., 2008). In macroscales, one would place a calibration target in the focal plane (or in the light sheet) to determine the relative position of the two cameras and the focal plane, but this is very difficult inside a closed micrometer-sized channel (Lindken et al., 2006). One solution is to make the calibration in air and then correct the parameters for the presence of the fluid and glass (Bown et al., 2006).

Lindken et al. (2006) used the self-calibration procedure described by Weineke (2005) and modified mapping function with a three-media model (Maas, 1992) to account for the air–glass–fluid transitions, which is usual with microchannels, from calibration measurements in air. Lindken et al. (2006) showed that the results had a root mean square error <10% of the expected value of the in-plane velocity component. The measurement area was 800 × 200 μm², with a spatial resolution of 44 × 44 × 15 μm³ and a scan separation of 22 μm. A drawback of this method is that this stereo-objective typically has a low

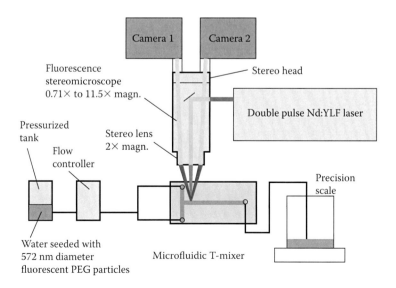

FIGURE 2.35
Schematic of a stereoscopic μPIV system. (From Lindken, R., Westerweel, J., and Wieneke, B., *Exp. Fluids*, 41, 161–171, 2006. With permission.)

NA (0.14–0.28) and a large depth of focus, which can reduce the spatial resolution in the z-direction (depth-wise direction) and the accuracy. It is also found that the accuracy of the correlation-based PIV technique is limited by the degree of overlap of the two focal planes in the stereomicroscope (Bown et al., 2006).

Bown et al. (2006) used a super-resolution PTV algorithm by Keane et al. (1995) to improve the spatial resolution by circumventing misalignment issues of the two focal planes in the stereomicroscope. This algorithm does not require exact alignment of the two focal planes, which enhances the accuracy of the out-of-plane velocity measurement. The measurements show that the spatial resolution was $10 \times 10 \times 10$ μm with a measurement volume of $900 \times 720 \times 45$ μm³.

Hagsäter et al. (2008) presented a stereoscopic viewing configuration for stereoscopic μPIV using millimeter-sized mirrors located between the microscope objective lens and the sample. This setup allows the angle between the views to be made larger than what is typically given by a stereomicroscope lens, which is a limiting factor for the accuracy of the stereoscopic μPIV method.

2.5.6.2 Defocusing μPIV/μPTV

Another technique for the measurements of three-dimensional velocity components of microscale flows is defocusing μPIV. Yoon and Kim (2006) presented three-dimensional velocity measurements of microflows in a backward-facing step of a 50-μm-deep micro-channel using a three-pinhole defocusing mask. The principle of the defocusing technique was initially described by Willert and Gharib (1992). The main principle is that the distance between the triangle vertices of the images increases as the particle position moves away in the depth-wise direction from the focusing plane (reference plane), and that distance is calculated to find three-dimensional velocity components (see Figure 2.36). For example, Yoon and Kim (2006) used fluorescent particles fixed on a glass plate for calibration (see Figure 2.37). The top glass plate was fixed, and the bottom glass plate with the particles was movable. The fluid was then injected into the gap between the two glass plates, and the several particle images were captured. The size of the triangular pattern caused by the mask increased proportionally to the distance from the reference plane and therefore the depth-wise positions can be calculated from the distance between the triangle vertices of the triangular pattern. The macroscale equations previously derived assuming that the lens plane is located at the same position as the pinhole mask plane cannot be used for microscale applications and need to be modified. In fact, the distance between the lens plane and the pinhole mask plane is not small.

Figure 2.38 shows the schematic of a defocusing μPIV setup using a three-pinhole mask where the three-pinhole mask is positioned on an objective lens. The setup consists of a CCD camera, a microscope equipped with a long pass filter cube and a pinhole mask, a laser light source, and a syringe pump. Some of the drawbacks of this method are low light intensity caused by the pinholes and relatively small number of velocity vectors. The volume measured was 768 μm long, 388 μm wide, and 50 μm deep, and the spatial resolution in a vector volume was 5 μm in x and y and 1 μm in z (Yoon and Kim, 2006).

Pereira et al. (2007) developed a high-speed, three-dimensional micro-defocusing digital PIV system. They addressed calibration difficulties of system developed by Yoon and Kim (2006) and proposed a practical solution. A microvolume of 400×300 μm² with a depth of 150 μm with submicron resolution has been mapped with a 20× objective lens.

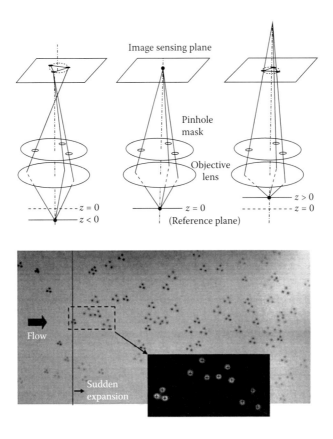

FIGURE 2.36
(Top) Schematic of the basic defocusing concept. (Bottom) Instantaneous image with inverted intensities and the peaks shown within the dashed region. (From Yoon, S.Y., and Kim, K.C., *Meas. Sci. Technol.*, 17, 2897–2905, 2006. With permission.)

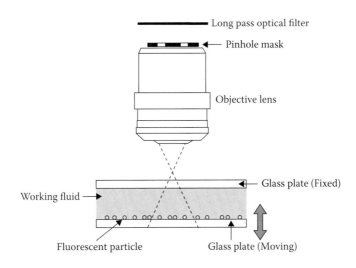

FIGURE 2.37
Calibration methods of a defocusing μPIV. (From Yoon, S.Y., and Kim, K.C., *Meas. Sci. Technol.*, 17, 2897–2905, 2006. With permission.)

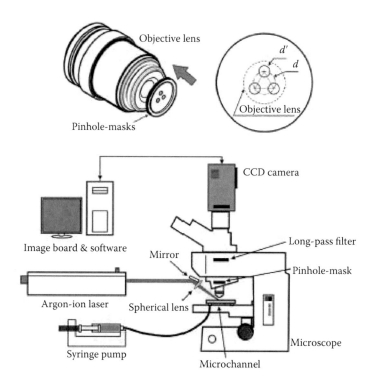

FIGURE 2.38
Schematic diagram of the experimental setup using a three-pinhole defocusing mask. (From Yoon, S.Y., and Kim, K.C., *Meas. Sci. Technol.*, 17, 2897–2905, 2006. With permission.)

Peterson et al. (2008) introduced three-dimensional particle tracking using μPIV system. In this method, the three-dimensional position of a particle was identified, and the three velocity components of the particle were calculated on the basis of the information encoded in the ring structure of the particle image. Although this technique requires a set of calibration images of a particle at various known distances from the focal plane, it can be easily used in most existing μPIV systems without any additional hardware and details of the optical train are not required.

2.5.6.3 Holographic μPIV/μPTV

Holography consists of two steps: recording of the laser light-field scattered by the seed particles and reconstruction of this field from the hologram (Arroyo and Hinsch, 2008). Figure 2.39 shows the experimental setup for digital holographic PTV measurements of flow in a microtube (Kim and Lee, 2007) and a schematic of the digital in-line holographic PTV (Lee and Kim, 2009). A light source is collimated directly onto an imaging plane as a reference wave. The superposition of the forward scattering of tracer particles in the flow and the unaffected reference wave creates interference patterns on the imaging plane of a photographic film or digital camera. The interference pattern is directly captured as a hologram image, and the interference pattern (the hologram) stores the amplitude and phase of a light wave. The pattern is used to find the original wave field by illuminating it with a replica of the reference wave (Hinsch, 2002). The three-dimensional velocity information of the tracer particles in the measurement volume can be calculated from the numerical

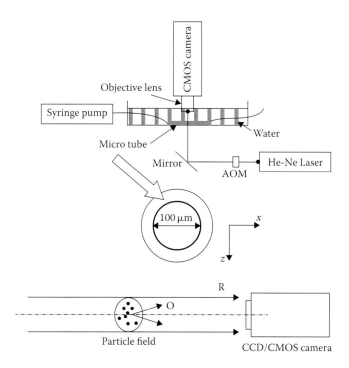

FIGURE 2.39
(Top) Experimental setup for digital holographic PTV measurements of flow in a microtube. (From Kim, S., and Lee, S.J. *J. Micromech. Microeng.*, 17, 2157–2162, 2007. With permission.) (Bottom) Schematic diagram of the digital in-line holographic PTV. (From Lee, S.J., and Kim, S. *Microfluid Nanofluidics.*, 6, 577–588, 2009. With permission.)

reconstruction of a hologram and a subsequent PTV analysis. The numerical reconstruction is based on the Huygen–Fresnel principle and the FFT (Lee and Kim, 2009).

Yang and Chuang (2005) used a hybrid multiplexing holographic velocimetry for three-component three-dimensional measurements in a 550- × 26-μm chamber using a concise cross-correlation and a photopolymer plate. They compared it with μPIV measurements, and they found that the photopolymer-based velocimetry can be used for microflow investigation but its accuracy needed to be improved. The resolution of measurements was approximately $30 \times 30 \times 2.6$ μm (Bown et al., 2006).

Kim and Lee (2007) developed a microdigital holographic PTV system for three-dimensional measurement of a microtube. The system can capture 4000 hologram fringe images directly for 1 s, and the temporal evolution of water flow in a microtube of 100 μm inner diameter is obtained successfully using this microdigital holographic PTV system. The measurement volume was $350 \times 100 \times 100$ μm³.

Ooms et al. (2009) presented measurements of a three-dimensional flow in a T-shaped micromixer by means of digital holographic microscopy (conventional in-line recording system with an added magnifying optical element). They illustrated four streamwise vortices in the micromixer outlet and investigated accuracy and resolution of the holographic measurements for the three-dimensional flows. They also discussed several parameters that affect the performance of digital holography, dynamic spatial range, and dynamic velocity range. The dynamic spatial range and dynamic velocity range obtained were larger than 20 and 30, respectively.

References

Adrian RJ. (1991). Particle-imaging techniques for experimental fluid mechanics. *Annu Rev Fluid Mech.* 23:261–304.

Adrian RJ. (1996). *Bibliography of Particle Image Velocimetry Using Imaging Methods: 1917–1995.* Urbana (IL): University of Illinois at Urbana-Champaign.

Adrian RJ, Yao CS. (1985). Pulsed laser technique application to liquid and gaseous flows and the scattering power of seed materials. *Appl Opt.* 24:44–52.

Amabile M, Dunsmuir J, Lazillotto AM, Len T. (1996). Applications of x-ray micro-imaging, visualization and motion analysis techniques to fluidic microsystems. Technical Digest of the IEEE Solid State Sensor and Actuator Workshop, Hilton Head Island, SC, June 3–6; pp. 123–6.

Anderson JM, Streitlien K, Barrett DS, Triantafyllou MS. (1998). Oscillating foils of high propulsive efficiency. *J Fluid Mech.* 360:41–72.

Arroyo MP, Hinsch KD. (2008). Recent developments of PIV towards 3D measurements. In: Schröder A, Willert CE, editors. *Particle Image Velocimetry: New Developments and Recent Applications.* New York: Springer; pp. 127–54.

Atkinson B, Brocklebank MP, Card CCH, Smith JM. (1969). Low Reynolds number developing flows. *AIChE J.* 15:548–53.

Beavers GS, Sparrow EM, Magnuson RA. (1970). Experiments on hydrodynamically developing flow in rectangular ducts of arbitrary aspect ratio. *Int J Heat Mass Transf.* 13(4):689–702.

Bendat JS, Piersol JG. (1986). *Random Data: Analysis and Measurement Procedures.* New York: Wiley.

Beskok A, Karniadakis GE, Trimmer W. (1996). Rarefaction and compressibility. *J Fluid Eng.* 118:448–56.

Billy F, David L, Pineau G. (2004). Single pixel resolution correlation applied to unsteady flow measurements. *Meas Sci Technol.* 15:1039–45.

Bown MR, MacInnes JM, Allen RWK, Zimmerman WBJ. (2006). Three-dimensional, three-component velocity measurements using stereoscopic micro-PIV and PTV. *Meas Sci Technol.* 17:2175–85.

Born M, Wolf E. (1997). *Principles of Optics.* New York: Pergamon Press.

Charmarthy P, Garimella SV, Wereley ST. (2009). Non-intrusive temperature measurement using microscale visualization techniques. *Exp Fluids.* 47:159–70.

Chen R-Y. (1973). Flow in the entrance region at low Reynolds numbers. *J Fluid Eng.* 95:153–8.

Chen Z, Milner TE, Dave D, Nelson JS. (1997). Optical Doppler tomographic imaging of fluid flow velocity in highly scattering media. *Opt Lett.* 22:64–6.

Chung K, Cho JK, Park E, Breedveld V, Lu H. (2009). Three-dimensional in situ temperature measurement in microsystems using Brownian motion of nanoparticles. *Anal Chem.* 81:991–9.

Chuang H-S, Wereley ST. (2007). *In vitro* wall shear stress measurements for microfluid flows by using second-order SPE micro-PIV. 2007 ASME International Mechanical Engineering Congress and Exposition, Seattle, WA, November 11–15, IMECE 2007-41171.

Compton DA, Eaton JK. (1996). A high-resolution laser Doppler anemometer for three-dimensional turbulent boundary layers. *Exp Fluids.* 22:111–7.

Cowen EA, Monismith SG. (1997). A Hybrid Digital Particle Tracking Velocimetry Technique. *Exp Fluids.* 22:199–211.

Cowen EA, Chang KA, Liao Q. (2001). A single-camera coupled PTV-LIF technique. *Exp Fluids.* 31:63–73.

Cummings EB. (2001). An image processing and optimal nonlinear filtering technique for PIV of microflows. *Exp Fluids.* 29(Suppl):42–50.

Dahm WJA, Su LK, Southerland KB. (1992). A scalar imaging velocimetry technique for fully resolved four-dimensional vector velocity field measurements in turbulent flows. *Phys Fluid.* 4(10):2191–206.

Devasenathipathy S, Santiago JG, Wereley ST, Meinhart CD, Takehara K. (2003). Particle imaging techniques for microfabricated fluidic systems. *Exp Fluids.* 34:504–14.

Einstein A. (1905). On the movement of small particles suspended in a stationary liquid demanded by the molecular-kinetic theory of heat. In: *Theory of Brownian Movement.* New York: Dover; pp. 1–18.

Fleming DP, Sparrow EM. (1969). Flow in the hydrodynamic entrance region of ducts of arbitrary cross section. *J Heat Transfer*. 91(3):345–54.

Fox RW, McDonald AT. (1998). *Introduction to Fluid Mechanics*. 5th ed. New York: Wiley.

Friedmann M, Gillis J, Liron N. (1968). Laminar flow in a pipe at low and moderate Reynolds numbers. *Appl Sci Res*. 19:426–38.

Gharib M., Modarress, D., Fourgette, D., Wilson, D. (2002). Optical microsensors for fluid flow diagnostics. 40th AIAA, Aerospace Sciences Meeting and Exhibit, Reno, NV; January 14–17, 2002. AIAA Paper No.: 2002-0252.

Gomez R, Bashir R, Greng A, Bhunia M, Ladishch J, Wereley S. (2001). Microfluidic biochip for impedance spectroscopy of biological species. *Biomed Microdevices*. 3(3):201–9.

Guezennec YG, Brodkey RS, Trigui N, Kent JC. (1994). Algorithms for fully automated 3-dimensional particle tracking velocimetry. *Exp Fluids*. 17:209–19.

Gui L, Merzkirch W, Shu JZ. (1997). Evaluation of low image density PIV recordings with the MQD method and application to the flow in a liquid bridge. *J Flow Visual Image Process*. 4:333–43.

Gui L, Wereley ST, Lee SY. (2002). Digital filters for reducing background noise in micro PIV measurements. Proceedings of the 11th International Symposium on the Application of Laser Techniques to Fluid Mechanics, Lisbon, Portugal; July 8–11, 2002. Paper No.: 12.4.

Hagsäter SM, Westergaard CH, Bruus H, Kutter JP. (2008). A compact viewing configuration for stereoscopic micro-PIV utilizing mm-sized mirrors. *Exp Fluids*. 45:1015–21.

Han LS. (1983). Hydrodynamic entrance lengths for incompressible laminar flow in rectangular ducts. *J Appl Mech*. 27:273–7.

Han G, Breuer KS. (2001). Infrared PIV for measurement of fluid and solid motion inside opaque silicon microdevices. Proceedings of 4th International Symposium on Particle Image Velocimetry, Göttingen, Germany; September 2001. Paper No.: 1146.

Hierck BP, van der Heiden K, Poelma C, Westerweel J, Poelmann RE. (2008). Fluid shear stress and inner curve remodeling of the embryonic heart. *Sci World J*. (revision submitted).

Hinsch KD. (2002). Holographic particle image velocimetry. *Meas Sci Technol*. 13:R61–72.

Hitt DL, Lowe ML, Newcomer R. (1995). Application of optical flow techniques to flow velocimetry. *Phys Fluids*. 7(1):6–8.

Hitt DL, Lowe ML, Tindra JR, Watters JM. (1996). A new method for blood velocimetry in the microcirculation. *Microcirculation*. 3(3):259–63.

Hohreiter V, Wereley ST, Olsen M, Chung J. (2002). Cross-correlation analysis for temperature measurement. *Meas Sci Technol*. 13:1072–8.

Hove J, Koster R, Forouhar A, Acevedo-Bolton G, Fraser S, Gharib M. (2003). Intracardiac fluid forces are an essential epigenetic factor for embryonic cardiogenesis. *Nature*. 421:172–7.

Inoué S, Spring KR. (1997). *Video Microscopy*. 2nd ed. New York: Plenum Press.

Jang J, Wereley ST. (2006). Effective heights and tangential momentum accommodation coefficients of gaseous slip flows in deep reactive ion etching rectangular microchannels. *J Micromech Microeng*. 16:493–504.

Jin S, Huang P, Park J, Yoo JY, Breuer KS. (2003). Near-surface velocimetry using evanescent wave illumination. ASME International Mechanical Engineering Congress & Exposition, Washington, DC; November 16–21, 2003. Paper No.: IMECE2003-44015.

Jones KD, Dohring CM, Platzer MF. (1998). Experimental and computational investigation of the Knoller–Betz effect. *AIAA J*. 36:1240–6.

Keane RD, Adrian RJ, Zhang Y. (1995). Super-resolution particle imaging velocimetry. *Meas Sci Technol*. 6:754–68.

Kim S, Lee SJ. (2007). Measurement of 3D laminar flow inside a micro tube using micro digital holographic particle tracking velocimetry. *J Micromech Microeng*. 17:2157–62.

Kim YH, Wereley ST, Chun CH. (2004). Phase-resolved flow field produced by a vibrating cantilever plate between two endplates. *Phys Fluids*. 16:145–62.

Koochesfahani MM, Cohn RK, Gendrich CP, Nocera DG. (1997). Molecular tagging diagnostics for the study of kinematics and mixing in liquid phase flows. In: Adrian RJ, et al., editors. *Developments in Laser Techniques in Fluid Mechanics*. New York: Springer-Verlag; pp. 125–34.

Koutsiaris AG, Mathioulakis DS, Tsangaris S. (1999). Microscope PIV for velocity-field measurement of particle suspensions flowing inside glass capillaries. *Meas Sci Technol.* 10:1037–46.

Kumar A, Gorti V, Shang H, Lee GU, Yip NK, Wereley ST. (2008). Optical diffusometry techniques and applications in biological agent detection. *J Fluid Eng.* 130:111401–8.

Lee SJ, Kim S. (2009). Advanced particle-based velocimetry techniques for microscale flows. *Microfluid Nanofluidics.* 6:577–88.

Lee S-Y, Jang J, Wereley ST. (2008). Effects of planar inlet plenums on the hydrodynamically developing flows in rectangular microchannels of complementary aspect ratios. *Microfluid Nanofluidics.* 5:1–12.

Lee SY, Wereley ST, Gui LC, Qu WL, Mudawar I. (2002). Microchannel flow measurement using micro particle image velocimetry. Proceedings of ASME/IMECE, New Orleans, LA; November 2002. Paper No: 2002-33682.

Lempert WR, Jiang N, Sethwram S, Samimy M. (2001). Molecular tagging velocimetry measurements in supersonic micro nozzles. Proceedings of 39th AIAA Aerospace Sciences Meeting and Exhibit, Reno, NV; January 2001. Paper No.: 2001-2044.

Leu TS, Lanzillotto AM, Amabile M, Wildes R. (1997). Analysis of fluidic and mechanical motions in MEMS by using high speed x-ray micro-imaging techniques. Proceedings of Transducers '97, 9th International Conference on Solid-State Sensors and Actuators, Chicago, IL; June 16–19, 1997. p. 149–50.

Lindken R, Westerweel J, Wieneke B. (2006). Stereoscopic micro particle image velocimetry. *Exp Fluids.* 41:161–71.

Liu D, Garimella SV, Wereley ST. (2005). Infrared micro-particle image velocimetry of fluid flow in silicon-based microdevices. *Exp Fluids.* 38:385–92.

Maynes D, Webb AR. (2002). Velocity profile characterization in sub-millimeter diameter tubes using molecular tagging velocimetry. *Exp Fluids.* 32:3–15.

Meinhart CD, Wereley ST, Santiago JG. (1999a). PIV measurements of a microchannel flow. *Exp Fluids.* 27:414–9.

Meinhart CD, Wereley ST, Santiago JG. (1999b). A PIV algorithm for estimating time-averaged velocity fields. *J Fluid Eng.* 122:285–9.

Meinhart CD, Zhang H. (2000). The flow structure inside a microfabricated inkjet printer head. *J Microelectromech Syst.* 9:67–75.

Meinhart CD, Wereley ST, Santiago JG. (2000a). Micron-resolution velocimetry techniques. In: Adrian RJ, et al., editors. *Laser Techniques Applied to Fluid Mechanics.* New York: Springer-Verlag; pp. 57–70.

Meinhart CD, Wereley ST, Gray MHB. (2000b). Volume illumination for two-dimensional particle image velocimetry. *Meas Sci Technol.* 11:809–14.

Melling, A. Seeding Gas Flows for Laser Anemometry, Advanced Instrumentation for Aero Engine Components, Proceedings of the Propulsion and Energetics Panel 67th Symposium, Philadelphia, Pennsylvania, 19–23 May 1986, AD-A182 954, p8-1–8-11.

Olsen MG, Adrian RJ. (2000a). Out-of-focus effects on particle image visibility and correlation in particle image velocimetry. *Exp Fluids.* Suppl:166–74.

Olsen MG, Adrian RJ. (2000b). Brownian motion and correlation in particle image velocimetry. *Opt Laser Technol.* 32:621–7.

Ohmi K, Li HY. (2000). Particle-tracking velocimetry with new algorithm. *Meas Sci Technol.* 11:603–16.

Ooms TA, Lindken R, Westerweel J. (2009). Digital holographic microscopy applied to measurement of a flow in a T-shaped micromixer. *Exp Fluids* [Internet]. doi: 10.1007/s00348-009-0683-9.

Paul PH, Garguilo MG, Rakestraw DJ. (1998). Imaging of pressure- and electrokinetically driven flows through open capillaries. *Anal Chem.* 70:2459–67.

Pereira F, Lu J, Castaño-Graff E, Gharib M. (2007). Microscale 3D flow mapping with μDDPIV. *Exp Fluids.* 42:589–99.

Peterson SD, Chuang H-S, Wereley ST. (2008). Three-dimensional particle tracking using micro-particle image velocimetry hardware. *Meas Sci Technol.* 19:115406.

Prasad AK, Adrian RJ, Landreth CC, Offutt PW. (1992). Effect of resolution on the speed and accuracy of particle image velocimetry interrogation. *Exp Fluids.* 13:105–16.

Probstein RF. (1994). *Physicochemical Hydrodynamics: An Introduction.* New York: Wiley.

Raffel M, Willert C, Kompenhans J. (1998). *Particle Image Velocimetry: A Practical Guide*. New York: Springer.

Santiago JG, Wereley ST, Meinhart CD, Beebe DJ, Adrian RJ. (1998). A particle image velocimetry system for microfluidics. *Exp Fluids*. 25:316–9.

Shah RK, London AL. (1978). *Laminar Flow Forced Convection in Ducts—A Source Book for Compact Heat Exchanger Analytical Data, Supplement 1 to Advances in Heat Transfer Series*. New York: Academic Press.

Sparrow EM, Hixon CW, Shavit G. (1967). Experiments on laminar flow development in rectangular ducts. *J Basic Eng*. 89:116–24.

Sparrow EM, Anderson CE. (1977). Effect of upstream flow processes on hydrodynamic development in a duct. *J Fluids Eng*. 99(3):556–60.

Stitou A, Riethmuller ML. (2001). Extension of PIV to super resolution using PTV. *Meas Sci Technol*. 12:1398–403.

Sugii Y, Nishio S, Okamoto K. (2002). *In vivo* PIV measurements of red blood cell velocity field in microvessels considering mesentery motion. *Physiol Meas*. 23:403–16.

Takehara K, Adrian RJ, Etoh GT, Christensen KT. (2000). A Kalman tracer for super resolution PIV. *Exp Fluids*. 29:s34–41.

Tieu AK, Mackenzie MR, Li EB. (1995). Measurements in microscopic flow with a solid-state LDA. *Exp Fluids*. 19:293–4.

Tsukada K, Minamitani H, Sekizuka E, Oshio C. (2000). Image correlation method for measuring blood flow velocity in microcirculation: correlation 'window' simulation and *in vivo* image analysis. *Physiol Meas*. 21:459–71.

Vennemann P, Kiger KT, Lindken R, Groenendijk BCW, Stekelenburgde Vos S, Ten Hagen TLM, et al. (2006). *In vivo* micro particle image velocimetry measurements of blood-plasma in the embryonic avian heart. *J Biomech*. 39:1191–200.

Vrentas JS. Duda JL, Bargeron KG. (1996). Effect of axial diffusion of vorticity on flow development in circular conduits: part I. numerical solutions. *AIChE J*. 15:837–44.

Wereley ST, Gui LC. (2001). PIV measurement in a four-roll-mill flow with a central difference image correction (CDIC) method. Proceedings of 4th International Symposium on Particle Image Velocimetry, Göttingen, Germany; September 2001. Paper No.: 1027.

Wereley ST, Gui L. (2003). A correlation-based central difference image correction (CDIC) method and application in a four-roll mill flow PIV measurement. *Exp Fluids*. 34:42–51.

Wereley ST, Gui L, Meinhart CD. (2001). Flow measurement techniques for the microfrontier. Proceedings of the 39th AIAA Aerospace Sciences Meeting and Exhibit, Reno, NV; January 8–11, 2001. Paper No.: 2001-0243.

Wereley ST, Gui L, Meinhart CD. (2002b). Advanced algorithms for microscale particle image velocimetry. *AIAA J*. 40:1047–55.

Wereley ST, Lee SY, Gui LC. (2002a). Entrance length and turbulence transition in microchannels. American Physics Society, Division of Fluid Dynamics Annual Meeting, Dallas, TX; November 2002a.

Wereley ST, Meinhart CD. (2001). Adaptive second-order accurate particle image velocimetry. *Exp Fluids*. 31:258–68.

Wereley ST, Meinhart CD, Santiago JG, Adrian RJ. (1998). Velocimetry for MEMS applications. Proceedings of the Micro-Electro-Mechanical Systems, DSC. 66, ASME; pp. 453–9.

Westerweel J, Geelhoed PF, Lindken R. (2004). Single-pixel resolution ensemble correlation for micro-PIV applications. *Exp Fluids*. 37:375–84.

Wiesendanger R. (1994). *Scanning Probe Microscopy and Spectroscopy: Methods and Applications*. Cambridge, England: Cambridge University Press.

Yang C-T, Chuang H-S. (2005). Measurement of a microchamber flow by using a hybrid multiplexing holographic velocimetry. *Exp Fluids*. 39:385–96.

Yoon SY, Kim KC. (2006). 3D particle position and 3D velocity field measurement in a microvolume via the defocusing concept. *Meas Sci Technol*. 17:2897–905.

Zettner CM, Yoda M. (2003). Particle velocity field measurements in a near-wall flow using evanescent wave illumination. *Exp Fluids*. 34:115–21.

Zhao Y. (2003). Design and Characterization of Micro and Molecular Flow Sensors. M.S.M.E Thesis. West Lafayette, IN: Purdue University.

3

Near-Surface Particle-Tracking Velocimetry

Peter Huang, Jeffrey S. Guasto, and Kenneth S. Breuer

CONTENTS

3.1 Introduction

The advent of microfluidics in the late 1990s brought about a new frontier in fluid mechanics. Since the introduction of the first microfluidic device, these miniaturized fluidic manipulation systems have been regarded as one of the most promising technologies for the twenty-first century. In particular, investigations into its application in biotechnology have been the most intense. Examples of such applications include immunosensors [1], reagent mixing [2], content sorter [3], and drug delivery [4]. Microfluidic devices are very attractive in biotechnology over conventional technology because they require small sample volume and produce rapid results. In addition, the use of colloids for self-assembly processes [5] and the medical application of nanoparticles have recently become areas of research interest [6]. The idea of a laboratory on a chip spawned an industry that strives to miniaturize and popularize the ability to detect, process, and analyze biological and chemical specimens on smaller, less-expensive microfluidic-based platforms. As the device dimension shrinks, bulk properties of the fluid medium become less important while a thorough understanding of interfacial and near-wall fluid–solid interactions become vital to the advancement of these technologies. Indeed, for chemical reactions that take place at solid surfaces, the high surface-area-to-volume ratio characteristic of microfluidics offers a much higher efficiency than its large-scale counterpart [7]. On the flip side, the high surface-area-to-volume ratio also means that near-surface phenomena will have a much larger impact on the bulk of the fluid content. An example of such near-surface phenomena is the fluidic slip on the channel walls and its influence on flow pattern and velocity [8]. Thus, a strong grasp of the fluidic and colloidal dynamics near a solid boundary is critical in designing and analyzing microfluidic devices.

Current fabrication technology of small-scale fluidic devices and application of microscopy techniques to fluid mechanics allow us to quantitivly characterize new and interesting near-surface physical phenomena critical to micro- and nanofluidics. Under most circumstances, the solid boundary is rigid and inert such that its physical and structural changes due to fluidic forces are nonexistent. Thus, the majority of important surface-induced physical phenomena occur in the near-surface region of the fluid phase and can be categorized into two groups: (1) changes in the fluid mechanical characteristics due to the presence of the solid surface and (2) interactions between the dissolved molecules, the suspended particulates, and the solid surface. Examples of physical phenomena in the former group include electrokinetic flow [9], slip flow [10], and surface chemistry directed flow [11,12], whereas particle or cell adhesion [13] and detachment [14], increased hydrodynamic drag [15], electrostatic interactions [16], and particle depletion layers [17] are effects of the latter group.

With so much interest in near-surface phenomena, researchers have developed various techniques to study them. Optical microscopy has been widely used to observe interactions in the micrometer scale. However, as fabrication technology advances, the definition of "near surface" has also evolved from the micrometer to the nanometer scale. Traditional optical techniques are no longer sufficient now because the visible wavelength limits the probing resolution to ~0.5 μm. An optical technique demonstrated to overcome this obstacle is evanescent wave microscopy or, when combined with fluorescence microscopy, total internal reflection fluorescence (TIRF) microscopy [18]. The principle of total internal reflection has been known for more than a thousand years, since the time of the Persian scientist Ibn Sahl [19]. It is most commonly associated with Rene Descartes and Willebrord Snellius (Snell) after whom the common law of refraction is named. The evanescent wave propagation in the less dense optical medium was first described by Isaac Newton and later formalized in Maxwell's theory of electromagnetic wave propagation. However, the adoption of the evanescent wave as a means to achieve localized illumination rose to widespread use in the life sciences and biological physics community, where researchers realized that the near-surface illumination provided by the evanescent field can be used to probe cellular structure, kinetics, diffusion, and dynamics with unprecedented spatial resolution. Since the 1970s, the TIRF microscopy technique has been used to measure chemical kinetics, surface diffusion, molecular conformation of adsorbates, cell development during culturing, visualization of cell structures and dynamics, and single-molecule visualization and spectroscopy [20–24].

Surprisingly, a long time had passed before physical scientists finally caught up with the merits of evanescent wave imaging. Beginning in the 1990s, several research groups started studying near-wall colloidal dynamics by observing the light scattered by micron-sized particles inside evanescent wave field. Notable achievements include successful and accurate measurements of gravitational attraction, double-layer repulsion, hindered diffusion, van der Waals forces, optical forces, depletion and steric interactions, and particle surface charges [17,25–32]. One application of particular relevance to our discussion is that of Prieve and coworkers [30–32], who used the evanescent field as a means to measure the behavior of micron-sized colloidal particles in close proximity to a solid surface. Although this was not strictly velocimetry, they did track the statistical motion of particles in the evanescent field to back out the contributions of Brownian motion, gravitational sedimentation, and electrostatic surface interactions. However, evanescent wave scattering microscopy could not further advance to the nanoscale because scattering by nanometer-sized particles is weak and thus limited the minimum particle size for evanescent wave scattering microscopy to approximately 1 µm. Therefore, without fluorescence, experimental investigations of near-surface phenomena would be restricted to at least 1 µm away from the solid boundary.

The advantage of the TIRF microscopy technique, in contrast to evanescent wave light scattering microscopy, lies in its ability to produce extremely confined illumination and submicron imaging depths and resolutions at a dielectric interface by reflecting an electromagnetic wave off of the interface. An extremely high sensitivity is achieved by imaging fluorescent dyes or particles and illuminating only those fluorophores within the first few hundred nanometers of the interface (Figure 3.1). Because no extraneous, out-of-focus fluorescence is excited, there is little background noise as demonstrated by the TIRF image of the 200-nm-diameter colloidal particles in Figure 3.2. In addition, because the illumination intensity decreases monotonically away from the interface, it is possible to infer an object's distance from the interface through intensity.

Particle-based velocimetry has long been used in flow visualization and measurement [33]. It is based on an intuitive and for most part correct assumption that the seeding tracer particles are carried by the fluid surrounding them and therefore their translational velocities must be that of the local fluid elements. Therefore, fluid velocities can be inferred from

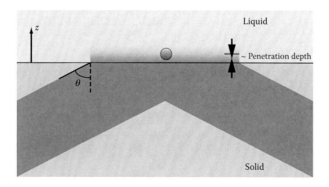

FIGURE 3.1
A schematic of TIRF microscopy. An illumination beam is brought to the liquid–solid interface at an incident angle, θ, greater than the critical angle predicted by Snell's law. As a result, total internal reflection occurs at the solid–liquid interface, and an evanescent field is created in the liquid phase. The evanescent energy then illuminates the encapsulated fluorophores inside a colloidal particle in the close vicinity of the interface.

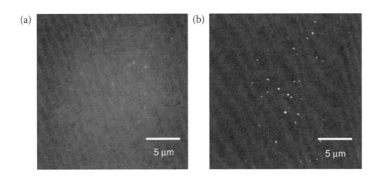

FIGURE 3.2
Sample images of (a) conventional bright-field illumination versus and (b) TIRF illumination for 200-nm particles.

apparent velocities of the tracer particles calculated on the basis of displacements of the tracer particles and the time between successive particle imaging. When particle-based velocimetry methods were adopted to study microfluidics, submicron fluorescent tracer particles were used to minimize light scattering and imaging noise while attaining spatial resolutions of tens of nanometers [34]. The first concerted effort to use TIRF microscopy with particle image velocimetry (PIV) was reported by Zettner and Yoda [35], who demonstrated prism-coupled TIRF to measure the motion of tracer particles within the electric double layer of an electroosmotic flow in a microchannel. Although ground breaking, the resolution of this approach was somewhat limited by the relatively poor spatial and temporal performance of the camera system used. More recently, TIRF microscopy has been integrated with improved particle velocimetry techniques and termed total internal reflection velocimetry [36] and multilayer nano-PIV (nPIV) [37]. These methods have been used to measure the dynamics of significantly smaller scales (10–300 nm), and applications have included the characterization of electroosmotic flows [38], slip flows [39,40], hindered diffusion [41,42], near-wall shear flows [43–47], and quantum dot (QD) tracer particles [45,48–52]. It is believed that evanescent-wave-based near-surface particle-tracking velocimetry (PTV) will become a workhorse in near-surface nanofluidic, colloidal, and molecular dynamic investigations as technology strives further toward smaller and smaller-scale systems.

The last 10 years have seen a number of near-surface PTV-based experimental results, each with different approaches, advantages, and disadvantages, as will be discussed during the remainder of the chapter. In this chapter, we provide a thorough review of the established fundamentals and recent development of near-surface PTV to the reader. We first discuss the theories, measurement designs, and experimental procedures that are essential to successful near-surface PTV for nanofluidics. We follow that with a discussion of experimental studies reported in the literature and state-of-the-art development of evanescent-wave-based velocimetry techniques. We then conclude with the future directions and perceived potentials of near-surface PTV in nanotechnology.

3.2 Theoretical Considerations

In this section, we present theoretical considerations most closely relevant to conducting near-surface PTV. They include evanescent wave illumination, fluorescent particle intensity variations, hindered Brownian motion, near-wall shear effects, and particle distribution.

3.2.1 Evanescent Wave Illumination

When an electromagnetic plane wave (light) in a dielectric medium of refractive index, n_1, is incident upon an interface of a different dielectric material with a lower index of refraction, n_2, at an angle, θ, greater than the critical angle predicted by Snell's law such that $\theta > \theta_{cr} = \sin^{-1}(n_2/n_1)$, total internal reflection occurs at the interface between the two media as illustrated in Figures 3.3 and 3.4.

Although all of the incident energy is reflected, the full solution of Maxwell's equations predicts that in the less dense medium there exists an electromagnetic field whose intensity decays exponentially away from the two-medium interface. This electromagnetic field, termed "evanescent waves" or "evanescent field," propagates parallel to the interface and has a decay length, δ, on the order of the wavelength of the illuminating light, λ. Furthermore, photons are not actually reflected at the interface but rather tunnel into the low-index material (a process called "optical tunneling"). As a result, the reflected beam of light is shifted along the interface by a small amount ($\Delta x \approx 2\delta \tan \theta$), which is known as the Goos–Haenchen shift [53].

The full details about this solution of Maxwell's equations are outlined elsewhere [22]. Only the basic results relevant to evanescent wave microscopy are presented below, specifically the intensity distribution in the lower optical density material. The solution

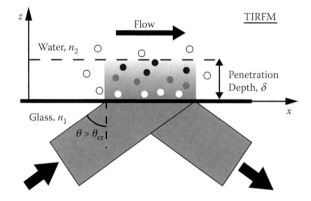

FIGURE 3.3
A schematic of evanescent wave illumination.

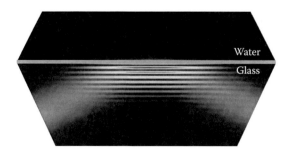

FIGURE 3.4
COMSOL simulation of total internal reflection in Figure 3.3. Plotted in the figure is time-averaged total electromagnetic energy density in the vicinity of the glass–water interface. Because the refractive indices of glass and water are 1.515 and 1.33, respectively, total internal reflection occurs at $\theta > \theta_{cr} = 61.39°$. In this figure, the incident angle of the incoming Gaussian illumination beam is $\theta = 64.54°$, and the illumination wavelength is 514 nm.

presented here assumes an infinite plane wave incident on the interface, which is a good approximation to a Gaussian laser beam typically used in practice. The intensity has the exponential form

$$I(z) = I_0 e^{-z/\delta}, \tag{3.1}$$

where z is the coordinate normal to the interface into the low-index medium, I_0 is the wall intensity, and the decay length, δ, is given by

$$\delta = \frac{\lambda}{4\pi n_1} [\sin^2 \theta - n^2]^{-1/2}, \tag{3.2}$$

and $n = n_2/n_1 < 1$. In a typical system with a glass substrate ($n_1 = 1.515$), water as the working fluid ($n_2 = 1.33$), and an argon ion laser for illumination ($\lambda = 514$ nm), a penetration depth of approximately $\delta = 128$ nm can be produced with an incident angle of $\theta = 64.54°$ (Figure 3.5). The polarization of the incident beam does not affect the penetration depth, but it does affect the amplitude of the evanescent field. For plane waves incident on the interface with intensity, I_1, in the dense medium, the amplitude of the field in the less dense medium, I_0, is given by

$$I_0^{\parallel} = I_1^{\parallel} \frac{4\cos^2 \theta (2\sin^2 \theta - n^2)}{n^4 \cos^2 \theta + \sin^2 \theta - n^2}, \tag{3.3}$$

$$I_0^{\perp} = I_1^{\perp} \frac{4\cos^2 \theta}{1 - n^2}, \tag{3.4}$$

for incident waves parallel and perpendicular to the plane of incidence, respectively, as shown in Figure 3.6. Both polarizations yield a wall intensity significantly greater than the

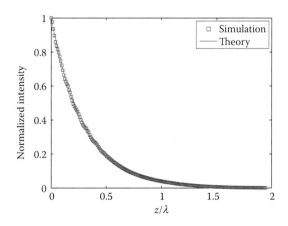

FIGURE 3.5
The exponential intensity decay of evanescent field in Figure 3.4. There exists a close agreement between the numerical solution of Maxwell's equations (COMSOL simulation) and theoretical calculations (Equations 3.1 and 3.2).

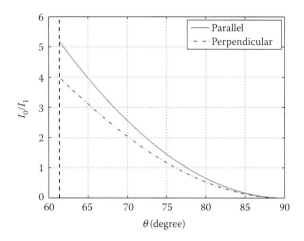

FIGURE 3.6
Evanescent field wall intensity as a function of incident angle for both parallel and perpendicular polarizations at a typical glass–water interface.

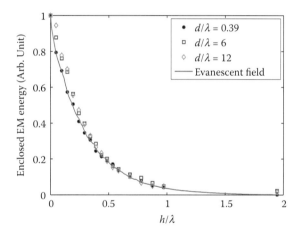

FIGURE 3.7
Enclosed electromagnetic (EM) energy inside suspended particles when illuminated by evanescent waves. The enclosed EM energies are obtained from COMSOL simulations of spherical polystyrene particles at various diameters, d, and gap sizes where $h = z - d/2$ is the shortest distance between the particle surface and the glass substrate ($n = 1.515$). A particle would be touching the substrate if $h = 0$. The suspending liquid of consideration is water ($n = 1.33$). The illumination wavelength is $\lambda = 514$ nm. The enclosed EM energies are normalized by the total enclosed EM energy inside a particle when $h = 0$. The evanescent field intensity decay curve is obtained from Figure 3.4.

incident radiation, with the parallel polarization being 25% greater than the perpendicular polarization.

Using the TIRF microscopy, many researchers have exploited the monotonic decay of the evanescent field to map the intensity of fluorescent dye molecules or particles to their distances from the fluid/solid interface [39,41,42,46]. It is intuitive to assume that for a particle that has fluorophores embedded throughout its whole volume, its fluorescent intensity will be proportional to the amount of evanescent electromagnetic energy entering its spherical shape. In Figure 3.7, the amounts of electromagnetic energies enclosed inside particles of various sizes in evanescent fields are found to be in close agreement with the local intensities of the illuminating evanescent waves. It can, therefore, be inferred that the emission intensities of fluorescent particles can be used to determine the distances between the particles and the fluid/solid interface. Still practical applications of such intensity–position correlation require additional experimental calibrations (Section 3.3.3).

3.2.2 Fluorescent Nanoparticle Intensity Variation

When applying the intensity–distance correlation described in the previous section to an ensemble of nanometer-sized particles that are typically used in fluid mechanics and colloid dynamics measurements, one must consider the polydispersity of the particles and the variation of emission intensity with particle size. All commercially available polystyrene and latex nanoparticles are manufactured with a finite-size distribution where the particle radius is specified by a mean value a_0 and a coefficient of variation up to 20%. Several researchers have attempted to compensate for this variation statistically when making ensemble-averaged measurements of fluorescent nanoparticles with TIRF [39,47]. Most manufacturers impregnate the volume of the polymer particles with fluorescent dye and thus it is often assumed that the light intensity emitted by a particle is proportional to its volume. For instance, Huang et al. [39] proposed that the intensity of a given particle, I^P, of radius a at a distance h from the interface is

$$I^P(z,a) = I_0^P \left(\frac{a}{a_0} \right)^3 \exp\left[-\frac{z-a}{\delta} \right], \tag{3.5}$$

where I_0^P is the intensity of a particle with a radius a_0 and δ is the penetration depth of the evanescent field.

Below, we quote results from Chew [54] for dipole radiation inside dielectric spheres to support the claim that particle intensity is proportional to volume and demonstrate the limits of this assumption for larger particles. Consider a dielectric sphere of radius, a, permittivity, ε_1, permeability, μ_1, and index of refraction, $n_1 = \sqrt{\mu_1 \varepsilon_1}$, inside of a second infinite dielectric medium with ε_2, μ_2, and n_2. The radiation from an emitting dipole with free space wavelength, λ_0, will have momentum vectors, $k_{1,2} = 2\pi n_{1,2}/\lambda_0$, and subsequently, $\rho_{1,2} = k_{1,2}a$. The power emitted by a dipole is proportional to the dipole transition rate, $R^{\perp,\|}$, for perpendicular and parallel polarizations. These relations are provided in Chew [54] and are normalized by the transition rates for dipoles contained in an infinite medium 1, $R^{\perp,\|}/R_0^{\perp,\|}$. For a distribution of dipoles, $c(\vec{r})$, located within the sphere, the volume-averaged emission is

$$\left\langle \frac{R^{\perp,\|}}{R_0^{\perp,\|}} \right\rangle = \frac{\int \left(R^{\perp,\|}/R_0^{\perp,\|} \right) c(\vec{r}) \, d^3\vec{r}}{\int c(\vec{r}) \, d^3\vec{r}}. \tag{3.6}$$

The volume-averaged emission for randomly oriented dipoles, R/R_0, with a uniform concentration distribution, $c(\vec{r}) = c_0$, is

$$\left\langle \frac{R}{R_0} \right\rangle = \frac{1}{3} \left\langle \frac{R^\perp}{R_0^\perp} + 2\frac{R^\|}{R_0^\|} \right\rangle = 2H \sum_{n=1}^{\infty} \left[\frac{J_n}{|D_n|^2} + \frac{GK_n}{|D_n'|^2} \right], \tag{3.7}$$

where

$$H = \sqrt{\frac{\mu_1 \varepsilon_1 \varepsilon_2}{\mu_2}} \left(\frac{9\varepsilon_1}{4\rho_1^5} \right),$$

$$G = \frac{\mu_1 \mu_2}{\varepsilon_1 \varepsilon_2},$$

$$K_n = \frac{\rho_1^3}{2} [j_n^2(\rho_1) - j_{n+1}(\rho_1) j_{n-1}(\rho_1)],$$

$$J_n = K_{n-1} - n\rho_1 j_n^2(\rho_1),$$

$$D_n = \varepsilon_1 j_n(\rho_1) [\rho_2 h_n^{(1)}(\rho_2)]' - \varepsilon_2 h_n^{(1)}(\rho_2) [\rho_1 j_n(\rho_1)]',$$

$$D_n' = \mu_1 j_n(\rho_1) [\rho_2 h_n^{(1)}(\rho_2)]' - \mu_2 h_n^{(1)}(\rho_2) [\rho_1 j_n(\rho_1)]'.$$

Spherical Bessel functions of the first kind are denoted by j_n, and spherical Hankel functions of the first kind are denoted by $h_n^{(1)}$. The terms D_n and D_n' are the same denominators of the Mie scattering coefficients [55]. In the Rayleigh limit ($ka \ll 1$), the transition rates become independent of polarization and simplify greatly to

$$\left\langle \frac{R}{R_0} \right\rangle = \frac{9}{((\varepsilon_1/\varepsilon_2) + 2)^2} \sqrt{\frac{\varepsilon_2 \mu_2^3}{\varepsilon_1 \mu_1^3}}. \tag{3.8}$$

Figure 3.8 shows the normalized mean emission rates (power) for both the Rayleigh limit and the full solution proposed by Chew [54] as a function of the particle radius, a, for a polystyrene particle ($n_1 = 1.59$) immersed in water ($n_2 = 1.33$) with an emission wavelength $\lambda_0 = 600$ nm. For particles with radius $a \leq 200$ nm, the Rayleigh limit is a good approximation to the full solution.

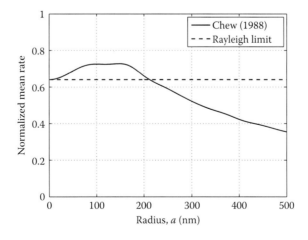

FIGURE 3.8
Volume-averaged emission rate for uniformly distributed, randomly oriented radiating dipoles with emission wavelength $\lambda_0 = 600$ nm, within a polystyrene sphere ($n_1 = 1.59$) immersed in water ($n_2 = 1.33$).

The total power, $\dot{E}$, emitted by a particle is the volume-averaged emission rate, $\langle R/R_0 \rangle$, scaled by the volume of a given particle

$$\dot{E} = \frac{4}{3}\pi a^3 \left\langle \frac{R}{R_0} \right\rangle. \tag{3.9}$$

The total power emitted by a particle is shown in Figure 3.9 normalized by the emission of a particle with radius $a = 500$ nm. Because all particle radii considered here are subwavelength ($a < \lambda_0$), the Rayleigh limit is a descent approximation. For particles with radii $a \lesssim 125$ nm, the Rayleigh approximation follows the full solution quite closely as seen in the inset of Figure 3.9. Thus, the total power scales with the particle volume for subwavelength particle at or near the Rayleigh limit, which partially vindicates the approximation made in Equation 3.5. Further validation of Equation 3.5 can be achieved through verification of the uniformity of excitation in both the plane wave and the evanescent wave excitation cases.

3.2.3 Hindered Brownian Motion

The Brownian motion of small particles due to molecular fluctuations is generally well understood [56] and can be significant in magnitude for nanoparticles commonly used for near-surface PTV. The random thermal forcing of the particles is damped by the hydrodynamic drag resulting from the surrounding solvent molecules, and the particle's motion can be described as a diffusion process [57]:

$$\frac{\partial p(\vec{r},t)}{\partial t} = \nabla \cdot (D(\vec{r})\nabla p(\vec{r},t)), \tag{3.10}$$

where p is the probability of finding a particle at a given location, $\vec{r}$, at time, t, and D is the diffusion coefficient. For an isolated spherical particle that is significantly larger

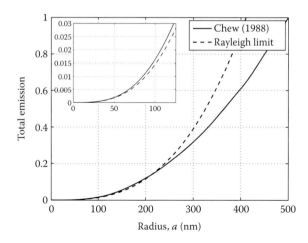

FIGURE 3.9
Total emission (power) for uniformly distributed, randomly oriented radiating dipoles with emission wavelength, $\lambda_0 = 600$ nm, within a polystyrene sphere ($n_1 = 1.59$) immersed in water ($n_2 = 1.33$). For particles approaching the Rayleigh limit with radii, $a \leq 125$ nm, the emitted power scales with the particle volume.

than the surrounding solvent molecules, the diffusion coefficient is constant and isotropic, and it is described by the Stokes–Einstein relation [58]

$$D_0 = \frac{k_B T}{\xi} = \frac{k_B T}{6\pi\mu a},$$ (3.11)

where k_B is Boltzmann's constant, T is the absolute temperature, ξ is the drag coefficient, μ is the dynamic viscosity of the solvent, and a is the particle radius. In this case, the solution to Equation 3.10 subject to the condition that $p(\vec{r}, t = t_0) = \delta(\vec{r} - \vec{r}_0)$ becomes

$$p(\vec{r},t) = \frac{1}{8(\pi D_0 \Delta t)^{3/2}} \exp\left[-\frac{|\vec{r}-\vec{r}_0|^2}{4 D_0 \Delta t}\right],$$ (3.12)

where $\Delta t = t - t_0$ [59].

When an isolated particle in a quiescent fluid is in the vicinity of a solid boundary, its Brownian motion is hindered anisotropically because of an increase in hydrodynamic drag. Several theoretical studies have accurately captured this effect for various regimes of particle-wall separation distance [15,60–62]. The hindered diffusion coefficient in the wall-parallel direction, D_x, is described by

$$\frac{D_x}{D_0} = 1 - \frac{9}{16}\left(\frac{z}{a}\right)^{-1} + \frac{1}{8}\left(\frac{z}{a}\right)^{-3} - \frac{45}{256}\left(\frac{z}{a}\right)^{-4} - \frac{1}{16}\left(\frac{z}{a}\right)^{-5} + O\left(\frac{z}{a}\right)^{-6},$$ (3.13)

where z is the particle center distance to the wall. This is a direct result of the drag force on a moving particle near a stationary wall in a quiescent fluid calculated by the "method of reflections," which is accurate far from the wall, $z/a > 2$ [63]. A better approximation for small particle-wall separation distances results from an asymptotic solution for the drag force on the basis of lubrication theory for $z/a < 2$ [61]. Under these assumptions, the corresponding normalized diffusion coefficient is

$$\frac{D_x}{D_0} = -\frac{2[\ln((z/a)-1)-0.9543]}{[\ln((z/a)-1)]^2 - 4.325\ln((z/a)-1)+1.591}.$$ (3.14)

The relative hindered diffusion coefficient for a particle diffusing in the wall-normal direction, D_z, is described by

$$\frac{D_z}{D_0} = \left\{\frac{4}{3}\sinh\alpha \sum_{n=1}^{\infty} \frac{n(n+1)}{(2n-1)(2n+3)} \right.$$
$$\left.\left[\frac{2\sinh(2n+1)\alpha + (2n+1)\sinh 2\alpha}{4\sinh^2(n+(1/2))\alpha - (2n+1)^2\sinh^2\alpha} - 1\right]\right\}^{-1},$$ (3.15)

where $\alpha = \cosh^{-1}(z/a)$. This equation results from an exact solution of the force experienced by a particle for motions perpendicular to a stationary wall in a quiescent fluid [60]. The wall-parallel and wall-normal hindered diffusivities described in Equations 3.13 through

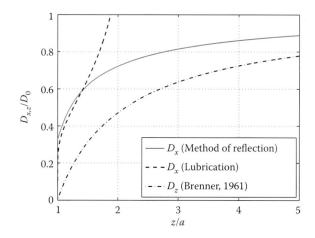

FIGURE 3.10
Hindered diffusion coefficients in the wall-parallel, D_x, and wall-normal, D_z, directions for a neutrally buoyant spherical particle near a solid boundary.

3.15 are plotted in Figure 3.10. We also note that Equation 3.15 has been shown to be well approximated by

$$\frac{D_z}{D_0} = \frac{6((z/a)-1)^2 + 2((z/a)-1)}{6((z/a)-1)^2 + 9((z/a)-1) + 2}, \tag{3.16}$$

which is convenient for fast computation [28]. These theoretical results have been verified over different length scales by various researchers including several evanescent wave illumination studies [28,30,42,64–67]. Deviations from the bulk diffusivity become noticeable for particle-wall separation distances of order 1. For instance, the wall-parallel diffusivity drops to one half of its bulk value when $z/a \approx 1.2$, whereas the wall-normal diffusivity drops to one half at $z/a \approx 2.1$. The implications of hindered Brownian motion on near-surface PTV have been intensely investigated recently and are further discussed in Section 3.4.4.

3.2.4 Near-Wall Shear Effects

The motion of an incompressible fluid obeys the continuity condition (conservation of mass)

$$\nabla \cdot \vec{u} = 0, \tag{3.17}$$

where the velocity field, $\vec{u}$, is divergence free. The Navier–Stokes equation governs the motion of a viscous fluid and in the case of an incompressible fluid is

$$\rho\left(\frac{\partial \vec{u}}{\partial t} + \vec{u} \cdot \nabla \vec{u}\right) = -\nabla P + u\nabla^2 \vec{u}, \tag{3.18}$$

where ρ is the fluid density and P is the dynamic pressure. For the small Reynolds numbers ($Re \ll 1$) typical of microfluidic devices, Equation 3.18 greatly reduces in complexity to Stokes' equation [68]:

$$\nabla P = u \nabla^2 \vec{u}. \tag{3.19}$$

Most microfabrication techniques produce microchannels with approximately rectangular cross sections. Thus, a useful result from Equation 3.19 is the solution for the velocity profile in a rectangular duct with height, d, and width, w, subject to the no-slip boundary condition ($\vec{u} = 0$) at the walls. The laminar, unidirectional flow occurs in the pressure gradient direction with velocity, u_x, described below [59]:

$$u_x(y,z) = \frac{1}{2\mu}\left(-\frac{\partial P}{\partial x}\right)\left[\frac{d^2}{4} - z^2 + \frac{8}{d}\sum_{n=1}^{\infty}\frac{(-1)^n \cos(mz)\cosh(my)}{m^3 \cosh(mw/2)}\right],$$
$$m = \frac{\pi(2n-1)}{d}, \tag{3.20}$$

where the pressure gradient and volumetric flow rate, Q, are related by

$$Q = \frac{wd^3}{12\mu}\left(-\frac{\partial P}{\partial x}\right)\left[1 - \frac{192d}{\pi^5 w}\sum_{n=1,3,5,\ldots}^{\infty}\frac{1}{n^5}\frac{1-\exp(-n\pi w/d)}{1+\exp(-n\pi w/d)}\right]. \tag{3.21}$$

It is well known that rigid particles tend to rotate in shear, and in the special case of a sphere near a planar wall, additional hydrodynamic drag slows the particle's translational velocity below that of the local fluid velocity [15,62]. For wide microchannels ($w \gg d$) in the very near-wall region ($h \ll d$), the nearly parabolic velocity profile can be approximated by a linear shear flow

$$u \approx zS, \tag{3.22}$$

where S is the shear rate. The wall-parallel drag force experienced by a neutrally buoyant, free particle with radius a and a distance z between its center and the wall in a linear shear flow results in a particle translational velocity, v, that is different from the fluid velocity at the particle center's plane. For large z/a, the particle's translational velocity can be estimated by the method of reflections [62]. This translational velocity, normalized by the local fluid velocity at the particle's center, is

$$\frac{v}{zS} \simeq 1 - \frac{5}{16}\left(\frac{z}{a}\right)^{-3}. \tag{3.23}$$

For small particle-wall separation distances (small z/a), an asymptotic solution on the basis of lubrication theory has also been established as

$$\frac{v}{zS} \simeq \frac{0.7431}{0.6376 - 0.2\ln((z/a)-1)}, \tag{3.24}$$

which is also normalized by the unperturbed fluid velocity at the particle's center [62]. The method of reflections solution and asymptotic lubrication solution from Equations 3.23 and 3.24 are shown in Figure 3.11. In addition, Pierres et al. [69] used a cubic approximation to segment the solutions for intermediate values of z/a:

$$\frac{v}{zS} \simeq \left(\frac{z}{a}\right)^{-1} \exp\left\{0.68902 + 0.54756\left[\ln\left(\frac{z}{a}-1\right)\right]\right.$$
$$\left. + 0.072332\left[\ln\left(\frac{z}{a}-1\right)\right]^2 + 0.0037644\left[\ln\left(\frac{z}{a}-1\right)\right]^3\right\}. \tag{3.25}$$

Particle rotation can also induce a lift force, which tends to make the particles migrate away from the wall [70]. Obviously, such lift force can potentially lead to biased sampling of local fluid velocities by the tracer particles during near-surface PTV measurements and should be cautiously treated when designing experiments and analyzing results. The subject of lift forces acting on a small sphere in a wall-bounded linear shear flow has been thoroughly studied by Cherukat and McLaughlin [71]. Here we will present only the theory that applies to the flow and colloidal conditions commonly encountered in near-surface PTV experiments. Suppose that a free-rotating rigid sphere of radius a is in a Newtonian incompressible fluid of kinematic viscosity ν and is in the vicinity of a solid wall. In the presence of a linear shear flow, this free-rotating sphere can travel at a velocity U_{sph} that is different from the fluid velocity, U_G, of the shear plane located at its center [62] because of shear-induced particle rotation described previously. We can define a characteristic Reynolds number

$$\text{Re}_\alpha = \frac{U_s \alpha}{\nu}, \tag{3.26}$$

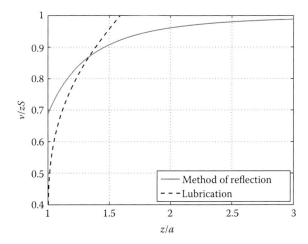

FIGURE 3.11
Particle velocity normalized by the local fluid velocity in a near-wall shear flow given by the method of reflection and lubrication approximation solutions from Goldman et al. [62].

on the basis of the velocity difference, $U_s = U_{sph} - U_G$. A second characteristic Reynolds number on the basis of shear rate can be defined as

$$Re_\beta = \frac{Ga^2}{v},$$ (3.27)

where G is the wall shear rate. In this geometry, the wall is considered as located in the "inner region" of flow around the particle if $Re_\alpha \ll \Omega$ and $Re_\beta \ll \Omega^2$, where $\Omega \equiv a/(z-a)$. For near-wall PTV using nanoparticles, $Re_\alpha \sim Re_\beta \lesssim 10^{-4}$ while $\Omega \sim O(1)$; thus, the inner region theory of lift force applies.

For a flat wall located in the inner region of flow around a free-rotating particle, the lift force, F_L, which is perpendicular to the wall, is scaled by [71]

$$F_L \sim Re_\alpha \cdot I_\Omega,$$ (3.28)

where I_Ω is a coefficient that can be numerically estimated by

$$\begin{aligned}
I_\Omega = &[1.7631 + 0.3561\Omega - 1.1837\Omega^2 + 0.845163\Omega^3] \\
&- \left[\frac{3.21439}{\Omega} + 2.6760 + 0.8248\Omega - 0.4616\Omega^2\right]\left(\frac{Re_\beta}{Re_\alpha}\right) \\
&+ [1.8081 + 0.879585\Omega - 1.9009\Omega^2 + 0.98149\Omega^3]\left(\frac{Re_\beta}{Re_\alpha}\right)^2.
\end{aligned}$$ (3.29)

Again, for near-surface PTV using nanoparticles, $I_\Omega \lesssim O(10^2)$. Therefore,

$$F_L \sim Re_\alpha \cdot I_\Omega \lesssim (10^{-4})(10^2) \ll 1,$$ (3.30)

and the lift force acting on near-wall particles is insignificant and can be neglected for most practical cases.

3.2.5 Near-Wall Particle Concentration

Electrostatic forces arise from the Coulombic interactions between charged bodies such as polystyrene tracer particles and glass immersed in water. When immersed in an ionic solution, these forces are moderated by the formation of an ionic double layer on their surfaces, which screen the charge. The characteristic length scale of these forces is given by the Debye length,

$$\kappa^{-1} = \sqrt{\frac{\varepsilon_f \varepsilon_0 k_B T}{2ce^2}},$$ (3.31)

where ε_0 is the permittivity of free space, ε_f is the relative permittivity of the fluid, e is the elementary charge of an electron, and c is the concentration of ions in solution [72]. In the case of a plane–sphere geometry for like-charged objects (e.g., a spherical polystyrene particle and a flat glass substrate), the immobile substrate can exert a repulsive force on the particle and is quantified by the potential energy of the interaction:

$$U^{el}(z) = B_{ps}e^{-\kappa(z-a)}. \tag{3.32}$$

The magnitude of the electrostatic potential is given by

$$B_{ps} = 4\pi\varepsilon_f\varepsilon_0 a\left(\frac{k_B T}{e}\right)^2\left(\frac{\hat{\psi}_p + 4\gamma\Omega\kappa a}{1+\Omega\kappa a}\right)\left[4\tanh\left(\frac{\hat{\psi}_s}{4}\right)\right], \tag{3.33}$$

where $\gamma = \tanh(\hat{\psi}_p/4)$, $\Omega = (\hat{\psi}_p - 4\gamma)/2\gamma^3$, $\hat{\psi}_p = \hat{\psi}_p e/k_B T$, and $\hat{\psi}_s = \hat{\psi}_s e/k_B T$ [73]. ψ_p and ψ_s represent the electric potentials of the particle and the substrate, respectively.

Contributions from attractive, short-ranged van der Waals forces, which originate from multipole dispersion interactions [74], should also be considered when the particle-wall separation is on the order of 10 nm. The potential due to van der Waals interactions for a plane–sphere geometry is given by

$$U^{vdw}(z) = -\frac{A_{ps}}{6}\left[\frac{a}{z-a} + \frac{a}{z+a} + \ln\left(\frac{z-a}{z+a}\right)\right], \tag{3.34}$$

where A_{ps} is the Hamaker constant [75, 76]. The gravitational potential can also be important for large or severely density mismatched particles. The gravitational potential of a buoyant particle in a fluid is given by

$$U^g(z) = \frac{4}{3}\pi a^3(\rho_s - \rho_f)g(z - a), \tag{3.35}$$

where ρ_s and ρ_f are the densities of the sphere and fluid, respectively, and g is the acceleration due to gravity.

Finally, optical forces due to electric field gradients from the illuminating light can trap or push colloidal particles [77]. Below, we present an order of magnitude estimation for the potential of a dielectric particle in a weak illuminating evanescent field typically found in evanescent-wave-based near-surface PTV. Following Novotny and Hecht [78], the force on a dipole is given by

$$\vec{F} = (\vec{\mu}\cdot\nabla)\vec{E} = (\alpha\vec{E}\cdot\nabla)\vec{E} = \alpha\nabla\left|\vec{E}\right|^2. \tag{3.36}$$

The dipole moment, $\vec{\mu}$, polarizability, α, and electric field magnitude are given by the following:

$$\vec{\mu} = \alpha\vec{E}, \tag{3.37}$$

$$\alpha = 4\pi\varepsilon_0 a_0^3\frac{n_p^2 - n_0^2}{n_p^2 + 2n_0^2}, \tag{3.38}$$

$$I(z) = I_0 e^{-z/\delta} = \frac{1}{2} c \varepsilon_0 n_0 |\vec{E}|^2, \tag{3.39}$$

where n_0 is the index of the surrounding medium, n_p is the index of the particle, and c is the speed of light in a vacuum. Combining the above expressions, we can write an approximation to the potential of a particle near an interface due to an evanescent field:

$$U^{\mathrm{opt}} \approx -\alpha |\vec{E}|^2 = -\frac{8\pi a_0^3}{cn_0} \frac{n_p^2 - n_0^2}{n_p^2 + 2n_0^2} I_0 e^{-z/\delta}, \tag{3.40}$$

which is an attractive force. However, for strongly light-absorbing particles such as the semiconductor materials found in QDs, this optical force can be repulsive and more detailed analyses should be carefully carried out.

The equilibrium distribution for an ensemble of noninteracting, suspended Brownian particles in an external potential has been given by a Boltzmann distribution [79]. For a brief discussion, we follow Doi and Edwards [79] and consider a one-dimensional distribution below. Fick's law of diffusion describes the flux, j, of a material

$$j(z,t) = -D \frac{\partial p(z,t)}{\partial z}, \tag{3.41}$$

where D is the diffusion coefficient and p is the continuous probability function of finding a particle at a location z in the wall-normal coordinate at time t. In the presence of an external potential, $U(z)$, particles experience an additional force

$$F_z = -\frac{\partial U}{\partial z}. \tag{3.42}$$

Fick's law (Equation 3.41) is modified to reflect the additional flux induced by this force

$$j(z,t) = -D \frac{\partial p}{\partial z} - \frac{p}{\xi} \frac{\partial U}{\partial z}, \tag{3.43}$$

where the drag coefficient, ξ, is related to the diffusion coefficient, $D = k_B T / \xi$. In the steady state, the net flux vanishes, $j \to 0$, and the solution of Equation 3.43 leads to the Boltzmann distribution

$$p(z) = \frac{\exp[-U(z)/k_B T]}{\int_{z_1}^{z_2} \exp[-U(z)/k_B T] dz} = p_0 e^{-U/k_B T}, \tag{3.44}$$

where p_0 is a normalization constant [30] from all particles in the range $z_1 \le z \le z_2$ and U is the total potential energy given by the sum of all potentials experienced by the particle (electrostatic, van der Waals, etc.). An example illustrating the nonuniform particle concentrations in the near-wall region is shown in Figure 3.12 for a 500-nm-diameter polystyrene particle in water near a glass substrate with a 10-nm Debye length. In this case,

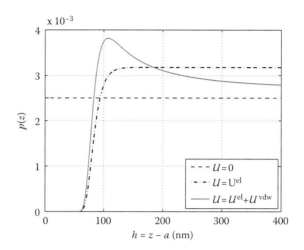

FIGURE 3.12
Near-wall particle concentration profiles for a 500-nm-diameter polystyrene particle in water near a glass substrate with the effect of electrostatic (10-nm Debye length) and van der Waals interactions.

electrostatic and van der Waals forces dominate, clearly forming a depletion layer within approximately 100 nm of the wall. The implications of the presence of a near-surface particle depletion layer to velocimetry accuracy have recently been investigated [44,47].

3.3 Experimental Procedures

In this section, we discuss in detail the experimental procedures of conducting successful evanescent-wave-based near-surface PTV, including selection of experimental materials, sample preparation, optical and imaging setup, measurement calibration, and the particle-tracking algorithm.

3.3.1 Materials and Preparations

As discussed before, creation of evanescent waves inside a microfluidic or nanofluidic channel requires the solid substrate or the channel wall to have higher optical density than the flowing fluid. That is, the substrate must have an index of refraction higher than that of the fluid. Furthermore, the substrate must be transparent to both the illumination and the fluorescence emission wavelengths for high-precision imaging. Examples of solid materials that satisfy these conditions include glass ($n = 1.47$–1.65), quartz ($n = 1.55$), polymethyl methacrylate ($n = 1.49$), and other types of clear plastics. Among these, glass is the most common choice as it is chemically inert, physically robust, and optically transparent to all visible light wavelengths (350–700 nm). Surface roughness of less than 10 nm is found to be not impeding the creation of evanescent waves. However, surface waviness presents a more critical issue as the local illumination incident angle could significantly deviate from the predicted value and thus changes the properties of the created evanescent waves. It should also be noted that thin-film chemical coatings of subwavelength thickness on the substrate surface do not prevent creation of evanescent waves. Examples of coatings used in near-surface PTV includes octadecyltrichlorosilane [39,40] and P-selectin glycoprotein ligand-1 (PSGL-1) [47] self-assembled monolayers.

Selection of the experimental fluid is based on the following criteria: (1) lower refractive index than that of the solid substrate, (2) lack of chemical reactions with the solid substrate or surface coatings, (3) availability of chemically compatible tracer particles, (4) and desired physical properties such as density, viscosity, and polarity. Air and various inert gases have the lowest refractive indices possible among fluids ($n \approx 1$), but creating submicron-sized aerosol tracer particles presents a difficult challenge. Water ($n = 1.33$) is the most common choice of fluid for its chemical stability and compatibility with biochemical reagents. Other organic and inorganic solvents such as hexane ($n = 1.375$) and ethanol ($n = 1.36$) are also potential candidates.

A wide range of micron-sized and nanometer-sized tracer particles are commercially available for near-surface PTV. For light-scattering-based experiments, metallic, glass, and quartz particles should be considered as they are stronger scatterers of evanescent waves. For fluorescence-based measurements, tracer particle candidates include fluorescent polystyrene and latex particles, fluorescently tagged macromolecules such as dextran and DNA, and semiconductor materials such as QDs. Properties of tracer particles such as density, average size and size variations, deformation tendency, potential affinity to substrate, coagulation tendencies, chemical compatibility with fluid, and fluorescence quantum efficiency and wavelength should be carefully evaluated before and during experimentation. In general, particles that have fluorophores embedded throughout its whole volume is preferred for maximum imaging signals. Density mismatch between the tracer particles and the fluid can lead to buoyancy and sedimentation that cause velocimetry measurement bias. These problems can be avoided if the chosen type of tracer particles satisfies the condition,

$$\frac{4\pi a^4 g \left| \rho_p - \rho_f \right|}{3 k_B T} \ll 1, \tag{3.45}$$

where a is average particle radius, g is gravitational acceleration, ρ_p is particle density, ρ_f is fluid density, k_B is Boltzmann constant, and T is experimental fluid temperature. Coagulation of tracer particles also presents serious problems for particle identifications and intensity-based three-dimensional positioning and should be avoided as much as possible. Simple sonication of particle suspension is usually quite effective in breaking up particle clumps. Finally, the tracer particle seeding density of the measurement suspension should be moderately low to avoid tracking ambiguity between frames of images and assure velocimetry accuracy. A good rule of thumb is that the average spacing (in pixels) between adjacent tracer particles in the acquired images should be at least five times larger than the average particle size (also in pixels) of the same images.

3.3.2 Evanescent Wave Microscopy Setup

The basic components of an evanescent wave imaging system include a light source, conditioning optics, specimen or microfluidic device, fluorescence emission imaging optics, and a camera. In reported experimental setups, light sources have included both continuous-wave (CW) lasers (argon ion, helium neon) and pulsed lasers (Nd:YAG) because they produce collimated, narrow wavelength-band illumination beams. Noncoherent sources (arc-lamps) are not common because they require band-pass filters for wavelength selection and the light beams produced cannot be as perfectly collimated. However, commercial versions of lamp-based evanescent wave microscopes are available for qualitative imaging because they are more economical than laser-based systems. Conditioning optics are used to create the angle of incidence necessary for total internal reflection and are of two types: prism based and objective based [21]. Prism-based evanescent imaging systems are typically laboratory built and cost-effective. A prism is placed in contact with the sample substrate, and the illumination light beam is coupled into by inserting an immersion medium in between. The illumination beam is focused through the prism onto the substrate at an angle greater than the critical angle such that the substrate then becomes a waveguide where evanescent waves are generated along its surface (Figure 3.13a). An air- or a water-immersion, long working-distance objective is often used for imaging to prevent decoupling of the guided wave from the substrate into the objective. Detailed prism and microscope configurations can be found in Axelrod [21]. In contrast, objective-based evanescent wave imaging is used exclusively with fluorescence and requires a high numerical aperture objective (NA > 1.4) to achieve the large incident angles required for total internal reflection (Figure 3.13b). These objectives are usually high-magnification (60× ≤ M ≤ 100×) and oil immersion objectives. In this method, a collimated illumination light beam is focused onto the back focal plane (BFP) of the objective and translated off the optical axis of the objective to create the required large incident angle. The emitted fluorescence of tracer particles is collected by the objective and recorded by a camera as in typical

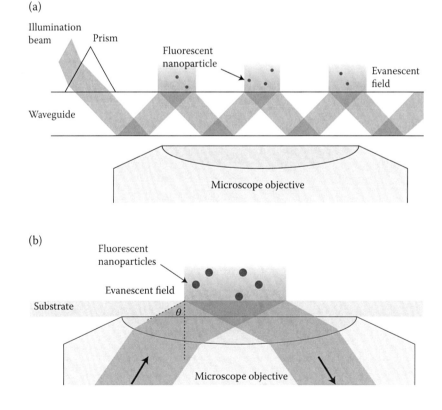

FIGURE 3.13
Two types of evanescent wave illumination: (a) prism-based setup and (b) objective-based setup.

fluorescence microscopy. To prevent the returning excitation light from being recorded by the camera, spectral filtering with dichroic mirrors and filters is used.

Another advantage of evanescent wave imaging to note is that its imaging depth provides significantly greater imaging resolution than the diffraction-limited depth of field (DOF) of the objective under bright-field illumination. The diffraction-limited DOF is calculated by

$$\text{DOF} = \frac{\lambda n_1}{\text{NA}^2} + \frac{n_1}{M \times \text{NA}} e, \tag{3.46}$$

where e is the smallest distance that can be resolved by the detector [80]. Even for a high-magnification and large NA microscope objective, the DOF is typically at least 600 nm and is thus unable to achieve submicron resolution.

An example of an objective TIRF microscope system is shown in Figure 3.14. An illumination beam produced by a laser is first regulated by a power attenuator–half-wave plate-polarizing beam splitter combination to achieve the desired power level. This step down in power is especially critical to high-power laser beams produced by pulsed lasers as their high-energy density can easily damage the optical components inside a microscope objective. A portion of the beam energy is diverted to an energy meter to monitor the laser stability. The beam is then "cleaned up" by a spatial filter (concave lens–10-μm pinhole–concave lens combination) before being directed through an NA1.45 100X oil-immersion microscope objective at an angle that creates evanescent waves inside a microchannel. Fluorescent images of near-surface tracer particles are captured by the same microscope objective and screened by a dichroic mirror and a barrier filter before being projected onto an intensified charge-coupled device (CCD) camera, capable of recording extremely low-intensity events. A TTL pulse generator is used to synchronize laser firing and intensified CCD image acquisitions to ensure precise control imaging timing. The energy of the

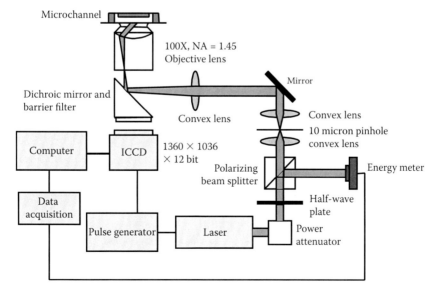

FIGURE 3.14
A schematic of objective-based TIRF microscope setup.

illuminating laser beam can also be recorded simultaneously with each image acquisition to account for illuminating energy fluctuation, if necessary.

Recent research literature has shown that objective-based TIRF microscopy systems are becoming much more common for experimental microfluidic and nanofluidic investigations. Extremely high signal-to-noise ratio images can be produced with proper alignment and conditioning of the incident beam and good control over the fluorescence imaging. Here, we discuss general methodologies for alignment and beam conditioning and supply the relevant details to reconstruct a proven system with an inverted epifluorescent microscope. The procedure, which follows, is valid for both pulsed and CW lasers. Caution should always be exercised when working with high-power laser. Proper eye protection should be worn at all times, and power should be kept to a minimum during alignment to prevent injury or damage to equipment.

To begin, a Coherent Innova CW Argon Ion laser capable of several hundred milliwatts of output power in both the green (514 nm) and the blue (488 nm) is used as the light source. The microscope is a Nikon TE2000-U with two epifluorescence filter turrets. The lower turret accommodates the mercury lamp, and the upper turret can be accessed from the rear of the microscope for free space optical alignment. The components of the conditioning and alignment optics are shown in Figure 3.15, which are categorized in several subsystems: power control, power meter, spatial filter, shifting prism, periscope, incident beam angle control, and reflected beam monitor. Not all of these systems are necessary for TIRF imaging (optional elements will be pointed out), but each system should be aligned in turn working from the laser to the microscope. We will discuss each system and its purpose.

The laser should always be operated near maximum power for the best thermal stability. As a safety mechanism and control, a mechanical beam chopper is placed directly in front of the laser aperture (Figure 3.15a). It is used as a beam stop for warm-up and can be controlled electronically for periodic modulation of the beam if desired. Next, the already

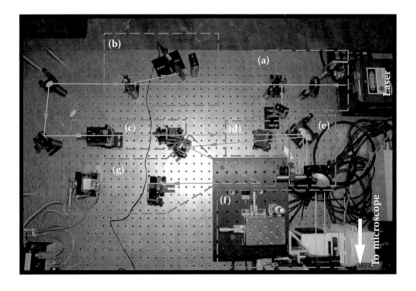

FIGURE 3.15

Schematic of the beam conditioning and manipulation optics for an objective-based TIRF microscopy system. The components are broken down into several subsystems: (a) power control, (b) power meter, (c) spatial filter, (d) shifting prism, (e) periscope, (f) incident beam angle control, and (g) reflected beam monitor.

vertically polarized laser passes through a half-wave plate to rotate the polarization to an arbitrary angle. The wave plate in combination with the polarizing beam splitter that follows allows one to continuously vary the evanescent wave intensity of the TIRF microscopy system while maintaining maximum operating power at the laser. By rotating the polarization, the vertical component is allowed to propagate along its original path through the beam splitter while the horizontally polarized light (excess power) is dumped to a beam stop toward the interior of the optical table. Next, the useful component of the beam is split again for power measurement (Figure 3.15b). With a fairly sensitive meter, a few hundred microwatts is sufficient for accurate measurement without being wasteful of the excitation power.

Two mirrors are now used to redirect the beam toward the back of the microscope and align the beam conveniently along the optical table bolt pattern using two irises (not shown). After referencing the beam to the table, a spatial filter is implemented to obtain the TEM00 mode and to expand the laser beam diameter (Figure 3.15c). The spatial filter movement containing an objective lens (f = 8 mm) and pinhole (~20 μm) is first aligned to be colinear with the beam, as shown in Figure 3.16a. When properly aligned, the spatial filter movement should produce a diverging, concentric ring pattern that is symmetric and bright (Figure 3.16b) while maintaining the beam propagation to the original trajectory along the optical table. To collimate and expand the beam, a lens ($f ≈ 20$ cm) is placed roughly one focal length away from the pinhole. The focal length of the lens and divergence angle of the expanding beam will determine the final beam diameter, which is approximately 1 cm in our case. Next, an adjustable iris is placed close to the collimating lens, with sufficient space in between for further adjustment of the lens (Figure 3.16a). The iris is aligned to block all of the rings from the diverging beam by narrowing the opening of the iris to the first minimum of the concentric ring pattern. If successful, one should now be left with a very "clean" Gaussian beam spot as shown in Figure 3.16c. Finally, fine-tune the position of the collimating lens such that the beam is collimated and again maintains the original trajectory along the bolt pattern of the optical table. One can determine if the beam is collimated by measuring the beam diameter just after the collimating lens and

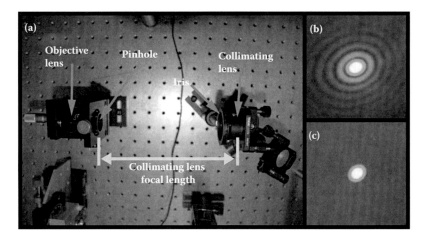

FIGURE 3.16
(See color insert.) Spatial filter schematic and resulting laser beam modes: (a) spatial filter components and orientation, (b) concentric rings produced by diffraction through the spatial filter pinhole, and (c) resulting Gaussian beam spot produced by clipping the concentric rings with an iris.

subsequently projecting the beam on a wall several meters away to ensure that the beam diameter remains the same.

The shifting prism (Figure 3.15d) is actually one of the final elements to be placed in the beam path and is optional. For now, we proceed with aligning the beam to the microscope's optical axis. First, an epifluorescence filter set to be used with the TIRF system is placed in the upper turret of the microscope. Next, one should prepare the objective-housing nosepiece of the microscope to have an empty slot, a mirror fixed atop a second empty slot with the reflective side facing down into the microscope, a slot occupied by a TIRF objective (100×, 1.45 NA or 60×, 1.49 NA in our case), and a last slot occupied by a low-magnification air objective (~10×). The TIRF objective should be adjusted to the correct operating height by placing and focusing on a sample of dried particles on the microscope stage. After focusing, the sample is removed while not disturbing operating height and maintaining the same plane of focus for the rest of the alignment procedure. A target mark should be made on the ceiling directly along the optical axis of the microscope. As an alternative, a semitransparent optical element (diffuser glass) with a cross-hair can be fitted to the empty slot of the nosepiece.

The nosepiece should now be rotated to the empty slot. Two large 5-cm-diameter periscope mirrors (Figure 3.15e) are placed at the rear of the microscope such that the laser beam is directed into the microscope, reflected off the dichroic mirror, and projected through the empty slot of the nosepiece and onto the ceiling. Large mirrors are chosen for the periscope to capture the reflected beam because the incident and reflected beams will not travel on the same axis once the system is shifted into the TIRF mode later. In the mean time, the goal of the current task is to align the laser beam to be colinear with the microscope's optical axis. The next step is to rotate the nosepiece to the up-side-down mirror position. At this time, one should see a reflected beam exit the rear of the microscope. By using only one of the periscope mirrors, the incident and reflected beams are to be aligned to become colinear at the rear of the microscope. Next, the nosepiece is rotated back to the open position while the incident beam is redirected to the target mark on the ceiling using the second periscope mirror. This process should be repeated until the optical axes are colinear and the incident beam lands on the ceiling target without any further periscope adjustments needed. The optional shifting prism (Figure 3.15d), which is simply a rectangular solid piece of glass, can now be inserted between the collimating lens of the spatial filter and the periscope. The prism is adjusted and rotated until the incident beam hits the target mark on the ceiling and the optical axes of the incident and reflected beams are again colinear.

The incident beam angle control (Figure 3.15f), which consists of a large 5-cm lens ($f \approx 30$ cm) on a two-axis rotational lens mount on a three-axis translational stage, should be placed between the periscope and microscope at approximately one focal distance away from the objective's BFP as shown in Figure 3.17a. Again, the large lens is used to accommodate the shifted reflection of the incident beam in the TIRF mode. By placing a closed iris adjacent to the lens, one can align the optical axis of the lens to be colinear with the existing beam path. The iris is then opened, and the rotational mount is adjusted to align the beam to the marked target on the ceiling through the open slot in the nosepiece. Subsequently, the iris is again closed and the lens is translated on the plane perpendicular to the laser beam optical axis to align the beam through the center of the closed iris (do not translate along the optical axis). This operation is repeated until no further adjustments are necessary. Upon completion of the repetitive steps, the nosepiece is rotated to the low-magnification objective to repeat alignment of the focusing lens (low-magnification objective alignment is optional). The nosepiece is then rotated again to the high-magnification

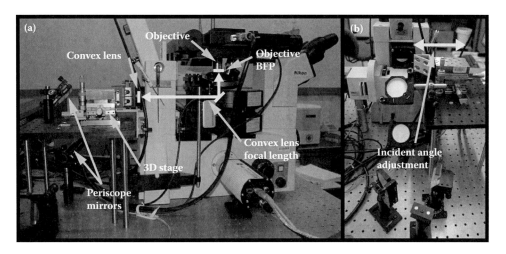

FIGURE 3.17
(See color insert.) Schematic of the periscope and beam angle control lens orientation in relation to the microscope: (a) by focusing the beam onto the BFP of the objective, a collimated beam emerges from the microscope objective and (b) the proper translational adjustment direction for manipulating the incident angle into TIRF mode.

TIRF objective. At this point, one should place a clean slide with dry fluorescent particles on the microscope stage and verify that the objective is still in focus (this is *extremely* important). Once verified, alignment of the beam angle control lens should be done once more until the optical axes are once again colinear. The final step is to collimate the beam emerging from the objective lens by ensuring that the focal plane of the beam angle control lens and the objective's BFP coincide. This can be achieved by translating the beam angle control lens along the optical axis of the beam to minimize the spot projected on the ceiling. If all steps are completed perfectly, this spot will appear Gaussian and symmetric.

The incident angle is controlled and calibrated by translating the beam shifting lens perpendicular to the optical axis (Figure 3.17b) and measuring the angle of the beam emerging from the objective and extrapolating the incident angle as a function of the translation stage position through a least-squares fit [81]. Once this relationship is obtained, a sample drop of fluid containing fluorescent tracer particles is placed on the slide and the beam angle is adjusted until evanescent waves are created in the fluid phase (again, remember to wear proper eye protection here). The experimentally determined angle should be compared with the predicted TIRF angle to verify the calibration. Last, the shifting prism can be rotated to make the evanescent wave spot at the center of the eyepiece field of view. Although this final adjustment is optional, if it is performed, the incident angle should be recalibrated. The final and optional system for the evanescent wave microscopy system is the reflected beam monitoring system (Figure 3.15g), which can be used to monitor changes in the total internal reflection conditions if necessary.

3.3.3 Fluorescent Particle Intensity and Particle Position

As mentioned before, the monotonic decay of the evanescent field has been exploited to map the intensities of fluorescent tracer particles to their distances from the fluid–solid interface [39,41,42,46]. Using this information, one can use a calibrated ratiometric fluorescence intensity to track particle motions three dimensionally. Although this method sounds theoretically feasible, successful use of this technique in practice requires precise knowledge of the illumination beam incident angle and a solution of Maxwell's equation for an evanescent field in a three-medium system (substrate, fluid, and tracer particles), which can be difficult to express explicitly. However, an experimental method can be devised to obtain a ratiometric relation between particle emission intensity and its distance to the glass surface such as that shown in Figure 3.4. In our TIRF calibration, we attached individual fluorescent nanoparticles to polished fine tips of graphite rods, which were translated perpendicularly through the evanescent field to the glass substrate with a 0.4-nm precision translation stage (MadCity Nano-OP25). Multiple images of the attached particles were captured at translational increments of 20 nm. The intensity values of the imaged particles were averaged and fitted to a two-dimensional Gaussian function to find their center intensities. The process was repeated several times with different particles. This procedure produces an intensity–distance correlation such as one shown in Figure 3.18:

$$I = Ae^{-(z-a)/\delta_T} + B, \qquad (3.47)$$

where z is the distance between a particle's center to the substrate surface and a is the particle radius. The δ_T is the TIRF decay length constant that, along with A and B, is determined by a least-square exponential fit. As predicted by Figure 3.4 and experimentally demonstrated in Figure 3.18, the particle intensity decay is very close to the decay of the evanescent wave intensity.

Figure 3.4 also predicts that emission intensities of fluorophore-embedded, micron-sized particles illuminated by evanescent waves also decay exponentially as a function of distance from the substrate, although the sizes of these particles are significantly larger than the evanescent field penetration depth. This unique characteristic exists because frustration of the evanescent field [82] by the dielectric particles can excite fluorophores well beyond the evanescent field at several microns from the interface. This effect produces radially asymmetric particle images, which also decrease in intensity with distance from the wall, as shown in Figure 3.19. To quantify the fluorescence intensity of micron-sized particles as a function of distance from the interface, one can again attach individual particles to

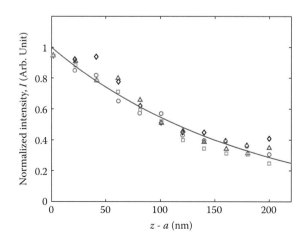

FIGURE 3.18
Fluorescent particle intensity as a function of its distance to the glass surface. The particles used here are 100 nm in radius. The solid line is a least-square exponential fit to the data.

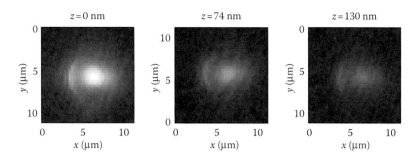

FIGURE 3.19

Characteristic images of a 6-μm particle at various distances from a glass–water interface created because of frustration of the evanescent wave by the fluorescent dielectric particles.

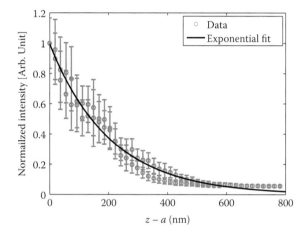

FIGURE 3.20

Mean, integrated fluorescence emission intensity of individual 6-μm particles as a function of distance from a glass–water interface. The intensity variance is due to both thermal motion of the particle and stage noise.

the tip of an opaque micropipette connected to a one-dimensional nanoprecision stage. The particle is traversed perpendicular to the substrate through the evanescent field while multiple images of the particle at each position are taken to correlate its intensity with position. From Figure 3.4, one can predict that the form of the integrated particle intensity, I, also decays exponentially with distance from the surface (Equation 3.18). A sample fit for the intensity decay of 6-μm fluorescent tracer particles is shown in Figure 3.20. The yielded decay length, $\delta_T = 204$ nm, is found to be nearly identical to the evanescent field penetration depth measured independently. Other experimental and computational investigations for the scattering intensity of similar-sized particles in an evanescent field [32, 81] also found similar intensity decaying results.

3.3.4 Particle-Tracking Velocimetry

With a TIRF microscopy system and an intensity–particle position correlation function in place, quantitative analysis of near-wall particle motions can be used to examine near-surface micro- and nanofluidics by using one of several velocimetry methods. Micro-PIV [83] and nPIV [35] infer the most probable displacement of a fluid element from the cross-correlation peak between two sequential image segments taken in time. There are several shortcomings to this approach in near-wall studies. First, the high velocity gradient near the wall cannot be easily resolved directly. Second, because particles in the near-wall evanescent field are brighter than the ones that are farther away, the cross-correlation method weights slower moving particles close to the wall more heavily, thus biasing the mean velocity. Third, near-surface microfluidic and nanofluidic investigations using particle-based velocimetry typically have low Reynolds numbers, Re, and Peclet numbers, Pe, of order unity. They are defined as

$$Re \equiv \frac{\rho V a}{\mu}, \tag{3.48}$$

and

$$Pe \equiv \frac{V a}{D_0}, \tag{3.49}$$

where a is the particle radius, V is the mean local velocity, ρ is the fluid density, and D_0 is the Stokes–Einstein diffusivity defined in Equation 3.11. The high levels of diffusion of small Brownian particles used in velocimetry tend to degrade the sharpness of the cross-correlation peak and introduce additional uncertainty. Fourth, no particle depth information is provided by PIV methods because all information regarding each particle's intensity is lost during cross-correlation analysis. Finally, the loss of particle intensity information during cross-correlation analysis means that one could not directly measure the near-wall particle concentration profile and would have to assume the concentration distribution of the tracer particles in the near-wall region most commonly as a uniform distribution. As discussed in Section 3.2.5, such an assumption can significantly deviate from the actual concentration profile and lead to analysis bias.

Because of the statistical nature of colloidal dynamics, one method to unmask the true physics hiding behind the randomness of the Brownian motion is individual particle tracking [84]. As shown in Section 3.3.3, the intensities of micron-sized particles and nanoparticles decay exponentially away from the fluid–solid interface with decay lengths similar to that of the illuminating evanescent waves [32,39,42,46]. This makes it possible to discern the height of a particle from the substrate surface on the basis of its intensity and has been applied to track particle motions three dimensionally [42,47]. Although there have been some attempts to evaluate cross-correlations over multiple particle layers [37] within the evanescent field, many challenges still remain. Tracking individual particles to resolve near-surface velocities remains a much more direct method to investigate near-wall dynamics. We discuss the most common algorithms for tracking near-wall particles with details to help the readers develop their own image analysis codes.

In PTV, bright particles with intensities above a predetermined threshold value are first identified in a series of images. To track particle motions, all particle locations from two or more successive video frames must be identified to good accuracy, most commonly through identification of the particle center positions. For micron-sized particles or larger, this is typically done by finding their intensity centroids through weighted-function particle image analysis. For subwavelength particles, center positions are found by fitting a two-dimensional Gaussian distribution to the imaged diffraction-limited spots of the tracer particles [78,85]. A two-dimensional Gaussian curve fit closely approximates the actual point-spread function of nanoparticle intensities near their peaks and allows one to locate the particle center coordinates with subpixel resolution. At this point, it is also critical to distinguish real particles from noise signals as noise tracking will unnecessarily corrupt the motion statistics obtained. This is typically accomplished by building an additional abnormality detection algorithm into the particle identification code. Examples include detecting the particle shapes and sizes on the basis of their images [36].

It is important to point out that the determination of an intensity threshold during particle identification sets an *observation depth*, as illustrated in Figure 3.21. The lower bound of the observation range is the particle radius, representing a particle in contact with the channel surface. As the particle moves farther from the wall and thus into a region of lower evanescent wave illumination intensity, the emitted intensity falls. Thus, the intensity threshold chosen sets an upper bound on the observation depth, which is different from the decay length of the evanescent field.

Next, particles identified are matched between frames to track their trajectories. In some cases (e.g., in fast-moving flows), only two frames (image pairs) may be available at a time for matching because of limitations in camera acquisition speed. The time duration between image acquisitions should be set such that most tracer particles translate between 5 and 10 pixels for highest velocimetry accuracy. If the tracer concentration is dilute, the nearest neighbor matching is simple and very effective [45]. The center position of an identified particle in frame 1 is first identified and noted. A search is then started in frame 2, around the center position of the identified particle in frame 1, until the nearest neighbor is found in frame 2. The two particle center positions then become a matching pair and are considered as the locations of a single tracer particle at different times. The distance between the center positions is now used to infer the displacement of the local fluid or Brownian motion of the particle between image acquisitions. Here, noise-detection algorithms can also be inserted to improve velocimetry accuracy. One can first make an educated guess of the largest distance that a tracer particle can travel between image acquisition and use this distance as the radial limit of nearest neighbor search. If the nearest neighbor search done in frame 2 for a particle identified in frame 1 is beyond this set limit, one can safely assume that this "particle" is probably misidentified and is most likely a noise rather than an actual tracer. Second, if more than two signals in frame 2 can be matched to a particle identified in frame 1 using the above criteria, it is advised to discard displacement information provided by this particle. This is because the probabilistic nature of the Brownian motion makes it impossible to resolve this matching ambiguity with any kind of certainty. It should be noted that the image acquisitions, particle identifications, and tracking processes should be repeated for a large number of times until the number of tracked trajectories becomes a

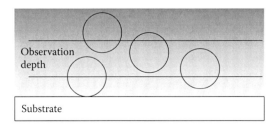

FIGURE 3.21
Schematic of near-wall particles moving near the surface illustrating the observation depth.

statistically sample size. It is recommended that at least 1000 successful trackings should be obtained for each experimental condition.

Clearly many of the tracking ambiguities and unwanted photon noises can be avoided if the tracer particle density is low. Therefore, the tracer particle density should be kept minimal while still capturing a good number of successful trackings from each image pair. For less-than-perfect tracer particle seeding conditions, several variations of the basic tracking algorithm have been proposed. For large Peclet numbers (less significant particle Brownian motions), a multiple matching method may be useful [86]. However, for small Peclet numbers or highly concentrated particle solutions, a statistical tracking method [49] or a neural network matching algorithm [41,87] might be advantageous. For a multiple-frame image sequence using window shifting and predictor corrector methods is beneficial [87], but only for large Peclet number.

Once the ensemble particle displacements are measured, the individual intensities of matched particles can provide information about the particle distance to the substrate surface as mentioned previously. Colloidal particles typically have a large diameter size variation (3%–20%) that can bias the interpretation of intensity to distance. Because particle intensity can also be a function of particle size as discussed previous studies [54,55,88] and in Section 3.2.2, a large particle far from the wall can appear to have the same intensity as that of a small particle near the wall [39,47]. Caution must be exercised when inferring the distance of the particle from the substrate based on its intensity.

Figure 3.2 compares particle images of bright-field illumination and TIRF illumination. A large amount of background noise is observed in the case of bright-field illumination. These background noises are attributed to the fluorescent light emitting from out-of-focus tracer particles in the bulk of the fluid and can lead to difficulty in particle identification and trajectory tracking. TIRF illumination, on the other hand, eliminates much of the background noise because the evanescent wave illumination is restricted to the near-surface region only and particles in the bulk fluid are not illuminated. This characteristic allows easy detection of only particles that are close to the channel surface and thus significantly improve PTV accuracies.

Using the geometric scale and the applied time separation between image acquisitions, a velocity vector can be calculated from each successful particle tracking. Figure 3.22 shows an example of a collection of velocity vectors obtained for a single shear rate. The Brownian motion is particularly strong because of the particle's small size. Alternatively, if one desires to investigate the three-dimensional translational motions of the tracer particles, one can use the calibration Equation 3.47 with the fitted peak particle intensities, I, to obtain the instantaneous position of each particle relative to the substrate, z, and subsequently track the three-dimensional trajectories of these particle over time.

A straightforward two-dimensional Gaussian fit used in nanoparticle identification has been demonstrated to be very accurate and is currently the gold standard in near-surface PTV. However, this method does not provide as accurate center positions for micron-sized particles, given the oddly shaped, asymmetric particle images as shown in Figure 3.19. An altered version of particle identification and center positioning algorithm has been developed to improve particle-tracking analysis accuracy [47]. In this method, particle locations are first coarsely identified by intensity thresholding and matched to new positions in subsequent images by a nearest neighbor search. Because of the oddly shaped, asymmetric particle images, a cross-correlation-tracking algorithm was used to refine the particle center locations [85]. In the next step, Gaussian fitting to the peak of the correlation map is performed and can now yield an accuracy of approximately 0.1 pixels (28.3 nm) in the wall-parallel directions, x and y. After that, overall intensity of an individual particle is

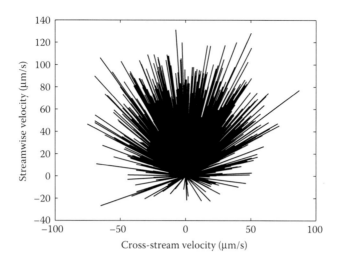

FIGURE 3.22
Distribution of particle velocity vectors of 200-nm particles for a near-wall shear flow of shear rate = 469 s^{-1}.

obtained from integration of all its local pixel intensities and normalized by the Gaussian shape of the illuminating beam as measured by an aqueous solution of rhodamine B dye. Finally, the relative particle position in the wall-normal direction, z, is computed by inverting Equation 3.47, where A is given by the largest intensity of a given particle along its trajectory (i.e., closest position to the wall). The resolution in the wall-normal direction is estimated to be on the order of 10 nm because of intensity variation resulting from particle diffusion, laser fluctuation, and camera noise [47]. Additional uncertainty can result from nonuniformity in the cover slip, nonuniformity in surface coating thickness, and illumination light intensity variation over the field of view.

Here we present our experimental tracking results for adhesion dynamics of micron-sized particles to demonstrate the tracking algorithm's effectiveness. P-selectin-coated 6-μm fluorescent particles and PSGL-1-coated microfluidic channels are used as mechanical models for investigating adhesion characteristics of leukocytes in pressure-driven flows inside blood vessels. As is typical in flow-chamber-based assays, tethering adhesion can be detected by arresting events in the particles' in-plane motion [89]. Figure 3.23a shows a segment of such a trajectory, where the plateaus signal that the particle is arrested. The instantaneous particle displacements (Figure 3.23b) show much more detail of this process. There are distinctly different features among the various binding events as indicated by the position fluctuations in the plateaus. This likely indicates the strength of an arresting event due to variations in the number of tethers that combat the Brownian motion. Although we have demonstrated this technique for a single particle here, from statistical averaging of such measurements, reaction rate constants for off-times can be computed by binning the lengths of the arresting times [90]. Furthermore, a histogram of the particle displacements shows a strongly bimodal distribution (Figure 3.24), where the system transitions stochastically between the free and the tethered states. There is also evidence of a third intermediate mode that may indicate a steady rolling velocity with $\Delta x \sim 100$ nm.

Particle motion in the wall-normal direction, z, tends to be more complex than in the wall-parallel case as shown for a segment of a single particle trajectory in Figure 3.25. Tethering events are clearly visible in the wall-normal trajectory by sharp transitions between plateau regions, especially in comparison with Figure 3.23. The plateaus correspond to times when

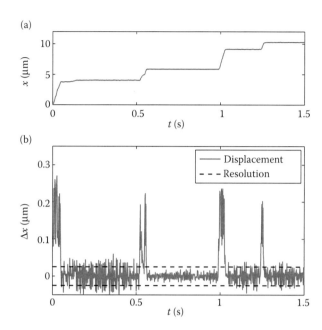

FIGURE 3.23
Time trace of the wall-parallel motion in the flow direction for a single particle, illustrating the tethering dynamics near the substrate: (a) particle trajectory and (b) particle displacement

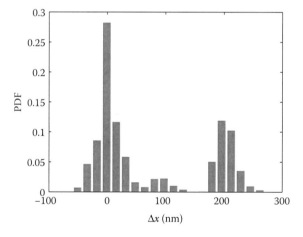

FIGURE 3.24
A histogram of the wall-parallel particle displacements for the trajectory of a typical particle. The bimodal nature of the histogram demonstrates the fraction of time that the particle is tethered versus free.

the particle is temporarily arrested by a tether. The statistical variation of the particle's height during tethering events contains additional information about the tether stiffness, which may also be examined.

We end Section 3.3 with some words on system evaluation and testing. If one follows the procedures on constructing a TIRF microscope system, setting up an intensity–position calibration component using a nanometer-precision translation stage, and developing a PTV software to conduct near-surface PTV, it would then be necessary to test and evaluate

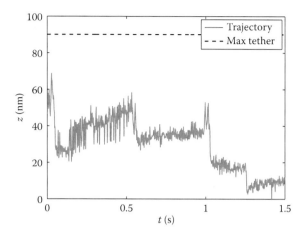

FIGURE 3.25
Time trace of the vertical position for a single particle, where the transitions between tethered and free states are clearly detectable by jumps in the wall-normal position, z (interpreted from the particle intensity), compared with the horizontal displacements. The dashed line shows the typical maximum bond length of approximately 92 nm for the P-selectin/PSGL-1 system.

the performance of this homemade system, both on functionality and on accuracy. In our opinion, the simplest evaluation experiment that one can conduct is an experimental verification of the Brownian motion. The tracer particles, in quiescent fluid, will undergo hindered Brownian motion in the vicinity of a solid substrate. Because the theory of hindered Brownian motion is quite well established (see Section 3.2.3), quantitative observations of the tracer micro- or nanoparticles using near-surface PTV can be easily compared with theories to determine if the experimental setup and the analysis software are truly functioning with the expected precision and accuracy. Many of the images and results shown in this section should provide sufficient examples for one to make a proper evaluation.

3.4 Recent Developments and Applications

Near-surface particle tracking has made tremendous progress over the past several years due in large part to increased interest in nanofluidics and a demand for higher resolution diagnostic techniques. In this section, we discuss the advances in tracer particles, imaging systems, and velocimetry algorithms that have elevated near-surface velocimetry to its current level of precision and flexibility. In addition, we discuss several recent applications and advantages that make this method an appealing measurement technique for micro- and nanofluidics, soft condensed matter physics, and biophysics fields. Near-surface tracking uses a wide variety of tracer particles (for both fluorescence and scattering imaging) with sizes ranging from as large as 10 μm to just a few nanometers where the choice in tracer particle is determined by the application. Advances in nanofabrication have lead to the development of single, uniform fluorescent particles just tens of atoms in diameter. Ultrahigh NA microscope objectives in combination with high-speed image intensifiers and cameras now provide unparalleled light collection and imaging speed capabilities. Micro- and nanoscale particle velocimetry algorithms have largely been adopted from their macroscale counterparts but have evolved to address the unique challenges of near-wall physics (large velocity gradients, highly diffusive tracer particles, etc.). Several recent applications have included electroosmotic flows, slip flows, near-surface temperature measurement, QD tracking, hindered diffusion, and velocity profile measurements.

3.4.1 Tracer Particles

With the many advances in micro- and nanofabrication techniques, there is a wide variety of commercially available tracer particles for almost any application ranging in size from several microns to several nanometers and even the molecular level. Typically, though, the size of a tracer particle is chosen for a particular application in near-wall tracking velocimetry where either the particle dynamics or the fluid dynamics are of interest. Although larger particles (>500 nm) may be imaged by scattering, fluorescence imaging is often the only way to image diffraction-limited particles. Some applications require that particles be coated with bioproteins for conjugation to surfaces or other particles.

Large tracer particles (>1 µm) are often used to study hydrodynamic and electrostatic forces in plane–sphere geometries as shown in Figure 3.26. Interesting behavior is often captured for short-ranged forces (~200 nm) or when the gap between the sphere and plate is much smaller than the particle radius. Scattering imaging has been used extensively in the past, which has the advantage of avoiding photobleaching and allowing extended observation times [30]. Fluorescence imaging of large particles has become popular, which uses lower illumination intensity than scattering and thus avoids unwanted optical forces that can bias measurements. Here applications have included the measurement of spatially resolved, anisotropic diffusion coefficients [42]. In addition to these typical geometries, larger particles have also been used to model leukocyte adhesion dynamics where they served as surrogate white blood cells to convey bioproteins [47].

The most common tracer particles for an array of applications in near-wall particle tracking are fluorescent polystyrene nanoparticles ranging in size from 40 to 300 nm. Such particles have served as tracers to measure slip velocities and near-wall velocity profiles [39,46]. Over the past 5 years, semiconductor nanocrystals or QDs have attracted increasing attention as

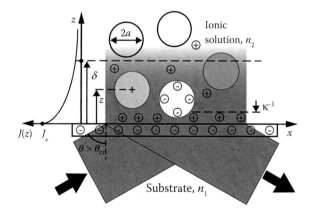

FIGURE 3.26
Typical geometry of evanescent wave illumination, where a plane monochromatic wave is incident on a dielectric interface at an angle greater than the critical angle, $\theta \geq \theta_{cr}$. The resulting evanescent field intensity, $J(z)$, has a decay length, δ, on the order of 100–200 nm. Ionic solutions screen the electrostatic forces between charged particles and surfaces with a length scale characterized by the Debye length, κ^{-1}.

extremely small, bright, and robust tracer particles for near-wall particle tracking. QDs are single fluorophores with reasonable quantum efficiencies and fluorescence lifetimes similar to conventional fluorophores, but with significantly higher resistance to photobleaching [91]. They exhibit several qualities beneficial to nanoscale velocimetry including small diameters ranging from 5 to 20 nm, a narrow and finely tunable emission wavelength, and even temperature sensing abilities [51]. However, single QDs have a significantly lower emission intensity compared with much larger polystyrene particles containing several thousand fluorophores. In addition, their emission can fluctuate randomly (Figure 3.27), which is known as "fluorescence intermittency" or "blinking" [92]. The first practical demonstrations using QDs in aqueous solutions for velocimetry used either extremely dilute solutions [48] or statistical velocimetry algorithms [49] to negotiate the necessarily long exposure times and high diffusivities. Applications include near-wall velocity bias measurements, high-speed imaging, and simultaneous temperature and velocity measurements. Hybrid particles consisting of 50 nm of polystyrene particles conjugated with a series of QDs also show promise as small yet extremely bright tracer particles [93].

3.4.2 Imaging Systems

Although the basic list of components for evanescent wave imaging systems (light source, conditioning optics, specimen or microfluidic device, fluorescence emission imaging optics, and a camera) have remained unchanged for some time, vast improvements in the quality and implementation of those components have translated into marked scientific achievements. Laser sources have diminished significantly in size with increasing stability and tunability. The conditioning optics used to create the angle of incidence necessary for total internal reflection are still primarily prism based or objective based with widely varying components [21]. Prism-based systems are still typically home built, with no significant recent advances in technology. In contrast, objective-based imaging has benefited greatly from advances in high-magnification, high NA oil-immersion objectives. Most recently, extremely large NAs (>1.49) have become commercially available to minimize vignetting

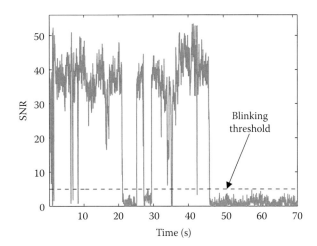

FIGURE 3.27
Sample intensity time trace for single, immobilized QDs under continuous illumination. An intensity greater than the threshold of signal-to-noise ratio = 5 designates blinking on-times from off-times.

and provide for larger illumination spots and also larger incident angles, which allows for extremely small penetration depths.

A wide variety of cameras are also suitable for near-wall particle tracking depending on the sensitivity and speed requirements dictated by the tracer particles and application. Sensitive, low-noise CCD cameras are a typical choice for imaging small tracer particles by scattering or fluorescence. For nanometer-sized particles, intensifiers are often used (integrated or external) to amplify light levels. This is useful not only for low-intensity particles but also for minimizing exposure times, t_e, to capture fast dynamics and to avoid particle image streaking. CCD cameras produce low-noise, high-resolution images but suffer from the disadvantage of slow frame rates (10–100 Hz) because of readout time. This problem can be circumvented somewhat by acquiring image pairs (PIV-type imaging) rather than triggering the camera at a constant rate. This method has been shown to improve velocity resolution by capturing extremely fast tracer particle velocities but does not provide for the Lagrangian particle tracking that is truly desirable.

Most recently, multistage, high-speed image intensifiers have been used to amplify extreme low-intensity single QD images captured by high-speed CMOS camera sensors allowing for frame rates more than 5 kHz [50]. Although CMOS cameras are less sensitive with lower spatial resolution than CCD cameras, they provide desirable high frame rates. Two key intensifier features that are necessary for imaging single molecules and fluorophores at high speeds are (1) a two-stage microchannel plate to amplify small numbers of photons to a sufficient level for detection by the CMOS sensor and (2) a fast decay phosphor screen (P24, 6-μs decay) to prevent ghost images. Multistage image intensifiers have an inherently low resolution because of imperfect alignment of finite-resolution microchannel plates. Although three-stage intensifiers are available, two-stage systems produce sufficient light amplification while maintaining the image resolution.

3.4.3 Tracking Algorithms

Near-wall particle velocimetry algorithms have largely been adopted from macroscale algorithms and are of two types: (i) PIV methods and (ii) PTV. PIV methods use cross-correlation techniques to determine the most probable displacement for a grouping of tracer particles suspended in fluid [94]. This yields an instantaneous snapshot of the spatially resolved velocity field (Eulerian description). Conversely, PTV algorithms identify distinct individual particles and follow their positions in time (Lagrangian description) [87]. Both techniques have been adapted to suit micro- and nanofluidics, and here, in particular, we discuss the various benefits and costs in relation to near-surface velocimetry.

3.4.3.1 PIV Methods

Micro-PIV techniques are used to measure fluid velocity fields with length scales $L < 1$ mm [83]. The DOF of the imaging objective defines a measurement plane with a typical thickness of 500 nm [80]. nPIV uses the same cross-correlation analysis techniques as its microscale counterpart but instead uses the penetration depth of the evanescent wave intensity to define an imaging plane with $100 \leq \delta \leq 200$ nm of the surface [35]. This provides obvious advantages over micro-PIV including a more defined imaging plane and significantly better signal-to-noise ratio images. This method was taken one step farther by exploiting the monotonic intensity decay of the evanescent wave in the wall-normal direction. Several groups have demonstrated that the intensities of a wide range of particles diameters decay exponentially away from the fluid–solid interface with similar decay lengths to the penetration depth [32,39,46]. This makes it possible to segregate particles at various ranges from the surface, on the basis of their intensity, providing a three-dimensional, two-component measurement of the near-surface velocity field [37,46].

Although PIV techniques are becoming highly developed, there are several shortcomings to this approach in near-wall studies. Currently, the wall-normal resolution has only course-grained discretization. In addition, because particles near the wall are brighter because of the evanescent field gradient, the cross-correlation method weights slower moving particles close to the wall more heavily introducing possible measurement biases in the velocity. Small Brownian particles in low Reynolds number flows typically have Peclet numbers of order unity. These high levels of diffusion and Brownian motion tend to degrade the sharpness of the cross-correlation peak, and consequently, several hundred images are required to sufficiently average and smooth the correlation map. This would drastically reduce PIV's advantages in providing time-resolved measurements. In addition, PIV methods were originally developed to measure fluid velocities under the assumption that tracer particles closely follow the fluid, but as tracer particles become smaller ($\lesssim 2$ μm radius), this assumption starts to break down because of the Brownian motion. This random motion is a direct manifestation of the thermal fluctuations in the solvent medium. Averaging over this motion masks important statistical fluctuations that can be used to measure complex physical phenomena that only occur within several nanometers of the wall (electrostatic, van der Waals, etc.).

3.4.3.2 PTV Methods

The statistical nature of colloidal dynamics makes particle tracking a natural fit [95], and near-surface measurements are no exception [36]. In PTV, all particle locations from two or more successive video frames can be identified to subpixel accuracy (typically ≥0.1 pixels) using their intensity centroids for large particles [85] or by fitting a two-dimensional Gaussian distribution to the diffraction-limited spot for subwavelength particles [78,85]. Once identified, particles are matched to one another between frames to link their positions into trajectories. The specific algorithm to achieve this task depends on several factors including tracer particle concentration, diffusivity, velocity, and number of consecutive video frames available for tracking.

With sensitive but slow CCD cameras (10 Hz), image pairs or PIV-type imaging is often used to improve the range of measurable velocities and capture fast-moving particles. If the tracer concentration is dilute, then nearest neighbor matching is simple and effective [45]. Also, when dealing with fast uniform velocities, window shifting may be used. When the interparticle distance is small compared with the particle displacements and Peclet number is large, a multiple matching method may be useful [86]. However, for small Peclet numbers or highly concentrated particle solutions, a statistical tracking method [49] is advantageous. Both these methods rely on similar principles. A number of possible particle displacements between two frames are computed to yield a statistical ensemble of displacements, a fraction of which are physical whereas the rest are artificial. On the basis of the type of flow, some assumptions can be made about the distribution of unphysical trackings and thus they can be statistically subtracted from the distribution. However, one significant drawback to these statistical techniques is that they do not yield any useful information about individual tracer particle tracks; only the *distribution* of displacements is obtained in a meaningful way. We also note that neural network matching algorithms have been used with some success for both image pair and multiframe particle tracking [41,87].

Several advantages are gained in the case that multiple-frame image sequences are available. Predictive methods or minimum acceleration methods [87] are useful for large Peclet number flows. The history of a given tracer particle (position, velocity, and acceleration) is used to predict its position some time in the future. The particle in the following frame most closely matching the predicted position is then added as a link in the trajectory. If several matches are plausible, additional frames into the future may be examined to judge the validity of a possible match. When the Peclet number is small, as in the case of QDs, the tracer particle concentration is often reduced to avoid mismatching particles, which again allows for simple nearest neighbor matching. To avoid mismatches between particles (and noise), multiple-frame nearest neighbor tracking can help [48]. For extremely small Peclet numbers, there is no substitute for fast imaging frequencies. Recently, high-speed intensified imaging has been incorporated into evanescent wave microscopy systems for the purpose of velocimetry via particle-tracking techniques with QD tracers [50], at frame rates in excess of 5 kHz for tracking single fluorophores. The fast imaging reduces the interframe displacement of the tracers significantly below the interparticle distance, thus allowing a return to simple, reliable interframe matching techniques (i.e., nearest neighbor).

Another useful feature of particle tracking is that with proper calibration, tracer particle intensity can be used to determine the particle's distance from the surface [30,39,40,46], yielding three-dimensional, three-component particle tracks. In some cases, this method of tracking can provide a resolution of several nanometers.

Finally, although PTV provides the statistics of particle motion more directly compared with PIV, many problems remain to be solved especially as tracer particles continually decrease in size. Given the extremely small image plane thickness for near-surface tracking (100–300 nm), highly mobile tracer particles may easily diffuse parallel to the velocity gradient and out of the imaging depth, which is a phenomenon known as "dropout." The opposite process can sporadically bring tracers into the imaging depth creating false particle tracks. Similarly, in the case of QDs, fluorescence intermittency or blinking can in effect cause an optical dropout; however, this phenomenon is usually insignificant compared with the physical dropout [83]. Large particle concentrations or mobilities can cause confusion for matching algorithms, and near-surface velocity gradients create dispersion that is only recently becoming well understood [43,44]. In the case of three-dimensional, near-wall particle tracking, nonuniform fluorescent particle sizes and temporally fluctuating particle intensities can also bias measurements.

3.4.4 Measurement Applications and Advantages

Evanescent wave microscopy and near-surface imaging have been used by biophysics researchers since the 1970s [22]. During the 1990s, several groups began studying near-wall colloidal dynamics by observing the light scattered by micron-sized particles in the evanescent field [27,30]. More recently, evanescent wave microscopy has been integrated with the well-established particle velocimetry techniques of microfluidics [35,36]. Typically, these methods have been used to measure the dynamics of small colloidal particles (10–300 nm), and applications have included the characterization of electroosmotic flows [38], slip flows [39,40,96], hindered diffusion [41,42], near-wall shear flows [43–47], and QD tracer particles [45,48,51,52] (Figure 3.28).

3.4.4.1 Near-Surface Flows

Several of the first near-surface particle-tracking experiments were investigations of two well-known but poorly understood surface flows: electroosmotic flows and slip flows. Until recently, there were no direct experimental measurements of electroosmotic flows

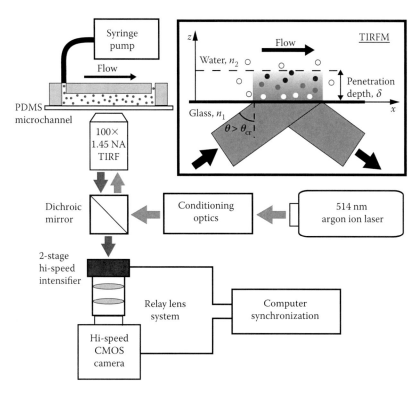

FIGURE 3.28
Schematic of the experimental setup and high-speed evanescent wave (TIRF) microscopy imaging system.

within the electric double layer approximately 100 nm from the fluid–solid interface. nPIV techniques were used to measure two wall-parallel velocity components in electroosmotic flows (EOF) within 100 nm of the wall. Analytical and numerical studies suggesting uniform flow near the wall were verified using nPIV, demonstrating that the electric double layer is much smaller than 100 nm as predicted [38]. Also, the microscopic limits of the no-slip boundary condition between a liquid and a solid have been the source of much debate in recent years, and this assumption has been challenged by recent experimental results and molecular dynamic simulations. Experimental studies have reported a wide range of slip lengths, ranging from micrometers to tens of nanometers or smaller (including no-slip) [97–104]. Molecular dynamics simulations, on the other hand, suggest small slip lengths, mostly less than 100 nm [105–110]. Several researchers have confirmed these simulations using near-surface particle tracking [39,40,52,96], and they showed that hydrophilic surfaces show minimal slip to within measurement accuracy. Hydrophobic surfaces do appear to introduce a discernible but small slip length of approximately 10–50 nm [39,40].

3.4.4.2 Temperature Measurements

Simultaneous, noninvasive thermal and velocimetry diagnostic methods have many potential applications in fields such as DNA amplification (polymerase chain reaction) and heat transfer in microelectromechanical systems [111,112]. Two viable methods of microscale optical temperature measurement have been successfully demonstrated including (i) laser-induced fluorescence thermometry, which exploits the change in emission intensity of laser dye with changing temperature [113,114], and (ii) PIV thermometry, which uses the random motion of tracer particles to estimate temperature [115]. More recently, these techniques were demonstrated using near-surface tracking of QD tracer particles within approximately 200 nm of a liquid–solid interface [51]. Because QDs also exhibit temperature-sensitive emission intensity (–1.1% K^{-1}) [116] and increased Brownian motion with increasing temperature, both velocity and temperature may be measured simultaneously.

3.4.4.3 Nanoparticle Tracking

There is a constant demand for increased resolution in micro- and nanofluidic diagnostics, and near-surface tracking has been applied to measure nanoparticle dynamics at a smaller scale. Recently, interest has peaked in the use of semiconductor nanocrystals or QDs (3–25 nm in diameter) as nanofluidic flow tracers [45,48,50–52,93,117]. Their quantum efficiency is comparable with typical fluorescent molecules, and they are considerably more resistant to photobleaching [91]. However, their small size means that QDs are significantly less intense than tracer particles measuring several hundred nanometers in diameter containing thousands of fluorescent molecules, and their small diameters yield high diffusivity, making imaging and tracking extremely difficult. Previously, single QD dynamics were only realized in elevated viscosity solvents [117], but the high sensitivity and low-noise imaging provided by evanescent wave microscopy made measurements in aqueous solutions possible [48]. Since that time, QDs and near-surface tracking have been used in the measurement of temperature [51], velocity profiles [52], and dispersion-related velocity bias [45, 50]. Most recently, the integration of two-stage, high-speed intensifiers and CMOS cameras has provided frame rates of more than 5 kHz and ability to measure velocities of nearly 1 cm/s within approximately 200 nm of the liquid–solid interface of microchannel.

3.4.4.4 Three-Dimensional Measurements, Velocity Profiles, and Hindered Brownian Motion

When attempting to measure the mean velocity or velocity profile near a solid boundary through particle-based imaging, the concentration distribution of particles in the wall-normal direction must be established to properly weight the spatially varying velocity. In many previous studies, the concentration distribution has been assumed to be uniform, which is almost never the case [35,36]. The equilibrium concentration of colloidal particles in the wall-normal direction from a liquid–solid interface is given by the Boltzmann distribution

$$p(z) = p_0 \exp[-U(z)/k_{\mathrm{B}}T], \qquad (3.50)$$

where p_0 is a normalization constant [30] and U is the total potential energy of a particle. In the absence of any forces between particles and the surface, the potential energy is zero, leading to a uniform particle concentration distribution. However, electrostatic, van der Waals, optical, and gravitational forces can create nonuniform potentials between the particle and the wall, thus leading to nonuniform concentration distributions as discussed in Section 3.2.5. The formation of a depletion layer near the wall can, thus, skew the inferred mean fluid velocity to higher values [30,44,46,47]. Measurements of particle distance to the wall have allowed for estimates of the wall-parallel velocity profile in Poiseuille flows within a few hundred nanometers of the liquid–solid interface [46,52] and proper weighting of ensemble-averaged near-wall measurements [39,44].

When the distance $h = z - a$ between a spherical particle of radius a and a solid boundary becomes sufficiently small ($h/a \sim 1$), hydrodynamic interactions between the particle and the wall hinder the Brownian movement of the particle. Such effects are critical to fundamental near-wall measurements and the accuracy of microvelocimetry techniques, which rely on the accurate measurement of micro- and nanoparticle displacements to infer fluid velocity. Near-surface particle tracking of fluorescent particles has been used to determine the three-dimensional anisotropic hindered diffusion coefficients for particle gap sizes $h/a \sim 1$ with 200-nm-diameter particles [41] and $h/a \ll 1$ with 3-μm-diameter particles [42]. Figure 3.29 shows a comparison between the experimental results of near-surface tracking by Huang and Breuer [42] and several theoretical approximations.

3.4.4.5 Velocimetry Bias

Recently, much interest has centered around diffusion-induced velocity bias, which is a result of dispersion stemming from diffusion of tracer particles parallel to velocity gradient and a bias imposed by the presence of the wall [43,44]. Small Brownian fluctuations in the wall-normal direction result in large stream-wise displacements. This phenomenon has been predicted by both Langevin simulation [44] and integration of the Fokker–Plank equation [43,118]. Experimental verification has come by way of near-surface particle tracking using nanoparticles [44] and QDs [45,50]. Figure 3.30 shows the effects of this phenomenon on ensemble-averaged tracer velocities as a function of interframe time, where the error can vary by as much as ±20%. Figure 3.31 further reveals that the presence of the wall and the associated hindered particle mobility can induce asymmetric particle velocity distribution, in violation of an assumption commonly made in particle-based velocimetry analysis, and thus lead to measurement bias [39,44,81]. Some of the reported studies in diffusion- and shear-induced velocimetry bias have offered analytical formula and protocols for retrieving the physically accurate flow velocities from flawed near-surface PTV data [43,44].

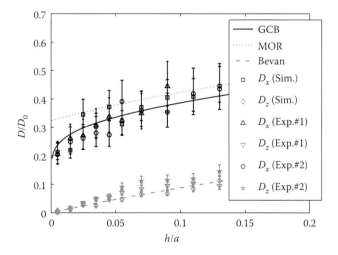

FIGURE 3.29

Normalized hindered diffusion coefficients for the wall-parallel, D_x/D_0, and wall-normal directions, D_z/D_0, near a fluid–solid interface as a function of nondimensional gap size. "GCB," "MOR," and "Bevan" represent asymptotic solution of Goldman et al. [61], method of reflection solution [63], and the Bevan approximation [28], respectively. "Exp." represents experimental data, whereas "Sim." means data obtained from Brownian dynamics simulation. Each error bar represents the 95% confidence interval of measurement.

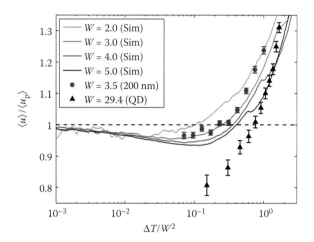

FIGURE 3.30
The ensemble-averaged stream-wise velocity, $\langle u \rangle$, for a Langevin simulation, 200 nm nanoparticles and QD tracers within a nondimensional observation depth, W, which varies with nondimensional interframe time, ΔT, as predicted by the results from a Langevin simulation of Brownian tracer particles in a near-wall shear flow. The variation is due to dispersion effects and described as diffusion-induced velocity bias in the context of velocimetry. The velocity is scaled by the ensemble-averaged velocity of non-Brownian tracer particles, $\langle u_p \rangle$, and the interframe time is appropriately scaled by the observation depth.

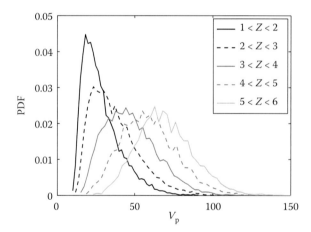

FIGURE 3.31
Apparent velocity (V_p) probability density function (PDF) of particles at various depths of observation. $Z = z/a$, where z is the distance between particle center and the wall and a is the particle radius. All apparent velocity distributions are obtained at Peclet number Pe = 10. Particles that start off farther away from the surface move faster because they are carried by fluids at higher velocity planes, and their distributions are more symmetric due to less influence of the wall and hindered Brownian motion.

3.5 Conclusions

In this chapter, we have not only outlined the basic concepts and underlying physics associated with near-wall PTV using TIRF microscopy but also highlighted the practical issues associated with the design, assembly, and operation of a TIRF velocimetry system for micro- and nanoscale fluid measurements. As is often the case with new diagnostic methods, the first few experiments reported are exciting but often only suggestive—they reveal the promise of a new technique but expose more questions than they provide answers. This pattern has certainly been true in the history of near-wall velocimetry. However, in the past decade, the technique has matured considerably, and many, certainly not all, of these questions have been identified and in some cases answered. As of today, many of the issues associated with the operation of a TIRF velocimetry system and the analysis of the resultant data have been optimized, algorithms have been developed to track particles, and many of the issues that make interpretation difficult have been identified and explained, so as to make TIRF microscopy a useful and quantitative approach for practical microfluidic measurements.

To be sure, many challenges still remain to be addressed. Accurate determination of the wall-normal position of tracer particles remains difficult and is hindered by the difficulty in discriminating between the particle size variations and the position of the particle in the evanescent field. This will improve as particle manufacturing techniques improve, and with the adoption of even more advanced optical methods that use, for example, interferometry or other phase- and polarization-sensitive methods. Another challenge is the ability to identify and track ensembles of particles whose thermal motions may be orders of magnitude larger than the local fluid velocity. This will become easier as imaging systems continue to improve and with the further development of statistical methods for extracting particle displacements. Lastly, the physics of near-wall flows and of the motion of particles in proximity to the liquid–solid interface is extraordinarily subtle, and these complexities continue to generate results that are often unexpected and require explanation. These explanations will take time as we continue to sort out the relative roles of each force and phenomenon before we arrive at a unified understanding of near-wall fluidic flows.

References

1. Bange A, Halsall HB, Heineman WR. Microfluidic immunosensor systems. *Biosens Bioelectron*. 2005;20:2488–503.
2. Sia SK, Whitesides GM. Microfluidic devices fabricated in poly(dimethylsiloxane) for biological studies. *Electrophoresis*. 2003;24:3563–76.
3. Kruger J, Singh K, O'Neill A, Jackson C, Morrison A, O'Brien P. Development of a microfluidic device for fluorescence activated cell sorting. *J Micromech Microeng*. 2002;12:486–94.
4. Saltzman WM, Olbricht WL. Building drug delivery into tissue engineering. *Nat Rev*. 2002;1:177–86.
5. Bigioni TP, Lin X-M, Nguyen TT, Corwin EI, Witten TA, Jaeger HM. Kinetically driven self assembly of highly ordered nanoparticlemonolayers. *Nat Mater*. 2005;5:265–70.
6. Salata OV. Applications of nanoparticles in biology and medicine. *J Nanobiotechnology*, 2, 2004.
7. Beebe DJ, Mensing GA, Walker GM. Physics and applications of microfluidics in biology. *Annu Rev Biomed Eng*. 2002;4:261–86.
8. Westin KJA, Breuer KS, Choi C-H, Huang P, Cao Z, Caswell B, et al. Liquid transport properties in sub-micron channel flows. Proceedings of 2001 ASME International Mechanical Engineering Congress and Exposition; 2001.
9. Karniadakis GE, Beskok A. *Micro Flows: Fundamentals and Simulation*. New York: Springer; 2002.

10. Lauga E, Brenner MP, Stone HA. Microfluidics: the no-slip boundary condition. In: Foss J, Tropea C, Yarin A, editors. *Handbook of Experimental Fluid Dynamics*. New York: Springer; 2007.

11. Zhao B, Moore JS, Beebe DJ. Surface-directed liquid flow inside microchannels. *Science*. 2001;291:1023–6.

12. Zhao B, Moore JS, Beebe DJ. Principles of surface-directed liquid flow in microfluidic channels. *Anal Chem*. 2002;74:4259–68.

13. Adamczyk Z, Jaszczolt K, Siwek B, Weronski P. Irreversible adsorption of particles at random-site surfaces. *J Chem Phys*. 2004;120:11155–62.

14. Chang K-C, Hammer DA. Influence of direction and type of applied force on the detachment of macromolecularly-bound particles from surfaces. *Langmuir*. 1996;12:2271–82.

15. Chaoui M, Feuillebois F. Creeping flow around a sphere in a shear flow close to a wall. *Q J Mechanics Appl Math*. 2003;56:381–410.

16. Wit PJ, Poortinga A, Noordmans J, van der Mei HC, Busscher HJ. Deposition of polystyrene particles in a parallel plate flow chamber under attractive and repulsive electrostatic conditions. *Langmuir*. 1999;15:2620–6.

17. Kok PJAH, Kazarian SG, Briscoe BJ, Lawrence CJ. Effects of particle size on near-wall depletion in mono-dispersed colloidal suspensions. *J Colloid Interface Sci*. 2004;280:511–7.

18. Axelrod D. Total internal reflection fluorescence microscopy. *Methods Cell Biol*. 1989;30:245–70.

19. Rashed R. A pioneer in anaclastics: Ibn Sahl on burning mirrors and lenses. *Isis*. 1990;81:464–91.

20. Tiatt CR, Anderson GP, Ligler FS. Evanescent wave fluorescence biosensors. *Biosens Bioelectron*. 2005;20:2470–87.

21. Axelrod D. Total internal reflection fluorescence microscopy in cell biology. *Traffic*. 2001;2:764–74.

22. Axelrod D, Burghardt TP, Thompson NT. Total internal reflection fluorescence. *Annu Rev Biophys Bioeng*. 1984;13:247–68.

23. Thompson NL, Langerholm BC. Total internal reflection fluorescence: applications in cellular biophysics. *Curr Opin Biotechnol*. 1997;8:58–64.

24. Toomre D, Manstein DJ. Lighting up the cell surface with evanescent wave microscopy. *Trends Cell Biol*. 2001;11:298–303.

25. von Grunberg HH, Helden L, Leiderer P, Bechinger C. Measurement of surface charge densities on Brownian particles using total internal reflection microscopy. *J Chem Phys*. 2001;114:10094–104.

26. Flicker SG, Tipa JL, Bike SG. Quantifying double-layer repulsion between a colloidal sphere and a glass plate using total internal reflection microscopy. *J Colloid Interface Sc*. 1993;158:317–25.

27. Bike SG. Measuring colloidal forces using evanescent wave scattering. *Curr Opin Colloid Interface Sci*. 2000;5:144–50.

28. Bevan MA, Prieve DC. Hindered diffusion of colloidal particles very near to a wall: revisited. *J Chem Phys*. 2000;113:1228–36.

29. Kun R, Fendler JH. Use of attenuated total internal reflection-Fourier transform infrared spectroscopy to investigate the adsorption of and interactions between charged latex particles. *J Phys Chem*. 2004;108:3462–8.

30. Prieve DC. Measurement of colloidal forces with TIRM. *Adv Colloid Interface Sci*. 1999;82:93–125.

31. Prieve DC, Frej NA. Total internal reflection microscopy: a tool for measuring colloidal forces. *Langmuir*. 1990;6:396–403.

32. Prieve DC, Walz JY. Scattering of an evanescent surface wave by a microscopic dielectric sphere. *Appl Opt*. 1993;32:1629–41.

33. Buchhave P. Particle image velocimetry. In: Lading L, Wigley G, Buchhave P, editors. *Optical Diagnostics for Flow Processes*. New York: Plenum Press; 1994, pp. 247–70.

34. Wereley ST, Meinhart CD. Micron-resolution particle image velocimetry. In Breuer K, editor. *Microscale Diagnostic Techniques*. New York: Springer; 2005, pp. 51–112.

35. Zettner CM, Yoda M. Particle velocity field measurements in a near-wall flow using evanescent wave illumination. *Exp Fluids*. 2003;34:115–21.

36. Jin S, Huang P, Park J, Yoo JY, Breuer KS. Near-surface velocimetry using evanescent wave illumination. *Exp Fluids*. 2004;37:825–33.
37. Li H, Sadr R, Yoda M. Multilayer nano-particle image velocimetry. *Exp Fluids*. 2006;41:185–94.
38. Sadr R, Yoda M, Zheng Z, Conlisk AT. An experimental study of electro-osmotic flow in rectangular microchannels. *J Fluid Mech*. 2004;506:357–67.
39. Huang P, Guasto JS, Breuer KS. Direct measurement of slip velocities using three-dimensional total internal reflection velocimetry. *J Fluid Mech*. 2006;566:447–64.
40. Huang P, Breuer KS. Direct measurement of slip length in electrolyte solutions. *Phys Fluids*. 2007;19:028104.
41. Kihm KD, Banerjee A, Choi CK, Takagi T. Near-wall hindered Brownian diffusion of nanoparticles examined by three-dimensional ratiometric total internal reflection fluorescence microscopy (3-D R-TIRFM). *Exp Fluids*. 2004;37:811–24.
42. Huang P, Breuer KS. Direct measurement of anisotropic near-wall hindered diffusion using total internal reflection velocimetry. *Phys Rev E Stat Nonlin Soft Matter Phys*. 2007;76:046307.
43. Sadr R, Hohenegger C, Li H, Mucha PJ, Yoda M. Diffusion-induced bias in near-wall velocimetry. *J Fluid Mech*. 2007;577:443–56.
44. Huang P, Guasto JS, Breuer KS. The effects of hindered mobility and depletion of particles in near-wall shear flows and the implications for nano-velocimetry. *J Fluid Mech*. 2009;637:241–65.
45. Pouya S, Koochesfahani MM, Greytak AB, Bawendi MG, Nocera D. Experimental evidence of diffusion-induced bias in near-wall velocimetry using quantum dot measurements. *Exp Fluids*. 2008;44:1035–8.
46. Li H, Yoda M. Multilayer nano-particle image velocimetry (MnPIV) in microscale Poiseuille flows. *Meas Sci Technol*. 2008;19:075402.
47. Guasto JS. *Micro- and Nano-scale Colloidal Dynamics Near Surfaces*. PhD thesis. Providence, RI: Brown University; 2008.
48. Pouya S, Koochesfahani M, Snee P, Bawendi M, Nocera D. Single quantum dot (QD) imaging of fluid flow near surfaces. *Exp Fluids*. 2005;39:784–6.
49. Guasto JS, Huang P, Breuer KS. Statistical particle tracking velocimetry using molecular and quantum dot tracer particles. *Exp Fluids*. 2006;41:869–80.
50. Guasto JS, Breuer KS. High-speed quantum dot tracking and velocimetry using evanescent wave illumination. *Exp Fluids*. 2009;47(6):1059–66.
51. Guasto JS, Breuer KS. Simultaneous, ensemble-averaged measurement of near-wall temperature and velocity in steady micro-flows using single quantum dot tracking. *Exp Fluids*. 2008;45:157–66.
52. Lasne D, Maali A, Amarouchene Y, Cognet L, Lounis B, Kellay H. Velocity profiles of water flowing past solid glass surfaces using fluorescent nanoparticles and molecules as velocity probes. *Phys Rev Lett*. 2008;100:214502.
53. Kasap SO. *Optoelectronics and Photonics: Principles and Practices*. Upper Saddle River, NJ: Prentice Hall; 2001. p. 1–49.
54. Chew H. Radiation and lifetimes of atoms inside dielectric particles. *Phys Rev A*. 1988;38:3410–6.
55. Chew H, McNulty PJ, Kerker M. Model for Raman and fluorescent scattering by molecule embedded in small particles. *Phys Rev A*. 1976;13:396–404.
56. Brown R. A brief account of microscopical observations made in the months of June, July and August, 1827, on the particles contained in the pollen of plants; and on the general existence of active molecules in organic and inorganic bodies. *Philos Mag*. 1828;4:161–73.
57. Einstein A. Ber die von der molekularkinetischen theorie der wrme geforderte bewegung von in ruhenden flssigkeiten suspendierten teilchen. *Ann Phys*. 1905;17:549–60.
58. Cussler EL. *Diffusion: Mass Transfer in Fluid Systems*. New York: Cambridge University Press, 1997.
59. McQuarrie DA. *Statistical Mechanics*, chapter 17, pp. 379–401. University Science Books, 2000.
60. Brenner H. The slow motion of a sphere through a viscous fluid towards a plane wall. *Chem Eng Sci*. 1961;16:242–51.

61. Goldman AJ, Cox RG, Brenner H. Slow viscous motion of a sphere parallel to a plane wall—I: motion through a quiescent fluid. *Chem Eng Sci.* 1967;22:637–51.
62. Goldman AJ, Cox RG, Brenner H. Slow viscous motion of a sphere parallel to a plane wall—II: Couette flow. *Chem Eng Sci.* 1967;22:653–60.
63. Happel J, Brenner H. *Low Reynolds Number Hydrodynamics: With Special Applications to Particulate Media.* Springer; 1983.
64. Frej NA, Prieve DC. Hindered diffusion of a single sphere very near a wall in a nonuniform force field. *J Chem Phys.* 1993;98:7552–64.
65. Lin B, Yu J, Rice SA. Direct measurements of constrained Brownian motion of an isolated sphere between two walls. *Phys Rev E Stat Nonlin Soft Matter Phys.* 2000;62:3909–19.
66. Oetama RJ, Walz JY. Simultaneous investigation of sedimentation and diffusion of a single colloidal particle near an interface. *J Chem Phys.* 2006;124:164713.
67. Banerjee A, Kihm KD. Experimental verification of near-wall hindered diffusion for the brownian motion of nanoparticles using evanescent wave microscopy. *Phys Rev E Stat Nonlin Soft Matter Phys.* 2005;72:042101.
68. Dean WM. *Analysis of Transport Phenomena.* New York: Oxford University Press; 1998.
69. Pierres A, Benoliel A-M, Zhu C, Bongrand P. Diffusion of microspheres in shear flow near a wall: use to measure binding rates between attached molecules. *Biophys J.* 2001;81:25–42.
70. King MR, Leighton DT Jr. Measurement of the inertial lift on a moving sphere in contact with a plane wall in a shear flow. *Phys Fluids.* 1997;9:1248–55.
71. Cherukat P, McLaughlin JB. The inertial lift on a rigid sphere in a linear shear flow field near a flat wall. *J Fluid Mech.* 1994;263:1–18.
72. Jones RAL. *Soft Condensed Matter.* New York: Oxford University Press; 2004.
73. Oberholzer MR, Wagner NJ, Lenhoff AM. Grand canonical Brownian dynamics simulation of colloidal adsorption. *J Chem Phys.* 1997;107:9157–67.
74. Israelachvili JN, Tabor D. The measurement of van der Waals dispersion forces in the range 1.5 to 130 nm. *Proc R Soc Lond A.* 1972;331:19–38.
75. Hamaker HC. The London van der Waals attraction between spherical particles. *Physica (Amsterdam).* 1937;4:1058–72.
76. Parsegian VA. *Van der Waals Forces: A Handbook for Biologists, Chemists, Engineers and Physicists.* New York: Cambridge University Press; 2006.
77. Ashkin A, Dziedzic JM, Bjorkholm JE, Chu S. Observation of a single-beam gradient force optical trap for dielectric particles. *Opt Lett.* 1986;11:288–90.
78. Novotny L, Hecht B. *Principles of Nano-Optics.* New York: Cambridge University Press; 2006.
79. Doi M, Edwards SF. *The Theory of Polymer Dynamics.* New York: Oxford University Press; 1986, pp. 46–50.
80. Inoue S, Spring KR. *Video Microscopy: The Fundamentals.* 2nd ed. New York: Plenum Press; 1997.
81. Huang P. *Near-surface slip flow and hindered colloidal diffusion at the nano-scale.* PhD thesis. Providence, RI: Brown University; 2006.
82. Zhu S, Yu AW, Hawley D, Roy R. Frustrated total internal reflection: a demonstration and review. *Am J Phys.* 1986;57:601–7.
83. Santiago JG, Wereley ST, Meinhart CD, Beebe DJ, Adrian RJ. A particle image velocimetry system for microfluidics. *Exp Fluids.* 1998;25:316–9.
84. Crocker JC, Grier DG. When like charges attract: the effects of geometrical confinement on long-range colloidal interactions. *Phys Rev Lett.* 1996;77:1897–900.
85. Cheezum MK, Walker WF, Guilford WH. Quantitative comparison of algorithms for tracking single fluorescent particles. *Biophys J.* 2001;81:2378–88.
86. Breedveld V, van den Ende D, Acrivos ATA. The measurement of the shear-induced particle and fluid tracer diffusivities in concentrated suspensions by a novel method. *J Fluid Mech.* 1998;375:297–318.
87. Ouellette NT, Xu H, Bodenschatz E. A quantitative study of three-dimensional Lagrangian particle tracking algorithms. *Exp Fluids.* 2006;40:301–13.

88. Schniepp H, Sandoghdar V. Spontaneous emission of europium ions embedded in dielectric nanospheres. *Phys Rev Lett.* 2002;89:257403.

89. Schmidt BJ, Huang P, Breuer KS, Lawrence MB. Catch strip assay for the relative assessment of two-dimensional protein association kinetics. *Anal Chem.* 2008;80:944–50.

90. Robert P, Benoliel A-M, Bongrand P. What is the biological relevance of the specific bond properties revealed by single-molecule studies? *J Mol Recognit.* 2007;20:432–47.

91. Bruchez M, Moronne M, Gin P, Weiss S, Alivisatos AP. Semiconductor nanocrystals as fluorescent biological labels. *Science.* 1998;281:2013–6.

92. Nirmal M, Dabbousi BO, Bawendi MG, Macklin JJ, Trautman JK, Harris TD, et al. Fluorescence intermittency in single cadmium selenide nanocrystals. *Nature.* 1996;383:802–4.

93. Freudenthal PE, Pommer M, Meinhart CD, Piorek BD. Quantum nanospheres for sub-micron particle image velocimetry. *Exp Fluids.* 2007;43:525–43.

94. Adrian RJ. Particle imaging techniques for experimental fluid mechanics. *Ann Rev Fluid Mech.* 1991;23:261–304.

95. Crocker JC, Grier DG. Methods of digital video microscopy for colloidal studies. *J Colloid Interface Sc.* 1996;179:298–310.

96. Bouzigues CI, Tabeling P, Bocquet L. Nanofluidics in the Debye layer at hydrophilic and hydrophobic surfaces. *Phys Rev Lett.* 2008;101:114503.

97. Choi C-H, Westin JA, Breuer KS. Apparent slip flows in hydrophilic and hydrophobic microchannels. *Phys Fluids.* 2003;15:2897–902.

98. Zhu Y, Granick S. Limits of the hydrodynamic no-slip boundary condition. *Phys Rev Lett.* 2002;88:106102.

99. Neto C, Craig VSJ, Williams DRM. Evidence of shear-dependent boundary slip in Newtonian liquids. *Eur Phys J E Soft Matter.* 2003;12:S71–4.

100. Cottin-Bizonne C, Cross B, Steinberger A, Charlaiz E. Boundary slip on smooth hydrophobic surfaces: intrinsic effects and possible artifacts. *Phys Rev Lett.* 2005;94:056102.

101. Pit R, Hervet H, Leger L. Direct experimental evidence of slip in hexadecane: solid interface. *Phys Rev Lett.* 2000;85:980–3.

102. Tretheway DC, Meinhart CD. Apparent fluid slip at hydrophobic microchannel walls. *Phys Fluids.* 2002;14:L9–12.

103. Joseph P, Tabeling P. Direct measurement of the apparent slip length. *Phys Rev E Stat Nonlin Soft Matter Phys.* 2005;71:035303(R).

104. Lumma D, Best A, Gansen A, Feuillebois F, Radler JO, Vinogradova OI. Flow profile near a wall measured by double-focus fluorescence cross-correlation. *Phys Rev E Stat Nonlin Soft Matter Phys.* 2003;67:056313.

105. Thompson PA, Troian SM. A general boundary condition for liquid flow at solid surfaces. *Nature.* 1997; 389:360–2.

106. Barrat J-L, Bocquet L. Large slip effect at a nonwetting fluid–solid interface. *Phys Rev Lett.* 1999;82:4671–4.

107. Cieplak M, Koplik J, Banavar JR. Boundary conditions at a fluid–solid surface. *Phys Rev Lett.* 2001;86:803–6.

108. Galea TM, Attard P. Molecular dynamics study of the effect of atomic roughness on the slip length at the fluid–solid boundary during shear flow. *Langmuir.* 2004;20:3477–82.

109. Nagayama G, Cheng P. Effects of interface wettability on microscale flow by molecular dynamics simulation. *Int J Heat Mass Transf.* 2004;47:501–13.

110. Cottin-Bizonne C, Barrat J-L, Bocquet L, Charlaiz E. Low-friction flows of liquid at nanopatterned interfaces. *Nat Mater.* 2005;2:237–40.

111. Liu J, Enzelberger M, Quake S. A nanoliter rotary device for polymerase chain reaction. *Electrophoresis.* 2002;23:1531–6.

112. Thome JR. Boiling in microchannels: a review of experiment and theory. *Int J Heat Fluid Flow.* 2004;25:128–39.

113. Ross D, Gaitan M, Locascio LE. Temperature measurement in microfluidic systems using a temperature-dependent fluorescence dye. *Anal Chem.* 2001;73:4117–23.

114. Kim HJ, Kihm KD, Allen JS. Examination of ratiometric laser induced fluorescence thermometry for microscale spatial measurement resolution. *Int J Heat Mass Transf.* 2003;46:3967–74.
115. Olsen MG, Adrian RJ. Brownian motion and correlation in particle image velocimetry. *Opt Laser Technol.* 2000;32:621–7.
116. Liu T-C, Huang Z-L, Wang H-Q, Wang J-H, Li X-Q, Zhao Y-D, et al. Temperature-dependent photoluminescence of water-soluble quantum dots for a bioprobe. *Anal Chim Acta.* 2006;559:120–3.
117. Bausch AR, Weitz DA. Tracking the dynamics of single quantum dots: beating the optical resolution twice. *J Nanopart Res.* 2002;4:477–81.
118. Sadr R, Li H, Yoda M. Impact of hindered Brownian diffusion on the accuracy of particle-image velocimetry using evanescent-wave illumination. *Exp Fluids.* 2005;38:90–8.

4

Finite-Volume Method for Numerical Simulation: Fundamentals

Pradip Dutta and Suman Chakraborty

CONTENTS

4.1 Introduction

The fluid flow, heat transfer, and other related physical phenomena in any system are modeled by applying conservation laws of mass, momentum, energy, and species to obtain the required governing differential equations. These equations, usually in the form of partial differential equations, define the physics of the problem. For example, the momentum equations express the conservation of linear momentum; the energy equation expresses the conservation of total energy. A complete description of the problem comprises

- Governing differential equations
- Geometrical details
- Initial and boundary conditions
- Material properties

An analytical solution may be possible only for some simplified cases. In many cases, however, the complicated and/or coupled nature of the governing differential equations makes it difficult, if not impossible, to solve them analytically. Keeping that in view, it has always been a challenging task for the research community to develop efficient numerical techniques for solving the transport equations. However, in most problems of practical interest, computational techniques are necessary. These computational techniques generally belong to the field of computational fluid dynamics (CFD) or heat transfer.

Over the years, CFD has emerged as a powerful tool to solve engineering and scientific problems related to fluid flow and other transport processes. This ranges from solving flow past a supersonic airplane, flow in an internal combustion engine, flow in an oil reservoir, atmospheric flows, and flow in turbomachines to such esoteric applications as geophysical and astrophysical fluid dynamics.

The development of advanced CFD tools and the increase in computing power that roughly doubles every 2 years has contributed immensely to the feasibility of solving realistic engineering problems. CFD has gradually evolved to become a cost-effective alternative to experimentation with respect to engineering design in many cases. For instance, in the aerospace industry, the lead time in design and development of aircrafts has been considerably reduced by using CFD. The cost-effectiveness is expected to improve further, as computing costs decrease.

The physical laws that are traditionally used for describing the conservation principles in fluid dynamics and transport processes are commonly expressed in terms of mathematical statements that are valid within the constraints of continuum hypothesis. Following the continuum hypothesis, equations describing the transport features may be formulated disregarding the details of underlying molecular behavior by expressing the governing transport equations in terms of macroscopic properties such as velocity, pressure density, temperature, and concentration and their suitable spatial and temporal derivatives. The consequent mathematical expressions may be conceived to be representing the statistically averaged behavior of a sufficiently large number of molecules, disregarding their discreteness. These considerations, however, may cease to be appropriate for cases in which discreteness of the underlying molecular picture becomes important and continuum considerations cannot be invoked. Such cases include gas flows in high Knudsen number regimes, liquid flow features over sub-nanometer length scales in ultranarrow confinements, and so forth. Nevertheless, in several problems concerning microfluidics, the standard continuum considerations may still succeed, with possible modifications in boundary conditions to accommodate special effects such as interfacial slip. With such considerations, many problems in microfluidics can be analyzed using conventional CFD techniques.

The objective of this chapter is to give an introduction to the basics of CFD and how it can be applied to analyze such problems. Among the different methods available for CFD, the finite-volume method (FVM) is presented in detail because it is the most popular method for fluid flow and heat transfer computations. Most commercially available codes (such as FLUENT [www.fluent.com], FLOTHERM [www.flomerics.com], CFX [www.software.aeat.com/cfx], and STAR-CD [www.cd-adapco.com]) are based on this method. Even if someone uses such a commercial code for solving a problem in microfluidics, a good background on the fundamentals of CFD is desirable for speedy and correct implementation of the code for the particular application. The material presented in this chapter includes various CFD techniques (Section 4.3), basics of FVM for computing diffusion and convection–diffusion transport problems (Sections 4.4 and 4.5, respectively), an introduction to solving the flow field using FVM (Section 4.6), a discussion on discretization using unstructured meshes (Section 4.7), and a discussion on the issues of stability of the numerical scheme (Section 4.8). Some future directions of research are also indicated in the concluding section (Section 4.9).

4.2 Elements of a CFD Code

Typically, a CFD code has the following three main components:

(1) In preprocessor, in which the geometry of the problem is defined, that is, the computational domain. The grid is generated and the domain (the work field) is divided into mesh, grids, control volumes, elements, and so forth, as the case may be. The nature of gridding usually depends on the method selected to solve the problem. This part of the program also deals with providing material properties and physical constants and applying the appropriate boundary conditions at the periphery and the initial conditions dealing with definition of the geometry (computational domain), grid generation (mesh, grids, control volumes, elements), material properties, physical constants, boundary conditions, and initial conditions.

(2) In solver part, approximation of the unknown flow variable by simple functions is made. Discretization is done in which approximate functions are substituted in the governing equation to obtain algebraic equations in terms of the unknown

nodal values of the field variables. The algebraic equations, thus, obtained by discretization are solved to obtain the solution of field variables in the domain.

(3) In the postprocessor, the result obtained in the solver part is used to obtain contour plots, vector plots, three-dimensional plots, and so on. Particle tracking is also done in this session. Most modern CFD packages have their own postprocessors. Alternatively, the output data for the field variables obtained from the solver is formatted according to the input requirements of standard postprocessing packages such as MATLAB or TECPLOT.

4.3 Various CFD Techniques

Various CFD techniques are available for solving problems in heat and fluid flow in systems. The techniques differ from each other in the manner in which the differential equations are approximated by a system of algebraic equations for the variables at some set of discrete locations in space and time. This process is called *discretization*. Some of the most popular discretization techniques include the finite-difference method (FDM), finite-element method (FEM), and FVM (also known as the control volume method). Each type of method yields the same solution if the grid is sufficiently fine. However, some methods are more suitable to some classes of problems than others. Fundamentals of CFD analysis using the above techniques can be found in several textbooks (Patankar, 1980; Ferziger and Peric, 1999). Only a brief description is presented below.

4.3.1 Finite-Difference Method

FDM is the oldest method for numerical solution of PDEs, introduced by Euler in the eighteenth century. In this method, the derivatives in the governing differential equation are approximated using truncated Taylor series expansions for the purpose of discretization. The starting point of formulating a finite-difference scheme is to discretize the domain into a number of grid points and to express the derivatives at those grid points in terms of the nodal values of the function itself. The grid points can be numbered according to a number of free indices (depending on the dimensionality of the problem), provided the grid system is a structured one. The result is one algebraic equation per grid node, in which the variable value at that node and a certain number of neighbor nodes appear as unknowns.

FDM is the easiest method of discretization to use for simple geometries. In principle, this method can be applied to any grid type, including structured grids. On structured grids, FDM is very simple to apply and effective in obtaining. Higher order schemes on regular grids. The disadvantages of FDMs are that the conservation is not enforced unless special care is taken and they are restricted to simple geometries.

4.3.2 Finite-Element Method

In FEM, the domain is broken into a set of discrete volumes or finite elements. The shape of the elements can be arbitrary. In two dimensions, these are usually triangles or quadrilaterals, whereas in three dimensions, tetrahedral or hexahedral elements are often used. In FEM, the equations are multiplied by a weight function before they are integrated over the entire domain in an effort to minimize the error in an integral sense (Zienkiewicz et al., 2000). The solution is approximated by simple piecewise functions (shape functions) valid on elements to describe the local variations of the unknown flow variables (or scalars). Such a function can be constructed from its values at the corners of the elements. This approximation is then substituted into the weighted integral of the governing equation, and a residual is defined to measure the deviation from the actual solution. The algebraic equations to be solved are derived by requiring the derivative of the integral with respect to each nodal value to be zero; this corresponds to minimization of the residual.

4.3.3 Finite-Volume Method

FVM was originally developed as a special finite-difference formulation in a conservative form, where the entire solution domain was divided into a number of control volumes (CVs). In the structured grid approach developed by Patankar (1980), the control volumes were created using grid lines of a coordinate system. FVM begins with a formal integration of the governing equations of fluid flow over all the control volumes of the solution domain. At the centroid of each CV lies a computational node at which the variable values are to be calculated. Interpolation is used to express variable values at the CV surface in terms of the nodal values. As a result, one obtains an algebraic equation for each CV, in which a number of neighbor nodal values appear. FVM approach is perhaps the simplest to understand and to program. All terms that need be approximated have physical meaning, and this is the reason why FVM is popular. FVM, which was originally formulated with structured grids, is now successfully adapted for unstructured grid systems (Mathur and Murthy, 1997), thus making it suitable for complex geometries as well. Because the discretization is performed using an integral form of the governing equations, conservation is always preserved. Moreover, the discretization process in FVM is directly related to the physics of the problem unlike variational formulation or functionals used in FEM, which have no obvious physical interpretation of the problems involving transport phenomena. It is this combination of ease of formulation of a flow problem and the geometric flexibility in the choice of grids, along with the flexibility in defining the discrete flow variables, that makes FVM suitable for solving physical problems involving fluid flow, heat, and mass transfer. Most commercially available codes are based on this method.

4.4 FVM for Diffusion Problems

The generalized transport equation for any scalar, ϕ, is

$$\frac{\partial(\rho\phi)}{\partial t} + \text{div}(\rho\phi u) = \text{div}\left(\Gamma \text{ grad } \phi\right) + S_\phi \tag{4.1}$$

where Γ is the diffusion coefficient and S is the source term. The equations of mass, momentum, and energy conservation, for example, can all be cast in the above general form, with the expressions for ϕ, Γ (general diffusion coefficient), and S_ϕ (general source term) differing between equations (Table 4.1). In Table 4.1, μ denotes viscosity; p, pressure; $\vec{F}$, the body force per unit volume; T, the temperature; k, thermal conductivity; c_p, specific

TABLE 4.1

General Form of Conservation Equations

Equation	ϕ	Γ	S_ϕ
Continuity	1	0	0
Momentum	$\vec{V}$	μ	$-\nabla p + \nabla\left(\dfrac{\mu}{3}\nabla\cdot\vec{V}\right) + \vec{F}$
Energy	T	k/c_p	$\dfrac{\beta}{c_p}T\dfrac{Dp}{Dt} + \dfrac{\mu\Phi}{c_p} + \dfrac{\dot{Q}}{c_p}$

heat; β, volumetric expansion coefficient, Φ, viscous dissipation function; and $\dot{Q}$, rate of heat generation per unit volume.

For pure diffusion and steady state, the first two terms on the left-hand side of Equation 4.1 vanish, and the resulting equation is

$$\text{div}(\Gamma\,\text{grad}\,\phi) + S_\phi = 0 \tag{4.2}$$

As a first step in an FVM, integration of Equation 4.2 over a control volume gives

$$\int_{CV}\text{div}\left(\Gamma\,\text{grad}\,\phi\right)dV + \int_{CV}S_\phi dV = \int_A n\cdot\left(\Gamma\,\text{grad}\,\phi\right)dA + \int_{CV}S_\phi dV = 0 \tag{4.3}$$

By carrying out the above integration over a control volume, appropriate approximation equations are introduced, which lead to discretized equations. For simplicity, the starting point chosen is the integration process over a one-dimensional control volume, through which most of the issues related to discretization, property interpolation, and solution techniques can be illustrated, and comparison with analytical solution is possible for some cases. Subsequently, the method is extended to two- or three-dimensional diffusion problems.

4.4.1 FVM for One-Dimensional Steady-State Diffusion

Consider the steady-state diffusion of a property ϕ in a one-dimensional domain shown in Figure 4.1.

The process is governed by the following equation:

$$\frac{d}{dx}\left(\Gamma\frac{d\phi}{dx}\right) + S = 0 \tag{4.4}$$

The first step is the grid generation process that, in the case of FVM, involves dividing the domain into discrete control volumes or cells, as shown in Figure 4.2. A representative nodal point corresponding to a control volume is denoted by P. In a one-dimensional domain, the two neighboring nodes to the west and east of P are denoted by W and E, respectively. The west side control volume face is denoted by "w," and the east side control volume face is denoted by "e." The distances between nodes W and P and between nodes P and E are denoted by $(\delta x)_w$ and $(\delta x)_e$, respectively. The control volume width is Δx.

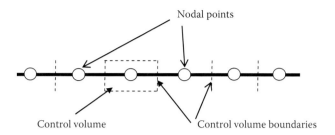

FIGURE 4.1
One-dimensional domain showing nodal points and control volume.

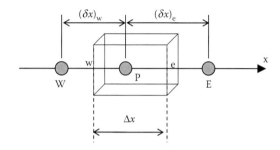

FIGURE 4.2
Grid system of a one-dimensional computational domain.

The next important step is the discretization process, which involves integration of the governing equation (4.4) over a control volume. The integration process yields an equation representing the balance of diffusive flux across the control volume. For the control volume defined in Figure 4.2, the diffusive flux of ϕ leaving the east face minus the diffusive flux of ϕ entering the west face is equal to the generation of ϕ. Mathematically, this can be expressed as

$$\int_{\Delta V} \frac{d}{dx}\left(\Gamma \frac{d\phi}{dx}\right) dV + \int_{\Delta V} S_\phi dV = \left(\Gamma A \frac{d\phi}{dx}\right)_e - \left(\Gamma A \frac{d\phi}{dx}\right)_w + \bar{S}\Delta V = 0 \tag{4.5}$$

where A is the face area of the control volume, ΔV is the volume, and $\bar{S}$ is the average value of source S over the control volume. A key approximation in the finite-volume approach is that the scalar ϕ and the material properties are assumed to be constants and are solely represented by the values specified at the nodal point of the control volume. To calculate the diffusive flux at an interface, the gradient of the scalar ϕ at that location needs to be evaluated, for which an approximate distribution of ϕ between the two adjacent nodal points is used. Linear approximation (Figure 4.3) is the simplest way of calculating interface values and the gradients and is called *central difference*. The gradients are calculated as

$$\left(\frac{d\phi}{dx}\right)_w = \left(\frac{\phi_P - \phi_W}{(\delta x)_w}\right) \quad \text{and} \quad \left(\frac{d\phi}{dx}\right)_e = \left(\frac{\phi_E - \phi_P}{\delta x_e}\right) \tag{4.6}$$

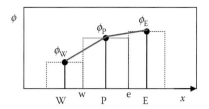

FIGURE 4.3
Linear variation of temperature between adjacent grid points.

The diffusive flux terms are

$$\left(\Gamma A \frac{d\phi}{dx}\right)_w = \Gamma_w A_w \left(\frac{\phi_P - \phi_W}{(\delta x)_w}\right) \quad \text{and} \quad \left(\Gamma A \frac{d\phi}{dx}\right)_e = \Gamma_e A_e \left(\frac{\phi_E - \phi_P}{\delta x_e}\right) \tag{4.7}$$

As the aim is to find a solution to the discretized equations in a linear framework, a non-linear source term must be linearized. The source term is expressed in a linear form (considering a piecewise constant profile of ϕ within each control volume):

$$\bar{S}\Delta V = S_C + S_P \phi_P \tag{4.8}$$

where S_C is the constant part of the source term. For a source term that is a nonlinear function of ϕ, a linear component is represented in the second term of Equation 4.8, whereas any nonlinearity is integrated with S_C, making use of existing values of ϕ at the nodal points.

By substituting Equations 4.6 through 4.8 in Equation 4.5, the discretized equation takes the form:

$$\left(\Gamma A \frac{d\phi}{dx}\right)_e - \left(\Gamma A \frac{d\phi}{dx}\right)_w + \bar{S}\Delta V = \Gamma_e A_e \left(\frac{\phi_E - \phi_P}{(\delta x)_e}\right) - \Gamma_w A_w \left(\frac{\phi_P - \phi_W}{(\delta x)_w}\right) + S_C + S_P \phi_P = 0 \tag{4.9}$$

Upon rearranging,

$$\left(\frac{\Gamma_e}{(\delta x)_e} A_e + \frac{\Gamma_w}{(\delta x)_w} A_w - S_P\right)\phi_P = \left(\frac{\Gamma_w}{(\delta x)_w} A_w\right)\phi_W + \left(\frac{\Gamma_e}{(\delta x)_e} A_e\right)\phi_E + S_C \tag{4.10}$$

or

$$a_P \phi_P = a_W \phi_W + a_E \phi_E + S_C \tag{4.11}$$

with the coefficients a's being

$$a_W = \frac{\Gamma_w}{(\delta x)_w} A_w, \quad a_E = \frac{\Gamma_e}{(\delta x)_e} A_e, \quad \text{and} \quad a_P = a_W + a_E - S_P \tag{4.12}$$

The discretized equations must be set up at each of the nodal points to solve a problem.

4.4.1.1 Some Notes on Discretization Rules

For numerical stability and to yield realistic solutions, there are certain rules to be followed for discretization (Patankar, 1980). The foremost among them is the flux consistency at control volume faces: when an interface is common to two adjacent control volumes, the flux across it must be represented by the same expression in the discretization equations for the two control volumes. The second rule is that all coefficients (a's) must always be positive (essentially, all coefficients must be of the same sign; by convention that sign is taken as positive). A negative coefficient amid positive coefficients of the other neighbors implies a negative influence of a neighbor, which is physically unrealistic. The third rule involves linearization of the source term. When the source term is linearized as $S = S_C + S_P \phi_P$, the coefficient S_P must always be less than or equal to zero. Effectively, this ensures satisfaction of the second rule that all coefficients of the discretized equations must be positive.

This principle of flux consistency is also used to determine interfacial diffusivity. Diffusivities and other physical properties are specified at the nodal points, whereas flux has to be calculated at a control volume interface where the value of the property is not specified explicitly. For a method of evaluating the interfacial diffusivity, consider a heat conduction problem shown schematically in Figure 4.4.

The first guess of interfacial conductivity at e is a linearly interpolated value:

$$k_e = f_e k_P + (1 - f_e) k_E \tag{4.13}$$

where

$$f_e = \frac{(\delta x)_{e^+}}{(\delta x)_e} \tag{4.14}$$

The interfacial heat flux may be calculated as

$$q_e = \frac{k_e (T_P - T_E)}{(\delta x)_e} = \frac{(T_P - T_E)}{(\delta x)_e / k_e} \tag{4.15}$$

Two different ways of calculating the same interfacial flux are

$$q_e = \frac{k_P (T_P - T_e)}{(\delta x)_{e^-}} \quad \text{and} \quad q_e = \frac{k_E (T_e - T_E)}{(\delta x)_{e^+}} \tag{4.16}$$

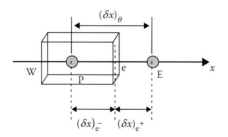

FIGURE 4.4
Schematic for evaluating interfacial diffusivity.

Equivalently,

$$q_e = \frac{(T_P - T_E)}{((\delta x)_{e^-}/k_P) + ((\delta x)_{e^+}/k_E)} \tag{4.17}$$

$$\frac{1}{k_e} = \frac{f_e}{k_E} + \frac{(1 - f_e)}{k_P}. \tag{4.18}$$

For $(\delta x)_{e^+} = (\delta x)_{e^-}$, $f_e = 0.5$. Accordingly, the expression for k_e becomes:
$k_e = 2k_P k_E/(k_P + k_E)$, which is actually the Harmonic mean of the two adjacent nodal conductivity values. This is different from the value obtained by linear interpolation (i.e., $k_e = (k_P + k_E)/2$), which is the arithmetic mean instead.

For control volumes that are adjacent to the domain boundaries, the general discretized equation is modified to incorporate boundary conditions. The resulting system of linear algebraic equations is solved to obtain the distribution of the property ϕ at nodal points.

4.4.1.2 Treatment of Boundary Conditions

Boundary conditions generally come in three different forms: (1) the Dirichlet type, as a specified value of the dependent variable (say, temperature); (2) the Neumann type, as a specified flux condition (i.e., specified gradient of the dependent variable; e.g., specified temperature gradient); and (3) the mixed type, as a value of the dependent variable specified as a function of its flux (e.g., convective boundary condition at an interface). In case 1, nothing special needs to be done for discretization, as the boundary node with specified (fixed) temperature is not solved for. In case 2, however, the boundary temperature (here we take heat conduction problem as a generic example) is unknown and hence needs to be solved for. The discretization technique described above uses the value of two neighboring nodes (one on either side of the unknown node) and hence cannot be applied directly in this case. Special methods of discretizing boundary conditions are described below.

Let q_B be the heat flux prescribed at the left boundary of a domain, as shown in Figure 4.5. One obvious method of discretization is by performing energy balance with respect to the boundary control volume:

$$q_B - q_i + \bar{S}\Delta x = 0. \tag{4.19}$$

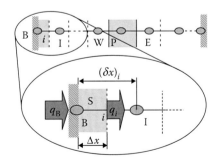

FIGURE 4.5
Discretization of a boundary flux.

Expressing the heat flux q_i using a forward difference technique and linearizing the source term,

$$q_B - \frac{k_i(T_B - T_I)}{(\delta x)_i} + (S_C + S_P T_B)\Delta x = 0 \tag{4.20}$$

If q_B is specified as a constant, then

$$a_B T_B = a_I T_I + b \tag{4.21}$$

where $a_B = a_I - S_P \Delta x$; $a_I = k_I/(\delta x)_i$; and $b = S_C \Delta x + q_B$.

If q_B is specified via a heat transfer coefficient h and the temperature of the surrounding fluid T_f (Boundary condition of type 3), then

$$q_B = h(T_f - T_B) \tag{4.22}$$

The corresponding discretization equation is

$$a_B T_B = a_I T_I + b \tag{4.23}$$

where $a_B = a_I - S_P \Delta x + h$; $a_I = k_I/(\delta x)_i$; $b = S_C \Delta x + h T_f$.

The method of discretizing the boundary heat flux described above mimics the actual diffusion process from the boundary and hence it often takes a number of iterations and convergence can be slow. A faster convergence can be achieved if the boundary heat flux is replaced by an equivalent volumetric heat source term at the boundary control volume and the boundary is treated as adiabatic simultaneously.

4.4.1.3 Solution of Linear Algebraic Equations

The set of linear algebraic equations resulting from a particular discretization technique needs to be solved computationally. The linear framework of discretization allows standard solution techniques for a system of linear equations. There are two families of solution techniques for linear algebraic equations: direct methods and iterative methods. Examples of direct methods include Cramer's rule matrix inversion and Gaussian elimination. In this class of solution technique, the simultaneous storage of all N^2 coefficients of the set of N equations with N unknowns in the core memory is required. On the other hand, iterative (or indirect) methods use repeated application of a relatively straightforward algorithm resulting in a final convergence of solution. Popular example include the Jacobi and the Gauss–Seidel point-by-point iteration method. In this method, only nonzero coefficients of the equations need to be stored in core memory. Because the systems are often very large, up to 100,000 or 1 million equations, iterative methods are generally much more economical than direct methods.

4.4.1.4 The Tridiagonal Matrix Algorithm

For one-dimensional problems, the dependent variable (say temperature) at a grid point is a sole function of the temperature of up to two other grid points, which are essentially its neighbors. Thus, the coefficient matrix is tridiagonal in nature. In such cases, a computationally efficient algorithm (i.e., the tridiagonal matrix algorithm [TDMA]) may be devised by storing the coefficients in a one-dimensional array, instead of a two-dimensional array. Although the TDMA (Thomas, 1949) is actually a direct method for one-dimensional situation, it can be applied iteratively, in a line-by-line fashion, to solve multidimensional problems and is widely used in CFD programs.

To begin with an illustration of the TDMA, one may consider a template-discretized equation in the following form:

$$a_P T_P = a_E T_E + a_W T_W + b \tag{4.24}$$

With the change in notation shown in Figure 4.6,

$$a_i T_i = b_i T_{i+1} + c_i T_{i-1} + d_i \tag{4.25}$$

Equation 4.25 is written for each i, with $i = 1, 2, \ldots, N$, as follows:

$$
\begin{aligned}
i &= 1 & a_1 T_1 &= b_1 T_2 + d_1 \\
i &= 2 & a_2 T_2 &= b_2 T_3 + c_2 T_1 + d_2 \\
i &= 3 & a_3 T_3 &= b_3 T_4 + c_3 T_2 + d_3 \\
&\ \ \vdots & &\ \ \vdots \\
i &= N-1 & a_{N-1} T_{N-1} &= b_{N-1} T_N + c_{N-1} T_{N-2} + d_{N-1} \\
i &= N & a_N T_N &= c_N T_{N-1} + d_N
\end{aligned}
$$

One may note here that for boundaries, $c_1 = 0$, $b_N = 0$. With the above setup of discretized equations, the TDMA algorithm is executed as given below.

Given the original relation, a relationship of the following form is sought:

$$T_i = P_i T_{i+1} + Q_i. \tag{4.26}$$

Supposing one already has $T_{i-1} = P_{i-1} T_i + Q_{i-1}$, one may substitute for T_i in Equation 4.25 and obtain

$$a_i T_i = b_i T_{i+1} + c_i \left(P_{i-1} T_i + Q_{i-1} \right) + d_i \tag{4.27}$$

Upon rearranging,

$$T_i = \left(\frac{b_i}{a_i - c_i P_{i-1}} \right) T_{i+1} + \left(\frac{d_i + c_i Q_{i-1}}{a_i - c_i P_{i-1}} \right) \tag{4.28}$$

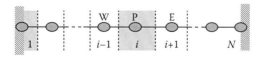

FIGURE 4.6
One-dimensional grid system.

Comparing with coefficients in Equation 4.26, one obtains

$$P_i = \frac{b_i}{a_i - c_i P_{i-1}}; \quad Q_i = \frac{d_i + c_i Q_{i-1}}{a_i - c_i P_{i-1}} \tag{4.29}$$

This is a recursive relation, and one needs to start somewhere. For $i = 1$,

$$a_1 T_1 = b_1 T_2 + d_1$$
$$P_1 = b_1 / a_1$$
$$Q_1 = d_1 / a_1$$

Using $T_i = P_i T_{i+1} + Q_i$, one may calculate $P_1 = b_1/a_1$ and $Q_1 = d_1/a_1$. Subsequently, one may use recurrence relations to get P_i and Q_i for $i = 2, 3, \ldots, N$. Noting that at the boundary, $b_N = 0$, one gets $P_N = 0$ and $T_N = Q_N$. By setting $T_N = Q_N$, one may use recurrence relations to get $T_i = P_i T_{i+1} + Q_i$ for $i = N - 1, N - 2, \ldots, 3, 2, 1$. Finally, one may obtain $T_{N-1}, T_{N-2}, \ldots, T_3, T_2, T_1$. Because of a simple linear storage, the complexity of this algorithm is $O(N)$.

4.4.2 FVM for One-Dimensional Unsteady-State Diffusion

Equation 4.4 represents the one-dimensional steady-state diffusion equation for any general scalar ϕ. Consider, now, the unsteady-state diffusion in the context of heat transfer, in which the temperature, T, is the scalar. The corresponding partial differential equation is:

$$\rho c \frac{\partial T}{\partial t} = \frac{\partial}{\partial x}\left(k \frac{\partial T}{\partial x}\right) + S \tag{4.30}$$

The term on the left-hand side of Equation 4.30 is the storage term, arising out of accumulation/depletion of heat in the domain under consideration. Note that Equation 4.30 is a partial differential equation as a result of an extra independent variable, time (t). The corresponding grid system is shown in Figure 4.7.

To obtain a discretized equation at the nodal point P of the control volume, integration of the governing Equation 4.30 is required to be performed with respect to time and

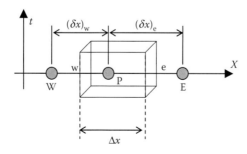

FIGURE 4.7
Grid system of an unsteady one-dimensional computational domain.

space. Integration over the control volume and over a time interval gives

$$\int_{t}^{t+\Delta t}\left(\int_{CV}\left(\rho c\frac{\partial T}{\partial t}\right)dV\right)dt = \int_{t}^{t+\Delta t}\left(\int_{CV}\frac{\partial}{\partial x}\left(k\frac{\partial T}{\partial x}\right)dV\right)dt + \int_{t}^{t+\Delta t}\left(\int_{CV}SdV\right)dt \qquad (4.31)$$

Rewritten,

$$\int_{w}^{e}\left(\int_{t}^{t+\Delta t}\rho c\frac{\partial T}{\partial t}dt\right)dV = \int_{t}^{t+\Delta t}\left(\left(kA\frac{\partial T}{\partial x}\right)_{e} - \left(kA\frac{\partial T}{\partial x}\right)_{w}\right)dt + \int_{t}^{t+\Delta t}(\overline{S}\Delta V)dt \qquad (4.32)$$

If the temperature at a node is assumed to prevail over the whole control volume, applying the central difference scheme, one obtains

$$\rho c\left(T_{P}^{new}-T_{P}^{old}\right)\Delta V = \int_{t}^{t+\Delta t}\left(\left(k_{e}A\frac{T_{E}-T_{P}}{\delta x_{e}}\right)-\left(k_{w}A\frac{T_{P}-T_{W}}{\delta x_{w}}\right)\right)dt + \int_{t}^{t+\Delta t}(\overline{S}\Delta V)dt \qquad (4.33)$$

Now, an assumption is made about the variation of T_P, T_E, and T_W with time. By generalizing the approach by means of a weighting parameter f between 0 and 1,

$$\int_{t}^{t+\Delta t}\phi_{P}dt = \phi_{P}\Delta t = \left[f\phi_{P}^{new}-(1-f)\phi_{P}^{old}\right]\Delta t \qquad (4.34)$$

Repeating the same operation for points E and W,

$$\rho c\left(\frac{T_{P}^{new}-T_{P}^{old}}{\Delta t}\right)\Delta x = f\left[\left(k_{e}\frac{T_{E}^{new}-T_{P}^{new}}{\delta x_{e}}\right)-\left(k_{w}\frac{T_{P}^{new}-T_{W}^{new}}{\delta x_{w}}\right)\right]$$
$$+(1-f)\left[\left(k_{e}\frac{T_{E}^{old}-T_{P}^{old}}{\delta x_{e}}\right)-\left(k_{w}\frac{T_{P}^{old}-T_{W}^{old}}{\delta x_{w}}\right)\right]+\overline{S}\Delta x \qquad (4.35)$$

Upon rearranging, dropping the superscript "new," and casting the equation into the standard form,

$$a_{P}T_{P} = a_{W}[fT_{W}+(1-f)T_{W}^{old}]+a_{E}[fT_{E}+(1-f)T_{E}^{old}]+[a_{P}^{0}-(1-f)a_{W}-(1-f)a_{E}]T_{P}^{old}+b$$

$$(4.36)$$

where

$$a_{P} = \theta(a_{W}+a_{E})+a_{P}^{0}; \quad a_{P}^{0} = \rho c\frac{\Delta x}{\Delta t}; \quad a_{W} = \frac{k_{w}}{\delta x_{w}}; \quad a_{E} = \frac{k_{e}}{\delta x_{e}}; \quad b = \overline{S}\Delta x \qquad (4.37)$$

The time integration scheme would depend on the choice of the parameter f. When $f = 0$, the resulting scheme is "explicit"; when $0 < f \le 1$, the resulting scheme is "implicit"; when $f = 1$, the resulting scheme is "fully implicit"; and when $f = 1/2$, the resulting scheme is the

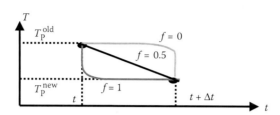

FIGURE 4.8
Variation of T within the time interval Δt for different schemes.

Crank–Nicolson scheme (Crank and Nicolson, 1947). The variation of T within the time interval Δt for the different schemes is shown in Figure 4.8.

4.4.2.1 Explicit Scheme

Linearizing the source term as $b = S_u + S_p T_P^{old}$ and setting $f = 0$ in Equation 4.36, the explicit discretization becomes

$$a_P T_P = a_W T_W^{old} + a_E T_E^{old} + [a_P^0 - (a_W + a_E)]T_P^{old} + S_u \tag{4.38}$$

where

$$a_P = a_P^0; \quad a_P^0 = \rho c \frac{\Delta x}{\Delta t}; \quad a_W = \frac{k_w}{\delta x_w}; \quad \text{and} \quad a_E = \frac{k_e}{\delta x_e} \tag{4.39}$$

The above scheme is based on backward difference, and its Taylor series truncation error accuracy is first order with respect to time. For stability, all coefficients must be positive in the discretized equation. Hence,

$$a_P^0 - (a_W + a_E - S_P) > 0, \quad \rho c \frac{\Delta x}{\Delta t} - \left(\frac{k_w}{\delta x_w} + \frac{k_e}{\delta x_e} \right) > 0, \quad \rho c \frac{\Delta x}{\Delta t} > \frac{2k}{\Delta x}, \quad \text{or} \quad \Delta t < \rho c \frac{(\Delta x)^2}{2k} \tag{4.40}$$

The above limitation on time step suggests that the explicit scheme becomes very expensive to improve spatial accuracy. Hence, this method is generally not recommended for general transient problems. Nevertheless, provided that the time-step size is chosen with care, the explicit scheme described above is efficient for simple conduction calculations.

4.4.2.2 The Crank–Nicolson Scheme

Setting $f = 0.5$ in Equation 4.36, the Crank–Nicolson discretization becomes

$$a_P T_P = a_E \left(\frac{T_E + T_E^{old}}{2} \right) + a_W \left(\frac{T_W + T_W^{old}}{2} \right) + \left[a_P^0 - \frac{a_E}{2} - \frac{a_W}{2} \right] T_P^0 + b \tag{4.41}$$

where

$$a_P = \frac{1}{2}(a_E + a_W) + a_P^0 - \frac{1}{2}S_P; \quad a_P^0 = \rho c \frac{\Delta x}{\Delta t}; \quad a_W = \frac{k_w}{\delta x_w}; \quad a_E = \frac{k_e}{\delta x_e}; \quad \text{and} \quad b = S_u + \frac{1}{2}S_P T_P^{old}$$

(4.42)

The above method is implicit, and simultaneous equations for all node points need to be solved at each time step. For stability, all coefficients must be positive in the discretized equation, requiring

$$a_P^0 > \frac{a_E + a_W}{2} \quad \text{or} \quad \Delta t < \rho c \frac{(\Delta x)^2}{k}$$

(4.43)

The Crank–Nicolson scheme is only slightly less restrictive than the explicit method. It is based on central difference, and hence it is second-order accurate in time.

4.4.2.3 The Fully Implicit Scheme

Setting $f = 1$ in Equation 4.36, the fully implicit discretization becomes

$$a_P T_P = a_E T_E + a_W T_W + a_P^0 T_P^{old}$$

(4.44)

where

$$a_P = a_P^0 + a_E + a_W - S_P; \quad a_P^0 = \rho c \frac{\Delta x}{\Delta t}; \quad a_W = \frac{k_w}{\delta x_w}; \quad \text{and} \quad a_E = \frac{k_e}{\delta x_e}$$

(4.45)

A system of algebraic equations must be solved at each time level. The accuracy of the scheme is first order in time. The time marching procedure starts with a given initial field of the scalar ϕ^0. The system is solved after selecting time step Δt. For the implicit scheme, all coefficients are positive, which makes it unconditionally stable for any size of time step. Hence, the implicit method is recommended for general-purpose transient calculations because of its robustness and unconditional stability.

4.4.3 Implicit Scheme for Unsteady Multidimensional Diffusion Problems

A two-dimensional unsteady thermal conduction (diffusion) equation is considered:

$$\rho c \frac{\partial T}{\partial t} = \frac{\partial}{\partial x}\left(k \frac{\partial T}{\partial x}\right) + \frac{\partial}{\partial y}\left(k \frac{\partial T}{\partial y}\right) + S$$

(4.46)

The corresponding grid system is shown in Figure 4.9. When the equation is formally integrated over the control volume and over the time step, the discretized unsteady two-dimensional heat conduction for implicit scheme (with $f = 1$) equation becomes

$$a_P T_P = a_E T_E + a_W T_W + a_N T_N + a_S T_S + b$$

(4.47)

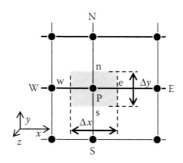

FIGURE 4.9
Grid system for two-dimensional diffusion process.

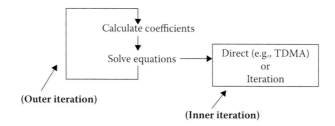

FIGURE 4.10
The levels of iteration in a solution algorithm.

where

$$a_E = \frac{k_e \Delta y}{(\delta x)_e}; \quad a_W = \frac{k_w \Delta y}{(\delta x)_w}; \quad a_N = \frac{k_n \Delta x}{(\delta y)_n}; \quad a_S = \frac{k_s \Delta x}{(\delta y)_s}; \quad a_P^{old} = \frac{\rho c \Delta x \Delta y}{\Delta t}$$

$$b = S_C \Delta x \Delta y + a_P^{old} T_P^{old} \tag{4.48}$$

$$a_P = a_E + a_W + a_N + a_S - S_P \Delta x \Delta y.$$

4.4.4 Solution of Multidimensional Equations

The direct method of TDMA that works conveniently with one-dimensional systems cannot be easily extended to two or three dimensions. For multidimensional problems, the matrix no longer remains tridiagonal. For two- and three-dimensional problems, the matrix becomes penta- and hepta-diagonal, respectively. Solving such problems numerically using these methods becomes computationally very expensive. Hence, iterative methods are usually chosen as these methods can take care of the inherently nonlinear nature of fluid flow problems. It may be noted here that this can lead to two levels of iterations, as shown in Figure 4.10.

The outer level of iteration can take care of nonlinearity arising out of properties, which are functions of the dependent variable ϕ, because of coupled nature of equations, whereas the inner iterations solve a particular set of linear algebraic equations.

Two of the most popular iterative methods used as part of inner iterations are Gauss–Siedel point-by-point scheme and line-by-line TDMA, which are elaborated below.

4.4.4.1 The Gauss–Siedel Point-by-Point Scheme

First, the generalized discretization equation is considered:

$$a_P T_P = \sum a_{nb} T_{nb} + b \tag{4.49}$$

or

$$T_P = \frac{\sum a_{nb} T_{nb}^* + b}{a_P} \tag{4.50}$$

where T_{nb}^* is the current estimate of temperature on any grid point

$$T_P = \frac{\sum a_{nb} T_{nb}^* + b}{a_P} \tag{4.51}$$

This can be solved if T_{nb}^*s are known.

Procedure:

   (i) Start with an initial guess temperature distribution;
   (ii) Visit a grid point, substitute latest estimate for T_{nb}^*, get new T;
   (iii) Visit all points in this manner;
   (iv) Iterate until convergence.

4.4.4.2 The Line-by-Line TDMA Method

As the matrix becomes sparse in the case of multidimensional problems, TDMA cannot be applied to all the grid points (with unknown temperature) simultaneously. However, TDMA can be applied to a particular line with unknown temperature, by considering the temperatures of the grid points on the neighboring lines as known values (using existing values). Considering a two-dimensional system shown in Figure 4.11, the procedure for a line-by-line TDMA is elaborated below.

Procedure:

   (i) Choose a line (unknown temperature);
   (ii) Substitute for Ts along neighboring lines from the current estimates;

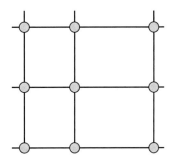

FIGURE 4.11
Two-dimensional grid system with vertical lines highlighted for solving by TDMA.

(iii) Solve for Ts along the line by TDMA;

(iv) Repeat for all lines;

(v) Repeat the whole process by alternating the line direction if necessary.

4.5 FVM for Convection–Diffusion Problems

The generalized convection–diffusion equation can be written in the following form:

$$\frac{\partial(\rho\phi)}{\partial t} + \frac{\partial J_i}{\partial x_i} = S \tag{4.52}$$

where J_i is the total flux including convection and diffusion and is given by

$$J_i = \rho u_i \phi - \Gamma\left(\frac{\partial\phi}{\partial x_i}\right) \tag{4.53}$$

The velocity field associated with convection should also satisfy the continuity equation.

4.5.1 One-Dimensional Convection–Diffusion Problems

Consider the one-dimensional domain shown in Figure 4.7. For steady-state problem without any source term, the governing equation can be written as

$$\frac{d}{dx}(\rho u \phi) = \frac{d}{dx}\left(\Gamma\frac{d\phi}{dx}\right) \tag{4.54}$$

where u is the velocity in the x-direction. In the same domain, the mass conservation equation becomes

$$\frac{d}{dx}(\rho u) = 0 \quad \text{or} \quad \rho u = \text{const} \tag{4.55}$$

Integrating Equation 4.54 over the control volume shown in Figure 4.7, one obtains

$$\underbrace{(\rho u \phi)_e - (\rho u \phi)_w}_{\text{net convection}} = \underbrace{\left(\Gamma\frac{d\phi}{dx}\right)_e - \left(\Gamma\frac{d\phi}{dx}\right)_w}_{\text{net diffusion}} \tag{4.56}$$

4.5.1.1 *Central Difference Scheme*

Using a central difference scheme and assuming that an interface is located midway between two adjacent grid points, the interfacial values of the scalar ϕ can be expressed as

$$\phi_e = \frac{1}{2}(\phi_E + \phi_P)$$
$$\phi_w = \frac{1}{2}(\phi_P + \phi_W). \tag{4.57}$$

The discretized form of Equation 4.56, thus, can be written as

$$\frac{1}{2}(\phi_E + \phi_P)(\rho u)_e - \frac{1}{2}(\phi_P + \phi_W)(\rho u)_w = \frac{\Gamma_e(\phi_E - \phi_P)}{(\delta x)_e} - \frac{\Gamma_w(\phi_P - \phi_W)}{(\delta x)_w} \tag{4.58}$$

Rearranging the discretization equation in the usual form, one obtains

$$a_P \phi_P = a_E \phi_E + a_W \phi_W \tag{4.59}$$

where

$$
\begin{aligned}
a_E &= \frac{\Gamma_e}{\delta x} - \frac{\rho u_e}{2} = D_e - \frac{F_e}{2} \\
a_W &= \frac{\Gamma_w}{\delta x} + \frac{\rho u_w}{2} = D_w + \frac{F_w}{2} \\
a_P &= \frac{\Gamma_e}{\delta x} + \frac{\rho u_e}{2} + \frac{\Gamma_w}{\delta x} - \frac{\rho u_w}{2} = a_E + a_W + (F_e - F_w)
\end{aligned}
\tag{4.60}
$$

In Equation 4.60, $F = \rho u$ represents the advection strength, whereas $D = \Gamma/\delta x$ is the diffusion strength.

4.5.1.2 Upwind Scheme

The central difference scheme is found to be suitable for flows in which diffusion strength is high compared with the advection strength, as far as flux of the scalar ϕ is concerned. The relevant nondimensional number representing the relative strength of the two transport mechanisms is called the *Peclet number* ($P = \rho u L/\Gamma$), where L is a representative length scale and Γ is the diffusivity. The central difference scheme assumes equal influence of the ϕ values of two adjacent grid points in determining the convective flux at the interface of the control volume between those grid points. This assumption will be valid only for low Peclet number flows. For very high Peclet number flows, the influence of the upstream node will be dominant at the interface, and this is the basis of the upwind method. In this scheme, the value of ϕ at an interface is equal to that at the node on the *upstream* side of the interface. In other words,

$$
\begin{aligned}
\phi_e &= \phi_P \quad \text{if } F_e > 0 \\
\phi_e &= \phi_E \quad \text{if } F_e < 0
\end{aligned}
\tag{4.61}
$$

Accordingly, the advective flux at the interface e can be written as

$$(\rho u \phi)_e = \phi_P \text{Max}[\rho u_e, 0] - \phi_E \text{Max}[-\rho u_e, 0] \tag{4.62}$$

Defining ϕ_w in a similar manner, the advective flux at the interface w can be written as

$$(\rho u \phi)_w = \phi_W \text{Max}[\rho u_w, 0] - \phi_P \text{Max}[-\rho u_w, 0] \tag{4.63}$$

The discretized equation is then written in the standard form as

$$a_P \phi_P = a_E \phi_E + a_W \phi_W \tag{4.64}$$

where the coefficients are expressed as

$$
\begin{aligned}
a_E &= \frac{\Gamma_e}{\delta x} + \langle -\rho u_e, 0 \rangle = D_e + \langle -F_e, 0 \rangle \\
a_W &= \frac{\Gamma_w}{\delta x} + \langle \rho u_w, 0 \rangle = D_w + \langle F_w, 0 \rangle \\
a_P &= \frac{\Gamma_e}{\delta x} + \langle \rho u_e, 0 \rangle + \frac{\Gamma_w}{\delta x} + \langle -\rho u_w, 0 \rangle \\
&= a_E + a_W + (\rho u_e - \rho u_w)
\end{aligned}
\tag{4.65}
$$

4.5.1.3 Exponential Scheme

The upwind scheme described above is ideal for high Peclet number flows. However, to cover the entire range of Peclet number, the analytical solution for variation of ϕ over the length of a one-dimensional control volume can be observed. The steady-state one-dimensional transport equation can be written as

$$\frac{d}{dx}(\rho u \phi) = \frac{d}{dx}\left(\Gamma \frac{d\phi}{dx}\right) \tag{4.66}$$

The boundary conditions are

$$
\begin{aligned}
x &= 0 \quad \phi = \phi_0 \\
x &= L \quad \phi = \phi_L
\end{aligned}
\tag{4.67}
$$

The solution is

$$\frac{\phi - \phi_0}{\phi_L - \phi_0} = \frac{\exp(P(x/L)) - 1}{\exp(P) - 1} \tag{4.68}$$

A graphical representation of the behavior of Equation 4.68 is depicted in Figure 4.12.

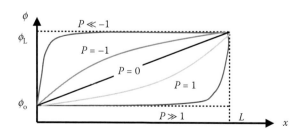

FIGURE 4.12
Variation of ϕ along control volume length for various Peclet numbers (exact solution).

In the exponential scheme, Equation 4.68 is used as the profile assumption for ϕ between successive grid points. Considering the two neighboring grid points to be located at $x = 0$ and $x = L$, respectively, in a local coordinate system where the origin corresponds to the face at which the flux of ϕ needs to be calculated, the above may lead to a discretization equation of the form of Equation 4.64, where

$$a_E = \frac{F_e}{\exp(F_e/D_e) - 1}$$

$$a_W = \frac{F_w \exp(F_w/D_w)}{\exp(F_w/D_w) - 1} \tag{4.69}$$

$$a_P = a_E + a_W + (F_e - D_w)$$

Here

$$\text{Pe} = \frac{(\rho u)_e (\delta x)_e}{\Gamma_e} = \frac{F_e}{D_e} \tag{4.70}$$

The exact solution gives the variation of ϕ along the control volume length for any Peclet number. However, it is inconvenient to use an analytical expression in the context of a numerical solution, as an accurate representation of the expression in a computer program will be computationally expensive. Hence, approximate representations of ϕ variations have emerged, which can be implemented in a more practical sense. The hybrid and the power law schemes are two such examples described subsequently.

4.5.1.4 Hybrid Scheme

The hybrid difference scheme (Spalding, 1972) is based on a combination of central and upwind difference schemes. The central difference scheme is used for small Peclet numbers (Pe < 2), whereas the upwind scheme is used for large Peclet numbers (Pe ≥ 2). The hybrid scheme is essentially a piecewise linear interpolation of the exponential scheme. A summary of the scheme is

$$\text{For Pe} < -2, \qquad \frac{a_E}{D_e} = -\text{Pe}$$

$$\text{For} -2 \leq \text{Pe} \leq 2, \qquad \frac{a_E}{D_e} = 1 - \frac{\text{Pe}}{2} \tag{4.71}$$

$$\text{For Pe} > 2, \qquad \frac{a_E}{D_e} = 0$$

4.5.1.5 Power Law Scheme

The power law difference scheme of Patankar (1980) is a more accurate approximation to the one-dimensional exact solution than the hybrid scheme. In the hybrid scheme, the influence of diffusion is switched off for Pe > 2. In the power law scheme, diffusion is set to zero when cell Pe exceeds 10. If $0 < \text{Pe} < 10$, the flux is evaluated by using a polynomial expression. In summary, the scheme is

$$
\begin{aligned}
&\text{For Pe} < -10, && \frac{a_E}{D_e} = -\text{Pe} \\
&\text{For } -10 \le \text{Pe} < 0, && \frac{a_E}{D_e} = (1+0.1\text{Pe})^5 - \text{Pe} \\
&\text{For } 0 \le \text{Pe} \le 10, && \frac{a_E}{D_e} = (1-0.1\text{Pe})^5 \\
&\text{For Pe} > 10, && \frac{a_E}{D_e} = 0
\end{aligned} \tag{4.72}
$$

4.5.2 Convection–Diffusion Equation for Two-Dimensional Problems

To derive the discretized equations in two dimensions, consider a typical control volume around the grid point P as shown in Figure 4.8. The integration of Equation 4.52 over this control volume would give

$$
\left(\rho_P \phi_P - \rho_P^0 \phi_P^0\right)\left(\frac{\Delta V}{\Delta t}\right) + J_e A_e - J_w A_w + J_n A_n - J_s A_s = (S_C + S_P \phi_P)\,\Delta V \tag{4.73}
$$

where ρ_P^0 and ϕ_P^0 denote the old values of ρ_P and ϕ_P at the beginning of the time step Δt, J is the total fluxes across the control volume faces, $(S_C + S_P \phi_P)$ is the source term in the linearized form, ΔV is the volume of the control volume, and A is the area of the control volume faces. The details of the derivation of Equation 4.73 can be found in Patankar (1980) and will not be repeated here for the sake of brevity.

The final set of discretized equations, derived from the continuity equation and Equation 4.73, is as follows:

$$
a_P \phi_P = a_E \phi_E + a_E \phi_E + a_N \phi_N + a_S \phi_S + b \tag{4.74}
$$

where

$$
a_E = D_e A\left(|\text{Pe}|\right) + max\{-F_e, 0\} \tag{4.75}
$$

$$
a_w = D_W A(|P_w|) + max\{F_w, 0\} \tag{4.76}
$$

$$
a_N = D_n A(|P_n|) + max\{-F_n, 0\} \tag{4.77}
$$

$$
a_s = D_s A(|P_s|) + max\{F_s, 0\} \tag{4.78}
$$

$$
a_P^0 = \rho_P^0 \frac{\Delta V}{\Delta t} \tag{4.79}
$$

$$
b = S_C \Delta V + a_P^0 \phi_P^0 \tag{4.80}
$$

$$
a_P = a_E + a_W + a_S + a_N + a_P^0 - S_P \Delta V \tag{4.81}
$$

TABLE 4.2

$A(|P|)$ for Various Convection–Diffusion
Discretization Schemes

Scheme	$A(|P|)$
Central difference	$1 - 0.5|P|$
Upwind	1
Exponential	$|P|/(\exp|P| - 1)$
Hybrid	$\max(0, 1 - 0.5|P|)$
Power law	$\max(0, (1 - 0.5|P|)5)$

F represents the flow rates at the control volume faces (e.g., $F_e = (\rho u)_e A_e$) and D_e is diffusion conductance of the link PE and is calculated as $D_e = A_e [(\delta_{e-}/\Gamma_P) + (\delta_{e+}/\Gamma_E)]^{-1}$. The definition of F and D at other interfaces can be constructed in the similar manner.

The function $A(|P|)$ depends on the scheme of interpolation to compute the cell face values for discretizing the convective flux crossing a cell face. Appropriate expressions of the same, for different schemes, are given in Table 4.2. The simplest scheme is central difference, which computes the cell face value using a linear interpolation, which is the same technique used for calculating the cell face value for evaluating the diffusion flux. However, the central difference scheme is suitable only for low Peclet number systems, which is characterized by strong diffusive flux compared with convection. For high Peclet number systems, the upwind scheme is more suitable, as the interpolated cell face value for convection is taken to be the same as that of the upstream node. However, such a scheme tends to overpredict diffusion when Peclet number is low. Hybrid and power law schemes, on the other hand, suitably adapt to the local flow conditions, ranging from very low to very high Peclet numbers. Before proceeding further, it may be instructive to reiterate and summarize some of the important features of the convection–diffusion discretization schemes as introduced in the preceding discussions.

1. The central difference scheme is not suitable for flows in which the advection strength is higher than the diffusion strength and hence can give rise to unrealistic solutions for cell-based Peclet numbers greater than 2. The upwind scheme is a possible remedy, but this scheme tends to overpredict diffusion effects.

2. Although the exponential scheme gives an accurate estimation of the profile variation of the dependent variable considering locally one-dimensional convection–diffusion process, it involves expensive computation of exponential terms.

3. The hybrid scheme is essentially a linear interpolation of the exponential scheme and therefore it represents the actual physical behavior more closely. However, the influence of diffusion is switched off for Pe > 2.

4. The power law scheme overcomes the above-mentioned drawback of the hybrid scheme in the sense that it is essentially a curve fitting of the exponential scheme, and therefore it mimics the physical behavior very closely in a one-dimensional convection–diffusion situation. It switches off the diffusion effects only beyond a cell-based Peclet number of 10, unlike the case of upwind scheme that does so beyond a cell-based Peclet number of 2.

5. The upwind scheme is usually associated with "numerical diffusion" or "false diffusion," which essentially implies overprediction of diffusion flux. However,

an alternative viewpoint is that the false diffusion can be attributed to the gridding system and consequent interpolation technique to calculate gradient at an interface, and it may not depend so much on the particular convection–diffusion scheme chosen. This effect becomes more pronounced when the local flow direction is somewhat oblique with respect to the grid lines that form the discretized space. Certain higher order difference schemes (an example will be illustrated in the subsequent subsection) have been suggested as alternative remedies to overcome false diffusion effects.

4.5.3 Higher-Order Difference Schemes

Higher-order schemes involve more neighbor points and reduce the discretization errors by bringing in a wider influence. A popular higher-order scheme is the quadratic upwind interpolation for convective kinetics (QUICK) scheme (Leonard, 1979). For a one-dimensional domain shown in Figure 4.13, the ϕ value at a control volume face is approximated using a three-point quadratic function passing through the two surrounding nodes, and an additional node on the upstream side.

Let $u_e > 0$, $u_w > 0$. Then $\phi_w = \phi_w\,(\phi_P, \phi_W, \phi_{WW})$ and $\phi_e = \phi_e\,(\phi_P, \phi_E, \phi_W)$. For uniform grid, a quadratic interpolation gives

$$\phi_w = (6/8)\,\phi_W + (3/8)\,\phi_P - (1/8)\,\phi_{WW} \tag{4.82}$$

$$\phi_e = (6/8)\,\phi_P + (3/8)\,\phi_E - (1/8)\,\phi_W \tag{4.83}$$

For the diffusion terms, one uses the gradient of the appropriate parabola. The discretized equation takes the standard form:

$$a_P\phi_P = a_E\phi_E + a_W\phi_W + a_{WW}\phi_{WW} \tag{4.84}$$

where

$$a_E = D_e - \frac{3}{8}F_e; \quad a_W = D_w + \frac{6}{8}F_w - \frac{1}{8}F_e; \quad a_{WW} = -\frac{1}{8}F_w$$
$$a_P = a_E + a_W + a_{WW} + (F_e - F_w) \tag{4.85}$$

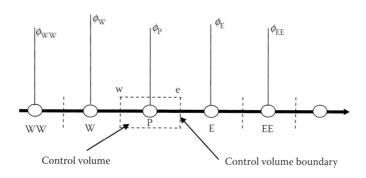

FIGURE 4.13
The quadratic upwind interpolation for convective kinetics interpolating scheme.

4.6 Solution of Velocity Field Using a Pressure-Based Algorithm

Consider the equations governing a two-dimensional laminar steady flow:

- *x*-momentum equation

$$\frac{\partial}{\partial x}(\rho u u) + \frac{\partial}{\partial y}(\rho v u) = -\frac{\partial p}{\partial x} + \frac{\partial}{\partial x}\left(\mu \frac{\partial u}{\partial x}\right) + \frac{\partial}{\partial y}\left(\mu \frac{\partial u}{\partial y}\right) + S_u \tag{4.86}$$

- *y*-momentum equation

$$\frac{\partial}{\partial x}(\rho u v) + \frac{\partial}{\partial y}(\rho v v) = -\frac{\partial p}{\partial y} + \frac{\partial}{\partial x}\left(\mu \frac{\partial v}{\partial x}\right) + \frac{\partial}{\partial y}\left(\mu \frac{\partial v}{\partial y}\right) + S_v \tag{4.87}$$

- continuity equation

$$\frac{\partial}{\partial x}(\rho u) + \frac{\partial}{\partial y}(\rho v) = 0 \tag{4.88}$$

The momentum conservation equations are looked into separately because of their coupling with the continuity equation. Also, difficulties of these equations are that (i) the convective terms of the momentum equation contain nonlinear quantities and (ii) all three equations are coupled. The most complex issue to resolve is the role played by the pressure. The pressure gradient terms in the momentum equations cannot be treated as source terms, as pressure is also one of the variables to be solved for. In case of compressible flow, a separate equation of state can relate the pressure with the density. However, for incompressible flow, the coupling between pressure and density is rather weak. Hence, an equation of state will not have much role to play in solving the momentum conservation equations. In this case, the continuity equation is used determine the solution, in the sense that if the correct pressure field is applied in the momentum equations, the resulting velocity field should satisfy continuity. Hence, unlike density-based algorithms used in compressible flow calculations, pressure-based iterative algorithms are more appropriate for incompressible flows. Such iterative techniques can also take care of the problems associated with the inherent nonlinearities in the equation set. A popular technique is the well-known semi-implicit method for pressure-linked equations (SIMPLE) algorithm of Patankar and Spalding (1972).

4.6.1 Concept of Staggered Grid

FVM starts with the discretization of the flow domain and of the relevant transport equations. Consider a typical control volume around the grid point P as shown in Figure 4.9. It seems logical to define the velocities at the same locations as the scalar variables such as pressure, temperature, and so forth. If the linear interpolation is used, the pressure gradient in the *u*-momentum equation becomes

$$\frac{\partial p}{\partial x} = \frac{p_e - p_w}{\delta x} = \frac{((p_E + p_P)/2) - ((p_P + p_W)/2)}{\delta x} = \frac{p_E - p_W}{2\delta x} \tag{4.89}$$

The corresponding pressure gradient in the y-momentum equation is $(\partial p/\partial y) = (p_N - p_S/2\delta y$. For such discretization of the pressure gradient term, a highly nonuniform pressure field can act like a uniform field in the discretized momentum equations. The remedy is to use a "staggered grid" for the velocity components. In the staggered grid system, the scalar variables are evaluated, such as pressure, density, temperature, and so forth, at the usual nodal points, whereas the velocity components are calculated on staggered grids centered about the cell faces (Figures 4.14 and 4.15). A new system of notation is introduced on the basis of a numbering of grid lines and cell faces. The control volumes for u and v are different from the scalar control volumes. The pressure gradient terms are given by

$$\frac{\partial p}{\partial x} = \frac{p_E - p_W}{\delta x_u} \quad \text{and} \quad \frac{\partial p}{\partial y} = \frac{p_N - p_S}{\delta y_v} \tag{4.90}$$

where δx_u is the width of the u-control volume and δy_v is the width of the v-control volume.

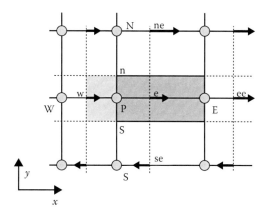

FIGURE 4.14
Staggered grid system for u-momentum equation, showing control volume centered about the location e.

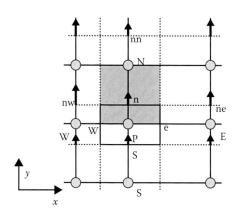

FIGURE 4.15
Staggered grid system for v-momentum equation, showing control volume centered about the location n.

It can be seen from Figure 4.14 that the pressure force acting on the velocity control volume can be calculated directly as $(P_P - P_E) \times$ area of control volume face. Then the discretization equation of the x-momentum equation can be written as

$$a_e u_e = \Sigma a_{nb} u_{nb} + b + (P_P - P_E) A_e \qquad (4.91a)$$

where the coefficients a_e, a_{nb}, and b are identical to those as considered for discretizing the convection–diffusion problems. The discretized equation for the control volume around v_n can be obtained in the similar way.

$$a_n v_n = \Sigma a_{nb} v_{nb} + b + (P_P - P_N) A_n \qquad (4.91b)$$

Here it should be noticed that the pressure gradient is not included in the source term. If pressure field is known, these equations can be solved similar to the one explained earlier for convection–diffusion equation.

4.6.2 Velocity and Pressure Correction Equations

It is obvious that unless correct pressure field is used, the resulting velocity field will not satisfy the continuity equation. Thus, with a guessed pressure field P^*, the solution leads to imperfect velocity fields u^* and v^*.

$$a_e u_e^* = \Sigma a_{nb} u_{nb}^* + b + (P_P^* - P_E^*) A_e \qquad (4.92a)$$
$$a_n v_n^* = \Sigma a_{nb} v_{nb}^* + b + (P_P^* - P_N^*) A_n \qquad (4.92b)$$

These velocity fields will not satisfy continuity equation, which necessitates a pressure correction as $P = P^* + P'$. The velocity components will respond to this pressure correction. The corresponding velocity corrections can be introduced as $u = u^* + u'$ and $v = v^* + v'$. Subtracting Equation 4.92 from Equation 4.91,

$$a_e u_e' = \Sigma a_{nb} u_{nb}' + (PP' - P_E') A_e \qquad (4.93)$$

An important simplification is now made by approximating Equation 4.93 as $a_e u_e' = (PP' - P_E')A_e$, by dropping the term $\Sigma a_{nb} u_{nb}'$. The justification of this simplification will be given subsequently. The resulting velocity correction formulae are as follows:

$$u_e = u_e^* + d_e (PP' - P_E') \qquad (4.94a)$$
$$v_n = v_n^* + d_n (PP' - P_E') \qquad (4.94b)$$

where $d_e = A_e/a_e$ and $d_n = A_n/a_n$. The continuity equation is used for the pressure correction. The continuity equation (Equation 4.88) can be discretized for the main control volume shown in Figure 4.9 as follows:

$$(\rho A u)_w - (\rho A u)_e + (\rho A v)_s - (\rho A v)_n - (\rho_P - \rho_P^o)\left(\frac{\Delta V}{\Delta t}\right) = 0 \qquad (4.95)$$

Now, by substituting u and v velocities from Equation 4.94 into the continuity Equation 4.95, the discretized equation for pressure correction can be derived as follows:

$$a'_P P'_P = a_E P'_E + a_W P'_W + a_N P'_N + a_S P'_S + b \qquad (4.96)$$

where

$$a_E = \rho_e d_e A_e \qquad (4.97a)$$

$$a_W = \rho_w d_w A_w \qquad (4.97b)$$

$$a_N = \rho_n d_n A_n \qquad (4.97c)$$

$$a_S = \rho_s d_s A_s \qquad (4.97d)$$

$$b = (\rho A u^*)_w - (\rho A u^*)_e + (\rho A v^*)_s - (\rho A v^*)_n - (\rho_P - \rho_P^0)\frac{\Delta V}{\Delta t} \qquad (4.97e)$$

The density ρ at the interface can be calculated using some convenient interpolation from the neighboring grid point densities. It should be noted that the term b stands for the mass source (or mass imbalance) implied by the starred velocity field. Through the pressure correction equation, the velocity and pressure fields are guided toward a solution that satisfies both the continuity and the momentum equations. Starting with any guessed pressure field and solving for the momentum conservation equations, there is no guarantee that the resulting velocity field will satisfy the continuity equation. The pressure correction equation corrects the pressure and velocity fields to ensure that the resulting field completely nullifies this mass imbalance.

It is also important to mention in this context that although the fluid flow solution algorithms were originally introduced for staggered grids, it is not a compulsion that staggered grids only need to be used for such purposes. Use of staggered grids essentially eliminates the needs (and hence also the associated troubles and possible errors incurred) of pressure interpolation at the control volume faces. However, if one uses higher order and more effective pressure interpolation schemes at the control volume faces, then it becomes possible to use the same grid points for velocity and pressure computation. Such a grid arrangement is called a *colocated grid*, as is often used for flow computations in CFD codes with both structured and unstructured meshes.

4.6.3 Solution Algorithms

Having established the basic framework with all the equations needed for obtaining the velocity components and pressure, some algorithms suitable for solving the momentum conservation equations can now be formulated. The solution algorithms are iterative in nature, and they differ from each other in the starting point (e.g., start with guessed pressure field or velocity field) and in other procedures followed.

4.6.4 The SIMPLE Algorithm

The SIMPLE algorithm stands for semi-implicit method for pressure-linked equations. In this method, one starts with a guessed pressure field and follows an iterative procedure using the velocity and pressure correction equations as formulated above. The procedure can be outlined as follows:

- Guess the pressure field P^*
- Solve the momentum equations to obtain u^*, v^*, w^*
- Solve the P' equation (continuity)
- Correct pressure field $P = P^* + P'$
- Calculate the velocity corrections u', v', w'
- Correct velocity field $u = u^* + u', v = v^* + v', w = w^* + w'$
- Treat the corrected pressure P as a new guess P^* and go to the second step. Iterate until a converged solution is obtained.

4.6.5 Discussions

1. It may be noted that because the pressure correction is being solved for rather that the pressure itself, the omission of the terms $\sum a_{nb} u_{nb}'$ in Equation 4.93 has no effect on the final converged solution. In the final iteration, u^*, v^*, and P^* satisfy the momentum equations, and the b term in the p' equation is zero. As discussed earlier, the p' equation does not generate any corrections, and $u = u^*, v = v^*$, and $P = P^*$ hold true. Thus, the p' equation has no effect on the final converged solution. Only the discretized momentum and continuity equations determine the final solution.

2. On the other hand, the dropping of the $\sum a_{nb} u_{nb}'$ terms will affect the path of convergence and its rate. The x-velocity correction in Equation 4.93 is a function of both the velocity correction term $\sum a_{nb} u_{nb}'$ and the pressure correction term. The term $\sum a_{nb} u_{nb}'$ brings in the influence of the neighboring cells in the overall correction scheme, and by dropping it, the entire burden of correction falls on the pressure correction term. This results in overcorrection of pressure and may lead to solution divergence, unless underrelaxation is used. Typically, the corrected pressure is expressed as $P = P^* + \alpha_P P'$, where α_P is an underrelaxation factor (usually set to about 0.8).

3. With regard to boundary conditions for pressure, two common boundary conditions are considered here: the given normal velocity and the given static pressure. For given pressure at the boundary, the value of P' at that boundary will be zero. If the normal velocity u_e is given at a boundary that also coincides with the boundary control volume face, the flow rate across the boundary should be expressed in terms of u_e and not u^*. In other words, a_E will be zero in the P' equation.

4. For incompressible flows, density is not a function of pressure. For such cases, with the given velocity boundary conditions on all boundaries, the level of pressure in the domain is not set. Differences in pressure are unique, but the absolute pressure values themselves are not. It can be verified that P and $P + C$ are solutions to the governing differential equations.

4.6.5.1 The SIMPLER Algorithm (SIMPLE–Revised)

One disadvantage of using SIMPLE algorithm is the rate of convergence. The approximation introduced by the omission of the $\sum a_{nb} u_{nb}'$ terms in setting up the pressure correction equation results in overprediction of pressure correction and the consequent requirement of underrelaxation. On the other hand, if the pressure correction equation is used for correcting the velocities and the pressure field is improved by some other means, a more efficient algorithm results. The following revised algorithm (SIMPLER), introduced by Patankar, has a better convergence rate compared with SIMPLER (Patankar, 1980).

Another disadvantage of SIMPLE algorithm is that it starts with a guessed pressure field, which cannot always be given intuitively by the user. On the other hand, a trial velocity field is easier to provide, as most users have a better feel of approximate velocity distribution for a given problem.

The development of this algorithm starts with the establishment of a discretized pressure equation, as follows.

4.6.5.2 Pressure Equation

The equation for obtaining the pressure field can be derived as follows. The momentum Equation 4.91 can be written as

$$u_e = \hat{u}_e + d_e(P_P - P_E) \tag{4.98a}$$

where

$$\hat{u}_e = \frac{\sum a_{nb} u_{nb} + b}{a_e}$$

Similarly,

$$v_n = \hat{v}_n + d_n(P_P - P_N) \tag{4.98b}$$

Now, by substituting these velocities into the discretized continuity equation, the pressure equation can be derived as

$$a_P P_P = a_E P_E + a_W P_W + a_N P_N + a_S P_S + b \tag{4.99}$$

where a_P, a_E, a_W, a_N, and a_S may be derived in a manner analogous to the derivation of coefficients appearing in Equation 4.96 and

$$b = (\rho A \hat{u})_w - (\rho A \hat{u})_e - (\rho A \hat{v})_s + (\rho A \hat{v})_n - (\rho_P - \rho_P^o)\frac{\Delta V}{\Delta t} \tag{4.100}$$

4.6.5.3 Summary of the Overall Calculation Procedure

The above set of discretized equations is used for solving the momentum equation through an iterative procedure. The procedure followed is the SIMPLER scheme suggested by Patankar (1980), and the sequence of operation is summarized as follows:

1. Guess the velocity field.
2. Compute $\hat{u}$ and $\hat{v}$ from Equation 4.98a.

3. Solve the pressure equation (Equation 4.99) using the guessed field and obtain the pressure.

4. Solve the momentum equations using the pressure field just computed to obtain u^* and v^*.

5. Compute the mass source term b from Equation 4.97e and then solve the pressure correction equation (Equation 4.96).

6. Correct u^* *and* v^* using Equation 4.94. The pressure is not corrected.

7. Return to step 2 and repeat the procedure until converged solution is obtained.

4.6.5.4 The SIMPLEC Algorithm

The SIMPLE-corrected (SIMPLEC) algorithm attempts to overcome the primary drawback of the SIMPLE algorithm arising from the omission of the $\Sigma a_{nb}\, u'_{nb}$ terms in pressure correction equation. Instead of ignoring the contribution from the neighboring cells completely, the SIMPLEC algorithm (Van Doormal and Raithby, 1984) attempts to approximate the neighbor corrections by using the correction as

$$\sum a_{nb} u'_{nb} = u'_e \sum a_{nb} \tag{4.101}$$

Thus, the velocity corrections take the form

$$\left(a_e - \sum a_{nb}\right) u'_e = A_e (P'_P - P'_E) \tag{4.102}$$

In this manner, d_e is redefined as

$$d_e = \frac{A_e}{a_e - \sum a_{nb}} \tag{4.103}$$

Other than the above changes, the SIMPLEC algorithm is similar to the SIMPLE algorithm, with the exception that the pressure correction need not be underrelaxed. The convergence of the SIMPLEC algorithm has been shown to be faster than that of the SIMPLE, and it is also less computationally intensive compared with the SIMPLER algorithm. Several other variants of the SIMPLE algorithm have also been subsequently introduced in the literature, which is not discussed here for the sake of brevity. For example, Issa (1986) introduced the pressure implicit with splitting of operators algorithm for the noniterative computation of unsteady compressible flows.

4.7 Introduction to FVM with Unstructured Meshes

For long, the FVM was considered to be unsuitable for dealing with irregular geometry as the method was initially implemented with the help of structured grids. The FVM with body-fitted coordinate systems permitted the use of irregular geometry to some extent, but the method became too tedious for highly irregular geometries. Subsequently, the

FVM was adapted to accommodate the features of unstructured meshes (Mathur and Murthy, 1997), which gave the method the geometric flexibility usually enjoyed by the finite-element method. A fundamental difference between structured and the unstructured meshes is that the latter has the provision of a variable number of neighboring cell vertices, unlike the former that has a fixed number of neighboring cell vertices for all internal cells. The basic principle of FVM, however, is retained in the sense that the discretized equation is obtained through a balance of fluxes across the cell interfaces. However, in this case, the fluxes calculated at the cell faces in unstructured meshes need not necessarily be aligned with the direction joining the neighboring grid points, and they need to be resolved into "orthogonal" and "nonorthogonal" components.

The basic principles of discretization methodology for unstructured meshes can be illustrated by considering a triangular mesh as shown in Figure 4.16.

An illustration of the unstructured FVM is done through integration and subsequent discretization of the steady-state x-momentum equation. The integration is the triangular control volume shown in Figure 4.16, to yield

$$\underbrace{\int_{CV} \nabla \cdot \left(\rho \vec{V} u\right) d\forall}_{\text{term 1}} = \underbrace{-\int_{CV} \frac{\partial p}{\partial x} d\forall}_{\text{term 2}} + \underbrace{\int_{CV} \nabla \cdot (\mu \nabla u) d\forall}_{\text{term 3}}. \tag{4.104}$$

Divergence theorem is applied on each term appearing in Equation 4.104, converting the volume integrals into equivalent surface integrals. The resulting expressions become

$$\text{term } 1 = \int_{CV} \nabla \cdot (\rho \vec{V} u)\, d\forall = \int_{CS} (\rho \vec{V} u) \cdot d\vec{A} = \overset{\text{number of edges (3)}}{\sum_{i=1}} (\rho u_f)_i (\vec{V} \cdot d\vec{A})_i$$

or

$$\text{term } 1 = \sum_{i=1}^{3} (\rho u_f)_i U_{f,i} = \sum_{i=1}^{3} F_i U_{f,i}, \quad \text{say} \tag{4.105}$$

where $\vec{U} = \vec{V} \cdot d\vec{A} = u dA_x + v dA_y$ is the contravariant velocity vector, the subscript "f" refers to a cell face, and F represents the strength of advection. Similarly, the second term in

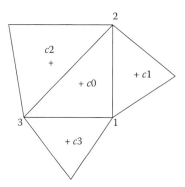

FIGURE 4.16
A triangular control volume (with cell center at $c0$), along with three of its neighboring control volumes. Points $c0$, $c1$, $c2$, and $c3$ are cell centers, and 1, 2, and 3 are cell vertices.

Equation 4.104 becomes

$$\text{term } 2 = \int_{CV} -\frac{\partial p}{\partial x}\,d\forall = \int_{CS} -(p\hat{i})\cdot d\vec{A} = -\sum_{i=1}^{3}(pdA_x)_i$$

where $\hat{i}$ is a unit vector along positive x-direction.

In the context of the control volume shown in Figure 4.16, the second term can be written as,

$$\text{term } 2 = -\sum_{i=1}^{3} p_i \Delta y_i = -[(p_{c1}-p_{c0})\Delta y_1 + (p_{c2}-p_{c0})\Delta y_2 + (p_{c3}-p_{c0})\Delta y_3] \tag{4.106}$$

In Equation 4.106, Δy_i represents the difference in the y-coordinates of the two vertices of the edge "i." Here, a convention is followed with respect to the order of vertices (positive oriented, in this case) taken for the evaluation of this parameter, for all edges. Following this convention, the third term is expressed as

$$\text{term } 3 = \int_{CV} \nabla \cdot (\mu \nabla u)\,d\forall = \int_{CS} (\mu \nabla u)\cdot d\vec{A} = \sum_{i=1}^{3}\mu_i\left[\frac{\partial u}{\partial x}\Delta y - \frac{\partial u}{\partial y}\Delta x\right]_i \tag{4.107}$$

The discretization of the partial derivatives appearing in Equation 4.107 can be performed by choosing auxiliary control volumes surrounding an edge "i". As an illustration, for evaluation of the integrals over the edge 1–2, the term $(\partial u/\partial x)_i$ can be evaluated by constructing an auxiliary control volume with the bounding edges as 1–c1–2–c0–1, as follows: $(\partial u/\partial x)_i = \int_{A_i}(\partial u/\partial x)\,dA/A_i = \int_C u\,dy/A_i$, where C is a closed contour that forms the surface of the auxiliary control volume. This contour integral can be numerically approximated by employing the Trapezoidal rule (for example), so that one can write $(\partial u/\partial x)_1 = [(u_{c1}-u_{c0})\Delta y_1 + (u_1-u_2)\Delta y_{c1-c0}]/A_1$, where $\Delta y_1 = y_2 - y_1$ and $\Delta y_{c1-c0} = y_{c1} - y_{c0}$. Analogous considerations can be made for the other integrals to be evaluated for term 3. The discretized form becomes

$$\text{term } 3 = \underbrace{\sum_{i=1}^{3}D_i(u_{ci}-u_{c0})}_{\text{orthogonal diffusion flux}} + \underbrace{D_{n1}(u_1-u_2) + D_{n2}(u_2-u_3) + D_{n3}(u_3-u_1)}_{\text{nonorthogonal diffusion flux}} \tag{4.108}$$

where $D_i = (\mu_i/2A_i)\,[(\Delta x)^2 + (\Delta y)^2]_i$ and $D_{ni} = (\mu_i/2A_i)\,[(\Delta x)_i(\Delta x)_{ci-c0} + (\Delta y)_i(\Delta y)_{ci-c0}]$.

The orthogonal components can be treated in the same manner as in the structured grid formulation, whereas the nonorthogonal components can be combined with the source term. With the help of interpolation functions relating the variables at the vertices with those at the cell centers, the three terms of the x-momentum equation can be combined to yield the following discretized equation:

$$a_P u_{c0} = A_1 u_{c1} + A_2 u_{c2} + A_3 u_{c3} + S \tag{4.109}$$

where $A_i = [D_i A(|\text{Pe}|_i) + \max(\pm F_i, 0)]$, $A_P = A_1 + A_2 + A_3$, and $A(|\text{Pe}|_i)$ depends on the chosen convection–diffusion discretization scheme (Table 4.2).

4.8 Issues of Consistency, Stability, and Convergence

The success of a discretization scheme in obtaining an "acceptable solution" is often qualified with the aid of performance parameters such as consistency, stability, and convergence (Ferziger and Peric, 1996). Consistency of a numerical scheme implies that the set of discretized algebraic equations perfectly represents the original governing differential equation in the limit as the grid size tends to zero. In other words, the truncation error must vanish as grid sizes become infinitesimally small. To obtain an estimate of the truncation error, let us consider two different grid spacings (say h and $2h$), and let the numerical solutions be ϕ_h and ϕ_{2h}, respectively. If the grids are sufficiently fine, the truncation error is proportional to the leading order term in the Taylor series, that is, $T.E \approx \alpha h^p + H$, where α depends on the derivatives at a given point but is independent of h, and H stands for the higher order terms. Accordingly, the exact solution can be expressed in terms of the numerical solutions as follows:

$$\phi_{\text{exact}} = \phi_h + \alpha h^p + H = \phi_{2h} + \alpha(2h)^p + H \tag{4.110}$$

Accordingly, the truncation error for the grid size of h can be estimated as

$$(T.E)_h \approx \frac{\phi_h - \phi_{2h}}{2^p - 1} \tag{4.111}$$

Stability of a numerical scheme is another important consideration for the successful performance of a discretization scheme. A numerical solution scheme is said to be stable if it does not amplify the numerical errors (essentially, "round off errors") that appear as the numerical solution proceeds. For unsteady problems, stability of a numerical method implies that the solution remains bounded provided that the solution of the original governing differential equation is also bounded. For solving linear problems, a conventional method of evaluating the stability of a solution is by using the von Neumann analysis. For illustration, the stability analysis corresponding to an explicit scheme is described below. Let us consider a one-dimensional diffusion-type problem without any source term (analogous to a heat conduction problem without heat generation) with constant physical properties. Let us also consider a uniform grid system. As a minor modification of Equation 4.38, the discretized equation for this case may be written as

$$a_P \phi_P = a_E \phi_E^0 + a_W \phi_W^0 + (a_P^0 - a_E - a_W)\phi_P^0 \tag{4.112}$$

where $a_E = \Gamma_e/\delta x_e$, $a_W = \Gamma_w/\delta x_w$, $a_P^0 = \rho \Delta x/\Delta t$, and $a_P = a_P^0$. Let Φ represent an exact solution of the discretized equations (using infinitesimally small grid size), such that the numerical error can be quantified as $\varepsilon = \phi - \Phi$. Substituting this in Equation 4.112 and noting that Φ itself satisfies this equation exactly, one obtains

$$a_P \varepsilon_P = a_E \varepsilon_E^0 + a_W \varepsilon_W^0 + (a_P^0 - a_E - a_W)\varepsilon_P^0 \tag{4.113}$$

To assess whether the round-off error gets amplified with iterations, one may expand the error in an infinite series as

$$\varepsilon(x,t) = \sum_m e^{\sigma_m t} e^{i\lambda_m x}, \quad m = 0,1,2,...$$ (4.114)

where σ_m is either real or complex $\lambda_m = m\pi/L$, and $m = 0,1,2,...$, with L being the length of the problem domain. If σ_m is real and greater than zero, then the error gets amplified with time, whereas if σ_m is real and less than zero, then the error dampens. If σ_m is complex (with an imaginary component), the solution is oscillatory. The nature of evolution of the error can be estimated from the amplification parameter, $a = \varepsilon(x, t + \Delta t)/\varepsilon(x,t)$. If a is greater than unity, then the error grows with the iterations, whereas if a is less than unity, then the error is damped out with the iterations so that the scheme can be predicted as stable. In order to examine the underlying consequences, one may utilize the linearity of the problem and just examine the stability consequences of a single error term in Equation 4.114. Accordingly, one gets

$$a = e^{\sigma_m \Delta t} = 1 - \frac{4\Gamma\Delta t}{(\Delta x)^2}$$ (4.115)

Equation 4.115 gives rise to the conditional stability criterion of the explicit scheme, as

$$\Delta t \le \frac{\rho(\Delta x)^2}{2\Gamma}$$ (4.116)

Analogous stability analyses can be executed for the other time-discretization schemes (such as the fully implicit scheme and the Crank-Nicolson scheme). One may also note that although the von Neumann stability analysis yields a limiting time-step estimate to keep the round-off errors bounded, it does not preclude the occurrence of "bounded" but "unphysical" solutions.

4.9 Future Directions of Research

Algorithms in CFD have now parts and parcels of problem solving and design practices in thermal-fluid sciences. However, there are several remaining challenges to the CFD analysts, requiring significant research attention and endeavors. Some such issues are mentioned below.

1. With CFD analysis being applied to probe systems with highly refined spatiotemporal scales, the issue of massive scale parallelization of CFD algorithm becomes an important consideration. As such, there are still several unsolved problems in flow computation because of challenges not originating from the standard flow modeling considerations but mainly because of challenges from the perspectives of their implementations within the constraints of the state-of-the-art computational

resources. Addressing this whole issue requires an effective agglomeration of advancements in computer science (typically, high-performance computing) with the CFD algorithms.

2. In several applications (including microfluidics), the physics encompasses multiple length scales and time scales. For example, to understand the interfacial phenomena close to the fluid/solid boundaries in nanochannels, one may essentially need to resolve the underlying phenomena up to the molecular level (noncontinuum considerations). On the other hand, in the bulk fluid, continuum considerations may still be applicable. One may, thus, adopt a hybrid solution strategy with multiple scale considerations so that a certain part of the problem domain may be analyzed using molecular/mesoscopic approaches, whereas the remaining part may be addressed using continuum approaches. This essentially requires an effective patching of an overlapping region between the subdomains under consideration so that matching conditions can be obtained. Establishing efficient hybrid simulation strategies with coupled molecular- and continuum-based discretizations is definitely one of the key areas in ongoing and future research.

References

Crank J, Nicolson P. A practical method for numerical evaluation of solutions of partial differential equations of the heat conduction type. *Math Proc Cambridge Phil Soc*. 1947;43:50–67.

Ferziger JH, Peric M. *Computational Methods for Fluid Dynamics*. Springer: New York; 1996.

Leonard BP. A stable and accurate convective modeling procedure based on quadratic upstream interpolation. *Comput Methods in Appl Mech Eng*. 1979;19:59–98.

Issa RI. Solution of the implicitly discretized fluid flow equations by operator splitting. *J Comput Phys*. 1986;62:40–65.

Mathur SR, Murthy, JY. A pressure-based method for unstructured meshes. *Numer Heat Transf B*. 1997;31:195–215.

Patankar SV. *Numerical Heat Transfer and Fluid Flow*. New York: Hemisphere Publishing Corporation; 1980.

Patankar SV, Spalding DB. A calculation procedure for heat, mass and momentum transfer in three-dimensional parabolic flows. *Int J Heat Mass Transf*. 1972;15:1787–806.

Spalding DB. A novel finite difference formulation for differential expressions involving both first and second derivatives. *Int J Numer Methods Eng*. 1972;4:551–9.

Thomas LH. *Elliptic Problems in Linear Differential Equations over a Network*. Watson Science Computer Lab Report. New York: Columbia University; 1949.

Van Doormal JP, Raithby GD. Enhancements of the SIMPLE method for predicting incompressible fluid flows. *Numer Heat Transf*. 1984;7:147–63.

Zienkiewicz C, Taylor RL, Zhu JZ. The finite element method its basis and fundamentals. Burlington, MA: Elsevier Butterworth-Heinemann; 2000.

5

Level-Set Method for Microscale Flows

Y.F. Yap, J.C. Chai, T.N. Wong, and N.T. Nguyen

CONTENTS

5.1 Introduction

5.1.1 Purpose

This chapter focuses on presenting a numerical method for two-phase flows in microscale systems using the level-set (LS) method of Osher and Sethian (1988). The interface is characterized using an LS function that is modeled as an additional scalar equation as part of the finite-volume method (Patankar, 1980) used to model various transport phenomena. The main elements of the calculation procedure were presented by Yap et al. (2006). This chapter presents much of the same material in a condensed form. As a result, at times, readers are directed to the above-mentioned study for more details. Nevertheless, this chapter is more or less self-contained.

This chapter does not require the reader to have prior knowledge of two-phase flows modeling. It will start from the basic set of governing equations and formulate a solution procedure for two-phase flows.

5.1.2 Scope and Limitations

As the LS method, as presented by the authors, forms the basis for this chapter, a comprehensive review of all literature dealing with the subject will not be attempted here. This chapter focuses on the *method* of solution rather than on the different ways to handle various physical models. The method of choice (this reflects the authors' personal choice) important to the LS method will be discussed.

In the interest of providing a complete discussion of the LS method as implemented by the authors, it is possible that many significant studies might be omitted from this chapter. However, any such omissions are unintentional and do not imply any judgment as to the quality and usefulness of the studies.

Although admissible in the current approach, phase change between the two phases is not discussed in this chapter. Interested readers are referred to Yap et al. (2005), where phase change using the current approach was presented.

5.1.3 Outline

The rest of the chapter is divided into four sections. Section 5.2 gives a presentation of the conservation equations governing a two-phase flow. This is followed by Section 5.3 on the treatment of the interface between the two phases under the framework of an LS method. Two essential ingredients of a successful LS method, that is, redistancing and mass correction, are presented with considerably detailed Sections 5.3.2 and 5.3.3, respectively. In Section 5.4, the relevant boundary conditions are briefly discussed. The general solution procedure is outlined in Section 5.5. Three two-phase flow problems are presented in Section 5.6 to showcase the method. Finally, some concluding remarks are given.

5.2 Governing Equations

Consider a two-phase flow shown in Figure 5.1. The two immiscible phases are separated by the interface *i*. In the single-fluid approach, only one set of conservation equations governing the flow of both phases is required. This set of equations will be presented in this section.

5.2.1 The Continuity and Momentum Equations

The continuity equation using a single-fluid formulation is given by

$$\frac{\partial \rho}{\partial t} + \nabla (\rho \vec{u}) = 0, \tag{5.1}$$

where ρ is the density appropriate for the phase occupying a particular spatial location at a given instant of time, t is the time, and $\vec{u}$ is the velocity vector. Similarly, the momentum

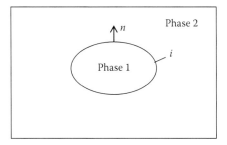

FIGURE 5.1
Schematic of a two-phase flow.

equation can be written as (Scriven, 1960; Edward et al., 1991)

$$\frac{\partial(\rho\vec{u})}{\partial t} + \nabla(\rho\vec{u}\vec{u}) = -\nabla p + \nabla\left[\mu\left(\nabla\vec{u} + \nabla\vec{u}^{T}\right)\right] + \vec{f}_{i}, \tag{5.2}$$

where μ is the viscosity appropriate for the phase occupying the particular spatial location at a given instant of time, p is the pressure, and $\vec{f}_{i}$ is the interfacial force (per unit volume basis) at the interface i. The interfacial force $\vec{f}_{i}$ can be written as

$$\vec{f}_{i} = (\Pi_{1} - \Pi_{2})\hat{n}\ \delta\{\hat{n}\cdot(\vec{x} - \vec{x}_{i})\}, \tag{5.3}$$

where Π is the stress tensor, $\vec{x}$ is the position vector of a point in the solution domain, $\vec{x}_{i}$ is the position vector on the interface i, $\hat{n}\cdot(\vec{x} - \vec{x}_{i})$ is the *shortest* normal distance between $\vec{x}$ and interface i, and $\delta\{\hat{n}\cdot(\vec{x} - \vec{x}_{i})\}$ is the Dirac delta function on the basis of the shortest normal distance to the interface. We defer the exact definition of the Dirac delta function to a later section.

We shall model the interfacial force, Equation 5.3, using the continuum surface force (CSF) model of Brackbill et al. (1992). The interfacial force is reformulated as an equivalent body force acting at the interface i as

$$\vec{f}_{i} = \left[-\kappa\sigma\hat{n} + (\hat{n}\times\nabla\sigma)\times\hat{n}\right]\delta\{\hat{n}\cdot(\vec{x} - \vec{x}_{i})\}, \tag{5.4}$$

where κ is the mean curvature and σ is the surface tension force per unit length. The first and the second terms on the right side of Equation 5.4 represent the capillary and the Marangoni effects, respectively. The capillary force is proportional to local mean curvature of the interface κ. The Marangoni force is induced by a surface tension gradient $\nabla\sigma$. Although the capillary force acts perpendicular to the interface i, the Marangoni force acts tangentially along the interface i. The mean curvature is defined as

$$\kappa \equiv \nabla\cdot\hat{n}. \tag{5.5}$$

Similar to the Dirac delta function, we defer the discussion on the surface normal $\hat{n}$ to a later section.

At this point, we have written the continuity and momentum equations for a two-phase flow problem. The surface tension (capillary and Marangoni) forces are written using the CSF approach. However, we have not discussed how interface between the two phases is captured and thus the properties, the Dirac delta function, and the normal are modeled. As indicated by the title of the chapter, we shall model the evolution of the interface using the LS method of Osher and Sethian (1988), which we shall discuss next.

5.3 The Level-Set Method

5.3.1 Background

We shall first use a single-phase flow to examine an important concept used in the LS method for two-phase flow.

Figure 5.2a shows four concentric circles, each of which encompasses a *fixed* mass of fluid. These circles are labeled using a scalar function ξ whose physical meaning is yet to be determined. These four pockets of fluid are convected by an underlying flow field. Figure 5.2b shows the same four curves, noncircular now; again each encompasses the same mass of fluid with the corresponding original circular curves after some time. The most important concept here is that we can track a fixed mass of fluid by following the evolution of an appropriately defined scalar function ξ. We shall now extend this concept to two-phase flow problems.

Figure 5.3a shows the same four circles as those shown in Figure 5.2a. Now, the space inside three of the four circles, namely, ξ_1, ξ_2, and ξ_3, is occupied by one fluid. The area outside of the third circle, namely, ξ_3, is occupied by the other fluid. With time, these circles deform into different shapes as shown in Figure 5.3b. The interface between the two fluids is still denoted by ξ_3.

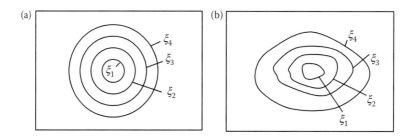

FIGURE 5.2
Evolution of four fixed mass of circular fluid pockets with time: (a) initial and (b) at a later time.

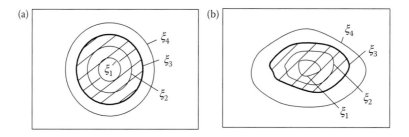

FIGURE 5.3
Evolution of an interface of a two-phase flow problem: (a) initial and (b) at a later time.

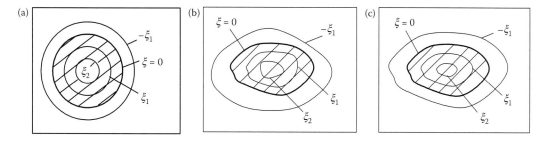

FIGURE 5.4
Evolution of the LS function at (a) $t = 0$, (b) $t_1 > 0$, and (c) t_1 after redistancing.

The choice of ξ is shown in Figure 5.4a, where constant ξ forms concentric circles. We shall set ξ to be zero right on the interface. The constant ξ of the remaining lines is set to the shortest normal distance from the interface i, that is, ξ is a *distance function*. To differentiate the two fluids, the ξ of the two fluids will be assigned different signs. ξ, shown in Figure 5.4a, is called the *level-set* (LS) function and generically described as

$$\xi(\bar{x}, t) = \pm d, \tag{5.6}$$

where the constant d is the shortest normal distance from the interface i for each line of constant ξ and the sign $\pm$ differentiates the two phases. In particular, the interface is described by $\xi(\bar{x}, t) = 0$. Because the LS function is a distance function, it satisfies

$$|\nabla \xi| = 1. \tag{5.7}$$

Using the chain rule, we can differentiate Equation 5.6 and rewrite it as

$$\frac{\partial \xi}{\partial t} + \frac{\partial \xi}{\partial x}\frac{dx}{dt} + \frac{\partial \xi}{\partial y}\frac{dy}{dt} = 0 \tag{5.8}$$

or

$$\frac{\partial \xi}{\partial t} + u\frac{\partial \xi}{\partial x} + v\frac{\partial \xi}{\partial y} = 0 \tag{5.9}$$

We shall track the interface between two fluids using Equation 5.9. Because the driving velocity field is generally nonuniform and discretization schemes are not exact, the lines of constant ξ are no longer a distance function at the end of a time step. The numerical error of the discretization schemes has a more serious setback. The lines of constant ξ are not convected exactly. This has a serious implication, that is, the zero LS line will generally not be the actual interface at the end of a time step. In particular, the mass of the fluid contained within the interface i is artificially increased or decreased because of the numerical error of the discretization schemes. This mass gain or loss of the fluid contained within the interface i increases cumulatively over all the time steps and results in a substantial mass imbalance. Thus, we need to perform redistancing to ensure ξ remains as a distance function. Second, mass correction is needed to ensure that mass is conserved. These two issues are discussed next.

5.3.2 Redistancing

As seen from Figure 5.4b, lines of constant ξ do not remain as a distance function. We shall perform redistancing to ensure that ξ remains as a distance function. Redistancing changes the LS function from Figure 5.4b to Figure 5.4c and is carried out by solving for the "steady-state" solution of a new distance function ψ (Sussman et al., 1994). This new function is

$$\frac{\partial \psi}{\partial \bar{t}} = \overline{\text{sign}}(\xi)(1 - |\nabla \psi|). \tag{5.10}$$

In Equation 5.10, $\bar{t}$ is a pseudotime for ψ, and $\overline{\text{sign}}(\xi)$ is a modified sign function given by Peng et al. (1999) as

$$\overline{\text{sign}}(\xi) = \frac{\xi}{\sqrt{\xi^2 + |\nabla \xi|^2 \Delta^2}}, \tag{5.11}$$

where Δ is the smallest control volume size. Equation 5.10 is subjected to the following initial condition:

$$\psi(\bar{x}, \bar{t} = 0) = \xi(\bar{x}, t). \tag{5.12}$$

Note that $\xi(\bar{x}, t)$ is the currently available values of the LS function. Obviously, the steady-state solution of Equation 5.10 or $\psi(\bar{x}, \bar{t} \to \infty)$ implies that $|\nabla \psi| = 1$, thus ensuring ψ as a distance function. The initial condition of Equation 5.10 ensures that ψ is zero at the same spatial location and thus possesses the same interface. The LS function $\xi(\bar{x}, t)$ is then replaced by the steady-state of $\psi(\bar{x}, \bar{t} \to \infty)$. This turns the LS function $\xi(\bar{x}, t)$ into a distance function with the same zero LS curve.

If the exact solution to Equation 5.9 is available, then the location of the interface ($\xi = 0$ curve) is exact and mass is conserved. However, because of discretization errors, the $\xi = 0$ curve obtained numerically is *not* exact and thus mass is *not* conserved. This is the major weakness of the LS method. Because of this error, mass correction must be performed to conserve mass.

5.3.3 Mass Correction

A mass correction scheme selects one of the fluids as the reference fluid and conserves the mass of the reference fluid. This in turn conserves the mass of the other fluid automatically. The basic concept involves adding mass to the reference fluid when there is mass loss and removing mass from the reference fluid when there is mass surplus.

For the purpose of discussion and without loss of generality, the sign of the distance function of the reference fluid ξ_{ref} is taken to be positive. Figure 5.5 shows the two possible situations where mass corrections are needed. Figure 5.5a shows a situation where there is mass loss during the solution process. Mass must be added to ensure mass of the reference fluid is conserved. The reverse is shown in Figure 5.5b. Mass must be removed to ensure mass of the reference fluid is conserved. In both cases, the mass correction term must shift the interfaces from the dotted lines to the solid lines. In Figure 5.5a, because of the addition of mass of the reference fluid, the values of the distance function (for the *whole* solution domain) are larger after mass correction. In Figure 5.5b, the values of the distance function

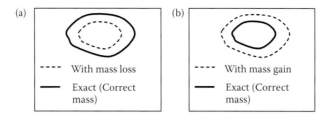

FIGURE 5.5
Mass conservation in the LS method (a) mass addition and (b) mass removal.

(again for the whole solution domain) decrease when mass is removed to ensure the conservation of the mass of the reference fluid. To accomplish this, a global mass correction (GMC) procedure (Yap et al., 2006) is used. The steady-state solution of another distance function ψ' is obtained through

$$\frac{\partial \psi'}{\partial t'} = M_{cor},$$ (5.13)

where t' is a pseudotime, which can be different from $\bar{t}$, and M_{cor} is the mass correction term. A dimensionless mass correction factor is used to ensure mass conservation. This is written as

$$M_{cor} = \frac{M_d - M_c}{M_d},$$ (5.14)

where M_d is the desired mass of the reference fluid and M_c is the current mass of the reference fluid. In the absence of injection, the desired mass is the original mass. The desired mass increases when there is injection of the reference fluid but decreases when there is removal of the reference fluid. Because the sign of the distance function of the reference fluid is a matter of choice, it is possible that the sign of this fluid is chosen to be negative. In such situation, the reverse of the above is observed. The mass correction term is, thus,

$$M_{cor} = \frac{M_c - M_d}{M_d}.$$ (5.15)

Depending on the choice of the Heaviside function of the reference phase, the mass of the reference phase can be calculated using

$$M = \begin{cases} \sum \rho_{ref} H \Delta V & H_{ref} = 1 \\ \sum \rho_{ref} (1 - H) \Delta V & H_{ref} = 0 \end{cases}.$$ (5.16)

The summation is performed over the whole solution domain. The mass correction given by Equation 5.13 (1) reduces to zero once mass is conserved, (2) modifies the distance functions (by a constant) for the whole domain, and (3) can be applied to all density ratios including when $\rho_1/\rho_2 = 1$. For condition 2, the present procedure corrects the distance function by adding a uniform correction factor globally, thus maintaining the LS function as a distance function. This uniform correct may seem too crude. However, tests have shown that it corrects the mass correctly. This will be discussed further in a later section.

5.3.4 Heaviside Function and Fluid Properties

The Heaviside function H is defined as

$$H \equiv \begin{cases} 0 & \xi < 0 \\ 1 & \xi \geq 0 \end{cases}. \tag{5.17}$$

However, the properties will change abruptly at the interface with Equation 5.17. A smeared Heaviside function is introduced to facilitate the calculation of the fluid properties. It is given by Chang et al. (1996) as

$$H(\xi) = \begin{cases} 0 & \xi < -\varepsilon \\ \dfrac{\xi + \varepsilon}{2\varepsilon} + \dfrac{1}{2\pi}\sin\left[\dfrac{\pi\xi}{\varepsilon}\right] & |\xi| \leq \varepsilon. \\ 1 & \xi > \varepsilon \end{cases} \tag{5.18}$$

The Heaviside function is smoothed over a finite thickness of 2ε. The parameter ε is related to the grid size and is usually taken as a factor of the grid spacing (one and a half control volume thickness). With the Heaviside function H defined, the properties of the fluid are calculated conveniently using

$$\alpha = (1 - H)\alpha_1 + H\alpha_2, \tag{5.19}$$

where α can be the density, viscosity, or other properties of interest. The subscripts 1 and 2 refer to phase 1 and phase 2, respectively.

5.3.5 Dirac Delta Function

The Dirac delta function, regularized over a distance of 2ε, is defined as (Peskin, 1977)

$$D(\xi) = \begin{cases} \dfrac{1 + \cos(\pi\xi/\varepsilon)}{2\varepsilon} & |\xi| < \varepsilon \\ 0 & \text{otherwise} \end{cases}. \tag{5.20}$$

5.3.6 Unit Surface Normal

The unit vector $\nabla\xi/|\nabla\xi|$ points in the direction of increasing ξ. The definition of the unit surface normal in Figure 5.1 depends on the sign of ξ adopted for phase 2. If ξ is positive for phase 2, the unit surface normal is given as

$$\hat{n} = \frac{\nabla\xi}{|\nabla\xi|}. \tag{5.21}$$

If ξ is negative for phase 2, the unit surface normal is given as

$$\hat{n} = -\frac{\nabla\xi}{|\nabla\xi|}. \tag{5.22}$$

These can be combined as

$$\hat{n} = \mathrm{sign}\left(\xi_{\mathrm{phase}\,2}\right)\frac{\nabla\xi}{|\nabla\xi|},\tag{5.23}$$

where $\mathrm{sign}(\xi_{\mathrm{phase}\,2})$ returns the sign of ξ adopted for phase 2.

5.4 Boundary Conditions

Generally, no slip condition is used at all solid surfaces. If there is inflow into the domain, the normal and tangential velocities at the inlet are specified. At the outlet, the gradient of the tangential velocity is set to zero, while the normal velocity is calculated to ensure mass conservation. For the LS functions, $|\nabla\xi| = 1$ is imposed at all boundaries.

5.5 Solution Procedure

The continuity equation (Equation 5.1), the momentum equation (Equation 5.2), and the LS equations (Equations 5.9, 5.10, and 5.13) are special cases of a general transport equation

$$\rho\frac{\partial\phi}{\partial t} + \rho u_j\frac{\partial\phi}{\partial x_j} = \frac{\partial}{\partial x_j}\left(\Gamma\frac{\partial\phi}{\partial x_j}\right) + S,\tag{5.24}$$

where ϕ, ρ, Γ, and S are the dependent variable, the density, the diffusion coefficient, and the source term, respectively. The finite-volume method of Patankar (1980) is used to solve the transport equation given in Equation 5.24. The fully implicit scheme is used to discretize the transient term. The power-law scheme is used to model the combined convection–diffusion effect in the momentum equations. To resolve the velocity–pressure coupling, the semi-implicit method for pressure-linked equations–revised algorithm is used. The LS function, Equation 5.9, is a transient, convective transport of a scalar function, ξ, in this case, with zero diffusion. Higher order scheme, such as the curved-line advection method (CLAM) (Van Leer, 1974), must be used to ensure accurate transport of the zero LS function. The resulting algebraic equations are solved using the TriDiagonal Matrix Algorithm.

The solution procedure can be summarized as follows:

1. Specify the locations of the interface at time $t = 0$.
2. Calculate the LS function for all nodes using Equation 5.6.
3. Specify the properties for all nodes using Equations 5.18 and 5.19.
4. Solve the continuity and momentum equations given by Equations 5.1 and 5.2 at $t + \Delta t$.
5. Solve for ξ from Equation 5.9 using the velocities obtained in step 4.

6. Solve for the steady-state ψ from Equation 5.10 using the values of ξ from step 5 as the initial values.

7. Solve for the steady-state ψ' from Equation 5.13 using the values of ψ from step 6 as the initial values.

8. Set $\xi(\bar{x}, t + \Delta t) = \psi'(\bar{x})$.

9. Repeat steps 3 through 8 for all time steps.

5.6 Examples

Three two-phase flow problems will be presented next to *showcase* the method. These are evolutions of a bubble carried by a primary fluid in a (1) straight channel (Figure 5.6a), (2) double-bend channel (Figure 5.7a), and (3) constricted channel (Figure 5.8a). These examples have all the essential features of a microscale two-phase flow, that is, strong surface tension effect but negligible gravitational effect.

5.6.1 Evolution of a Bubble Flowing in a Straight Channel

Figure 5.6a shows a bubble of phase 1 suspended in a channel filled with phase 2. The channel is of length $L = 2$ and height $W = 1$. The bubble has a radius of $r_c = 0.15$ and is initially located at $(x_c, y_c) = (0.4, 0.5)$. The densities and viscosities for phase 1 are 2 and 2 and for phase 2 are 1 and 1, respectively. Initially, both phases are at rest. A stream of phase 2 then flows into the channel with a uniform velocity of $u_{in} = 0.01$ at the inlet. As phase 2 flows, it carries the bubble with it. The bubble evolves gradually under the action of both the hydrodynamic and surface tension forces.

Figure 5.6b shows the evolution of a bubble in a channel with different coefficients of surface tension, namely, $\sigma = 0$, 0.01, and 0.10. These, obtained with a mesh of 100 × 50 control volumes (CVs) and a time step size of $\Delta t = 0.1$, are grid-size-independent solutions when GMC is used. When the coefficient of surface tension is set to zero, there is no surface tension force acting at the interface. Thus, the hydrodynamic force is the only force acting on the bubble. It distorts the bubble into a "bullet-like" shape. Generally, a bullet-like shape possesses a smaller drag force. Therefore, the hydrodynamic force reduces the drag force on the bubble by deforming it. The surface tension force acts to minimize the interfacial area. In this case, it strives to retain the bubble in its original shape as a circle. Therefore, the hydrodynamic and surface tension forces have opposite effects. When the bubble is deformed (e.g., by hydrodynamic force), surface tension force acts to restore the bubble to its original circular shape. As σ increases, smaller deformation is observed. There is no noticeable deformation when $\sigma = 0.10$. Figure 5.6b also shows comparisons between the present solutions (LS) with that of the VOF. Although not shown here, the VOF solutions are grid independent. The present solutions ($\sigma = 0$ and 0.1) are in good agreement with that of VOF.

5.6.2 Evolution of a Bubble through a Double-Bend Channel

Figure 5.7a shows a bubble of phase 1 suspended in a double-bend channel filled with phase 2. The detailed geometry of the double-bend channel is given in the figure.

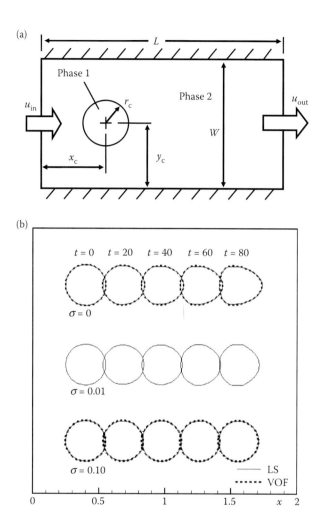

FIGURE 5.6
The evolution of a bubble flowing in a straight channel. (a) Schematic of a bubble flowing in a straight channel. (b) Effects of surface tension.

The bubble has a radius of $r_c = 0.20$ and is initially suspended at $(x_c, y_c) = (0.3, 0.8)$. The densities and viscosities for phase 1 are 2 and 2 and for phase 2 are 1 and 1, respectively. Initially, both phases are at rest. A stream of phase 2 then flows into the channel with a uniform velocity of $u_{in} = 0.01$ at the inlet. The bubble is carried by the phase 2 stream through the double-bend channel. Large deformation of the bubble is expected as the bubble negotiates the bends.

The following solutions were obtained on a mesh of 360×200 CVs with $\Delta t = 0.10$, which is sufficient to resolve the relevant physics of the problem. The solutions for $\sigma = 0$ obtained with and without GMC are shown in Figures 5.7b and 5.7c, respectively. Without GMC, the tail of the bubble is shorter at $t = 110$, an indication of mass loss. The present solution with GMC is compared with that of the VOF in Figure 5.7d. The present solution agrees well with that of the VOF.

Figure 5.7e shows the fluid velocity profiles at a few selected locations for $t = 110$. The fluid properties of the two phases do not differ much. Therefore, this velocity field would

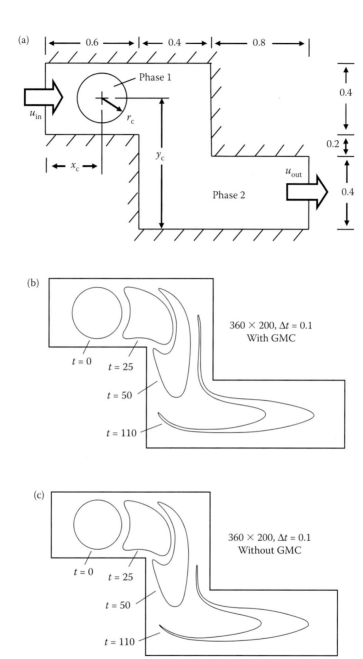

FIGURE 5.7
Evolution of a bubble flowing in a double-bend channel. (a) Schematic of a bubble flowing through a double-bend channel; (b) with GMC when $\sigma = 0$; (c) without GMC when $\sigma = 0$; (d) comparison with the VOF method when $\sigma = 0$; (e) velocity profiles at $t = 110$ when $\sigma = 0$; (f) $\sigma = 0.01$; (g) $\sigma = 0.05$; (h) with and without GMC when $\sigma = 0.01$; (i) mass errors.

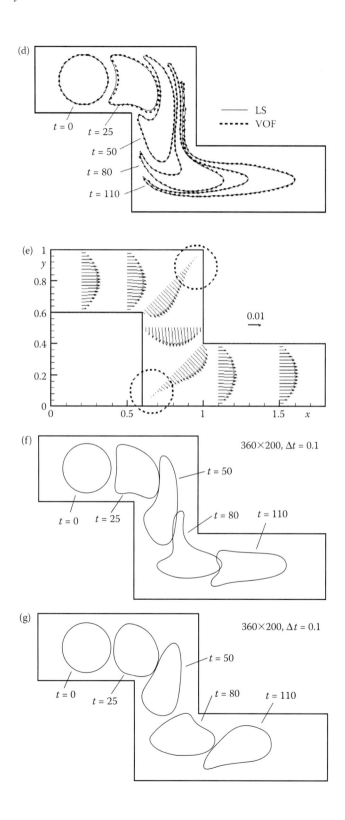

FIGURE 5.7
Continued

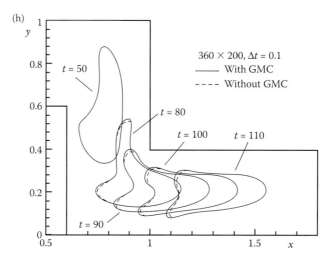

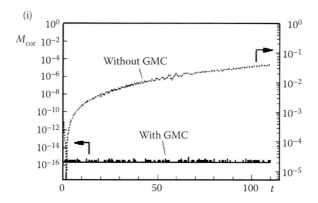

FIGURE 5.7
Continued

be qualitatively similar to that of other times even if the bubble is at different locations. There are two regions of low velocity, that is, regions near the upper corner of the first bend and the lower corner of the second bend. These regions are circled using dotted line for clarity. In each of these corners, a small region of recirculation is detected. Let the bubble be separated qualitatively into two portions, a portion near the upper corner and another portion consisting of the rest of the bubble. As the bubble flows through the first bend, the portion of the bubble near the upper wall moves with a much lower velocity. The rest of the bubble experiences a higher velocity. Therefore, the portion of the bubble near the upper corner will be increasingly "lagged" behind. This creates a long and thin tail. This process repeats as the bubble flows through the second bend, forming the second tail. To capture the formation and subsequent development of these tails properly, a very fine mesh is required.

Figures 5.7f and 5.7g show the effect of surface tension on the bubble evolution where the surface tensions are set to $\sigma = 0.01$ and 0.05, respectively. When $\sigma = 0.01$, one tail forms as the bubble flows through the first bend. This tail is shorter and thicker compared with the case without surface tension ($\sigma = 0$). Such a shape gives a smaller interfacial area, consistent with the effect of surface tension that minimizes interfacial area. The same applies to the second tail. When the surface tension coefficient is increased to $\sigma = 0.05$, surface tension effect becomes more profound. For a smaller σ, the first tail tends to form around $t = 50$. This tendency is suppressed by the surface tension force to the extent that no obvious tail is formed at $t = 80$. At $t = 110$, a small thick tail develops after the second bend. It is expected if further computations are made, this tail will become shorter and thicker. A portion of Figure 5.7f is enlarged in Figure 5.7h, showing the effect of GMC. The prediction without GMC is obviously different at the two blunt tails. Careful examination of Figure 5.7h reveals that the size of the bubble is smaller for the case without GMC. A parameter called mass error is introduced to examine the mass losses. It is given by

$$M_{\text{err}} = \left| \frac{M_{\text{d}} - M_{\text{c}}}{M_{\text{d}}} \right|. \tag{5.25}$$

The mass error represents the percentage of mass surplus or loss. Figure 5.7i shows the mass errors with GMC and without GMC. For the case without GMC, the mass error increases with time. This is an indication of either mass surplus or loss of the reference phase (phase 1 in this case). The bubble losses around 5% of its original mass at $t = 110$. On the basis of the trend of the mass error, more mass loss is expected as time passes. When GMC is used, the mass error of the reference phase remains well in the order of 10^{-16}, an indication of "perfect" mass conservation.

5.6.3 Evolution of a Bubble through a Constricted Channel

Figure 5.8a shows a bubble of phase 1 suspended in a constricted channel filled with phase 2. The channel is of length 2 units. It has a sudden contraction followed by a sudden expansion. At the inlet, the channel height is 1.0 unit. Then, the height contracts abruptly to $W = 0.2$ to form a smaller channel of length $L = 0.7$, after which it expands to its initial height. For ease of discussion, this smaller channel is henceforth referred to as the small channel. The details of the channel geometry are given in Figure 5.8a. The bubble has a radius of $r_c = 0.20$ and is initially suspended at $(x_c, y_c) = (0.3, 0.5)$. The densities and viscosities for phase 1 are 2 and 2 and for phase 2 are 1 and 1, respectively. Initially, both phases are at rest. A stream of phase 2 then flows into the channel with a uniform velocity of $u_{\text{in}} = 0.01$ at the inlet. The bubble is carried by the stream and squeezed through the small channel eventually. Large deformation of the bubble is expected as the bubble has a diameter larger than the height of the small channel.

The grid-independent solutions, on a mesh of 100×50 CVs with $\Delta t = 0.10$, are shown in Figures 5.8b and 5.8c, respectively, for $\sigma = 0$ and 0.05. Instead of the whole computational domain, only the portion of the small channel with the bubble, which is of interest, is shown. For these computations, GMC is used. The present solutions are compared with that of VOF. The present solutions are in good agreement with that of VOF both qualitatively and quantitatively.

Generally, the fluid flows faster in the small channel. A lower pressure is developed in the small channel, sucking the bubble into the small channel. This creates a cusp at the leading edge of the bubble. No slip condition is enforced at all walls, including those of the small channel. A typical velocity profile across the small channel for $t = 5$ is shown in Figure 5.8b. In the middle of the small channel, the fluid particles flow faster than those near the wall. Because the cusp is driven by a larger velocity at the middle of the channel,

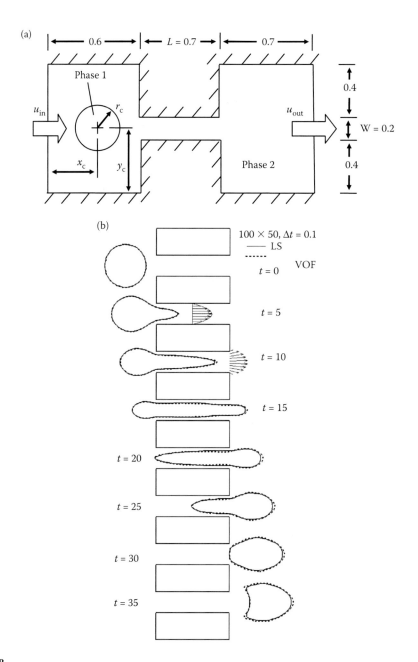

FIGURE 5.8
The evolution of a bubble squeezed through a constricted channel. (a) Schematic of a bubble flowing through a constricted channel, (b) $\sigma = 0$, (c) $\sigma = 0.05$, (d) bubble interface and velocity vectors at $t = 25$ for $\sigma = 0$, (e) mass errors for with and without GMC for $\sigma = 0$.

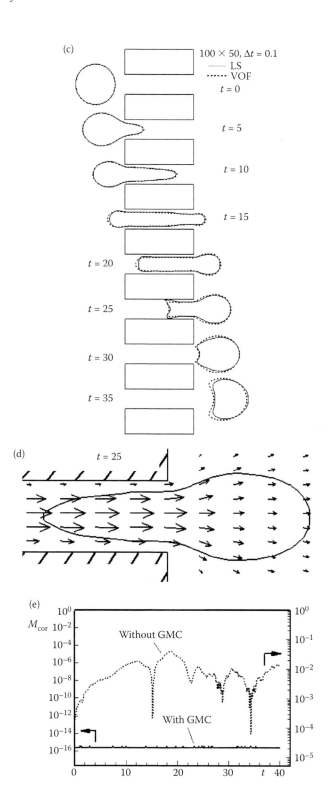

FIGURE 5.8
Continued

it becomes longer and sharper. Once in the small channel, the deformed bubble is squeezed by the main stream through the small channel. At the exit of the small channel, the flow expands because there is a sudden increase in cross-sectional area. At this point, the velocity component v is no longer negligible.

From $t = 10$ to $t = 25$, mass (phase 1) is transferred from the "tail" of the bubble to the leading cusp. This is shown in Figure 5.8d, which depicts an enlarged portion of Figure 5.8b at $t = 25$. Velocity vectors are also presented for the ease of explanation. The velocity in the "tail" is larger than that in the leading cusp. This suggests that mass is transferred from the "tail" and accumulated in the leading cusp. As a result, the leading cusp swells. This process is accompanied by shrinkage of the "tail" as it is pushed by the main stream. The "tail" disappears completely at $t = 30$ as the bubble emerges from the small channel. The emerged bubble is further deformed by the hydrodynamic force downstream of the flow field.

The effect of surface tension on the bubble evolution is shown in Figure 5.8c. The surface tension force acts to smooth interface with high curvature. In the absence of surface tension, both the leading cusp and the "tail" of the bubble are of large curvature, that is, pointed. When surface tension is present, the pointed nature of both the cusp and "tail" is suppressed.

Figure 5.8e shows the mass errors with GMC and without GMC for $\sigma = 0$. Without GMC, the mass error remains finite, in the order of 10^{-2} to 10^{-3}. There is either mass surplus or mass loss. With GMC, the mass error reduces to machine zero (10^{-16}). The same observation was made for nonzero surface tension. This shows that GMC is a necessity to ensure mass conservation.

5.7 Concluding Remarks

In this chapter, a single-fluid formulation for a two-phase flow problem under the framework of an LS method is presented. In the model, surface tension is accounted for by using the CSF model. The mass loss or gain problem of the LS method is greatly alleviated with the use of a mass correction procedure. The approach presented is showcased via three two-phase flow problems. The results have been cross-examined with those of the VOF method where good agreements are attained.

References

Brackbill JU, Kothe DB, Zemach C. (1992). A continuum method for modelling surface tension. *J Comput Phys*. 100:335–54.

Chang YC, Hou TY, Merriman B, Osher S. (1996). A level set formulation of Eulerian interface capturing methods for incompressible fluid flows. *J Comput Phys*. 124:449–64.

Edwards DA, Brenner H, Wasan DT. (1991). *Interfacial Transport Processes and Rheology*. Boston: Butterworth-Heinemann.

Osher S, Sethian JA. (1988). Fronts propagating with curvature-dependent speed: algorithms based on Hamilton–Jacobi formulations. *J Comput Phys*. 79:12–49.

Patankar SV. (1980). *Numerical Heat Transfer and Fluid Flow*. New York: Hemisphere Publisher.

Peng D, Merriman B, Osher S, Zhao H, Kang M. (1999). A PDE-based fast local level-set method. *J Comput Phys*. 155:410–38.

Peskin CS. (1977). Numerical analysis of blood flow in the heart. *J Comput Phys*. 25:220–52.

Scriven LE. (1960). Dynamics of a fluid interface. *Chem Eng Sci*. 12:98–108.

Sussman M, Smereka P, Osher S. (1994). A level set approach for computing solutions to incompressible two-phase flow. *J Comput Phys*. 114:146–59.

Van Leer B. (1974). Towards the ultimate conservative difference scheme. II. Monotonicity and conservation combined in a second order scheme. *J Comput Phys*. 14:361–70.

Yap YF, Chai JC, Toh KC, Wong TN, Lam YC. (2005). Numerical modeling of unidirectional stratified flow with and without phase change. *Int J Heat Mass Transf*. 48:477–86.

Yap YF, Chai JC, Wong TN, Toh KC, Zhang HY. (2006). A global mass correction scheme for the level-set method. *Numer Heat Transf B*. 50(5):455–72.

6

Characterization of Chaotic Stirring and Mixing Using Numerical Tools

Shizhi Qian, Bayram Celik, and Ali Beskok

CONTENTS

6.1 Introduction

Over the past decade, there has been a growing interest in the development of lab-on-a-chip (LOC) devices for biodetection, biotechnology, chemical, and biological reactors or even medical, pharmaceutical, and environmental monitoring [1–24]. LOC is a minute chemical processing plant, where common laboratory procedures ranging from filtration and mixing to separation and detection are carried out in the palm of the hand. Resembling electronic circuit boards, these integrated devices contain a network of microchannels and reservoirs that dilute the sample, separate it into multiple routes for parallel analysis, mix it with target-specific reagents, propel it from one part of the device to the other, and detect the presence of chemical or biological targets. This whole set of precise and reproducible operation is done automatically in a single platform, which ultimately results in high data quality and reduces the need for trained personnel. This technology, therefore, has the potential of revolutionizing various bioanalytical applications, which require handling tiny amounts of samples in a rapid and cheap manner, and automatically analyzing them through multiple parallel procedures. To meet these very demanding criteria, rapid mixing of fluids must be achieved within the LOC devices that are used in biochemical analysis, drug delivery, point-of-care testing, crystallization, protein and RNA folding, and sequencing or synthesis of nucleic acids.

Mixing is the process of homogenization of species distribution as a result of stirring and diffusion. Stirring brings the constituents to close proximity, whereas diffusion homogenizes through "blending of the constituents." Using this simplified definition of mixing, stirring is indicative of *flow kinematics* that is often determined by the mixer's geometry and flow conditions, which are primarily described by Reynolds number (Re), defined as the ratio of inertial and viscous forces (other dimensionless groups can also exist on the basis of the specific mixer design). The effects of diffusion are determined by the species that is being mixed, which can be characterized as a function of the Schmidt number (Sc), defined as the ratio of momentum and mass diffusivities. Although the characteristic lengths associated with LOC devices are very small—typically in the order of 100 μm—in the case of large molecules, diffusion alone does not allow for sufficiently fast mixing. For example, at room temperature, myosin's diffusion coefficient in water is approximately 10^{-11} m^2/s, and the time constant for the diffusion along a length of 100 μm is thus intolerably large, approximately 10^3 s. Therefore, mixing several fluids in reactors at the micron scale is not as easy as it might seem at first sight. Because the Reynolds numbers of flows in microdevices are usually very small [i.e., Re~$O(1)$], the flows are laminar, and mixing enhancement cannot be reached by making use of turbulence-like flow patterns. To achieve reasonable speed and yield of chemical reactions and bioassays, micromixers must be necessarily integrated into the chips. The device integration step may bring several limitations regarding the flow kinematics used in the mixer design, where simply increasing the flow rate (or Re) to mix different species may not be compatible with the upstream and downstream components of the LOC device. Given these limitations, one needs to determine the kinematically favorable conditions for mixing by choosing the mixer geometry and flow conditions (Re and other relevant dimensionless parameters) and then ensure efficient mixing for various species by varying Sc under these predetermined kinematic conditions.

In recent years, numerous micromixers based on various operation principles have been developed for LOC applications. These technologies have been extensively reviewed by Nguyen and Wu [25], Qian and Duval [26], and Nguyen [27]. In general, micromixers can be categorized into passive and active ones. A mixer is said to be active when its functioning relies on the use of a time-variant external energy source such as that provided by applied pressure, electric field, and/or magnetic field, to quote only a few. This external energy source excludes the source used for propelling the fluid itself, regardless of its state of mixing. Unlike active mixers, passive mixers use the same power source (which is in most of the cases time-independent) to propel the fluid and typically take the benefit of irregular or asymmetric channel geometry to laminate the flowing fluids either in-plane or out-of-plane to promote chaotic advection. Each of these two types of micromixers has different capacity, mixing speed, and operating requirements. The lack of moving components makes passive mixers free of additional friction and wear effects, even if their intricate channel topologies are often hard to fabricate. Passive mixers are generally not adaptable to operate beyond their design conditions. Once a passive mixer is incorporated into an LOC device, it fulfills its function when fluids pass through it. In contrast, active mixers can be controlled externally, which makes them suitable and attractive for reconfigurable LOC systems that can perform several different functions, given different states of external controls.

Various types of micromixers have been designed, fabricated, and experimentally characterized using fluorescent dyes to measure the fluorescence intensity at various sections of these mixers. In these studies, the mixing efficiency was quantified using standard deviation of the fluorescence intensity from a perfect mix [28–33]. A remarkable amount of the experiments used a single type of dye (i.e., fixed Schmidt number, Sc), and the mixing length or mixing time was investigated as a function of the Peclet number (Pe $\equiv$ Sc $\times$ Re), which gives the forced convection to diffusion ratio of a system, by varying the Reynolds number. An important aspect of this approach that is overlooked is that varying Re while keeping Sc fixed changes the flow kinematics. Especially, beyond the Stokes flow regime, significant changes in flow kinematics can be achieved by varying the flow rate, which may lead to different stirring conditions. Therefore, such studies should be interpreted as attempts to identify the flow kinematics that enhances mixing. A fundamentally important, yet mostly underappreciated, aspect of mixing is the characterization of the stirrer under fixed flow kinematics but for mixing of different species. This approach requires varying the Sc to vary the Peclet number at fixed Re. Only this latter approach should be used to assess the chaotic nature of species mixing on the basis of fluorescent dye experiments and numerical simulations of the species transport equations. The reasons for this claim are substantiated in this chapter.

The main *objective* of this chapter is to provide an introductory review on characterization of chaotic stirrers using appropriate numerical tools. The rest of this chapter is organized as follows: The section "Routes to Chaotic Advection in LOC Devices" reviews the general routes to achieve chaotic advection and gives examples of chaotic stirrers. The section "Numerical Tools for Characterization of Chaotic Stirrers" describes computational tools used for quantification of chaos and mixing efficiency with their relevance to the experimental characterization. Finally, the chapter ends with concluding remarks.

6.2 Routes to Chaotic Advection in LOC Devices

Chaos was discovered and studied almost one century ago and has been mostly thought of in the context of turbulence. The concept of chaotic advection in laminar flows was introduced in the early 1980s by Aref [34]. Since then, a substantial number of investigators have demonstrated that chaotic advection occurs in a wide variety of laminar flows, ranging from creeping flow to potential flow, and in different flow systems including unsteady two-dimensional flow and both steady and time-dependent three-dimensional flows [35–39].

The idea underlying chaotic advection is the observation that a certain regular velocity field, $u(x,t)$, can produce fluid path lines $x(x_0, t)$, which uniformly fill the volume in an ergodic way. The motion of passive tracers is governed by the advection equation,

$$\dot{x} = u(x,t), \quad x(t=0) = x_0. \tag{6.1}$$

Hereafter, bold letters represent vectors. In such velocity fields, fluid elements that are originally close to one another trace paths that diverge rapidly (exponentially fast in the ideal case), so that the material is dispersed throughout the volume very efficiently. This typically leads to significantly fast mixing. Therefore, chaotic advection in LOC devices can provide the best possibility of achieving efficient and thorough mixing of fluids.

Because of the nature of the dynamical system, chaotic advection requires time-dependent flow in either simple two-dimensional geometries or complex three-dimensional geometries [36,39]. Typically, active chaotic micromixers that are actuated externally by time-dependent energy sources (i.e., pressure, electric, and/or magnetic fields) use time-dependent two-dimensional flow to achieve chaotic advection for mixing enhancement. On the other hand, passive chaotic micromixers typically use complex three-dimensional twisted conduits fabricated in various substrates such as silicon [40], polydimethylsiloxane [41], ceramic tape [42], or glass [40] to create three-dimensional steady flow velocity with a certain complexity to achieve chaotic advection. Typical examples of the aforementioned two routes to achieve chaotic advection and mixing in LOC devices are presented in the following.

6.2.1 Mixing with Chaotic Advection in Two-Dimensionality

Various active micromixers using two-dimensional time-dependent flow to achieve chaotic advection have been developed [25–27]. Because electro-osmosis is very attractive for manipulating fluids in LOC devices, a chaotic electro-osmotic stirrer developed by Qian and Bau [43] is described as an example to achieve chaotic advection and mixing by two-dimensional time-dependent electro-osmotic flow.

Most solid surfaces carry electrostatic charges because of broken bonds and surface charge traps. When a liquid containing a small amount of ions is brought into contact with such a solid boundary, counterions in the liquid are strongly attracted to the charged solid surface, forming an "immobile" compact layer at the solid–liquid interface. This layer is called the *Stern layer*. Outside the Stern layer, counterions and co-ions are accumulated following a Boltzmann statistics in a thin liquid layer, which is called diffuse layer. The electrical double layer (EDL), consisting of Stern and diffuse layer, typically has a thickness in the order of a few nanometers depending on the electrolyte concentration. Away from the EDL, the electrolyte is neutral, and local ionic concentrations equal bulk values. This ionic-charge separation within the EDL next to the solid wall causes either a positive or a negative potential difference across the EDL, denoted as zeta potential. In the presence of an external electric field, the counterions in the EDL are attracted to the oppositely charged electrode and drag the liquid along, and the induced fluid motion is called electro-osmotic flow. When the thickness of the EDL is much smaller than the conduit's dimensions, the electro-osmotic flow can be described by specifying a slip velocity, U_{HS}, on the wall:

$$U_{HS} = -\frac{\varepsilon \varepsilon_0 \zeta}{\mu} E, \qquad (6.2)$$

where ζ is the zeta potential, ε is the relative dielectric constant, ε_0 is the permittivity of the vacuum, μ is the dynamic viscosity of the medium, and E is the applied electric field. Equation 6.2 is valid for sufficiently thin double layers and is further strictly applicable for low values of the zeta potential, where the Debye–Hückel approximation is valid. Equation 6.2 shows that the magnitude and direction of the electro-osmotic velocity can be controlled by the magnitude and polarity of the zeta potential. Therefore, by selectively patterning negatively, neutrally, or positively charged material patches on the channel walls in LOC devices, a wide range of flow patterns in microfluidic channels can be created. In addition, the zeta potential of a channel wall can be dynamically modulated as a result of changes in the potential imposed to gate electrodes embedded beneath the solid–liquid interface [44–47]. One can pattern several individually controlled gate electrodes beneath the solid–liquid interface; complex flow field to achieve chaotic advection for mixing enhancement can be generated by spatial and temporal modulation of the zeta potentials along the channel walls [43,48].

The electro-osmotic chaotic stirrer developed by Qian and Bau [43] consists of a closed cavity ($|x| \leq L$ and $|y| \leq H$), with two electrodes mounted along the walls, $x = \pm L$, inducing an electric field, E, parallel to the x-axis. Four additional electrodes are embedded in the cavity's upper and lower walls. These electrodes are not in contact with the liquid and are

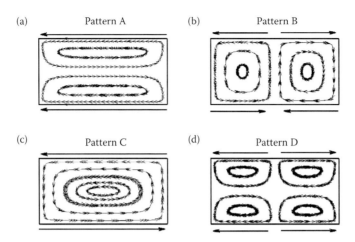

FIGURE 6.1
Two-dimensional flow patterns encountered in the electro-osmotic chaotic stirrer introduced by Qian and Bau. (From Qian, S., and Bau, H.H., *Appl. Math. Model.*, 29, 726–753, 2005. With permission.)

used to control the ζ potential at the liquid–solid interface. The cavity contains an electrolytic solution. Various two-dimensional flow patterns induced through the modulation of the zeta potentials along the top left, top right, bottom left, and bottom right walls are depicted in Figures 6.1a through 6.1d. These flows are, however, highly regular. In the absence of diffusion, trace particles will follow the streamlines, with no transport occurring transverse to the streamlines. To induce chaotic advection in the cavity, two different flow patterns, A and B, are alternated with a period of *T*. In other words, the flow field type A is maintained for a time interval $0 < t < T/2$ and then switched to the flow field type B for the time interval $T/2 < t < T$. This allows crossing of the streamlines between two separate half periods. Subsequently, the process is repeated, and chaotic advection is created for mixing enhancement at certain actuation conditions [43]. Further quantification of the chaotic strength and mixing in this stirrer using Lagrangian particle tracking-based methods, such as box-counting, Poincaré sections, Lyapunov exponents, and probability density of stretching as well as the species transport-equation-based mixing index calculations can be found in the study of Kim and Beskok [49]. We must note that this two-dimensional example uses Stokes flow regime, where Re ≪ 1. As a result, varying the Reynolds number does not greatly affect the flow patterns. Therefore, the flow kinematics for this case is described by the time modulations between the flow patterns A and B and could be described as a function of the Strouhal number.

6.2.2 Mixing with Chaotic Advection in Three-Dimensionality

To achieve chaotic advection in three-dimensional steady flow in a geometry that has certain complexity created by three-dimensionality, microconduits can be used. When all the conduits such as the zigzag and square-wave microconduit lie in the same plane, the symmetry of the flow field is preserved, and chaotic advection occurs only at high Reynolds numbers (i.e., Re > 80). When the Reynolds number is below a certain threshold value, the mixing process is governed by molecular diffusion [50,51]. Previous results have shown that micromixers consisting of twisted pairs of bends, with each pair forming a C-shape or an L-shape work well only at moderate Reynolds numbers and are not efficient at low

Reynolds numbers because the bend-induced vortices decay well before they may significantly stir the fluid [42].

In three-dimensionality, chaotic advection can be achieved by using passive micromixer designs, which can efficiently operate at limited kinematic conditions because of the availability of beneficial flow mechanisms, such as the Dean vortices, the expansion vortices, or the secondary flows. In the study of Nguyen [27], passive micromixers in which chaotic advection may take place are classified by the Reynolds number. A spiral channel-like design reported by Sudarsan and Ugaz [32] uses Dean vortices to introduce secondary flow in a 1 < Re < 10 range. Despite its relatively simple structure compared with the sequential lamination micromixers reported in Reference 33, Sudarsan and Ugaz's mixer is able to create transverse flow and thereby enhance mixing.

To demonstrate the effects of Dean vortices on laminar flows, we present numerical simulation results for a planar channel with a 90° turn in Figure 6.2. In the curved part of the channel, the fluid encounters interplay between the inertial and the centrifugal effects, which are along the channel and the radius of the conduit's curvature direction, respectively. Inertial to centrifugal effect ratio is characterized by the Dean number, De, which can be written in terms of the Reynolds number as

$$\text{De} = \text{Re}(D_\text{h}/2R)^{0.5}, \tag{6.3}$$

where D_h and R are the hydraulic diameter and the radius of the curved channel, respectively. Dean vortices arise in the flows above the critical Dean number of 150. In Figure 6.2, we plotted the streamlines initiated from seven equally spaced points along the diagonal of the channel entrance at De = 300. As can be seen from the figure, the streamlines represented by different colors are along the channel direction up to the curved section. Then, they are stretched and twisted inside the curved section because of the secondary flow induced by the Dean flow. To show the evolution of the Dean vortices, the planar vorticity

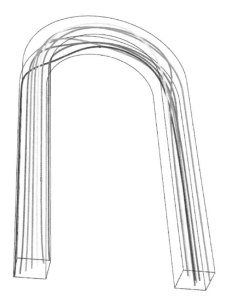

FIGURE 6.2
(See color insert.) The streamlines of the fluid particles released from seven equally spaced points along the diagonal of the channel entrance at De = 300.

contours at the curved sections' inlet, middle, and outlet are presented in Figure 6.3a–c, respectively. It is obvious that this three-dimensional streamline structure can be used to stretch the interface between two coflowing species and enhance mixing. In the case of a planar channel geometry consisting of repeated right- and left-turning curved sections, additional Dean vortices arise, and thereby, mixing on the basis of chaotic advection can take place.

Instead of using twisted complex three-dimensional geometry, ribs or grooves on the channel walls can lead to the formation of transverse velocity components at low Reynolds numbers [52–54] and thus create chaotic advection for mixing enhancement. The experimental results demonstrate that staggered herringbone mixer (SHM) works well in the range $0 < Re < 100$ [53]. The SHM mixing is more efficient than that achieved with similar microfluidic channels devoid of internal structures such as ribs and grooves. For example, The basic T-mixer (i.e., two coflowing species under pressure-driven flow in a straight microchannel) requires mixing lengths of approximately 1 and 10 m at $Pe = 10^4$ and 10^5, whereas the SHM mixer performs the same task within 1 and 1.5 cm only, respectively. The mixing efficiency of the SHM mixer can be further greatly improved with ribs or grooves placed on both the top and the bottom of the channel, which not only increases the driving force behind the lateral flow but also allows for the formation of advection patterns that cannot be created with structures on the bottom alone [54].

We must reemphasize that the flow kinematics in these mixers is a function of the flow rate and hence the Reynolds number. For example, in the case of mixers with bend-induced vortices, the Dean number is a function of Re, and experiments using fluorescent dyes with constant Sc but different flow rates (Re) should be interpreted as a search for kinematically favorable stirring conditions. Such results should be characterized as a function of the Reynolds number rather than the Peclet number because these do not reveal the behavior of the device for mixing species with varying mass diffusivities.

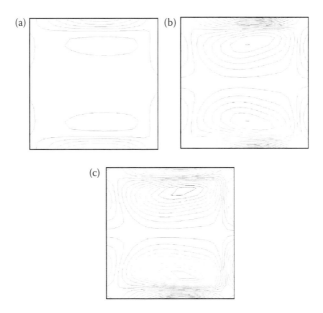

FIGURE 6.3
(See color insert.) Planar vorticity contours at the inlet (a), the middle (b), and the outlet (c) of the curved section of the channel.

6.3 Numerical Tools for Characterization of Chaotic Stirrers

Mathematical modeling and numerical simulations of mixing in LOC devices provide a convenient and fast method for optimizing the design and operation parameters of a chaotic stirrer, which otherwise would require enormous effort. In the numerical studies, appropriate tools to qualitatively/quantitatively characterize the stirrers are required. In the following, we focus on the appropriate applications of some diagnostic tools such as the Poincaré section, the finite time Lyapunov exponent (FTLE), and the mixing index to characterize chaotic stirrers. Using a continuous-flow chaotic stirrer developed by Kim and Beskok [55] as an example, the applications and limitations of the aforementioned tools to characterize the stirrer are illustrated.

Figure 6.4 shows the continuous-flow micromixer developed by Kim and Beskok [55], which consists of periodically repeating mixing blocks (the length and height of each block are $L = 4h$ and $H = 2h$, respectively) with zeta potential patterned surfaces (Figure 6.4a) that induce two-dimensional electro-osmotic flow (Figure 6.4b) under an axial electric field. A time-periodic flow can be generated by altering the axial electric field in the form of a cosine wave with a frequency of ω. In addition, a pressure-driven unidirectional flow in x-direction (see Figure 6.4c) is superposed to the electro-osmotic flow, with channel center-line velocity of U_0. The flow field in the mixer is governed by the following dimensionless Navier–Stokes equations:

$$\nabla \cdot \mathbf{u} = 0, \tag{6.4}$$

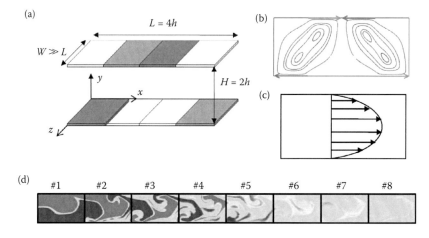

FIGURE 6.4
(See color insert.) Schematic of the continuous chaotic stirrer developed by Kim and Beskok. The stirrer consists of periodically repeating mixing bocks with zeta potential patterned surfaces (a), and an electric field parallel to the *x*-axis is externally applied resulting in (b) an electro-osmotic flow. Combining a unidirectional (*x*-direction) pressure-driven flow (c) with electro-osmotic flow under time-periodic external electric field (in the form of a cosine wave with a frequency ω), a two-dimensional time-periodic flow is induced to achieve chaotic stirring in the mixer. Two fluid streams colored with red and blue are pumped into the mixer from the left and are almost well mixed after eight repeating mixing blocks for Re = 0.01, St = $1/2\pi$, Pe = 1000, and $A = 0.8$ (d).

$$\frac{\partial \boldsymbol{u}}{\partial t} + (\boldsymbol{u} \cdot \nabla)\boldsymbol{u} = -\nabla p + \frac{1}{\text{Re}}\nabla^2 \boldsymbol{u}. \tag{6.5}$$

The dimensionless species concentration distribution (C) is described by the time dependent convection–diffusion equation:

$$\frac{\partial C}{\partial t} + (\boldsymbol{u} \cdot \nabla)C = \frac{1}{\text{Pe}}\nabla^2 C. \tag{6.6}$$

Using dimensional analysis, the mixer's performance can be shown to depend on the following dimensionless parameters:

$$\text{Re} = \frac{U_{\text{HS}}h}{v}; \quad St = \frac{\omega h}{U_{\text{HS}}}; \quad A = \frac{U_0}{U_{\text{HS}}}, \tag{6.7}$$

where Re and St are the Reynolds and the Strouhal numbers, and A is the ratio of the Poiseuille (U_0) and electro-osmotic flow velocities (U_{HS}), which is kept constant ($A = 0.8$). Each repeated pattern of the mixer, a mixing block, has an aspect ratio of $L/H = 2$. It is also essential to assume a quasi-steady flow, which further requires small Stokes numbers ($= \sqrt{\text{Re} \times St} \leq 0.2$). Figure 6.4d shows concentration contours obtained within eight mixing blocks for Re = 0.01, Pe = 1000, and St = $1/2\pi$, which shows rapid mixing between the red and blue streams, generating fully mixed (green contours) toward the mixer's exit.

Similar to the Qian and Bau's mixer presented in Reference 48, this mixer also uses electro-osmotic flow, which operates in the Stokes flow regime. As a result, the streamlines are rather insensitive to the variations in Reynolds number. Therefore, the flow kinematics in this system is determined mostly by the Strouhal number.

6.3.1 Poincaré Sections

Poincaré maps are often used to qualitatively characterize the quality of stirring over a wide range of operating parameters because the Poincaré mapping visualizes invariant manifolds and Kolmogorov–Arnold–Moser (KAM) curves on a Poincaré section [37]. For a closed mixer, a Poincaré section consists of a finite number of points (where the number of points is the period of the advection cycles), which show the sequential positions of a specific fluid particle, and is obtained by integrating the advection equation (Equation 6.1) and recording the position at the end of each advection cycle. However, for an open flow system such as a channel flow, obtaining images to display the dynamic states in a physical channel domain (x, y) through time-periodic projections is not straightforward because motions of passive tracer particles are not bounded within a limited physical domain.

Poincaré sections of a continuous-flow mixer can be obtained with the mapping method introduced by Niu and Lee [56]. We choose Poincaré section to be $x_n = n \times L$ for the trajectory of any initial point, where L is the length of the mixing block and $n = 1, 2, \ldots, N$. Particle trajectories will intersect boundaries of the mixing blocks successively at points $P_1, P_2, \ldots, P_N$. This mapping is defined as

$$P_{n+1} = \Phi_{\text{p}}(P_n), \tag{6.8}$$

where Φ_p is the Poincaré mapping. At point P_n, the vertical positions of passive tracer particles are recorded in y-axis of the Poincaré section, while time (t) increases to infinity with the repetition of mapping. To convert t into a periodic variable, we adopt a new variable α. Because the flow is periodic with a specified St, and 2π is a common factor in our definition of St, we define α = modular (t, 2π). The values of α are recorded in the horizontal axis of the Poincaré section. Thus, the mathematical expression of projection points on the Poincaré section is $P_n(\alpha_n, y_n)$.

Figure 6.5a depicts the Poincaré section of the continuous-flow stirrer when St = $1/4\pi$ and Re = 0.1. The Poincaré sections are obtained by numerically tracking four passive tracer particles initially located at (0.005, −0.5), (0.005, 0.0), (0.005, 0.5), and (0.005, 1.0) during 10^5 convective time scales (H/U_{HS}). A quasi-periodic motion of the passive tracer particle that is initially located at (0.005, 0.0) results in a regular formation separating the upper and lower halves of the Poincaré section. A zoomed image showing this KAM boundary is presented in Figure 6.5c. The passive tracer particles initially located at the upper and lower halves of the channel entry cannot pass this global barrier. In addition to this, there are two unstirred zones called *void zones* surrounded by a well-stirred zone (chaotic sea) at the bottom half of the Poincaré section. A zoomed image of these two void zones can be seen in Figure 6.5b.

Figure 6.6 depicts the Poincaré section when St = $3/4\pi$. All other parameters are the same as those in Figure 6.5. There are sinusoidal curves like regular global patterns formed by the passive tracer particles initially located at (0.005, 0.1) and (0.005, 0.0). Zoomed images of these two patterns are shown in Figure 6.6b and c, respectively. Similar to the previous case with St = $1/4\pi$, away from this area, there are two separate chaotic seas formed by the particles initially located at (0.005, 1.0) and (0.005, −0.5), respectively. Both chaotic

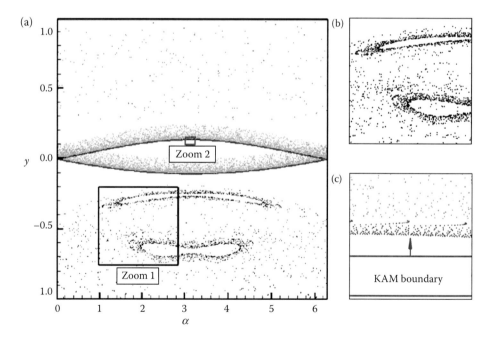

FIGURE 6.5
(See color insert.) Poincaré section for four particles initially located at (0.005, −0.5)—black, (0.005, 0.0)—red, (0.005, 0.5)—green, and (0.005, 1.0)—blue, at stirring conditions of St = $1/4\pi$, Re = 0.01, and A = 0.8 (a). The Poincaré sections in the zoomed area 1 with two void zones (b), and the global regular pattern in the zoomed area 2 (c).

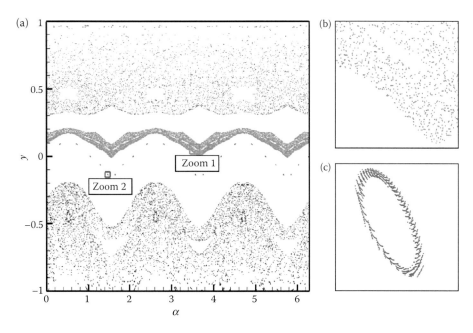

FIGURE 6.6
(See color insert.) Poincaré section for the same four particles in Figure 6.5 at stirring conditions of St = 3/4π, Re = 0.01, and *A* = 0.8 (a). The Poincaré sections in the zoomed areas 1 and 2 are shown, respectively, in panels b and c.

seas have several small islands that are relatively small compared with the one with lower St. Although the particles are tracked for quite a long time period, they cannot penetrate into the regular islands separated from the chaotic sea with KAM boundaries. These two Poincaré sections shown in Figures 6.5 and 6.6 qualitatively show that stirring at St = 1/4π and 3/4π is partially chaotic.

Figure 6.7 shows a featureless Poincaré section at St = 1/2π, and all other conditions are the same as those in Figures 6.5 and 6.6. The KAM boundaries and the quasi-periodic zones shown in the previous two Poincaré sections (Figures 6.5 and 6.6) are destroyed. By comparing Figures 6.5–6.7, the stirring quality less than St = 1/2π is much better than those less than 1/4π and 3/4π.

Before we finish this section, we would like to discuss the practical limitations of the Poincaré sections, which require Lagrangian particle tracking for extended times. In reality, Figures 6.5–6.7 are stroboscopic images of the same four particles passing through thousands of mixing block boundaries. This has two basic implications. First, numerical calculation of the Poincaré sections requires either analytical solutions or high-order accurate discretizations of the velocity field. Otherwise, the results may suffer from numerical diffusion and dispersion errors, and the KAM boundaries may not be identified accurately [48,49]. Second, it is experimentally difficult, if not impossible, to track particles (in three dimensions) beyond a certain distance allowed by the field of view of the microscopy technique. Also, the mixers always have a limited number of mixing blocks/sections. Therefore, the Lagrangian particle tracking results cannot be directly associated with the experiments. It may be possible to track many particles for a limited time or within a finite mixing length. However, one needs to observe these results in color to identify each particle correctly and to observe certain features of the Poincaré section. Despite these

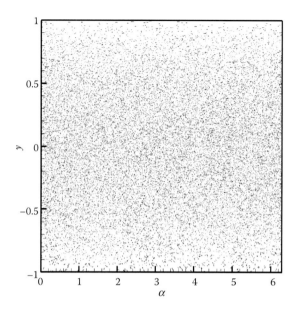

FIGURE 6.7
(See color insert.) Poincaré section for the same four particles in Figure 6.5 at stirring conditions of St = 1/2π, Re = 0.01, and A = 0.8.

limitations and challenges, the Poincaré sections can differentiate between the partially and the totally chaotic flows. Overall, one would expect better mixing with decreased regular flow zones. The basic problem is that the Poincaré section method is qualitative, and it cannot determine the chaotic strength between two fully chaotic systems, both resulting in featureless Poincaré sections.

6.3.2 Finite Time Lyapunov Exponent

To quantitatively characterize the chaotic strength in the case of featureless Poincaré sections, the Lyapunov exponent (LE) needs to be calculated [38]. Positive LE indicates the presence of chaos in the system, and higher LE values are considered as higher chaotic strength. Calculation of the LE requires extremely long time integration; therefore, FTLE, λ_{FTLE}, has been widely used in quantification of chaotic strength [49]. FTLE is calculated using the algorithm suggested by Sprott [57]. Briefly, arbitrary pairs of nearby points with an initial distance of $|dx|$ are chosen, and the distance at the time of Δt, $|dx(\Delta t)|$, is evaluated to calculate $\ln(|dx(\Delta t)/dx|$ by integrating the particles' paths. This process is repeated n times using the following mathematical expression:

$$\lambda_{FTLE} = \frac{1}{n}\sum\frac{1}{\nabla t}\ln\left(\frac{dx(\Delta t)}{dX}\right) = \frac{1}{n\Delta t}\sum\ln\left(\frac{dx((n+1)\Delta t)}{dx((n)\Delta t)}\right). \tag{6.9}$$

Figure 6.8 depicts the FTLE distribution obtained by tracking 4000 passive tracer particle pairs at the end of 10 time periods for the stirring case of Figure 6.7. The FTLE has a normal distribution with mean $\bar{\lambda}_{FTLE} = 0.1$ and standard deviation $\sigma = 0.03$. The positive FTLE is consistent with the chaotic flow, which is qualitatively observed from the Poincaré

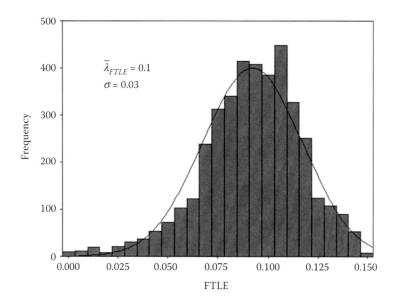

FIGURE 6.8
The FTLE distribution for mixing with St = $1/2\pi$, Re = 0.01, and $A = 0.8$ obtained by tracking 4000 passive particle pairs for 10 time periods.

section shown in Figure 6.7. Therefore, the mixing characteristics can be quantitatively determined by using FTLE as a complementary tool to the Poincaré section.

Before finishing this section, we would like to discuss the practical limitations of FTLE calculations. Similar to the Poincaré sections, FTLE requires Lagrangian particle tracking techniques, and therefore one needs to have highly accurate velocity field. The computational cost for the FTLE calculation per time step per particle is approximately twice that of the Poincaré section because the former uses particle pairs. In the example above, 4000 passive particle pairs have been tracked for 10 time periods. This resulted in dispersion of particles within the entire simulation domain consisting of 10 mixing blocks. Obviously, these are computationally expensive calculations, especially for continuous-flow mixer problems. For characterization of the mixing performance, one should use particle pairs that are uniformly distributed within the first mixing block at time zero and then follow their behavior in time, while calculating the FTLE values, and ensure that each FTLE converges to a corresponding constant value. After such calculation, one can analyze the mean FTLE value and its standard deviation. It is also possible to map these values to the initial particle position to observe the regular flow zones and KAM boundaries [55]. Because the FTLE calculations require utilization of particle pairs, motion of the primary and the satellite particles cannot be controlled in experiments. Therefore, the FTLE cannot be realized within an experimental setup.

6.3.3 Mixing Index

To quantify a mixer's performance, the mixing index (M) is defined as [58]:

$$M = \frac{\sigma}{C_\infty} = \frac{1}{C_\infty}\sqrt{(C_i^2) - (C_i)^2} \approx \sqrt{\frac{1}{N-1}\sum_{i=1}^{N}\left(\frac{C_i}{C_\infty} - 1\right)^2},$$

(6.10)

where C_i is the average concentration inside the ith section of a total of N interrogation areas and C_∞ is the concentration value corresponding to the perfect mix. The $<>$ symbol denotes averaging over the volume of a single mixing block. On the basis of the initial distribution of the species, a perfect mix would reach $C_\infty = 0.5$. According to the definition in Equation 6.9, a perfect mix results in $M = 0$. Hence, smaller values of M show better mixing. For the continuous-flow mixer considered here, the mixing index varies as a function of the channel length, and it can be used as a metric to assess the mixing efficiency. Using the inverse of the mixing index (M^{-1}) instead of the index itself is more preferable because $M^{-1} \rightarrow \infty$, whereas $M \rightarrow 0$. For example, $M^{-1} = 20$ corresponds to $(1 - \sigma/C_\infty) \times 100\% = 95\%$ mixing efficiency. Similarly, $M^{-1} = 10$ corresponds to 90% mixing.

To calculate M from experimental fluorescence images captured by a high-resolution charge-coupled device camera, the dimensionless concentration, C_i, in Equation 6.9 corresponds to the fluorescence intensity of a pixel in the fluorescence image. For a theoretical study, the dimensionless concentration is calculated from the species transport equation (Equation 6.5). For high Pe flows, the solution of scalar transport equation is quite challenging and requires high accuracy both in time and in space [24,49]. Mixing length (l_m) and mixing time (t_m) are used to assess the mixing efficiency for continuous-flow and closed mixers, respectively. However, Pe dependence of mixing length or time must be investigated by varying the Schmidt number (i.e., different molecular dyes) while keeping the Reynolds number constant.

Figure 6.9 depicts the dimensionless species concentration distribution in the stirrer with St = $1/4\pi$ (Figure 6.9a) and $1/2\pi$ (Figure 6.9b) at Re = 0.01 for Pe = 1000. Red and blue colors correspond, respectively, to the concentration values of two different species entering the mixer from the left end of the mixer with the concentration values of $C = 1$ and 0. The green areas correspond to well-mixed zones with $C_\infty = 0.5$. By comparing Figure 6.9a and b, the mixing efficiency of the flow with St = $1/2\pi$ is higher than that of the flow with St = $1/4\pi$ because the concentration distribution of the former at the exit of the mixer is almost uniform.

The mixing process can be characterized globally by evaluating its l_m–Pe behavior at a fixed kinematic condition (i.e., constant Re and St). For laminar convective/diffusive transport, mixing length typically varies as $l_m \propto Pe^{0.5}$. It is possible to reduce the mixing length drastically by inducing chaotic stirring, which results in $l_m \propto \ln(Pe)$ for fully chaotic and $l_m \propto Pe^\beta$ (with $\beta < 1$) for partially chaotic flows. Figure 6.10 depicts the dimensionless species concentration distribution in the continuous mixer when Pe = 500 (a), 1000 (b), and 2000 (c), whereas the kinematic condition is fixed at St = $1/2\pi$ and Re = 0.01. On the basis of the spatial species concentration distribution, the mixing index, M, is then calculated using Equation 6.9. Figure 6.11 depicts M^{-1} as a function of the dimensionless mixing length, l_m/h, under the same kinematic condition (St = $1/2\pi$ and Re = 0.01) for Pe = 500 (rectangles), 1000 (triangles), and 2000 (circles). The corresponding 95%, 93%, and 90% mixing efficiencies are also marked in Figure 6.11.

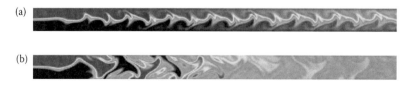

FIGURE 6.9
(See color insert.) Dimensionless species concentration distribution for mixing at Pe = 1000, Re = 0.01, and A = 0.8, with stirring at (a) St = $1/4\pi$ and (b) St = $1/2\pi$.

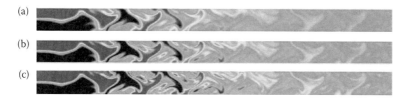

FIGURE 6.10
(See color insert.) Dimensionless species concentration distribution for Pe = 500 (a), 1000 (b), and 2000 (c) at St= $1/2\pi$, Re = 0.01, and A = 0.8.

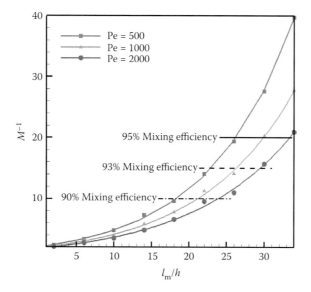

FIGURE 6.11
Mixing index inverse (M^{-1}) as a function of the normalized mixing length for Pe = 500, 1000, and 2000 at St = $1/2\pi$, Re = 0.01, and A = 0.8.

As a succinct and clear evidence of global chaos, the logarithmic relation between mixing length (l_m) and Pe should be investigated under the same kinematic condition. Figure 6.12 depicts the mixing length for 95%, 93%, and 90% mixing efficiency versus ln(Pe) under the kinematic condition of St = $1/2\pi$ and Re = 0.01. The linear relationship between l_m and ln(Pe) indicates fully chaotic flow in the mixer. Figure 6.12 can also be used to determine the length or number of mixing blocks of the mixer that is required to achieve a certain mixing efficiency.

Before finishing this section, we would like to discuss the practical aspects of mixing index calculations. The method requires utilization of the species transport equation along with a flow solver, which presently exists in most commercial software packages. Therefore, the mixing index can be the preferred methodology to characterize the mixing efficiency. One must be careful about accuracy of the used numerical solver in the case of long-time integration errors that have significant impacts on the results of the flow and species distribution. Although the color contour plots are often too forgiving, significant discrepancy in the M^{-1} values of high-order accurate solver from others is observed at high Pe values because of the high numerical diffusion of the latter. Finally, the mixing index is relevant to the experimental observations on the basis of mixing of fluorescent dyes. The decision for the desired M^{-1} value or the mixing efficiency can depend on the application.

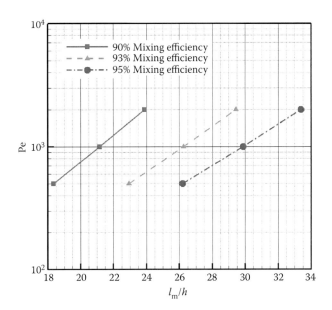

FIGURE 6.12
Peclet versus the normalized mixing length at 90%, 93%, and 95% mixing efficiencies for St = 1/2π, Re = 0.01, and A = 0.8 conditions.

For example, certain applications may require perfect molecular diffusion for a reaction to take place. Numerical modeling of this situation could be challenging as it would require very large M^{-1} values, requiring very long numerical integration times and excessive number of mixing blocks for open mixers.

6.4 Concluding Remarks

We outlined the conditions and benefits to obtain chaotic stirring in LOC devices. On the basis of Lagrangian particle tracking of passive tracers, the Poincaré sections provide qualitative detection of bad mixing zones, such as islands, and cannot differentiate the stirring performance after the Poincaré sections become featureless. When the Poincaré section is featureless, FTLE should be used to quantify the chaotic strength. Strictly positive FTLE values indicate fully chaotic flow in the mixer, and the average value of the FTLE quantifies the chaotic strength. However, the computations of the Poincaré section and FTLE are very expensive because they require both accurate time integration and accurate resolution of the flow field.

Alternatively, mixing index and scaling of l_m as a function of Pe can be used to characterize the mixing behavior under fixed kinematic conditions by varying the Sc. Spatial distribution of species concentration can be determined either by numerically solving the species transport equation or by performing imaging analysis of fluorescence images obtained from experiments using confocal microscopy with dye additions. The relationship l_m versus Pe should be evaluated under fixed kinematic conditions (i.e., fixed Re, etc.) to check whether the continuous-flow mixer is globally chaotic. Often $l_m \propto Pe^\beta$ with $\beta < 1$

is observed. Pure diffusion results in $\beta = 1$, whereas certain convective flows result in $\beta = 0.5$. For partially chaotic flows, the value of β depends on the extent of the regular flow regions, and $\beta \rightarrow 0$ as the regular flow zones diminish. For fully chaotic flows $l_m \propto \ln(\text{Pe})$ is observed. Similarly, the logarithmic relationship between the mixing time (t_m) and Pe should be investigated in a closed mixer at a fixed kinematic condition (i.e., fixed Re, etc.), where $t_m \propto \ln(\text{Pe})$ indicates fully chaotic flow in closed mixers [49].

Given these observations, we outline a methodology that could be used in the design and characterization of chaotic mixers using computational tools. First, the designer should determine the limitations induced by the LOC application, especially from the upstream and downstream device components. This should determine the range of flow rate and other conditions such as the level of desired mixing and Sc of the species to be mixed. Then the designer should choose either an active or a passive mixer system as well as a closed or an open mixer system. Such choices are often based on the designer's experience and the limitations induced by the LOC application. After these decisions, the designer should come up with a mixer geometry, which is expected to induce enhanced stirring. In the case of active mixers, crossing of the streamlines between different actuation instances is important to induce chaotic stirring. Such conditions can be checked using flow solvers. However, it is often not possible to pick the best kinematic condition for a given mixer geometry. For this reason, it may be beneficial to simultaneously run the species transport solver at a reasonably high Pe value while varying the Reynolds number and other design conditions, such as the St in the continuous-flow mixer example used here. Note that Pe $\gg$ Re is essential for this methodology, and such numerical iterations are hard to duplicate in actual experiments. On the basis of the numerical results, the best kinematic condition for mixing can be determined at a given (high) Pe value. However, characterization of mixing efficiency requires fixing the kinematic conditions (Re) and systematically varying Pe by varying Sc, the result of which will indicate l_m or t_m variation as a function of Pe, as outlined above.

Nomenclature

$A = U_0/U_{HS}$	Poiseuille to electro-osmotic flow ratio		R	curved channel radius
C	species concentration		Re	Reynolds number, $U_{HS}h/\nu$
C_i	average concentration inside the ith section of N		Sc	Schmidt number
C_∞	the concentration corresponding to the perfect mix		St	Strouhal number, $\omega h/U_{HS}$
De	Dean number, $\text{Re}(D_h/2R)$		T	period
D_h	hydraulic diameter		t	time
E	applied electric field		t_m	mixing time
H	height of the mixing block, $2h$		u	velocity vector
h	half of the channel height		U_0	centerline velocity
L	length of the mixing block		U_{HS}	electro-osmotic velocity
l_m	mixing length		$x(x_0, t)$	fluid path lines
M	mixing index		ω	angular frequency, $2\pi/T$
N	total number of interrogation areas		ν	kinematic viscosity
P	pressure		ζ	zeta potential

Pe	Peclet number, Sc × Re	σ	standard deviation
P_n	position of passive tracer particle n	λ_{FTLE}	finite time Lyapunov exponent
$/dx/$	initial distance of arbitrary pairs of nearby points	ε_0	permittivity of the vacuum
Φ_P	Poincaré mapping	μ	dynamic viscosity
ε	relative dielectric constant		

References

1. Jensen KF. Microchemical systems: Status, challenges, and opportunities. *AIChE J.* 1999; 45:2051–4.
2. Jain KK. *Biochips & Microarrays: Technology and Commercial Potential.* London: Informa Global Pharmaceutical Publications. 2000.
3. Langer R. Biomaterials: Status, challenges, and perspectives. *AIChE J.* 2000;46:1286–9.
4. Stone HA, Kim S. Microfluidics: Basic issues, applications, and challenges. *AIChE J.* 2001; 47:1250–4.
5. Chow AW. Lab-on-a-chip: Opportunities for chemical engineering. *AIChE J.* 2002;48:1590–5.
6. Verpoorte E. Microfluidic chips for clinical and forensic analysis. *Electrophoresis.* 2002; 23:677–712.
7. Huikko K, Kostiainen R, Kotiaho T. Introduction to micro-analytical systems: Bioanalytical and pharmaceutical applications. *Eur J Pharm Sci.* 2003;20:149–71.
8. Gardeniers H, Van den Berg A. Micro and nanofluidic devices for environmental and biomedical applications. *Int J Environ Anal Chem.* 2004;84:809–19.
9. Thilmany J. Think small. *EMBO Rep.* 2005;6:913–6.
10. Dittrich PS, Manz A. Lab-on-a-chip: Microfluidics in drug discovery. *Nature.* 2006;5:210–8.
11. Dittrich PS, Tachikawa K, Manz A. Micro Total Analysis Systems. Latest advancements and trends. *Anal Chem.* 2006;78:3887–907.
12. Crevillen AG, Hervas M, Lopez MA, Gonzalez MC, Escarpa A. Real sample analysis on microfluidic devices. *Talanta.* 2007;74:342–57.
13. Tanaka Y, Sato K, Shimizu T, Yamato M, Okano T, Kitamori T. Biological cells on microchips: New technologies and applications. *Biosens Bioelectron.* 2007;23:449–58.
14. Roman GT, Kennedy RT. Fully integrated microfluidic separation systems for biochemical analysis. *J Chromatogr A.* 2007;1168:170–88.
15. Kuswandi B, Nuriman J, Huskens J, Verboom W. Optical sensing systems for microfluidic devices: A review. *Anal Chim Acta.* 2007;601:141–55.
16. Abgrall P, Gue AM. Lab-on-chip technologies: making a microfluidic network and coupling it into a complete microsystem—a review. *J Micromech Microeng.* 2007;17:R15–49.
17. Borgatti M, Bianchi N, Mancini I, Feriotto G, Gambari R. New trends in non-invasive prenatal diagnosis: Applications of dielectrophoresis-based lab-on-a-chip platforms to the identification and manipulation of rare cells (Review). *Int J Mol Med.* 2008;21:3–12.
18. Yager P, Domingo GJ, Gerdes J. Point-of-care diagnostics for global health. *Ann Rev Biomed Eng.* 2008;10:107–44.
19. Hansen EH, Miro M. Interfacing microfluidic handling with spectroscopic detection for real-life applications via the lab-on-valve platform: A review. *Appl Spectrosc Rev.* 2008;43:335–57.
20. Chen P, Feng XJ, Du W, Liu BF. Microfluidic chips for cell sorting. *Front Biosci.* 2008; 13:2464–83.
21. Ong SE, Zhang S, Du HJ, Fu YQ. Fundamental principles and applications of microfluidic systems. *Front Biosci.* 2008;13:2757–73.

22. Sathuluri RR, Yamamura S, Tamiya E. Microsystems technology and biosensing. *Adv Biochem Eng Biotechnol*. 2008;109:285–350.

23. Hertzog DE. Microfluidic mixers for protein folding. PhD thesis. Stanford, CA: Stanford University; February 2006.

24. Boy DA, Gibou F, Pennathur S. Simulation tools for lab on a chip research: advantages, challenges, and thoughts for the future. *Lab Chip*. 2008;8:1424–31.

25. Nguyen NT, Wu ZG. Micromixers - A review. *J Micromech Microeng*. 2005;15:R1–16.

26. Qian S, Duval JFL. Mixers. In: *Comprehensive Microsystems*. Edited by Gianchandani YB, Tabata O, and Zappe H. London: Elsevier; 2007;2:323–374.

27. Nguyen NT. Micromixers: fundamentals, design and fabrication. In: *Micro & Nano Technologies Series*. New York: William Andrew; 2008.

28. Stroock AD, Dertinger SKW, Ajdari A, Mezic I, Stone HA, Whitesides GM. Chaotic mixer for microchannels. *Science*. 2002;295:647–51.

29. Sasaki N, Kitamori T, Kim HB. Electroosmotic micromixer for chemical processing in a microchannel. *Lab Chip*. 2004;6:550–4.

30. Simonnet C, Groisman A. Chaotic mixing in a steady flow in a microchannel. *Phys Rev Lett*. 2005;94:134501–1–134501–4.

31. Chang SH, Cho YH. Static micromixers using alternating whirls and lamination. *J Micromech Microeng*. 2005;15:1397–405.

32. Sudarsan AP, Ugaz VM. Fluid mixing in planar spiral microchannels. *Lab Chip*. 2006;6:74–82.

33. Xia HM, Shu C, Wan SYM, Chew YT. Influence of the Reynolds number on chaotic mixing in a spatially periodic micromixer and its characterization using dynamical system techniques. *J Micromech Microeng*. 2006;16:53–61.

34. Aref H. Stirring by chaotic advection. *J Fluid Mech*. 1984;143:1–21.

35. Aref H. Chaotic advection of fluid particles. *Proc R Soc Lond A*. 1990;333:273–88.

36. Aref H. The development of chaotic advection. *Phys Fluids*. 2002;14:1315–25.

37. Ottino JM. *The Kinematics of Mixing: Stretching, Chaos, and Transport*. Cambridge: Cambridge University Press; 1989.

38. Wiggins S, Ottino JM. Foundations of chaotic mixing. *Philos Transact A Math Phys Eng Sci*. 2004;362:937–70.

39. Stremler MA, Haselton FR, Aref H. Designing for chaos: applications of chaotic advection at the micro scale. *Philos Transact A Math Phys Eng Sci*. 2004;362:1019–36.

40. Liu RH, Stremler MA, Sharp KV, Olsen MG, Santiago JG, Adrian RJ, et al. A passive three-dimensional 'C-shape' helical micromixer. *J Microelectromech Syst*. 2000;9:190–7.

41. Stremler MA, Olsen MG, Adrian RJ, Aref H, Beebe DJ. Chaotic Mixing in Microfluidic Systems. Solid-State Sensor and Actuator Workshop, Hilton Head, SC; 2000 June 4–8.

42. Yi M, Bau HH. The kinematics of bend-induced mixing in microconduits. *Int J Heat Fluid Flow*. 2003;24:645–56.

43. Qian S, Bau HH. Theoretical investigation of electro-osmotic flows and chaotic stirring in rectangular cavities. *Appl Math Model*. 2005;29:726–53.

44. Lee CS, Blanchard WC, Wu CT. Direct control of the electroosmosis in capillary zone electrophoresis by using an external electric field. *Anal Chem*. 1990;62:1550–2.

45. Schasfoort RBM, Schlautmann S, Hendrikse J, Van den Berg A. Field-effect flow control for microfabricated fluidic networks. *Science*. 1999;286:942–5.

46. Lin A, Lee KH, Lee GB. Active mixing inside microchannels utilizing dynamic variation of gradient zeta potentials. *Electrophoresis*. 2005;26:4605–15.

47. Wu HY, Liu CH. A novel electrokinetic micromixer. *Sens Actuators A Phys*. 2005;118:107–15.

48. Qian S, Bau HH. A chaotic electroosmotic stirrer. *Anal Chem*. 2002;74:3616–25.

49. Kim HJ, Beskok A. Quantification of chaotic strength and mixing in a micro fluidic system. *J Micromech Microeng*. 2007;17:2197–210.

50. Mengaud V, Josserand J, Girault HH. Mixing processes in a zigzag microchannel: Finite element simulations and optical study. *Anal Chem*. 2002;74:4279–86.

51. Therriault D, White SR, Lewis JA. Chaotic mixing in three-dimensional microvascular networks fabricated by direct-write assembly. *Nat Mat*. 2003;2:265–71.
52. Johnson TJ, Ross D, Locascio LE. Rapid microfluidic mixing. *Anal Chem*. 2002;74:45–51.
53. Stroock AD, Dertinger SKW, Ajdari A, Mezic I, Stone HA, Whitesides GM. Chaotic mixer for microchannels. *Science*. 2002;295:647–51.
54. Howell PB, Mott DR, Fertig S, Kaplan CR, Golden JP, Oran ES, et al. A microfluidic mixer with grooves placed on the top and bottom of the channel. *Lab Chip*. 2005;5:524–30.
55. Kim HJ, Beskok A. Numerical modeling of chaotic mixing in electroosmotically stirred continuous flow mixers. *J Heat Transfer*. 2009;131(9):092403.
56. Niu X, Lee Y. Efficient spatial-temporal chaotic mixing in microchannels. *J Micromech Microeng*. 2003;13:454–62.
57. Sprott JC. *Chaos and Time-Series Analysis*. Oxford: Oxford University Press; 2003.
58. Antonsen TM, Fan Z, Ott E, Garcia-Lopez E. The role of chaotic orbits in the determination of power spectra of passive scalars. *Phys Fluids*. 1996;8:3094–104.

7

Lattice Boltzmann Method and Its Applications in Microfluidics

Junfeng Zhang

CONTENTS

7.1 Introduction

Microfluidics studies the behavior, control, and manipulation of small volumes of fluids, typically in the order of microliters, nanoliters, or even picoliters. Microfluidic devices have dimensions ranging from millimeters to micrometers. Because of the high surface-area-to-volume ratio of the devices, microscopic interactions and related interfacial phenomena (such as electrokinetics, surface wettability, and interfacial slip), which are typically neglected in macroscopic systems, have to be considered carefully in microfluidics. Traditional computational fluid dynamics (CFD) methods start from the continuum Navier–Stokes equations and solve then numerically by appropriate schemes. However, such an approach has difficulties in incorporating the microscopic interactions, which are crucial in many microfluidic circumstances, for example, the dynamics of wetting and solid–liquid interfacial slip. The continuum assumption may even break up in micro scopic situations, such as gaseous flows in microchannels. On the other hand, microscopic simulation approach, such as molecular dynamics (MD), studies the fluid behaviors by the evolution of individual molecules interacting with each other through intermolecular potentials. Here, the microscopic interactions can be represented well; however, the huge computation demand limits its applications to very small space and time scales [1].

Between these two approaches, there exists a representative mesoscopic approach, the lattice Boltzmann method (LBM) [2–4], which has experienced rapid development and attracted increasing interest during the past two decades in simulating complex fluid systems and other processes and phenomena. LBM models the fluid as fictitious particles, and such particles perform consecutive propagation and collision processes over a discrete lattice mesh. Because of its particulate nature and local dynamics, the LBM has several advantages over conventional CFD methods, especially in dealing with complex boundaries, incorporation of microscopic interactions, and parallel computation [3].

Historically, LBM originated from the lattice gas automata (LGA), which can be considered as a simplified, fictitious version of MD in which space, time, and particle velocities are all discrete. Each lattice node is connected to its neighbors by, for example, six lattice velocities in a hexagonal Frisch–Hasslacher–Pomeau (FHP) model [3,4]. There can be either one or zero particle at a lattice node moving along a lattice direction, according to a set of collision rules. Good collision rules should conserve the particle number (mass), momentum, and energy before and after the collision. In spite of its many successful applications, LGA suffers from several native defects, including the lack of Galilean invariance, the presence of statistical noise, and the exponential complexity for three-dimensional lattices [3]. The main motivation for the transition from LGA to LBM was the desire to remove statistical noise by replacing the Boolean particle number in a lattice direction with its ensemble average, the so-called *density distribution function*. Accompanying with this replacement, the discrete collision rules also have to be modified as a continuous function—the collision operator. In the LBM development, an important simplification is the approximation of the collision operator with the Bhatnagar–Gross–Krook (BGK) relaxation term. This lattice BGK model makes simulations more efficient and allows flexibility of the transport coefficients. Through a Chapman–Enskog analysis, one can recover the governing continuity and Navier–Stokes equations from the LBM algorithm [2]. In addition, the pressure field is also directly available from the density distributions, and hence there is no additional Poisson equation to be solved as in the traditional CFD methods. It has been shown that the LBM scheme can also be interpreted as a special discretized form of the continuous Boltzmann equation [5–7].

Simulating multiphase/multicomponent flows has always been a challenge because of the moving and deformable interfaces. More fundamentally, the interfaces between two bulk phases (e.g., liquid and vapor) originate from the specific interactions among molecules. Therefore, it is difficult to implement such microscopic interactions into the macroscopic Navier–Stokes equation. However, in LBM, the particle kinetics allows a relatively easy and consistent avenue to incorporate the underlying microscopic interactions. Several LBM multiphase/multicomponent models have been developed. Phase separations can be generated automatically from particle dynamics, and no special treatment is needed to manipulate the interfaces as in traditional CFD methods.

From a pure mathematical point of view, the general LBM can be considered as a numerical solver of the macroscopic momentum conservation equation, that is, the Navier–Stokes equation, which is a nonlinear, second-order, partial differential equation. For this reason, several LBM-like algorithms have also been proposed to solve differential equations governing other physical processes and phenomena, such as heat transfer, electrical field, diffusion processes, flows in porous materials, and shallow flows [4]. Such models inherit the advantages in programming and parallel computation, but with no physics involved.

In this chapter, we describe the basic LBM algorithm for general fluid dynamics and introduce LBM models for other processes and phenomena, which may be useful in microfluidics, such as multiphase/multicomponent flows, diffusion–convection, heat transfer, and electrical field. Representative LBM simulations are also reviewed to demonstrate their attractiveness in computational microfluidics.

7.2 LBM Algorithm and Boundary Conditions for Fluid Dynamics

Theoretically, the LBM method can be derived from the classical Boltzmann equation in statistical physics [8]

$$\frac{\partial f}{\partial t} + \xi \cdot \nabla f = \Omega(f), \tag{7.1}$$

where f is the single-particle distribution function, with $f(\mathbf{x}, \xi, t)\mathrm{d}\mathbf{x}\mathrm{d}\xi$ representing the probability to find a particle positioned in $[\mathbf{x}, \mathbf{x} + \mathrm{d}\mathbf{x}]$ and with a velocity in $[\xi, \xi + \mathrm{d}\xi]$ at time t [8]. ξ is the microscopic velocity of a fluid particle. On the right-hand side, Ω is the collision operator for incorporating the distribution function change due to particle–particle collisions. The macroscopic fluid density ρ and momentum $\rho\mathbf{u}$ are obtained from the moments of the density distribution functions with respect to the microscopic velocity ξ [6,8,9]:

$$\rho = \int f(\mathbf{x}, \xi, t)\, \mathrm{d}\xi, \quad \rho\mathbf{u} = \int f(\mathbf{x}, \xi, t)\xi\, \mathrm{d}\xi. \tag{7.2}$$

For simplicity, the collision operator Ω is often expressed as

$$\Omega = -\frac{f - f^{\mathrm{eq}}}{\tau}, \tag{7.3}$$

with the BGK approximation [10], which assumes that f relaxes toward an equilibrium distribution f^{eq} with a relaxation time τ. The equilibrium distribution f^{eq} is given by the Boltzmann–Maxwellian distribution function:

$$f^{\mathrm{eq}} = \frac{\rho}{(2\pi c_s^2)^{D/2}} \exp\left(-\frac{(\xi - \mathbf{u})^2}{2c_s^2}\right), \tag{7.4}$$

where D is the dimension of the space, and $c_s^2 = RT$ is the speed of sound, with R the gas constant and T the temperature. By integrating the Boltzmann equation (Equation 7.1) and by multiplying Equation 7.1 by ξ and then integrating, correct macroscopic continuity and momentum equations, respectively, can be derived from this statistical kinetic description [8].

   Although historically developed from LGA, LBM can also be considered as a particular discrete version of the Boltzmann theory with time, space, and momentum all discretized [5,6]. In LBM, a fluid is modeled as fictitious particles moving with several lattice velocities (momentum discretization) in a lattice domain (space discretization) at consequential time steps (time discretization). Typical D2Q9 (two-dimensional and nine-velocity) square and D3Q19 (three-dimensional and 19-velocity) cube lattice structures, which

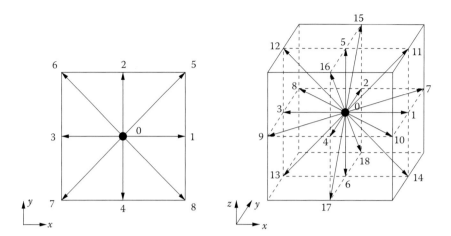

FIGURE 7.1
Schematics of the D2Q9 (left) and D3Q19 (right) lattice models. The edge length of both the D2Q9 square and D3Q19 cube is $2\Delta x/\Delta t$, and dashed lines show the midplanes of the D3Q19 cube. The lattice velocities are indicated by arrows starting from the square/cube center. Note that there is a zero lattice velocity $c_0 = 0$ in both models.

are commonly used in LBM simulations among many other options [11], are shown in Figure 7.1. For simplicity, here we take the D2Q9 model as an example for our discussion. Details of other models are similar and readily available in the literature.

The nine lattice velocities of the D2Q9 model are

$$\mathbf{c}_0 = 0,$$
$$\mathbf{c}_i = \left(\cos\frac{i-1}{2}\pi, \sin\frac{i-1}{2}\pi\right)\frac{\Delta x}{\Delta t}, \quad i = 1-4,$$
$$\mathbf{c}_i = \sqrt{2}\left(\cos\frac{2i-9}{4}\pi, \sin\frac{2i-9}{4}\pi\right)\frac{\Delta x}{\Delta t}, \quad i = 5-8,$$

(7.5)

where Δx and Δt are the lattice unit and the time step, respectively. The major variable in LBM is the density distribution $f_i(\mathbf{x}, t)$, indicating the number of particles moving with the ith lattice velocity at lattice site $\mathbf{x}$ and time t. Because $\mathbf{c}_0 = (0, 0)$, f_0 corresponds to the portion that does not move at all. After one time step Δt, $f_i(\mathbf{x}, t)$ will arrive at its neighboring lattice site $\mathbf{x} + \mathbf{c}_i\Delta t$ along the lattice velocity $\mathbf{c}_i$. This process is usually called the streaming or propagation process. Because there are other particles coming to this same site from different directions, collision among them will occur at this site, and the original particle numbers moving in each direction will be changed. After this collision, a new set of density distributions leave the collision site in different lattice velocities, and another propagation process starts. In an LBM simulation, such propagation and collision processes are repeated again and again till satisfactory results are achieved.

The above-mentioned dynamics can be expressed by the so-called *lattice Boltzmann equation*,

$$f_i(\mathbf{x} + \mathbf{c}_i\Delta t, t + \Delta t) - f_i(\mathbf{x}, t) = \Omega_i(f).$$

(7.6)

Here Ω_i is the collision operator that takes account of the distribution change after a collision. The left-hand side represents the propagation process. Applying the single relaxation time approximation, which was originally proposed by Bhatnagar, Gross, and Krook (BGK) for Boltzmann equation in continuum kinetic theory [10], the lattice BGK equation is written as

$$f_i(\mathbf{x} + \mathbf{c}_i \Delta t, t + \Delta t) - f_i(\mathbf{x}, t) = -\frac{f_i(\mathbf{x}, t) - f_i^{eq}(\mathbf{x}, t)}{\tau}, \tag{7.7}$$

where τ is a relaxation parameter toward the equilibrium distribution f_i^{eq}, which can be expressed as a discretization of the Maxwell–Boltzmann equilibrium distribution Equation 7.4:

$$\begin{aligned}
f_0^{eq} &= \rho \left(1 - \frac{2}{3c_s^2} \mathbf{u}^2 \right), \\
f_i^{eq} &= \rho \left[\frac{1}{9} + \frac{1}{3c_s^2} \mathbf{c}_i \cdot \mathbf{u} + \frac{1}{2c_s^4} (\mathbf{c}_i \cdot \mathbf{u})^2 - \frac{1}{6c_s^2} \mathbf{u}^2 \right], \quad i = 1 - 4, \\
f_i^{eq} &= \rho \left[\frac{1}{36} + \frac{1}{12c_s^2} \mathbf{c}_i \cdot \mathbf{u} + \frac{1}{8c_s^4} (\mathbf{c}_i \cdot \mathbf{u})^2 - \frac{1}{24c_s^2} \mathbf{u}^2 \right], \quad i = 5 - 8.
\end{aligned} \tag{7.8}$$

Here the fluid density ρ and velocity $\mathbf{u}$ can be readily obtained from the density distributions at each lattice site through

$$\rho = \sum_i f_i, \quad \rho \mathbf{u} = \sum_i f_i \mathbf{c}_i. \tag{7.9}$$

Here $c_s^2 = \Delta x^2 / 3\Delta t^2$. Through the Chapman–Enskog expansion, one can recover the macroscopic continuity and momentum (Navier–Stokes) equations from the above-defined LBM dynamics [12,13]:

$$\begin{aligned}
\frac{\partial \rho}{\partial t} + \nabla \cdot (\rho \mathbf{u}) &= 0, \\
\frac{\partial \mathbf{u}}{\partial t} + (\mathbf{u} \cdot \nabla) \mathbf{u} &= -\frac{1}{\rho} \nabla p + \nu \nabla^2 \mathbf{u},
\end{aligned} \tag{7.10}$$

with the kinematic viscosity ν given by

$$\nu = \frac{2\tau - 1}{6} \frac{\Delta x^2}{\Delta t} \tag{7.11}$$

and the pressure P expressed as

$$P = c_s^2 \rho. \tag{7.12}$$

To induce a fluid flow, an external force $\mathbf{F}$ can be included in the above LBM algorithm by adding an extra term to the collision operator in the right-hand side of Equation 7.7.

$$f_i(\mathbf{x} + \mathbf{c}_i \Delta t, t + \Delta t) - f_i(\mathbf{x}, t) = -\frac{f_i(\mathbf{x}, t) - f_i^{eq}(\mathbf{x}, t)}{\tau} + \frac{w_i \Delta t}{c_s^2} \mathbf{F} \cdot \mathbf{c}_i, \tag{7.13}$$

or by redefining an equilibrium velocity

$$\mathbf{u}^{eq} = \frac{\sum_i f_i \mathbf{c}_i + \Delta t \mathbf{F}}{\rho} \tag{7.14}$$

and using it in the calculation of equilibrium distribution f_i^{eq} in Equation 7.8 [14]. For our D2Q9 model, the weight factors are $w_0 = 4/9$, $w_{1-4} = 1/9$, and $w_{5-8} = 1/36$. After such modifications, the total fluid momentum increases by an amount of $\Delta t \mathbf{F}$ during a time step, whereas the total fluid density remains unchanged. It can be seen from the above description that the LBM dynamics is local (i.e., only the very neighboring lattice nodes are involved in updating the density distributions), and hence LBM algorithm is advantageous for parallel computations.

As with other numerical approaches, appropriate boundary conditions are important for meaningful simulations. Because the principal variables in LBM are the density distributions f_i, LBM boundary conditions are implemented through specifying the unknown f_i entering the simulation domain across boundaries to achieve desired macroscopic fluid properties such as velocity and pressure. This feature poses both conveniences and difficulties. For example, to model a no-slip boundary over a solid surface, one can simply reverse the particle directions toward the boundary back to their original locations, the so-called *bounce-back scheme*. Periodic boundaries are even easier to implement: all particles that leave the domain across a periodic boundary will re-enter the domain from the opposite side [3,4]. However, for the general pressure and velocity boundary conditions, there are usually more unknown incoming density distributions than the number of constraint equations from the boundary conditions (see Equation 7.9), that is, more unknowns to be determined from less constraints. To have a complete system of equations, assumptions are usually necessary, and inappropriate choices could lead to unphysical boundary effects and also hamper simulation stability. Tremendous efforts have been devoted to develop accurate and efficient boundary schemes for different situations, and detailed descriptions can be found in the literature [15–20].

There are several other, more complicated LBM schemes proposed for fluid dynamics. For example, a multiple-relaxation-times model can improve numerical stability and offers the option of adjusting the bulk and shear viscosities independently [4,21], whereas carefully designed off-grid lattice structures can be adopted for microscopic flows at finite Knudsen numbers [22,23]. Unstructured LBM is also available for complex geometry boundaries [24].

7.3 Multiphase/Multicomponent LBM Models

Multiphase fluids with droplets are commonly found in recent microfluidic applications, such as lab-on-a-chips, microelectromechanical systems (MEMS), fuel cells, and micro heat exchangers. The use of liquid droplets, which represent the smallest units of mass flux and can function as the smallest reactors in chemical and biochemical reactions, is considered as the next generation microfluidic technology, or the so-called *digital microfluidics*. In such systems for biological and biochemical analysis, interfacial tension and surface wettability are usually used for flow metering, activation, and control [25,26]. It can be seen from the pressure–density relationship Equation 7.10 that the fluid from the general LBM is an ideal fluid. To simulate a multiphase/multicomponent system, the above method must be modified to incorporate microscopic intermolecular interactions [14,27,28] or to impose the macroscopic equation of state [29,30]. Such a modification can be carried out either by applying an additional force term as in Equations 7.7 and 7.8 or by reformulating the equivalent density distribution given in Equation 7.2. Chin et al. [31] have described the latter equation of state approach as a top-down model because it imposes a macroscopic free energy functional without considering its molecular origin. On the other hand, the former microscopic interaction approach is considered a bottom–up model because the force term is related to pairwise interparticle interactions. New sets of density distributions are also introduced in multicomponent models to represent the individual components or their density differences [14,29,32,33].

Different from traditional CFD methods, the LBM multiphase/multicomponent algorithm is uniform throughout the entire domain, and phase separation as well as interface evolution can be obtained without front-capturing and front-tracking treatments. These models typically had been validated by means of stationary bubbles for the Laplace law of capillary and dynamic interfacial waves for the wavelength–frequency dispersion relations. Successful applications of the multiphase/multicomponent LBM models can be found in various complex fluid systems, including interface instability, bubble/droplet dynamics, wetting on solid surfaces, interfacial slip, and droplet electrohydrodynamic deformations.

A notable problem with these physics-originated models is that the density difference between different phases is usually limited when compared with realistic systems. For example, the water density is ~1000 times that of air. To address this point, researchers apply treatments similar to those in other traditional CFD methods, and several high-density-ratio models have been proposed [34–36]. Another point is that the interface thickness from these models is usually orders larger than the actual thickness. This is not surprising because several lattice units are needed to represent the property (e.g., density) change from one phase to another. Some models have also been proposed by treating the interface as an artificial boundary [37,38]. However, these models require extra sets of distribution functions or more complicated interface treatments, and the phase-separation process has less physical basis. Therefore, these approaches, although attractive from a practical point of view, blur the distinction between LBM and other traditional CFD methods.

7.4 LBM-Like Algorithms for Other Phenomena

Realizing that the LBM dynamics can be viewed as a numerical solver of differential equations (Navier–Stokes equation here), several LBM-like algorithms have been proposed to solve problems that can be described by similar differential equations. Such algorithms are designed so that the corresponding macroscopic differential equations can be recovered through a Chapman–Enskog analysis. However, we should be aware that these models are purely numerical solvers, and the density distributions in these models have no physical meaning as in the original LBM for fluid mechanics. As an example, an LBM-like model for electrical field is illustrated below.

Generally, the electric potential ϕ over space is governed by the Poisson equation

$$\nabla \cdot (\varepsilon \nabla \phi) = -\frac{\rho_e}{\varepsilon_0}, \tag{7.15}$$

where ε and ε_0 are the dielectric constant of the medium and the permittivity of the vacuum, respectively, and ρ_e is the net charge density, which can be related to the electric field through the Boltzmann distribution in an electrolyte [39]. To solve this equation in the framework of LBM, He and Li [40] proposed the following lattice equation for density distribution h_i

$$h_i(\mathbf{x} + \mathbf{c}_i \Delta t, t + \Delta t) - h_i(\mathbf{x}, t) = -\frac{h_i(\mathbf{x}, t) - h_i^{eq}(\mathbf{x}, t)}{\tau_\phi} + \frac{w_i \rho_e}{\varepsilon_0}. \tag{7.16}$$

The corresponding equilibrium distribution h_i^{eq} is related to the electrical potential ϕ via the lattice weight factor w_i by $h_i^{eq} = w_i \phi$, and the relaxation parameter τ_ϕ is determined from the dielectric constant ε with $\tau_\phi = 3\varepsilon + 0.5$. The integral electrical potential ϕ is just a sum of all potential distributions h_i, that is, $\phi = \Sigma_i h_i$. Again the Chapman–Enskog analysis can be applied to Equation 7.16, and the original Poisson equation (Equation 7.15) can be obtained.

Similar LBM-like schemes exist in the literature for other processes and phenomena, including convection-diffusion [41,42], heat transfer [43,44], and magnetic field [9,45]. A novel enthalpy-based thermal LBM model has also been proposed recently by Chatterjee [46]. These models might be useful for microfluidic simulations involving mixing, Joule heating, and magnetohydrodynamics (MHD). Interested readers can refer the above references and those therein.

7.5 Applications in Microfluidics

In recent years, LBM has been applied to study many microscale transport phenomena and processes, including gas flows, electro-osmotic flows, interfacial phenomena, microscopic biofluidics, and colloid suspensions. In the following, we briefly review some representative applications of LBM in these areas.

7.5.1 Microscopic Gaseous Flows

MEMS have experienced rapid progress in recent years for their applications in various areas, including automotive, aerospace, medical, and telecommunications. In such microscopic systems, gaseous flows are commonly involved, and their behaviors are important to the system performance [47,48]. The Knudsen number $K_n = \lambda/L$, the ratio of the gas mean free path λ to a characteristic length of the flow domain L, usually serves as a criterion in determining whether the continuum assumption is applicable to a fluid flow. By definition, the mean free path is the average distance traveled by a gas molecule between two subsequent collisions with other molecules. The mean free path is usually in a range of 20–200 nm. Several flow regimes can be classified according to the Knudsen number [47]. For flows in large domains with $K_n \leq 10^{-3}$, the general Navier–Stokes equation and no-slip boundary are reasonable to describe the flow behaviors, as we do in most macroscopic situations. As the flow domain becomes smaller, the Knudsen increases, and some rarefaction effects, such as velocity slip and temperature jumps on solid boundaries, have been observed. For the slip flow regime of $10^{-3} \leq K_n \leq 10^{-1}$, the Navier–Stokes equation is still applicable in the bulk region; however, slip boundary condition has to be considered over the solid boundaries. For the transition ($10^{-3} \le K_n \le 10^{-1}$) and free-molecule ($K_n > 10$) flows, the particulate and kinematic nature of the gas becomes dominant, and the Boltzmann equation is necessary to analyze such flows.

Microscopic gaseous flows in MEMS are usually in the slip and transition regimes. As LBM can be considered as a numerical solver of the Navier–Stokes equation and also as a discretized version of the Boltzmann equation, it appears that LBM should be an attractive choice for these microflows. Indeed, several attempts have been made since 2002. These models usually take the same LBM formulations. The boundary slip was achieved by using a variable relaxation parameter τ as a function of K_n [49–51], or a slip-allowed boundary condition [52,53], or both [54]. Although interesting results have been obtained in these studies, most of them were limited to simple-geometry structures such as two-dimensional channels and the slip regime with relatively low Knudsen numbers.

7.5.2 Droplet/Bubble Dynamics

With the multiphase/multicomponent models described above, LBM has been used extensively to study various situations with bubbles and droplets. For example, droplet/bubble deformation, breakup, and collision have been simulated and compared with experimental observations [55–59]. Moreover, the droplet formation processes by using a flow focusing device [60] or cross-junction microchannel [61] have been studied, which has demonstrated the potential usefulness of these LBM models in microfluidic system design.

7.5.3 Electrokinetic and Electrohydrodynamic Flows

Application of LBM to single-phase flows in microsystems is straightforward. To simulate electrokinetic flows, one needs to incorporate the electrical forces in the electrical double-layer region near the solid–liquid interface due to the net charge and external electrical field. In these studies, the general LBM is used for the fluid dynamics, and the electrical potential can be solved by an LBM-like algorithm [62,63]. For example, Tian et al. [39] have investigated the electro-osmotic flows in microchannels with heterogeneous surface potentials and found that local circulations can be produced near the heterogeneous region. This model has also been adopted to study the electro-osmotic flow near the body surface of an earthworm when the earthworm moves in moist soil [64]. Simulations show that moving vortices, which likely contribute to antisoil adhesion, can be formed near the earthworm body from the nonuniform and variational electrical force. In addition, the LBM-like algorithm for electrical potential can be combined with a multicomponent model to study the electrohydrodynamic interface deformation [65]. The simulated drop deformation and flow field were found in good agreement with other theoretical and experimental studies. Furthermore, Guo et al. [66] had simulated electro-osmotic flows with the effect of Joule heating considered by using three sets of density distribution functions for the highly coupled fluid flow, electrical field, and temperature field, respectively.

7.5.4 Magnetohydrodynamic Flows

Magnetic forces have been extensively used in various microfluidic applications, including micropumps, microvalves, and stirrers, and in manipulation of magnetic particles in microflows [67–69]. With the LBM models for magnetohydrodynamics (MHD) available [9,45], several MHD simulations of microfluidic systems have been reported. For example,

Sofonea et al. [70,71] had simulated the deformation of magnetic fluid drops (or gas bubbles in magnetic fluids) and the normal field instability of magnetic fluids under the action of an external magnetic field. Similar phenomena has also been investigated recently by Falcucci et al. [72]. Clime et al. [73] had simulated the splitting process of a magnetic droplet with a hydrophilic magnetic plug inside electrowetting-on-dielectric devices, and the simulation results had been in excellent agreement with experimental observations. Moreover, the dynamics of paramagnetic particles under rotating magnetic fields in a fluid has also been studied by Calhoun et al. [74] and Krishnamurthy et al. [75] for its potential application in microfluidic mixing. Another interesting area in which LBM methods could be useful is the electro-magnetohydrodynamic-based microfluidic devices [69], which involve electrical field, magnetic field, fluid mechanics, ion diffusion–convection, and possible heat transfer. However, so far no LBM simulations of these complex but important processes have been reported.

7.5.5 Surface Wettability and Dynamic Wetting

Because it is relatively easy to incorporate microscopic interactions in the LBM, the LBM provides an attractive alternative to study the wetting dynamics of a liquid on a surface. Simulations have shown that different surface wettabilities (contact angles) can be achieved from the solid–fluid interaction [30,76]. Dynamic contact angles have been derived from LBM simulations, which are in good agreement to a theoretical adsorption/desorption model [77]. Recently, superhydrophobic surfaces have attracted great interest for their potential in reducing flow resistance. Because of the geometrical and physical complexity of the system, traditional theoretical and numerical methods encounter severe challenges to study this phenomenon. Several LBM attempts have been performed [78,79], and the results are promising. For example, it has been found that the advancing contact angle increases and resistance to droplet sliding decreases with decreasing fractional solid area. Static and dynamic behaviors of droplets on a chemically heterogeneous surface have also been investigated [80–82]. Another interesting process that has been examined with LBM is the droplet impacting on solid substrates [83] or a liquid film [84]. Further LBM simulations could be valuable to improve our understanding about these important phenomena.

7.5.6 Liquid–Solid Interfacial Slip

Different from the boundary slip of gaseous flows, recent measurements indicate significant slip of liquid flows on solid surfaces, and this fact has also been confirmed by MD simulations. In general, both experimental and MD simulation results show that there is a strong relationship between the magnitude of slip and the surface hydrophobicity: larger slip is usually observed on more hydrophobic surfaces and no slip on hydrophilic surfaces. LBM has also been used to study this interesting phenomenon [85]. The fluid was modeled as a nonideal fluid, and surface hydrophobicity was adjusted by tuning the specific solid–fluid interactions. The slip magnitude can be characterized by the apparent slip length, which was found to increase with surface hydrophobicity (contact angles). For relatively hydrophilic surfaces, even negative slip lengths could result because of the strong solid–fluid interactions. These findings are in good agreement with other experimental and MD studies. Because such an interfacial slip is generated by direct variation of the solid–fluid interactions instead of an applied slip boundary, these results demonstrate that the underlying microscopic interactions have been represented well, and LBM could be useful in microfluidic studies.

7.6 Conclusion

This chapter briefly summarized the development, model, and microfluidic applications of LBM. The purpose of this chapter is to provide an introduction to potential users for microfluidics simulations. For this reason, rigorous theoretical derivations are not included, and many important applications (e.g., biofluidics and colloidal suspensions) are not discussed. For a more complete picture, readers are referred to other comprehensive reviews and books. Compared with other traditional CFD methods, LBM is still relatively young. The author believes that the relation between LBM and other CFD methods should be complementary, not competitory. There are many particular situations where LBM can be advantageous because of its physical and numerical features, especially in microfluidics.

Acknowledgment

This work was supported by the Natural Sciences and Engineering Research Council of Canada (NSERC).

References

1. Kadau K, Germann TC, Lomdahl PS. *Int J Mod Phys C*. 2006;17:1755.
2. Chen S, Doolen GD. *Annu Rev Fluid Mech*. 1998;30:329.
3. Succi S. *The Lattice Boltzmann Equation*. Oxford: Oxford University Press; 2001.
4. Sukop MC. Thorne DT Jr. *Lattice Boltzmann Modeling*. The Netherlands: Springer; 2006.
5. Abe T. *J Comput Phys*. 1997;131:241246.
6. He X, Luo L-S. *Phys Rev E*. 1997;55:6333.
7. Sterling JD, Chen S. *J Comput Phys*. 1996;123:196.
8. Cercignani C. *The Boltzmann Equation and Its Application*. New York: Springer; 1988.
9. Dellar PJ. *J Comput Phys*. 2002;179:95.
10. Bhatnagar P, Gross E, Krook M. *Phys Rev*. 1954;94:511.
11. Wolf-Gladrow DA. *Lattice-Gas Cellular Automata and Lattice Boltzmann Models: An Introduction*. New York: Springer-Verlag; 2000.
12. Buick JM, Greated CA. *Phys Rev E*. 2000;61:5307–20.
13. Latt J. PhD thesis. *Hydrodynamic Limit of Lattice Boltzmann Equations*. Geneva (Switzerland): University of Geneva; 2007.
14. Shan X, Chen H. *Phys Rev E*. 1993;47:1815.
15. Guo Z, Zheng C, Shi B. *Phys Fluids*. 2002;14:2007.
16. Ladd A. *J Fluid Mech*. 1994;271:285.
17. Ladd A. *J Fluid Mech*. 1994;271:311.
18. Le G, Zhang J. *Phys Rev E*. 2009;79:026701.
19. Mei R, Luo L-S, Shyy W. *J Stat Phys*. 1999;155:307.
20. Zhang J, Kwok DY. *Phys Rev E*. 2006;73:047702.
21. Du R, Shi B, Chen X. *Phys Lett A*. 2006;359:564.
22. Shan X, Yuan X, Chen H. *J Fluid Mech*. 2006;550:413.
23. Zhang R, Shan X, Chen H. *Phys Rev E*. 2006;74:046703.

24. Ubertini S, Succi S. *Prog Comput Fluid Dyn.* 2005;5:85.
25. Haeberle S, Zengerle R. *Lab Chip.* 2007;7:1094.
26. Madou M, Zoval J, Jia G, Kido H, Kim J, Kim N. *Annu Rev Biomed Eng.* 2006;8:601.
27. Sbragaglia M, Benzi R, Biferale L, Succi S, Sugiyama K, Toschi F. *Phys Rev E.* 2007;75:026702.
28. Zhang J, Tian F. *Europhys Lett.* 2008;81:66005.
29. Swift M, Orlandini E, Osborn W, Yeomans J. *Phys Rev E.* 1996;54:5041.
30. Zhang J, Li B, Kwok DY. *Phys Rev E.* 2004;69:032602.
31. Chin J, Boek E, Coveney P. *Philos Trans R Soc Lond A.* 2002;360:547.
32. Gunstensen A, Rothman D, Zaleski S, Zanetti G. *Phys Rev A.* 1991;43:4320.
33. Zhang J, Kwok DY. *Eur Phys J Spec Top.* 2009;171:45.
34. Inamuro T, Ogata T, Tajima S, Konishi N. *J Comput Phys.* 2004;198:628.
35. Lee T, Lin C-L. *J Comput Phys.* 2005;206:16.
36. Zheng HW, Shu C, Chew YT. *J Comput Phys.* 2006;218:353.
37. Ginzburg I, Steiner K. *J Comput Phys.* 2003;185:61.
38. Korner C, Thies M, Hofmann T, Thurey N, Rude U. *J Stat Phys.* 2005;121:179.
39. Tian F, Li B, Kwok DY. *Langmuir.* 2005;21:1126.
40. He X, Li N. *Comput Phys Commun.* 2000;129:158.
41. Joshi AS, Grew KN, Peracchio AA, Chiu WKS. *J Power Sources.* 2007;164:631.
42. Shi B, Deng B, Du R, Chen X. Comput Math Appl. 2008;55:1568.
43. He X, Chen S, Doolen GD. *J Comput Phys.* 1998;146:282.
44. Yan Y, Zu Y. *Int J Heat Mass Transf.* 2008;51:2519.
45. Breyiannis G, Valougeorgis D. *Phys Rev E.* 2004;69:065702.
46. Chatterjee D. *Europhys Lett.* 2009;86:14004.
47. Gad-el-Hak M. *J Fluids Eng.* 1999;121:5.
48. Ho CM, Tai YC. *Annu Rev Fluid Mech.* 1998;30:579.
49. Guo Z, Zhao T, Shi Y. *J Appl Phys.* 2006;99:074903.
50. Lim C, Shu C, Niu X, Chew Y. *Phys Fluids.* 2002;14:2299.
51. Nie X, Boolen GD, Chen S. *J Stat Phys.* 2002;107:279.
52. Lee T, Lin CL. *Phys Rev E.* 2005;71:046706.
53. Sofonea V, Sekerka RF. *J Comput Phys.* 2005;207:639.
54. Guo Z, Zheng C, Shi B. *Phys Rev E.* 2008;77:036707.
55. Premnath KN, Abraham J. *Phys Fluids.* 2005;17:122105.
56. Sakakibara B, Inamuro T. *Int J Heat Mass Transf.* 2008;51:3207.
57. Sehgal BR, Nourgaliev RR, Dinh TN. *Prog Nucl Energy.* 1999;34:471.
58. Xi H, Duncan C. *Phys Rev E.* 1999;59:3022.
59. Yang Z, Dinh TN, Nourgaliev RR, Sehgal BR. *Int J Therm Sci.* 2000;39:1.
60. Dupin MM, Halliday I, Care CM. *Phys Rev E.* 2006;73:055701.
61. Wu L, Tsutahara M, Ha LSKM. *Int J Multiphas Flow.* 2008;34:852.
62. Tian F, Kwok DY. *Langmuir.* 2005;21:2192.
63. Wang J, Wang M, Li Z. *J Colloid Interface Sci.* 2006;296:729.
64. Yan YY, Zu YQ, Ren LQ, Li JQ. *Proc Inst Mech Eng C J Mech Eng Sci.* 2007;221:1201.
65. Zhang J, Kwok DY. *J Comput Phys.* 2005;206:150.
66. Guo Z, Zhao T, Shi Y. *J Chem Phys.* 2005;122:144907.
67. Gijs MAM. *Microfluid Nanofluidics.* 2004;1:22.
68. Pamme N. *Lab Chip.* 2006;6:24.
69. Qian S, Bau H. *Mech Res Commun.* 2009;36:10.
70. Sofonea V, Fruh WG. *Eur Phys J B.* 2001;20:141.
71. Sofonea V, Fruh WG, Cristea A. J Magn Magn Mater. 2002;252:144.
72. Falcucci G, Chiatti G, Mohamad SSAA, Kuzmin A. *Phys Rev E.* 2009;79:056706.
73. Clime L, Brassard D, Veres T. *J Appl Phys.* 2009;105:07B517.
74. Calhoun R, Yadav A, Phelan P, Vuppu A, Garcia A, Hayes M. *Lab Chip.* 2006;6:247.
75. Krishnamurthy S, Yadav A, Phelan PE, Calhoun R, Vuppu AK, Hayes AAGMA. *Microfluid Nanofluidics.* 2008;5:33.

76. Lee T, Liu L. *Phys Rev E*. 2008;78:017702.
77. Zhang J, Kwok DY. *Langmuir*. 2004;20:8137.
78. Hyvaluoma J, Koponen A, Raiskinmaki P, Timonen J. *Eur Phys J E*. 2007;23:289.
79. Zhang J, Kwok DY. *Langmuir*. 2006;22:4998.
80. Dupuis A, Yeomans JM. *Pramana*. 2005;64:1019.
81. Zhang J, Kwok DY. *J Colloid Interface Sci*. 2005;282:434.
82. Zhang J, Li B, Kwok DY. *Eur Phys J Spec Top*. 2009;171:73.
83. Lunkad SF, Buwa VV, Nigam KDP. *Chem Eng Sci*. 2007;62:7214.
84. Shi Z, Yan Y, Yang F, Qian Y, Hu G. *J Hydrodyn*. 2008;20:267.
85. Zhang J, Kwok DY. *Phys Rev E*. 2004;70:056701.

Part II

Fabrication and Other Applications

SU-8 Photolithography and
Its Impact on Microfluidics

Rodrigo Martinez-Duarte and Marc J. Madou

CONTENTS

8.1 Introduction

Microfluidics is the science and technology of systems that process or manipulate small amounts (micro- to picoliters) of fluids, using channels with dimensions of tens to hundreds of micrometers (Whitesides, 2006). Microfluidics offers several advantages in a variety of fields including fuel cells (Dyer, 2002; Jankowski et al., 2002; Erdler et al., 2006; Nguyen et al., 2006; Morse, 2007; Kuriyama et al., 2008; Kjeang et al., 2009), forensics (Verpoorte, 2002), clinical diagnostics (Figeys et al., 2000; Mitchell, 2001; Verpoorte, 2002; Andersson et al., 2003; Verpoorte et al., 2003; Chung et al., 2007), biotechnology (Craighead, 2006), and drug discovery (Weigl et al., 2003; Dittrich and Manz, 2006). An insight into the origin, the present, and the future of this exciting field has been presented by Whitesides (2006). Other excellent reviews are those by Verpoorte et al. (2003), Gravesen et al. (1993), Reyes et al. (2002), Auroux et al. (2002), Dittrich et al. (2006), and Vilkner et al. (2004). The physics and scaling laws in fluidics have been detailed by Beebe et al. (2002), Hu et al. (2007), Janasek et al. (2006), Mijatovic et al. (2005), and Stone et al. (2004). A treatise on the different dimensionless numbers in microfluidics has been given by Squires and Quake (2005). Although the first reported implementation of microfluidics dates back to ink-jet printers at IBM in the late 1970s (Bassous et al., 1977; Petersen, 1979) and gas chromatographs at Stanford University in the early 1980s (Terry et al., 1979; Tuckerman et al., 1981; Zdeblick et al., 1986), a wider embrace of the technology did not come until the 1990s. In 1990, the term miniaturized total chemical analysis systems, or µ-TAS, was introduced by Manz and colleagues (1990) to refer to the integration of different laboratory steps into a single device, which they projected would range in size from few millimeters to tens of micrometers. In practice, today fluidic platforms might have individual features in that size range, but the whole system is considerably larger. Over the years, the term µ-TAS has been interchangeably used with the term lab-on-a-chip (LOC), although the latter is mostly used to refer to the use of µ-TAS in the healthcare field. LOC-based platforms are expected to effectively diminish the footprint, complexity, and cost of clinical diagnostics and other health-care-related platforms to enable the replacement of centralized, expensive laboratories by point-of-care, portable instruments. To conduct a total analysis, different capabilities must be incorporated into the chip to allow for sample pretreatment, sample separation, selective isolation, and amplification and for the sensing and detection of relevant phenomena. The advantages of such devices include their small size, improved sensitivity, low sample volume requirements, rapid analysis, potential disposability, and most importantly their ease of use that eliminates the need for skilled personnel to perform the assays. The expected impact of this technology on the improvement of global health promises to be highly significant, especially in developing countries where the lack of medical infrastructure is one of the main causes of high mortality rates (Chin et al., 2007). Examples of commercial µ-TAS include the i-STAT® portable clinical analyzer by Abbott Point of Care, Inc., the Piccolo® Xpress from Abaxis, Inc., and the Apolowako® from Wako Diagnostics, Inc.

Microfluidics today relies on the use of different fabrication materials such as silicon, glass, and polymers (Zhang et al., 2006). The first microfluidics devices were fabricated in silicon mainly because microfabrication techniques and materials at that time greatly depended on those used by the integrated circuit (IC) industry. For the ability to optically monitor chemical and biological assays, borosilicate glasses, Pyrex®, and quartz were soon added to the menu of fabrication materials. The very stable negatively charged surfaces of glasses (silanol groups) also allowed for the implementation of electro-osmotic flow, an electrically induced flow that scales well with miniaturization and offers important advantages over the more traditional flow injection systems (Iverson et al., 2008; Wang et al., 2009). Microfabrication techniques, including wet etching, chemical vapor deposition, and deep reactive ion etching, are commonly used to fabricate silicon-glass microfluidic devices. Significant breakthroughs in gas chromatography, capillary electrophoresis, and other fields have been achieved using these microfluidics devices (Gravesen et al., 1993; Reyes et al., 2002; Jensen, 2006). However, polymers are replacing silicon and glass as the fabrication material of choice for most applications. The wide variety of available polymer compositions, low cost of materials, and relatively inexpensive processing infrastructure enables tailoring the microfluidic substrates/structures to specific applications and allows for disposability (Becker et al., 2000, 2002; Fiorini et al., 2005; Bakajin et al., 2006; Stroock et al., 2006). Commonly used polymers include poly(dimethylsiloxane) (PDMS), poly(methyl methacrylate) (PMMA), polycarbonate, cyclic olefin polymers, polyimide, and epoxies such as SU-8. Other materials include parylene C, Zeonor 1020R® (a polyolefin), and polytetrafluoroethylene (PTFE or Teflon®). Polymer processing techniques include laser ablation, photolithography, hot embossing, casting, and injection. Perhaps the polymer used mostly in microfluidics research is PDMS, an elastomer that is easily cast. PDMS is an optically clear, generally inert, nontoxic, nonflammable, porous elastomer that allows for gas exchange, a useful property for cell culturing. It is a hydrophobic material that can be rendered hydrophilic by plasma oxidation (although for short times, only ~0.5 h). Although PDMS is impermeable to aqueous solvents, organic solvents can penetrate the matrix and cause the material to swell (Mata et al., 2005). PDMS microfluidic devices are commonly fabricated on the basis of soft lithography. The latter techniques involve casting PDMS from a master mold and yield affordable processing with fast turnaround times (Duffy et al., 1998b; Xia et al., 1998; Anderson et al., 2000). Despite all its advantages in a research setting, PDMS devices are not yet viewed as strong candidates for commercial applications. Commercial microfluidic platforms are expected to be made from materials such as polycarbonate or cyclic oleofin and to be fabricated with techniques such as hot embossing, injection molding, and roll-to-roll processing (Madou et al., 2000). A major problem with polymers as opposed to glasses is their rather low and/or unstable negative surface charges reflected in a difficult-reproduce zeta potential and nonuniform electro-osmotic flow. Different groups have tried to remedy this problem, for example, Locascio and colleagues used polyelectrolyte multilayers* (Barker et al., 2000a, 2000b; Liu et al., 2000) and laser ablation (Pugmire et al., 2002; Henry et al., 2002) to modify the surface of different polymers including polystyrene, poly(ethylene terephthalate) glycol, PMMA, polyvinyl chloride, polycarbonate, and PDMS. PDMS is commonly rendered hydrophilic using oxygen plasma, but other available methods for surface modification include ultraviolet (UV)/ozone oxidation (Berdichevsky et al., 2004) and surface coatings, for example, the use of a three-layer (biotinylated IgC, neutravidin, and biotinylated dextran) biotin–neutravidin sandwich (Linder et al., 2001). A review on the zeta potential of different polymer substrates is given by Kirby et al. (2004a, 2004b). Sikanen et al. (2005) characterized the zeta potential on SU-8 surfaces and observed an electro-osmotic flow equal to that for glass microchannels at pH ≥ 4.

Although SU-8 is currently the material of choice to fabricate casting molds for PDMS, the authors believe that SU-8 has not yet been fully exploited as a structural material for microfluidics. For example, although SU-8 offers several advantages for the fabrication of structures of high aspect ratio, its processing parameters require further characterization. In this chapter, we detail the photolithography process to fabricate SU-8 structures with dimensions ranging from millimeters to hundreds of nanometers. In the first section, the reader will get acquainted with the origin and properties of SU-8 as a material and why it is so useful in microfluidics. The section "SU-8, a Versatile Photoresist" details all the fabrication steps in the SU-8 photolithography process and presents optimization tips so that one might be able to better exploit the potential of this versatile material. As previously noted, we emphasize SU-8 photolithography that results in either better molds or permanent structural elements in better-functioning microfluidic devices.

8.2 SU-8, a Versatile Photoresist

To understand the photolithographic process better ,we start detailing the photoresist. The principal components of a photoresist are a polymer (base resin), a sensitizer, and a casting solvent. The polymer changes structure when exposed to electromagnetic radiation, the

---

* Multilayers are created by exposing a surface to alternative solutions of positively and negatively charged polyelectrolytes.

solvent allows for spin-coating uniform layers on a flat substrate, and the sensitizers control the photochemical reactions in the polymeric phase. Photoresists must meet several rigorous requirements: good adhesion, high sensitivity, high contrast, good etching resistance (wet or dry etching), good resolution, easy processing, high purity, long shelf life, minimal solvent use, low cost, and high glass transition temperature, T_g. Most resins, such as novolacs, used as a basis for photoresists are amorphous polymers that exhibit viscous flow with considerable molecular motion of the polymer chain segments at temperatures above the glass transition point. At temperatures below T_g, the motion of the segments is halted, and the polymer behaves as a glass rather than a rubber. If the T_g of a polymer is at or below room temperature, the polymer is considered a rubber; if it lies above room temperature, it is considered to be a glass.

SU-8 is an acid-catalyzed negative photoresist,[*] made by dissolving EPON® SU-8 resin (a registered trademark of Shell Chemical Company) in an organic solvent such as propylene glycol methyl ether acetate (PGMEA), cyclopentanone, or gamma-butyrolactone (GBL) and adding up to 10 wt% of triarylsulfonium hexafluoroantimonate salt as a photoinitiator. Commercial formulations also include 1%–5% propylene carbonate. In a chemically amplified resist such as SU-8, one photon produces a photoproduct that, in turn, causes hundreds of reactions to change the solubility of the film. Because each photolytic reaction results in an "amplification" via catalysis, this concept is dubbed "chemical amplification" (Ito, 1996). The viscosity of the photoresist and hence the range of thicknesses accessible are determined by the ratio of solvent to SU-8 resin. The EPON resist is a multifunctional, highly branched epoxy derivative that consists of bisphenol-A novolac glycidyl ether. On an average, a single molecule contains eight epoxy groups that explain the 8 in the name SU-8 (Figure 8.1). The material has become a major workhorse in miniaturization science because of its low UV absorption (up to thicknesses of 2 mm), high chemical and thermal resistance, and good mechanical properties that make it suitable as a structural material. For example, Abgrall et al. (2007) fabricated SU-8 microfluidic devices with different techniques, including the successive lamination and patterning of SU-8 layers on existent topographies. The use of SU-8 photoresists allows for the coating of thick layers (up to 500 µm) on a single spin coat, or thicker layers in multiple spin coatings, and high-aspect-ratio structures with nearly vertical side walls.

SU-8 cross-linking starts upon the irradiation of the photoresist. In the exposed areas, the photoinitiator decomposes to form hexafluoroantimonic acid that protonates the epoxides on the oligomer. These protonated oxonium ions are, in turn, able to react with neutral epoxides in a series of cross-linking reactions after application of heat. In other words, irradiation generates a low concentration of a strong acid that opens the epoxide rings and acts as a catalyst of the chemically amplified cross-linking process that gets further activated by the application of heat.

On the basis of discoveries in the late 1970s by Crivello and Lam at General Electric (Crivello et al., 1977, 1979, 1980; Crivello, 2000), scientists at IBM discovered that certain photoinitiators, such as onium salts, polymerize low-cost epoxy resins such as EPON-SU-8. Compositions of SU-8 photoresist were patented by IBM as far back as 1989 (Gelorme et al., 1989) and 1992 (Angelo et al., 1992). Originally, SU-8 was intended for printed circuit board and electron beam lithography (EBL), but it is now used in a wide variety of

---

[*] If the photoresist is of the type called *negative* (also *negative tone*), the photochemical reaction strengthens the polymer by random cross-linkage of main chains or pendant side chains, thus becoming less soluble. If the photoresist is of the type called *positive* (also *positive tone*), the photochemical reaction during exposure of a resist weakens the polymer by rupture or scission of the main and side polymer chains, and the exposed resist becomes more soluble in developing solutions. In other words, in negative photoresists light cross-links, whereas in positive ones light scissions.

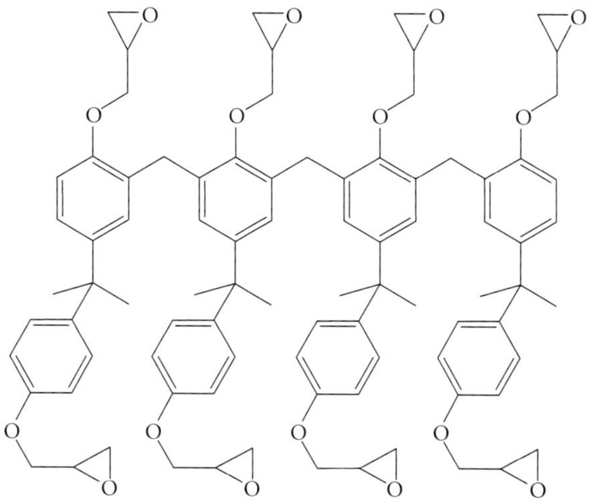

FIGURE 8.1
The SU-8 molecule.

other processes (Lee et al., 1995; Lorenz et al., 1997, 1998; Shaw et al., 1997). In view of the many advantages of SU-8 photoresists over available UV photoresists, including the fabrication of high-aspect-ratio microelectromechanical system (MEMS) features, different formulations of SU-8 photoresists began to be commercialized by MicroChem in 1996. Because of its aromatic functionality and highly cross-linked matrix, the SU-8 resist is thermally stable and chemically very inert. When fully polymerized, it withstands nitric acid, acetone, and even sodium hydroxide (NaOH) at 90°C, and it is more resistant to prolonged plasma etching and is better suited than PMMA as a mold for electroplating (Harris, 1976). The low molecular weight (~7000 ± 1000 Da) and multifunctional nature of an epoxy gives it a high cross-linking propensity, which reduces the solvent-induced swelling typically associated with negative resists. Very fine feature resolution, unprecedented for negative resists, was obtained, and as a result epoxy-based formulations are now used in the fabrication of high-resolution semiconductor devices and nanofluidics. For example, nanochannels can be fabricated using EBL for a variety of applications as suggested by Koller et al. (2009) and Gersborg-Hansen et al. (2006). Low molecular weight characteristics also translate into high contrast and high solubility. Because of its high solubility, very concentrated resist-casting formulations can be prepared. The increased concentration benefits thick film deposition (up to 500 μm in one spin-coating step) and planarization of extreme topographies. The high epoxy content promotes strong SU-8 adhesion to many types of substrates and makes the material highly sensitive to UV exposure. From the microfluidics point of view, strong adhesion to the substrate and chemical inertness of the SU-8 are very desirable. In contrast, the same two properties make resist stripping a very challenging problem for those applications where the resist must be removed such as in the IC industry. Stripping of SU-8 may be carried out with hot NMP (1-methyl-2-pyrrolidone), plasma, laser ablation, or simple burning (Dentinger et al., 2002).

Previously, we listed several benefits of the SU-8 photoresist. Here, we analyze some of the negative aspects such as the thermal mismatch between SU-8 and common substrates such as Si or glass, which produces stress and may cause film-cracking. Also, the absorption spectrum of SU-8 shows much higher absorption coefficients at shorter wavelengths and as a result, lithography, using a broadband light source, tends to result in overexposure at the surface of the resist layer and underexposure at the bottom of the resist layer. The exaggerated negative slope at the top of the resist structure's surface is often called *T-topping*. UV light shorter than 350 nm is strongly absorbed near the surface's creating locally more acid that diffuses sideways along the top surface. Selective filtration of the light source is often used to eliminate these undesirable shorter wavelengths (below 350 nm) and thus obtain better lithography results.

In Table 8.1, we list some of the most important properties of SU-8. Data sources for the properties SU-8 of listed here (and many more) can be found in Chollet (2009) and Guerin (2005).

The appearance of SU-8 is pale yellow to clear, and it has a faint to mild odor. The mechanical properties of cross-linked SU-8 listed in this table include a Young's modulus ranging from 4.02 to 4.95 GPa, a Poisson ratio of 0.22, and a friction coefficient of 0.19. The glass temperature, T_g, of the unexposed resin ranges from 50°C to 55°C but increases to >200°C when it is fully cross-linked. This allows for the use of SU-8 structural elements in microfluidics applications where heating is necessary, such as polymerase chain reaction assays (where the sample temperature must be increased to 98°C). The degradation temperature, T_d, is ~380°C. Other thermal properties listed here include a coefficient of thermal expansion (CTE) of 52×10^{-6}/K and a thermal conductivity of 0.2 W/m K. SU-8, the EPON resin not the photoresist, features a density of 1200 kg/m^3. Cross-linked SU-8 features a refractive index of 1.67–1.8 and a relative dielectric constant of 4–4.5 depending on the frequency. Unexposed SU-8 has a refractive index of 1.668 at 365 nm wavelength. As previously noted, the viscosity of the different commercial photoresists depends on the resin to solvent ratio. Table 8.2 lists the viscosity values, percentage of solids, and

TABLE 8.1

Selected Properties fo SU-8

Appearance	Pale yellow to clear
Odor	Faint to mild
Young's modulus	4.02–4.95 GPa
Poisson ratio	0.22
Friction coefficient	0.19
Glass temperature (T_g)	50°C–55°C, uncross-linked; >200°, cross-linked
Degradation temperature (T_d)	~380°C
Boiling point	204°C
Flash point	100°C
Autoignition temperature	455°C
CTE	52×10^{-6}/K
Thermal conductivity	0.2 W/m K
Specific heat	1500 J/kg K
Vapor pressure	0.3 mmHg at 20°C
Density (of EPON SU-8 resin)	1200 kg/m^3
Refractive index	1.668, uncross-linked; 1.67–1.8, cross-linked
Dielectric constant	4–4.5 ε_0
Electrical breakdown fields	~10^7 V/m
Resistivity	>10^8 Ω cm

TABLE 8.2

Viscosity, Percentage of Solids, and Density of Different SU-8 Photoresist Formulations Available from MicroChem

SU-8 XX	SU-8 (Casting Solvent: Gamma-Butyrolactone)			SU-8 2XXX.X	SU-8 2000 (Casting Solvent: Cyclopentanone)			SU-8 30XX	SU-8 3000 (Casting Solvent: Cyclopentanone)		
	Viscosity (cSt)	% Solid	Density (g/ml)		Viscosity (cSt)	% Solid	Density (g/ml)		Viscosity (cSt)	% Solid	Density (g/ml)
2	45	39.5	1.123	2000.5	2.49	14.3	1.070				
				2002	7.5	29	1.123				
5	290	52	1.164	2005	45	45	1.164	3005	65	50	1.075
				2007	140	52.5	1.175				
10	1050	59	1.187	2010	380	58	1.187	3010	340	60.4	1.106
				2015	1250	63.45	1.2				
25	2500	63	1.200	2025	4500	68.55	1.219	3025	4400	72.3	1.143
				2035	7000	69.95	1.227	3035	7400	74.4	1.147
50	12,250	69	1.219	2050	12,900	71.65	1.233	3050	12,000	75.5	1.153
				2075	22,000	73.45	1.236				
100	51,500	73.5	1.233	2100	45,000	75	1.237				
				2150	80,000	76.75	1.238				

density of all SU-8 photoresist formulations currently (2009) available from MicroChem and is intended to aid the reader in the process of choosing a formulation for his/her micro/nanofluidic application. The characterization of the surface properties of a material, such as electro-osmotic mobility and contact angle with aqueous-based media, is of extreme importance in microfluidics applications. It has been suggested that the SU-8 photoresist series from MicroChem possess similar electro-osmotic mobility properties to those of glass at pH ≥ 4 (Sikanen et al., 2005). However, the authors attribute this property to the photoinitiator contained in this formulations and not to the EPON SU-8 resist itself. The surface of SU-8 is hydrophobic, with a contact angle close to 90°, but it can be rendered hydrophilic by plasma treatment or by chemical modification of the bulk (Wu et al., 2003) or the surface (Nordstrom et al., 2004).

SU-8 photoresist formulations are commercially available from MicroChem (www. microchem.com), in Newton, Massachusetts, and more recently from Gersteltec (www. gersteltec.ch), in Pully, Switzerland. The latter also offers SU-8 with different color dyes, carbon nanotubes, silica, or silver nanoparticles incorporated. MicroChem sells three different series of the product: SU-8 series, using gamma-butyrolactone (GBL) as the casting solvent, and SU-8 2000 and SU-8 3000 series where GBL is replaced by cyclopentanone. Cyclopentanone is a faster drying, more polar solvent system that results in improved coating quality and increased process throughput (Shaw et al., 2003).

Other SU-8 composition changes in the research and development stage include the use of different photoacid generators to reduce internal stress (Ruhmann et al., 2001) and the formation of copolymers with hydrophilic epoxy molecules to render the resist hydrophilic (Wu et al., 2003).

8.3 SU-8: Photolithography

Photolithography refers to the use of light to pattern a substrate. Because of, in part, its heavy use by the IC industry, UV photolithography is the most widely used form of lithography. Other resist-patterning techniques include x-ray, electron, and ion-lithography as well as soft lithography, nanoimprinting, and so forth. SU-8 photolithography generally involves a set of basic processing steps: photoresist deposition, soft bake, exposure, postexposure treatment, and developing. Descumming and postbaking might also be part of the process. A detailed resume of all the possible SU-8 photolithography steps is presented in Table 8.3. When patterned at 365 nm, the wavelength at which the photoresist is the most sensitive, total absorption of the incident light in SU-8 is achieved at a depth of 2 mm. In principle, resist layers up to 2 mm thick can be structured (Bertsch et al., 1999a). Yang and Wang recently confirmed this astounding potential experimentally (Yang et al., 2005) by fabricating structures with aspect ratios more than 190 (for features with a thickness of 6 μm and a height of 1150 μm). Aspect ratios greater than 10 are routinely achieved with SU-8. Aspect ratios up to 40 for lines and trenches have been demonstrated in SU-8-based contact lithography (Lorentz et al., 1998; Ling et al., 2000; Williams et al., 2004b).

For a more thorough review of the fundamentals of photolithography and its use to pattern other resists, the reader is referred to Madou (2009). The reader is also encouraged to consult other recent SU-8 reviews by del Campo et al. (2007) and Abgrall et al. (2007). Table 8.3 details the process of SU-8 patterning.

TABLE 8.3

SU-8 Photolithography Processing Steps

Process Step	Process Description	Processing Parameters	Remarks
Photoresist deposition Un-cross-linked SU-8 Substrate	Photoresist is deposited on a clean substrate. Substrate materials include glass, Si, quartz, polymers, etc. Different deposition techniques exist but spin coating is the most common	Spin time, speed, and acceleration when using spin coating. Layer thickness is inversely proportional to spin time, speed, and acceleration	• Layer thicknesses from a few micrometers to hundreds of micrometers (~500 μm) are possible in a single coat • Long spin times at a given speed yield better layer uniformity • Accumulation of resist on the substrate edges during spin coating causes ridges (known as edge beads), which can be removed with commercial solutions or acetone • Lamination and casting are alternatives to spin coating to deposit layers thicker than 500 μm
Soft bake or prebake Solvent evaporation	Casting solvent is evaporated from the photoresist. SU-8 does not flow at room temperature after a proper soft bake. The glass transition temperature of SU-8 resist at this point, in the uncross-linked state, is 55°C	Temperature and time. Soft bake temperature is usually 95°C. Times can be as short as a couple of minutes (layers <5 μm thick) or as long as hours (layers >200 μm thick)	• The use of a hot plate is recommended to avoid *skin effects*. Baking the photoresist layer in convection ovens evaporates the solvent present on the top surface of the photoresist first, hardens the surface, and hinders the evaporation of solvent from the bulk of the photoresist • Elevated temperatures (>120°C) during soft bake can activate resist polymerization and reduce contrast • A sufficient amount of residual solvent in the soft-baked resist allows the polymer matrix to relax more and thus minimizes the residual stress, which otherwise can cause pattern debonding from the substrate at the end of the photolithography process
Exposure Energy Irradiation Photo mask	Irradiation from an energy source generates a low concentration of a strong acid that opens the epoxide rings of the resist and acts as a catalyst of the chemically amplified cross-linking reaction	Exposure dose. This value is also known as energy dose and is given by the product of the source power intensity and the exposure time. In practice, the power intensity of the source is usually fixed, and the energy dose can be varied by changing the exposure time	• Energy sources include extreme, deep, and near-UV lights, x-rays, and ion and electron beams • SU-8 is commonly exposed using light in the near-UV range [including the i-line (365 nm) and g-line (435 nm) of a mercury lamp]. In this case, the use of a filter to eliminate light wavelengths below 360 nm is recommended to minimize *T-topping* or an exaggerated negative slope of the structure walls • SU-8 is usually exposed through a 1:1 photomask in a contact or proximity exposure setup • Multilevel topographies can be fabricated in one exposure step using GTMs or software masks • Photoresist can be exposed through the substrate (known as back-side exposure or back exposure) if the substrate has a low UV-absorption coefficient. This approach yields structure walls with shallow, positive slopes that prove ideal in molding applications

continued

TABLE 8.3 (continued)

SU-8 Photolithography Processing Steps

Process Step	Process Description	Processing Parameters	Remarks
Postexposure bake Cross-linked SU-8	The cross-linking process activated during exposure gets further activated by the application of heat. SU-8 is fully polymerized after a proper postexposure bake. The glass transition temperature of cross-linked SU-8 is more than 200°C	Temperature and time. More than time, PEB temperature is of crucial importance. A two-step PEB (at 65°C and 95°C for different times) is recommended by MicroChem but temperatures between 55°C and 75°C and extended baking times have been suggested to minimize internal stresses in the pattern	• The precise control of PEB times and temperatures critically determines the quality of the final features • Insufficient PEB times and/or temperatures that are too low yield structures that are not completely cross-linked and can be attacked by the developer • Temperatures below 55°C do not polymerize the matrix completely regardless of the bake time • Thermal stresses increase with resist thickness and can cause cracking of the pattern's surface, structure bending, or complete peeling from the substrate in the worst cases • The use of a substrate material with a CTE similar to that of SU-8 can minimize thermal stresses in the pattern
Development	Unpolymerized SU-8 dissolves upon immersion in a developer agent such as PGMEA	Time and agitation rate. Development times range from few minutes (for layers thinner than 10 μm) to hours (for layers more than 400 μm thick). Agitation is usually done manually. In this case, an approximate value for the agitation rate is 60 Hz	• Constant agitation during development is recommended to constantly feed fresh developer to the resist pattern and decrease developing times • Care must be taken when developing high-aspect-ratio structures, as excessive agitation can cause mechanical breakage • Layers thinner than 2 μm are recommended to be developed by rinsing with PGMEA instead of batch immersion in the developer • Megasonic cleaning systems can be useful to develop densely packed high-aspect-ratio structures. Megasonic frequencies are higher than ultrasonic ones and offer a less violent cavitation that prevents pattern erosion and debonding from the substrate
Drying	This step removes the developer. Drying methods include nitrogen blowing, spinning, freeze-drying, and supercritical drying	Surface tension of the liquid where the SU-8 pattern is immersed before drying. The replacement of PGMEA by a liquid with lower surface tension is recommended for the prevention of structure collapse during the resist-drying process	• Isopropyl alcohol is the most commonly used liquid to replace the developer • Stiction forces developed in narrow gaps can be strong enough to bend and join SU-8 patterns together during drying, even when using isopropyl alcohol • As the gap between high-aspect-ratio structure decreases below a few micrometers, freeze-drying and supercritical drying methods must be used

Step	Description	Parameters	Comments
Descumming	A mild oxygen plasma is performed to remove unwanted resist left behind after drying. This step is not necessary for all applications and can be omitted	Oxygen pressure, polarizing voltage, and plasma exposure time. Short exposure (~1 min) of the sample to oxygen plasma created at an oxygen pressure of 200 mTorr and 200 W polarizing voltage is recommended	• SU-8 leaves a thin polymer film at the resist–substrate interface and on the pattern surface that is most likely due to the solid saturation of the employed developer • An alternative to oxygen plasma is a short dip immersion in acetone. SU-8 withstands acetone when fully polymerized; otherwise, the surface of the pattern will noticeably degrade • The use of a liquid phase during descumming requires a further drying step
Hard bake	This last bake step removes any casting solvent residue from the polymer matrix and anneals the film. This step is not necessary for all applications and can be omitted	Temperature and time. Hard baking temperatures frequently range from 120°C to 150°C. Baking times are usually longer than those used in the soft bake step	• Hard baking improves the hardness of the film • Improved hardness increases the resistance of the resist to subsequent etching steps • Fully cross-linked SU-8 has a glass transition temperature higher than 200°C, which makes resist reflow practically impossible
Removal	Removal of SU-8 from substrate. Necessary if SU-8 is used as a sacrificial material or if the substrate is to be reused	Structure thickness, amount of solvent present in the polymer matrix	• When using the photoresist as a sacrificial material, the primary consideration is complete removal of the photoresist without damaging the device under construction • A hard bake step is not recommended if SU-8 is intended to be removed • Dry stripping using oxygen plasma, or ashing, is recommended for a cleaner, environmentally friendly process

8.4 Resist Profiles—An Overview

Manipulation of resist profiles is one of the most important concerns of a lithography engineer. Depending on the final objective, one of the three resist profiles shown in Figure 8.2 is attempted. A reentrant, an undercut, or a reverse resist profile (resist sidewall >90°) is required for metal lift-off. Some authors confusingly call slopes >90° *overcut* (Moreau, 1987); most, including these authors, refer to this type of resist profile as an *undercut*. Shallow resist angles (<90°) enable continuous deposition of thin films over the resist sidewalls. A vertical (90° resist sidewall angle) slope is desirable when the resist is intended to act as a permanent structural element such as in microfludics and molding applications. For more details, refer to Madou (2009). A vertical slope is desired in most microfluidics devices to avoid a flow velocity gradient (other than the one introduced by possible surface effects) along the height of the channel. Furthermore, a rectangular channel cross section is easier to model than a trapezoidal one.

8.5 Choice of Substrate

Traditionally, SU-8 photolithography is conducted on rigid substrates such as silicon, quartz, and glass. In these cases, the CTE of the substrate is significantly different from

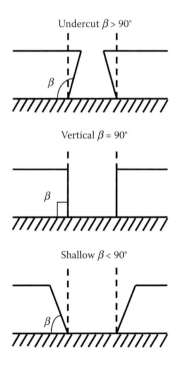

FIGURE 8.2
The three important resist profiles. A reentrant, undercut, or a reverse resist profile (resist sidewall $\beta > 90°$) is required for metal liftoff. A vertical ($\beta = 75°$–$90°$ resist sidewall angle) slope is desirable for a perfect fidelity transfer of the image on the mask to the resist. Shallow resist angles ($45° < \beta < 90°$) enable continuous deposition of thin films over the resist sidewalls.

that of SU-8 and results in patterns with built-in stresses that may cause film cracking and distorted sidewalls. Moreover, silicon, glass, and quartz substrates are expensive and lead to a fragile, brittle microfluidics device. An inexpensive, robust material with a CTE that is close to that of SU-8 is more desirable. In Table 8.4, we list different substrate materials and their properties together with the pertinent SU-8 attributes. Polymers offer a viable option to replace silicon and glass as the substrate material. Polyetheretherketone (PEEK) and PMMA most closely resemble SU-8, with PEEK exhibiting a higher operating temperature and improved solvent resistance than PMMA (Song et al., 2004). Other viable materials from Table 8.4 include polycarbonate, polyester, and polyimide. The use of flexible film rather than rigid disk substrates allows for peeling or release of the substrate from the SU-8 patterns. Using either a 70-µm-thick PET or a 127-µm-thick polyimide film and the peeling approach, our research team obtained less-stressed freestanding SU-8 parts (Martinez-Duarte, 2009; Martinez-Duarte et al., 2009). Thick photoresist layers (>50 µm) prove to be less brittle than thinner layers and facilitate their clean release from the film. The recommended thickness for the substrate film ranges from 50 to 130 µm. Too thin of a film tears easily and is difficult to manipulate. Films that are too thick are not as flexible and can induce sufficient stress on the structures during peeling, which causes their mechanical failure. Polymer films can be purchased in roll-to-roll sheets, and this may

TABLE 8.4

Properties of Materials Used as Substrates Compared with Those of SU-8

Material	(CTE) $(10^{-6}/K)$	Thermal Conductivity at 293 K (W/m K) (Silver = 429)	Density (g/cm^3)	Glass Temperature or Melting Point
Silicon (Hull, 1999)	2.6–4.442	1.56	2.33	1414°C (1687 K)
Silicon oxide film (MEMS Exchange, 2009)	0.55	1.1	2.2	1700°C
Glass (Giancoli, 1998)	0.52	1.1	2.4–2.8	550°C–600°C
Polycarbonate (Boedeker Plastics, 2009b)	70.2	0.2	1.3	145°C
Polyimide (PI), Kapton® (Kapton Polyimide Film, 2009 and Cirlex, 2009)	20–40	0.12	1.42	360°C–410°C (633–683 K)
Polyester (PET), Mylar® (Mylar Technical Information, 2009)	17.1	0.37	1.390	254°C (527 K)
PTFE (Teflon) (FLUOROTHERM Polymers, Inc., 2009)	99	2.94	2.2	327°C
PEEK (Boedeker Plastics, 2009a)	58	0.25	1.32	249°C
PDMS (2009)	310	0.15	0.965	–125°C (uncross-linked); ~100°C (cross-linked)
PMMA (Tangram Technology Ltd., 2005)	55	0.2	1.19	65°C for standard PMMA, 100°C for heat-stabilized types
SU-8 (Chollet, 2009)	50–52	0.2	1.2	50°C–55°C (323–328 K) not cross-linked; >200°C (>473 K) when cross-linked

open an avenue to continuous lithography of SU-8 patterns as desired (e.g., to fabricate disposable microfluidics devices for diagnostics or inexpensive molds).

Previous work on the use of polyester (Abgrall et al., 2006) or Kapton (Feng et al., 2003; Agirregabiria et al., 2005) films carrying SU-8 patterns still relied on an underling silicon or glass substrate to provide the needed rigidity for the successful application of the photoresist on these polymer films. In a novel approach, Martinez-Duarte (2009) uses aluminum support rings to hold and tighten polymer support films (drum-skin-like) to provide a rigid, planar surface ready for spin coating (see Figure 8.3). SU-8 has been patterned on both polyester and Kapton (2009) films in this manner. SU-8 processing on rigid PMMA, PTFE, PEEK, and polycarbonate disks has also been demonstrated (Song et al., 2004). For SU-8 processing on a given substrate, the chemical resistance of the substrate material to acids, bases, and solvents used in a particular SU-8 process must be investigated (e.g., when using polycarbonate and PMMA substrates, acetone should be avoided).

For many applications, the optical properties of the substrate must also be optimized. Materials with low-UV light absorption coefficient, such as glass and quartz, enable the exposure of the photoresist through the substrate, a technique referred to as back-side exposure or back exposure. Some polymers, polyester, for example, also allow for backside photopatterning. The use of back-side exposure in SU-8 photolithography yields SU-8 structures with a slight inward taper (shallow angle, <90°) that is highly advantageous in molding and other applications. For example, thick, high-aspect-ratio SU-8 molds with positive wall slopes (<90°) were fabricated by Martinez-Duarte (2009) using a transparent polyester film as substrate and by Peterman et al. (2003) using a Schott Borofloat® glass wafer. Kim et al. (2004) implemented backside SU-8 photolithography with a Pyrex glass substrate to fabricate tapered SU-8 pillars and later used them as templates to fabricate metallic microneedles for drug delivery. Sato et al. (2004a) and Yoon et al. (2006) combined inclined exposure (to be detailed below) and back exposure to pattern a micromesh to filter particles flowing in a microchannel. Another example is that presented by Lü et al. (2007) who used back-side exposure to fabricate SU-8 templates that were later used to electroplate nickel parts.

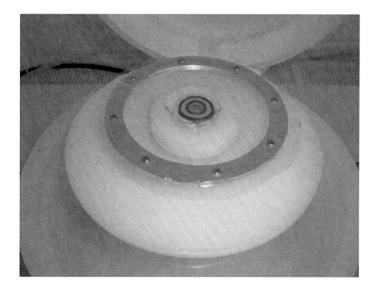

FIGURE 8.3
The use of a stretched polyester film on an aluminum ring as a substrate for SU-8 spin coating. Kapton film was also demonstrated successfully for the same purpose.

8.6 The Clean Room

To ensure industrially accepted yields, micro- and nanofabrication processes should be carried out in a *clean room*, a specially designed area where the size and the number of airborne particulates are highly controlled, together with the temperature ($\pm 0.1°F$), air pressure, humidity (from 0.5% to 5% RH), vibration, and lighting (yellow to avoid resist activation). Clean rooms are classified on the basis of the number and size of particles allowed per unit volume of air. The official international standard for clean room classification is ISO 14644–1 (from the International Organization for Standardization), which specifies the decimal logarithm of the maximum number of particles 0.1 μm or larger that are permitted per cubic meter of air. For example, in an ISO 1 clean room, the particle count must not exceed ten 0.1-μm particles (or larger) per cubic meter of air, whereas in an ISO 6 clean room, the 0.1-μm particle count per cubic meter of air can be as high as one million. Although the standard FED-STD-209E, using a cubic foot as the unit volume of air (a cubic meter roughly equals 35 cubic feet), is still widely used today, it was officially cancelled by the General Services Administration of the US Department of Commerce on November 29, 2001, and was superseded by ISO 14644–1. ISO-14644 is a series of nine documents and includes additional clean room requirements that FED-STD-209E does not have, such as design and construction, operations, separation enclosures, molecular contamination, and surface contamination. Table 8.5 illustrates the different ISO classes and their FED-STD-209E equivalent.

8.7 Substrate Cleaning and Priming

Substrate cleaning is the first and a very important step in any lithographic process, as the adhesion of the resist to the substrate could be severely compromised by the presence of impurities and residual coatings. Poor adhesion can result in leaks at the substrate/ SU-8 interface and compromise the performance of the microfluidic device. Contaminants

TABLE 8.5

ISO 14644–1 Clean Room Classification and FED STD 209E Equivalents

	Maximum Number of Particles per Cubic Meter						FED-STD-209E
Class	≥0.1 μm	≥0.2 μm	≥0.3 μm	≥0.5 μm	≥1 μm	≥5 μm	Equivalent
ISO 1	10	2					
ISO 2	100	24	10	4			
ISO 3	1000	237	102	35	8		Class 1
ISO 4	10,000	2370	1020	352	83		Class 10
ISO 5	100,000	23,700	10,200	3520	832	29	Class 100
ISO 6	1,000,000	237,700	102,000	35,200	8320	293	Class 1000
ISO 7				352,000	83,200	2930	Class 10,000
ISO 8				3,520,000	832,000	29,300	Class 100,000
ISO 9				35,200,000	8,320,000	293,000	Room air

TABLE 8.6

Some Common Clean Room Contaminant Sources

Location: a clean room near a refinery, smoke stack, sewage plant, and cement plant spells big trouble
Construction: the floor is an important source of contamination. Also, items such as light fixtures must be sealed, and room construction tolerances must be held very tight
Wafer handling: transfer box
Process equipment: never use fiber glass duct liner; always use 100% polyester filters, and eliminate all nonessential equipment
Chemicals: residual photoresist or organic coatings, metal corrosion
Attire: only proper attire and dressing in the anteroom
Electrostatic charge: clean room must have a conductive floor
Furniture: only clean room furniture
Stationary: use ballpoint pen instead of lead pencil; use only approved clean room paper
Operator: no eating, drinking, smoking, chewing gum, or makeup of any kind

include solvent stains (methyl alcohol, acetone, trichloroethylene, isopropyl alcohol, xylene, etc.) and airborne dust particles from operators, equipment, smoke, and so forth (see Table 8.6). As previously noted, SU-8 photolithography makes use of different kinds of substrates, and thus a cleaning procedure must be chosen accordingly. Wet immersion cleaning carried out using diluted hydrofluoric acid, Piranha (a mix of sulfuric acid and hydrogen peroxide at different ratios including 5:1 and 3:1. Note that this mixture is exothermic. When cool, it may be refreshed by very slowly adding more hydrogen peroxide), and RCA (a process developed by W. Kern that involves ammonium hydroxide, hydrogen peroxide, DI water, and hydrofluoric and hydrochloric acids at different stages*) can only be used with silicon or glass substrates. Milder, but not as effective, procedures such as DI water rinsing followed by solvent rinse can be used for the cleaning of certain polymer substrates. Other cleaning methods for substrates include ultrasonic agitation, polishing with abrasive compounds, and supercritical cleaning in which the liquid-solvative properties, gas-like diffusion, and viscosity of critical fluids, CO_2, for example, enable rapid penetration into crevices for a complete removal of organic and inorganic contaminants contained therein (McHardy et al., 1998; King et al., 2003). Gaseous cleaning methods include vapor cleaning, thermal treatment, (e.g., baking the substrate at 1000°C in vacuum or in oxygen), and plasma or glow discharge techniques (e.g., in Freon® with or without oxygen). In general, vapor phase cleaning methods use significantly less chemicals than wet immersion cleaning and because the US Environmental Protection Agency regulations are becoming the preferred methods. In the case of wet immersion cleaning, dehydration before resist deposition is recommended as humidity plays a crucial role in the adhesion of SU-8 to the substrate.

Poor adhesion of SU-8 photoresist to a given substrate is also due to partial wetting of the substrate surface and can occur even if a clean substrate is used. Modification of the substrate surface chemistry can significantly improve the adhesion of a photoresist to a substrate. Common adhesion promoters such as hexamethyldisilazane, AP300, and methacryloxy [propyl] trimethoxysilane, effectively prime the substrate surface for photoresist deposition. MicroChem's Omni-coat and Gersteltec's adhesion promoter are two choices optimized to improve SU-8 adhesion to the substrate. Adhesion promoters are not

---

* The prevalent RCA1 and RCA2 wet cleaning procedures are as follows: RCA1: add one part of NH_3 (25% aqueous solution) to five parts of DI water and heat to boiling and add one part of H_2O_2. Immerse the substrate for 10 min. This procedure removes organic dirt (resist). RCA2: add one part of HCl to six parts of DI water and heat to boiling and add one part of H_2O_2. Immerse the substrate for 10 min. This procedure removes metal ions.

always necessary, and the reader, to save cost and time, is encouraged to pattern SU-8 on the bare substrate first and characterize the results. SU-8 exhibits good adhesion to Si and Si/SiO$_2$ but poor adhesion to glass substrates. From the metals commonly used in microfabrication, Al offers the strongest adhesion for SU-8 followed by Ti and Au. The quality of adhesion of SU-8 to Cu and Cr has been reported to depend inversely on feature size, with smaller features exhibiting the best performance. From the metals investigated, Ni exhibits the lowest adhesion strength (Dai et al., 2005; Nordström et al., 2005; Barber et al., 2007). The choice of the casting solvent in a photoresist composition is also important. For example, the use of cyclopentanone instead of gamma butyrolactone produces a photoresist composition with lower surface tension that improves SU-8 spreading characteristics (Shaw et al., 2003).

8.8 Photoresist Deposition

Spin coating is the current method of choice to deposit SU-8 photoresist. A uniform layer ranging in thickness from a few to hundreds of micrometers can be deposited by simply tuning spin profile parameters such as time, speed, and acceleration and using an appropriate SU-8 photoresist formulation (thicker layers benefit from more viscous formulations). Successive spin coating and baking steps of individual SU-8 layers lead to a total thickness of up to 2 mm. Spin coating features three steps: dispensing, spreading, and coating. In the first step, the photoresist is dispensed onto a substrate held in place by a vacuum-actuated chuck in a resist spinner (Metz et al., 1992). Dispensing too much resist onto the substrate results in edge covering or run-out; too little resist may leave uncovered areas. The resist must be dispensed at the center of rotation of the substrate. Dispensing resist on an area other than the center of rotation leads to an imbalance on the centripetal forces acting on the resist puddle during spinning that causes nonuniform coatings. The shape of the substrate must also be taken into account to guarantee a uniform coating. Circular substrates benefit from dispensing in one single spot such that the resist puddle takes in a circular shape after dispensing. Rectangular substrates benefit instead from dispensing in several continuous spots (think of a line). A time after dispensing and before spreading is recommended for the photoresist to relax and eliminate air bubbles introduced during dispensing. Low rotation speeds (<100 rpm) during resist dispensing may also prove beneficial. The second step, spreading, uses a rotating speed of approximately 500 rpm for 10–20 s to spread the fluid over the substrate. Acceleration for this step is recommended to be approximately 100 rpm/s. After spreading the photoresist, higher rotating speeds are used to thin down the fluid near to its final desired thickness and completely coat the substrate. Typical values for this step include spin speeds in the range of 1000 to 4000 rpm, spin acceleration between 200 and 500 rpm/s, and spinning times from a few seconds to several minutes. The reader is referred to either MicroChem's or Gersteltec's processing data sheets for guidelines on how to optimize spin parameters and obtain a specific layer thickness, given a photoresist formulation. In general, the last number(s) of the resist formulation name (see Table 8.2) depicts the layer thickness that is obtained when spinning at 3000 rpm for 30 s. For example, the spin coating of MicroChem's SU-8 2010 at 3000 rpm for 30 s yields a 10-μm-thick layer; the use of SU-8 100 under the same parameters yields a 100-μm-thick layer. Layer uniformity is improved by using prolonged spin times (Chen et al., 2001), but this decreases the layer thickness. The final layer thickness is inversely proportional to spin speed, acceleration, and time but directly proportional to solution concentration and molecular weight (measured by intrinsic viscosity) of the photoresist. Inherent to spin coating is the formation of edge beads or photoresist ridges that can be approximately 10 times higher than the mean thickness of the rest of the substrate. This introduces air gaps between the mask and the surface of the photoresist that can lead to pattern broadening because of light diffraction and nonuniform exposure throughout the substrate. For SU-8 formulations with low viscosity (<7000 cSt), the edge bead is likely to disappear because of reflow of the photoresist after spin coating. For thicker formulations, commercial edge bead removal solutions such as MicroChem's edge bead removal solution can be used. Other alternatives include GBL and acetone. Edge bead removal is commonly integrated in the spin-coating process and is carried out immediately after the coating of the photoresist layer.

An empirical expression to predict the thickness of the spin-coated film as a function of its molecular weight and solution concentration has been given by Thompson (Thompson, 1994). For theoretical analysis and modeling of the rheology of the spin-coating process, the reader is referred to the works of Schwartz et al. (2004), Acrivos et al. (1960), Emslie et al. (1959), and Washo (1977).

Optimization of the photoresist coating process in terms of resist dispense rate, dispense volume, and spin speed presents a growing challenge. SU-8 photoresists currently average US$ 1000 per liter (2009) and make processing optimization necessary. The need for an alternative photoresist deposition technique arises as the amount of waste material generated by spin coating is high, with most of the resist solution (>95%) thrown off the substrate during the spin casting process (the waste resist must be disposed as a toxic material). For example, in spray coating, the substrates to be coated pass under a spray of photoresist solution. The spray system includes an ultrasonic spray nozzle that generates a distribution of droplets in the micrometer range.

Spray coating does not suffer from resist thickness variation caused by the centrifugal force because the resist droplets are supposed to stay where they are being deposited. Because of the same reason, the amount of photoresist wasted during spray coating is significantly less than that with spin coating. In a spray coating system, the resist is pushed out of a pressurized tank via a supply pipe to the spray head that has a defined aperture where the spray mist is formed. The shape of the spray pattern can be adjusted by a secondary air cushion, and undesired overspray can be minimized by keeping the spray gun close to the substrate (~5 cm). A major advantage of this technique is its ability to coat uniformly over nonuniform surfaces, making the technique appropriate for substrates with three-dimensional topographies such as those used in MEMS and microfluidics. Importantly, sprayed coatings do not have the internal stress forces that are common in spin-coated films. The process can be automated and may coat substrates double-sided. The disadvantages of spray coating include its inability to control the thickness of the deposited film as precisely as spin coating and that it is difficult to deposit layers thicker than 20 µm.

Other photoresist deposition methods include electrostatic spraying, a variant of spray coating, and roller, curtain, extrusion, dip, and meniscus coating. As with spray coating, these techniques are not as efficient when depositing layers thicker than tens of micrometers. Alternatives for the deposition of thick layers include casting (Lin et al., 2002), silkscreen printing, and lamination (Abgrall et al., 2008). Unfortunately, SU-8 dry laminated sheets are not commercially available yet and must be fabricated in-house. For more details on alternative techniques to spin coating for photoresist deposition, the reader is referred to Madou (2009).

8.9 Soft Baking or Prebaking

After coating, the photoresist still contains 4–24 wt% solvent, depending on the thickness of the layer. Layers thinner than 20 μm contain less than 10% solvent, whereas thicker layers, 100 μm or greater, can contain up to 24% solvent. The photoresist must be soft baked (also known as preexposure bake or prebake) for a given time in an oven or on a hot plate at temperatures ranging from 65°C to 100°C (Feng et al., 2003; Anhoj et al., 2006) to remove solvents and stress and to promote adhesion of the resist layer to the substrate. SU-8 soft bake is usually conducted at 95°C. Hot-plating the resist is faster, more controllable, and does not trap solvent as convection oven baking does. In convection ovens, the solvent at the top surface of the resist is evaporated first, and this can cause an impermeable resist skin, trapping the remaining solvent inside. Soft bake was considered a trivial step in the SU-8 photolithography process until Zhang et al. (2001) demonstrated a substantial increase in patterning resolution and overall device yield. Excessive or extended baking leads to a decrease in material hardness and an increase in crack density. The presence of cracks in the SU-8 surface introduces crevices that can distort the flow field or act as random physical traps for cells, DNA, and other objects of interest. Elevated soft bake temperatures (>120°C) result in thermal activation of the photoinitiator and reduced contrast, leading to line broadening and poor resolution. It has been suggested that the presence of a certain amount of solvent in the resist after the prebake increases the effective concentration of the polymerization catalyst during exposure, leading to improved sensitivity and yielding a better cross-linked (stronger) material (Anhoj et al., 2006). A sufficient amount of residual solvent in the soft-baked resist also allows the polymer matrix to relax more and thus minimizes the residual stress that otherwise can cause delamination of the final resist structures. However, excessive solvent in the polymer matrix can lead to bubbles and cause a stress gradient to form during postexposure baking. Solvent pockets in the final SU-8 structure create weak points in the matrix that compromise the mechanical integrity of the pattern (more details are presented below under hard bake).

The rate of solvent loss is determined by the solvent diffusion coefficient. This number increases with temperature and has been suggested to decrease exponentially with the amount of solvent present in the polymer–solvent system (Vrentas et al., 1977a, 1977b). Experimental data suggest that the bulk of th solvent contained in the photoresist evaporates within the first 5 min of the bake (Shaw et al., 2003). Because the amount of solvent in the resist exponentially decreases with baking time, the solvent evaporation rate also reduces exponentially as soft bake time increases. The difference in the evaporation rates of different solvents, for example, between gamma-butyrolactone and cyclopentanone, tends to disappear as the thickness of the layer increases (Shaw et al., 2003). Thick resists may benefit from a longer bake time to completely remove the solvent.

The resist must be allowed a period of relaxation after the soft bake and before exposure for the resist and substrate to cool down to room temperature. A disagreement on the optimal relaxation time exists; some authors suggest relaxation times of hours (e.g., Anhoj et al., 2006), whereas others, including the current authors, suggest a few minutes (<5 min) [e.g., Williams et al. (2004b) and our own team's practice—not published].

8.10 Exposure

After soft baking, the resist-coated substrates are transferred to an illumination or exposure system where they are aligned with the features on a photomask. For any lithographic technique to be of value, it must provide a very precise alignment technique capable of aligning the mask and substrate within a small fraction of the minimum feature size of the devices under construction. In the case of SU-8 photolithography, exposure is normally conducted in ontact or proximity systems that consist of a UV lamp illuminating the resist-coated substrate through a mask without any lenses between the two.* The purpose of the illumination systems is to deliver light with the proper intensity, directionality, spectral characteristics, and uniformity across the substrate, allowing a nearly perfect transfer or printing of the mask image onto the resist in the form of a latent image. The incident light intensity (in W/cm²) multiplied by the exposure time (in seconds) gives the incident energy (J/cm²) or dose, D, across the surface of the resist film. Radiation induces the generation of a strong acid that initiates polymerization in the exposed areas. The smaller the dose needed to transfer the mask features onto the resist layer with good resolution, the better the sensitivity. Pattern geometry, including area and thickness, has a strong influence on the required dose. Large-area thick patterns, commonly required in microfluidics, require a higher exposure dose than small-area shallow patterns.

The absolute minimum feature size in a miniature device, whether it involves a channel width, spacing between features, or contact dimension, is called the *critical dimension*. A critical dimension defines the overall resolution of a process, that is, the consistent ability to print a minimum size image, under conditions of reasonable manufacturing variation (Wolf and Tauber, 2000). Many aspects of the process, including hardware, photoresist, and processing considerations, can limit the resolution of photolithography. Hardware limitations that include diffraction and reflection of light (or scattering of charged particles in the case of charged-particle lithography or hard x-rays), lens aberrations, and mechanical stability of the system (vibrations) must all be minimized. The resist material properties that impact resolution include sensitivity, contrast, and energy absorption at different wavelengths.

Light wavelengths used in SU-8 photolithography range from the very short of extreme UV (10–14 nm) to deep UV (150–300 nm) to near UV (350–500 nm). Other energy sources amenable to induce SU-8 cross-linking have shorter wavelengths yet and include x-ray (0.01–10 nm) and electron (0.12 nm to 7 pm in the energy range 100 eV to 30 keV) and ion (2.8–0.16 pm also in the energy range 100 eV to 30 keV) beams. X-ray techniques enable the fabrication of SU-8 structures with very high aspect ratios (Bogdanov et al., 2000; Shew et al., 2004; Barber et al., 2005; Tan et al., 2006; Reznikova et al., 2008), whereas nanometer-sized features have been fabricated with EBL and ion-beam lithography[†] (van Kan et al., 1999; Aktary et al., 2003; Nallani et al., 2003; Pépin et al., 2004; Bilenberg et al., 2006; Robinson et al., 2006). It was shown that SU-8 is more sensitive to x-ray and electron beams than PMMA, the traditional material of choice.

Here, we focus on SU-8 patterning with light wavelengths in the near-UV range where one typically uses the g-line (435 nm) or i-line (365 nm) of a broadband mercury lamp. SU-8 photoresists feature a high absorption of energy of wavelengths that are less than 350 nm. One of the main reasons for this is that triarylsulfonium hexafluoroantimonate

---

* In contrast to projection lithography where the pattern is imaged onto the resist through a lens.
[†] Ion beams undergo less scattering than electron beams and achieve higher resolution.

salt, used as photoacid generator in common SU-8 photoresist formulations, has absorption bands at 231.5, 268.5, and 276 nm, as measured in methanol solution (Crivello et al., 1979). When SU-8 is exposed using a broadband mercury lamp, UV light shorter than 350 nm is strongly absorbed at the photoresist's top surface. This causes the creation of more acid that diffuses sideways and polymerizes a thin layer along the top surface of the resist film. This effect is commonly known as *T-topping* and is illustrated in Figure 8.4. Selective filtration of the light from the light source is required to eliminate the undesirable shorter wavelengths and to obtain a vertical wall that is desired in most microchannels. An easy and affordable way to implement a filter is by using a 50- to 100-μm layer of SU-8 placed in between the light source and the mask. Commercial high-pass filters with a cutout wavelength of 360 nm are also available. For example, Reznikova et al. (2005) used a 100-μm-thick SU-8 resist layer to filter exposure radiation at 334 nm, and Lee et al. (2003) reported using a commercial Hoya UV-34 filter to eliminate the T-topping.

Introduction of a filter between the photoresist surface and the light source attenuates the incident light intensity at the SU-8 film. It is necessary to take this attenuation factor, and any others introduced by elements in the optical pathway between the light source and the photoresist (Figure 8.5), into account and adjust the exposure dose accordingly. The following relation must be obeyed:

$$\text{Experimental dose} = \frac{\text{Recommended dose}}{t_1 \times t_2 \times t_3 \times \cdots \times t_n} \tag{8.1}$$

where t denotes UV transmission percentage of an element n in the optical pathway. The recommended dose is the value recommended in the SU-8 processing data sheets, whereas the experimental dose is the one implemented in practice. Underexposure causes the intended pattern to dissolve or to lift off during developing because the cross-linking acid concentration was not enough to fully polymerize the resist all the way to the underlying substrate. Overexposure leads to extreme T-topping and pattern broadening.

Besides T-topping, the Fresnel diffraction is also responsible for the decrease in resolution and pattern broadening at the resist–air interface of the structure (Chuang et al., 2002; Yang et al., 2005). SU-8 exposure is commonly carried out in a contact or proximity mode. In contact mode, the resist-coated substrate is brought into contact with the mask (Figure 8.5), whereas in proximity mode, the resist's top surface is 10–20 μm away from the mask.

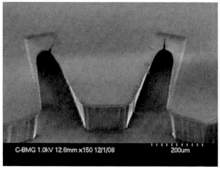 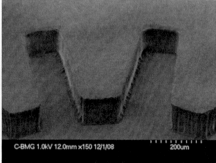

FIGURE 8.4
The effect of T-topping. Left: gear pattern showing T-topping. Right: same gear pattern but now T-topping is minimized by the use of an in-house fabricated filter (a 50 μm layer of SU-8 on glass).

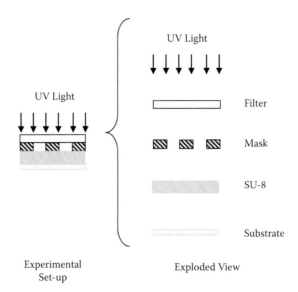

FIGURE 8.5
Optical pathway between light source and SU-8 surface in a common exposure setup.

The contact mode can be further classified as soft contact and hard contact depending on the amount of contact pressure between the resist and the mask, with higher pressures being applied in hard contact mode. Exposure in a contact or proximity setup results in a 1:1 image of the entire mask onto the photoresist. The three setups introduce air gaps, with hard contact minimizing the size of such gaps, between the mask and the resist surface that amplify light diffraction and cause pattern broadening because of the difference between the refractive indexes of air and SU-8. These gaps are often the result of an edge bead or the use of an accidentally inclined hot plate. Different measures can minimize the negative impact of air gaps. For example, the air gaps were filled with glycerol (refractive index $n = 1.473$) by Chuang et al. (2002), whereas Yang et al. (2005) used a type of refractive-index-matching liquids from Cargille Laboratories, in Cedar Grove, New Jersey, to exactly match the refractive index of SU-8 ($n = 1.668$). The use of the latter matching liquid allowed for the fabrication of structures with aspect ratio greater than 190:1. The reflection of light on certain substrates also degrades wall verticality, broadens the pattern, and adds another parameter to consider when setting the exposure time and intensity (Zhang et al., 2004).

8.11 Masks and Gray Scale Lithography

The stencil used to repeatedly generate a desired pattern on resist-coated substrates is called a mask. Like resists, masks can be positive or negative. A positive or a dark-field mask is a mask on which the pattern is clear with the background dark. A negative or a clear-field mask is a mask on which the pattern is dark with the background clear. SU-8 microfluidic devices are usually fabricated with negative or clear-field masks where the microchannels, reservoirs, sample inlets, and so forth, are denoted by dark patterns on a clear background. A light-field or a dark-field image, known as mask polarity, is then transferred to the surface. Masks can be further classified using a criteria similar to that described above for exposure setups. Masks that make direct physical contact with the substrate are called contact masks and degrade faster than noncontact, proximity masks, which are slightly raised, say 10–20 μm, above the substrate. However, diffraction effects are minimized by the use of contact masks. Contact masks are classified as hard contact and soft contact masks. In both cases, one brings the resist-coated wafer into contact with a mask, but by using more pressure, in a hard contact mask setup, the gap between the wafer and the mask can be further reduced. Contact masks, mainly the soft contact type, are still in use in research and development, in mask making itself, and in prototyping. Contact mask and proximity mask printing are collectively known as shadow printing. A more reliable method of masking, but significantly more expensive, is projection printing where, rather than placing a mask in direct contact with (or in proximity to) a substrate, the photomask is imaged by a high-resolution lens system onto the resist-coated substrate. In projection printing, the only limit to the mask's lifetime results from operator handling. Mask fabrication is less challenging in projection lithography because the imaging lens can reduce the mask pattern by 10:1, although reduction values of practical systems are 4:1 or 5:1 (Levinson, 2005). Diffraction can be minimized by the use of highly collimated light sources and/or collimator lenses. Projection lithography is the printing method used in the fabrication of devices based on very large scale integration, such as ICs.

In miniaturization science, one is often looking for low-cost and fast-turnaround methods to fabricate masks. Besides the traditional photomask—a nearly optically flat glass (transparent to near UV) or quartz plate (transparent to deep UV) with an absorber pattern metal layer (e.g., an 800-Å-thick chromium layer) generated by EBL—other less expensive approaches are common. These may involve in-house fabricated masks by manually drawing patterns on cut-and-peel masking films and photoreducing them. Alternatively, these may involve direct writing on a photoresist-coated plate with a laser-plotter (~2-μm resolution) (Arnone, 1992). Simpler yet, using a drawing program such as Canvas™ (ACD Systems, Ltd.), Freehand®, Illustrator® (Adobe Systems, Inc.), or L-Edit™ (Tanner Research, Inc.), a mask design can be created on a computer and saved as a PostScript® or Gerber file to be printed with a high-resolution printer on a transparent film (Duffy et al., 1998a). The transparency with the printed image may then be clamped between a presensitized chrome-covered mask plate and a blank plate to make a traditional photomask from it. After exposure and development, the exposed plate is put in a chrome etch for a few minutes to generate the desired metal pattern, and the remaining resist is stripped off. Simpler yet, the printed transparency may be attached to a blank quartz plate and used directly as a photolithography mask. The maximum resolution with transparency methods is currently around 7 μm (2009) and is highly dependent on the photoplotter used to print the transparency. Although transparency masks are significantly less expensive and have a fast-turnaround time, the quality of the exposed patterns, that is, wall roughness, and the resolution are obviously less than those obtained with photomasks patterned with EBL (where the resolution can be as high as 10 nm) (see Figure 8.6).

Photolithography, as described so far, constitutes a binary image transfer process—the developed pattern consists of regions with resist (1) and regions without resist (0). In contrast, in gray scale lithography, the partial exposure of a photoresist renders it soluble to a developer in proportion to the local exposure dose, and as a consequence, after development, the resist exhibits a surface relief or three-dimensional topography. This way, covered microchannels and complex fluid networks can be fabricated in a single exposure step to eliminate further bonding of a separate top cover needed to close a channel fabricated

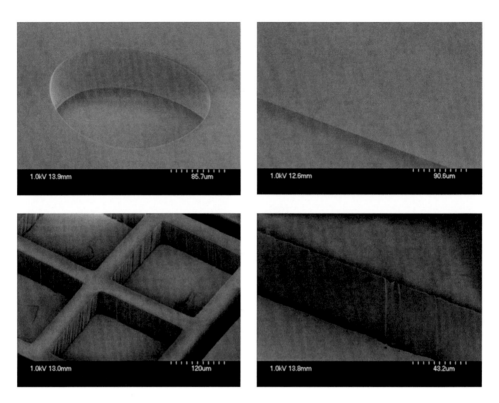

FIGURE 8.6
The quality of a pattern is far superior when using photomasks patterned with EBL (top) than when using transparency masks printed with a commercial photo-plotter (bottom).

with conventional SU-8 photolithography. Gray scale lithography has a great potential use in miniaturization science as it allows for the mass production of microstructures with varying topography. The possibility of creating profiled three-dimensional microstructures offers tremendous additional flexibility in the design of microfluidic, microelectronic, optoelectronic, and micromechanical components (Sure et al., 2003). A key part in the development of a gray scale process is the characterization of the resist thickness as a function of the optical density in the mask for a given lithographic process. It is also desirable to use photoresists that exhibit a low contrast to achieve a wide process window. Ideally, the resist response can be linearized to the optical density within the mask. Gray scale lithography can be achieved using gray-tone masks (GTMs) and software masks. Possible methods for making GTMs, or variable transmission masks, include magnetron sputtering of amorphous carbon onto a quartz substrate. Essentially any transmittance (T) desired in the $0\% < T < 100\%$ range can be achieved by controlling the film thickness (t) in the 200-nm $> t > 0$-nm range with subnanometer precision (Windt et al., 1999). Perhaps more elegantly, gray levels may be created by the density of dots that will appear as transparent holes in a chromium mask. These dots must be small enough not to be transferred onto the wafer because they are below the resolution limit of the exposure tool; otherwise, pixelation might occur. Another attractive way to fabricate a GTM is with high-energy beam-sensitive (HEBS) glass. HEBS glass turns dark upon exposure to an electron beam; the higher the electron dosage, the darker the glass turns. In HEBS glass, a top layer, a couple of microns thick, contains silver ions in the form of silver-alkali-halide $(AgX)_m (MX)_n$

complex nanocrystallites that are approximately 10 nm or less in size and are dispersed within cavities of the glass SiO_4 tetrahedron network. Chemical reduction of the silver ions produces opaque specks of silver atoms upon exposure to a high-energy electron beam (>10 kV) (Sure et al., 2003). Another example of a GTM was demonstrated by Hung et al. (2005). A positive resist is first patterned on a glass substrate and reflowed to obtain curved structures. These curved extrusions are then used as a GTM mask to fabricate cups in SU-8 using glycerol to fill the air gap between the GTM and the SU-8 surface. The use of microfluidics in GTM technology was demonstrated by Chen et al. (2003) using PDMS microchannels as photomasks. The microchannels are filled and emptied at will with light-absorbing liquids, that is dyes, to enable rapid reconfiguration to different gray scale patterns. The authors claim an unlimited number of gray scales given by variations in dye concentration and composition. A similar approach but without the use of microfluidics is the use of colored masks as demonstrated by Taff et al. (2006), where they first characterized the UV absorption of different colors, printed on a transparent film using a standard laser color printer, and then fabricated SU-8 holes of varying depths depending on the color used during exposure.

An alternative to GTMs is the use of maskless optical projection lithography techniques, where a physical mask is replaced by a *software mask*. One approach to make multilevel photoresist patterns directly with a software mask is by variable-dose electron beam writing, in which the electron dosage (the current multiplied by the dwell time) is varied across the resist surface (Stauffer et al., 1992). A laser writer can produce a similar topography but at a lower resolution (Yu et al., 2006). However, variable-dose electron beam and laser writing are serial, slow, and costly, making GTMs a better alternative if high-throughput production is required. A gray scale lithography technique that uses a software mask and yet allows for batch fabrication is based on the digital micromirror device (DMD) chip from Texas Instruments, Inc. and relies on the same spatial and temporal light modulation technology used in digital light processing projectors and high-definition televisions. A commercial instrument based on this chip is offered by Intelligent Micro Patterning LLC. Enormous simplification of lithography hardware is feasible by using the movable mirror arrays in a DMD chip to project images on the photoresist. This technique is capable of fabricating micromachined elements with any surface topography and can, just like EBL or laser writing, be used for implementing maskless binary and gray scale lithography. The maximum resolution of DMD-based maskless photolithography (1 μm in 2009) is far less than with EBL (~10 nm) or laser writers (<1.0 μm), but it is a parallel technique, and for many applications, microfluidics, for example, the lower resolution might not be an obstacle. Unfortunately, the maximum field of exposure is currently (2009) only a few square centimeters, and image stitching is necessary if a large-area pattern is to be fabricated. The unique capability of representing a gray scale is probably the most essential merit of this type of maskless lithography. When a mirror is switched on more frequently than switched off, it reflects a light gray pixel; a mirror that is switched off more frequently reflects a darker gray pixel. In this way, the mirrors in a DMD system can reflect pixels in up to 1024 shades of gray to convert the video or graphic signal entering the DMD chip into a highly detailed gray scale image. Examples of gray scale features obtained with this type of maskless lithography are shown in Figure 8.7.

Another interesting technique to fabricate gray scale structures includes the tilting and/or rotation of the substrate stage with respect to the UV source (Beuret et al., 1994; Han et al., 2004; Yoon et al., 2006) to fabricate pillars, V-grooves, cones, and other structures as illustrated in Figure 8.8a. One of the most astonishing applications is the fabrication of three-dimensional gratings embedded in a microchannel, as shown in Figure 8.8b

FIGURE 8.7
Examples of gray scale lithography obtained with an SF-100 maskless lithography system (Intelligent Micro Patterning, LLC) using SU-8 photoresist. (Courtesy of P. Dwivedi and A. Sharma at the Indian Institute of Technology in Kanpur, India. With permission.)

(Sato et al., 2004b). The combination of tilting and rotation is also capable of fabricating embedded channels in a single exposure step. The angle of the structure depends on both the inclination of the holding stage and the refractive indexes of the different materials in the optical path between the light and the photoresist. The elimination of air gaps is also desired in this setup to improve the maximum tilt angle of the structures (Hung et al., 2004).

Other gray scale exposure techniques include moving mask lithography (Hirai et al., 2007), holography (Deubel et al., 2004; Ullal et al., 2004), and stereolithography (Bertsch et al., 1999b).

8.12 Postexposure Bake

In the case of a chemically amplified resist, such as SU-8, the postexposure bake (PEB) is critical to complete the cross-linking of the polymer matrix. Although reactions induced by the catalyst formed during exposure take place at room temperature, their rate is significantly increased by baking at 60°C–100°C. The precise control of PEB times and temperatures critically determines the quality of the final features and the performance of the pattern as a mold or as a structural element for microfluidics. More than time, PEB temperature is

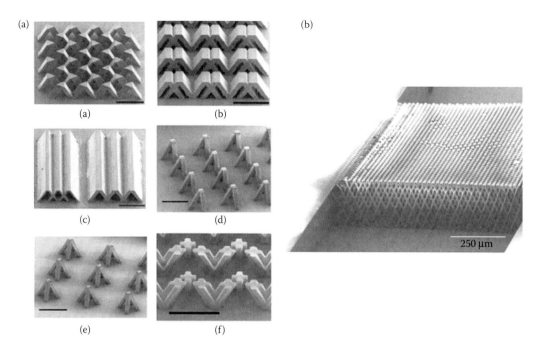

FIGURE 8.8
(a) Different geometries obtained with inclined exposure. (Reprinted from *Sensors and Actuators A Physical* 111, Han, M., W. Lee, S. Lee, and S. S. Lee, 3D microfabrication with inclined/rotated UV lithography, 14–20, Copyright (2004), with permission from Elsevier.) (b) A grating embedded in a microchannel. (Reprinted from *Sensors and Actuators A: Physical* 111(1), Sato, H., T. Kakinuma, J. S. Go, and S. Shoji, In-channel 3-D micromesh structures using maskless multi-angle exposures and their microfilter application, 87–93, Copyright (2004a), with permission from Elsevier.)

of crucial importance; longer times at lower temperatures are recommended. High PEB temperatures have been shown to induce a high amount of internal stress in the structure when using traditional substrates such as silicon and glass. The reason behind this is the significant difference between the coefficients of thermal expansion (CTE) of SU-8 and the substrate as discussed previously. Thermal stresses increase with resist thickness and can cause cracking of the pattern's surface, structure bending, or complete peeling from the substrate in the worst cases. These defects are most likely to be present in extended, large surface area features as the ones required in microfluidics and in those with sharp angles and can be minimized by choosing an alternative substrate, a polymer, one for example, or by reducing the pattern dimensions, if the application allows it. On the other hand, insufficient PEB times and/or temperatures that are too low yield structures that are not completely cross-linked and that can be attacked by the developer. This causes extremely high surface roughness or even complete dissolution of the photoresist. MicroChem's processing data sheets (MicroChem Corp.) recommend PEB to be carried out at 95°C, but temperatures between 55°C and 75°C and extended baking times have been suggested to minimize internal stresses in the pattern. Temperatures below 55°C do not polymerize the matrix completely regardless of the bake time (Li et al., 2005). Rapid heating and cooling of the resist should be avoided as swift temperature changes induce a significant amount of stress. Controlled cooling and heating rates are suggested, and the slower the temperature ramps, the better the results (Williams et al., 2004b). The current authors believe that the use of a convection oven benefits the PEB step as thermal energy is applied more uniformly across the surface of the polymer structure, but no studies have been performed to confirm this point. The replacement of silicon and glass substrates by those having a CTE similar to that of SU-8, for example, polymers, can prove to be very beneficial.

SU-8 microfluidic devices benefit from an optimal PEB to improve adhesion, reduce scumming (resist left behind after development), increase contrast and resist profile (higher edge-wall angle), and reduce the effects of standing waves in an SU-8 resist.

8.13 Development

Development is the dissolution of unpolymerized SU-8 that transforms the latent resist image, formed during exposure, into a relief topography that can serve as a mold or as a permanent structural element. In general, two main technologies are available for resist development: wet development and dry development. The latter is starting to replace wet development for some of the ultimate line-width resolution applications, but wet development is still widely used in a variety of applications, including SU-8 photolithography. Wet development can be based on at least three different types of exposure-induced changes: variation in molecular weight of the polymers (by cross-linking or by chain scission), reactivity change, and polarity change (Le Barny, 1987). The development of SU-8 patterns exploits the variation in molecular weight of the cross-linked polymer. During the development of SU-8, those areas that were not cross-linked dissolve upon immersion in PGMEA (sold by MicroChem as SU-8 developer). Constant agitation during development is recommended to constantly feed fresh developer to the resist pattern and decrease developing times, but care must be taken when developing high-aspect-ratio structures, as excessive agitation can cause mechanical breakage. In general, microfluidic devices do not have high aspect ratios but instead have a large-area surface that makes them more robust to mechanical breakage and shortens developing times. Agitation can be conducted manually or with a rotator and may be assisted by the application of thermal energy to the developer bath. An alternative to mechanical agitation is the use of a sonicator bath. The use of megasonic frequencies instead of ultrasonic ones proves beneficial in the development of high-aspect-ratio structures and patterns with large surface area such as those shown in Figure 8.8. An important distinction between the two sonication methods is that the higher megasonic frequencies (800–2000 kHz) do not cause the violent cavitation effects found with ultrasonic frequencies (15–400 kHz). This significantly reduces or eliminates cavitation erosion that can lead to pattern fracture and structure debonding from the substrate (Williams et al., 2004a).

The replacement of PGMEA by a liquid with lower surface tension is recommended for the prevention of structure collapse during the resist-drying process. This step is crucial when developing high-aspect-ratio structures (HARS) with high-aspect-ratio gaps in between them; that is, densely packed tall, narrow structures. An example of such structures in microfluidics is the fabrication of very dense arrays of tall pillars to act as physical filters or to increase the area of a functionalized surface in microfluidic applications, exploiting the specific binding of two entities (proteins, DNA, etc.). Stiction forces developed in narrow gaps are strong enough to bend and join SU-8 patterns together during drying as shown in Figure 8.9. More examples of this problem are given by Wang et al. (2005). Although rinsing with isopropyl alcohol (IPA) is the most common method to reduce the surface forces arising during drying, perfluorohexane rinse has also been suggested (Yamashita, 1996). As the gap between the HARS decreases below a few micrometers, other methods must be

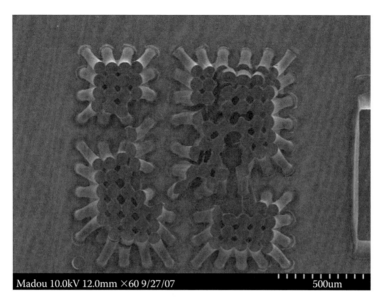

FIGURE 8.9
Example of SU-8 posts joined together by stiction forces.

used, such as freeze-drying (Tanaka et al., 1993a) and supercritical drying (Namatsu et al., 2000; Jincao et al., 2001). In a supercritical liquid, the surface tension becomes negligible and the capillary force that causes pattern stiction is nonexistent. One example of commercial supercritical drying systems is the Tousimis Samdri® line (www.tousimis.com). Theoretical modeling can take into account the Young's modulus of SU-8 and the immersion fluid's surface tension to optimize the cross section, aspect ratio, and packing density of the pattern to avoid stiction (Tanaka et al., 1993b; Jincao et al., 2001; Vora et al., 2005).

8.14 Descumming and Hard Baking

A mild oxygen plasma treatment, so-called *descumming*, removes unwanted resist left behind after development. Negative resists such as SU-8 leave a thin polymer film at the resist–substrate interface and on the pattern surface, which is most likely due to the solid saturation of the used developer (Zaouk, 2008) and can introduce an unexpected change in the surface properties of the SU-8 microfluidics device. An alternative to oxygen plasma is a short dip immersion in acetone. SU-8 withstands acetone when fully polymerized; otherwise, the surface of the pattern will noticeably degrade.

Hard baking removes any casting solvent residues and anneals a resist film that has been weakened either by developer penetration along the resist–substrate interface or by swelling of the resist. Hard baking also improves the hardness of the film and avoids solvent bursts during vacuum processing (Melai et al., 2008). Improved hardness increases the resistance of the resist to subsequent etching steps and makes it less prone to swelling caused by chemicals/buffers that can be used in microfluidic applications. Hard baking frequently occurs at higher temperatures (120°C–150°C) and for longer times than soft baking. Fully cross-linked SU-8 has a glass transition temperature, T_g, that exceeds 200°C, which makes reflow of the resist practically impossible. A hard bake induces further stress in the pattern and can be omitted from the photolithography process depending on the final application of the SU-8 structure.

8.15 Resist Stripping or Removal

Although this chapter is geared toward the use of SU-8 patterns as molds or structural elements for microfluidic and nanofluidic applications, the following section is an overview of available methods for SU-8 photoresist stripping. The reader may find it useful for the use of SU-8 as a sacrificial material or for the removal of SU-8 from substrates intended to be recycled.

Photoresist stripping, in slightly oversimplified terms, is organic polymer etching. When using the photoresist as sacrificial material, the primary consideration is complete removal of the photoresist without damaging the device under construction. SU-8 photoresist is effectively stripped off with a strong acid such as H_2SO_4 or an acid–oxidant combination such as H_2SO_4–Cr_2O_3 or Piranha (detailed under substrate cleaning), attacking the photoresist but not silicon, silicon oxide, or glass. Other liquid strippers include organic solvent strippers (such as Remover PG from MicroChem and SRGM-Red Stripper from Gersteltec), alkaline strippers (with or without oxidants), and even acetone. These strippers can be used if the postexposure bake is not too long or occurs at a low enough temperature such that solvent molecules are still present in the polymer matrix and act as weak points for the stripper to attack. A hard bake is not recommended if SU-8 is to be removed. High temperatures or long baking times cause the resist to develop a tough "skin," prevent attack from common strippers, and force a removal step using oxygen plasma.

Dry stripping or oxygen plasma stripping, also known as ashing, has become more popular as it poses fewer disposal problems with toxic, flammable, and dangerous chemicals. Wet stripping solutions lose potency in use, causing stripping rates to change with time. Accumulated contamination in solutions can be a source of particles, and liquid phase surface tension and mass transport tend to make photoresist removal difficult and uneven. Dry stripping is more controllable than liquid stripping, less corrosive with respect to metal features on the wafer, and more importantly, it leaves a cleaner surface under the right conditions. Finally, it does not cause the undercutting and broadening of photoresist features, which can be caused by wet strippers.

In solid–gas resist stripping, a volatile product forms either through reactive plasma stripping (e.g., with oxygen), gaseous chemical reactants (e.g., ozone), and radiation (UV) or a combination thereof (e.g., UV/ozone-assisted). Plasma stripping uses a low-pressure electrical discharge to split molecular oxygen (O_2) into its more reactive atomic form (O). This atomic oxygen converts an organic photoresist into a gaseous product that may be pumped away. This type of plasma stripping belongs to the category of chemical dry stripping and is isotropic in nature. In ozone strippers, ozone, at atmospheric pressure, attacks the resist. In UV/ozone stripping, UV helps to break bonds in the resist, paving the way for a more efficient attack by ozone. Ozone strippers have the advantage that no plasma damage can occur on the devices in the process. Reactive plasma stripping is currently the predominant commercial technology because of its high removal rate and throughput. Some different stripper configurations are barrel reactors, downstream strippers, and parallel plate systems.

8.16 Conclusion

The versatility of SU-8 has positioned it as one of the most important materials in polymer microfabrication. Microfluidics benefits from SU-8 photolithography in the batch fabrication of structures of high aspect ratio and/or large surface area, which can range in size from a few millimeters down to tens of nanometers. The good mechanical and excellent chemical properties of cross-linked SU-8 yield polymer microfluidics devices that can handle a variety of samples such as blood, urine, milk, and so forth, as well as buffers and cleaning agents at a wide range of flow pressures and working temperatures. SU-8 has a high transparency for light at wavelengths more than 400 nm and allows for the use of fluorescence and other visualization and detection techniques that are common in microfluidics. The recent commercial introduction of SU-8 with dyes and nanoparticles incorporated in it further enlarges the potential of this photoresist in microfluidics and other applications. SU-8 photolithography yields or supports the fabrication of research devices that are applied in a variety of fields including optics, precision mechanics, energy, and space. Furthermore, complex carbon microstructures can be derived by pyrolyzing SU-8 patterns, a technique known as carbon MEMS (Wang et al., 2005).

Although general guidelines for carrying out SU-8 photolithography exist, significant practical experience is still required to develop successful new SU-8 applications. A generic guideline for SU-8 processing is not really possible because the parameters must always be finely tweaked depending on the available equipment and facilities, the specific application, and most importantly, the geometry of the pattern. Even process parameters once thought trivial are still subjected to further optimization, for example, soft baking time, temperature, and method of baking. Perhaps the most important variable to control in the SU-8 structure manufacturing process is the internal stress. The introduction of thermally induced stresses significantly degrades the final product. Crack formation, bending, and debonding from the substrate can all be minimized by controlled thermal management that includes the use of appropriate heating and cooling ramps. However, baking techniques and procedures must be optimized to eliminate long heating ramps and relaxation times between processing steps. The use of polymer substrates with thermal properties similar to that of SU-8 proves beneficial to minimize thermal stresses induced during baking. Moreover, polymer substrates are less expensive and significantly less brittle when compared with silicon and glass. Alternatives to traditional SU-8 processing include the use of SU-8 dry films for lamination. This approach eliminates the casting solvent from the photoresist formulation, reduces process time significantly, especially when working with thick layers (>200 µm), and can make SU-8 photolithography a roll-to-roll process. The constant optimization of process methodologies, techniques, and equipment has enabled a steady improvement of SU-8 photolithography, and hopefully, this will eventually result in the application of SU-8 photolithography in a number of commercial devices rather than just in research tools.

Acknowledgments

The authors thank Rob Hardman from MicroChem for useful discussions regarding SU-8 formulations.

References

Abgrall P, Charlot S, Fulcrand R, Paul L, Boukabache A, Gué A. (2008). Low-stress fabrication of 3D polymer free standing structures using lamination of photosensitive films. *Microsyst Technol.* 14:1205–14.

Abgrall P, Conedera V, Camon H, Gue A, Nguyen N. (2007). SU-8 as a structural material for labs-on-chips and microelectromechanical systems. *Electrophoresis.* 28:4539–51.

Abgrall P, Lattes C, Conédéra V, Dollat X, Colin S, Gué AM. (2006). A novel fabrication method of flexible and monolithic 3D microfluidic structures using lamination of SU-8 films. *J Micromech Microeng.* 16:113–21.

Acrivos A, Shah MJ, Petersen EE. (1960). On the flow of a non-Newtonian liquid on a rotating disk. *J Appl Phys.* 31(6):963–8.

Agirregabiria M, Blanco FJ, Berganzo J, Arroyo MT, Fullaondo A, Mayora K, et al. (2005). Fabrication of SU-8 multilayer microstructures based on successive CMOS compatible adhesive bonding and releasing steps. *Lab Chip.* 5:545–52.

Aktary M, Jensen MO, Westra KL, Brett MJ, Freeman MR. (2003). High-resolution pattern generation using the epoxy novolak SU-8 2000 resist by electron beam lithography. *J Vac Sci Technol B.* 21(4):L5–7.

Anderson JR, Chiu DT, Jackman RJ, Cherniavskaya O, McDonald JC, Wu H, et al. (2000). Fabrication of topologically complex three-dimensional microfluidic systems in PDMS by rapid prototyping. *Anal Chem.* 72:3158–64.

Andersson H, van den Berg A. (2003). Microfluidic devices for cellomics: a review. *Sens Actuators B Chem.* 92(3):315–25.

Angelo R, Gelorme J, Kucynski J, Lawrence W, Pappas S, Simpson L. (1992). *Photocurable epoxy composition with sulfonium salt photoinitiator.* United States patent 5102772. April 7.

Anhoj TA, Jorgensen AM, Zauner DA, Hubner J. (2006). The effect of soft bake temperature on the polymerization of SU-8 photoresist. *J Micromech Microeng.* 16:1819–24.

Arnone C. (1992). The laser-plotter: a versatile lithographic tool for integrated optics and microelectronics. *Microelectron Eng.* 17:483–6.

Auroux P, Iossifidis D, Reyes DR, Manz A. (2002). Micro total analysis systems. II. Analytical standard operations and applications. *Anal Chem.* 74(12):2637–52.

Bakajin O, Fountain E, Morton K, Chou SY, Sturm JC, Austin RH. (2006). Materials aspects in micro- and nanofluidic systems applied to biology. *MRS Bull.* 31:108–13.

Barber RL, Ghantasala MK, Divan R, Vora KD, Harvery EC, Mancini DC. (2005). Optimisation of SU-8 processing parameters for deep x-ray lithography. *Microsyst Technol.* 11(4):303–10.

Barber R, Ghantasala MK, Divan R, Mancini DC, Harvey EC. (2007). Study of stress and adhesion strength in SU-8 resist layers on silicon substrate with different seed layers. *Journal of Micro/ Nanolithography, MEMS and MOEMS.* 6(3):033006–8.

Barker SLR, Ross D, Tarlov MJ, Gaitan M, Locascio LE. (2000a). Control of flow direction in microfluidic devices with polyelectrolyte multilayers. *Anal Chem.* 72(24):5925–9.

Barker SLR, Tarlove MJ, Canavan H, Hickman JJ, Locascio LE. (2000b). Plastic microfluidic devices modified with polyelectrolyte multilayers. *Anal Chem.* 72(20):4899–903.

Bassous E, Taub HH, Kuhn L. (1977). Ink jet printing nozzle arrays etched in silicon. *Appl Phys Lett.* 31:135–7.

Becker H, Gartner C. (2000). Polymer microfabrication methods for microfluidic analytical applications. *Electrophoresis.* 21(1):12–26.

Becker H, Locascio LE. (2002). Polymer microfluidic devices. *Talanta.* 56(2):267–87.

Beebe DJ, Mensing GA, Walker GM. (2002). Physics and applications of microfluidics in biology. *Ann Rev Biomed Eng.* 4:261–86.

Berdichevsky Y, Khandurina J, Guttman A, Lo Y. (2004). UV/ozone modification of poly(dimethylsiloxane) microfluidic channels. *Sens Actuators B Chem.* 97(2–3):402–8.

Bertsch A, Lorenz H, Renaud P. (1999a). 3D microfabrication by combining microstereolithography and thick resist UV lithography. *Sens Actuators A Phys.* 73:14–23.

Bertsch A, Lorenz H, Renaud P. (1999b). 3D microfabrication by combining microstereolithography and thick resist UV lithography. *Sens Actuators A Phys.* 73(1–2):14–23.

Beuret C, Racine G, Gobet J, Luthier R, de Rooji NF. (1994). Microfabrication of 3D multidirectional inclined structures by UV lithography and electroplating. Paper presented at 17th International Conference on MEMS, MEMS '94, Oiso, Japan.

Bilenberg B, Jacobsen S, Schmidt MS, Skjolding LHD, Shi P, Bøggild P, et al. (2006). High resolution 100 kV electron beam lithography in SU-8. *Microelectron Eng.* 83(4–9):1609–12.

Boedeker Plastics, Inc. (2009a). PEEK (PolyEtherEtherKetone) specifications [cited July 31, 2009]. Available from: http://www.boedeker.com/peek_p.htm.

Boedeker Plastics, Inc. (2009b). Polycarbonate specifications [cited July 31, 2009]. Available from: http://www.boedeker.com/polyc_p.htm.

Bogdanov AL, Peredkov SS. (2000). Use of SU-8 photoresist for very high aspect ratio x-ray lithography. *Microelectron Eng.* 53(1–4):493–6.

Chen C, Hirdes D, Folch A. (2003). Gray-scale photolithography using microfluidic photomasks. *Proc Natl Acad Sci USA.* 100(4):1499–504.

Chen R, Cheng C. (2001). Study of spin coating properties of SU-8 thick-layer photoresist. Paper presented at Advances in Resist Technology and Processing XVIII, Santa Clara, CA.

Chin CD, Linder V, Sia SK. (2007). Lab-on-a-chip devices for global health: past studies and future opportunities. *Lab Chip.* 7:41–57.

Chollet F. (2009). SU-8: thick photo-resist for MEMS [cited April 15, 2009]. Available from: http://memscyclopedia.org/su8.html.

Chuang Y, Tseng F, Lin W. (2002). Reduction of diffraction effect of UV exposure on SU-8 negative thick photoresist by air gap elimination. *Microsyst Technol.* 8(4–5):308–13.

Chung TD, Kim HC. (2007). Recent advances in miniaturized microfluidic flow cytometry for clinical use. *Electrophoresis.* 28(24):4511–20.

Cirlex. [cited 2009 Apr 15]. Available from: http://www.matweb.com.

Craighead H. (2006). Future lab-on-a-chip technologies for interrogating individual molecules. *Nature.* 442:387–93.

Crivello JV. (2000). The discovery and development of onium salt cationic photoinitiators. *J Polym Sci A Polym Chem.* 37(23):4241–54.

Crivello JV, Lam JHW. (1977). Diaryliodonium salts. A new class of photoinitiators for cationic polymerization. *Macromolecules.* 10(6):1307–15.

Crivello JV, Lam JHW. (1979). Photoinitiated cationic polymerization with triarylsulfonium salts. *J Polym Sci.* 17:977–99.

Crivello JV, Lam JHW. (1980). Complex triarylsulfonium salt photoinitiators. I. the identification, characterization, and syntheses of a new class of tryarylsulfonium salt photoinitiators. *J Polym Sci A Polym Chem.* 18:2677–95.

Dai W, Lian K, Wang W. (2005). A quantitative study on the adhesion property of cured SU-8 on various metallic surfaces. *Microsyst Technol.* 11:526–34.

del Campo A, Greiner C. (2007). SU:8 a photoresist for high-aspect-ratio and 3D submicron lithography. *J Micromech Microeng.* 17:R81–95.

Dentinger PM, Clift WM, Goods SH. (2002). Removal of SU-8 photoresist for thick film applications. *Microelectron Eng.* 61–62:993–1000.

Deubel M, von Freymann G, Wegener M, Pereira S, Busch K, Soukoulis CM. (2004). Direct laser writing of three-dimensional photonic-crystal templates for telecommunications. *Nat Mater.* 3(7):444–7.

Dittrich PS, Manz A. (2006). Lab-on-a-chip: microfluidics in drug discovery. *Nat Rev Drug Discov.* 5:210–8.

Dittrich PS, Tachikawa K, Manz A. (2006). Micro total analysis systems. latest advancements and trends. *Anal Chem.* 78(12):3887–907.

Duffy DC, McDonald JC, Schueller OJA, Whitesides GM. (1998a). Rapid prototyping of microfluidic systems in poly(dimethylsiloxane). *Anal Chem.* 70:4974–84.

Duffy DC, McDonald JC, Schueller OJA, Whitesides GM. (1998b). Rapid prototyping of microfluidic systems in poly(dimethylsiloxane). *Anal Chem.* 70(23):4974–84.

Dyer CK. (2002). Fuel cells for portable applications. *J Power Sources.* 106:31–4.

Emslie AG, Bonner FT, Peck LG. (1959). Flow of a viscous liquid on a rotating disk. *J Appl Phys.* 29(5):858–62.

Erdler G, Frank M, Lehmann M, Reinecke H, Muller C. (2006). Chip integrated fuel cell. *Sens Actuators A Phys.* 132:331–6.

Feng R, Farris RJ. (2003). Influence of processing conditions on the thermal and mechanical properties of SU8 negative photoresist coatings. *J Micromech Microeng.* 13:80–8.

Figeys D, Pinto D. (2000). Lab-on-a-chip: a revolution in biological and medical sciences. *Anal Chem.* 72(9):330 A, 335 A.

Fiorini GS, Chiu DT. (2005). Disposable microfluidic devices: fabrication, function, and application. *Biotechniques.* 38(3):429–46.

FLUOROTHERM Polymers, Inc. (2009). PTFE characteristics and behavior. [cited July 31, 2009]. Available from: http://www.fluorotherm.com/ptfe_properties.html.

Gelorme JD, Cox RJ, Gutierrez SAR. (1989). *Photoresist composition and printed circuit boards and packages made therewith.* Patent 4882245. November 21.

Gersborg-Hansen M, Kristensen A. (2006). Optofluidic third order distributed feedback dye laser. *Appl Phys Lett.* 89:103518–0,103518–3.

Giancoli DC. (1998). *Physics, Principles with Applications.* 5th ed. New Jersey: Prentice Hall.

Gravesen P, Branebjerg J, Jensen OS. (1993). Microfluidics—a review. *J Micromech Microeng.* 3:168–82.

Guerin LJ. (2005). The SU8 homepage [cited July 30, 2009]. Available from: http://www.geocities.com/guerinlj/.

Han M, Lee W, Lee S, Lee SS. (2004). 3D microfabrication with inclined/rotated UV lithography. *Sens Actuators A Phys.* 111:14–20.

Harris TW. (1976). *Chemical Milling.* Oxford: Clarendon Press.

Henry AC, Tutt TJ, Galloway M, Davidson YY, McWhorter S, Soper SA, McCarley RL. (2000). Surface modification of poly(methyl methacrylate) used in the fabrication of microanalytical devices. *Anal Chem.* 72(21):5331–7.

Henry AC, Waddell EA, Shreiner R, Locascio LE. (2002). Control of electroosmotic flow in laser-ablated and chemically modifiec hot imprinted poly(ethylene terephthalate glycol) microchannels. *Electrophoresis.* 23:791–8.

Hirai Y, Inamoto Y, Sugano K, Tsuchiya T, Tabata O. (2007). Moving mask UV lithography for three-dimensional structuring. *J Micromech Microeng.* 17:199–206.

Hu G, Li D. (2007). Multiscale phenomena in microfluidics and nanofluidics. *Chem Eng Sci.* 62(13):3443–54.

Hull R. (1999). *Properties of Crystalline Silicon.* London: The Institution of Engineering and Technology.

Hung K, Hu H, Tseng F. (2004). Application of 3D glycerol-compensated inclined-exposure technology to an integrated optical pick-up head. *J Micromech Microeng.* 14:975–83.

Hung K, Tseng F, Chou H. (2005). Application of 3D gray mask for the fabrication of curved SU-8 structures. *Microsyst Technol.* 11:365–9.

Ito H. (1996). Chemical amplification resists: history and development within IBM. *IBM J Res Dev.* 41:69–80.

Iverson BD, Garimella SV. (2008). Recent advances in microscale pumping technologies: a review and evaluation. *Microfluid Nanofluidics.* 5:145–74.

Janasek D, Franzke J, Manz A. (2006). Scaling and the design of miniaturized chemical-analysis systems. *Nature.* 442:374–80.

Jankowski AF, Hayes JP, Graff RT, Morse JD. (2002). Micro-fabricated thin film fuel cells for portable power requirements. *Mater Res Soc Symp Proc.* 730:93–8.

Jensen KF. (2006). Silicon-based microchemical systems: characteristics and applications. *MRS Bull.* 31:101–7.

Jincao Y, Matthews MA. (2001). Prevention of photoresist pattern collapse by using liquid carbon dioxide. *Ind Eng Chem Res*. 40(24):5858–60.

Kapton Polyimide Film. [cited 2009 Apr 15]. Available from: http://www2.dupont.com/Kapton/en_US/.

Kim K, Park DS, Lu HM, Che W, Kim K, Lee J, et al. (2004). A tapered hollow metallic microneedle array using backside exposure of SU-8. *J Micromech Microeng*. 14:597–603.

King JW, Williams LL. (2003). Utilization of critical fluids in processing semiconductors and their related materials. *Curr Opin Solid State Mater Sci*. 7(4–5):413–24.

Kirby BJ, Hasselbrink EF Jr. (2004a). Zeta potential of microfluidic substrates: I. Theory, experimental techniques, and effects on separations. *Electrophoresis*. 25:187–202.

Kirby BJ, Hasselbrink EF Jr. (2004b). Zeta potential of microfluidic substrates: II. Data for polymers. *Electrophoresis*. 25:203–13.

Kjeang E, Djilali N, Sinton D. (2009). Microfluidic fuel cells: a review. *J Power Sources*. 186:353–69.

Koller DM, Galler N, Ditlbacher H, Hohenau A, Leitner A, Aussenegg FR, et al. (2009). Direct fabrication of micro/nano fluidic channels by electron beam lithography. *Microelectron Eng*. 86(4–6):1314–6.

Kuriyama N, Kubota T, Okamura D, Suzuki T, Sasahara J. (2008). Design and fabrication of MEMS-based monolithic fuel cells. *Sens Actuators A Phys*. 145–146:354–62.

Le Barny P. (1987). *Molecular Engineering of Ultrathin Polymeric Films*. Stroeve P, Franses E, editors. New York: Elsevier; pp. 99–150.

Lee KY, LaBianca N, Rishton SA, Zolgharnain S, Gelorme JD, Shaw J, et al. (1995). Micromachining applications of a high resolution ultrathick photoresist. *J Vac Sci Technol B*. 13:3012–6.

Lee SJ, Shi W, Maciel P, Cha SW. Boise (ID); 2003.

Levinson HJ. (2005). *Principles of Lithography*. 2nd ed. Bellingham (WA): SPIE Publications.

Li B, Liu M, Chen Q. (2005). Low-stress ultra-thick SU-8 UV photolithography process for MEMS. *J Microlithogr Microfabr Microsyst*. 4(4):043008–1, 043008–6.

Lin C, Lee G, Chang B, Chang G. (2002). A new fabrication process for ultra-thick microfluidic microstructures utilizing SU-8 photoresist. *J Micromech Microeng*. 12:590–7.

Linder V, Verpoorte E, Thormann W, de Rooji NF, Sigrist H. (2001). Surface biopassivation of replicated poly(dimethylsiloxane) microfluidic channels and application to heterogeneous immunoreaction with on-chip fluorescence detection. *Anal Chem*. 73(17):4181–9.

Ling ZG, Lian K, Jian L. (2000). Improved patterning quality of SU-8 microstructures by optimizing the exposure parameters. Paper presented at Advances in Resist Technology and Processing XVIII, Santa Clara, CA.

Liu Y, Fanguy JC, Bledsoe JM, Henry CS. (2000). Dynamic coating using polyectrolyte multilayers for chemical control of electroosmotic flow in capillary electrophoresis microchips. *Anal Chem*. 72(24):5939–44.

Lorentz H, Despont M, Fahrni N, Brugger J, Vettiger P, Renaud P. (1998). High-aspect-ratio, ultrathick, negative-tone near-UV photoresist and its applications for MEMS. *Sens Actuators A Phys*. 64:33–9.

Lorenz H, Despont M, Fahrni M, LaBianca N, Vettiger P, Renaud P. (1997). SU-8: a low-cost negative resist for MEMS. *J Micromech Microeng*. 7:121.

Lü C, Yin X, Wang M. (2007). Fabrication of high aspect ratio metallic microstructures on ITO glass substrate using reverse-side exposure of SU-8. *Sens Actuators A Phys*. 136(1):412–6.

Madou MJ. (2009). *Manufacturing Techniques for Microfabrication and Nanotechnology*. 1st ed. Boca Raton (FL): CRC Press.

Madou M, Florkey J. (2000). From batch to continuous manufacturing of microbiomedical devices. *Chem Rev*. 100:2679–91.

Manz A, Graber N, Widmer HM. (1990). Miniaturized total chemical analysis systems: a novel concept for chemical sensing. *Sens Actuators*. B1:244–8.

Martinez-Duarte R. (2009). Fabrication of carbon micro molds. Master of Science in Mechanical and Aerospace Engineering. Irvine: University of California.

Martinez-Duarte R, Madou MJ, Kumar G, Schroers J. (2009). A novel method for amorphous metal micromolding using carbon MEMS. Paper presented at 15th International Conference on Solid-State Sensors, Actuators & Microsystems, Transducers 2009, Denver, CO.

Mata A, Fleischman AJ, Roy S. (2005). Characterization of polydimethylsiloxane (PDMS) properties for biomedical micro/nanosystems. *Biomed Microdevices*. 7(4):281–93.

McHardy J, Haber SP, editors. (1998). Supercritical fluid cleaning: Fundamentals, technology and applications. In: *Materials Science and Process Technology*. Westwood (NJ): Noyes Publications.

Melai J, Salm C, Wolters R, Schmitz J. (2008). Qualitative and quantitative characterization of outgassing from SU-8. *Microelectron Eng*. 86(4–6):761–4.

MEMS Exchange. (2009). Silicon dioxide (SiO_2), film [cited July 31, 2009]. Available from: http://www.memsnet.org/material/silicondioxidesio2film/.

Metz TE, Savage RN, Simmons HO. (1992). In situ control of photoresist coating processes. *Semiconductor Int*. 15:68–9.

MicroChem Corp. (2009). SU-8 2000 permanent epoxy negative photoresist processing guidelines. [cited April 12, 2009]. Available from: www.microchem.com.

Mijatovic D, Eijkel JCT, van den Berg A. (2005). Technologies for nanofluidic systems: top–down vs. bottom–up—a review. *Lab Chip*. 5:492–500.

Mitchell P. (2001). Microfluidics—downsizing large-scale biology. *Nat Biotechnol*. 19:717–21.

Moreau WM. (1987). *Semiconductor Lithography: Principles, Practices and Materials*. New York: Springer.

Morse JD. (2007). Micro-fuel cell power sources. *Int J Energy Res*. 31(6):576–602.

Mylar Technical Information. [cited 2009 Apr 16]. Available from: http://usa.dupontteijinfilms.com/informationcenter/technicalinfo.aspx.

Nallani AK, Park SW, Lee JB. (2003). Characterization of SU-8 as a photoresist for electron-beam lithography. Paper presented at Smart Sensors, Actuators, and MEMS, Maspalomas, Gran Canaria, Spain.

Namatsu H, Yamazaki K, Kurihara K. (2000). Supercritical resist dryer. *J Vac Sci Technol B*. 18(2):780–4.

Nguyen N, Chan SH. (2006). Micromachined polymer electrolyte membrane and direct methanol fuel cells—a review. *J Micromech Microeng*. 16:R1–12.

Nordstrom M, Marie R, Calleja M, Boisen A. (2004). Rendering SU-8 hydrophilic to facilitate use in micro channel fabrication. *J Micromech Microeng*. 14:1614–7.

Nordström M, Johansson A, Noguerón ES, Clausen B, Calleja M, Boisen A. (2005). Investigation of the bond strength between the photo-sensitive polymer SU-8 and gold. *Microelectron Eng*. 78–79:152–7.

PDMS (polydimethylsiloxane). [cited 2009 Jul 31]. Available from: http://web.mit.edu/6.777/www/matprops/pdms.htm.

Pépin A, Studer V, Decanini D, Chen Y. (2004). Exploring the high sensitivity of SU-8 resist for high resolution electron beam patterning. *Microelectron Eng*. 73–74:233–7.

Peterman MC, Huie P, Bloom DM, Fishman HA. (2003). Building thick photoresist structures from the bottom up. *J Micromech Microeng*. 13:380–2.

Petersen KE. (1979). Fabrication of an integrated, planar silicon ink-jet structure. *IEEE Trans Electron Devices*. ED-26(12):1918–20.

Pugmire DL, Waddell EA, Haasch R, Tarlov MJ, Locascio LE. (2002). Surface characterization of laser-ablated polymers used for microfluidics. *Anal Chem*. 74(4):871–8.

Reyes DR, Iossifidis D, Auroux P, Manz A. (2002). Micro total analysis systems. I. Introduction, theory and technology. *Anal Chem*. 74(12):2623–36.

Reznikova EF, Mohr J, Hein H. (2005). Deep photo-lithography characterization of SU-8 resist layers. *Microsyst Technol*. 11:282–91.

Reznikova E, Mohr J, Boerner M, Nazmov V, Jakobs P. (2008). Soft x-ray lithogaphy of high aspect ratio SU8 submicron structures. *Microsyst Technol*. 14(9–11):1683–8.

Robinson APG, Zaid HM, Gibbons FP, Palmer RE, Manickam M, Preece JA, et al. (2006). Chemically amplified molecular resists for electron beam lithography. *Microelectron Eng*. 83(4–9):1115–8.

Ruhmann R, Ahrens G, Schuetz A, Voskuhl J, Gruetzner G. (2001). Reduction of internal stress in SU-8-like negative tone photoresist for MEMS applications by chemical modification. Paper presented at Advances in Resist Technology and Processing XVIII, Santa Clara, CA.

Sato H, Kakinuma T, Go JS, Shoji S. (2004a). In-channel 3-D micromesh structures using maskless multi-angle exposures and their microfilter application. *Sens Actuators A Phys.* 111(1):87–92.

Schwartz LW, Valery R. (2004). Theoretical and numerical results for spin coating of viscous liquids. *Phys Fluids.* 16(3):569–84.

Shaw JM, Gelorme JD, LaBianca NC, Conley WE, Holmes SJ. (1997). Negative photoresists for optical lithography. *IBM J Res Dev.* 41:81–94.

Shaw M, Nawrocki D, Hurditch R, Johnson D. (2003). Improving the process capability of SU-8. *Microsyst Technol.* 10:1–6.

Shew B, Huang T, Liu K, Chou C. (2004). Oxygen quenching effect in ultra-deep x-ray lithography with SU-8 resist. *J Micromech Microeng.* 14:410–4.

Sikanen T, Tuomikoski S, Ketola RA, Kostiainen R, Franssila S, Kotiaho T. (2005). Characterization of SU-8 for electrokinetic microfluidic applications. *Lab Chip.* 5:888–96.

Song Y, Kumar CSSR, and Hormes J. (2004). Fabrication of an SU-8 based microfluidic reactor on a PEEK substrate sealed by a 'flexible semi-solid transfer' (FST) process. *J Micromech Microeng.* 14:932–40.

Squires TM, Quake SR. (2005). Microfluidics: fluid physics at the nanoliter scale. *Rev Mod Phys.* 77:977–1026.

Stauffer JM, Oppliger Y, Regnault P, Baraldi L, Gale MT. (1992). Electron beam writing of continuous resist profiles for optical applications. *J Vac Sci Technol B.* 10:2526.

Stone HA, Stroock AD, Ajdari A. (2004). Engineering flows in small devices: microfluidics toward a lab-on-a-chip. *Annu Rev Fluid Mech.* 36:318–411.

Stroock AD, Cabodi M. (2006). Microfluidic biomaterials. *MRS Bull.* 31:114–9.

Sure A, Dillon T, Murakowski J, Lin C, Pustai D, Prather DW. (2003). Fabrication and characterization of three-dimensional silicon tapers. *Optics Express.* 11:3555–61.

Taff J, Kashte Y, Spinella-Mamo V, Paranjape M. (2006). Fabricating multilevel SU-8 structures in a single photolithographic step using colored masking patterns. *J Vac Sci Technol A.* 24(3):742–6.

Tan TL, Wong D, Lee P, Rawat RS, Springham S, Patran A. (2006). Characterization of chemically amplified resist for x-ray lithography by fourier transform infrared spectroscopy. *Thin Solid Films.* 504(1–2):113–6.

Tanaka T, Mitsuaki M, Hiroaki O, Taro O. (1993a). Freeze-drying process to avoid resist pattern collapse. *Jpn J Appl Phys.* 32(12R):5813–4.

Tanaka T, Morigami M, Atoda N. (1993b). Mechanism of resist pattern collapse during development process. *Jpn J Appl Phys.* 32:6059–64.

Tangram Technology Ltd. (2005). Polymethyl methacrylate—PMMA (acrylic) polymer data sheet [cited July 31, 2009]. Available from: http://www.tangram.co.uk/TI-Polymer-PMMA.html.

Terry SC, Jerman JH, Angell JB. (1979). A gas chromatographic air analyzer fabricated on a silicon wafer. *IEEE Trans Electron Devices.* ED-26:1880–6.

Thompson LF. (1994). An introduction to lithography. In: Thompson LF, Willson CG, Bowden MJ, editors. *Introduction to Microlithography.* 2nd ed. Washington (DC): American Chemical Society; pp. 1–17.

Tuckerman DB, Pease RF. (1981). High-performance heat sinking for VLSI. *IEEE Electron Device Lett.* ED-2:126–9.

Ullal CK, Maldovan M, Thomas EL. (2004). Photonic crystals through holographic lithography: simple cubic, diamond-like, and gyroid-like structures. *Appl Phys Lett.* 84:5434–6.

van Kan JA, Sanchez JL, Xu B, Osipowicz T, Watt F. (1999). Micromachining using focused high energy ion beams: deep ion beam lithography. *Nucl Instrum Methods Phys Res B.* 148:1085–9.

Verpoorte E. (2002). Microfluidic chips for clinical and forensic analysis. *Electrophoresis.* 23:677–712.

Verpoorte E, De Rooji NF. (2003). Microfluidics meets MEMS. *Proc IEEE.* 91(6):930–53.

Vilkner T, Janasek D, Manz A. (2004). Micro total analysis systems. Recent developments. *Anal Chem.* 76(12):3373–86.

Vora KD, Shew BY, Harvey EC, Hayes JP, Peele AG. (2005). Specification of mechanical support structures to prevent SU-8 stiction in high aspect ratio structures. *J Micromech Microeng.* 15:978–83.

Vrentas JS, Duda JL. (1977a). Diffusion in polymer-solvent systems. I. Reexamination of the free-volume theory. *J Polym Sci B.* 15:403–16.

Vrentas JS, Duda JL. (1977b). Diffusion in polymer-solvent systems. II. A predictive theory for the dependence of diffusion coefficients on temperature, concentration, and molecular weight. *J Polym Sci B.* 15:417–39.

Wang C, Jia G, Taherabadi LH, Madou MJ. (2005). A novel method for the fabrication of high-aspect ratio C-MEMS structures. *J Microelectromech Syst.* 14(2):348–58.

Wang SC, Perso CE, Morris MD. (2000). Effects of alkaline hydrolysis and dynamic coating on the electroosmotic flow in polymeric microfabricated channels. *Anal Chem.* 72(7):1704–6.

Wang X, Cheng C, Wang S, Liu S. (2009). Electroosmotic pumps and their applications in microfluidic systems. *Microfluid Nanofluidics.* 6:145–62.

Washo BD. (1977). Rheology and modeling of the spin coating process. *IBM J Res Dev.* March:190–98.

Weigl BH, Bardell RL, Cabrera CR. (2003). Lab-on-a-chip for drug development. *Adv Drug Deliv Rev.* 55(3):349–77.

Whitesides GM. (2006). The origins and the future of microfluidics. *Nature.* 442:368–73.

Williams JD, Wang W. (2004a). Using megasonic development of SU-8 to yield ultra-high aspect ratio microstructures with UV lithography. *Microsyst Technol.* 10:694–8.

Williams JD, Wang W. (2004b). Study on the postbaking process and the effects on UV lithography of high aspect ratio SU-8 microstructures. *J Microlithogr Microfabr Microsyst.* 3:563–8.

Windt DL, Cirelli RA. (1999). Amorphous carbon films for use as both variable-transmission apertures and attenuated phase shift masks for deep ultraviolet lithography. *J Vac Sci Technol B.* 17:930–2.

Wolf S, Tauber RN. (2000). *Silicon Processing for the VLSI Era.* Sunset Beach: Lattice Press.

Wu C, Chen M, Tseng F. (2003). *SU-8 Hydrophilic Modification by Forming Copolymer with Hydrophilic Epoxy Molecule.* Squaw Valley, CA.

Xia Y, Whitesides GM. (1998). Soft lithography. *Annu Rev Mater Sci.* 28:153–84.

Yamashita Y. (1996). Sub-0.1 um patterning with high aspect ratio of 5 achieved by preventing pattern collapse. *Jpn J Appl Phys.* 35:2385–6.

Yang R, Wang W. (2005). A numerical and experimental study on gap compensation and wavelength selection in UV-lithography of ultra-high aspect ratio SU-8 microstructures. *Sens Actuators B Chem.* 110(2):279–88.

Yoon Y, Park J, Allen MG. (2006). Multidirectional UV lithography for complex 3-D MEMS structures. *J Microelectromech Syst.* 15(5):1121–30.

Yu H, Li B, Zhang X. (2006). Flexible fabrication of three-dimensional multi-layered microstructures using a scanning laser system. *Sens Actuators A Phys.* 125:553–64.

Zaouk RB. (2008). Carbon MEMS from the nanoscale to the macroscale: novel fabrication techniques and applications in electrochemistry. PhD in Mechanical and Aerospace Engineering. Irvine: University of California.

Zdeblick MJ, Bartha PW, Angell JA. (1986). Microminiature fluidic amplifier. In: *Technical Digest of the IEEE Solid-State Sensor and Actuator Workshop (Hilton HEad Island, SC).* New York: IEEE.

Zhang J, Chan-Park MB, Conner SR. (2004). Effect of exposure dose on the replication fidelity and profile of very high aspect ratio microchannels in SU-8. *Lab Chip.* 4:646–53.

Zhang J, Tan KL, Hong GD, Yang LJ, Gong HQ. (2001). Polymerization optimization of SU-8 photoresist and its applications in microfluidic systems and MEMS. *J Micromech Microeng.* 11:20–6.

Zhang X, Haswell SJ. (2006). Materials matter in microfluidic devices. *MRS Bull.* 31:95–9.

9

System Integration in Microfluidics

Morteza Ahmadi, John T.W. Yeow, and Mehdi Shahini

CONTENTS

9.1 Introduction

Microelectromechanical systems (MEMS) technology has been a promising technology since the invention of transistors, integrated circuits, and computer chips. However, many researchers believe that the next revolution will occur with the combination of MEMS technology and microfluidics to fabricate integrated microfluidic devices and systems (Beebe et al., 2002; Geschke et al., 2008).

Lab-on-a-chip microfluidic devices are fabricated for particular tasks, but there are common structures and functions in most of them. The components of microfluidic devices that are used most frequently include valves, pumps, reactors, detectors, and sensors. The standard processes in most of the microfluidic devices are sample loading, sample transfer, target separation, sample mixing, temperature measurements, fluorescence detection, and electrochemical sensing.

The first part of a microfluidic device that is in contact with the sample is the sample loading inlet. External or built-in pumps can be used for flowing the sample into the inlet; however, depending on the design and the material used, the sample can be loaded into the device by dipping part of the device into the fluidic sample. Such a chip requires specific tools or processes to move the sample through different steps on the chip. Favorable cells are separated from the rest of the sample and are lysed for DNA extraction. Polymerase chain reaction (PCR) is used for performing DNA amplification in a relatively short amount of time. This is performed such that the DNA population would be able to generate a reasonable detectable signal and an optimized signal-to-noise ratio. A single integrated chip that supports all the applications mentioned above has the advantage of reduced risk of sample loss or sample contamination (Heller and Guttman, 2002). As such, less amount of sample is required for the total analysis.

We start the next section by discussing the importance of system integration and summarizing the common components and processes in integrated microfluidic devices. Afterward, we provide a review on many devices that have used different integration techniques and tools toward specific targets in biomedical engineering. In particular, we focus on the lab-on-a-chip devices in which cell and DNA processing and analysis are performed on a single integrated microfluidic chip. The goal of this chapter is to provide the reader with information about the most important factors to be taken into consideration when either different parts of a single chip or different processes related to segregated chips are combined to make an integrated MEMS microfluidic device. Several examples of partially integrated and fully integrated microfluidic devices are given.

9.2 Integrated Microfluidics: Overview

Technological advances in microfluidics are leading toward more complex devices and applications. Fabrication of a fully functional microfluidic device capable of performing several analytical processes on a single chip requires a higher degree of integration.

9.2.1 The Significance of Integration in Microfluidics

Several microfuidic devices have been already fabricated with the goal of performing only a single task on a microliter volume of sample. The question is why we need to integrate the functions of different such devices on an integrated microfluidic chip while one can perform each task separately on different devices. Some advantages of an integrated microfluidic device over several nonintegrated ones are accuracy, contamination control, ease of use, and low cost.

It is easier to use a single integrated device rather than using several chips. Integrated devices are fabricated when several processes are required to be performed in series and/or in parallel. If the set of processes are in series, the sample is needed to be transferred from one stage to another. In a nonintegrated device, sample transfer can be done using a micropipette, for example. This increases the risk of sample contamination with unwanted agents. Also, more amount of sample is usually required as part of the sample might be wasted in transfer. In a set of parallel processes, an integrated system can be very accurate in controlling the amount of reaction agents and synchronizing the reaction time in different reactors. Although the degree of complexity increases with the integration, the cost of fabrication of an integrated microfluidic device is normally less than the total cost of the equivalent set of devices.

9.2.2 Common Components in Integrated Microfluidics

9.2.2.1 Valves

Valves are designed to control fluid flow and its direction in channels. An ideal valve is the one without any leakage or dead volume. It should also operate without any energy source and in a negligible response time. Depending on the application of the device, there are two types of valves used in microfluidics: active valves that are attached to an external energy source for operation and passive valves that do not need any external energy source to function. There are usually microscale actuators in active valves that control the open–close function of the valve. Two commonly used actuators in active valves are those working electromagnetically (Capanu et al., 2000) or by air pressure (Unger et al., 2000). An example of a passive valve is the one made of hydrogels. The volume of the hydrogel can change depending on a physical or chemical stimulus like pH of the fluid. Such volume change can act as a valve by either blocking or opening the fluid passage according to the pH of the fluid (Beebe et al., 2000). The valves selected for a system can be different depending on the fluid, the number of times that the valve has to function, and the available energy source.

9.2.2.2 Pumps

Pumps are used to provide the necessary force of motion for the fluid flowing in the device. The most important factors in choosing the right pump for an integrated system are its working liquid, maximum flow rate, pressure and pulsation, power consumption, reservoir capacity, and finally its production costs. Normally, the microfabricated pumps use similar actuation mechanisms as the active valves. The actuator controls the displacement of a membrane or a diaphragm placed on the pump chamber. Such displacement controls the flux into and out of the chamber. Other pump designs may benefit from those physical phenomena that are dominant in the microscale, for example, a pump that works with the creation of a bubble in the channel (Geng et al., 2001) or a pump based on osmotic force (Su et al., 2002).

9.2.2.3 Mixers

Mixers are required to produce a homogeneous mixture of two or more fluids. Diffusion is the main phenomenon used in the mixers in microfluidic devices. To increase the efficiency of the mixer, the diffusion distances have to be decreased or the interfacial area has to be increased. In addition to that, it is possible to use actuators and generate turbulence in the mixture to decrease the mixing time. For example, two electrodes can be fabricated in a chamber to generate an oscillating electric field. Such field introduces flow instabilities that lead to a chaotic behavior in the fluid and a decrease of the mixing time (Oddy et al., 2001).

9.2.2.4 Reactors

Microreactors are the parts of microfluidic systems in which biochemical reactions take place. They consist of a reaction chamber with at least two sample inputs and an output.

As such reactions may require controlled conditions, a microreactor is integrated with one or several sensors. If temperature change is important, a heater or a cooler and a thermal sensor, which can be connected together by a feedback circuit, are integrated with the microreactor. Depending on the reaction, other sensors such as a pressure sensor, a pH sensor, or a fluorescent detector might also be present, which are generally linked to valves or pumps through feedback circuits.

9.2.2.5 Detectors and Spectrometers

In addition to the applications of the sensors in controlling the reactions, detectors are required to measure the output of the system. In an integrated microfluidic device, the detectors normally have to be sensitive enough to small amounts of samples (usually in the microliter scale or less) and low concentrations. Detectors are used according to the application of the device and the physical/chemical quantities that need to be measured. Miniature detectors and spectrometers are required for on-chip cell or DNA detection and analysis. The basic idea in most of the on-chip detection techniques is that the DNA or cell samples are tagged by fluorescent solutions, and the samples fluoresce when irradiated with UV light. To collect the fluorescent emission, micro-optics can be integrated with microfluidics. For example, this is achieved by fabrication of optical waveguide micro-optics and microchannels in the same substrate as an integrated monolithic device (Mazurczyk et al., 2006). The optical waveguides are used to excite the fluorescence in the microchannels by directing a laser beam on the sample and collecting the emitted fluorescence. Optical waveguides are fabricated on soda lime glass plates by ion exchange techniques, and microchannels are produced on the same substrate using photolithography and wet etching methods. The integration process is summarized in Figure 9.1.

In another work, filtered silicon detector arrays have been combined with replica-molded elastomeric microchannels to fabricate a miniature fluorescent spectrometer in the visible and near-UV region (Figure 9.2) (Adams et al., 2003). Elastomers are transparent and provide hermetic seals to silicon dioxide. The external light source excites the sample with a wavelength 40–80 nm far from the emitted fluorescent light. A dielectric thin-film mirror can be grown on top of the complementary metal-oxide semiconductor (CMOS) as a filter. Such a filter blocks the external light and passes the emission fluorescent light to be detected by an imager. Without an efficient filter, the imager gets saturated by the excitation beam (Figure 9.3). To create a complete on-chip visible spectrometer, it is required that the light source be fabricated on the chip and be placed on top of the microfluidic channels that are created on the detector array. If LED arrays are used for this purpose, optical cavity filters need to be added in the design to narrow down the source wavelength to the excitation range.

Burns et al. (1998) have reported an integrated nanoliter DNA analysis device in which they have fabricated diode detectors on a silicon substrate, with a thick optical UV filter deposited on them. However, the fluorescence excitation light source is not incorporated in the device. This device is discussed further in the following sections.

9.2.3 Common Processes in Integrated Microfluidics

9.2.3.1 Cell Lysis

Extraction of subcellular materials such as DNA, RNA, and proteins is required for amplification, sample analysis, and detection. Cell lysis can be performed mechanically,

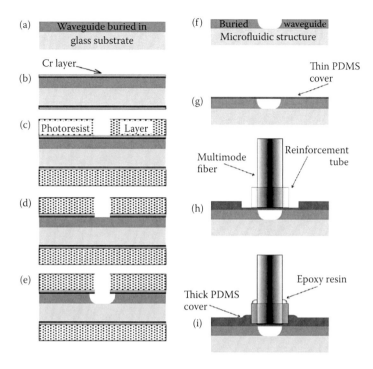

FIGURE 9.1
(a) Integrated waveguides; (b) Cr layer deposition; (c) photolithography; (d) a window is opened in Cr film; (e) microchannels are etched; (f) mask removal; (g) thin PDMS film covers the window; (h) structure reinforcement and fiber alignment; (i) more PDMS cover and gluing the fiber structure using epoxy resin. (From Mazurczyk, R., Vieillard, J., Bouchard, A., Hannes, B., and Krawczyk, S., A novel concept of the integrated fluorescence detection system and its application in a lab-on-a-chip microdevice, *Sensors and Actuators B: Chemical*, 118, 11–19, 2006. With permission.)

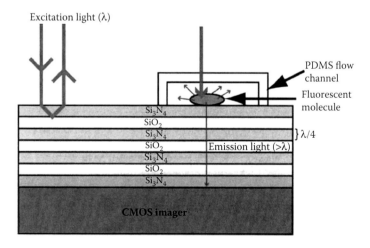

FIGURE 9.2
A commercial CMOS imager integrated in a microfluidic spectrometer chip. (From Adams, M. L., Enzelberger, M., Quake, S., and Scherer, A., Microfluidic integration on detector arrays for absorption and fluorescence micro-spectrometers, *Sensors and Actuators A*, 104, 25–31, 2003. With permission.)

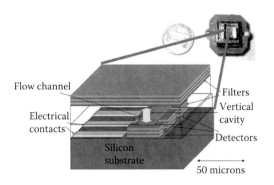

Flow channel
Electrical contacts
Silicon substrate
Filters
Vertical cavity
Detectors
50 microns

FIGURE 9.3

Dielectric filters deposited on the CMOS imager to block the pump wavelength and let the fluorescence signal pass with minimal loss. (From Adams, M. L., Enzelberger, M., Quake, S., and Scherer, A., Microfluidic integration on detector arrays for absorption and fluorescence micro-spectrometers, *Sensors and Actuators A*, 104, 25–31, 2003. With permission.)

electrically, or chemically. Osmotic shock or pressure and shear stress can cause the cell membrane to break mechanically. Sufficiently high electric field can also cause an irreversible cell lysis. In chemical cell lysis, detergents, solvents, or antibiotics can be used. In integrated microfluidic devices, electrical cell lysis is favored because of the simplicity of fabrication, the compatibility with the microscale size, and the control on the applied voltage. The drawback is the size of the high-voltage source required to generate the electric field, although there has been a lot of progress on fabrication of a miniaturized high-voltage power supply for applications in microfluidics (Erickson et al., 2004). In an electric field, a transmembrane potential is induced across the dielectric cell bilayer. If this induced potential is higher than 1 V, then the membrane becomes permeable (Lee & Tai, 1999). Although this process can be reversible in a low external electric field, using a high voltage can cause a permanent damage to the membrane.

9.2.3.2 Polymerase Chain Reaction

Nucleic acid molecules can be amplified *in vitro* by a molecular biological method called PCR. With its wide applications in biology, medical sciences, and chemistry, this method has been of interest for miniaturization and integration in microfluidic devices. In a PCR cycle, the double-stranded DNA is melted, the specific primers get bound to their target sites, and the primers are extended with the thermostable DNA polymerase (enzymes such as Taq). As such, the number of DNA molecules is doubled with each PCR cycle. In PCR, the temperature of the sample and reactants in reaction chambers and the timing are very important. Hence, it is necessary to have a complete temperature control in each reaction chamber. For PCR on a chip, mixers, heaters, coolers, and temperature sensors are required.

For temperature measurements, metal resistance temperature detectors are very popular if the required sensitivity is less than 1% K^{-1}. In particular, thin films of platinum or nickel–iron alloy are deposited and then patterned to form the sensor. For greater temperature sensitivity, ceramic or oxides are used to form thermistors. For integrated microfluidics based on silicon substrates, one can use the silicon resistance temperature sensor with the sensitivity of 0.8% K^{-1} or the diode sensor with the accuracy of 0.1 K (Geschke et al., 2008). Thermocouples based on wires of different metals are also very popular and easy to integrate. Other options for temperature sensing in PCR include fluorescence thermometry (Robinson et al., 2009) and liquid–crystal thermometry (Chaudhari et al., 1998).

PCR can be performed in a continuous flow at a high speed (Kopp et al., 1998). The device has thermostated temperature zones with DNA sample flowing from one to another and to the output. The process from input to output takes only 1.5 to 18.7 min. Giordano et al. (2001) have shown that PCR can be performed in a comparable time (4 min) using an infrared-mediated temperature control system.

9.2.3.3 Electrophoretic Separation

In samples containing more than one component, it is usually necessary to separate one or all of the components to perform different tasks on each. If the molecules are electrically charged, it is easy to separate them in an electric field based on their charge-to-mass ratios. This is called electrophoretic separation. Assuming two electrodes with opposite electric poles at the ends of a microchannel containing a solution full of free positively and negatively charged molecules, one can expect the order of arrival of the molecules to be (i) small, highly positively charged; (ii) larger, less positively charged; (iii) uncharged molecules; (iv) larger, less negatively charged; and (v) small, highly negatively charged. The separation process is very efficient and quick and can even occur in the range of milliseconds (Jacobson et al., 1998). For DNA molecules with close charge-to-mass ratio, the microchannel is filled with a polymer gel that assists the separation.

9.2.3.4 Fluorescence Detection

One of the techniques most frequently applied to measure physical and chemical quantities in microchannels is fluorescence. In this method, fluorescent dyes or beads are added to the system to label the sample. The excitation light is irradiated on the labeled target, the emitted light is collected, and the fluorescence intensity is measured. The spectrum of the applied incident beam on the fluorescent material matches the excitation energy levels of the fluorescent material. Also, the spectrum of the emitted light from the fluorescent material corresponds to the absorption bandwidth of the detector. To deliver the excitation beam and to read the fluorescence emission, micro-optics can be integrated with microfluidics (Mazurczyk et al., 2006). Temperature-sensitive fluorescent dyes are used to monitor the temperature increase of 1 K (Duhr et al., 2004). The fluorescence technique is also used in the study of flow profiles in microchannels. For example, diffusion of low-molecular-weight species across the interface of two aqueous solutions in a laminar flow in microchannels was studied by Ismagilov et al. (2000).

The drawback of the fluorescence technique is the necessity for the use of fluorescent materials and their effect on the sample and the complications because of the integration of mixers. One possibility is to read the fluorescent spectrum of the sample without tagging it with fluorescent dyes. There is a research on fluorescence detection of "unlabeled" basic proteins by means of deep UV at an excitation wavelength of 266 nm (Schulze et al., 2005). In this method, no fluorescent dye is added to the sample; however, the fluorescent detection setup is not integrated with the microchip. This is a major roadblock for many applications that require portability. For those with an integrated fluorescence detector, a coating layer is deposited between the detector and the fluorescence-emitting material such that the layer filters out the excitation light and does not let it reach the detector (Dandin et al., 2007). This coating may not be required if the detector is sensitive only to a narrow bandwidth around the emission wavelength and the excitation wavelength does not overlap with that bandwidth.

9.3 Integrated Microfluidic Devices

The promise of integration in microfluidics is to fabricate devices that, in the sample processing and analysis process, would not need external tools such as pumps, heaters, and lenses. The input of such devices would be microfluidic samples and chemical reagents, and the output could be the analyzed data stored on a memory or shown on a small display.

9.3.1 Integration of Cell Lysis, Capillary Electrophoresis, PCR, and Detection

An integrated DNA analysis system was fabricated by coupling silicon PCR microreactors and glass capillary electrophoresis (CE). The link is a photolithographically fabricated channel filled with hydroxyethylcellulose-sieving matrix. This integrated device was used for PCR amplification and CE separation of β-globin in less than 20 min (Woolley et al., 1996). A similar device with additional capability of handling multiple samples (up to four samples) was fabricated later by another group (Waters et al., 1998b) and was further developed by adding the cell lysis function to its capabilities (Waters et al., 1998a). In the latter device, cells are lysed, and DNAs are amplified by PCR. The products are separated by size in a sieving medium and detected using fluorescence technique.

One of the most complex integrated microfluidic devices has the capability of nanoliter DNA sample analysis (Burns et al., 1998; Mastrangelo et al., 1999). This device comprises injectors, a mixer, a heating chamber, temperature sensors, and a separation channel with fluorescence detectors, all integrated in one device. This chip is capable of moving, amplifying, separating, and detecting the *Mycobacterium tuberculosis* DNA samples. The diode detectors coated with UV filter are placed on the silicon substrate, and electrodes and heaters are fabricated on top of them and passivated with silicon dioxide and *p*-xylylene. An etched glass wafer shapes the fluidic chambers and capillaries. This device has a high degree of complexity in terms of integration and does not need external pumps, heaters, or lenses for full operation (Figure 9.4). The only external connections are input electronic control lines, data output lines, and an air pressure manifold.

In Table 9.1, we summarize and compare integrated microfluidic systems with different degrees of integration and complexity.

9.3.2 Polymer-Based Microfluidic Devices

Elastomers are used extensively in the fabrication of microfuidic devices. They are used to fabricate patterns and structures with feature sizes as small as 30 nm (Xia & Whitesides, 1998). Polydimethylsiloxane (PDMS) is a nontoxic biocompatible siloxane-based elastomer. PDMS is porous to certain materials, very flexible, and optically transparent at more than 300-nm wavelengths. Because of the porosity of PDMS, it is difficult to fabricate an efficient device working with small amounts of samples. To solve this problem, Parylene coating of PDMS is suggested (Shin et al., 2003). Parylene film reduces the moisture permeability of PDMS and increases its stability. PDMS is used in the fabrication of microchannels, valves, pumps, and seals (De Volder et al., 2007). PDMS shows exceptional valving/switching

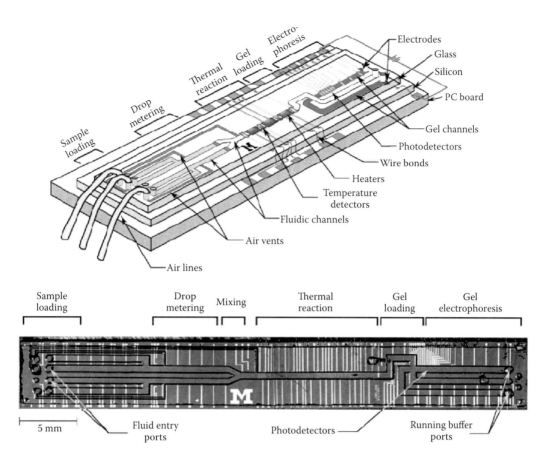

FIGURE 9.4
(Top) Schematic of integrated device fabricated by Burns et al. (1998). (Bottom) Optical micrograph of the device from above. The pressure manifold and buffer wells that fit over the entry holes at each end of the device are not shown.

capabilities, which make it a good material for fabrication of cell sorters (Fu et al., 1999, 2002). For on-chip detection applications, the microfluidic channels have to be transparent to excitation light, and PDMS has this advantage. It is also relatively easy to work with PDMS and use soft lithography techniques to fabricate microfluidic channels on a spectrometer chip (Eteshola & Balberg, 2004). The flexibility of PDMS is a key parameter in applications of this polymer in tunable gratings for pressure sensing in microfluidics (Hosokawa et al., 2002) and PDMS two-dimensional varifocal optical lenses integrated with microfluidic channels (Camou et al, 2003).

Zare's laboratory has developed a single-cell analysis chip (Huang et al., 2007), which is able to manipulate, lyse, label, separate, and quantify the protein content of a single cell using single-molecule fluorescence counting. The single cell is captured and lysed, and the lysate is chemically separated and analyzed. This provides a tool to understand the functioning of cells. The device is fabricated using multilayer soft lithography techniques, and the main material used in it is PDMS (Wheeler et al., 2003).

In addition to PDMS, there are several other elastomers and plastics that are used in microfluidics. One of the best examples of fully integrated microfluidic devices is the

TABLE 9.1

Integrated Microfluidic Systems

References	Integrated Functions	Integrated Components/Materials	Advantages	Disadvantages
Lee and Tai (1999)	Cell lysis	Silicon	A microsized device	Single task
Jacobson, Culbertson, Daler, and Ramsey (1998)	Electrophoresis	Glass	Resolve a binary mixture in 0.8 ms	Single task
Shin et al. (2003)	PCR	Parylene-coated PDMS	Disposable; volume of the PCR mixture as low as 2 μL	Single task
Kopp et al. (1998)	PCR amplification of a 176-base pair fragment from the DNA gyrase gene of *Neisseria gonorrhoeae*	Glass	Continuous flow; high speed (1.5–18.7 min)	Single task
Giordano et al. (2001)	PCR amplification of a 500-base-pair fragment of λ-phage DNA	Polyimide	High speed (amplification time <240 s)	Single task
Hong et al. (2001)	PCR, CE	PDMS glass	Disposable	Volume of the PCR mixture: 30–50 μL
Woolley et al. (1996)	PCR, CE	Double-sided polished silicon 100 wafer	High-speed PCR, total analysis in <20 min	Partial integration
Schulze et al. (2005)	Electrophoresis, separation, and detection of egg white chicken proteins	Fused silica chips coated with poly(vinyl alcohol)	Fluorescence detection of "unlabeled" drugs and proteins	Partial integration, 266-nm UV source and microscope not integrated on microchip
Mazurczyk et al. (2006)	Separation and detection of the mixture of Cy3 fluorescent dye and Cy3-tagged streptavidine	Soda lime glass plates	Integration of micro-optics (optical waveguides and fiber optics) with microfluidics	Extra size due to integrated optics
Waters et al. (1998a)	Cell lysis, PCR, electrophoretic sizing	Glass	Up to four samples at the same time	Not disposable; expensive
Burns et al. (1998)	Sample mixing and positioning, temperature-controlled reactions, electrophoretic separation, fluorescence detection	Glass and silicon	Nanoliter DNA sample; fully integrated	Light source not integrated; not disposable; expensive
Liu et al. (2004)	Sample preparation (cell capture, purification, and concentration), cell lysis, PCR, detection	Polycarbonate plastic substrate	Fully integrated; self-contained; no external pressure source, fluid storage, or mechanical pump	Not disposable; expensive

work by Liu et al. (2004). This fully automated device is capable of cell capture, concentration and purification, cell lysis, PCR, DNA hybridization, and detection. This device is very self-contained, and all the fabricated fluid control tools on it make it needless of any external pump or valve (Figure 9.5). Four pumps are fabricated in this chip: the first three pumps are electrochemical pumps and the fourth pump is a thermopneumatic pump. Chambers in this device are used for more than one function. For example, only one chamber is used for cell capture, preconcentration, cell lysis, and PCR. Although the fluidic parts are machined on a polycarbonate plastic substrate, the electrical controls are printed on a circuit board. There is also a sensor attached to the circuit board for measuring the electrochemical signal of the hybridized DNA. The plastic part, the circuit board, and the DNA sensor were attached together by double-sided adhesive tapes. The device was used to detect pathogenic bacteria (*Escherichia coli*) in rabbit whole-blood sample.

9.3.3 Paper-Based Microfluidic Devices

Martinez et al. (2007) and Bracher et al. (2009) have developed microfluidic devices based on paper. Paper is ubiquitous and inexpensive, and the fabrication process for making these devices is not complicated. Biological fluids can get defused in paper easily; therefore, such devices do not require external tools for sample loading. Such devices are fabricated in paper using photolithography and are functionalized with reagents for colorimetric assays (Figure 9.6). Calibration curves are then used to compare the color intensities in each assay. This system has been successful in the determination of glucose and protein concentration in artificial urine. Another advantage of using paper is the relatively easy integration process to make three-dimensional microfluidic devices. This is possible by stacking multiple layers of patterned papers on top of each other with double-sided adhesive tapes in between. Such three-dimensional microfluidic devices are capable of wicking microliter fluids from single input and distributing them into thousands of detection zones. Most integrated microfluidic devices require multilevel fabrication of electronics and microchannels using hard substrates such as silicon and glass, which is not a trivial task. However, three-dimensional stacking of patterned papers is not very complex and it is cheap. This makes it

Integrated fluidic chip

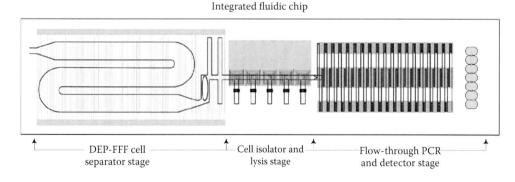

|←——— DEP-FFF cell ———→|←— Cell isolator and —→|←———Flow-through PCR———→|
separator stage   lysis stage   and detector stage

FIGURE 9.5
(Left) The plastic microfluidic chip including units for mixing, cell preconcentration, purification, lysis, PCR, and DNA microarray chamber. (Right) The printed circuit board comprising resistive heaters, control unit, and attached DNA electrochemical signal reader (eSensor). (From Liu, R. H., Yang, J., Lenigk, R., Bonanno, J., and Grodzinski, P., Self-contained, fully integrated biochip for sample preparation, polymerase chain reaction amplification, and DNA microarray detection, *Analytical Chemistry*, 76, 1824–1831, 2004. With permission.)

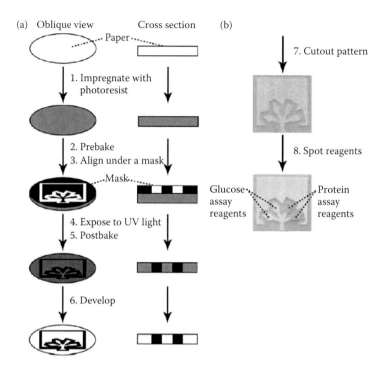

FIGURE 9.6
Fabrication of paper-based microfluidic devices. (a) Hydrophobic photoresist is used for patterning. (b) Applying reagents on the assay. (From Martinez, A. W., Phillips, S. T., Carrilho, E., Thomas, S. W., Sindi, H., and Whitesides, G. M., Simple telemedicine for developing regions: camera phones and paper-based microfluidic devices for real-time, off-site diagnosis, *Analytical Chemistry*, 80, 3699–3707, 2008. With permission.)

a good candidate for healthcare applications in places where there is a lack of medical laboratories and specialists. An example of how to integrate such a system is given in Figure 9.7. The efficiency of this integrated system has been tested by passing streams of fluids across each other multiple times, and the figure illustrates that there was no mixing. In Figure 9.7e, a channel that connects the top paper layer to the bottom one is shown. The cross section of orthogonal channels is also presented in Figure 9.7f.

9.4 Large-Scale Integrated Microfluidics

Many lab-on-a-chip devices comprise only a small number of microfluidic channels. However, like integrated circuits in microelectronics, large-scale integration is necessary for more complicated microfluidic devices. According to Thorsen et al. (2002), multiplexed control of leakproof and scalable monolithic microvalves is the key requirement for large-scale integration in microfluidics. Microfluidic multiplexers are combinatorial arrays of binary valve patterns and allow control of n fluid channels with only $2 \log_2 n$ control channels (Figure 9.8). For example, only 20 control channels are required to harbor 1024 flow channels.

To fabricate such a complex pattern containing two layers, multilayer soft lithography techniques can be used (Unger et al., 2000). Using this technique, elastomers can be applied

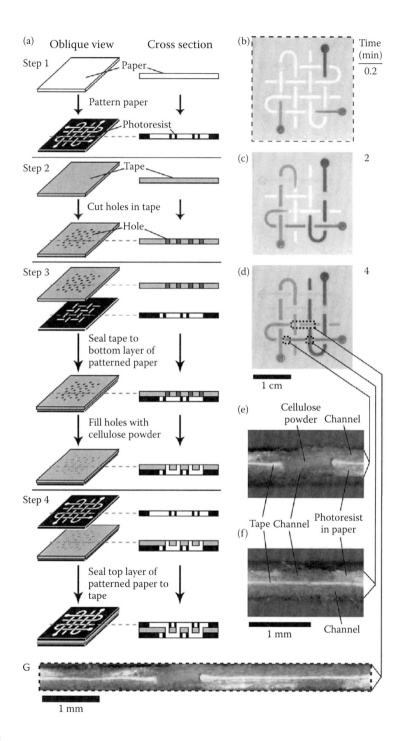

FIGURE 9.7

(a) Fabrication process of an integrated three-dimensional microfluidic assay. (b) Ten minutes after filling the sample reservoirs with red, yellow, green, and blue aqueous solutions of dyes. (From Martinez, A. W., Phillips, S. T., and Whitesides, G. M., Three-dimensional microfluidic devices fabricated in layered paper and tape. *PNAS*, 105, 19606–19611, 2008. With permission.)

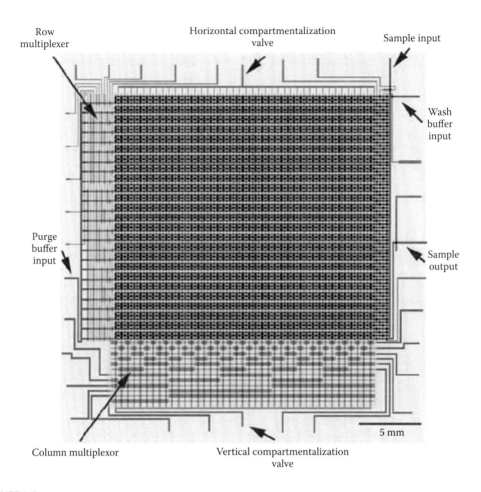

FIGURE 9.8
Large-scale integrated microfluidic chip with multiplexers. (From Thorsen, T., Maerkl, S. J., and Quake, S., Microfluidic large-scale integration, *Science*, 298, 580–584, 2002. With permission.)

to fabricate on–off valves, switching valves, and pumps. Elastomers have advantages over silicon, such as device areas reduced by two orders of magnitude, rapid prototyping, and ease of fabrication. The bottom layer contains the network of microfluidic flow channels and the top layer controls all channels required to actuate the valves. The intersection of each flow channel with a control channel is a valve. Steel pins inserted into holes punched through the silicone are used to introduce the fluid into the system. Silicone provides a tight seal around the pins up to 300 kPa. The multiplexers are controlled by external solenoid valves, which are the connections between the microfluidic chip and a computer. The chip contains an array of 25 × 40 number of ~250-pL volume chambers, each accessed by horizontal and vertical multiplexers. Two separate inputs and relevant outputs are made such that each chamber can contain water or color dye. There are three microchannels located in each row in parallel. To purge the fluid from a particular chamber, pressurized fluid is introduced in the purge buffer (water) input, and the row multiplexer leads the fluid toward the lower channel of the selected row. The column multiplexer opens the vertical valves of the chamber, and the pressurized fluid flows through the chamber and purges the contents of the chamber. Figure 9.9 illustrates how the microfluidic multiplexers work.

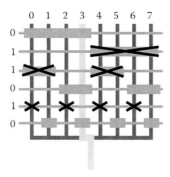

FIGURE 9.9
Vertical lines are microfluidic channels containing the sample. Horizontal lines represent the control channels. The intersection of the wide part of a control channel with a flow channel is a valve. Cross signs show that the valves are closed. In this diagram, all channels are closed except the fourth from left. (From Thorsen, T., Maerkl, S. J., and Quake, S., Microfluidic large-scale integration, *Science*, 298, 580–584, 2002. With permission.)

9.5 Digital Microfluidics

Digital microfluidics is a novel concept of integration on the basis of the behavior of discrete pico–microliter droplets. An array of electrodes can generate electric fields such that the droplets of sample and chemical reactants can be manipulated and controlled. Therefore, samples and chemical agents can be addressed individually in digital microfluidics. This is the key distinction between microchannel microfluidics and digital microfluidics. As digital microfluidics is mainly based on arrays of electrodes, the geometry of the fabricated devices and the physics of the system are different in comparison with the microchannel devices. For example, the sample droplet can be moved, merged, and split into smaller independent droplets. Droplets can be controlled on the basis of two different electrode configurations: single-plate format and two-plate format. In the single-plate format, the droplet is placed on a set of electrodes on a single substrate. In the two-plate format, the droplet is sandwiched between two substrates with electrodes fabricated on them. To avoid electrolysis and to have a better control on the sample, a thin film of dielectric material and a layer of hydrophobic material are coated on the electrodes. If electrodes are coated on flexible substrates, the device will be mechanically flexible (Abdelgawad et al., 2008). There has been a lot of progress in applications of digital microfluidics in the fabrication of integrated microfluidic devices; however, devices in this field have not advanced as much as microchannel devices. One of the best examples of such devices is a PCR chip based on digital microfluidics (Chang et al., 2006). In this device, sample and PCR reagent droplets are merged and mixed and then transferred to the on-chip heater. The fluorescent signal from amplified DNA is then measured.

9.6 Conclusion

In this chapter, we reviewed many integrated microfluidic devices. There are several devices made for the purpose of accomplishing a single task; however, system integration provides the bridge to the fabrication of devices that perform multitasking. A lab-on-a-chip device targeted for biomedical usage is usually required to handle biosampling, cell manipulation, cell lyses, PCR, and detection on a single chip. Such devices are fabricated using biocompatible materials and require small amounts of fluidic sample equal or less than microliter for functioning. If the techniques and materials used in the fabrication make the device cheap enough for mass production, the device is expected to be commercialized.

References

Abdelgawad M, Freire SL, Yang H, Wheeler AR. (2008). All-terrain droplet actuation. *Lab Chip*. 8:672–7.

Adams ML, Enzelberger M, Quake S, Scherer A. (2003). Microfluidic integration on detector arrays for absorption and fluorescence micro-spectrometers. *Sens Actuators A Phys*. 104:25–31.

Beebe DJ, Mensing GA, Walker GM. (2002). Physics and applications of microfluidics in biology. *Annu Rev Biomed Eng*. 4:261–86.

Beebe D, Moore JS, Bauer JM, Yu Q, Liu RH, Devadoss C, et al. (2000). Functional hydrogel structures for autonomous flow control inside microfluidic channels. *Nature*. 404:588–90.

Bracher PJ, Gupta M, Whitesides GM. (2009). Shaped films of ionotropic hydrogels fabricated using templates of patterned paper. *Adv Mater*. 21:445–50.

Burns MA, Johnson BN, Brahmasandra SN, Handique K, Webster JR, Krishnan M, et al. (1998). An integrated nanoliter DNA analysis device. *Science*. 282:484–7.

Camou S, Fujita H, Fujii T. (2003). PDMS 2D optical lens integrated with microfluidic channels: principle and characteristion. *Lab Chip*. 3:40.

Capanu M, Boyd J, Hesketh P. (2000). Design, fabrication, and testing of a bistable electromagnetically actuated microvalve. *J Microelectromech Syst*. 9:181–9.

Chang Y-H, Lee G-B, Huang F-C, Chen Y-Y, Lin J-L. (2006). Integrated polymerase chain reaction chips utilizing digital microfluidics. *Biomed Microdevices*. 8:215.

Chaudhari A, Woudenberg T, Albin M, Goodson K. (1998). Transient liquid crystal thermometry of microfabricated PCR vessel arrays. *J Microelectromech Syst*. 7:345.

Dandin M, Abshire P, Smela E. (2007). Optical filtering technologies for integrated fluorescence sensors. *Lab Chip*. 7:955.

De Volder M, Ceyssens F, Reynaerts D, Puers R. (2007). A PDMS lipseal for hydraulic and pneumatic microactuators. *J Micromech Microeng*. 17:1232.

Duhr S, Arduini S, Braun D. (2004). Thermophoresis of DNA determined by microfluidic fluorescence. *Eur Phys J E Soft Matter*. 15:277–86.

Erickson D, Sinton D, Li D. (2004). A miniaturized high-voltage integrated power supply for portable microfluidic applications. *Lab Chip*. 4:87–90.

Eteshola E, Balberg M. (2004). Microfluidic ELISA: on-chip flurescence imaging. *Biomed Microdevices*. 6:7–9.

Fu AY, Chou H-P, Spence C, Arnold F, Quake S. (2002). An integrated microfabricated cell sorter. *Anal Chem*. 74:2451–7.

Fu AY, Spence C, Scherer A, Arnold FH, Quake S. (1999). A microfabricated fluorescence-activated cell sorter. *Nat Biotechnol*. 17:1109–11.

Geng X, Yuan H, Oguz H, Prosperetti A. (2001). Bubble-based micropump for electrically conducting liquids . *J Micromech Microeng*. 11:270–6.

Geschke O, Klank H, Telleman P. (2008). *Microsystem Engineering of Lab-on-a-Chip Devices*. Weinheim: Wiley-VCH Verlag GmbH & Co.

Giordano B, Ferrance J, Swedberg S, Hühmer A, Landers J. (2001). Polymerase chain reaction in polymeric microchips: DNA amplification in less than 240 seconds. *Anal Biochem*. 291:124–32.

Heller MJ, Guttman A. (2002). *Integrated Microfabricated Biodevices: Advanced Technologies for Genomics, Drug Discovery, Bioanalysis, and Clinical Diagnostics.* New York: Marcel Dekker, Inc.

Hong J, Fujii T, Seki M, Yamamoto T, Endo I. (2001). Integration of gene amplification and capillary gel electrophoresis on a polydimethylsiloxane-glass hybrid microchip. *Electrophoresis.* 22:328–33.

Hosokawa K, Hanada K, Maeda R. (2002). A polydimethylsiloxane (PDMS) deformable diffraction grating for monitoring of local pressure in microfluidic devices. *J Micromech Microeng.* 12:1.

Huang B, Wu H, Bhaya D, Grossman A, Granier S, Kobilka BK, et al. (2007). Counting low-copy number proteins in a single chip. *Science.* 315:81–4.

Ismagilov RF, Stroock AD, Kenis PJ, Whitesides G, Stone HA. (2000). Experimental and theoretical scaling laws for transverse diffusive broadening in two-phase laminar flows in microchannels. *Appl Phys Lett.* 76:2376–9.

Jacobson S, Culbertson C, Daler J, Ramsey J. (1998). Microchip structures for submillisecond electrophoresis. *Anal Chem.* 70:3476–80.

Kopp MU, de Mello AJ, Manz A. (1998). Chemical amplification: continuous-flow PCR on a chip. *Science.* 280:1046–8.

Lee SW, Tai YC. (1999). A micro cell lysis device. *Sens Actuators A Phys.* 73:74–9.

Liu RH, Yang J, Lenigk R, Bonanno J, Grodzinski P. (2004). Self-contained, fully integrated biochip for sample preparation, polymerase chain reaction amplification, and DNA microarray detection. *Anal Chem.* 76:1824–31.

Martinez AW, Phillips ST, Whitesides GM. (2008a). Three-dimensional microfluidic devices fabricated in layered paper and tape. *Proc Natl Acad Sci USA.* 105:19606–11.

Martinez AW, Phillips ST, Butte MJ, Whitesides GM. (2007). Patterned paper as a platform for inexpensive, low-volume, portable bioassays. *Angew Chem Int Ed Engl.* 46:1318–20.

Martinez AW, Phillips ST, Carrilho E, Thomas SW, Sindi H, Whitesides GM. (2008b). Simple telemedicine for developing regions: camera phones and paper-based microfluidic devices for real-time, off-site diagnosis. *Anal Chem.* 80:3699–707.

Mastrangelo CH, Burns MA, Burke DT. (1999). Integrated microfabricated devices for genetic assays. Microprocesses and Nanotechnology Conference, Yokohama, Japan; pp. 1–2.

Mazurczyk R, Vieillard J, Bouchard A, Hannes B, Krawczyk S. (2006). A novel concept of the integrated fluorescence detection system and its application in a lab-on-a-chip microdevice. *Sens Actuators B Chem.* 118:11–9.

Oddy MH, Santiago JG, Mikkelsen JC. (2001). Electrokinetic instability micromixing. *Anal Chem.* 73:5822–32.

Robinson T, Schaerli Y, Wootton R, Hollfelder F, Dunsby C, Baldwin G, et al. (2009). Removal of background signals from fluorescence thermometry measurements in PDMS microchannels using fluorescence lifetime imaging. *Lab Chip.* 9:3437.

Schulze P, Ludwig M, Kohler F, Belder D. (2005). Deep UV laser-induced fluorescence detection of unlabeled drugs and proteins in microchip electrophoresis. *Anal Chem.* 77:1325–9.

Shin YS, Cho K, Lim SH, Chung S, Park S-J, Chung C, et al. (2003). PDMS-based micro PCR chip with Parylene coating. *J Micromech Microeng.* 13:768.

Su Y, Lin L, Pisano A. (2002). Water-powered osmotic microactuator. *J Microelectromech Syst.* 11:736–42.

Thorsen T, Maerkl SJ, Quake S. (2002). Microfluidic large-scale integration. *Science.* 298:580–4.

Unger MA, Chou H-P, Thorsen T, Scherer A, Quake S. (2000). Monolithic microfabricated valves and pumps by multilayer soft lithography. *Science.* 288:113–6.

Waters LC, Jacobson SC, Kroutchinina N, Khandurina J, Foote RS. (1998a). Microchip device for cell lysis, multiplex PCR amplification, and electrophoretic sizing. *Anal Chem.* 70:158–62.

Waters LC, Jacobson SC, Kroutchinina N, Khandurina J, Foote RS, Ramsey JM. (1998b). Multiple sample PCR amplification and electrophoretic analysis on a microchip. *Anal Chem.* 70:5172–6.

Wheeler AR, Throndset WR, Whelan R, Leach AM, Zare RN, Liao YH, et al. (2003). Microfluidic device for single-cell analysis. *Anal Chem.* 75:3581–6.

Woolley AT, Hadley D, Landre P, deMello AJ, Mathies RA. (1996). Functional integration of PCR amplification and capillary electrophoresis in a microfabricated DNA analysis device. *Anal Chem.* 68:4081–6.

Xia Y, Whitesides GM. (1998). Soft lithography. *Annu Rev Mater Sci.* 28:153.

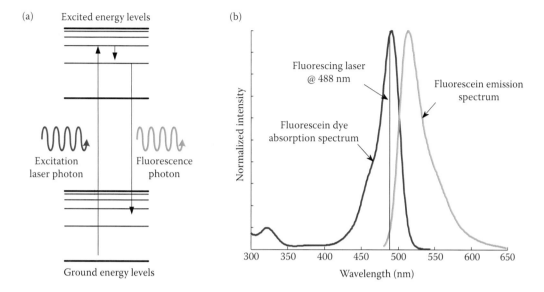

FIGURE 1.2
Schematics of the fluorescence process shown in (a) the excitation process transfer level and (b) the adsorption and emission characteristics of fluorescein.

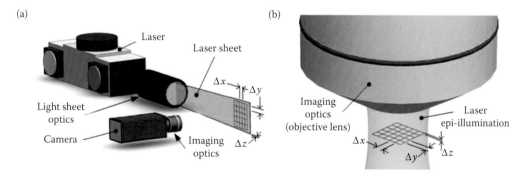

FIGURE 1.7
Schematics comparing the arrangement of hardware and defined region of interest and resolution for (a) macroscale PIV and (b) microscale PIV.

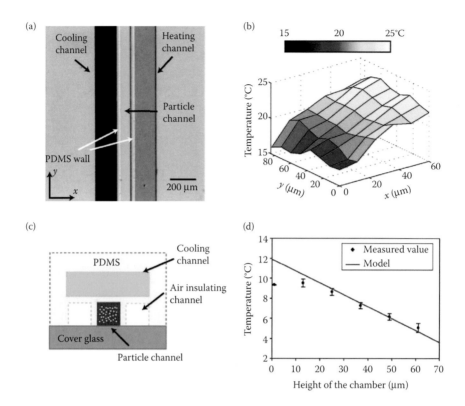

FIGURE 2.31
(a) Optical micrograph of the microdevice fabricated using soft lithography. (b) Temperature distribution measurement. Hot and cold streams were at ~35°C and ~5°C, respectively. (c) Cross-sectional view of microdevice to generate z-depth temperature distribution. (d) Measured temperature distribution in z-direction. (From Chung, K., Cho, J.K., Park, E., Breedveld, V., and Lu, H., *Anal. Chem.*, 81, 991–999, 2009. With permission.)

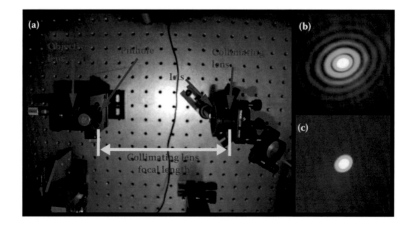

FIGURE 3.16
Spatial filter schematic and resulting laser beam modes: (a) spatial filter components and orientation, (b) concentric rings produced by diffraction through the spatial filter pinhole, and (c) resulting Gaussian beam spot produced by clipping the concentric rings with an iris.

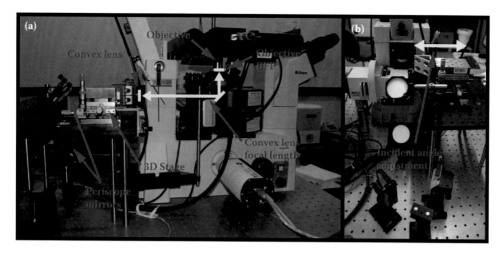

FIGURE 3.17
Schematic of the periscope and beam angle control lens orientation in relation to the microscope: (a) by focusing the beam onto the BFP of the objective, a collimated beam emerges from the microscope objective and (b) the proper translational adjustment direction for manipulating the incident angle into TIRF mode.

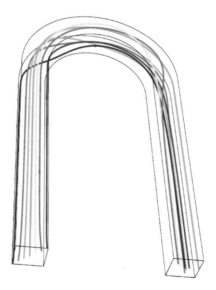

FIGURE 6.2
The streamlines of the fluid particles released from seven equally spaced point along the diagonal of the channel entrance at De = 300.

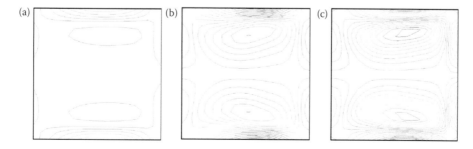

FIGURE 6.3
Planar vorticity contours at the inlet (a), the middle (b), and the outlet (c) of the curved section of channel.

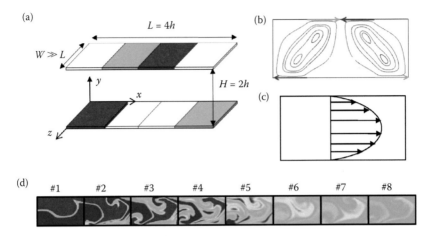

FIGURE 6.4
Schematics of the continuous chaotic stirrer developed by Kim and Beskok [55]. The stirrer consists of periodically repeating mixing bocks with zeta potential patterned surfaces (a) and an electric field parallel to the x-axis is externally applied resulting in (b) an electro-osmotic flow. Combining a unidirectional (x-direction) pressure-driven flow (c) with electro-osmotic flow under time-periodic external electric field (in the form of a cosine wave with a frequency ω), a two-dimensional time-periodic flow is induced to achieve chaotic stirring in the mixer. Two fluid streams colored with red and blue are pumped into the mixer from the left and are almost well mixed after eight repeating mixing blocks for Re = 0.01, St = $1/2\pi$, Pe = 1000, and A = 0.8 (d).

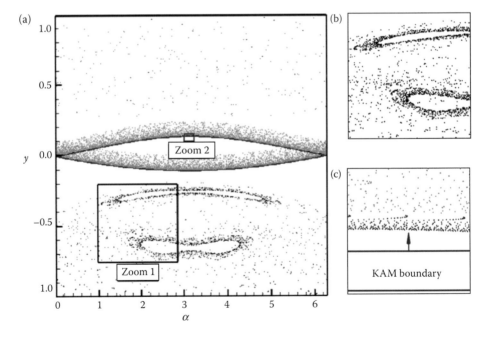

FIGURE 6.5
Poincaré section for four particles initially located at (0.005, –0.5)—black, (0.005, 0.0)—red, (0.005, 0.5)—green, and (0.005, 1.0)—blue, at stirring conditions of St = $1/4\pi$, Re = 0.01, and A = 0.8 (a). The Poincaré sections in the zoomed area 1 with two void zones (b), and the global regular pattern in the zoomed area 2 (c).

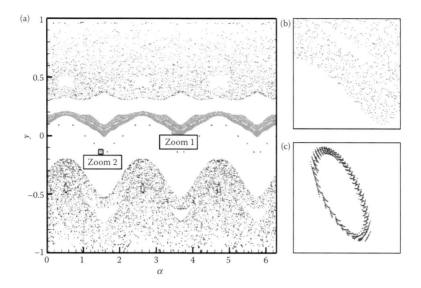

FIGURE 6.6
Poincaré section for the same four particles in Figure 6.5 at stirring conditions of St $= 3/4\pi$, Re $= 0.01$, and $A = 0.8$ (a). The Poincaré sections in the zoomed areas 1 and 2 are shown, respectively, in panels b and c.

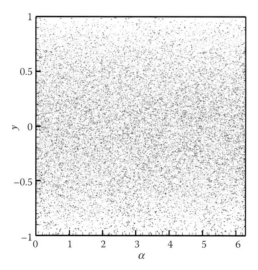

FIGURE 6.7
Poincaré section for the same four particles in Figure 6.5 at stirring conditions of St $= 1/2\pi$, Re $= 0.01$, and $A = 0.8$.

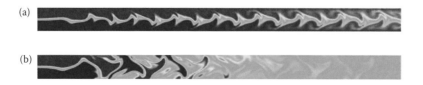

FIGURE 6.9
Dimensionless species concentration distribution for mixing at Pe $= 1000$, Re $= 0.01$, and $A = 0.8$, with stirring at (a) St $= 1/4\pi$ and (b) St $= 1/2\pi$.

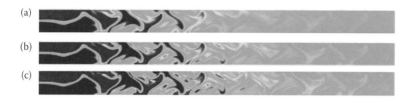

FIGURE 6.10
Dimensionless species concentration distribution for Pe = 500 (a), 1000 (b), and 2000 (c) at St = $1/2\pi$, Re = 0.01, and A = 0.8.

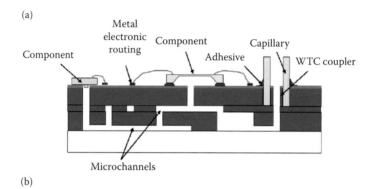

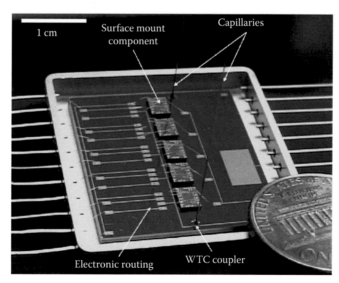

FIGURE 10.2
Modular microfluidic board approach: (a) illustration of example board technology with electronic and fluidic routing between surface mounted components and coupling to off-chip devices; (b) photograph of example microfluidic board in silicon with five mounted components (pressure sensors), embedded microchannels, surface metal electronic routing, and WTC interconnect to capillary tubing. (Adapted from B.L. Gray, D. Jaeggi, N.J. Mourlas, B.P. van Drieënhuizen, K.R. Williams, N.I. Maluf, G.T.A. Kovacs, Novel interconnection technologies for integrated microfluidic systems, *Sensors and Actuators A*, 77, 57–65, 1999.)

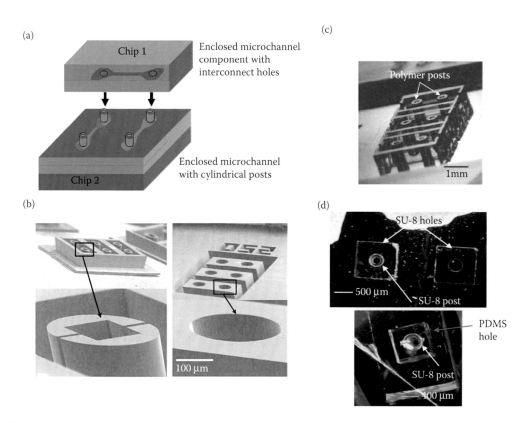

FIGURE 10.6
Examples of modular interconnect scheme on the basis of cylindrical peg-in-hole press fits: (a) overall concept (From Westwood, S., Jaffer, S., Gray, B.L. Enclosed SU-8 and PDMS microchannels with integrated interconnects for chip-to-chip and world-to-chip connections, *Journal of Micromechanics and Microengineering*, 18, 064014, 2008.); (b) example silicon mating joints with notched cylinders and holes requiring gasketing. As silicon is not compliant, the cylinders are notched to provide a spring force to hold the cylinders in the holes (Adapted from Gray, B.L., Collins, S.D., Smith, R.L., Interlocking mechanical and fluidic interconnections for microfluidic circuit boards, *Sensors and Actuators A Physics*, 112, 18–24, 2004.); (c) example polymer cylinders that do not require notches; and (d) example polymer mating joints on the basis of polymer compliancy press fits between holes and cylinders. (Panels c and d from Jaffer, S. and Gray, B.L., Polymer mechanically interlocking structures as interconnect for microfluidic systems, *Journal of Micromechanics and Microengineering*, 18, 035043, 2008.)

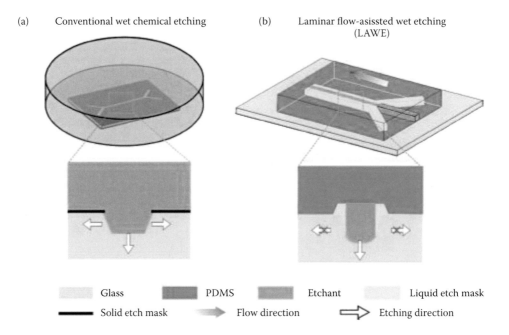

(a) Conventional wet chemical etching

(b) Laminar flow-asissted wet etching (LAWE)

Glass	PDMS	Etchant	Liquid etch mask
Solid etch mask	Flow direction	Etching direction	

FIGURE 11.5

Schematic representation of (a) conventional wet chemical etching where a solid mask is usually deposited on glass surface and (b) LAWE that require a PDMS layer with channels, which is sealed against a glass plate. (From Mu X, Liang Q, Hu P, Ren K, Wang Y, Luo G. Laminar flow used as liquid etch mask in wet chemical etching to generate glass microstructures with an improved aspect ratio. *Lab Chip*, 9, 2009, pp. 1994–1996. Reprinted with permission from The Royal Society of Chemistry.)

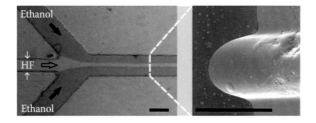

FIGURE 11.6

Microchannel with a central HF flow and two ethanol lateral flows at a velocity of 0.1 ms^{-1}. After a 13-min etch, the aspect ratio by LAWE is up to approximately 1. (From Mu X, Liang Q, Hu P, Ren K, Wang Y. Luo G. Laminar flow used as liquid etch mask in wet chemical etching to generate glass microstructures with an improved aspect ratio. *Lab Chip*, 9, 2009, pp. 1994–1996. Reprinted with permission from The Royal Society of Chemistry.)

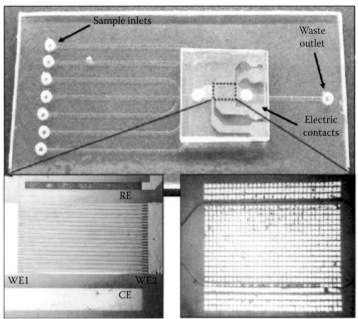

FIGURE 11.10
Photograph of the COC-based lab on a chip with a magnified view. The polymeric microfluidic was fabricated using the micropolymer injection molding technique with an electroplated Ni mold. (From Do J, Ahn CH. A polymer lab-on-a-chip for magnetic immunoassay with on-chip sampling and detection capabilities. *Lab Chip*, 8, 2008, pp. 542–549. Reprinted with permission from The Royal Society of Chemistry.)

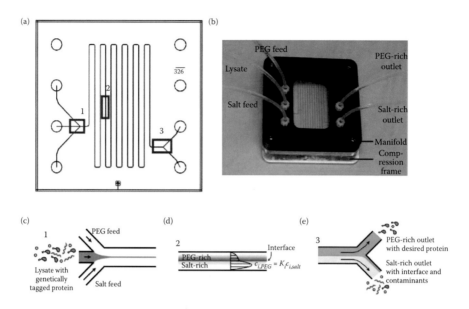

FIGURE 11.19

(a) Schematic representation of the microchannel design with three inlets, a serpentine channel with 326 mm in length and two outlets, all on a glass chip with 37 mm². (b) Photograph of the microfluidic system mounted in a Delrin manifold with inlets and outlets connections. (c) Zoom of region 1 of (a) that represents the cell lysate stream containing a genetically tagged protein of interest along with undisturbed contaminants hydrodynamically focused—between inlet stream containing PEG and salt potassium phosphate. (d) Zoom of region 2 of (a) that represents the farther downstream, laminar two-phase flow with a stable interface. Components from the cell lysate stream partition between the two phases, with the genetically tagged protein (green) strongly partitioning toward the PEG-rich phase. An approximate velocity profile is represented to show high flow velocity in the less viscous salt-rich phase. (e) End of the channel (zoom of region 3 of panel a), where the flow is split into two outlet streams. The tagged protein is concentrated in the PEG-rich outlet stream, whereas undesired proteins and slowly diffusing macromolecules are directed toward the salt-rich outlet along with the phase interface. (From Meagher RJ, Light YK, Singh, AK. Rapid, continuous purification of proteins in a microfluidic device using genetically-engineered partition tags. *Lab Chip*, 8, 2008, pp. 527–532. Reprinted with permission from The Royal Society of Chemistry.)

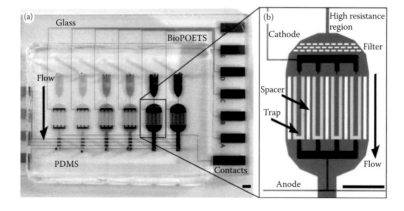

FIGURE 11.20

Integrated microfluidic cell culture and lysis on a chip: (a) top view of the chip composed by six separate devices (filled with colored dye for visualization); (b) magnified image of one chamber. (From Nevill JT, Cooper R, Dueck M, Breslauer DN, Lee LP. Integrated microfluidic cell culture and lysis on a chip. *Lab Chip*, 7, 2007, pp. 1689–1695. Reprinted with permission from The Royal Society of Chemistry.)

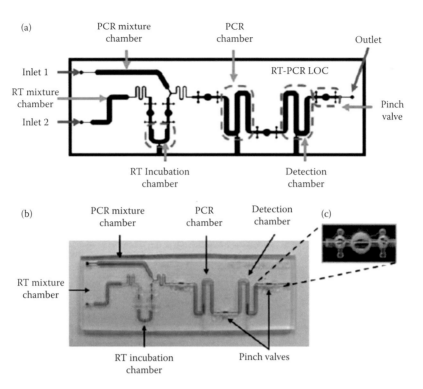

FIGURE 11.21
RT-PCR lab on a chip: (a) schematic representation; (b) photograph; (c) embedded micropinch valves. (From Lee SH, Kim SW, Kang JY, Ahn CH. A polymer lab-on-a-chip for transverse transcription (RT)-PCR based point-of-care clinical diagnostics. *Lab Chip* 8, 2008, pp. 2121–2127. Reprinted with permission from The Royal Society of Chemistry.)

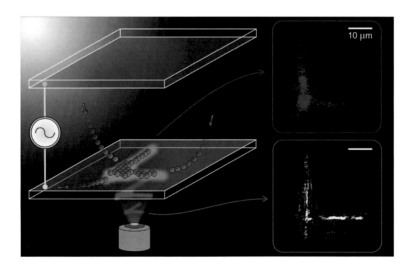

FIGURE 13.4
An example of the REP manipulation technique. An intense laser hologram is shone on an ITO surface, and the 690-nm red fluorescent PS beads take the shape of the hologram. (From Williams, S.J., Kumar, A., and Wereley, S.T., Electrokinetic patterning of colloidal particles with optical landscapes, *Lab on a Chip* 8(11), 1879–1882, 2008. Reproduced by permission of The Royal Society of Chemistry.)

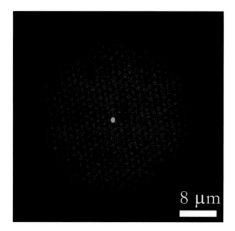

FIGURE 13.5
Hundreds of red fluorescent 1-μm particles form an REP aggregation on an electrode surface. The green dot in the middle shows the approximate location and size of the focused laser spot, which creates the REP aggregation.

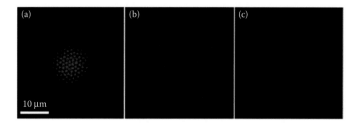

FIGURE 13.6
(a) An REP aggregation of 1.0-μm PS beads in a low-conductivity electrolyte solution, created by activating infrared illumination and electric field; (b) the laser is subsequently deactivated. The particles are scattered parallel to the electrode surfaces, indicating that (c) the electric field is subsequently deactivated. (From Williams, S.J., Kumar, A., Green, N.G., and Wereley, S.T., Optically induced electrokinetic concentration and sorting of colloids, *Journal of Micromechanics Microengineering* 20, 2010. Reproduced with permission from IOP Publishing.)

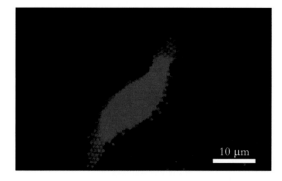

FIGURE 13.7
An "S"-shaped pattern is formed by 690-nm red fluorescent PS beads. The "S" pattern resulted from a line optical illumination, which was being slowly rotated.

10

Fluidic Interconnects for Microfluidics: Chip to Chip and World to Chip

Bonnie L. Gray

CONTENTS

10.1 Introduction

From the miniaturization of laboratory instrumentation for bedside rapid diagnosis of diseases to bioanalytical instrumentation for study of individual cells or molecules in new ways using micromachined structures, microfluidics is well established as an exciting new area of research with great promise and an ever-growing application base. Microfluidic components and systems are needed whenever the manipulation of microliter or (usually) nanoliter amounts of a fluid sample is required or when miniaturization of a fluidic system is needed. Such miniaturization can be required for a wide variety of reasons, including the desire to minimize sample or reagent volumes, the types of techniques benefiting from smaller scales, or the drive to shrink laboratory equipment for operation in the field using portable or even handheld instruments. Applications include implantable monitoring and drug delivery devices such as glucose sensors and insulin pumps for diabetes, miniaturized polymerase chain reaction (PCR) systems with fast thermal cycling of small liquid volumes, and other lab-on-a-chip devices for performing complete laboratory analysis on a sample using a portable instrument.

However, miniaturization presents new challenges as plumbing requirements also shrink, and new methods and tools must be developed for fluid manipulation, interfacing, and packaging. Microfluidic systems can be very complex, requiring the interfacing of many different passive and active components such as microchannels, valves, pumps, filters, and chemical reaction chambers as well as detection systems and other sensors operating on electrical, mechanical, or optical principles. Most research in microfluidics has centered around the optimization of particular components or subsystems using well-known or newly developed principles, for example, PCR chambers for increasing the quantity of DNA in a sample, electrode-based biosensors for determining the amount of a biological component in a sample, DNA microarrays for performing multiple DNA tests simultaneously, or microneedles for painless biofluid sampling and drug delivery. However, as systems become more complex, there is an increasing focus on efficient and cost-effective schemes for interfacing between microfluidic components and between these microfluidic systems and macroscale devices. Such microfluidic interconnect technology has been widely recognized as crucial to the practical realization and commercial success of complex microinstrumentation for bedside diagnostics and lab-on-a-chip systems, with increasing research every year to solve these problems generically and also case-by-case for particular applications. Many microfluidic interconnects have arisen from the need to connect to particular devices or to package-specific devices (Velten et al., 2005), rather than a generic approach to microfluidic interconnects, which seeks to standardize the field. However, researchers are increasingly investigating microfluidic interconnects as a separate research topic, with a view toward achieving generic structures and standardization, similar to that obtained for macroscale plumbing or microelectronic packaging.

This chapter covers the primary methodologies used to interconnect microfluidic devices to each other (device-to-device [DTD] or chip-to-chip [CTC] interconnect) and to interconnect microfluidic devices to external devices (world-to-chip [WTC] interconnect) as well as key research findings and example systems from academia and industry, which demonstrate these methods in practical systems. First, some general concepts and considerations of requirements, design, and materials are discussed. Next, interconnect and packaging methods including integrated, embedded, and modular systems are compared. Finally, irreversible and reversible WTC devices, both conventional devices and devices for specialized applications (e.g., nanofluidic and high-pressure interconnects), are discussed. The chapter concludes with progress in generic standardization methods and a view toward future research in this exciting microsystems packaging area.

10.1.1 General Concepts

Microfluidic systems are often composed of a number of devices that perform different functions, such as reagent mixers, sample separation devices, and sensors for analyte detection. These devices must be interconnected together to perform complex functions and must be connected to off-chip devices for sample and reagent introduction and waste removal. DTD or CTC interconnects fall into several main categories, with overlap existing not only between them but also in the methods of accomplishing WTC interconnection. Indeed, the CTC interconnect can be considered a special type of the DTD interconnect in which multiple substrates or chips are used with one or more devices located on each chip. DTD and CTC interconnects can be accomplished in several ways, including integration of all devices into a single fabrication process flow; embedding multiple devices in a (usually polymer) package; stacking devices, chips, or substrates in multilayers; modularly packaging discrete devices using a circuit board or multichip module (MCM) approach; or accomplishing CTC direct connection using tubes or other specially designed interconnect structures. For many components and systems, a combination of different types of integrated and/or modular approaches may result in the most easily packaged and best-performing devices. For example, a handheld instrument developed by Sandia National Laboratories features both system-to-system modular and stacked modules (Renzi et al., 2005). Other researchers combine integrated and modular techniques, and it is common for an industry to use cartridges and one or more other methods. Similarly, many methods of WTC interconnect have been investigated, including simple wells for fluid sample introduction, adhesively bonded tubes or other permanently sealed macro–micro interfaces, tubes that are not permanently sealed but can be attached mechanically, and socket-type approaches. In addition, specialized WTC interconnects are emerging for applications that include high-pressure and nanofluidic applications.

Before a detailed discussion of different microfluidic interconnects and practical examples, a few general concepts and considerations are presented in greater detail. These general concepts are necessary to understand the design and comparative operation of the wide variety of available interconnects.

10.1.2 Theoretical and Experimental Considerations

For any microfluidic interconnect to function properly in a microfluidic system, it must meet a few general criteria. First, it must be capable of operating at the system fluidic pressures and flow rates without failure (e.g., leakage). The maximum pressure that an interconnect is capable of withstanding before failure is known as the *pressurization limit*. The interconnect device must be able to withstand the pressure across it because of major and minor pressure drops in the interconnect but more importantly must withstand the internal pressure of the microfluidic system with respect to the outside environment. In general, higher pressurization limits are desired, but as long as the microfluidic interconnect meets the requirements of the application, the limit is considered to be adequate. The pressurization limit may also be given by the term *fail pressure*.

Conversely, the amount of *pressure drop* in the interconnect structure itself should be low, to keep the overall system pressure low, especially if many microfluidic interconnects must be used. This translates to the microfluidic resistance of the interconnect being as low as possible, so that there is lower pressure drop across the microfluidic interconnect for a given flow rate, as flow rate is usually the variable of interest in microfluidic systems. *Fluidic resistance* is defined as the relation between the pressure drop and the flow rate of a fluidic system. It is determined by the geometry of the system and is usually constant over a given range of pressure and flow rate, thus giving a linear relation between pressure and flow rate, much like electrical resistance that gives such a relation between voltage and current (e.g., Nguyen and Wereley, 2002; Westwood et al., 2008). Individual interconnect resistance is generally very low compared with the microfluidic system using the interconnects. This is due to the facts that the fluid path length within the interconnect is usually comparatively small, and although sudden contractions and expansions may introduce minor head loss in pressure, the effect is minimal compared with the larger pressure drop in the greater system due to laminar flow (Jaffer and Gray, 2008).

Another basic characteristic of an interconnect structure is its potential *dead volume*. The dead volume of an interconnect is an area of zero or very low flow, such as that occurring at sudden expansions or contractions or around sharp corners. Such volumes are problematic as they tend to trap analyte and bubbles, causing problems with cross-contamination, minimum amount of sample that can be used as the dead volume must be filled although it is not used, and uneven pressure and flow distributions. In general, it is desirable to minimize the dead volume of a microfluidic interconnect, especially if multiple interconnects are to be used as each additional interconnect's dead volume has a cumulative effect on system performance. Dead volume can be measured directly but is usually calculated on the basis of simulation results.

Microfluidic interconnects may be used individually or may be part of some sort of one or two-dimensional arrays. In the latter case, the potential *microfluidic interconnect density* may be of interest as it indicates how many interconnections can be made in a finite space. As microfluidic systems become more complex and a higher number of fluidic interconnections are required, this density becomes increasingly important. A few groups have considered multiple interconnect effects (e.g., Gray et al., 2008; Saarela et al., 2006), but the field is wide open for study.

WTC interconnects must be capable of interfacing not only to on-chip microchannels but also to external devices usually through some intermediary tubing that can be further coupled to macroscale tubings or devices. WTC interconnects often go by a second name, *microfluidic couplers*, indicating to some degree the change in scale that these structures provide. Another term macro–micro interface, is also indicative of the change in scale and is more general.

Another concept of importance in the literature, most notably with respect to WTC interconnects, is the concept of permanent versus reworkable interconnects. Permanent interconnects are usually those that are formed via a bond that cannot be easily removed, such as interconnects being integrated into the overall system flow or interconnects realized by direct material-to-material bonding. Such direct bonding may include high-temperature fusion bonding, anodic bonding (for glass and silicon), or adhesive bonding (glues such as epoxy or other curable material). Reworkable interconnects are those where the attachment is not permanent, relying, for example, on some spring-loaded mechanism or material compliance for direct substrate-to-substrate attachment, or screws in a jig for stacked systems to hold the two halves of the interconnect together. Reworkable interconnects often use a gasket, which is a compliant material used to aid in mating two hard surfaces such that they form a leak-tight interface. A popular type of gasket is a simple torus or an *o-ring*. This differs from a *ferrule*, which is a name for a type of metal or plastic object (usually a ring) that is used for fastening, joining, or reinforcement.

Testing of a microfluidic interconnect can range for very basic tests (e.g., pressurization limit) to much more complicated ones. In addition to the pressurization limit, another common test is the pullout force test. In this test, one half of the interconnect (e.g., a tube sitting in a WTC interconnect structure) is attached to a known force, such as a materials tester or voice coil actuator (Gray et al., 2008), and the mechanical force needed to pull apart the interconnect is measured. External devices typically used for microfluidic testing include the following: a pump (often a syringe-type pump) or high voltage for electro-osmotic pumping, external pressure and flow sensors, and a microscope to capture video imaging in microchannels. External plumbing fixtures include the following: small-bore capillary tubing (typically a few tens to hundreds of microns in diameter) fabricated in a variety of metals (e.g., steel hypodermic tubing), fused silica or quartz, or polymers (e.g., polyetheretherketone [PEEK]); luer fittings such those used for syringes; and small-scale fittings developed for liquid chromatography and other biological applications (e.g., Micro-Fingertights™ available from Upchurch Scientific, now known as IDEX Health & Science).

10.1.3 Materials Considerations

The materials used for microfluidics interconnects are usually the same as, or at the very least compatible with, the materials of which the microfluidic device is fabricated. For WTC interconnects, fabrication materials must also be compatible with the external tubing material (e.g., steel, silica, PEEK) that is usually connected to macroscale off-chip devices often through macroscale interconnects. Typical materials used for microfluidic devices include silicon, metal, glass, and an ever-increasing variety of polymers. Some excellent reviews of microfluidics are already available, which include a materials overview (e.g., Becker and Gärtner, 2008; Fiorini and Chiu, 2005; Velten et al., 2005; Whitesides, 2006) in addition to coverage in a number of textbooks (e.g., Nguyen and Wereley, 2002), so only a brief discussion is included here.

Silicon and silicon-based materials (e.g., silicon nitride, silicon dioxide) are widely used for microelectromechanical systems (MEMS). They are also used for microfluidics applications, but the trend is away from silicon toward other materials such as polymers (Whitesides, 2006). Still, the use of silicon as a microfluidic material certainly cannot be discounted, especially as many other devices such as sensors and electronics are still often produced using silicon, and new methods such as deep reactive ion etching (DRIE) and other silicon processing techniques are constantly under development for applications in microfluidics. Glass has been used extensively for microfluidic systems because of its optical properties such as transparency and compatibility with biological materials. Glass can be single crystal quartz, although far more used are glasses such as fused silica, Pyrex™, and those used for microscope slides. Metals and alloys are often used for microfluidics, although usually as thin or thick films rather than as a substrate, and include aluminum, silver, gold, platinum, halides, and many others.

A very large number of polymers have been used for microfluidic systems (see, e.g., Becker and Gärtner, 2008). Some of the more common polymers include those that can be photopatterned, such as a photoresist, epoxy-based SU-8, and polyimide. SU-8 has been especially important in the development of microfluidic systems. SU-8 is an epoxy-based negative-tone near-UV photoresist that was originally developed by IBM (Lorenz et al., 1997). A photoresist is called negative when the areas exposed to certain wavelengths of light are cross-linked because of energy imparted by the light and thus remain on the substrate during development while the unexposed (uncross-linked) areas are developed away. SU-8 is sensitive to wavelengths below 400 nm and will absorb light shorter than 360 nm. Exposure to near UV-light causes a photoacid generator in the SU-8 formulation to become a strong acid, which catalyzes the thermal cross-linking of SU-8. The manufacturer of SU-8, Microchem Corporation, distributes many formulations of SU-8, denoted by a numerical suffix ranging from 2 to 100, which denotes the estimated thickness of the resulting SU-8 film in micrometers when spun at 3000 RPM. SU-8 enables fabrication of high-aspect-ratio structures. By choosing the correct SU-8 formulation and controlling the spin speed, various thicknesses of SU-8 can be spun onto the substrate, starting from 2 μm to greater than 200 μm in a single coating, or even thicker using casting (e.g., Ueda et al., 2008). SU-8 consists of EPON Resin SU-8 (Shell Chemical) dissolved in a solvent (gamma-butyrolactone). SU-8 has been widely used in microfluidics as a structural material and as a mold for micro-patterning other polymers using soft lithography (Xia and Whitesides, 1998).

A wide variety of other polymer materials have been used for microfluidics. Although many classification methods and characterizations for polymer materials exist, a few definitions help in the understanding of the differences between commonly used materials. A thermosetting polymer is one that irreversibly cures, generally after it has been poured in a precured state into a mold master. The curing process transforms the resin into a plastic or elastomer by a cross-linking process. Energy (e.g., heat; hence, thermoset) and/or catalysts that cause the molecular chains to react at chemically active sites and cause cross-linking to occur are added. The cross-linking process forms a solid molecule with a larger molecular weight. One of the most commonly used polymers in microfluidics is the thermosetting polymer polydimethylsiloxane (PDMS) (Whitesides, 2006), an *elastomer* than can be easily micromolded with features as small as those in the submicron range (Xia and Whitesides, 1998) and bonded using various surface treatments (Bhattacharya et al., 2005).

A thermoplastic is a polymer that turns to a liquid when heated and freezes to a very glassy state when cooled. Most thermoplastics are high-molecular-weight polymers. Unlike thermosetting polymers, thermoplastic polymers can be remelted and remolded, usually using an embossing, or stamping, process under heat. Thermosetting polymers commonly used for microfluidics include polymethylmethacrylate (PMMA), polycarbonate, and polystyrene. For further review of micropatterning processes associated with both thermosetting and thermoplastic polymers common to the field of microfluidics, see Becker and Locascio (2002) or Becker and Gärtner (2008).

Many microfluidic interconnects rely upon the relative *compliance* of interfacing materials for a fluid-tight seal. Compliant materials will strain much more under stress than a noncompliant or stiff material. An example where compliant and stiffer materials are used together is the use of compliant polymer microfluidic gaskets between stiffer/harder functional material layers or devices such as silicon for a leak-free seal (e.g., Gonzalez et al., 1998; Jaffer and Gray, 2008; Verpoorte et al., 1994).

10.1.4 Application Requirements

A very large number of microfluidic systems applications exist. For a review of some of the more prevalent examples of microfluidic systems, the reader is referred to Erickson and Li (2004). These microfluidic systems may have application-specific requirements for their interconnect that must be considered in developing DTD, CTC, and WTC interconnects, which include typical pressure and flow rate requirements for the application, which can range from a few Pascal and microliters per hour to tens of megapascals and millimeters/minute; number of interconnects required, from a single interconnect to dozens or hundreds of DTD or WTC interconnects; acceptable maximum dead volume; biocompatibility, cleaning, chemical resistance, and/or antifouling materials-related requirements; minimum operating temperatures; requirements for other interconnects, such as optical or electronic interconnects; ease of testing; reliability and lifetime requirements; and many others that must be considered along with system design and development.

10.2 Microdevice Interconnect Technologies

In the macroscopic world, fluidic devices are attached together using various plumbing fixtures and readily interfaced to electronic, mechanical, and optical devices using wires, mechanical fasteners, and lenses. Fluidic components (e.g., pumps) may also be fabricated using a variety of mechanical attachments, gaskets, plumbing parts, and molded plastic or machined metal housing, using various integrated (all in the same plastic molding) or modular approaches. The microscale has similar problems, although on the basis of the difference in design, materials, and applications, the approaches to solving them may be very different.

This section discusses the two main types of microfluidic interconnects: integrated DTD interconnect where devices are designed from the ground to be integrated together and CTC interconnects that interface devices on different substrates together using a wide variety of interconnect structures and methods. WTC interconnects that interface on-chip devices to macroscale devices will be the focus of "WTC Interconnects." Each of the methods presented has been demonstrated in practical microfluidic systems, and each method has strong and weak points that may be dependent on the application. Although these classifications provide an organized method to discuss practical interconnect structures, there is a great deal of overlap between these methods, for example, between stacked methods and integrated methods that use multiply bonded substrates, or tube-based CTC and WTC interconnect structures. Many practical/commercial microfluidic systems use a combination of integrated and/or modular approaches and different types of each approach (integrated and modular). In these cases, portions of the system, such as capillary electrophoresis microchannels with integrated electrodes, or microsensors with integrated pumps and valves, may be combined with a modular approach toward the optics or other microfluidic elements. Such combinations of integrated and modular systems can be very effective, resulting in the "best of all worlds" for a microfluidic system. It should also be noted that although practical examples are given, this does not necessarily mean that the examples are exemplary only of the technology being discussed, as many examples feature more than one type of approach and, especially, may feature one or more of the WTC interfaces discussed in the section "WTC Interconnects."

10.2.1 On-Chip/Integrated DTD Interconnection

Integrated approaches seek to use a single fabrication process for the entire device or system, including all components, interconnect, and packaging. Although using a single process simplifies device interconnection, it can be at the expense of device performance, or it can limit the devices and interconnects that are possible because of materials or process constraints. Thus, components cannot be individually optimized using different material platforms or technologies, although interfacing problems between different materials can be minimized.

For integrated processes, a DTD interconnect is generally accomplished by providing a passive network of horizontal and vertical microfluidic pipes either in or on the substrate surface that interface to the individual devices fabricated using the integrated process flow. These microfluidic channels may be fabricated in the substrate itself via etching or photo-patterning of a polymer, or on top of the devices using subsequent thin or thick film layers that are patterned on top of the substrate containing devices. One early technology used printed circuit boards for an integrated microfluidic pH regulator (Läritz and Pagel, 2000).

Surface micromachining is a commonly used integrated process for MEMS, which has also been demonstrated for microfluidics. In surface micromachining, multiple layers of a structural material (e.g., thin-film polysilicon) are usually separated by layers of another sacrificial material (e.g., silicon dioxide) that is removed at the end of the fabrication process, resulting in free-standing structures in the structural material (e.g., using hydrofluoric acid to remove the *sacrificial* oxide). In addition to mechanical structures, electrodes and other electronics can be fabricated in the same (usually silicon) substrate. Although widely used for more traditional MEMS devices, for example, accelerometers and resonating sensors, integrated approaches can also be used for microfluidics, provided components and interconnect can be realized in the chosen technology. Surface micromachining has been used for integrated microfluidic systems. However, most surface micromachining processes involve deposition and patterning of thin films (limited to several microns in thickness) located only on the substrate surface, and use of silicon-based and other thin-film surface micromachining for microfluidics is limited in application. One example is the development of an integrated closed-loop drug delivery system at the University of California at Berkeley, with integrated microneedles, fluid manipulation devices, and analysis devices, all fabricated using the same surface micromachining process (Zahn et al., 2001) (see Figure 10.1). Other microneedles fabricated via an integrated microfluidic process have also been demonstrated, but for neural prosthesis applications (Retterer et al., 2004). Another surface micromachined microfluidics application was presented by Galambos and Benavides (2004). In this work, the microfluidic channels and level-to-level interconnects were formed in thin-film polysilicon. As with many surface micromachined microfluidics based on thin films, extensive packaging for WTC interfacing was required, which was complicated by the fragile nature of the microfluidic channels. Passive microfluidic channels were demonstrated and tested.

Polymer-based sacrificial processes have also been demonstrated. In an early work, an integrated microfluidic interconnect was accomplished by patterning of polymer microfluidic capillaries on top of a silicon wafer that could potentially contain already fabricated devices

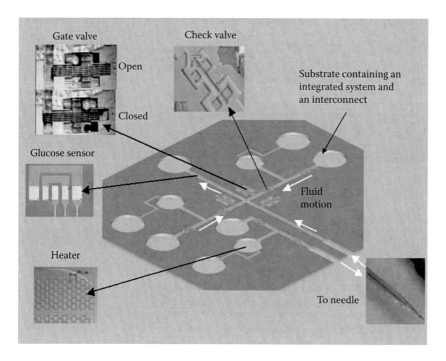

FIGURE 10.1
Example of an integrated system with fluidic devices and a fluidic interconnect fabricated simultaneously using surface micromachining. (From J.D. Zahn, A.A. Deshmukh, A.P. Papavasiliou, A.P. Pisano, D. Liepmann, An integrated device for the continuous sampling and analysis of biological fluids, Proceedings of 2001 ASME International Mechanical Engineering Congress and Exposition, New York, 6 pp., Nov. 2001. With permission.)

(Man et al., 1997). This IC-compatible microfluidic process used three polymers in a sacrificial process to produce polyimide/*p*-xylene microchannels ranging in height from 0.5 to 100 μm and several micrometers long, with walls comprising up to 40-μm-thick polyimide.

Another integrated method using polymers uses PDMS ports, through-hole silicon wafer vias, and backside parylene jumpers for user-defined microchannel routing defined by the jumper pattern (Shaikh et al., 2003).

Several new surface micromachining processes use thick-film polymers (e.g., the photopatternable thick-film polymer SU-8 or direct-write polymers) for fabrication of channels, wells, and other passive microfluidics that are integrated with electronics or optical detection devices in the substrate. SU-8 has been combined with complementary metal-oxide semiconductor (CMOS) MEMS using a transfer process for CMOS-compatible integrated microfluidics with multiple vertically interconnected microchannel levels (Peng et al., 2006). Example devices were demonstrated with integrated sensors fabricated in silicon and covered with photopatterned and transfer-bonded SU-8 layers, with WTC interfacing (simple inlet/outlet holes) integrated into the top channel level. SU-8 has also been fabricated directly on CMOS photodiodes for microfluidic wells and coverslip assembly structures (Young et al., 2003). Direct-write fabrication of microfluidic channels on CMOS-compatible chips has been demonstrated for integrated microfluidic systems. In such cases (e.g., Ghafar-Zadeh et al., 2007), polymer microfluidics is directly patterned on top of a chip containing nonmicrofluidic integrated electronics and/or sensors. The polymer (ink) is dispensed via a small nozzle using an *x-y-z* stage. The resolution capabilities of such systems are usually adequate for many microfluidics applications.

Other processes that may be considered as integrated rely on sequential bonding and patterning of multiple bulk layers rather than film deposition and patterning. One popular approach involves patterning all microchannels and microfluidic structures for one layer using, for example, DRIE of silicon, alternated with silicon fusion bonding of patterned or unpatterned wafers that can then be further structured in the integrated process (e.g., Gray et al., 1999). As with surface micromachining, such approaches can be limited to the type of structures that can be realized and are usually limited to a set number of materials. Approaches that rely on sequential bonding and patterning of multiple bulk layers rather than film deposition and patterning have been used successfully. A variation of this approach is, of course, to pattern all the layers individually and then bond them together, an approach that can border on modular dependence on the specific device or system in question and how the different layers are used. Surface-activated bonding of multiple layers of patterned PDMS with through holes has been used to build up a multi-level microchannel system (e.g., Jo et al., 2000). PDMS-to-PDMS bonding is accomplished by first treating the surfaces to be bonded, for example, with oxygen plasma or chemical treatments such as HCl (Bhattacharya et al., 2005).

Centrifugal microfluidics is a type of integrated approach in which microfluidic devices are integrated into a disc using embossing methods similar to compact disc technology, with pumping of fluid on the disc accomplished via centrifugal force produced due to spinning of the disc (e.g., Lai et al., 2004). Passive structured barrier valves are used, whereby fluid stays trapped closer toward the center of the disc until the centrifugal force is increased by spinning the disc faster, and fluid bursts out of the valve into the next chamber located at a farther radius from the disc center. This integrated interconnect platform has been adapted by companies such as Gyros AB (Sweden; www.gyros.com) and other researchers (e.g., Haeberle and Zengerle, 2007) for applications in immunoassays and other lab-on-a-chip applications, for example, blood analysis.

The development of actuation schemes and interconnect technologies may make integrated polymer microfluidic systems more attractive, although more research is needed for these components as well as for the integration of the polymers with other materials to realize truly integrated systems.

10.2.2 CTC Interconnection

Although many successful integrated systems have been demonstrated, CTC systems offer increased benefits such as developing and optimizing components independently before attaching them and allowing potential reconfiguration and replacement of individual modules rather than having to refabricate a completely new on-chip system when a change is introduced or an individual device becomes nonfunctional. CTC approaches have used a variety of techniques adapted to or developed for the microscale, including the following: microfluidic circuit boards with attachable components; gasketed stacked modules where each layer performs a specific function, either material based or device based; embedding of multiple devices in a single, usually polymer, package; a quick-connect modular assembly using a specially developed interfacial structure between components; and cartridge-based modularity for interfacing reusable control and detection circuitry to disposable microfluidic cartridges. Many patents exist for companies that have developed CTC modular schemes for their systems, especially for the overall scheme (e.g., Brennen and van de Goor, 2003; Frye-Mason et al., 2004; O'Connor et al., 2002). We now examine a number of the more popular methods, discuss their overlap, and present practical systems that demonstrate these CTC approaches.

10.2.2.1 Embedded Devices and Modules

Most of the earliest packaging of individual microsensors centered around a concept taken directly from microelectronics: the embedding of the device in a molded polymer package with a simple electronic (e.g., dual in-line package pins or surface mount bumps) and/or optic (e.g., transparent windows for LEDs) or fluidic (e.g., access holes for pressure sensors) interconnect as needed. This concept has been extended to microfluidic systems whereby multiple components are embedded into a relatively inexpensive polymer package at the same time. The most commonly used polymer in academia is PDMS as it is readily molded, easy to work with, and is in use by many researchers who use soft lithography, or molding against a photoresist master, as a primary method of fabrication of microdevices and systems. However, thermoplastic polymers, for example, PMMA and polycarbonate, are used for embedded packaging by industry and other researchers with access to techniques such as injection molding.

A good example of modular embedding can be found in a work from Lawrence Livermore Laboratories that is perhaps even better known for its novel WTC screw-in-place couplers (Krulevitch et al., 2002). Here, a glass capillary electrophoresis chip and a small-scale pump are embedded in a singular PDMS package. Early valve manifolds (Verlee et al., 1996) also featured an embedded device approach, although this could also be considered a microfluidic board because of the valve placement being located on the surface.

A major drawback of the embedding of components in a package is that the assembly is usually permanent. Several other devices successfully combine microchannels with optics for fluorescence-activated cell sorting, a closed-loop system where cells that are tagged with different fluorescent markers can be sorted according to color (Wolff et al., 2003). Fiber optics or light-emitting diodes and detectors are embedded in a polymer package containing microchannels. Other examples have been found in the industry, such as the iSTAT whole-blood gas analyzer (Abbott; www.abbottpointofcare.com/istat/), a handheld plastic device that manipulates blood samples and is embedded with simple sensors.

10.2.2.2 Microfluidic Circuit Board

One of the most straightforward and popular modular microfluidic approaches is the microfluidic circuit board or microfluidic board. These interconnection technologies seek to mimic the electronic printed circuit board in terms of ease of use, reliability, and versatility. In microfluidic board technologies, the substrate contains passive microfluidic channels and reservoirs, with active components (e.g., sensors, valves) mounted on top (e.g., Gray et al., 1999) (see Figure 10.2), similar to electronic components mounted on a printed circuit board or MCM in surface mount technologies. Electronic connections

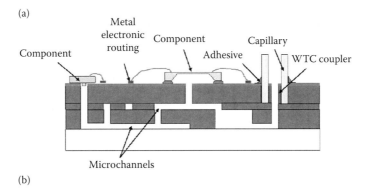

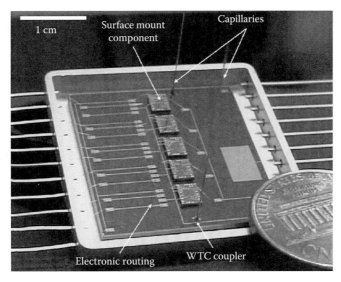

FIGURE 10.2
(See color insert.) Modular microfluidic board approach: (a) illustration of a board technology with electronic and fluidic routing between surface-mounted components and coupling to off-chip devices; (b) photograph of a microfluidic board in silicon with five mounted components (pressure sensors), embedded microchannels, surface metal electronic routing, and a WTC interconnect to a capillary tubing. (Adapted from B.L. Gray, D. Jaeggi, N.J. Mourlas, B.P. van Drieënhuizen, K.R. Williams, N.I. Maluf, G.T.A. Kovacs, Novel interconnection technologies for integrated microfluidic systems, *Sensors and Actuators A*, 77, 57–65, 1999.)

between components are provided on top of the board, whereas fluidic interconnection is usually provided in the board material itself. Microfluidic circuit boards can be made from a variety of materials, including silicon, glass, and polymers, for example, PDMS, epoxy printed electronic circuit boards, or PMMA. Microfluidic boards can be as simple as a single packaged component or as complex as a full lab-on-a-chip analysis system, with size and complexity constrained mainly by the mounting technique (e.g., bonding, adhesive, or reworkable interface) and via-hole fabrication method. The alternative terms *microfluidic MCM* or *microfluidic bench* are often used instead, but the concept of having a larger board upon which are mounted interconnected chips, either permanently or reversibly, remains the same.

Early microfluidic boards were modeled directly based on the electronic circuit board approach and usually featured a permanent interconnect attachment. These early boards were based on silicon (Gray et al., 1999), epoxy printed electronic circuit board materials (Lammerink et al., 1996), other polymers (LeClair et al., 1999), or glass (Schabmueller et al., 1999). At least one of them featured stereolithography for fabrication of high-aspect-ratio microstructures (Tse et al., 2003). Other more recent examples of microfluidic boards or MCM with multiple interconnect devices are as follows: an approach using passive user-definable custom chips with a breadboard containing active devices (Shaikh et al., 2005); a layer-based interconnect chip with multiple interconnected components forming a different layer (Miserendino and Tai, 2008); a similar approach with multiple interconnected devices on a microfluidic bench, with both electronic and fluidic interconnection (Chang et al., 2009); a spring-loaded cylinder-in-hole attachment approach using high-aspect-ratio silicon etching that can be used for either a microfluidic circuit board or direct component-to-component attachment and is reversible (Gray et al., 2004); and thermal-fluidic I/O interconnects between a board and chips which also features electronic solder bump technology for an electronic interconnect (Dang et al., 2006).

10.2.2.3 Stacked Multilevel Modules

Microfluidic systems composed of modules, where each module forms a layer that is stacked on previously layered modules, is a simple solution for interfacing components fabricated using different materials or for producing components from alternating passive (e.g., microchannels) and active (e.g., sensors, actuators for valves and pumps) layers. Each module (layer) performs a specific function and can be interfaced to the modules above and below it electrically, optically, or fluidically, provided suitable electronic and optical interfaces and fluidic gaskets are used (see Figure 10.3). The modules are usually compressed together in a jig, with modular replacement requiring complete disassembly of all modules. Stacked modules can be made of many different materials, including silicon, glass, and both thermoplastic and thermosetting polymers, provided compliant polymer gaskets (e.g., PDMS) are used to interface layers for fluid-tight seals. Other technologies border on integrated ones, with multilayers that are permanently fixed.

Many researchers have used stacked multilayers to realize microfluidic systems where layer modules perform different system functions. Verpoorte et al. (1994) produced one of the first microfluidic systems on the basis of stacked silicon modules with polymer gaskets forming interfacial layers. Since then, many other researchers have successfully used this concept for the assembly of even more complex microfluidic devices and systems. Some examples are systems with integrated WTC ports fabricated via stereolithography with a combination of microfluidic, electronic, and optical functionality (Han et al., 2003, 2005); a stacked technology with battery and external connection socket (Ucok et al., 2003); a

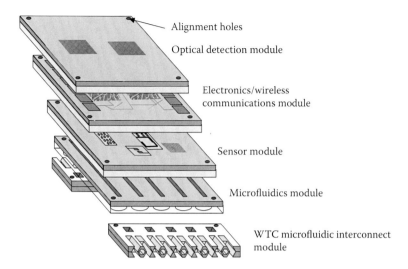

Alignment holes

Optical detection module

Electronics/wireless communications module

Sensor module

Microfluidics module

WTC microfluidic interconnect module

FIGURE 10.3
Illustration of stacked-module approach, in which the microfluidic system is composed of stacked modules, each performing specific functions and aligned and assembled together vertically in a jig using alignment holes and, for example, pegs or screws. Although not shown, often gaskets or other interface layers are inserted between multiple fluidic or other functional layers.

microdialysis device (Martin et al., 1999); a simple system that consists of layered microfluidics on an interconnect/clamping block and an instrument platform (Datta et al., 2006); a plug-together three-dimensional packaging technology based on Match-X interfacing (now called Easykit; www.easy-kit.de/EasyKit) demonstrated with an example valve (Schindler-Saefkow et al., 2004); a reconfigurable microfluidic chip platform for electrokinetics (Dalton and Kaler, 2007); and an approach to packaging specifically for bioanalytical applications (Schueneman et al., 2004).

A slightly modified version of this approach involves stacking of layers based more on materials requirements than on module function for opto-microfluidic fluorescence detection, with some devices also using an embedded approach (Kim et al., 2006). A modular cell concentrator and genetic sample preparation system was also developed using stacked layers, where the modules were stacked next to each other on a base with an o-ring and tube-type interconnects rather than in the typical vertical fashion (Grodzinski et al., 2003).

Other approaches consist of three-dimensional stacks of individual components that also happen to be layer-like modules, but these types of approaches are discussed in the section "Modular Direct CTC Techniques", although it should be noted that there is a significant overlap of these approaches with the stacked module approach.

10.2.2.4 Cartridge-Based Techniques

The use of disposable, usually hard polymer cartridges containing microfluidics that are interfaced temporarily to reusable components, for example, optics or control electronics, is an assembly approach that is widely used in industry for microfluidic systems involving DNA chips and biofluid analysis. In this approach, the cartridge containing the fluidic sample is slid into the instrument by the user, and after desired tests are performed on the sample, the sample is then slid out and disposed such that the sample does not interfere with the reusable and more expensive parts of the instrument. This also avoids washing steps, as the cartridge is the only part of the machine in contact with the liquid and is disposable. Provision for smooth motion and lock-in-place mechanisms between the disposable and the reusable portions of the instrument must be included in the system design. Two industrial examples of the many modular cartridge-based systems are a bead array counter for multiplex detection of nucleic acid sequences (Tamanaha et al., 2002) and a PCR chip (Oh et al., 2005), although there are many, many others. It should be noted that in the latter reference (Oh et al., 2005), the authors classify their technique as a WTC technology, of which it most closely resembles a socket-type WTC interface.

Figure 10.4 shows a relatively simple example of a cartridge-based integrated high-speed screening array, in which photodiodes are located under each well for fluorescence detection, but the wells themselves are located on a disposable cartridge that is slid into place using on-chip polymer guides (Young et al., 2003). Other examples include DNA chip cartridges and blood sample analysis cartridges.

10.2.2.5 Modular Direct CTC Techniques

Methods to efficiently interconnect components that have been fabricated separately, often using vastly different techniques and materials specific to each device, are more complex but may be required for full microfluidic systems with optimized performance. Microfluidic components or modules can be directly interconnected either via aligned CTC bonding methods or through specially designed interconnect structures that may be integrated with individual components and then used to join them together. Such modular techniques offer the most versatility in system design and reworkability but are often the hardest to realize as proper placement and assembly of components can become a significant bottleneck.

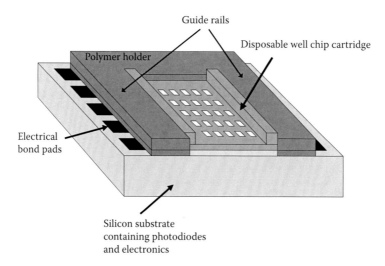

FIGURE 10.4

Illustration of a microfluidic cartridge concept where a disposable cartridge is interfaced to reusable instrumentation: a disposable well array aligned over reusable photodiode array using on-chip alignment structures (polymer holder guide rails). The chip is used for integrated high-speed screening by monitoring fluorescence changes as chemical reactions proceed inside the wells. (Adapted from I.T. Young, R. Moerman, R.L.R. van den Doel, V. Iordanov, A. Kroon, H.R.C. Dietrich, G.W.K. van Dede, A. Bossche, B.L. Gray, L. Sarro, P.W. Verbeek, L.J. van Vliet, Monitoring enzymatic reactions in nanolitre wells, *Journal of Microscopy* 212, 254–263, 2003.)

In many modular approaches and techniques, alignment keys and mechanical joints may facilitate microassembly, joint strength, or reworkability. Although this may also be true for the previously discussed microfluidic circuit boards and stacks, alignment keys are most often used in modular techniques in which the components or modules are freely joined. Alignment keys can be used for bonding of two substrates, parts, or components such that they line up in some useful manner (e.g., a through hole lines up with the end of a microchannel). Alignment keys are similar to those used for photolithography processes, although they may be larger to facilitate manual or automated assembly. One example of passive alignment keys used for a microfluidic interconnect is found in the study of You et al. (2006). Another interesting method of module-to-module attachment involves placing polymer modules side by side and then attaching them along the edges rather than the more conventional horizontal surface (Igata et al., 2002). The connection is accomplished by flowing an ultraviolet-light cross-linkable adhesive into the interconnect boundary (which is marked with alignment keys) and then curing it. The process is "reversible" or reworkable, in that soaking the multichip system in dichloromethane separates the chips.

Other structures can be used to facilitate the actual joining of parts. Such structures may be based on mechanical press fits, including simple fits of one part into another, and structures based on micro-peg-in-hole or beam-spring assembly. The inserted part is held in contact with the mating part via friction and the spring force from either the beams or the material compliancy holding the parts together. Prediction of how well a snap-together interface will work is based on mechanics (e.g., beam equations) and contact modeling, with mechanical characteristics of materials, such as Young's modulus and yield strength, of paramount importance. It should be noted that polymer characteristics can be heavily dependent on processing, which complicates prediction of behavior based on structure. A cylindrical peg inserted into a compliant hole can provide a microfluidic interconnect between two microchannel-containing substrates (Jaffer and Gray, 2008). Such a structure mechanically holds the substrates together and also provides a fluid-tight interconnection.

Quick connects were developed in silicon (Gonzalez et al., 1998) to aid in modular manual assembly. Both finger-joint interconnects perpendicular to the wafer surface as well as peg-in-hole interconnects parallel to the silicon substrate surface (see Figure 10.5) were realized using wafer sawing and/or KOH anisotropic etching along crystal planes. As these interconnects were developed in silicon, polymer gaskets were needed, complicating the fabrication process. Micro-peg-in-hole assembly has been investigated for use in microfluidics, in which a cylinder is used instead of a simple peg and is assembled into a fluidic hole making a fluid-tight connection (see Figure 10.6). Fluid-tight interconnects in both DRIE silicon (Gray et al., 2004) and polymers (Jaffer and Gray, 2008; Westwood et al., 2008) have been developed using this concept, which are used to combine microchanne-containing substrates. The polymer interconnects represent a significant improvement over their silicon counterparts, by requiring no additional gaskets and being compatible with not only silicon and glass but also polymer substrates, provided suitable bonding/embedding techniques are used. Polymer interconnects are also superior in that the restoring spring force used to hold each cylinder in its mating hole is the result of material compliancy rather than rigid beams and provides a more uniform joining force (Jaffer and Gray, 2008).

A three-dimensional direct component-to-component approach also allows stacking of modules (Morrissey et al., 1998), with demonstration of a micropump and sensor units. A combined fluidic and electronic direct module-to-module interconnect technique can be found in Pepper et al. (2007) for micromachined sensor packaging.

A tube-based interconnect attaches chips or modules together using an intermediary of a separate tubing that attaches two integrated mating structures, one on each of the two

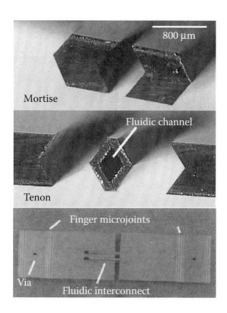

FIGURE 10.5

Example of a modular interconnect structure: quick-connect reusable interconnects in silicon on the basis of hexagonal mortise and tenon concept. The two haves of the interconnect slide together for interconnection between channels on two different substrates, but a gasket must be used for a fluid-tight seal. (Reprinted from *Sensors and Actuators A*, 66, C. Gonzalez, R.L. Smith, D.G. Howitt, S.D. Collins, MicroJoinery: concept, definition, and application to microsystem development, 315–322, Copyright 1998 with permission from Elsevier.)

components to be joined together. This differs only from other CTC methods in which the tube-type interconnect features in that tubes are separate entities, that is, they are separate tubes that fit into holes or over other cylinders on the microfluidic chips. In general, the tubes are short interconnectors (Friedrich et al., 2005; Perozziello et al., 2008), but they can be long, similar to ribbon cables (Ilic et al., 2002; Man et al., 1997).

In additional to pegs and holes, other mechanical joints for assembling microfluidic components and systems often involve insertion of one part into another. The aforementioned cartridges are a simple example, and structures must be included both on the cartridge and inside the instrument to allow smooth insertion. Such structures can include micro-rails to aid such a sliding assembly on a chip (Young et al., 2003).

10.3 WTC Interconnects

A major issue with many microfluidic interconnect technologies is the coupling of the fluidic circuits to the macroscopic world. Such WTC interconnects allow fluid to be pumped to the chip and waste to be removed from the chip in a controlled manner. Coupling between on-chip channels and off-chip devices is a scale-matching problem that seems universal in the field of microfluidics. Whether a researcher wishes to study the flow through a new micromixer or manipulate and analyze cells for lab-on-a-chip, most researchers eventually come across the WTC interconnect problem. As a result, a large number of solutions have been developed for providing fluid flow between on-chip microfluidic systems and off-chip devices. A few example approaches and specific devices are discussed here.

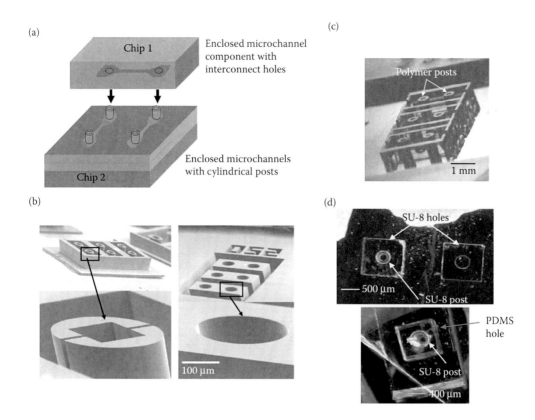

FIGURE 10.6
(See color insert.) Examples of modular interconnect scheme on the basis of cylindrical peg-in-hole press fits: (a) overall concept (From Westwood, S., Jaffer, S., Gray, B.L. Enclosed SU-8 and PDMS microchannels with integrated interconnects for chip-to-chip and world-to-chip connections, *Journal of Micromechanics and Microengineering*, 18, 064014, 2008.); (b) example silicon mating joints with notched cylinders and holes requiring gasketing. As silicon is not compliant, the cylinders are notched to provide a spring force to hold the cylinders in the holes (Adapted from Gray, B.L., Collins, S.D., Smith, R.L., Interlocking mechanical and fluidic interconnections for microfluidic circuit boards, *Sensors and Actuators A Physics*, 112, 18–24, 2004.); (c) example polymer cylinders that do not require notches; and (d) example polymer mating joints on the basis of polymer compliancy press fits between holes and cylinders. (Panels c and d from Jaffer, S. and Gray, B.L., Polymer mechanically interlocking structures as interconnect for microfluidic systems, *Journal of Micromechanics and Microengineering*, 18, 035043, 2008.)

Couplers can be integrated into the fabrication process or assembled to the device as a separate component. Microfluidic WTC interconnects must satisfy a number of different criteria, including being reliable (Han and Frazier, 2005), mechanically robust, leak-free, and simple to assemble, and must have a minimal dead volume. Several excellent reviews of microfluidic WTC interconnects already exist either as stand-alone reviews or as part of an article on wider packaging, and the reader is referred to these for an additional perspective (see, e.g., Fredrickson and Fan, 2004; Rameshan and Ghaffarian, 2000; Wang and Shaikh, 2008).

10.3.1 Sample Loading via Wells

The simplest WTC "interconnect" consists of a simple loading well or reservoir from which the sample is pumped into the other parts of the chip using electro-osmotic flow, capillary force, or differential pressure. The reservoir is open to the ambient environment at one side and interfaces to the microchannels in the microfluidic chip on the other side. Although such an interface is very simple to fabricate and operate and can be realized in virtually any substrate material, it may require priming depending on the type of fluid flow and must be watched by the user to make sure that the reservoir does not deplete or evaporate. The well-based sample loading method has been used for a very long time (see early works of Harrison et al., 1992, 1993) and has been, and still is, a popular method with research groups (e.g., Chiu et al., 2000; Moorthy et al., 2004; Seong et al., 2003) and industry (e.g., Boone et al., 2002; Bouse et al., 2001).

10.3.2 WTC Interconnects on the Basis of Tubes-and-Holes (Permanent and Reworkable)

Another commonly used WTC interconnect method uses tubes and holes, or tubes and tubes, that interface with one another. Although many different designs have been demonstrated, in general these WTC interconnects may be either permanently attached using an adhesive or bonding or may be reworkable using some sort of spring-loaded connection or compliant-material interface or ferrule. A few examples are discussed below.

10.3.2.1 Permanently Attached

Many glued WTC interconnects have been demonstrated, for example, as in Figure 10.7a, in which a length of a capillary tube (usually made of metal, fused silica glass, or polymer) is simply placed in an on-chip hole, and adhesive is used to hold the tube in place and provide a fluid-tight seal to the underlying microchannels. Macroscale tubing may then be attached via the permanently attached tube interface or the capillary tube used directly for interfacing to macroscale components. The interface hole is usually located at the end of an on-chip microchannel, and the fluidic coupler often consists of a capillary tube glued into an insertion channel that is etched into silicon or other substrate (see Figure 10.7b). Many other similar WTC interconnects have been demonstrated using solder ferrules and a metal tubing for interfacing to silicon microchannels (Murphy et al., 2007), quartz glass capillaries attached to PMMA microfluidic chips using UV-curable adhesive (Wang et al., 2005), Swagelok-compatible glass tubes directly bonded to silicon microchannels using high-temperature bonding (Mogulkoc et al., 2008; Unnikrishnan et al., 2009), and ulvrasonic bonding of PMMA interconnects to PMMA microfluidic chips (Ng et al., 2009).

The coupler shown in Figure 10.7a does not have an accurately fitted insertion channel, resulting in difficult handling and leakage of the adhesive into the microchannels, which is a common problem with adhesively bonded tubes for WTC interconnects. The coupler in Figure 10.7b solves the first problem but not the second. WTC couplers can also fabricated by more geometrically versatile techniques, such as DRIE of silicon, yielding higher density insertion channels that accurately fit the outer diameter of the capillary tubing (Gray et al., 1999) (see Figure 10.7c and d). Such couplers are still sealed with an adhesive, but integrated sidewalls around the tube prevent leakage of the adhesive into the microchannels.

WTC interconnects based on permanent bonding of a tube to the microfluidic chip have also been demonstrated, where the tubes are parallel rather than perpendicular to the microchip surface (e.g., Hartmann et al., 2008). In addition to WTC fluidic interconnects, optical interconnection was similarly accomplished using optical fibers instead of a tubing on the same chip.

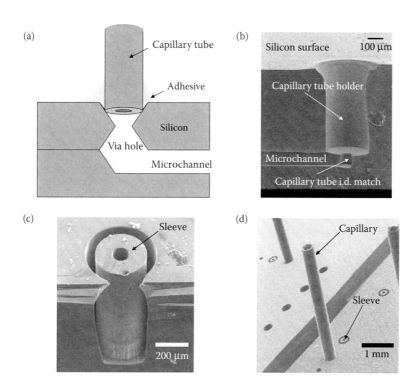

FIGURE 10.7
Examples of permanently bonded WTC interconnects between on-chip and off-chip devices: (a) illustration of a capillary tube glued in a microchannel fabricated by wet anisotropic etching of silicon; (b) capillary tubing WTC coupler with holes etched in silicon such that the inner and outer diameters of the capillary tubing match the holes in the silicon; (c and d) "sleeve"-type coupler that fits the inner and outer diameter of the capillary tubing and prevents adhesive seeping into the microchannel. (Panels b, c, and d are adapted from *Sensors and Actuators A*, 77(1), B.L. Gray, D. Jaeggi, N.J. Mourlas, B.P. van Drieënhuizen, K.R. Williams, N.I. Maluf, G.T.A. Kovacs, Novel interconnection technologies for integrated microfluidic systems, 57–65, Copyright 1999, with permission from Elsevier.)

10.3.2.2 Reworkable WTC Interconnects

Although permanently bonded tubes may provide adequate WTC interconnect to microchannels, reworkability of the interface is desirable from the standpoint of prototyping, ease of use, and reliability (e.g., permanently attached tubes sticking perpendicularly out of a chip surface tend to break, especially in transport). This section discusses reworkable interfaces on the basis of (1) press fits, in which a tube is mated to a hole or another tube structure that may contain a spring-fit mechanism or simply rely on the material's compliance with the interface to hold the tube in place and (2) screw-type interfaces, in which the tubes can be securely screwed in place but are usually comparatively large because of fabrication methods and which allow manual assembly, thereby resulting in reduced interconnect density.

10.3.2.2.1 WTC Interconnects on the Basis of Press Fits

Many reversible press-fit couplers have been developed, most of which are based on several basic designs. Press fits use the spring and/or elastic forces of the coupling parts to seal the fluidic connection, which is adequate, provided a high-pressure and/or very high density coupling is not required. Press-fit-type couplers can be designed such that the tubing fits on a coupler parallel to the on-chip microchannels or such that the tubing couples perpendicularly to the microchannel-containing chip surface. Examples include injected-molded plastic couplers, soft molded polymer couplers, and compression-molded couplers that can be used to hold and seal a fused silica tube or a metal capillary tube in place.

Press-fit WTC interconnects (see Figure 10.8) include those in which a tube is placed into a holder-type mechanism on the microchip surface, which may simply consist of a soft material such as PDMS or another polymer interfaced to a more rigid tube or in which the more compliant material forms a gasket or a wedge for the rigid tube (e.g., Chiou and Lee, 2004; Christensen et al., 2005; Gray et al., 1999; Han et al., 2007; Li and Chen, 2003; Westwood et al., 2008). A heat-shrink tubing may also be used for a reworkable press-fit WTC interconnect (Pan et al., 2006). Multiple WTC interconnects can also be made (Saarela et al., 2006). Other press-fit interconnects involve attaching a microfabricated interconnect structure directly to the tube, which then mates with on-chip structures (e.g., Liu et al., 2004; Meng et al., 2001; Puntambekar and Ahn, 2002).

Another successful design involves arrays of microports with a press-fit between a soft tube and a more rigid microstructure on the microchip surface, such as the hexagonal ports shown in Figure 10.9 (Gray et al., 2002) on the basis of prior work (Gonzalez et al., 1998). These ports have been fabricated parallel to the microchannel flow using KOH anisotropic

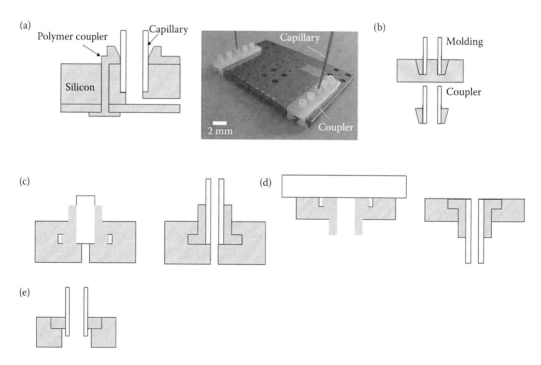

FIGURE 10.8

Examples of press-fit WTC interconnect methods: (a) Combination of silicon and injected-molded polymer coupler. (Adapted from *Sensors and Actuators A*, 77(1), B.L. Gray, D. Jaeggi, N.J. Mourlas, B.P. van Drieënhuizen, K.R. Williams, N.I. Maluf, G.T.A. Kovacs, Novel interconnection technologies for integrated microfluidic systems, 57–65, Copyright 1999, with permission from Elsevier.) (b) Illustration of a PDMS coupler molded to the tip of a capillary tube, similar to that developed by Meng et al. (2001). (c and d) Illustration of flanged thermoplastic tube couplers created by, for example, thermocompression molding, similar to that developed by Puntambekar and Ahn 2002. (e) Commonly used integrated o-ring coupler, where the o-ring may also be simply part of the compliant polymer (e.g., PDMS) chip.

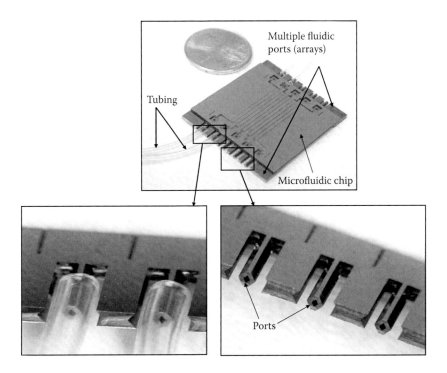

FIGURE 10.9
Structure similar to that shown in Figure 10.5, which is used instead as a soft-tubing press-fit WTC interconnect and placed in an array. (Adapted from B.L. Gray, D.K. Lieu, S.D. Collins, R.L. Smith, A.I. Barakat, Microchannel-based platform for the study of endothelial cell shape and function, *Biomedical Microdevices* 4: 9–16, 2002.)

etching, in an array of five ports. These ports are usable for low-pressure flows. At higher pressure flows, an adhesive such as room-temperature vulcanizing silicone must be used in addition to the press-fit. A similar port-type approach, in which the tubes are instead perpendicular to the substrate surface, has also been implemented using press-fits. For example, fluid ports have been fabricated by stereolithography, which is integrated with microfluidic chip fabrication (Han et al., 2005).

10.3.2.2.2 WTC Interconnects on the Basis of Screw-Type Connectors

For macroscale systems, screw-type interfaces, whereby a compression fit is introduced via a twisting (screw) force, are a popular fluidic interconnect method. On the microscale, such interfaces are complicated to fabricate. Fabrication of the screw interface typically limits the size scale and thus the interconnect density. As a result, many of the WTC interconnects presented in this section may be more properly considered to be milliscale rather than microscale, although they may be used to interface with on-chip microchannels.

Screw-type tubing interfaces may be bonded to on-chip microchannels (e.g., Bradley et al., 2007) or fabricated together with microchannels in an integrated process such as injection molding (e.g., Mair et al., 2006).

Other screw-in-place WTC interconnects include a very clever design (shown in Figure 10.10), whereby the screwing motion clamps down harder on the specially fabricated PEEK tubing (Krulevitch et al., 2002). Another interesting design involved electrochemical discharge machining (ECDM) to fabricate threaded vias and molds for mating plastic

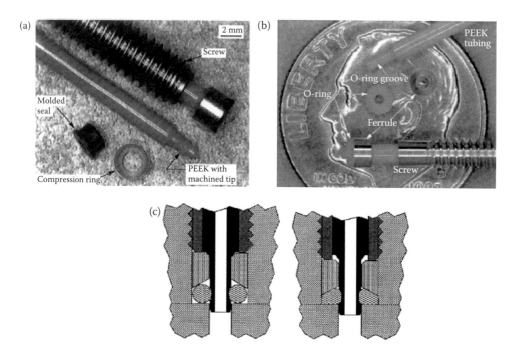

FIGURE 10.10
Screw-in-place WTC interconnect: (a) components that make up the interconnect structure; (b) assembled interconnect; and (c) interconnect operating principle, where the connection is made by finger tightening the screw in the hole and compressing a gasket seal. (Reprinted with kind permission from Springer Science + BusinessMedia: *Biomedical Microdevices*, Polymer-based packaging platform for hybrid microfluidic systems, 4(4), 2002, 301–308, P. Krulevitch, W. Benett, J. Hamilton, M. Maghribi, K. Rose.)

screws to glass microfluidics (see Figure 10.11) (Lee et al., 2004). Because microfabrication by ECDM was used, the size of these molded interconnects was comparatively small.

10.3.3 Socket-Type Approaches

Socket-type approaches cover any sort of device where the chip sits in or is enclosed in a device that performs interface adaptation to an external tubing or components. Often a combination of fluidic and electronic interfacing, or even optical interfacing, is provided although this may be simply a clear window to the device for viewing and optical detection. Both CMC Microsystems and Cascade Microtech (Cascade, www.cmicro.com/products/integrated-systems/microfluidics/l-series-microfluidics-systems; Mallard, 2008) have developed generic sockets that can be used for electrical and fluidic interfacing, with optical interfacing available, and have made these technologies available for use or purchase. In research labs, socket-type approaches are very popular as they are easy to use once developed, although the lead time for socket development may be high unless all chips in the given research laboratory have electrical and fluidic connections located in the same places and all are of the same size or can be adapted to fit into the existing socket.

A socket-like approach was used for an integrated surface micromachining microfluidics technology that was mentioned previously (Galambos and Benevides, 2004). A device featuring a socket-like interface with built-in valves was developed for PCR applications, which has been mentioned previously (Oh et al., 2005; Oh et al., 2005).

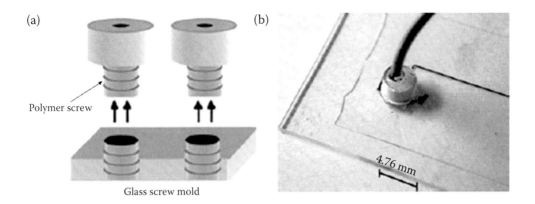

FIGURE 10.11
WTC interconnect using ECDM to fabricate threaded vias and molds for mating plastic screws to glass microflu-idics: (a) molding of polymer screws in glass mold and removal; (b) operation as a WTC coupler for a microfluidic device in glass, where the interconnect has been screwed into place in the threaded glass using a molded poly-mer screw. (Reprinted with permission from E.S. Lee, D. Howard, E. Liang, S.D. Collins, R.L. Smith, Removable tubing interconnects for glass-based micro-fluidic systems made using ECDM, *Journal of Micromechanics and Microengineering*, 14:535–541, 2005; permission granted by Rosemary L. Smith.)

Many other socket WTC interconnects have been developed by other research groups. One modular socket device was developed for capillary electrophoresis, in which the chip was loaded into the socket and covered and sealed in place with a hinged lid structure (Yang and Maeda, 2003). The socket provided both electronic and fluidic interconnect between the chip and the off-chip devices. The fluidic connections were accomplished by silicone tubes press-fit against holes on the microfluidic chip, thus providing a malleable gasket. Other socket-type WTC interfaces include lids containing microfluidic couplers that mate with microfluidic channels (Armani et al., 1999; Korivi and Jiang, 2007; Ueda et al., 2008). Another interesting socket-type approach involves the interconnection of a very large number of microfluidic channels via an array of WTC couplers located in "inter-connection blocks" (Sabourin et al., 2009; Yan et al., 2000).

10.3.4 WTC Interconnects for Special Applications

10.3.4.1 Microfluidic to Nanofluidic

Although many of the WTC interconnects thus far discussed may have relatively small dead volume compared with their size and may successfully interface macroscale equip-ment to microchannels, they may not be adequate for interconnection to nanosized chan-nels for applications such as DNA analysis. An analyte may become trapped in the dead volume of the WTC interconnects thus far discussed, and the size scale of the WTC inter-connects may not provide an effective interface to nanochannels for applications such as DNA linearizing. Several groups working in these areas have proposed solutions to WTC interconnects to nanochannels. In general, these techniques involve interfacing first to microchannels (even if only briefly) and then to the nanochannels.

The need for very sensitive DNA detection has resulted in new designs in microfluidic WTC interconnects for the efficient and gentle sequential delivery of biopolymers into device channels without fouling. One group (Humphreys et al., 2009) has used a planar interconnect for tubing, whereby the tubing is inserted edge-on into a silicon microchannel that has a slope defined by DRIE. This sloped WTC interconnect has very little dead volume and provides a gently sloping interface for delivery of DNA strands. Other researchers have used gradient nanostructures and membranes to stretch the DNA before its entry into nanochannels and/or gating operations or other fluid manipulation techniques (e.g., Cao et al., 2002; Chatterjee et al., 2005; Flachsbart et al., 2006; Shannon et al., 2005).

10.3.4.2 High-Pressure WTC Interconnects

For most biomedical and biological lab-on-a-chip applications, the pressures in the overall system are usually fairly low, with the flow rate being the variable of interest, especially in many cell research applications (Gray et al., 2002). However, for high-pressure chromatography, automotive applications, or energy and power applications, high-pressure and/or high-temperature operation of a microfluidic device, including its microfluidic interconnects, may be required.

Several research groups have investigated high-pressure and/or high-temperature WTC interconnects. Most involve adhesive or permanent bonding of the tubes that are attached to holes in the chip surface. One high-pressure or high-temperature WTC interconnect uses both epoxy and a reinforcing sleeve for Teflon capillary tubes in a WTC interconnect to silicon microchannels with a Pyrex™ cap (Pattekar and Kothare, 2003). Another involves Kovar™ tubes bonded to silicon microchannels via glass that is heated to high temperatures (>1000°C for 15 min) to achieve a bond with the glass preform (Peles et al., 2004), in high-pressure and high-temperature WTC interconnects for microengines and microrockets.

Hypodermic needles have been used for attachment to thermoplastic microfluidic chips via high-force friction fits and threaded interfaces (Chen et al., 2009). A socket-type WTC interconnect has also been demonstrated for high-pressure glass microreactor chips connected to fused silica capillaries (Nittis et al., 2001).

Although most WTC interconnects, even those designed for high-pressure applications, are generally formed perpendicular to the chip surface, a high-pressure interconnect can be realized for glass reactor chips simply by making the WTC tube insertion into the edge of the chip (in plane) and using an adhesive to seal it (Tiggelaar et al., 2007).

10.4 Interface Standardization and the Future

Many DTD, CTC, and WTC schemes that are applicable to both specific devices or systems have been developed or designed from the ground to be generically applicable with a design paradigm of interface standardization. Although many examples have previously been discussed, in this section we note the work by other researchers that specifically seeks interface standardization. We end the section with a discussion on emerging technologies.

Researchers have developed microfluidic assembly blocks with a goal toward making user-definable systems on the basis of generic building block structures that include WTC tubing ports, passive channels in generic geometries, chambers, connectors, culture beds for biological cells, and valves (Rhee and Burns, 2008). The generic blocks can be made in PDMS very quickly. Figure 10.12 shows some example blocks and illustration of how they might be connected. Unfortunately, modular connection was not fully developed, and the blocks had to be permanently bonded using uncured polymer, adhesive, or conventional

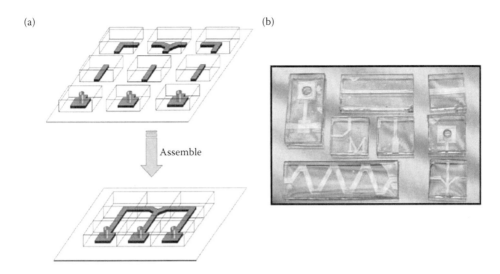

FIGURE 10.12
Example of generic microfluidic systems approach: (a) schematic of the basic concept where generic module blocks are placed on a glass slide, ordered as needed and (b) an example fabricated blocks. (From M. Rhee, M.A. Burns, Microfluidic assembly blocks, *Lab on a Chip* 8:1365–1373, August 2008. Reprinted with permission of The Royal Society of Chemistry.)

plasma surface treatments in order to achieve strong microfluidic seals. In essence, the blocks are just laid next to each other and bonded for a CTC interconnect, with a WTC interconnect of a conventional tube-in-PDMS hole type, with bonding of the tube for added WTC seal strength. Still, the project is one of the few projects that have sought a generic method of microfluidic system development.

Shaikh et al. (2005) have also tried to develop modular, customizable microfluidics. Rather than modules laid side by side and stacked on glass or other substrate materials, they have taken a more microfluidic circuit board or MCM approach. Here, the interconnect between two modules, one containing generic active devices and one, user-definable, containing passive devices such as interconnect channels, is accomplished through fluid ports that appear to be based on tube-type CTC interconnects of unknown reconfigurability. The researchers describe each module being developed in an integrated fashion, with modularity being introduced only in the interconnection between the fluidic bread-board containing active components and a user-definable passive module. Yuen (2008) has proposed "plug-and-play" microfluidics on the basis of a modular microfluidic circuit board approach with a CTC interconnect via basic H-shaped microchannel inserts with miniaturized luer fittings fabricated via stereolithography, which are a version of tube-type CTC fittings but with added geometry for better attachment because of the stereolithography methods. Pressurization limits of 51.1 psi were demonstrated, along with the two example systems. Although the channel sizes were fairly large by microfluidics standards (635 µm in diameter), the interconnections did appear to be somewhat reversible. Another approach uses a compliance mismatch between cylindrical pegs and holes for a multiple-chip interconnect of generic microchannel-containing modules (Westwood et al., 2008).

In addition to standardization needs for conventional microfluidic systems, other emerging microfluidic platforms and specific applications also require interconnect technology development. Both electrowetting-based droplet microfluidic systems (e.g., Fair, 2007;

Kanagasabapathi and Kaler, 2007) and systems featuring large-scale integration (Thorsen et al., 2002) require new interconnect paradigms. Systems for specialized applications, such as those that work with non-Newtonian fluids such as blood or highly viscous fluids, or those that are designed for harsh environments also require different considerations for packaging and interconnect technologies. Furthermore, systems that move fluid using electrokinetic principles may have unique problems at the interface boundaries of materials.

Microfluidic interconnect, both between devices/components and to external devices, is an exciting field that progresses and expands every year, expanding from a few references that could be cited by the author in articles in the late 1990s to the multiple pages of references for this chapter. Although a bottleneck to commercial acceptance of microfluidic interconnects has been widely recognized by many leaders in the field, microfluidic interconnects will continue to expand in importance as systems become increasingly closer to realization and as methods of interconnecting them easily and quickly become a selling feature.

References

Armani D, Liu C, Aluru N. (1999). Re-configurable fluid circuits by PDMS elastomer micromachining. Twelfth IEEE International Conference on Micro Electro Mechanical Systems (MEMS), Orlando, FL; January 17–21. pp. 222–7.

Becker H, Locascio LE. (2002). Polymer microfluidic devices (review). *Talanta*. 56:267–87.

Becker H, Gärtner C. (2008). Polymer microfabrication technologies for microfluidic systems. *Anal Bioanal Chem*. January;390(1):89–111.

Bhattacharya S, Datta A, Berg JM, Gangopadhyay S. (2005). Studies on surface wettability of poly(dimethyl) siloxane (PDMS) and glass under oxygen-plasma treatment and correlation with bond strength. *J Microelectromech Syst*. June;14(3):590–7.

Boone TD, Fan ZH, Hooper HH, Ricco AJ, Tan H, Williams SJ. (2002). Plastic advances microfluidic devices. *Anal Chem*. 74:78–86A.

Bousse L, Mouradian S, Minalla A, Yee H, Williams K, Dubrow R. (2001). Protein sizing on a microchip. *Anal Chem*. 73:1207–12.

Bradley WW, Gordon MH, Mock PEJ. (2007). World to chip sample introduction. IEEE Region 5 Technical Conference, Fayetteville; April, pp. 424–31.

Brennen R, van de Goor AAAM. (2003). Integrated microfluidics and electronic components. United States patent US 6,632,400 B1. October 14.

Cao H, Tegenfeldt JO, Austin RH, Chou SY. (2002). Gradient nanostructures for interfacing microfluidics and nanofluidics. *Appl Phys Lett*. October;81(16):3058–60.

Chang S, Suk SD, Cho Y-H. (2009). Characterization of a multi-chip microelectrofluidic bench for modular fluidic and electric interconnections. *Sens Actuators B Chem*. 140:342–8.

Chatterjee AN, Cannon DM, Gatimu EN, Sweedler JV, Aluru NR, Bohn PW. (2005). Modeling and simulation of ionic currents in three-dimensional microfluidic devices with nanofluidic interconnects. *J Nanopart Res*. 7:507–16.

Chen CF, Liu J, Hromada LOP, Tsao CW, Chang CC, DeVoe DL. (2009). High-pressure needle interface for thermoplastic microfluidics. *Lab Chip*. 9:50–5.

Chiou C-H, Lee G-B. (2004). Minimal dead-volume connectors for microfluidics using PDMS casting techniques. *J Micromech Microeng*. 14:1484–90.

Chiu DT, Jeon NL, Huang S, Kane RS, Wargo CJ, Choi IS, et al. (2000). Patterned deposition of cells and proteins onto surfaces by using three-dimensional microfluidic systems. *Proc Natl Acad Sci USA*. 97:2408–13.

Christensen AM, Chang-Yen DA, Gale BK. (2005). Characterization of interconnects used in PDMS microfluidic systems. *J Micromech Microeng*. 15:928–34.

Dalton C, Kaler KVIS. (2007). A cost effective re-configurable electrokinatic microfluidic chip platform. *Sens Actuators B Chem*. 123:628–35.

Dang B, Bakir MS, Meindl JD. (2006). Integrated thermo-fluidic I/O interconnects for an on-chip microchannel heat sink. *IEEE Electron Device Lett*. February;27(2):117119.

Datta P, Hammacher J, Pease M, Gurung S, Goettert J. (2006). Development of an integrated polymer microfluidic stack. *J Phys Conf Ser*. 34:853–8.

Erickson D, Li D. (2004). Integrated microfluidic systems. *Anal Chim Acta*. 507:11–26.

Fair RB. (2007). Digital microfluidics: is a true lab-on-a-chip possible? *Microfluid Nanofluidics*. 3(3):245–81.

Fiorini GS, Chiu DT. (2005). Disposable microfluidic devices: fabrication, function, and application. *Biotechniques*. 38:429–46.

Flachsbart BR, Wong K, Iannacone JM, Abante EN, Vlach RL, Rauchfuss PA, et al. (2006). Design and fabrication of a multilayered polymeric microfluidic chip with nanofluidic interconnects via adhesive contact printing. *Lab Chip*. 6:667–74.

Fredrickson CK, Fan ZH. (2004). Macro-to-micro interfaces for microfluidic devices. *Lab Chip*. 4:526–33.

Friedrich CR, Avula RRK, Gugale S. (2005). A fluid microconnector seal for packaging applications. *J Micromech Microeng*. 15:1115–24.

Frye-Mason GC, Martinez D, Manginelli RP, Heller EJ, Chanchani R. (2004). Method for making electro-fluidic connections in microfluidic devices. United States patent US 6,772,513 B1. August 10.

Galambos P, Benavides G. (2004). Electrical and fluidic packaging of surface micromachined electro-microfluidic devices. *Proc SPIE*. 4177:186–200.

Galambos P, Eaton WP, Shul R, Willison CG, Sniegowski JJ, Miller SL. (November 1999). Surface micromachine microfluidics: design, fabrication, packaging, and characterization. ASME Winter Annual Meeting, Nashville, Tennessee.

Ghafar-Zedeh E, Sawan M, Therriault D. (2007). Novel direct-write CMOS-based laboratory-on-chip: design, assembly, and experimental results. *Sens Actuators A Phys*. 134:27–36.

Gonzalez C, Collins SD, Smith RL. (1998). Fluidic interconnects for modular assembly of chemical microsystems. *Sens Actuators B Chem*. 49:40–5.

Gonzalez C, Smith RL, Howitt DG, Collins SD. (1998). MicroJoinery: concept, definition, and application to microsystem development. *Sens Actuators A Phys*. 66:315–22.

Gray BL, Collins SD, Smith RL. (2004). Interlocking mechanical and fluidic interconnections for microfluidic circuit boards. *Sens Actuators A Phys*. 112:18–24.

Gray BL, Jaeggi D, Mourlas NJ, van Drieënhuizen BP, Williams KR, Maluf NI, et al. (1999). Novel interconnection technologies for integrated microfluidic systems. *Sens Actuators A Phys*. 77(1):57–65.

Gray BL, Jaffer S, Sahota D, Westwood SM. (June 2008). Mechanical and fluidic characterization of microfluidic interconnects for lab on a chip applications. The 14th IEEE International Mixed-Signals, Sensors, and Systems Test Workshop, Vancouver; 5 p.

Gray BL, Lieu DK, Collins SD, Smith RL, Barakat AI. (2002). Microchannel-based platform for the study of endothelial cell shape and function. *Biomed Microdevices*. 4:9–16.

Grodzinski P, Yang J, Liu RH, Ward MD. (2003). A modular microfluidic system for cell pre-concentration and genetic sample preparation. *Biomed Microdevices*. 5(4):303–10.

Haeberle S, Zengerle R. (2007). Microfluidic platforms for lab-on-a-chip applications. *Lab Chip*. 7:1094–110.

Han A, Graff M, Wang O, Frazier AB. (2005). An approach to multi-layer microfluidic systems with integrated electrical, optical, and mechanical functionality. *IEEE Sens J*. 5(1):82–9.

Han A, Wang O, Graff M, Mohanty SK, Edwards TL, Han K-H, et al. (2003). Multi-layer plastic/glass microfluidic systems containing electrical and mechanical functionality. *Lab Chip*. 3:150–7.

Han K-H, Frazier AB. (2005). Reliability aspects of packaging and integration technology for microfluidic systems. *IEEE Trans Device Mater Reliab*. September;5(3):452–7.

Han K-H, McConnell RD, Easley CJ, Bienvenue JM, Ferrance JP, Landers JP, et al. (2007). An active miocrofludic system packaging technology. *Sens Actuators B Chem*. 122:337–46.

Harrison DJ, Manz A, Fan Z, Luedi H, Widmer HM. (1992). Capillary electrophoresis and sample injection systems integrated on a planar glass chip. *Anal Chem.* 64:1926–32.

Harrison DJ, Fluri K, Seiler K, Fan Z, Effenhauser CS, Manz A. (1993). *Science.* 261:895–7.

Hartmann DM, Nevill JT, Pettigrew KI, Votaw G, Kung P-J, Crenshaw HC. (2008). A low-cost, manufacturable method for fabricating capillary and optical fiber interconnects for microfluidic devices. *Lab Chip.* 8:609–16.

Humphreys T, Andersson J, Södervall U, Melvin T. (2009). World-to-chip interconnects for efficient loading of genomic DNA into microfluidic channels. *J Micromech Microeng.* 19:105024.

Igata E, Aryndell M, Morgan H, Cooper JM. (2002). Interconnected reversible lab-on-a-chip technology. *Lab Chip.* 2:65–9.

Ilic B, Czaplewski D, Zalalutdinov M, Schmidt B, Craighead HG. (2002). Fabrication of flexible polymer tubes for micro and nanofluidic applications. *J Vac Sci Technol B.* November/December;20(6):2459–65.

Jaffer S, Gray BL. (2008). Polymer mechanically interlocking structures as interconnect for microfluidic systems. *J Micromech Microeng.* March;18:035043.

Jo BH, van Lerberghe LM, Motsegood KM, Beebe DJ. (2000). Three-dimensional micro-channel fabrication in polydimethylsiloxane (PDMS) elastomer. *J Microelectromech Syst.* 9(1):76–81.

Kanagasabapathi TT, Kaler KVIS. (2007). Surface microfluidics—high-speed DEP liquid actuation on planar substrates and critical factors in reliable actuation. *J Micromech Microeng.* 17:743–52.

Kim Y-H, Shin K-S, Kang J-Y, Yang E-G, Pack K-K, Seo D-S, et al. (2006). Poly(dimethylsiloxane)-based packaging technique for microchip fluorescence detection system applications. *J Microelectromech Syst.* 15(5):1152–8.

Korvini NS, Jiang L. (2007). A generic chip-to-world fluidic interconnect system for microfluidic devices. 39th IEEE Southeastern Symposium on System Theory, Macon, GA; March, pp. 176–80.

Krulevitch P, Benett W, Hamilton J, Maghribi M, Rose K. (2002). Polymer-based packaging platform for hybrid microfluidic systems. *Biomed Microdevices.* 4(4):301–8.

Lai S, Wang S, Luo J, Lee LJ, Yang ST, Madou MJ. (2004). Design of a compact disk-like microfluidic platform for enzyme-linked immunosorbent assay. *Anal Chem.* 76(7):1832–7.

Lammerink TSJ, Spiering VL, Elwenspoek M, Fluitman JHJ, van den Berg A. (February 1996). Modular concept for fluid handling systems: a demonstrator micro analysis system. IEEE Proceedings of the 9th International Workshop on Micro Electro Mechanical Systems (MEMS '96), San Diego. pp. 389–94.

Läritz C, Pagel L. (2000). A microfluidic pH-regulation system based on printed circuit board technology. *Sens Actuators A Phys.* 84(9):230–5.

LeClair T, Harper A, Graham S, Ackley D. (1999). Flip chip interconnection of DNA chip devices. *Proc SPIE.* V3582. pp. 732–6.

Lee ES, Hoawrd D, Liang E, Collins SD, Smith RL. (2005). Removable tubing interconnects for glass-based micro-fluidic systems made using ECDM. *J Micromech Microeng.* 14:535–41.

Li S, Chen S. (2003). Polydimethylsiloxane fluidic interconnects for microfluidic systems. *IEEE Trans Adv Packag.* August;26(3):242–7.

Liu T, Masood S, Iovenitti P, Harvey E. (2004). Development of surface connectors for microfluidic systems. *Proc SPIE.* 5276:498–506.

Lorenz H, Despont M, Fahrni N, LaBianca N, Renaud P, Vettiger P. (1997). Su-8: a low-cost negative resist for MEMS. *J Micromech Microeng.* 7:121–4.

Mair DA, Geiger E, Pisano AP, Fréchet JMJ, Svec F. (2006). Injection molded microfluidic chips featuring integrated interconnects. *Lab Chip.* 6:1346–54.

Mallard R. (June 2008). A modular platform for proof-of-concept development of microsystems based on microfluidic technology. The 14th IEEE International Mixed-Signals, Sensors, and Systems Test Workshop, Vancouver.

Man PF, Jones DK, Mastrangelo CH. (1997). Microfluidic plastic interconnects for multi-bioanalysis chip modules. *Proc SPIE.* 3224:196–200.

Martin PM, Matson DW, Bennett WD, Lin Y, Hammerstrom DJ. (1999). Laminated plastic microfluidic components for biological and chemical systems. *J Vac Sci Technol A*. July/August;17(4):2264–9.

Meng E, Wu S, Tai Y-C. (2001). Silicon couplers for microfluidic applications. *Fresenius J Anal Chem*. 371:270–5.

Miserendino S, Tai Y-C. (2008). Modular microfluidic interconnect using photodefinable silicone microgaskets and MEMS O-rings. *Sens Actuators A Phys*. 143:7–13.

Mogulkoc B, Jansen H, Ter Brake M, Elwenspoek M. (September 2008). Borosilicate glass (Duran) tubes as micro-fluidic interconnects. Proceedings of the 19th micromechanics Europe conference, Aachen, Germany; pp. 28–30.

Moorthy J, Mensing GA, Kim D, Mohanty S, Eddington DT, Tepp WH, et al. (2004). Microfluidic tectonics platform: a colormetric, disposable, botulism toxin enzyme-linked immunoabsorbant assay system. *Electrophoresis*. 25:1705–13.

Morrissey A, Kelly G, Alderman J. (1998). Low-stress 3D packaging of a microsystem. *Sens Actuators A Phys*. 68:404–8.

Murphy ER, Inoue T, Sahoo HR, Zaborenko N, Jensen KF. (2007). Solder-based chip-to-tube and chip-to-chip packaging for microfluidic devices. *Lab Chip*. 7:1309–14.

Ng SH, Wang ZF, de Rooij NF. (2009). Microfluidic connectors by ultrasonic welding. *Microelectron Eng*. 86:1354–7.

Nguyen NT, Wereley ST. (2002). Fabrication techniques for microfluidics. In: Nguyen NT, Wereley ST, editors. *Fundamentals and Applications of Microfluidics*. Boston: Artech House; pp. 124–7.

Nittis V, Fortt R, Legger CH, de Mello AJ. (2001). A high-pressure interconnect for chemical microsystem applications. *Lab Chip*. 1:148–52.

O'Conner SD, Karp CD, Dantsker E, Pezzuto M. (2002). Modular microfluidic systems. United States patent US2002/0124896 A1. September 12.

Oh KW, Park C, Namkoong K. (2005). A world-to-chip microfluidic interconnection technology with dual functions of sample injection and sealing for a multi-chamber micro PCR chip. IEEE 18th International Conference on Micro Electro Mechanical Systems (MEMS 2005), Miami; pp. 714–7.

Oh KW, Park C, Namkoong K, Kim J, Ock K-S, Kim S, et al. (2005). World-to-chip microfluidic interface with built-in valves for multi-chamber chip-based PCR assays. *Lab Chip*. 5:845–50.

Pan T, Baldi A, Ziaie B. (2006). A reworkable adhesive-free interconnection technology for microfluidic systems. *J Microelectromech Syst*. February;15(1):267–72.

Pattekar AV, Kothare MV. (2003). Noval microfluidic interconnectors for high temperature and pressure applications. *J Micromech Microeng*. 13:337–45.

Peles Y, Srikar VT, Harrison TS, Protz C, Mracek A, Spearing SM. (2004). Fluidic packaging of microengine and mcirorocket devices for high-pressure and high-temperature operation. *J Microelectromech Syst*. February;13(1):31–40.

Peng Z-C, Ling Z-G, Tondra M, Liu C-G, Zhang M, Lian K, et al. (2006). CMOS compatible integration of three-dimensional microfluidic systems based on low-temperature transfer of SU-8 films. *J Microelectromech Syst*. June;15(3):708–16.

Pepper M, Palsandram NS, Zheng P. (2007). Interconnecting fluidic packages and interfaces for micromachined sensors. *Sens Actuators A Phys*. 134:278–85.

Pepper M, Palasandram NS, Zhang P, Lee M, Cho HJ. (2007). Interconnecting fluidic packages and interfaces for micromachined sensors. *Sens Actuators A Phys*. 134:278–85.

Perozziello G, Bundgaard F, Geschke O. (2008). Fluidic interconnections for microfluidic systems: a new integrated fluidic interconnection allowing plug'n'play functionality. *Sens Actuators B Chem*. 130:947–53.

Puntambekar A, Ahn CH. (2002). Self-aligning microfluidic interconnects for glass- and plastic-based microfluidic systems. *J Micromech Microeng*. 12:35–40.

Ramesheshan R, Ghaffarian R. (2000). Challenges in interconnection and packaging of microelectromechanical systems. Proceedings of the IEEE Electronic Components and Technology Conference, Las Vegas, NV; pp. 666–75.

Retterer ST, Smith KL, Bjornsson CS, Neeves KB, Spence AJH, Turner JN, et al. (2004). Model neural prosthesis with integrated microfluidics: a potential intervention strategy for controlling reactive cell and tissue response. *IEEE Trans Biomed Eng*. November;51(11):2063–73.

Renzi RF, Stamps J, Horn BA, Ferko S, VanderNoot VA, West JAA, et al. (2005). Hand-held microanalytical instrument for chip-based electrophoretic separations of proteins. *Anal Chem*. 77:435–41.

Rhee M, Burns MA. (2008). Microfluidic assembly blocks. *Lab Chip*. August;8:1365–73.

Saarela V, Franssila S, Tuomikoski S, Marttila S, Östman P, Sikanen T, et al. (2006). Re-usable multi-inlet PDMS fluidic connector. *Sens Actuators B Chem*. 114:552–7.

Sabourin D, Snakenborg D, Dufva M. (2009). Interconnection blocks: a method for providing reusable, rapid, multiple, aligned and planar microfluidic interconnections. *J Micromech Microeng*. 19:035021.

Schabmueller CGJ, Koch M, Evans AGR, Brunnschweiler A. (1999). Design and fabrication of a microfluidic circuit board. *J Micromech Microeng*. 9:176–9.

Schindler-Saefkow F, Jam KA, Großer V, Reichl H. (June 2004). A 3D-package technology for fluidic applications based on Match-X. Proceedings of Actuator 2004, Bremen.

Schuenemann M, Thomson D, Atkins M, Garst S, Yussuf A, Soloman M, et al. (2004). Packaging of disposable chips for bioanakytical applications. Proceedings of the IEEE Electronic Components and Technology Conference; pp. 853–61.

Seong GH, Heo J, Crooks RM. (2003). Measurements of enzyme kinetics using a continuous flow microfluidics system. *Anal Chem*. 75:3161–67.

Shaikh K, Ryu KS, Fan Z, Liu C. (October 2003). Fabrication of through-wafer fluid interconnects with low dead volume and integrated back-plane fluid jumpers. 7th International Conference on Miniaturized Chemical and Bioechemical Analysis Systems (Micro-TAS), Squaw Valley; pp. 1053–6.

Shaikh KA, Ryu KS, Goluch ED, Nam J-M, Liu J, Thaxton CS, et al. (2005). A modular microfluidic architecture for integrated biochemical analysis. *Proc Natl Acad Sci USA*. July;102(28):9745–50.

Shannon MA, Flachsbart BR, Iannacone JM, Wong K, Cannon DM, Fa K, et al. (May 2005). Nanofluidic interconnects within a multilayer microfluidic chip for attomolar biochemical analysis and molecular manipulation. Proceedings on the 3rd Annual International IEEE EMBS Special Topic Conference on Microtechnologies in Medicine and Biology, Kahuku, Hawaii; pp. 257–9.

Tamanaha CR, Whitman LJ, Colton RJ. (2002). Hybrid macro-micro fluidics system for a chip-based biosensor. *J Micromech Microeng*. 12:N7–17.

Thorsen T, Maerkl SJ, Quake SR. (2002). Microfluidic large-scale integration. *Science*. 298:580–4.

Tiggelaar RM, Benito-López F, Hermes DC, Rathgen H, Egberink RJM, Mugele FG, et al. (2007). Fabrication, mechanical testing and application of high pressure glass microreactor chips. *Chem Eng J*. 131:163–70.

Tse LA, Hesketh PJ, Rosen DW, Gole JL. (2003). Stereolithography on silicon for microfluidics and microsensor packaging. *Microsyst Technol*. 9:319–23.

Ucok AB, Giachino JM, Najafi K. (June 2003). Compact, modular assembly and packaging of multi-substrate microsystems. 12th International Conference on Solid State Sensors, Actuators, and Microsystems, Boston, 4A3.6; pp. 1877–8.

Ueda T, Gray BL, Chen Y, Li P. (2008). Flexible enclosure for fluidic sealing of microcomponents. *Proc SPIE*. January; 6886–OP.

Unnikrishnan S, Jansen HV, Berenschot JW, Mogulkoc B, Elwenspoek MC. (2009). Microfluidics within a Swagelok: a MEMS-on-tube assembly. *Lab Chip*. 9(13):1966–9.

Velten T, Ruf HH, Barrow D, Aspragathos N, Lazarou P, Jung E, et al. (2005). Packaging of Bio-MEMS: strategies, technologies, and applications. *IEEE Trans Adv Packag*. 28(4):533–46.

Verlee D, Alcock A, Clark G, Huang TM, Kantor S, Nemcek T, et al. (1996). Fluid circuit technology: integrated interconnect technology for miniature fluidic devices. Proceedings of the Solid-State Sensor and Actuator Workshop '96, Hilton Head; June 3–6. pp. 9–14.

Verpoorte EMJ, van der Schoot BH, Jeanneret S, Manz A, Widmer HM, de Rooij NF. (1994). Three-dimensional micro flow manifolds for miniaturized chemical analysis systems. *J Micromech Microeng*. 4:246–56.

Wang X, Shaikh KA. (2008). Interfacing microfluidic devices with the macro world. In: Tian W-C, Finehout E, editors. *Microfluidics for Biological Applications*. New York: Springer; pp. 93–115.

Wang X, Wang X, Yao G, Liu C, Wang L. (2005). Automatic micro-bonding technology of capillaries with adhesives. 6th International IEEE Conference on Electronic Packaging Technology, 0–7803-9449–6.

Westwood S, Jaffer S, Gray BL. (2008). Enclosed SU-8 and PDMS microchannels with integrated interconnects for chip-to-chip and world-to-chip connections. *J Micromech Microeng*. June;18(6):064014.

Whitesides GM. (2006). The origins and the future of microfluidics. *Nature*. 442:368–73.

Wolff A, Perch-Nielsen I, Larsen U, Friis P, Goranovic G, Poulsen C, et al. (2003). Integrating advanced functionality in a microfabricated high-throughput fluorescent-activated cell sorter. *Lab Chip*. 3:22–7.

Xia Y, Whitesides GM. (1998). Soft lithography. *Annu Rev Mater Sci*. 28:153–84.

Yan KY, Smith RL, Collins SD. (2000). Fluidic microchannel arrays for the electrophoretic separation and detection of bioanalytes using electrochemiluminescence. *Biomed Microdevices*. 2(3):221–9.

Yang Z, Maeda R. (2003). Socket with built-in valves for the interconnection of microfluidic chips to macro constituents. *J Chromatogr A*. 1013:29–33.

You BH, Chen P-C, Guy J, Datta P, Nikitopoulos DE, Soper SA, et al. (November 2006). Passive alignment structures in modular, polymer microfluidic devices. Proceedings of the ASME International Mechanical Engineering Congress and Exposition, Chicago, IMECE2006–16100.

Young IT, Moerman R, van den Doel RLR, Iordanov V, Kroon A, Dietrich HRC, et al. (2003). Monitoring enzymatic reactions in nanolitre wells. *J Microsc*. 212(3):254–63.

Yuen PK. (2008). SmartBuild—a truly plug-n-play modular microfluidic system. *Lab Chip*. August;8:1374–8.

Zahn JD, Deshmukh AA, Papavasiliou AP, Pisano AP, Liepmann D. (November 2001). An integrated device for the continuous sampling and analysis of biological fluids. Proceedings of 2001 ASME International Mechanical Engineering Congress and Exposition, New York; 6 p.

11

Micro Total Analysis Systems

V.F. Cardoso and G. Minas

CONTENTS

11.1 Introduction

During the last few decades, a large variety of micro total analysis systems (also called μTAS, lab on a chip, or miniaturized analysis systems) have been developed for applications such as medicine, drug discovery, forensic science, biowarfare defense, and environment monitoring. These systems have been designed to operate as analytical miniaturized devices capable of performing all the required processes of a sample analysis. Actually, they should be able to manipulate and monitor microliter (10^{-6}) to attoliter (10^{-18}) of fluid volumes (Figure 11.1) as well as incorporate into a single and fully automated platform all the elements of a chemical analysis system since sample preparation, liquid handling, detection, and readout electronics. Such microdevices have many advantages over macrosystems: (1) reduced size; (2) reduced materials consumption and hence a reduction in waste production; (3) improved analytical performance by high-speed and accurate analysis (due to short diffusion distances, fast heating, high surface-to-volume ratios); (4) systems compactness, once several functionalities such as parallel detection of many compounds can be performed in a single chip; (5) low cost due to mass production and inexpensive materials that nowadays can be used to enable disposable microchips fabrication; (6) lower power consumption; and (7) portability and easy handling, which leads to the use of microdevices for point-of-care applications where proper laboratory access is not

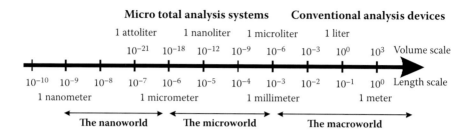

FIGURE 11.1
The length-scale technology.

available or in situ and rapid analysis results are required. Despite these advantages, at the small scale, the dominant physical and chemical effects are different from the ones at macroscale, and consequently they can be more complex to processes in µTAS than in conventional equipments. Because of the small dimensions and typically low-volume flow rate involved, the liquid handling system is restricted to the laminar flow regime (typically with Reynolds number ≤ 1). For many applications that rely on mixing, these properties could slow the mixing time. Therefore, considerable research efforts are being focused on the design of micropumps and micromixer systems that can be incorporated on a µTAS for fast transport and mixing. Additionally, detection principles may not always scale down in a positive way that could lead to low signal-to-noise ratios [1–3]. Indeed, the major criteria that lead the choice of a detection scheme are the properties of the analyte (optical, electrical, electrochemical, and physical), the composition of the sample, and the required detection limit. Other specific criteria can be the price, ease of use and implementation, and as far as µTAS devices are concerned, the ease of fabrication and the possibility for miniaturization. In applications where the µTAS is discarded after a single use, the cost of the device is very important and can even limit the choice of the detection scheme.

11.2 Materials and Micromanufacturing

The precision design required to obtain complex structures and to create increasingly smaller systems that are precise, accurate, and highly functional (transport, mixing, separation, manipulation, detection, etc.) is increasing almost as rapidly as the number of potential applications for these miniaturized devices, and it is dictated by the availability of materials and technology to manipulate them. Indeed, the ability to obtain fully integrated-automated and multilevel systems with functional apparatus, such as three-dimensional channels, functionalized surfaces, microelectromechanical systems (MEMS), or other components, is crucial to this relatively new field of microdevices to achieve its full potential. In this sense, new materials have been introduced and studied for this purpose, and some techniques have been modified and developed to obtain µTAS with new and enhanced features. The materials used most frequently in the manufacture of µTAS can be divided into three main groups: silicon, glass, and polymers. The first miniaturized analysis system has been fabricated mostly in silicon using photolithography and chemical etching technique. It is a gas chromatography system based on a sample injection valve and a 1.5-m-long separating capillary column. The device was able to separate and analyze a simple gas mixture [4]. Silicon has been a material widely used in the manufacture of µTAS because of its well-known properties and its wide use in the microelectronics and integrated circuit industry. The use of glass has also been reported because of its attractive properties such as its biocompatibility, its resistance to some chemicals, and its optical transparency, features that are suitable for many biological applications. Recently, polymers have been introduced and are slowly replacing silicon and glass, mainly when the target is the fabrication of miniaturized systems with low-cost production, good biocompatibility, and good resistance to chemicals. Moreover, polymers allow obtaining a multiplicity of surface properties because of their easy processing and manufacturing techniques [5–7]. This section provides an overview of these materials and several manufacturing methods currently available, emphasizing polymers manufacturing.

11.2.1 Silicon and Glass Manufacturing

Silicon was first identified by Antoine Lavoisier in 1787 and prepared by Jöns Jacob Berzelius in 1824. In the universe, silicon is one of the most abundant elements, after hydrogen, helium, neon, oxygen, nitrogen, and carbon. The earth's crust is composed of 25.7% silicon (the second most abundant element after oxygen). Silicon is a hard solid with a crystal structure similar to that of diamond, and its chemical reactions are similar to that of carbon [8].

In nature, silicon does not exist alone but combined with other elements. It is possible to find silica (silicon oxide) when combined with oxygen or silicates when combined with oxygen and other compounds (e.g., aluminum, magnesium, calcium, potassium, and iron). Silicon is the main component of glass, cement, ceramics, and silicones, which are plastic substances often confused with silicon. Because of its electronic structure, silicon is extremely important and widely used in the electronics industry as a semiconductor. For that, high-purity silicon is doped with other precise and very small amounts of boron, gallium, phosphorus, or arsenic to create basic materials for the construction of computer chips, transistors, switches, and various other special electronic devices [9]. For these reasons, many different techniques for processing and manipulating silicon and silicon-based materials that have been developed for the microelectronics industry have been subsequently adapted and optimized for the µTAS industry.

Glass is a hard and amorphous solid that has been manufactured for human use since 12,000 BCE. Examples of such materials include, but are not limited to, soda lime, borosilicate glass, acrylic glass, quartz, or Pyrex (as Corning 7740). Glass plays an important role in science and industry; indeed, it is frequently included in the microfluidic system of a µTAS. This is due to its intrinsic properties, such as its transparency (which allows for optical detection or visualization), its low autofluorescence, its excellent insulation property, and its chemical resistance. These properties make its use very interesting in many biological applications. However, its amorphous nature limits the acquisition of isotropic structures, so glass microfabrication techniques are not as advanced as silicon and polymer fabrication techniques. Consequently, etching in glass is usually achieved by bulk micromachining using the isotropic wet etch technique, which is more expensive and more complex [10].

11.2.1.1 Surface Micromachining

Surface micromachining, as its name suggests, produces structures above the surface using a succession of thin-film deposition and selective etching patterning [11]. Consequently, three-dimensional structures are created through sequential deposition, patterning, and selective etching of thin films onto the surface of a substrate. For that, two different materials are deposited alternately: the so-called sacrificial material and structural material. Consequently, the process requires a compatible set of these two materials and chemical etchants with high etch selectivity to remove the sacrificial material without affecting the structural material. As the name suggests, the sacrificial material is used to create gaps and spaces during the process, and therefore it will not be present in the final device. It should etch quickly during the release process, it should present good mechanical properties, and it should be stable during all fabrication processes. In turn, the structural material gives rise to the final structure of the device; thus, it should provide appropriate mechanical and chemical properties [12].

The surface micromachining process is illustrated in Figure 11.2. It starts with a substrate (Figure 11.2a), such as a silicon wafer or other substrate, that provides support for the manufacturing process. Subsequently, the sacrificial layer is deposited (Figure 11.2b); next it is covered by a photomask for patterning the device features in this sacrificial layer, and it is exposed to light (Figure 11.2c). The unwanted regions of the exposed layer are then removed by a specific chemical etchant (Figure 11.2d). After that, a thin film of the structural material is deposited, occupying the spacers that have been obtained with the etching of the sacrificial layer (Figure 11.2e), patterned (Figure 11.2f), and etched (Figure 11.2g). After all layers are completed, all the sacrificial material is removed (Figure 11.2h), the process being completed and the final structure obtained.

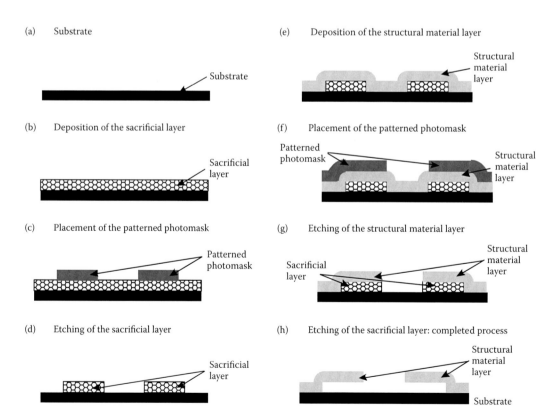

FIGURE 11.2
Process flow of surface micromachining.

TABLE 11.1

Materials Used for Surface Micromachining

Substrates	Sacrificial	Structural
Silicon	Silicon dioxide	Polysilicon
Glasses	Phosphosilicate glass	Silicon nitride
Plastics	Silicon nitride	Silicon carbide
Metals	Polymers	Metals
	Metals	Polymers
		Composites

Many variations of this technique can be used depending on the materials for the sacrificial, structural, and masking layers. Each material is chosen according to its mechanical, electrical, and/or chemical properties. Table 11.1 provides various materials, in addition to silicon, used for surface micromachining.

Silicon dioxide as a sacrificial material and polysilicon as a structural material are a combination frequently used for the fabrication of μTAS [12–14]. For this case, the final release process, that is, the silicon dioxide etch, is performed by washing the wafer in hydrofluoric acid. This etchant quickly removes the silicon dioxide while leaving the polysilicon undisturbed. Other materials have been tried, but the majority of the materials used in surface micromachined devices use common microelectronic materials because of the low cost and availability of the fabrication equipment.

Okandan et al. [15] presented examples of microfluidic systems (namely, pumps, valves, and a cell manipulator) fabricated using surface micromachining (SUMMIT™ technology). Polysilicon was used as the mechanical material and silicon dioxide as the sacrificial material. Additional layers of silicon nitride were used to enhance this technology and to enable microfluidic and BioMEMS applications. They provide functionality by allowing the fabrication of transparent microfluidic channels and insulation. Lin et al. [16] demonstrated the feasibility of a surface micromachined cell force transducer to measure forces generated by living heart muscle cells. Galambos et al. [17] proved the potential advantages of the surface micromachined technology for the fabrication of microfluidic devices and systems. For that, two different types of microfluidic channels, silicon nitride and polysilicon, were fabricated and studied. The channels were characterized for leak and flow rate tests in function of pressure measurements. This research indicated that the challenge in the development of surface micromachined microfluidic structures, that is, the difficulties to introduce the fluids into the very small channels that have been produced, the packaging problems, and the difficulties in device and system characterization, can be overcome, and the microfluidic structures can be successfully used. A similar study on the basis of the characterization of silicon micromachined devices was performed by Bien et al. [18]. They presented an automated measurement technique for fluid flow characterization of a surface micromachined valve. The system presented a pressure resolution of 2 Pa and has been used to measure liquid flow rates to less than 0.5 μL/min. Another interesting study was conducted by Han et al. [19]. They fabricated nanofluidic devices with channels down to 10 nm deep, 200 nm wide and up to 8 cm long, using surface micromachining technology. Different materials, such as silicon nitride, polysilicon, and silicon dioxide, were combined, making some modifications to both the silicon and the glass substrate fabrication procedures. The channels were integrated with specific elements of μTAS for electrokinetic separations.

11.2.1.2 Bulk Micromachining

Bulk micromachining represents a totally different process from surface micromachining. Whereas surface micromachining produces structures using a succession of thin-film deposition and selective etching patterning, the purpose of bulk micromachining is to selectively etch inside a substrate to produce structures. Table 11.2 gives the main features of the different substrates used for bulk micromachining. Although the plastic substrate is not favorable to metallization, both the ceramic and the glass substrates are difficult to machine. Consequently, the wafer most commonly used as substrate is silicon, due to its greater flexibility concerning design and manufacturing, well characterization, and ready availability. Despite the high cost of silicon in comparison with other substrates (e.g., plastics), if produced in large scale, the cost becomes negligible [20].

The used etching approaches seemingly require quite aggressive chemistries. However, they can be compatible with on-chip circuit platform and even micrometer-scale mechanisms if the overall process flow is correctly projected [20]. Consequently, bulk micromachining can be used to combine in the same silicon substrate microfluidics structures and electronic circuits (complementary metal-oxide semiconductor [CMOS]) to obtain a fully automated-integrated μ-TAS. A generalized comparison of various etchants is provided by Kovacs et al. [20]. The reaction that occurs between the etchant and the substrate relies on the oxidation of the silicon substrate to form compounds that can be physically removed.

The silicon micromachining technique can be divided according to the phase of the reactants, namely, *wet etching* that uses liquid etchant and *dry etching* that uses gaseous etchants (both vapor and plasma). In turn, the obtained etch geometry with these processes can be isotropic (the etching proceeds at the same rate, regardless of direction) or anisotropic (the etching proceeds at different rates in different directions of the substrate). As referred previously, for glass manufacturing, isotropic wet etching using hydrofluoric acid (HF)-based solutions is the most commonly used technique.

Possible bulk micromachined etching profiles are illustrated in Figure 11.3. The geometry of Figure 11.3a corresponds to an isotropic profile, whereas the geometries of Figures 11.3b, 11.3c, and 11.3d correspond to anisotropic profiles. The substrate, the etchant, the mask materials that are used to define the etched regions, the time of the etching process, and many other factors should be studied well to obtain the desired geometry structure.

11.2.1.2.1 Wet Etching

Wet etching technique consists of a removal process of specific parts of a substrate usually patterned by photoresist masks covering the substrate. This step can be achieved by submerging or by spraying the substrate with a chemical solution. During the diffusion of the etchant in the substrate (not covered by mask), the products formed by the reaction could also be removed by diffusion. Consequently, it is necessary to find a mask that will not dissolve or, at least, that will etch much slower than the material of the substrate to be patterned. The process of etching can be characterized by three parameters: (1) the etch

TABLE 11.2

Materials Used for Bulk Micromachining

Substrate	Cost	Metallization	Machinability
Silicon	+	++	++
Glass	−	++	−
Ceramic	+/−	+	−
Plastic/polymer	−	−	+

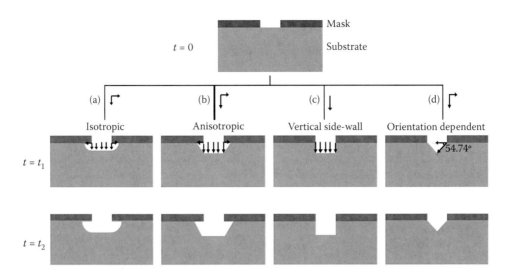

FIGURE 11.3
Illustration of possible bulk micromachined etching profiles.

rate, which provides the material thickness etched per unit time; (2) the etch selectivity, defined as the ability of the etchant to remove parts of the substrate without affecting the mask or other materials that can be present; and (3) the etch uniformity, which depends on the homogeneous property of the liquid etchant [21].

Because of the high etch rate, low cost, low complexity, and availability of masking materials, isotropic wet etching is extensively used (both for silicon and glass materials). However, anisotropic wet etching has been increasingly used in the microfluidic field because of the several geometries that can be obtained, opening doors to the development of new mechanisms (like MEMS).

Coltro et al. [22] proposed a simple, fast, innovative, and inexpensive masking technology on the basis of toner layers for wet chemical etching to fabricate glass microstructures. The microfabrication process is illustrated in Figure 11.4, and it resulted in isotropic etched channels using HF solution, with a trapezoid-like shape. The use of this technology to produce silicon and quartz microdevices is also preferred.

Isotropic wet etching was also used by Aziz et al. [23] to fabricate smooth biconcave conical silicon microneedles. For that, the work was focused on the characterization of isotropic wet etchant composition, in which a mixture of HF and nitric acid (HNO_3) was used. This study concluded that microneedles with a smooth etched surface can be obtained using HNA solution with 60–70 wt% of HNO_3.

The fabrication process of silicon microneedles was also studied by Wilke et al. [24]. However, in this case, the used fabrication technique was anisotropic wet etching using potassium hydroxide (KOH). This article presented several studies about the influence of the mask design and the processing environment, such as the etching parameters and the etching bath conditions on the formation of the silicon microneedle structures as well as its reproducibility and the mechanical stiffness of the fabricated microneedles. The authors had successfully produced silicon microneedles with very accurate needle heights and good reliability, after several optimization procedures. An interesting study on the basis of anisotropic wet etching was also presented by Mu et al. [25] and consisted in the use of three laminar distinct flows in glass microchannels to create microstructures with aspect ratio higher than 0.5 (Figure 11.5). This method was called as laminar-flow-assisted

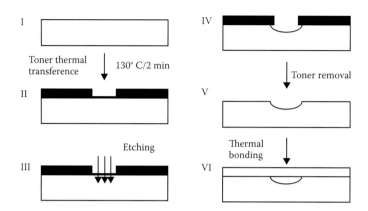

FIGURE 11.4
Toner-mediated lithography process: I, cleaned glass surface; II, thermal transference of the toner onto the glass surface; III, wet chemical etching; IV, channel profile before toner removal; V, toner removal; VI, thermal bonding. (From Coltro WKT, Piccin E, Fracassi da Silva JA, Lucio do Lago C, Carrilho E. A toner mediated lithographic technology for rapid prototyping of glass microchannels. *Lab Chip*, 7, 2007, pp. 931–934. Reprinted with permission from The Royal Society of Chemistry.)

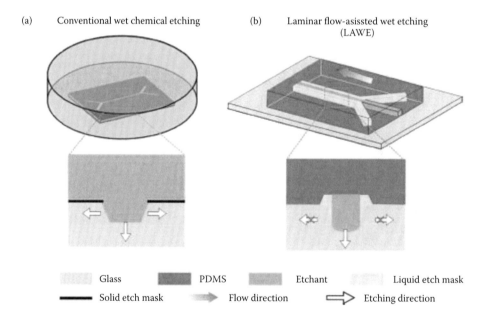

FIGURE 11.5
(See color insert.) Schematic representation of (a) conventional wet chemical etching where a solid mask is usually deposited on glass surface and (b) LAWE that require a PDMS layer with channels, which is sealed against a glass plate. (From Mu X, Liang Q, Hu P, Ren K, Wang Y, Luo G. Laminar flow used as liquid etch mask in wet chemical etching to generate glass microstructures with an improved aspect ratio. *Lab Chip*, 9, 2009, pp. 1994–1996. Reprinted with permission from The Royal Society of Chemistry.)

wet etching (LAWE). The two lateral flows were used as masks to protect sidewalls from the central etchant flow. Figure 11.6 demonstrates the ability of LAWE to generate three-dimensional glass microstructures with aspect ratio around 1 using HF as etchant and ethanol as liquid etch mask. This technique presented several advantages when compared with other etch masks in the solid phase, such as the improved aspect ratio of glass

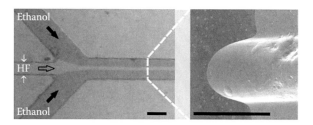

FIGURE 11.6
(See color insert.) Microchannel with a central HF flow and two ethanol lateral flows at a velocity of 0.1 ms^{-1}. After a 13-min etch, the aspect ratio by LAWE is up to approximately 1. (From Mu X, Liang Q, Hu P, Ren K, Wang Y. Luo G. Laminar flow used as liquid etch mask in wet chemical etching to generate glass microstructures with an improved aspect ratio. *Lab Chip*, 9, 2009, pp. 1994–1996. Reprinted with permission from The Royal Society of Chemistry.)

channel. Indeed, because of the isotropic nature of glass, the aspect ratio of its structure, obtained until then, was always lower than 0.5.

Wet etching micromachining provides a higher degree of selectivity, and it is faster than dry etching micromachining. Moreover, it presents a relatively planar etching surface, high repeatability, and a controllable etch rate by the etchant concentration. However, despite being a simple technology for obtaining good results, it is necessary to find the combination of the etchant and the mask material for a suitable application.

11.2.1.2.2 Dry Etching

As referred previously, the dry etching technology does not use any liquid solution to etch material from a substrate. In this case, the material is sputtered or dissolved using reactive ions contained in a gaseous or plasma phase, generating only volatile by-products in the process [26]. Dry etching can be split into three separate classes, which are obtained by adjusting the nature of the plasma containing the ions:

- Chemical etching: technique similar to wet etching. It is based on the chemical reaction between reactive compounds contained in a gas or plasma that is moving to the surface of the substrate by the action of an electrical field. Consequently, the products formed by the reaction volatilize and create holes and, consequently, the required structure.

- Physical etching: by the action of an electrical field, ions, electrons, or photons are bombarded, reaching the substrate and pulling atoms off its surface. This technique is anisotropic, and it has some disadvantages, including low selectivity, low etch rate, and usually surface damage.

- Combination of both physical and chemical etching: reactive ion etching (RIE) represents an example of dry etching technique and is based on the combination of both chemical and physical processes, allowing isotropic and anisotropic etching. In this case, a reaction chamber is formed by two parallel plates acting as electrodes: the anode and the cathode. The substrate is positioned in the middle. To create structures, masks are placed on the top of the substrate to protect specific regions from etching, exposing only the area to be etched. Then, a chemical reactive plasma containing ions (positively and negatively charged) is accelerated to the substrate by the action of a radiofrequency power source applied between the two electrodes, attacking the material. The chemical etching process occurs when the ions reach the substrate, reacting with it and forming gaseous materials. The physical etching part occurs when the ions have an energy so high that can etch atoms of the substrate. In this way, through a good optimization of the process, it is possible to control the isotropic etching because of the chemical process and the anisotropic etching because of the physical process to obtain the desired structure [27–29]. Because of its excellent process control, that is, etch-rate homogeneity, etch profile, and selectivity, this is the most commonly used technique in micro- and nanomanufacturing. This technique allows obtaining very thin thicknesses, in the order of 100 nm. A subclass of the RIE technique is the so-called deep RIE (DRIE). This technique has been increasingly studied because of its excellent ability to obtain structures with extremely high aspect ratios [26,30]. Ogawa et al. [31] fabricated a liquid pumping device using vibrating silicon-on-insulator microchannel walls 200 and 500 μm wide using DRIE. Ruan et al. [32] present new methods for improving the quality of silicon DRIE procedure using PECVD oxide layer and a thermal oxide layer. Marty et al. [33] used DRIE technology to create deep trenches in silicon substrate with an aspect ratio up to 107.

11.2.1.3 Comparison between Surface and Bulk Micromachining

Bulk micromachining presented a quicker advance in the market because it was easier to make a device that would work as projected. Surface micromachining showed more technological and compatibility problems, but now, surface micromachined devices are seen on the market. The choice takes into account the required application, the type of structure that should be made, and the availability of the equipment for the fabrication [34] (Table 11.3).

Moreover, surface micromachining is more expensive than bulk micromachining once it requires more fabrication steps. However, it is able to create more sophisticated mechanical elements such as micropumps, microvalves, and microactuators, which increase the functionality of the μTAS.

11.2.2 Polymers Manufacturing

As mentioned previously, polymer-based microfluidic devices have assumed the leading role as substrate materials because of their many advantages compared with silicon and

TABLE 11.3

Comparisons between Surface and Bulk Micromachining

Property	Surface	Bulk
Processing complexity	0/+	0
Lateral dimensions	100–500 μm	3–5 mm
Vertical dimensions	0.5–2 μm	100–500 μm
Mechanical quality	++	0/+
Mass	++	0
Capacitance	++	−
Chip area	−	++

Source: Reprinted with permission from French PJ, Sarro, PM. Surface versus bulk micromachining: the contest for suitable applications. *J. Micromech. Microeng.*, 8, 1998, pp. 45–53.

glasses, in particular, the low cost, the easier manufacturing procedures to produce complex three-dimensional structures, the high resources available, and above all, the fact that many polymers do not interact with chemical solutions [35]. Some of the principal properties of polymeric materials compared with silicon and glass are provided in Reference [35].

One of the most commonly used polymers in the µTAS field, poly(dimethylsiloxane) (PDMS), demonstrated those advantages. A study by the PubMed database indicates that the number of publications with the search query "PDMS" increased 30 times in 10 years, although many of them are also macroscopic applications [36]. However, many other polymers have also been studied and used to fabricate µTAS. For this reason and because of the diversity of existing polymers (thermoplastics, thermosets, and elastomers), various techniques of manufacturing have been studied, developed, and enhanced to process these materials and also to provide other functionalities for the development of enhanced microfluidic devices with properties that were previously impossible to achieve.

These techniques are often divided into two major groups: replication micromachining and direct micromachining. The replication methods have the potential for high-volume production, low cost, and consequently the opportunity to create disposable and/or point-of-care microdevices. They are based on the repetitive replication of a mold structure and are typically used for the fabrication of commercial microdevices. Hot embossing, injection molding, and casting are the techniques most commonly used in replication micromanufacturing, and these will be reported and compared in the next section. The direct methods manufacture each device individually, which limits the fabrication time each device and, consequently, the production volume. They do not need previous mastering, and their techniques are usually used for prototyping. Laser micromachining, stereolithography, and photopolymerization are the direct micromachining techniques that will be described in Section 11.2.2.2 [37,38].

11.2.2.1 Replication Micromachining

11.2.2.1.1 Hot Embossing

Hot embossing, also called compression molding, has become established in the last 10 years [38] and has been used for the fabrication of microfluidic devices, BioMEMS, biosensors, microoptics devices, and consequently µTAS. It uses a process that consists of three major steps (Figure 11.7). First, the mold is heated just above the polymer glass transition temperature (Figure 11.7a). Then, the mold is pressed under vacuum to the polymer, applying a specific embossing pressure, thereby imposing its design upon the polymer (Figure 11.7b). A vacuum chamber is used to remove air bubbles trapped at small features and to remove water vapor from the polymer. Finally, the system is cooled below the glass transition temperature to harden the polymer (Figure 11.7c). The result is an inverted replica of the mold (Figure 11.7d).

A wide variety of polymers (including polymethylmethacrylate [PMMA], polycarbonate [PC], polyvinylchloride, and cyclic olefin copolymers [COC]) have been successfully hot embossed at the microscale size (and below) to create structures with excellent reproducibility [38]. Narasimhan and Papautsky [39] have embossed microchannels in PMMA with aspect ratios up to 2, with depths from 5 to 250 µm, and with widths more than 40 µm. Figure 11.8 shows PMMA microchannels obtained in this study.

This technique presents an easy manufacturing process at a low cost and obtains high aspect ratio features (with feature sizes smaller than 100 nm [40]). However, it has a relatively long time cycle compared with injection molding. Moreover, and as with any other

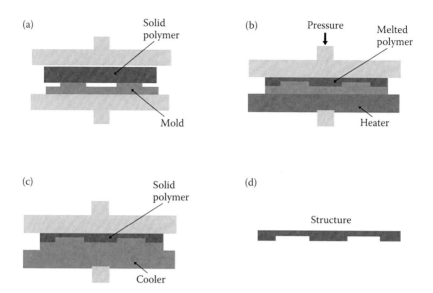

FIGURE 11.7
Process flow for hot embossing: (a) arrangement of the system; (b) the polymer is heated and pressed to acquire the mold structure; (c) the polymer is cooled to harden; (d) final polymeric structure.

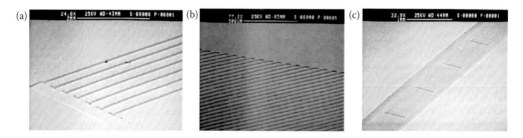

FIGURE 11.8
Scanning electron microscope of PMMA microchannels with: (a) 90 μm deep containing an array of 300-μm-wide features and a 500-μm center-to-center spacing; (b) 5 μm containing an array of 40-μm-wide features and 90-μm center-to-center spacing; (c) orthogonal three-dimensional microchannels with 1-mm wide, 110-μm deep rectangular features in the bottom surface of the microchannel with 20-μm thick, 500-μm long, and 1-mm center-to-center spacing. (From Narasimhan J, Papautsky I. Polymer embossing tools for rapid prototyping of plastic microfluidic devices. *J. Microchem. Microeng.*, 14, 2004, 96–103. With permission.)

manufacturing technique, the process should be optimized depending on the material used and the desired structure.

11.2.2.1.2 Injection Molding

Injection molding was first designed in 1930, and it is still a technique widely used, particularly in micromanufacturing. In this technique, the mold that creates the microstructure is heated at a temperature higher than the softening glass transition temperature of the polymer, and hence, the polymer is injected into the mold. Because of the high temperature normally required, metal molds are usually used.

During the process, the polymeric material is fed into the hopper. In the heated chamber, a screw carries the material toward the injection port of the molding tool. Throughout this transport, the polymer melts and reaches the tool in a liquid form, with the melt temperature depending on the polymer. The injection takes place under high pressure into the mold, which contains the microstructured mold inserted within. Then, the mold is cooled to the hardening point of the polymer, and then the molded microstructure can be removed (Figure 11.9) [38,41,42].

Many polymers such as PMMA, PC, polysulfone, polyoxymethylene, polypropylene, polyamide, or high-density polyethylene can be used in injection molding [6,38]. Liou and Chen [43] studied the effect of various injection molding parameters to obtain micro- and submicrostructures with high aspect ratios using PMMA, polypropylene, and high-density polyethylene polymers. They concluded that the mold temperature is crucial for the success of the injection molding to obtain micro- and sub-microwalls. Moreover, the injection pressure and the time also affect the aspect ratio of these walls. Do and Ahn [44] presented a polymeric lab on a chip fabricated by injection molding for magnetic bead-based immunoassay. The device consists of a magnetic bead-based separator, an interdigitated array microelectrode, and a microfluidic system, all fully integrated into a COC-based lab on a chip. A photograph of the polymeric microfluidics is shown in Figure 11.10. Each microchannel has a width of 150 μm and height of 60 μm [44].

The advantages of this technique include high production rates, design flexibility (complex geometry and fine features, as small as several nanometers, can be fabricated because of the high pressures that can be applied), repeatability (within its tolerance) due to the relatively low cycle times, possibility to make extremely high volumes of pieces with a single mold, possibility to process a wide range of materials, availability of a large variety of equipment suppliers, and automation solutions for large-volume manufacturing. However, it requires a high initial investment in equipment, it is difficult to accurately predict the microdevice fabrication cost, and the molds are expensive and complex (thus, low-volume production, less than 1000 parts, is not recommended for this technique). Despite being a well-known technique, a complex process of optimization is necessary, depending on the desired microstructure [6,42].

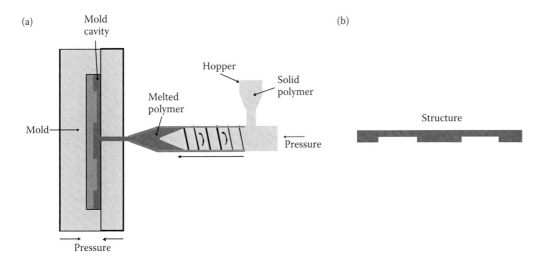

FIGURE 11.9
Process flow for injection molding: (a) arrangement of the system; (b) polymeric structure.

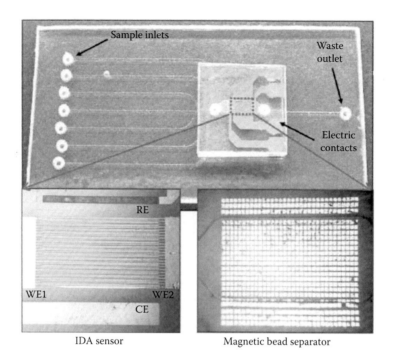

IDA sensor   Magnetic bead separator

FIGURE 11.10
(See color insert.) Photograph of the COC-based lab on a chip with a magnified view. The polymeric microfluidic was fabricated using the micropolymer injection molding technique with an electroplated Ni mold. (From Do J, Ahn CH. A polymer lab-on-a-chip for magnetic immunoassay with on-chip sampling and detection capabilities. *Lab Chip*, 8, 2008, pp. 542–549. Reprinted with permission from The Royal Society of Chemistry.)

11.2.2.1.3 Casting

Compared with hot embossing and injection molding that use heating and cooling cycles to produce their structures, casting uses chemical processes to harden the polymer. In this technique, a material is introduced in the mold in its liquid state, which later solidifies, acquiring the shape of the mold (Figure 11.11). For that, the material should be composed of a mixture of two compounds, the so-called prepolymer: a base and an initiator (crosslinker). This mixture, now a fluid, is spread in the mold surface until it attains the mold shape. Subsequent exposure to heat or ultraviolet (UV) light cold-sets the mixture through cross-linking (polymerization), increasing its viscosity by several orders of magnitude. Then, the structure can be removed [36,38].

Elastomers are very popular in the fabrication of microstructures using casting because they form hermetic, reversible seals to smooth surfaces by adhesion, like silicon and glass [38]. PDMS is the elastomer most frequently used for casting microstructures. The drawbacks are the full mixing, the distribution of the two compounds, and the chemical reaction (cross-linking) that must take place throughout the bulk to obtain a uniform strength. This can only be accomplished through a comparatively long cycle time. Nevertheless, stuctures on the scale of a few nanometers can be simply replicated.

PDMS casting was used by Narasimhan and Papautsky [39] to fabricate hot embossing tools, which were then used to fabricate the PMMA-based microchannels shown in Figure 11.8 (see Section 11.2.2.1.1). For that, a negative photoresist SU-8 or positive

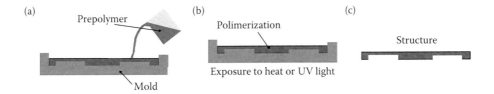

FIGURE 11.11
Process flow for casting: (a) introduction of the fluid polymer inside the mold; (b) polymerization process; (c) polymeric structure.

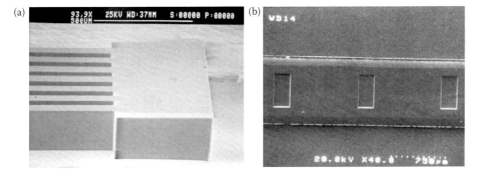

FIGURE 11.12
Scanning electron microscope of PDMS embossing tools obtained by casting manufacturing: (a) with a thickness of 90 μm and an array of 150-μm-wide features with a 250-μm center-to-center spacing; (b) orthogonal three-dimensional PDMS mold with a 1-mm wide and a 110-μm-thick microchannel; the rectangular indents in the microchannel are 20 μm thick, 500 μm wide, and 250 μm long, with a 1-mm center-to-center spacing. (From Narasimhan J, Papautsky I. Polymer embossing tools for rapid prototyping of plastic microfluidic devices. *J. Microchem. Microeng.*, 14, 2004, pp. 96–103. With permission.)

photoresist AZ4620 on silicon was used for molding during PDMS casting. Figure 11.12 shows a scanning electron microscope micrograph of the PDMS mold obtained [39].

11.2.2.1.4 Comparision between Replication Micromachining Methods

The distinct behavior of the three replication methods reported previously was summarized in [36]. For commercial use, injection molding is still the most adequate method because of the possibility of high-volume production that contributes significantly to the reduction of the cost per device, despite the high initial costs required. Moreover, it enables the fabrication of a complex three-dimensional geometry in a short cycle time.

11.2.2.2 Direct Micromachining

11.2.2.2.1 Laser Micromachining

In this technique, structures are obtained by applying an intense laser beam of UV or infrared (IR) radiation to the polymer (Figure 11.13). In both cases, a high-intensity laser beam is focused onto the material, and the concentrated energy of the beam evaporates the material at the focal point [36].

The irradiated polymer is decomposed by the generated laser heat, leaving a void in the polymer. Consequently, the wavelength used and the power of the radiation affect the removal mechanism.

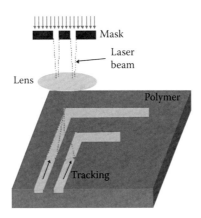

FIGURE 11.13
Schematic representation of laser micromachining.

The most conventional polymers that are frequently used in laser micromachining studies are PMMA, polyimide, poly(tetrafluoroethylene), polyethylene, PC, and polyamide.

The micromachining precision of this technique depends mainly on the quality of the optical system, on the energy distribution of the beam, and on the used laser wavelength [38,45]. The roughness of the surface is of the order of 200 to 500 nm. Although UV laser with masks is the more frequently used process for obtaining the structures, it is also possible to obtain these structures by direct writing systems. When a mask is used, the feature size depends mainly on the quality of the mask. Cheng et al. [46] studied the influence of the laser power, the motion speed, and the number of laser beam steps used to create trenches. In this study, they also overcame the general problem of rough surfaces after laser micromachining by using a simple thermal annealing process.

11.2.2.2.2 Stereolithography

Stereolithography is suited for rapid prototyping applications. Many approaches of stereolithography can be found in the literature [37,38]. In general, stereolithography is a technology that focuses a laser beam onto the free surface of a photosensitive liquid to induce polymerization of the liquid in that region and to transform it to a polymerized solid. The motion of the table inside the photopolymer, relative to the focal point of the laser, forms a structure [38]. Normally, the laser focus is scanned in the x–y direction while the table is moved in the z direction, consequently building up a layer-by-layer structure (Figure 11.14).

The resolution of the structure depends on the laser wavelength, on the scanning system used, and on the dispersion energy in the photosensitive liquid. Moreover, the thickness of the formed layer is determined by the penetration depth of the laser radiation into the liquid. Three-dimensional structures can be formed by spreading fresh liquid on top of the created structure, for example, on top of each two-dimensional cross-sectional layer and after its laser exposure. The second layer is created in the same way and so on. Depending on the technique used, for example, the combination of stereolithography and photolithography, it is possible to fabricate complex three-dimensional structures, such as integrated fluidic systems. Problems of leakage and expensive packaging can be solved with stereolithography. However, this technique is complex to use, it is necessary to get sophisticated control of the laser beam movement, and the structure is slowly built up by volume elements, which allows only restricted device volumes. On the other hand, no masks or other

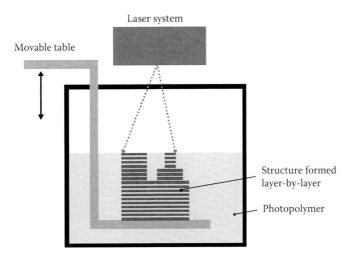

FIGURE 11.14
Schematic representation of stereolithography.

steps are involved, and the structure is directly created. The precision of this technique is of the order of a few micrometers [36,38].

Sun et al. [47] applied projection microstereolithography using a digital micromirror dynamic mask to obtain complex three-dimensional microstructures.

11.2.2.2.3 Photopolymerization

Photopolymerization or polymer lithography is based on the removal of specific areas of a thin polymer film. For that, the polymer must have the peculiarity to interact with energetic radiation, such as UV light. The radiation may break chemical bonds or lead to other types of chemical change within the polymer material. A mask is used to absorb or reflect the radiation light in specific regions, allowing radiation to reach the places where the microstructure (microwalls for creating microchannels and microwells) should be formed (Figure 11.15). Subsequent suction and flushing remove the unexposed polymer [48].

Photoresists are mostly used to fabricate microstructures, like the negative epoxy SU-8 [49]. The exposed and subsequently thermally cross-linked portions of the SU-8 film are rendered insoluble to liquid developers. SU-8 has excellent imaging characteristics and is capable of producing very high aspect ratio structures. Moreover, SU-8 is almost transparent above 360 nm, which makes it ideally suited for imaging near vertical sidewalls in very thick films. In addition, SU-8 can be used to form structures as well as the sacrificial layer forming the mold.

Williams [50] studied the effect of the postbacking and UV lithography process to obtain high aspect ratio of SU-8 microstructures. In this study, the author proved that microstructures with aspect ratio up to 40:1 and thickness between 1 and 1.5 mm can be obtained. Moreover, Chuang et al. [51] proved that the use of glycerol between the mask and the substrate reduces significantly the UV light diffraction, which greatly increases the sidewall straightness of high aspect ratio structures. Zhang et al. [52] studied in detail the effect of the exposure dose on the replication accuracy fidelity and profile of SU-8 microchannels with very high aspect ratio. Figure 11.16 shows the effect of the exposure time on the channel width and profile. They concluded, experimentally, that longer exposure time results in smaller (i.e., less steep) side wall angles and narrower microchannels. The authors also

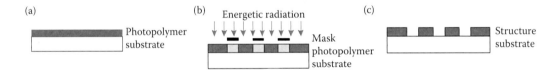

FIGURE 11.15
Schematic representation of photopolymerization: (a) deposition of the photopolymer on the substrate; (b) photopolymerization; (c) polymeric structure.

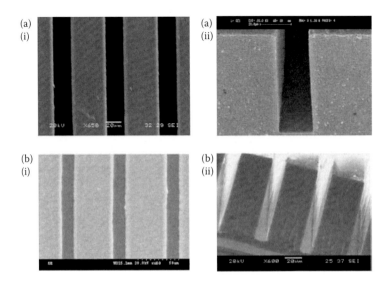

FIGURE 11.16
(i) Top view and (ii) cross section of microchannels exposed for (a) 35 s and (b) 90 s. The thickness of the SU-8 was 107.2 ± 1.9 μm, and the air gap was 40–50 μm. (Zhang J, Chan-Park B, Conner SR. Effect of exposure dose on the replication fidelity and profile of very high aspect ratio microchannels in SU-8. *Lab Chip*, 4, 2004, pp. 646–653. Reprinted with permission from The Royal Society of Chemistry.)

presented a new diffraction–refraction–reflection model to predict the microchannel width and sidewall angle (Figure 11.17) as a function of the exposure dose.

Electron beam or x-ray radiation can also be used to manufacture polymeric structures, such as SU-8 or PMMA. These methods have the advantage of featuring comparatively short processing times, once no dissolution steps have to take place. Moreover, completely sealed cavities can be fabricated by these techniques [38].

11.3 Applications

The field of μTAS is a quickly developing area with interesting studies that have a huge growth. The common goal is reflected in overcoming the limitations presented by such devices preserving their advantages, which consequently enable the development of new microdevices with their functionality, autonomy, point of care, and performances increased, which was unthinkable until now. Since 2000, thousands of excellent articles

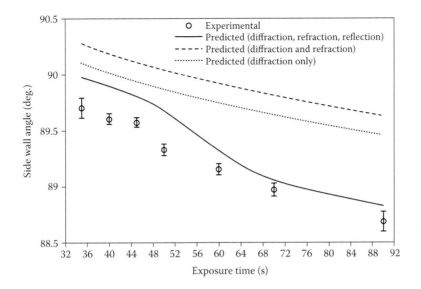

FIGURE 11.17
Effect of the exposure time on the side wall angle of microchannels. (From Zhang J, Chan-Park B, Conner SR. Effect of exposure dose on the replication fidelity and profile of very high aspect ratio microchannels in SU-8. *Lab Chip*, 4, 2004, pp. 646–653. Reprinted with permission from The Royal Society of Chemistry.)

associated with µTAS were reported by online keyword search. Beyond the many existing applications in the µTAS field, this chapter will emphasize some of the most cited and recent articles in the application fields of DNA, proteins, cells, and clinical diagnostics. These are the most rapidly growing fields of scientific research and technology development that attract more and more researchers because of their applicability to the health care, forensic, and environmental sectors. These areas are, nowadays, the most dynamic and where novelty is a strategic and operational imperative. Indeed, µTAS not only brings exciting opportunities to study new phenomena that are unique at these small scales, enabling the detection and manipulation of single molecules and cells otherwise undetectable but also provides better services at a lower price in those fields.

11.3.1 DNA Studies

Bader et al. [53] presented a silicon-based micromachined device that can transport DNA. It is based on a Brownian ratchet that rectifies the Brownian motion of microparticles and consequently allows the application of the device to be used as a pump or separation component for DNA or other any charged species. Ueda et al. [54] proved that a separation channel of only 6 mm is adequate for the separation of triplet repeat DNA fragments and DNA molecular markers. The separation occurred in just 12 s in a microfabricated capillary electrophoresis (CE), which is 18 times faster than the same separation carried out by the 24.5-cm conventional CE instrument. For that, a laser-induced fluorescence detection system was used. Wang [55] described and studied various biosensors and gene chips to allow fast, sensitive, and selective detection of DNA hybridization. Stelzle et al. [56] presented a possible on-chip electrophoretic biomolecules accumulation system that can be applied to sample preconditioning in lab-on-a-chip devices for DNA and proteins samples analysis. The system was demonstrated by the accumulation of DNA oligomers and streptavidin in aqueous solution through the application of an electrical field in an array of planar focusing electrodes. Lenigk et al. [57] demonstrated the feasibility and advantageousness of silicon wafers as support for use in the DNA microarrays field. Like that, the integration of CMOS silicon-based components (such as polymerase chain reaction [PCR] microreactors and CE) into a single chip is provided. Park and Swerdlow [58] presented a high-sensibility CE through a sample pretreatment device for the concentration and purification of a crude DNA sample in a flowing stream, providing both enhanced signal and resolution, compared with conventional CE. Niemeyer et al. [59] used self-assembled DNA–protein conjugates to obtain a very sensitive antigen detection and compared its advantages with other possible conjugates methods. Moreover, it allowed the reduction of the cost and the time of the analysis, and the authors referred its suitable use in µTAS. Kartalov and Quake [60] fabricated the first fully integrated microfluidic system in PDMS that combines fluorescence, surface chemistry, and microfluidic methods with fluorescent studies of DNA. They demonstrated the system for 4 base pairs of DNA sequencing. Yamashita et al. [61] developed a novel microfluidic analysis method based on a serpentine channel structure and fluorescence detection for simple sequence-selective DNA and single base mutation detection with a high accuracy. Moreover, specific detection of probe-bound DNA can also be done by the combination of laminar stream and laminar secondary flow in the microchannel. Taylor et al. [62] developed a rapid and novel on-chip method to analyze mitochondrial DNA. Its viability was pointed out in the characterization of the mitochondrial demographics of a single cell (homogeneous and heterogeneous population) by measurements of fluorescence. Simion et al. [63] fabricated a silicon microdevice for rapid testing of DNA analysis on the basis of PCR technique with biomolecules attached. It includes a microreactor and heating resistance to assure the necessary condition to PCR reaction. A transparent glass is used to cover the reactor, providing optical monitoring. Guiducci et al. [64] used an electrical capacitive measurement circuit to form an integrated biosensor prototype for direct, high-sensitive, specific, and reproducible detection of DNA sequences. The biosensor allowed its integration in standard CMOS technology. Vanderhoeven et al. [65] enhanced DNA microarray using a continuous and discontinuous rotating microchamber obtaining a significant larger hybridization intensity, about two to three factors compared with a continuous rotation. Lao and Hsing [66] referred the possibility to incorporate in a µTAS a miniaturized electrical field flow fractionation and a segmented electrode operated under a pulsed voltage for the manipulation and separation of DNA molecules. Blazej et al. [67] integrated on a single microchip the thermocycling amplification of DNA fragments as well as their separation by electrophoresis, that is, the Sanger DNA sequencing using only 1 fmol of DNA sample. They managed to sequence 556 continuous bases with 99% accuracy, which showed its great potential for sequencing human and other complex genomes. Huang et al. [68] used a chemical compatible elastomer, the perfluoropolyether, to fabricate a microfluidic DNA synthesizer using phosphoramidite reagents and conventional solvents. They referred the possibility of its use for screening siRNA sequences, creating DNA nanostructures, and for DNA computing. Li et al. [69] presented through both theoretical and experiment analysis the efficiency of a PMMA microchip on the basis of electroelution and optimal channel configuration for selective extraction of size-fractioned DNA during electrophoresis. Mohamed et al. [70] improved the efficiency of the electrophoresis technique through the development of an on-chip cellulose acetate-based biopolymer membrane. This was due to the rejection of the heme passage across the membrane, which inhibits PCR. Consequently, an improvement of detection and characterization of DNA from blood samples was achieved. Yea et al. [71] used fluorescence resonance energy transfer to increase the speed and detection sensitivity of DNA hybridization in a

PDMS microfluidic channel. This study is relevant in the detection of genetic diseases and gene expression profiling. Marie et al. [72] implemented a simple and direct method to fabricate microarrays in microsystems using SU-8 photoresist. For that, they successfully used direct coupling DNA to functionalized SU-8 surfaces, which allows a capture probe density of approximately 6–13 fmol/mm^2 determined by hybridization assays with fluorescent-labeled target molecules. Consequently, its use in genotyping was referred as a potential application. Yeung and Hsing [73] used functionalized magnetic particles for extracting a living cell's genomic DNA from other lysate mixtures in a silicon-based microchip. Yeung and Hsing [73] developed an efficient DNA extraction and real-time detection in one chamber through a combination of laser irradiation and magnetic beads. This allows optimal and rapid detection of pathogens present in biological samples because of the efficient and rapid DNA extraction. They proved this fact by lysing *Escherichia coli*, Gram-positive bacterial cells (*Streptococcus mutans* and *Staphylococcus epidermidis*), and also hepatitis B virus mixed in human serum [74]. An integrated staggered herringbone mixer on a microfluidic system was used by Lund-Olesen et al. [75] to improve the hybridization of fluorescent-labeled DNA in solution with probe DNA on magnetic beads that were immobilized on channel sidewalls in a magnetic bead separator. This technique can be a solution for many diffusion-limited reactions that take place on the surface in microfluidic devices. Li et al. [76,77] developed a MEMS local heat shock device and the corresponding microfluidic system for effective DNA transformation. The results showed an improvement of 200% in transformation efficiency compared with conventional methodology. This technology demonstrated its potential application on gene cloning/therapy and protein expression applications. Theoretical and experimental studies of the DNA transport in a pressure-driven postarray flow were performed by Teclemariam et al. [78]. They concluded that an appropriate design of postarray geometry in a lab-on-a-chip device can be used to control DNA conformation and to guide the location of hooking events. Chang et al. [79] presented a different detection method for DNA-strand studies. For that, they successfully demonstrated the feasibility of a nanogap-based electrical detection DNA biochip with different sizes of gold nanoparticles with DNA immobilized. An electroacoustic miniaturized DNA biosensor with a high detection level was developed by Gamby et al. [80]. It is based on the capacity of the resonance response evolution, that is, on the polarization of the dielectric material layer (polyethylene terephthalate) induced by the DNA hybridization that affected the electromechanical coupling established in the dielectric layer. Kumemura et al. [81] presented a simple and efficient way to isolate long single DNA molecules by electrophoresis force. They extended and oriented each molecule by alternate current (AC) dielectrophoresis, and finally they captured them using aluminum-integrated electrodes in a microfluidic environment. Brennan et al. [82] developed an integrated optical platform on the basis of standard resonant mirror waveguide detection for DNA hybridization and detection. Fluorescence detection and quantification of DNA molecules using a microdevice on the basis of a thin-film hydrogenated amorphous silicon *p-i-n* photodiode was developed by Pimentel et al. [83]. The detection limit of the presented device was approximately 5 nM in solution. Cretich et al. [84] used amino-modified DNA fragments covalently bound in a copolymer (DMA-NAS-MAPS) deposited in a PDMS slide for bacterial genotyping studies. Cheong et al. [85] used optothermal properties of gold nanoparticles for pathogen detection using a one-step real-time PCR system. These nanoparticles have the peculiarity of using near IR energy to generate heat that consequently causes pathogen lysis. In this way, the DNA was extracted from the cell and transferred to a PCR system. Lo and Ugaz [86] used a new alternative detection method by introducing an automated whole-gel scanning detection system. This allows an improvement on microchip-based gel electrophoresis of DNA to be continuously and in detail monitored along an entire microchannel. In this way, fundamental physical parameters associated with DNA migration phenomena can be rapidly and accurately measured in a single experiment. Terao et al. [87] reported a method for on-site single-molecule manipulation of DNA molecules in the order of mega base pairs, using optically driven microstructures and electro-osmotic flow to extended DNA without fragmentation, under a video microscope. In addition, this system allows the manipulation of any desired part of the targeted DNA in the microscope view. A microfluidic cartridge that integrates reservoir and valves has been optimized for lab-on-a-chip application and biotesting for DNA/RNA extraction from a human blood sample by Xie et al. [88]. The microchip was tried and sufficiently quantified from this cartridge for subsequent PCR amplification and detection. A real-time portable DNA analysis system on the basis of an integrated PCR-CE microdevice for forensic human identification was reported by Liu et al. [89]. This system was tested in a mock crime scene and this proved its feasibility to amplify and correctly type three blood strain samples. York et al. [90] presented a biosensing nano-device to detect single, sequence-specific target DNA molecules through the association of an immobilized F1-ATPase motor and functionalized gold nanorod sensors. This mechanism presented a sensitivity limit of one zeptomole. Chang et al. [91] presented an enhanced method for DNA molecules extraction from extreme low concentration in a sample solution of a fluidic microsystem on the basis of the immobilization of inorganic layered double hydroxides on the PC substrate. Malic et al. [92] reported a biochip platform functionalization using electrowetting-on-dielectric digital microfluidics for surface plasmon resonance imaging detection of DNA hybridization. A continuous-flow PCR of single-copy DNA in water-in-oil droplets was integrated in a high-throughput microfluidic device. The high efficiency demonstrated allows amplification from a single molecule of DNA per droplet [93]. The combination of electro-osmotic pumping and the extraction/elution of DNA coupled with gel-support reagent in a silica monolith support was reported by Oakley et al. [94]. Its benefits were proved on both the DNA extraction (with 65% efficiency) and the gene amplification using the PCR process. Micro- and nanotechnology was used by Hubálek et al. [95] for the implementation on a chip of rapid virus detection. For that, superparamagnetic nanoparticles were modified to provide easy separation and electrochemical method detection of viral nucleic acids. The method demonstrated its reliability for the HIV and H5N1 viruses. Lien et al. [96] reported an automatic magnetic bead-based microfluidic platform for rapid leukocytes purification and genomic DNA extraction for fast analysis of genes. Mahalanabis et al. [97] developed a disposable microfluidic chip for cell lysis and for DNA extraction of Gram-positive and Gram-negative bacteria (the tenth leading cause of death in the United States) from a microliter of whole blood (Figure 11.18). The obtained detection limits were 10^2 CFU/mL and 10^3–10^4 CFU/mL for Gram-negative bacteria and Gram-positive bacteria, respectively. These results proved to be comparable with the standard benchtop bacteria and viral lysis and nucleic acid extraction kits.

Manz et al. [98–101] published a series of review articles regarding advances within the µTAS field, including DNA applications. Other important review articles can be found in references [102–109].

11.3.2 Protein Studies

A fully integrated microfluidic biochemical detection system with magnetic-bead-based immunoassay, used as both immobilization surfaces and biomolecule carriers, was produced by Chol et al. [110]. The feasibility of the device was done by the capability of protein

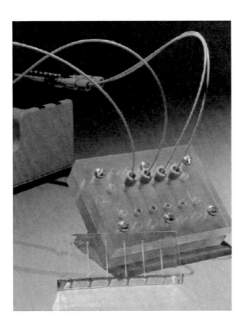

FIGURE 11.18
Photograph of the microfluidic chip showing the pump and the syringes for nucleic acid extraction. (From Mahalanabis M, Al-Muayad H, Kulinski MD, Altman D, Klapperich CM. Cell lysis and DNA extraction of gram-positive and gram-negative bacteria from whole blood in a disposable microfluidic chip. *Lab Chip*, 9, 2009, pp. 2811–2817. Reprinted with permission from The Royal Society of Chemistry.)

sampling to capture target antigens that can also be applied for the detection of a wide variety of biological materials (like DNA analysis). Shilling et al. [111] reported a microfluidic system for the lysis of cells and analysis of their constituent proteins. In this article, the detection, extraction, and quantification of a large intracellular enzyme, β-galactosidase, of bacterial cells were demonstrated using a fluorogenic enzyme assay and a numerical model. Hamamoto et al. [112] developed an integrated PDMS–glass hybrid microreactor array for protein research. The article includes the design and fabrication of all the components necessary for high-throughput cell-free protein synthesis, namely, the temperature control chip, the reaction chamber chip, and the surface treatment for preventing protein absorption. A multidimensional microfabricated device was fabricated by Slentz et al. [113] for protein proteolysis and electrochromatographic separation of histidine-containing peptide. The microfabricated frits that retained particulates down to 3-μm size were an important reason for the success of the device. Huber et al. [114] developed a microfluidic device with an active 4-nm-thick polymer film that can be thermally switched to adsorb and desorb protein monolayers in less than 1 s. This system proved that the protein traps selectively adsorb either myoglobin or bovine serum albumin (BSA) from a mixture of both. The authors referred that an array of such devices can be used with success for the extraction of particular proteins with a high selectivity. Carlo et al. [115] showed an effective extraction of intracellular protein from defibrinated sheep blood using a highly effective and reagentless mechanical cell lysing device integrated in microfluidic channels. The system functionality consisted of microfluidic filters with nanostructured barbs. The efficiency of the protein extraction was optically demonstrated. Bouaidat et al. [116] presented a process technology that can be used in biochip, BioMEMS, cell and protein studies, and high-throughput screening applications. This technology is based on cell adhesion and on protein selective adsorption using a nonadhesive plasma polymerization system that involves microscale patterning of the polyethylene oxide-like ppCrow coating. The system was tested for both human cells (HeLa) and fluorescently labeled proteins (fluorescein isothiocyanate BSA). Proteolytic digestion of proteins is crucial for protein studies in microdevices. Jin et al. [117] reported an electro-osmotic flow for electrokinetically pumping proteins to a proteolytic system, thus providing a simple way for integrating protein digestion into an electrophoretic μTAS for protein analysis and characterization. A notable ultraminiaturized functional protein capture assay was fabricated by Lynk et al. [118]. The microspot-based nanoarrays assay covers less than 1/1000th of the surface area of a conventional microarray spot. The significant decrease of the device size and, consequently, the sample volume allowed measurements in a single optical microscope image or with novel label-free methods, like atomic force microscopy detection. A technique for patterning active proteins on a glass substrate using perforated PDMS sheet-sieve was developed by Atsuta et al. [119]. They used the F_1-ATPase rotary motion to prove that the activity of the proteins does not change during the patterning process. Moreover, the system was improved by selective patterning to obtain a microarray of different active proteins. A total serum protein N-glycome profiling was successfully performed in a CE microfluidics platform by Callwaert et al. [120]. They used channels with a 11.5 cm effective length and 4% of linear polyacrylamide as the separation matrix to profile the N-glycans in human serum in 12 min with a good resolution. Koskinen et al. [121] combined in a unique and novel assay platform the dry etching bioaffinity reagents and the ArcDIA TPX™ binding assay technique. In this case, the performance of the new platform was studied for human α-fetoprotein. The results obtained showed the same performance of a corresponding wet-chemistry assay method, but with the particularity to enable the detection of bioaffinity reactions using miniature volumes. The design and fabrication of a fluidic device with two fluidic distinct separations for purification of toxic and nontoxic proteins was reported by Mia et al. [122]. For that, two pH-controllable linkers were used. The success of this technique proved its applicability to small-scale massively parallel proteomic separations. Kai et al. [123] developed and demonstrated the feasibility of an innovative, disposable, fast (detection time less than 10 min), and highly sensitive (detection limit of 1 ng/mL) protein lab on a chip for the detection of a prostate-specific antigen for the clinical diagnosis of prostate cancer. It is based on a COC substrate and uses a chemiluminescence-based detection method in a sandwich immunoassay. A high concentration of the biomarker C-reactive protein (CRP) in serum is directly correlated with cardiovascular disease or oral infections. Despite the several existing methods for its quantification from blood sample, Christodoulides et al. [124] developed a noninvasive microchip assay system on the basis of the electronic taste chip approach for measuring CRP protein in a saliva sample. Luo et al. [125] developed a highly specific immunoassay system for goat antihuman IgG using gold nanoparticles and microfluidic technology with a detection sensitivity of 10 ng/mL. A multiplex measurement of seven tumor markers (AFP, ferritin, carcinoembryonic antigen [CEA], hCG-â, CA 15–3, CA 125, and CA 19–9) using an array of immunosensing electrochemical sensors, composed of different immobilized antigens, was developed by Wilson and Nie [126]. The biosensor showed an excellent precision, accuracy, and performance comparable with commercially available single-analyte assays. Many important proteins are difficult or even impossible to study by current genomic and proteomic techniques. Moreover, in a living cell, the gene expression occurs stochastically, which makes difficult the observation of this event. However, Cai et al. [127] opened that possibility with the presentation of a microfluidic system that permits real-time observation of the expression of β-galactosidase in living *E. coli* cells. Lönnberg and Carlsson [128] developed a novel lab-on-a-chip technique for quantitative isoform profile determination using a porous monolith chip. Results discriminated transferrin isoforms differing by only 0.1 pH units in isoelectric point and specifically quantified the separated transferrin isoforms for every 42 µm along the separation zone in less than 15 min. An integrated clinic-proteomic chip presents several specifications for its management and analysis platform. Tsiknakis et al. [129] reported all the functional specifications, such as engineering aspect, challenges, and architectures for creating that microdevice environment, which monitors, documents, and interprets the digital data from the analyses. Li et al. [130] developed a lab on a chip for protein analysis on the basis of piezoelectric copolymer (PVDF-TrFE) diaphragm arrays. These diaphragms present a hydrophobic surface and consequently act as a natural biosensor to capture proteins of interest. A microthermodevice was developed by Arata et al. [131] for permitting real-time observations of the denaturation dynamics of fluorescent proteins at the millisecond scale. This was due to both on-chip microheater and microcontainer, which performed temperature control of the target within 2 ms under the microscope. Consequently, this technology can be a powerful apparatus in molecular biology. A simple and fast method on the basis of two printed V-shaped microchannels in mirror image orientation and separated by a 100-µm-wide toner gap was fabricated by Yu et al. [132] for rapid concentration and purification of proteins. By applying a high electrical field between the two channels, nanofissures were formed. Consequently, upon the application of a specific electrical field across its junction, negatively charged proteins are concentrated at the anode side of the nanofissures. A disadvantage arises from the positively charged analytes that cannot be concentrated despite their size; hence, the charge analytes have an important role in the concentration process [132]. Wang et al. [133] used hydrophobic polymer nanoparticles to develop a one-step affinity purification system for recombinant proteins. The system is composed of polyhydroxyalkanoate granule-associated protein phasing as a hydrophobic affinity tag, a pH-inducible self-cleaving intein and polyhydroxyalkanoate nanoparticles. Several proteins (RGFP, MBP, and β-galactosidase) were effectively purified in their active forms using this protein purification system, demonstrating its potential to be used for many other proteins [133]. Meagher et al. [134] reported a rapid, automated microscale process to isolate and to purify specific proteins from a submicroliter volume of cell lysate (in this specific study, *E. coli*) on the basis of a polyethylene glycol (PEG) salt aqueous two-phase microfluidic extraction process (Figures 11.19a and 11.19b). The protein of interest was genetically modified to express a hydrophobic partition tag, which strongly biases partitioning into the PEG-rich phase. When the cell lysate was introduced into the microfluidic device (Figure 11.19c), the protein of interest was extracted into one phase, whereas DNA or other nontagged proteins stayed on the other phase (Figure 11.19d) and were discarded (Figure 11.19e).

The system demonstrated its capability of removing 85% of contaminating proteins as well as unwanted nucleic acids and cell debris. Consequently, that two-phase extraction system can form the core component of an integrated µTAS with cell culture, lysis, purification, and analysis [134].

A possible new tool for future tissue engineering, cell adhesion, and antibody patterning investigations can be done using protein micropatterning because of its high contrast and stability. Huang et al. [135] concluded this patterning neutravidin on glass and quartz substrates using deep UV irradiation. Le Nel et al. [136] reported a microreactor for proteinase-K-mediated protein digestion as a step for the fabrication of a fully integrated microdevice. The objective was to detect pathological prion protein. The system was used to examine the digestion of prion protein in brain tissues, and the results showed that the complete proteolysis of normal protein was achieved in only 3 min. Koc et al. [137] showed

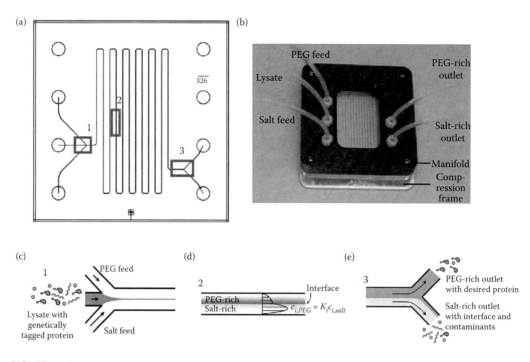

FIGURE 11.19

(See color insert.) (a) Schematic representation of the microchannel design with three inlets, a serpentine channel with 326 mm in length and two outlets, all on a glass chip with 37 mm². (b) Photograph of the microfluidic system mounted in a Delrin manifold with inlet and outlet connections. (c) Zoom of region 1 of (a) that represents the cell lysate stream containing a genetically tagged protein of interest along with undisturbed contaminants hydrodynamically focused—between inlet stream containing PEG and salt potassium phosphate. (d) Zoom of region 2 of (a) that represents the farther downstream, laminar two-phase flow with a stable interface. Components from the cell lysate stream partition between the two phases, with the genetically tagged protein (green) strongly partitioning toward the PEG-rich phase. An approximate velocity profile is represented to show high flow velocity in the less viscous salt-rich phase. (e) End of the channel (zoom of region 3 of panel a), where the flow is split into two outlet streams. The tagged protein is concentrated in the PEG-rich outlet stream, whereas undesired proteins and slowly diffusing macromolecules are directed toward the salt-rich outlet along with the phase interface. (From Meagher RJ, Light YK, Singh, AK. Rapid, continuous purification of proteins in a microfluidic device using genetically-engineered partition tags. *Lab Chip*, 8, 2008, pp. 527–532. Reprinted with permission from The Royal Society of Chemistry.)

how nanoscale superhydrophobic surfaces can be used to reduce protein adsorption taking place in the absence of fluid flow, but mainly they have demonstrated the considerable reduction of a large amount of adsorbed protein under flow conditions by increasing the desorption rate. This study was done to overcome the common problem of wall adsorption in microfluidic devices, mostly when protein solutions are used. Sedgwick et al. [138] developed an integrated lab on a chip that allowed the isolation, electroporation, and lysis of single cells. Dielectrophoresis within a lab-on-a-chip device containing sawtooth microelectrodes was used to capture a green fluorescent protein-labeled action, resulting from the A431 human epithelial carcinoma cell expression. Confocal fluorescent microscopy was used to monitor protein release. Self-interaction chromatography on a chip was studied and successfully used by Deshpande et al. [139] for the rapid screening of protein–protein interaction, avoiding the use of resins, beads, or particles normally used for chromatography columns. Therefore, this platform can be used for the design of protein crystallizations with a reduction of expensive proteins and experimentation time. Khnouf et al. [140] reported a polystyrene microfluidic device that uses passive pumping for *in vitro* (cell-free) protein expression. It is composed of 192 microchannels, each of which is connected to two wells positioned in a 384-well microplate format. Luciferase expression was demonstrated with an improvement of approximately five times and using a luminescence assay. Moreover, the device showed several benefits compared with a protein expression instrument commercially available [140]. A digital microfluidic approach on the basis of the manipulation of discrete microdroplets on an open array of electrodes was used by Luk and Wheeler [141] for rapid, reproducible, and automated proteomic sample processing, including reduction, alkylation, and enzymatic digestion. Vasina et al. [142] developed an online, free-access biomolecular adsorption database, which allowed the estimation of the amount of adsorbed protein, the thickness of the adsorbed protein layer, and the surface of the protein-covered surfaces on the basis of some input variables. This study is very important for protein adsorption at solid–liquid interfaces that can occur in many applications, including in μTAS, because of its particular relevance to the design of microfluidics devices. This database presented a predictive error of 5% or less. An application of microfluidics to protein extraction by precipitation was reported by Jebrail and Wheeler [143]. To extract and purify the proteins from heterogeneous fluids, a digital microfluidics method was used, and the results obtained showed a performance comparable with the conventional manual technique, combined with the advantage of miniaturization. A PDMS microchip integrated with 10-nm PC nanopore membranes was studied and fabricated by Wu and Steckl [144] to obtain high-speed protein accumulators. This can increase the efficiency in the identification of protein biomarkers with a very low concentration in target biofluids. The excellent results obtained by this research group (approximately 10^5–10^6 accumulation factor and an accumulation rate as high as 5000/s, 10 times higher than most results reported to date, for fluorescein-labeled human serum albumin solutions) were due to the double-sided injection control of electrokinetic fluid flow in the sample channels. Manz et al. published a series of review articles regarding the advances in the μTAS field, including protein applications [87–90]. Many other review articles can be found in references [145–153].

11.3.3 Cell Studies

An integrated microfabricated cell sorter was developed by Fu et al. [154] using multilayer soft lithography. To obtain coordinated and automated cell sorting, the system incorporated input and output wells, switch valves, and peristaltic pumps. The system demonstrated the sorting and the recovery of *E. coli* cells on chip as well as several advantages compared with the conventional fluorescence-activated cell sorters. A standard CMOS microsystem that incorporates a moving dielectrophoresis-cage approach to detect and individually manipulate more than 10,000 cells in parallel fashion was developed by Manaresi et al. [155]. The device was optimized for handling eukaryotic cells in the range of 20–30 μm, for which a unique electrode cage permitted individual manipulation. Another similar study was reported by Medoro et al. [156]. A printed circuit board prototype was used to trap, to concentrate, and to quantify polystyrene microbeads. In this case, the electronic detection and manipulation microsystem was successfully repeated with *Saccharomyces cerevisiae*. Di Carlo et al. [157] reported a simple and effective mechanical cell lysis system integrated in microfluidic channels on the basis of a microfluidic filter region with nanostructure barbs and a membrane-impermeable dye. This resulted in biomolecule accessibility for bioassays. Seger et al. [158] reported a microfluidic system that allows the accurate handling of particles such as biological cells in suspensions using an electrical field of aligned top and bottom electrodes in channels (negative dielectrophoresis barriers) combined with pressure-regulated fluid flows. The performance of this platform was studied in cell immersion with a second reagent and a novel type of flow-through cell dipping. This study can be useful in the fields of impedance spectroscopy flow cytometry and single-cell electroporation [158]. A MEMS-on-CMOS microsystem was designed by Liu et al. [159] for long-term measurements on arrays of single electrically active cells (more specifically to encage, to culture, and to monitor cells). Closable vials on the basis of a MEMS process were developed on a custom bioamplifier CMOS chip. The culture of bovine aortic smooth muscle cells was used both to test the biocompatibility of the microstructures and also to verify that the cell electrical signals can be recorded using the bioamplifier circuit [159]. A microfluidic device for continuous partitioning of relatively large cells was performed by Yamada et al. [160] using an aqueous two-phase laminar flow system. Several designs of the microchannel were studied, and the authors concluded that a microchannel with a pinched segment is highly advantageous because of the high partitioning efficiency even in high flow rate conditions. Inglis et al. [161] reported a microfluidic device for continuous cell-by-cell separation from a flow stream by selectively tagging with magnetic nanoparticles. An array of magnetic features was used to induce a lateral force on streaming tagged cells. The device was effectively verified by the separation of leukocytes from whole human blood. Gao et al. [162] developed a microfluidic system for the analysis of single biological cells including single-cell injection, lysis, separation, and detection of intracellular constituents by CE separation with laser-induced fluorescence detection using the docked-cell-electrolyzing approach. The device was demonstrated through the determination of glutathione in a human erythrocyte cell. Rhee et al. [163] described and tested a plasma-based dry etching approach that allows the micropatterning of biomaterials for selective attachment of cells on the substrates and enables culturing of patterned cells inside a microfluidic device (specifically for neuroscience and cell biology research). This technique was used to pattern cell-adhesive and nonadhesive areas on the glass and polystyrene substrates. Moreover, the authors concluded that it can also be used with protein coating and other organic or inorganic substrates. A novel method for measuring the effective refractive index of a single living cell in real time was reported by Liang et al. [164]. The microdevice integrates laser diode, microlenses, microfluidic channels, and the measurement system into a monolithic chip. The experimental results showed a high accuracy with an error of 0.25% when testing five types of cancerous cells, and consequently the device will offer an efficient approach for diseases or cancer diagnosis. Tourovskaia et al. [165] reported a microfluidic perfusion platform that allows long-term culture of muscle cells and supports muscle cell differentiation. Is this case, the process of differentiation from myoblasts to myotubes was studied. Surface modification techniques were used for selective cell attachment in the microfluidic device, and the process of cell fusion was geometrically controlled to obtain isolated single myotubes. The results showed that there are no differences between microfluidic and traditional cultures. Chung et al. [166] described a gradient-generating microfluidic system for optimizing proliferation and differentiation of neural stem cells in culture. The system was studied in the differentiation of human neural stem cells from the developing cerebral cortex into astrocytes using time-lapse microscopy and immunocytochemistry. Irimia and Toner [167] used microstructure membranes that are pneumatically actuated to conduct sample preparation procedures for suspensions of eukaryotic cells in a microfluidic device. Sathuluti et al. [168] developed a compartmental microfluidic device for a high-throughput multiplexed single-cell sorting (up to 1250 cells <100 s) and analysis system. This platform can also be studied and used to perform rapid and sensitive bioassays. Xing et al. [169] developed a sandwich flow microfluidic biochip for blood cell lysis using chemical lysing reagents. Experimental results demonstrated that the blood cells can be rapidly lysed in a few minutes when the flow rates of the lysing reagent are significantly faster than that of the cell sample. A rapid heterogeneous liver-cell patterning *via* the chip design of an enhanced field-induced dielectrophoresis trap inside a microfluidic chip was reported by Ho et al. [170]. The *in vitro* reconstruction of the heterogeneous lobule-mimetic radial pattern was demonstrated with good cell feasibility after cell patterning. A microfluidic system for dielectrophoresis-based continuous fractionation and purification of sample suspensions of biological cells using a planar microelectrode array structure was reported by Li et al. [171]. Huang et al. presented a fully integrated microfluidic system capable of automating cell culture. It comprises microheaters, a microtemperature sensor, micropumps, microvalves, microchannels, a cell culture area, and several reservoirs. The capability of this device was successfully demonstrated by a typical cell culturing process for human lung cancer cells [172]. Tai et al. [173] developed a microchip for cell transportation, separation, and collection using dielectrophoresis forces in an automated-microfluidic system that includes pneumatic micropumps and microvalves. Experimental results demonstrated that the viable and nonviable cells (human lung cancer cell and A549-luc-C8) can be successfully separated and collected. Nuclei can also be collected after cell lysis using the same platform [173]. Zhu et al. [174] studied the cell loss during sample transportation from macro- to microcomponents in integrated microfluidic devices. This loss can considerably deteriorate the cell detection sensitivity. However, possible solutions, such as a hydrodynamic focusing method, were also reported to reduce the cell loss. Lob et al. [175] reported an automated screening platform that provides real-time monitoring data of cell parameters, such as metabolic and morphological changes, important in many fields such as pharmacology and toxicology. Choi et al. [176] presented a hydrophoretic method for continuous blood separation using a microfluidic system that comprises slanted and filtration obstacles. Concerning the isolation of white blood cells from red blood cells, the system isolated white cells with an enrichment ratio of 210-fold at a throughput of $0.0004\,s^{-1}$ and within a filtration time of approximately only 0.3 s. Immune-magnetophoresis was presented by Kim et al. [177] as a novel technique for cell sorting. The immune affinity was used for the microcell sorter, and a high-gradient magnetic field was used for the separation of target T lymphocytes from biological suspensions. Despite its advantages, such as its continuous operation and fully automated one-step cell sorting, the system exhibited a lot of cell/particle loss because of the chip roughness surface [177]. Nevill et al. [178] proposed an integrated microfluidic system for cell culture and electrochemical lysis by hydroxide, without the need of multiple reagents or manual procedures. The system demonstrated the capability of culturing HeLa, MCF-7m Jurkat, and CHO-K1 cells and, consequently, lysing the cells, exposing both protein and DNA for being immunodetected and PCR amplified. Dielectrophoretic guiding of biological cells was studied by Christensen et al. [179] using a microfluidic system and baker's yeast (*S. cerevisiae*) as a model organism. The flow rate, the different medium conductivities, and the applied frequencies for the guiding efficiency were investigated. The microsystem was fabricated using the UV lithography technique for electrode fabrication and laser ablation for channel structuring. Nevill et al. [180] developed an integrated microfluidic cell culture and electrochemical lysis on a chip for automated cell analysis and sample preparation (Figure 11.20). Experimental results proved that both the proteins and the DNA release, after the lysis of the cells, maintained significant biological viability and, consequently, can be immunodetected and PCR amplified.

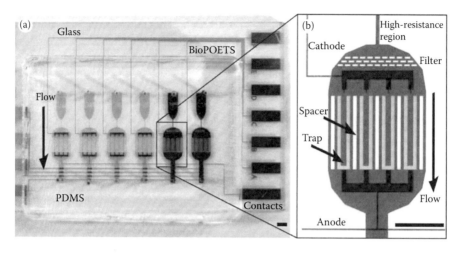

FIGURE 11.20
(See color insert.) Integrated microfluidic cell culture and lysis on a chip: (a) top view of the chip composed of six separate devices (filled with colored dye for visualization); (b) magnified image of one chamber. (From Nevill JT, Cooper R, Dueck M, Breslauer DN, Lee LP. Integrated microfluidic cell culture and lysis on a chip. *Lab Chip*, 7, 2007, pp. 1689–1695. Reprinted with permission from The Royal Society of Chemistry.)

Quinto-Su et al. [181] studied the dynamics of cell lysis produced by the application of a pulsed laser microbeam within a microfluidic channel using time-resolved bright-field and fluorescence imaging. Despite the deformation of the PDMS microfluidic channel walls, the initial release and dispersion of the cell contents occurred on the submicrosecond time scale. Flueckiger and Cheung [182] reported an integrated microfluidic system for cell culture and characterization. Hydrogels were used to trap and release cells for incubation, but they can also be used for nutrient and waste transport because of the porous properties of these gels. Moreover, the gel can present adhesive properties for adhesion of specific cells. Electrodes were also used for temperature control, electrokinetic cell handling, and dielectric characterization. The activation of single T cells in a microfluidic system was successfully presented by Kierschbaum et al. [183]. The system comprises microfluidic channels and microelectrodes for performing dielectrophoresis operation and for supplying optical access. Experimental results showed that after overnight cultivation, 76% of the processed cells remained vital. Ji et al. [184] studied four types of silicon-based microfilter: weir, pillar, cross-flow, and membrane, for sorting white blood cells and red blood cells in a whole-blood sample. Experimental results showed that the cross-flow microfilter is the most efficient, demonstrating a whole-blood handling capacity of <300 µL, with an average efficiency of >70%–80% in trapping white blood cells and passing red blood cells. Cimetta et al. [185] discussed the use of microbioreactor arrays for the control of cellular environments when factors that control the differentiation of human embryonic stem cells (hESCs) must be studied. Estes et al. [186] presented an on-chip magnetic system for cell sorting on the basis of multiple surface markers. The system integrates an electroplated array of Ni/Fe permalloy for the separation of cell substitutes. It was characterized and optimized for studying the flow rates and the incubation times for obtaining the best combination of high specific and low nonspecific cell capture. Khanna et al. [187] used nanocrystalline diamond microspikes to increase the efficiency of mechanical cell lysis. For that, B16-F10 murine melanoma cells were used. They verified a 400% increase in fluorescence reading using microspikes in only one chamber wall, significantly more than that without spikes. The microspikes puncture the cell membranes on collision, which enhance the efficiency of the cell lysis. Villa-Diaz et al. [188] reported a microfluidic system for the culture of individual hESC colonies. Moreover, the microsystem provided delivery-specific treatment by using patterned laminar flow to target specific regions of a single hESC colony and the potential to differentiate into three germ cell lineage. Consequently, it can perform in situ treatment and analysis with the ability to obtain cell samples from a specific part of a colony. The selective lyse of a single cell (without damaging the nucleus) within a group of cells was done by Lin and Lee [189] using an optically induced electric field at a specific position generating a transmembrane potential and the consequent lysis of the cell. Hsieh et al. [190] presented a very interesting platform that integrates all the functions necessary to a real-time, automatic, high-throughput, and high-magnification *in vitro* microscopic observation of cultured cells and their response to drug delivery on the basis of a unique ultrathin culture chamber (180 μm thick). It is mainly composed of two modules that include a cell culture and a microtemperature control that allows maintaining the cell culture microenvironment at 37°C. The platform was successfully tested using an oral cancer line OC-2 and nanodrugs on the cancer cells on the basis of quantum dots conjugated with the anticancer drug epirubicin [190]. Modak et al. [191] reported a theoretical study on the basis of a Eulerian–Lagrangian approach about the use of magnetic microspheres for cell separation in a T-microfluidic channel. An electrochemical impedance spectroscopy (EIS) cell chip was developed by Primiceri et al. [192] for real-time monitoring of cell growth and adhesion with complementary optical microscopy measurements. This application can be used to count cells or to evaluate cell morphology and changes as a consequence of different treatments. Green et al. [193] presented a theoretical and experimental study regarding the influence of the microchannel geometry on cell adhesion in microfluidic devices. The results showed that channels with curved turns provide more uniform and predictable cell adhesion and, consequently, are more suitable for cell-capture applications compared with channels with sharp turns. Manz et al. published a series of review articles regarding the advances in the μTAS field, including cell studies and applications [98–101]. Other important review articles can be found in references [194–202].

11.3.4 Clinical Diagnostics Studies

While clinical diagnostics comprises DNA, protein, and cell studies, it can be studied as an important area by itself, once its target is focused on the development of a complete μTAS for a specific application, where all the processes are integrated. That is why this section provides an overview of the most recently developed devices in that area.

Gascoyne et al. [203,204] used dielectrophoresis for the isolation of malaria-infected cells from blood. Infected cells lose internal ions when challenged by suspension in a low-conductivity medium because of the increased ionic permeability caused by the malarial pathogen. Consequently, the dielectric difference between infected and uninfected cells can be exploited by dielectrophoresis. Several parameters were studied by Ferrance et al. [205] for the development of a complete μTAS for Duchenne muscular dystrophy diagnosis. Microchip extraction of genomic DNA from whole blood using a novel-gel matrix was also reported. Moreover, an IR-mediated PCR amplification of a β-globin fragment from genomic DNA followed by electrophoresis analysis on a single integrated microsystem was also presented. Srinivasan et al. [206,207] reported a fully integrated and reconfigurable microdroplet-based digital lab on a chip for clinical diagnostics on human physiological fluids. The microdroplets were manipulated using the electrowetting effect that was proven to be compatible with body fluids. The feasibility of the system was performed and proved by a colorimetric enzymatic glucose assay. Ahn et al. [208] developed a disposable plastic lab on a chip for clinical diagnostics applications, including point-of-care testing. It comprises a smart passive microfluidics with embedded on-chip power sources and an integrated biosensor array. The measurements of partial oxygen concentration, glucose, and lactate level in human blood were successfully done using this biochip. Minas et al. [209] presented a fully integrated lab on a chip for biological fluid analysis on the basis of spectrophotometric measurements using a regular white light source. The system comprises all the components necessary for the analysis: the microfluidic system containing the microfluidic channels, the optical Fabry–Perot filtering system to filter the white light at a specific wavelength, which depends on the molecule of the biological fluids that is being analyzed, and the detection and readout electronics. The feasibility of the system was successfully demonstrated in the quantitative measurement of uric acid and total protein in human urine samples. The same group is improving the device using acoustic microagitation to accelerate the mix and the reaction time [210]. A lab-on-a-chip system for measurement of an important inflammatory marker, the CRP, in saliva, on the basis of the electronic taste chip approach, was presented by Christodoulides et al. [211]. Moreover, several advantages of this method in contrast to the ELISA measurement (commonly used in clinical laboratories to measure the concentration of this protein in serum) were presented. Experimental results demonstrated that this methodology has sufficient sensitivity and selectivity to detect CRP levels across the entire pathophysiological range for this protein in saliva samples. Huang et al. [212] proposed two BioMEMS platforms for fast diagnosis of infectious diseases. The first is based on a molecular biology technique for DNA/RNA-based diagnostic, and its performance was proved using RNA-based detection for type 2 dengue virus. The other is based on ELISA methods, and its performance was demonstrated for hepatitis C virus and syphilis serum tests. Guo et al. [213] reported an automatic lab on a chip for the clinical diagnosis of HIV. It is based on laser-induced fluorescence measurements for real-time monitoring of the signal after the sample preparation, reaction, and signal amplification, all features incorporated on the chip. Tsukagoshi et al. [214] developed a µTAS that incorporates the following three steps: immune reaction for high selectivity, formation, and transportation of the sample using electrophoresis and chemiluminescence detection for high selectivity. Human serum albumin and immunosuppressive acidic protein as a cancer marker in human serum were successfully detected and quantified with suitable selectivity and sensitivity. Liao et al. [215] presented a portable reverse transcription PCR (RT-PCR) system for amplification of specific nucleic acid and for detection of RNA-based viruses. Similar to the system presented by Huan et al. [212], the system is based on MEMS techniques and includes a microtemperature and a microfluidic control. Experimental results of type 2 dengue virus and enterovirus 71 (EV71) confirmed the capability of the RT-PCR system [215]. A microchip platform for point-of-care and real-time nucleic acid sequence-based amplification for detection of artificial human papillomavirus and SiHa cell line samples was developed by Gulliksen et al. [216]. The system is based on a COC microchip. Hofmann et al. [217] reported a fully automated microfluidic system for the quantitative measurement of human serum albumin (HSA) in urine and, consequently, the determination of microalbuminuria. The detection method is by fluorescence on the basis of the emission of a strong light at 620 nm when HSA reacts with albumin blue 580. The excitation source is an organic light-emitting diode with an emission peak at 540 nm. Experimental results showed that HSA concentrations down to 10 mg/L can be detected within a linear range from 10 to 100 mg/L [217]. A microfluidic device for monitoring the progression of the HIV infection by counting and sorting CD4 cells was presented by Sun et al. [218]. Compared with the device described by Guo et al. [213], this system uses electro-osmosis as a driving mechanism and a laser diode as a light source. Results demonstrated that the system can provide very high accuracy for evaluating the patients' HIV progression stage. Liao et al. [219] reported an integrated lab on a chip for rapid identification of bacterial pathogens in urine. The detection system comprises an amperometric detection mechanism on the basis of a DNA sandwich between capture and detector probe pairs and an oxidoreductase enzymatic transducer attached with a bacterial 16S rRNA. The system pointed out 98% sensibility for Gram-negative bacterial urinary pathogen detection. Moreover, the identification of the genotypic species was reached within 45 min, less than half a day needed for standard bacterial culture methods. Weigum et al. [220] presented a lab on a chip with an integrated sensor that uses an embedded track-etched membrane for the detection of an oral cancer biomarker. In this case, the epidermal growth factor receptor was used as the biomarker for the early detection of oral malignancies. Experimental results revealed the capability to characterize epidermal growth factor receptor overexpression in less than one-tenth the time of conventional methods. Herr et al. presented an integrated portable microfluidic device for the quantification of potential biomarkers of periodontal disease in saliva samples. It integrates all the components for a portable and self-contained device, like pretreatment of the samples, optical elements, and electronics for data acquisition. The device demonstrated a rapid measurement (<10 min) with low sample volume and a good sensitivity (nanomolar to picomolar). The feasibility of the device was pointed out measuring collagen-cleaving enzyme matrix metalloproteinase-8 in saliva from healthy and periodontal patients and comparing the results with conventional measurement systems [221,222]. A lab-on-a-chip biosensor developed by Blanes et al. [223] features the integration of immobilized enzyme reactor to CE. In this study, glucose, more than 100 μM, was quantified by packing particles modified with glucose oxidase at the end of the capillary channel. The performance of this device was also demonstrated by the separation and detection of noradrenaline. King et al. [224] reported a high-throughput microfluidic experimental platform to obtain real-time characterization of gene expression in living cells. For that, the platform comprises microfluidic arrays with a quantitative imaging of fluorescent protein transcriptional reporters in live cells. The dynamics of hepatocyte inflammatory responses was used to test the system and revealed that the results can be obtained in hours compared with months by conventional methods. Ortiz et al. [225] presented a disposable microfluidic cartridge on the basis of packaging and integration of MEMs mass sensor devices , according to a circular diaphragm resonator principle. Moreover, a biofunctionalizing polymer biosensor was deposited on the MEMS device, allowing the detection of cancer markers at low levels. Tan et al. [226] developed a lab on a chip for early detection of a trace amount of the nerve agent sarin (200 nM) in a small volume of blood. Sarin is a highly toxic phosphonate that can be used in military and terrorist attacks. It acts as a cholinesterase inhibitor and disrupts neuromuscular transmission. The system includes all the steps necessary for the analysis, namely, the nerve gas regeneration reactor, the cell lysis and filtering, the removal of fluoride ions, the inhibition reactor, and the optical detection. Zhang et al. [227] developed a very efficient lab on a chip for high-speed separation of blood cells from whole cells. This separation is the first step for subsequent blood analysis in clinical diagnosis. The merit of this device comes, in part, from the multiple separation mechanism embedded: the two centrifugal forces and the Coriolis force. Experimental results demonstrated a separation efficiency of 99% for diluted blood of 6% hematocrit. A microdevice for the isolation of specific T-cell subsets from a small blood volume and consequent detection of two cell-secreted cytokines, interleukin-2 and interferon-γ, was developed by Zhu et al. [228]. Mitogenic activation of the isolated T cell followed by immunofluorescent staining and microarray scanner was used to quantify the concentration of cytokines. A fully disposable and point-of-care RT-PCR microfluidic device with embedded micropinch valves on the basis of chemiluminescence assays was developed by Lee et al. [229]. Its application was the rapid detection of HIV and, consequently, early diagnosis of AIDS. Sieben et al. [230] integrated in a microfluidic chip the fluorescence in situ hybridization technique, which allows to significantly decrease the cost per test, compared with traditional clinical settings. Indeed, the on-chip fluorescence in situ hybridization allows the use of 20 times less reagent, reducing dramatically the analysis time as well as the support equipment needed. This technique allows the detection of chromosomal patterns that reveal relevant clinical information. Having in mind those advantages, Meagher et al. [231] reported a microfluidic portable platform-based electrophoretic immunoassay for the rapid detection (<20 min) of biotoxins in minimal sample body fluids (<10 µL). The system presented an appreciable sensitivity and dynamic range (micromolar to picomolar) for the biotoxins ricin, Shiga toxin I, and staphylococcal enterotoxin B, and it can also be applied for detection of proteinaceous biomarkers on many other diseases. He et al. [232] fabricated a lab on a chip for cancer cell detection in body fluids. The device was successfully used for the detection of mixtures of breast cancer cells MCF-7. After only 4 min, the fluorescent signals were about seven times stronger than when using a conventional glass slide. Rosenbluth et al. [233] developed a high-throughput microsystem on the basis of the biophysical flow cytometry technique to quantify differences in blood cell deformability. More precisely, the microsystem demonstrated its ability to detect distinct differences between normal and aberrant blood cells that cause hematologic complications, such as microvascular occlusion (in this article, sepsis and leukostasis were also studied). The deformability of cells was also studied by Bao et al. [234] using a microfluidic-based electroporative flow cytometry that combined electroporation with flow cytometry. Experimental results pointed out the high potential of the device for mechanistic studies of cytoskeleton dynamics and clinical application such as diagnosis and staging of cancer. Do et al. [235] developed a functional disposable and point-of-care polymeric lab-on-a-chip device that includes a portable analyzer with multianalyte detection capability for rapid and reliable quantifications of metabolic parameters in human whole blood. Glucose, lactate, and partial oxygen were successfully measured using a small amount of human blood sample (3.5 µL) in a short time (100 s). A lab on a chip was also reported by Do and Ahn [236] for magnetic-bead-based immunoassay with on-chip sampling and detection capabilities that can be used as a point-of-care system in hazardous agent detection, food inspection, or clinical diagnostics. An electrowetting-based digital microfluidic system for point-of-care analysis was developed by Sista et al. [237]. The performance of its magnetic-bead-based immunoassays (cardiac troponin I) in the microfluidic cartridge, using whole-blood samples, was performed in less than 8 min. The same system was successfully tested in a 40-cycle real-time PCR with a duration time of 12 min and in the sample preparation for bacterial infectious disease pathogen. Tang and Xia [238] developed a new immunosensor and immunoassay method for the detection of CEA in serum samples (associated with tumors and the developing fetus) suitable to be incorporated in a lab-on-a-chip device. The method consists in the immobilization of CEA antigen onto nanogold-fuctionalized carbon paste interface, and the results demonstrated a good reproducibility, accuracy, selectivity, and stability. A cell concentration microdevice on the basis of immunomagnetic beads and on-chip pumps was developed by Beyor et al. [239] for pathogen detection of dilute samples. The system pointed out a detection limit of 2 CFU/µL and maximum capture efficiency greater than 70%. A microchip CE with external contactless conductivity detection was developed by Kubán and Hauser [240] for the quantitative measurement of inorganic ions and lithium in serum and urine samples. The detection limits obtained with this microchip were 1 µM for K^+, 1.5 µM for Ca^{2+}, 3 µM for Na^+, 1.75 µM for Mg^{2+}, and 7.5 µM for Li^+. Moreover, the separation and the determination of small inorganic cations (NH_4^+, K^+, Na^+, Ca^{2+}, Mg^{2+}) and anions (Cl^-, NO_3^-, SO_4^{2-}, phosphate) in blood serum and urine samples were possible in one typical electrolyte solution by simply switching the high voltage from positive to negative polarity. The separation and the determination were performed in 35 and 90 s, respectively. Hattersley et al. [241] presented an experimental methodology on the basis of microfluidic technology in which pseudo *in vivo* tissue studies can be achieved under *in vitro* conditions. The results revealed a viable and functional state for more than 70 h for liver tissue with that methodology. Moreover, the disaggregation of tissue samples into individual cells was also performed and allowed the subsequent cell analysis. Consequently, this methodology can be very useful in biological and clinical studies. Kalatzis et al. [242] reported a point-of-care lab-on-a-chip platform on the basis of genomic microarrays of HLA typing for the early prognosis, diagnosis, and monitoring of a large number of autoimmune diseases. It allows the monitoring of immune system status and the supervision of chronic multiple sclerosis and rheumatoid arthritis, both autoimmune diseases. Marchand et al. [243] reported a DNA-chip-based integrated on a plastic disposable card for clinical diagnostic applications. In this article, the authors reported the study of an optoelectronic DNA chip and electrically activated embedded pyrotechnic microvalves with closing/opening functions for fluid control. However, they also suggested the design of a complete and fully automated card that will integrate all the electrical handling and DNA chip reading. Darain et al. [244] developed a disposable immunosensor microsystem for the quantification of myoglobin that allows the early evaluation of acute myocardial infection. The system was based on the immobilization of the antibody: antimyoglobin on a polystyrene substrate, and fluorescence technique was used as the detection method. The experimental results demonstrated a linear immunosensor response for myoglobin concentration between 20 and 230 ng/mL and a detection limit of 16 ng/mL, which is appreciably lower compared with the clinical cutoff value for myoglobin in healthy patients. An automated and fully integrated micro-ELISA system was fabricated by Ohashi et al. [245] for rapid, precise, and sensitive allergy diagnosis. The system is based on a BSA–biotin–avidin linker and polystyrene microbeads for the immobilization of allergens. The system pointed out a good detection limit of 2 ng/mL for IgE and a good correlation with the CAP conventional method, with the advantage of a significant reduction in sample volumes (1/10) and in the analysis time (1/20). A biochip readout system for point-of-care diagnostics on the basis of total internal reflection fluorescence was developed by Brandenburg et al. [246]. The applicability of this technique for immunoassays was testified by the limit detection of 1 ng/mL obtained for the inflammation parameter CRP, which is comparable with commercial laser scanner systems. Zang et al. presented a droplet-in-oil microfluidic PCR system for DNA methylation analysis. The system was successfully tested for the analysis of two tumor suppressor promoters (p15 and TMS1), showing consistent results with standard methylation analysis protocols. Moreover, multiple DNA methylation analyses can be performed in an array-based platform [247]. Javanmard et al. [248] developed a technique on the basis of the electrical detection and quantification of specific protein biomarkers, using protein-functionalized microfluidic channels. The technique was used to identify anti-hCG, a biomarker useful for female infertility diagnosis, and the results showed a limit detection of 1 ng/mL and a dynamic range of three orders of magnitude, in less than 1 h. Lee et al. [249] reported a portable and point-of-care lab on a chip (Figure 11.21) used as a portable analyzer for the early and rapid detection of HIV. The analyzer consists of a noncontact IR-based temperature control system for RT-PCR and of a chemiluminescence detection system.

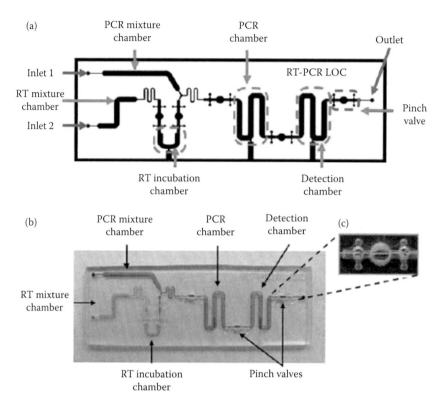

FIGURE 11.21
(See color insert.) RT-PCR lab on a chip: (a) schematic representation; (b) photograph; (c) embedded micropinch valves. (From Lee SH, Kim SW, Kang JY, Ahn CH. A polymer lab-on-a-chip for transverse transcription (RT)-PCR based point-of-care clinical diagnostics. *Lab Chip* 8, 2008, pp. 2121–2127. Reprinted with permission from The Royal Society of Chemistry.)

Manz et al. [98–101] published a series of review articles regarding advances in the µTAS field, including clinical diagnostics applications. Other important review articles can be found in references [250–259].

11.4 Conclusions

Researches in µTAS open up new possibilities for the miniaturization of conventional (bio) chemical analysis systems. However, this field has just begun, and a great deal of work needs to be done for the field to gain maturity and to be more than an active field of academic research. During the last few years, the number of publications associated within this research field has grown exponentially, many of them to overcome the limitations intrinsic to miniaturization. Indeed, the resolution of these problems requires imagination and a genuine interdisciplinary effort to integrate in a single chip the microfluidic, detection, and readout systems. Microfabrication technologies for polymers are probably the most promising fabrication techniques for the development of new µTASs, as it becomes possible to form both micro-optical and microfluidic elements in single devices by these processes. For more demanding applications, in which, for example, the optical signal is weak, some increase in complexity is inevitable. The challenge for the next generation of cheap and disposable μTASs will be the improvement of the sensitivity, so that these devices can accomplish, for example, in the optical field, low-light tasks. The final objective will be the development of multifunctional and commercial microdevices that are portable, autonomous (to be used successfully by any nonspecialist professional), at low-cost fabrication, with low-cost analysis, and achieving the same or better sensitivity as the commercially available equipment. Maybe it will not be too long until we will all be able to have our own μTAS at home: you feel sick one day and maybe you can know what is wrong and quantify it by simply loading urine, saliva, or whole blood into a μTAS device.

References

1. Reyes DR, Iossifidis D, Auroux PA, Manz A. Micro total analysis systems. I. Introduction, theory and technology. *Anal Chem*. 2002;74:2623–36.
2. Lee SJ, Lee SY. Micro total analysis system (μ-TAS) in biotechnology. *Appl Microbiol Biotechnol*. 2004;64(3):289–99.
3. Haeberle S, Zengerle R. Microfluidic platforms for lab-on-a-chip applications. *Lab Chip*. 2007;7:1094–110.
4. Terry SC, Jerman JH, Angell JB. A gas chromatographic air analyzer fabricated on a silicon wafer. *IEEE Trans Electron Devices*. 1979;26:18801886.
5. Jackson MJ. *Microfabrication and Nanomanufacturing*. Boca Raton: CRC Press, Taylor & Francis; 2006.
6. Becker H, Locascio LE. Polymer microfluidic devices. *Talanta*. 2002;56:267–87.
7. Becker H, Gärtner C. polymer microfabrication for microfluidic systems. *Anal Bioanal Chem*. 2008;390:89–111.
8. Colas A, Curtis J. Silicone biomaterials: history and chemistry & medical applications of silicones. In: *Biomaterials Science*. 2nd ed.; London: Elsevier Academic Press 2004, pp. 80–6.
9. Lécuyer C, Brock DC. The materiality of microelectronics. *Hist Technol*. 2006;22(3):301–25.
10. Stjernström M, Roeraade J. Method for fabrication on microfluidic systems in glass. *J Micromech Microeng*. 1998;8:33–8.
11. French PJ; Sarro PM. Surface versus bulk micromachining: the contest for suitable applications. *J Micromech Microeng*. 1998;8:45–53.
12. Linder C, Paratte L, Grétillat M-A, Jaecklin VP, Rooij NF. Surface micromachining. *J Micromech Microeng*. 1992;2:122–32.
13. Bustillo JM, Howe RT, Muller RS. Surface micromachining for microelectromechanical systems. *Proc IEEE*. 1998;86(8):1552–4.
14. Jang WI, Choi CA, Jun CH, Kim YT, Esashi M. Surface micromachined thermally driven micropump. *Sens Actuators A Phys*. 2004;115:151–8.
15. Okandan M, Galambos P, Mani SS, Jakubczak JF. Development of surface micromachining technologies for microfluidic and BioMEMS. *Proc SPIE*. 2000;4560:133–9.
16. Lin G, Pister KSJ, Roos KP. Surface micromachined polysilicon heat cell force transducer. *J Microelectromech Syst*. 2000;9(1):9–17.
17. Galambos P, Eaton WP, Shul R, Willison CG, Sniegowski JJ, Miller SL, et al. Surface micromachined microfluidics: design, fabrication, packaging, and characterization. ASME '99 MEMS; 1999. p. 441.
18. Bien DCS, Mitchell SJN, Gamble HS. Characterization of microfluidic devices. *Proceedings of the IEEE International Conference on Microelectronic Test Structures*, Vol. 15; 2002. pp. 211–215.

19. Han A, De Rooij NF, Staufer U. Design and fabrication of nanofluidic devices by surface micromachining. *Nanotechnology*, 2006;17:2498–503.
20. Kovacs GTA, Maluf NI, Peterson KE. Bulk micromachining of silicon. *Proc IEEE*. 1998;86(8): 1536–51.
21. Nguyen N-T, Wereley ST. *Fundamentals and Applications of Microfluidics*. 2nd ed. Norwood, MA Artech House; 2006.
22. Coltro WKT, Piccin E, Fracassi da Silva JA, Lucio do Lago C, Carrilho E. A toner-mediated lithographic technology for rapid prototyping of glass microchannels. *Lab Chip*. 2007;7:931–4.
23. Aziz NAA, Bais B, Hamzah A, Majlis BY. Characterization of HNA etchant for silicon microneedles array fabrication. *Proceedings of ICSE. The IEEE International Conference of Semiconductor Electronics*; 2008. pp. 203–6.
24. Wilke N, Mulcahy A, Ye S-R, Morrissey A. Process optimization and characterization of silicon microneedles fabricated by etch technology. *Microelectronics J*. 2005;36:650–6.
25. Mu X, Liang Q, Hu P, Ren K, Wang Y, Luo G. Laminar flow used as liquid etch mask in wet chemical etching to generate glass microstructures with an improved aspect ratio. *Lab Chip*. 2009;9:1994–6.
26. Olson A. Valve-less diffuser micropumps [dissertation]. Royal Institute of Technology Stockholm; 1998.
27. Janseny H, Gardeniers H, Boer M, Elwenspoek M, Fluitman J. A survey on the reactive ion etching of silicon in microtechnology. *J Micromech Microeng*. 1996;6:14–28.
28. Ghiocel DL. Development and characterization of a RIE process for anisotropic trenches in silicon. *24th Annual Microelectronic Engineering Conference*; 2006. pp. 54–9.
29. Boer MJ, Tjerkstra RW, Berenschot JW, Jansen HV, Burger GJ, Gardeniers JGE, et al. Micromachining of buried micro channels in silicon. *J Microelectromech Syst*. 2000;9(1):94–103.
30. Aziz NA, Majlis BY. Fabrication study of solid microneedles array using HNA. *IEEE International Conference on Semiconductor Electronics. ICSE '06*; 2006. pp. 20–4.
31. Ogawa J, Kanno I, Kotera H, Wasa K, Suzuki T. Development of liquid pumping devices using vibrating microchannel walls. *Sens Actuators A Phys*. 2009;152:211–8.
32. Ruan Y, Ren T, Liu L, Zhang D. Methods to protect silicon microstructures from the damages in deep reactive ion etching. *EDST* A18;2007:101–5.
33. Marty F, Rousseau L, Saadany B, Mercier B, Français O, Mita Y, Bourouina T. Advanced etching of silicon based on deep reactive ion etching for silicon high aspect ratio microstructures and three-dimensional micro- and nanostructures. *Microelectronics J*. 2005;36:673–7.
34. French PJ, Sarro PM. Surface versus bulk micromachining: the contest for suitable applications. *J Micromech Microeng*. 1998;8:45–53.
35. Becker H, Gärtner C. Polymer microfabrication technologies for microfluidic systems. *Anal Bioanal Chem*. 2008;380:89–111.
36. Available from: url: www.ncbi.nlm.nih.gov/sites/entrez. Cited June 2009.
37. Becker H, Gärtner C. Polymer microfabrication technologies for microfluidic systems. *Anal Bioanal Chem*. 2008;390:89–111.
38. Becker H, Gärtner C. Polymer microfabrication methods for microfluidic analytical applications. *Electrophoresis*. 2000;21:12–26.
39. Narasimhan J, Papautsky I. Polymer embossing tools for rapid prototyping of plastic microfluidic devices. *J Microchem Microeng*. 2004;14:96–103.
40. Schift H, David C, Gabriel M, Gobrecht J, Heyderman LJ, Kaiser W, et al. Nanoreplication in polymers using hot embossing and injection molding. *Microelectron Eng*. 2000;53(1–4):171–4.
41. Edwards TL, Mohanty SK, Edwards RK, Thomas CL, Frazier AB. Rapid micromold tooling for injection molding microfluidic components. *Sens Mater*, 2002;14(3):167–78.
42. Thiriez A, Gutowski T. An environment analysis of injection molding. *Proceedings of the IEEE International Symposium on Electronics and the Environment*; 2006, pp. 195–200.
43. Liou AC, Chen RH. Injection molding of polymer micro- and su-micro structures with high-aspect ratios. *Int J Adv Manuf Technol*. 2006;28:1097–103.
44. Do J, Ahn CH. A polymer lab-on-a-chip for magnetic immunoassay with on-chip sampling and detection capabilities. *Lab Chip*, 2008;8:542–9.

45. Becker H, Locascio LE. Polymer microfluidic devices. *Talanta*. 2002;56:267–87.

46. Cheng JY, Wei CW, Hsu KH, Young TH. Direct-write laser micromachining and universal surface modification of PMMA for device development. *Sens Actuators B Chem*. 2004;99:186–96.

47. Sun C, Fang N, Wu DM, Zhang X. Projection micro-stereolithography using digital micromirror dynamic mask. *Sens Actuators A Phys*. 2005;121:113–20.

48. Beebe DJ, Mensing G, Moorthy J, Khoury CM, Pearce TM. Alternative approaches to microfluidic systems design, construction and operation. *Micro Total Analysis Systems*, Proceedings 5th µTAS Symposium, Monterey, CA, 2001 Oct 21–25. Dordrecht (the Netherlands): Kluwer Academic Publishers; 2001. pp. 453–5.

49. Gelorme JD, Cox RJ, Gutierrez SAR, inventors; IBM, assignee. Photoresist composition and printed circuit boards and packages made therewith. United States patent 4882245; 1989.

50. Williams JD, Wang W. Study on the postbaking process and the effects on UV lithography of high aspect ratio SU-8 microstructures. *J Microlithogr Microfabrication Microsyst*. 2004;3(4):563–8.

51. Chuang YJ, Tseng FG, Lin WK. Reduction of diffraction effect of UV exposure on SU-8 negative thick photoresist by air gap elimination. *Microsyst Technol*. 2002;8:308–13.

52. Zhang J, Chan-Park B, Conner SR. Effect of exposure dose on the replication fidelity and profile of very high aspect ratio microchannels in SU-8. *Lab Chip*. 2004;4:646–53.

53. Bader JS, Hammond RW, Henck SA, Deem MW, McDermott GA, Bustillo JM, et al. DNA transport by micromachined Brownian ratchet device. *Proc Natl Acad Sci U S A*. 1999;96:13165–9.

54. Ueda M, Kiba Y, Abe H, Arai A, Nakanishi H, Baba Y. Fast separation of oligonucleotide and triplet repeat DNA on a microfabricated capillary electrophoresis device and capillary electrophoresis. *Electrophoresis*. 2000;21:176–80.

55. Wang J. From DNA biosensors to gene chips. *Nucleic Acids Res*. 2000;28(16):3011–6.

56. Stelzle M, Dürr M, Cieplik M, Nisch W. On-chip electrophoretic accumulation of DNA oligomers and streptavidin. *Fresenius J Anal Chem*. 2001;371:112–9.

57. Lenigk R, Carles M, Ip NY, Sucher NJ. Surface characterization of a silicon-chip-based DNA microarrays. *Langmuir*. 2001;17:2497–501.

58. Park SR, Swerdlow H. Concentration of DNA in a flow stream for high sensibility capillary electrophoresis. *Anal Chem*. 2003;75:4467–74.

59. Niemeyer CM, Wacker R, Adler M. Combination of DNA-directed immobilization and immune-PCR: very sensitive antigen detection by means of self-assembled DNA–protein conjugated. *Nucleic Acids Res*. 2003;31(16):1–7.

60. Kartalov EP, Quake SR. Microfluidic device reads up for four consecutive base pairs in DNA sequencing-by-synthesis. *Nucleic Acids Res*. 2004;32(9):2873–9.

61. Yamashita K, Yamaguchi Y, Miyazaki M, Nakamura H, Shimizu H, Maeda H. Sequence-selective DNA detection using multiple laminar streams: a novel microfluidic analysis methods. *Lab Chip*. 2004;4:1–3.

62. Taylor P, Manage DP, Helmle KE, Zheng Y, Glerum DM, Backhouse CJ. Analysis of mitochondrial DNA in microfluidic systems. *J Chromatogr B*. 2005;822:78–84.

63. Simon M, Kleps I, Ignat T, Condac E, Craciunoiu F, Angelescu A, et al. *A19 International Semiconductor Conference*, Vol. 1; 2005. pp. 235–8.

64. Guiducci C, Stagni C, Zuccheri G, Bogliolo A, Benini L, Samori B, et al. DNA detection by integrated electronics. *Biosens Bioelectron*. 2004;19:781–7.

65. Vanderhoeven J, Pappaert K, Dutta B, Hummelen PV, Desmet G. DNA microarray enhancement using a continuously and discontinuously rotating microchamber. *Anal Chem*. 2005;77:4474–80.

66. Lao AIK, Hsing I-M. Flow-based and sieving matrix-free DNA differentiation by miniaturized flied flow fractionation device. *Lab Chip*. 2005;5:687–90.

67. Blazej RG, Kumaresan P, Mathies RA. Microfabricated bioprocessor for integrated nanoliter-scale Sanger DNA sequencing. *Proc Natl Acad Sci U S A*. 2006;103(19):7240–5.

68. Huang Y, Castrataro P, Lee CC, Quake SR. Solvent resistant microfluidic DNA synthesizer. *Lab Chip*. 2007;7:24–6.

69. Li G, Ran R, Zhao H, Liu K, Zhao J. Design of a PMMA chip for selective extraction of size-fractioned DNA. *IEEE International Conference on NanoMicro Engineered and Molecular Systems*; 2006. pp. 105–9.

70. Mohamed H, Russo AP, Szarowski DH, McDonnell E, Lepk LA, Spencer MG, et al. *J Chromatogr A*. 2006;1111:214–9.

71. Yea KH, Lee S, Choo J, Oh CH, Lee S. Fast and sensitive analysis of DNA hybridization in a PDMS micro-fluidic channel using fluorescent resonance energy transfer. *Chem Commun*. 2006;14:1509–11.

72. Marie R, Schmid S, Johansson A, Ejsing L, Nordström M, Häfliger D, et al. Immobilization of DNA to polymerized SU-8 photoresist. *Biosens Bioelectron*. 2006;21:1327–32.

73. Yeung SW, Hsing IM. Manipulation and extraction of genomic DNA from cell lysate by functionalizes magnetic particles for lab on a chi application. *Biosens Bioelectron*. 2006;21:989–97.

74. Lee JG, Cheong KH, Huh N, Kim S, Choi JW, Ko C. Microchip-based one step DNA extraction and real-time PCR in one chamber for rapid pathogen identification. *Lab Chip*. 2006;6:886–95.

75. Lund-Olesen T, Dufva M, Hansen MF. Capture of DNA in microfluidic channel using magnetic beads: increasing capture efficiency with integrated microfluidic mixer. *J Magn Magn Mater*. 2007;311:396–400.

76. Li S, Anderson LM, Lin L, Yang H. DNA transformation by local heat shock. *IEEE 20th International Micro Electro Mechanical Systems*, 2007. MEMS; 2007. pp. 533–6.

77. Li S, Anderson LM, Yang JM, Lin L, Yang H. DNA transformation via local heat shock. *Appl Phys Lett*. 2007;911–3.

78. Teclemariam NP, Beck VA, Shaqfeh ESG, Muller SJ. Dynamics of DNA polymers in post arrays: comparison of single molecule experiments and simulations. *Macromolecules*. 2007;40:3848–59.

79. Chang TL, Lee YW, Chen CC, Ko FH. Effect of different gold nanoparticle sizes to build an electrical detection DNA between nanogap electrodes. *Microelectron Eng*. 2007;84:1698–701.

80. Gamby J, Lazerges M, Pernelle C, Perrot H, Girault H, Tribollet B. Electroacoustic miniaturized DNA-biosensor. *Lab Chip*. 2007;7:1607–9.

81. Kumemura M, Collard D, Yamahata C, Sakaki N, Hashiguchi G, Fujita H. Single DNA molecule isolation and trapping in microfluidic device. *Chem Phys Chem*. 2007;8:1875–80.

82. Brennan D, Lambkin P, Moore EJ, Galvin P. An integrated optofluidic platform for DNA hybridization and detection. *IEEE Sens J*. 2008;8(5):536–42.

83. Pimentel AC, Prazeres DMF, Chu V, Conde JP. Fluorescence detection of DNA using an amorphous silicon *p-i-n* photodiode. *J Appl Phys*. 2008;104:1–10.

84. Cretich M, Sedini V, Damin F, Di Carlo G, Oldani C, Chiari M. Functionalization of poly(dimethylsiloxane) by chemisorption of copolymers: DNA microarrays for pathogen detection. *Sens Actuators B Chem*. 2008;132:258–64.

85. Cheong KH, Yi DK, Lee JG, Park JM, Kim MJ. Edel JB, Ko C. Gold nanoparticles for one step DNA extraction and real-time PCR of pathogens in a single chamber. *Lab Chip*. 2008;8:810–3.

86. Lo RC, Ugaz VM. Microchip DNA electrophoresis with automated whole-gel detection. *Lab Chip*. 2008;8:2135–3145.

87. Terao K, Washizu M, Oana H. On-site manipulation of single chromosomal DNA molecules by using optically driven microstructures. *Lab Chip*. 2008;8:1280–4.

88. Xie L, Chew PM, Chong SC, Wai LC, Lau J. Optimization of a microfluidic cartridge for lab-on-a-chip (LOC) application and bio-testing for DNA/RNA extraction. *IEEE Electronic Components and Technology Conference*; 2008. pp. 1310–6.

89. Liu P, Yeung SHI, Crenshaw KA, Crouse CA, Scherer JR, Mathies RA. Real-time forensic DNA analysis at a crime scene using a portable microchip analyzer. *Forensic Sci Int Genet*. 2008;2:301–9.

90. York J, Spetzler D, Xiong F, Frasch WD. Single-molecule detection of DNA via sequence-specific links between F_1-ATPase motors and gold nanorod sensors. *Lab Chip*. 2008;8:415–9.

91. Chang CH, Chen JK, Chang FC. Specific DNA extraction through fluid channels with immobilization of layered double hydroxides on polycarbonate surface. *Sens Actuators B Chem*. 2008;133:327–32.

92. Malic LM, Veres T, Tabrizian M. Biochip functionalization using electrowetting-on-dielectric digital microfluidics for surface plasmon resonance imaging detection of DNA hybridization. *Biosens Bioelectron*. 2009;24:2218–24.

93. Schaerli Y, Wootton RC, Robinson T, Stein V, Dunsby C, Neil MAA, et al. Continuous-flow polymerase chain reaction of single-copy DNA in microfluidic microdroplets. *Anal Chem*. 2009;81:302–6.

94. Oakley JA, Shaw KJ, Docker PT, Dyer CE, Greenman, Greenway GM, et al. Development of a bi-functional silica monolith for electro-osmotic pumping and DNA clean-up/extraction using gel-supported reagents in a microfluidic device. *Lab Chip*. 2009;9:1596–1600.

95. Hubálek J, Adam V, Kizek R. New approach in rapid viruses detection and its implementation on a chip. *IEEE International Conference on eHealth, Telemedicine, and Social Medicine*; 2009. pp. 108–12.

96. Lien KY, Liu CJ, Lin YC, Kuo PL, Lee GB. Extraction of genomic DNA and detection of single nucleotide polymorphism genotyping utilizing an integrated magnetic bead-based microfluidic platform. *Microfluid Nanofluidics*. 2009;6:539–55.

97. Mahalanabis M, Al-Muayad H, Kulinski MD, Altman D, Klapperich CM. Cell lysis and DNA extraction of Gram-positive and Gram-negative bacteria from whole blood in a disposable microfluidic chip. *Lab Chip*. 2009;9:2811–7.

98. Auroux PA, Iossifidis D, Reyes DR, Manz A. Micro total analysis systems. II. Analytical standard operations and applications. *Anal Chem*. 2002;74:2637–52.

99. Vilkner T, Janasek D, Manz A. Micro total analysis systems. Recent developments. *Anal Chem*. 2004;76:3373–86.

100. Dittrich PS, Tachikawa K, Manz A. Micro total analysis systems. Latest advancements and trends. *Anal Chem*. 2006;76:3887–907.

101. West J, Becker M, Tombrink S, Manz A. Micro total analysis systems: latest achievements. *Anal Chem*. 2008;80:4403–19.

102. Ugaz VM, Elms RD, Lo RC, Shaikh FA, Burns M. Microfabricated electrophoresis systems for DNA sequencing and genotyping applications: current technology and future directions. *Philos Trans R Soc Lond A*. 2004;362:1105–29.

103. Liu S, Guttman A. Electrophoresis microchips for DNA analysis. *Trends Analyt Chem*. 2004;23:422–31.

104. Zhang C, Xu J, Ma W, Zheng W. PCR microfluidic devices for DNA amplification. *Biotechnol Adv*. 2006;24:243–84.

105. Sun Y, Kwok YC. Polymeric microfluidic system for DNA analysis. *Anal Chim Acta*. 2006;556:80–96.

106. Lee TMH, Hsing IM. DNA-based bioanalytical microsystems for handheld device applications. *Anal Chim Acta*. 2006;556:26–37.

107. Bianchessi M, Burgarella S, Cereda M. Point-of-care systems for rapid DNA quantification in oncology. *Tumori*. 2008;94:216–24.

108. Teles FRR, Fonseca LP. Trends in DNA sensors. *Talanta*. 2008;77:606–23.

109. Zhang YH, Pinar O. Microfluidic DNA amplification—a review. *Anal Chim Acta*. 2009;62(2):115–25.

110. Chol JW, Oh KW, Thomas JH, Helneman WR, Halsall HB, Nevin JH, et al. An integrated microfluidic biochemical detection system for protein analysis with magnetic bead-based sampling capabilities. *Lab Chip*. 2002;2:27–30.

111. Shilling EA, Amholz AE, Yager P. Cell lysis and protein extraction in a microfluidic device with detection by a fluorogenic enzyme assay. *Anal Chem*. 2002;74:1798–804.

112. Hamamoto T, Nojima T, Fujii T. PDMSUglass hybrid microreactor array with embedded temperature control device. Application to cell-free protein synthesis. *Lab Chip*. 2002;2:197–202.

113. Slentz BE, Penner NA, Regnier FE. Protein proteolysis and the multi-dimensional electrochromatographic separation of histidine-containing peptide fragments on a chip. *J Chromatogr A*. 2003;984:97–107.

114. Huber DL, Manginell RP, Samara MA, Kim BI, Bunker BC. Programmed adsorption and release of proteins in a microfluidic device. *Science*. 2003;301:352–4.

115. Carlo DD, Jeong KH, Lee LP. Reagentless mechanical cell lysis by nanoscale barbs in microchannels for sample preparation. *Lab Chip*. 2003;3:287–91.

116. Bouaidat S, Berendsen C, Thomsen P, Petersen SG, Wolff A, Jonsmann J. Micro patterning of cell and protein non-adhesive plasma polymerized coatings for biochip applications. *Lab Chip*. 2004;4:632–7.

117. Jin LJ, Ferrance J, Sanders JC, Landers JP. A microchip-based proteolytic digestion system driven by electroosmotic pumping. *Lab Chip*. 2003;3:11–18.

118. Lynk M, Mosher C, Huff J, Nettikadan S, Johnson J, Henderson E. Functional protein nanoarrays for biomarker profiling. *Proteomics*. 2004;4:1695–702.

119. Atsuta K, Noji H, Takeuchi S. Micro patterning of active proteins with perforated PDMS sheets (PDMS sieve). *Lab Chip*. 2004;4:333–6.

120. Callwaert N, Contreras R, Mitnik-Gankin L, Carey L, Matsudaira P, Ehrlich D. Total serum protein N-glycome profiling on a capillary electrophoresis-microfluidics platform. *Electrophoresis*. 2004;25:3128–31.

121. Koskinen JO, Meltola NJ, Soini E, Soini A. A lab-on-a-chip compatible bioaffinity assay method for human α-fetoprotein. *Lab Chip*. 2005;5:1408–11.

122. Mia J, Wu W, Spielmann T, Belfort M, Derbyshire V, Belfort G. Single-step affinity purification of toxic and non-toxic proteins on a fluidics platform. *Lab Chip*. 2005;5:248–53.

123. Kai J, Thaiti S, Ahn C. An ultra-high sensitive protein lab-on-a-chip on polymer for prostate cancer diagnostics. *Tranducers '05. The 13th International Conference on Solid-State Sensors, Actuators and Microsystem*; 2005. pp. 1653–5.

124. Christodoulides N, Mohanty S, Miller CS, Lahgub MC, Floriano PN, Dharshan P, et al. Application of microchip assay system for the measurement of C-reactive protein in human saliva. *Lab Chip*. 2005;5:261–9.

125. Luo C, Fu Q, Li H, Xu L, Sun M, Ouyang Q, et al. PDMS microfludic device for optical detection of protein immunoassay using gold nanoparticles. *Lab Chip*. 2005;5:726–9.

126. Wilson MS, Nie W. Multiplex measurement of seven tumor markers using an electrochemical protein chip. *Anal Chem*. 2006;78:6676–483.

127. Cai L, Friedman N, Xie XS. Stochastic protein expression in individual cells at the single molecule level. *Nature*. 2006;440:358–62.

128. Lönnberg M, Carlsson J. Lab-on-a-chip technology for determination of protein isoform profiles. *J Chromatogr A*. 2006;1127:185–2.

129. Tsiknakis M, Grangeat P, Binz PA, Potamias G, Lisacek F, Gerfault L, et al. Functional specifications of an integrated proteomics information management and analysis platform. *29th Annual International Conference of the IEEE EMBS*; 2007. pp. 6064–8.

130. Li C, Wu PM, Browne A, Lee S, Ahn CH. Hot-embossed piezoelectric polymer micro-diaphragm arrays integrated with lab-on-a-chip for protein analysis. *IEEE Sensors Conference*; 2007. pp. 462–5.

131. Arata HF, Gillot F, Nojima T, Fuji T, Fujita H. Millisecond denaturation dynamics of fluorescent proteins revealed by femtoliter container on micro-thermodevice. *Lab Chip*. 2008;8:1436–40.

132. Yu H, Lu Y, Zhou YG, Wang FB, He FY, Xia XH. A simple, disposable microfluidic device for rapid protein concentration and purification via direct-printing. *Lab Chip*. 2008;8:1496–501.

133. Wang Z, Wu H, Chen J, Zhang J, Yao Y, Chen GQ. A novel self-cleaving phasin tag for purification of recombinant proteins based on hydrophobic polyhydroxyalkanoate nanoparticles. *Lab Chip*. 2008;8:1957–62.

134. Meagher RJ, Light YK, Singh AK. Rapid, continuous purification of proteins in a microfluidic device using genetically-engineered partition tags. *Lab Chip*. 2008;8:527–32.

135. Huang YM, Uppalapati M, Hancock WO, Jackson TN. Neutravidin micropatterning by deep UV irradiation. *Lab Chip*. 2008;8:1745–7.

136. Le Nel A, Minc N, Smadja C, Slovakova M, Bilkova Z, Peyrin JM, et al. Controlled proteolysis of normal and pathological prion protein in a microfluidic chip. *Lab Chip*. 2008;8:294–301.

137. Koc Y, de Mello AJ, McHale G, Newton MI, Roach P, Shirtcliffe NJ. Nano-scale superhydrophobicity: suppression of protein adsorption and promotion of flow-induced detachment. *Lab Chip*. 2008;8:582–6.

138. Sedgwick H, Caron F, Monaghan PB, Kolch W, Cooper JM. Lab-on-a-chip technologies for proteomic analysis from isolated cells. *J R Soc Interface*. 2008;5:123–30.

139. Deshpande K, Ahamed T, van der Wielen LAM, ter Horst JH, Jansens PJ, Ottens M. Protein self-interaction chromatography on a microchip. *Lab Chip*. 2009;9:600–5.

140. Khnouf R, Beebe DJ, Fan ZH. Cell-free protein expression in a microchannel array with passive pumping. *Lab Chip*. 2009;9:56–61.

141. Luk VN, Wheeler AR. A digital microfluidic approach to proteomic sample processing. *Anal Chem*. 2009;81:4524–30.

142. Vasina EN, Paszek E, Nicolau DV, Nicolau DV Jr, Nicolau DV. The BAD project: data mining, database and prediction of protein adsorption on surfaces. *Lab Chip*. 2009;9:891–900.

143. Jebrail MJ, Wheeler AR. Digital microfluidic method for protein extraction by precipitation. *Anal Chem*. 2009;81:330–5.

144. Wu D, Steckl AJ. High speed nanofluidic protein accumulator. *Lab Chip*. 2009;9:1890–6.

145. Lion N, Rohner TC, Dayon L, Arnaud IL, Damoc E, Youhnovski N, et al. Microfluidic systems in proteomics. *Electrophoresis*. 2003;24:3533–62.

146. Schasfoort RBM. Proteomics-on-a-chip: the challenge to couple lab-on-a-chip unit operations. *Expert Rev Proteomics*. 2004;1:123–32.

147. Erickson D, Li D. Integrated microfluidic devices. *Anal Chim Acta*. 2004;507:11–26.

148. Joos T. Protein microarray technology. *Expert Rev Proteomics*. 2004;1:1–3.

149. Freire SLS, Wheeler R. Proteome-on-a-chip: mirage, or on the horizon? *Lab Chip*. 2006;6:1415–23.

150. Urban PL, Goodall DM, Bruce NC. Enzymatic microreactors in chemical analysis and kinetic studies. *Biotechnol Adv*. 2006;24:42–57.

151. Chin CD, Linder V, Sia SK. Lab-on-a-chip devices for global health: past studies and future opportunities. *Lab Chip*. 2007;7:41–57.

152. Haeberle S, Zengerle R. Microfluidic platforms for lab-on-a-chip applications. *Lab Chip*. 2007;7:1094–110.

153. Fonslow BR, Yates III JR. Capillary electrophoresis applied to proteomic analysis. *J Sep Sci*. 2009;32:1175–88.

154. Fu AY, Chou HP, Spence C, Arnold FH, Quake, SR. An integrated microfabricated cell sorter. *Anal Chem*. 2002;74:2451–7.

155. Manaresi N, Romani A, Medoro G, Altomare L, Leonardi A, Tartagni M, et al. A CMOS chip for individual manipulation and detection. *IEEE International Solid-State Circuits Conference*, Session 11; 2003.

156. Medoro G, Manaresi N, Leonardi A, Altomare L, Tartagni M, Guerrieri R. A lab-on-a-chip for cell detection and manipulation. *IEEE Sens J*. 2003;3(3):317–25.

157. Di Carlo D, Jeong KH, Lee LP. Reagentless mechanical cell lysis by nanoscale barbs in microchannels for sample preparation. *Lab Chip*. 2003;3:287–91.

158. Seger U, Gaward S, Johann R, Bertsch A, Renaud P. Cell immersion and cell dipping in microfluidic devices. *Lab Chip*. 2004;4:148–51.

159. Liu Y, Smela E, Nelson NM, Abshire P. Cell-lab chip: a CMOS-based microsystem for culturing and monitoring cells. *26th Annual International Conference of the IEEE EMBS*; 2004. pp. 2534–7.

160. Yamada M, Kasim V, Nakashima M, Edahiro J, Seki M. Continuous cell partitioning using an aqueous two-phase flow system in microfluidic devices. *Biotechnol Bioeng*. 2004;88(4):489–94.

161. Inglis D, Riehn R, Austin RH, Sturm JC. Continuous microfluidic immunomagnetic cell separation. *Appl Phys Lett*. 2004;85(21):5093–6.

162. Gao J, Yin XF, Fang ZL. Integration of single cell injection, cell lysis, separation and detection of intracellular constituents on a microfluidic chip. *Lab Chip*. 2004;4:47–52.

163. Rhee SW, Taylor AM, Tu CH, Crinns DH, Cotman W, Jeon NL. Patterned cell culture inside microfluidic devices. *Lab Chip*. 2005;5:102–7.
164. Liang XJ, Liu AQ, Zhang XM, Yap PH, Ayi TC, Yoon HS. Determination of refractive index for single living cell using integrated biochip. *Transducers '05—13th International Conference on Solid-State Sensors, Actuators and Microsystems*; 2005. pp. 1712–5.
165. Tourovskaia A, Figueroa-Masot X, Folch A. Differentiation-on-a-chip: a microfluidic platform for long-term cell culture studies. *Lab Chip*. 2005;5:14–9.
166. Chung BG, Flanagan LA, Rhee SW, Schwartz PH, Lee AP, Monuki ES, et al. Human neural stem cell growth and differentiation in a gradient-generating microfluidic device. *Lab Chip*. 2005;5:401–6.
167. Irimia D, Toner M. Cell handling using microstructured membranes. *Lab Chip*. 2006;6:345–52.
168. Sathuluti RR, Kitamura M, Yamamura S, Tamiya E. Development of high-throughput compartmental microfluidic devices for multiplexed single-cell sorting, manipulation and analysis. *5th IEEE Conference on Sensors*; 2006. pp. 642–5.
169. Xing C, Dafu C, Changchun L, Hahoyuan C. Microfluidic biochip for blood cell lysis. *Chin J Anal Chem*. 2006;34(11):1656–60.
170. Ho CT, Lin RZ, Chand WY, Chang HY, Liu CH. Rapid heterogeneous liver-cell on-chip patterning via the enhanced field-induced dielectrophoresis trap. *Lab Chip*. 2006;6:724–34.
171. Li Y, Dalton C, Crabtree HJ, Nilsson G, Kaler KVIS. Continuous dielectrophoretic cell separation microfluidic device. *Lab Chip*. 2007;7:239–48.
172. Huang CW, Huang SB, Lee GB. A microfluidic system for automatic cell culture. *Transducers and Eurosensors'07—The International Conference on Solid-State Sensors, Actuators and Microsystems*; 2007. pp. 783–6.
173. Tai CH, Hsiung SK, Chen CY, Tsai ML, Lee GB. Automatic microfluidic platform for cell separation and nucleus collection. *Biomed Microdevices*. 2007;9:533–43.
174. Zhu L, Peh XL, Ji HM Teo CY, Feng HH, Liu WT. Cell loss in integrated microfluidic device. *Biomed Microdevices*. 2007;9:745–50.
175. Lob V, Geisler T, Brischwein M, Uhl R, Wolf B. Automated live cell screening system based on a 24-wellmicroplate with integrated micro fluidics. *Med Biol Eng Comput*, 2007;45:1023–8.
176. Cloi S, Song S, Choi C, Park JK. Continuous blood cell separation by hydrophoretic filtration. *Lab Chip*. 2007;7:1532–8.
177. Kim J, Steinfeld U, Lee HH, Seidel H. Development of a novel micro b immune-magnetophoresis cell sorter. *IEEE Sensors Conference*; 2007. pp. 1081–4.
178. Nevill JT, Cooper R, Dueck M, Breslauer DN, Lee LP. Integrated microfluidic cell culture and lysis on a chip. *Lab Chip*. 2007;7:1689–95.
179. Christensen TB, Pedersen CM, Bang DD, Wolff A. Sample preparation by cell guiding using negative dielectrophoresis. *Microelectron Eng*. 2007;84:1690–3.
180. Nevill JT, Cooper R, Dueck M, Breslauer DN, Lee LP. Integrated microfluidic cell culture and lysis on a chip. *Lab Chip*. 2007;7:1689–95.
181. Quinto-Su PA, Lai HH, Yoon HH, Sims CE, Allbritton NL, Venugopalan V. Examination of laser microbeam cell lysis in a PDMS microfluidic channel using time-resolved imaging. *Lab Chip*. 2008;8:408–14.
182. Flueckiger J, Cheung KC. Integrated microfluidic systems for cell culture and characterization. *IMS3TW—IEEE 14th International Mixed-Signals, Sensors, and Systems Test Workshop*; 2008, pp. 1–6.
183. Kierschbaum M, Jaeger MS, Schenkel T, Breinig T, Meywehans A, Duschl C. T cell activation on a single-cell level in dielectrophoresis-based microfluidic devices. *J Chromatogr A*. 2008;1202:83–9.
184. Ji HM, Samper V, Chen Y, Heng CK, Lim TM, Yobas L. Silicon-based microfilters for whole blood cell separation. *Biomed Microdevices*. 2008;10:251–7.
185. Cimetta E, Figallo E, Cannizzaro C, Elvassore N, Navakovic GV. Micro-bioreactor arrays for controlling cellular environments: design principles for human embryonic stem cell applications. *Methods*. 2009;47:81–9.

186. Estes MD, Do J, Ahn CH. On chip cell separator using magnetic bead-based enrichment and depletion of various surface markers. *Biomed Microdevices*. 2009;11:509–15.
187. Khanna P, Ramachandran N, Yang J, Wang J, Kumar A, Jaroszeski M, et al. Nanocrystalline diamond microspikes increase the efficiency of ultrasonic cell lysis in a microfluidic lab-on-a-chip. *Diam Relat Mater*. 2009;18:606–10.
188. Villa-Diaz LG, Torisawa US, Uchida T, Ding J, Nogueira-de-Souza NC, O'Shea KS, et al. Microfluidic culture of single human embryonic stem cell colonies. *Lab Chip*. 2009;9:1749–55.
189. Lin YH, Lee GB. An optically induced cell lysis device using dielectrophoresis. *Appl Phys Lett*. 2009;94:1–3.
190. Hsieh CC, Huang SB, Wu PC, Shieh DR, Lee GB. A microfluidic cell culture platform for real-time cellular imaging. *Biomed Microdevices*. 2009;11:903–13.
191. Modak N, Datta A, Ganguly R. Cell separation in a microfluidic channel using magnetic microspheres. *Microfluid Nanofluidics*. 2009;6:647–60.
192. Primiceri E, Chiriacò MS, Ionescu RE, D'Amone E, Cingolani R, Rinaldi R, et al. Development of EIS cell chips and their application for cell analysis. *Microelectron Eng*. 2009;86:1477–80.
193. Green JV, Kniazeva T, Abedi M, Sokhey DS, Taslim ME, Murthy SK. Effect of channel geometry on cell adhesion in microfluidic devices. *Lab Chip*. 2009;9:677–85.
194. Sanders GHW, Manz A. Chip-based microsystems for genomic and proteomic analysis. *Anal Chem*. 2000;19(6):364–78.
195. Park TH, Shuler ML. Integration of cell culture and microfabrication technology. *Biotechnol Prog*. 2003;19:243–53.
196. Voldman J. Electrical forces for microscale cell manipulation. *Annu Rev Biomed Eng*. 2006;8:425–54.
197. Medoro G, Guerrieri R, Manaresi N, Nastruzzi C, Gambaru R. Lab on a chip for live-cell manipulation. *IEEE Des Test Comput*. 2007:26–36.
198. Brown RB, Audet J. Current techniques for single-cell lysis. *J R Soc Interface*. 2008;5:131–8.
199. Kim SM, Lee SH, Suh KY. Cell research with physically modified microfluidic channels: a review. *Lab Chip*. 2008;8:1015–23.
200. Chao TC, Ros A. Microfluidic single-cell analysis of intracellular compounds. *J R Soc Interface*. 2008;5:139–50.
201. Doshi R, Day PJR. Scrapheap challenge and the single cell. *Lab Chip*. 2008;8:1774–8.
202. Liu C, Stakenborg T, Peeters S, Lagae L. Cell manipulation with magnetic particles toward microfluidic cytometry. *J Appl Phys*. 2009;105:1–10.
203. Gascoyne P, Mahidol C, Ruchirawat M, Satayavivad J, Watcharasit P, Becker F. Microsample preparation by dielectrophoresis: isolation of malaria. *Lab Chip*. 2002;2:70–5.
204. Gascoyne P, Satayavivad J, Ruchirawat M. Microfluidic approaches to malaria detection. *Acta Trop*. 2004;89:357–69.
205. Ferrance JP, Wu Q, Giordano B, Hernandez C, Knok Y, Snow K, et al. Developments toward a complete micro-total analysis system for Duchenne muscular dystrophy diagnosis. *Anal Chim Acta*. 2003;500:223–36.
206. Srinivasan V, Pamula VK, Fair RB. An integrated digital microfluidic lab-on-a-chip for clinical diagnostics on human physiological fluids. *Lab Chip*. 2004;4:310–5.
207. Srinivasan V, Pamula VK, Fair RB. Croplet-based microfluidic lab-on-a-chip for glucose detection. *Anal Chim Acta*. 2004;507:145–50.
208. Ahn CH, Choi JW, Beaucage G, Nevin JH, Lee JB, Puntambekar A, et al. Disposable smart lab on a chip for point-of-care clinical diagnostics. *Proc IEEE*. 2004;92(1):154–73.
209. Minas G, Woffenbuttel RF, Correia JH. A lab-on-a-chip for spectrophotometric analysis of biological fluids. *Lab Chip*. 2005;5:1303–9.
210. Cardoso VF, Catarino SO, Serrado Nunes J, Rebouta L, Rocha JG, Lanceros-Méndez S, et al. Lab-on-a-chip with β-PVDF based acoustic microagitation. *IEEE Trans Biomed Eng*. In press.
211. Christodoulides N, Mahanty S, Miller CS, Langub MC, Florian PN, Dharshan P, et al. Application of microchip assay system for the measurement of C-reactive protein in human saliva. *Lab Chip*. 2005;5:261–9.

212. Huang FC, Lia CS, Wang CH, Lee GB. Automatic Bio-MEMS platforms for fast disease diagnosis. *IEEE International Conference on Robotics and Biomimetics*; 2005. pp. 10–15.
213. Guo WP, Ma XM, Zang Y. Clinical laboratories on a chip for human immunodeficiency virus assay. *IEEE Engineering in Medicine and Biology 27th Annual Conference*; 2005. pp. 1274–7.
214. Tsukagoshi K, Jinno N, Nakajima. Development of a micro total analysis system incorporating chemiluminescence detection and application to detection of cancer markers. *Anal Chim.* 2005;77:1684–8.
215. Liao CS, Lee GB, Liu HS, Hsieh TM, Luo CH. Miniature RT-PCR system for diagnosis of RNA-based viruses. *Nucleic Acids Res.* 2005;33(18):1–7.
216. Gulliksen A, Solli LA, Drese KS, Sörensen O, Karlsen F, Rogne H, et al. Parallel nanoliter detection of cancer markers using polymer microchips. *Lab Chip.* 2005;5:416–20.
217. Hofmann O, Wang X, de Mello JC, Bradley DDC, deMello AJ. Towards microalbuminuria determination on a disposable diagnostic microchip with integrated fluorescence detection based on thin-film organic light emitting diodes. *Lab Chip.* 2005;5:863–8.
218. Sun Y, Liu AQ, Yap PH, Ayi TC. A self-contained microfluidic cell sorting system for HIV diagnosis. *Transducers—The 13th International Conference on Solid-State Sensors, Actuators and Microsystem*; 2005. pp. 1708–11.
219. Liao JC, Ma Y, Gau V, Mastali M, Sun CP, Li Y, et al. A point-of-care micro-laboratory for direct pathogen identification in body fluids. *1st IEEE International Conference on Nano/Micro Engineered and Molecular Systems*; 2006. pp. 1109–12.
220. Weigum SE, Floriano PN, Christodoulides N, McDevitt JT. Cell-based sensor for analysis of EGFR biomarker expression in oral cancer. *Lab Chip.* 2007;7:995–1003.
221. Herr AE, Hatch AV, Throckmorton DJ, Tran HM, Brennan JS, Giannobile WV, et al. Microfluidic immunoassays as rapid saliva-based clinical diagnostics. *Proc Natl Acad Sci USA.* 2007;104(13):5268–73.
222. Herr AE, Hatch AV, Giannobile WV, Throckmorton DJ, Tran HM, Brennan JS, et al. Integrated microfluidic platform for oral diagnostics. *Ann N Y Acad Sci.* 2008:1–13.
223. Blanes L, Mora MF, do Lago CL, Ayon A, García CD. Lab-on-a-chip biosensor for glucose based on a packed immobilized enzyme reactor. *Electroanalysis.* 2007;19(23):2451–6.
224. King KR, Wang S, Irimia D, Jayaraman A, Toner M, Yarmush ML. A high-throughput microfluidic real-time gene expression living cell array. *Lab Chip.* 2007;7:77–85.
225. Ortiz P, Keegan N, Spoors J, Hedley J, Harris A, Burdess J, et al. A hybrid microfluidic system for cancer diagnosis based on MEMS biosensors. *Biomedical Circuits and Systems Conference. IEEE*; 2008. pp. 337–40.
226. Tan HY, Loke WK, Tan YT, Nguyen NT. A lab-on-a-chip for detection of nerve agent sarin in blood. *Lab Chip.* 2008;8:885–91.
227. Zhang J, Guo Q, Liu M, Yang J. A lab-on-CD prototype for high-speed blood separation. *J Micromech Microeng.* 2008;18:1–6.
228. Zhu H, Stybayeva G, Macal M, Ramanculov E, George MD, Dandekar S, et al. A microdevice for multiplexed detection of T-cell-secreted cytokines. *Lab Chip.* 2008;8:2197–205.
229. Lee SH, Kim SW, Kang JY, Ahn CH. A polymer lab-on-a-chip for reverse transcription (RT)-PCR based point-of-care clinical diagnostics. *Lab Chip.* 2008;8:2121–7.
230. Sieben VJ, Debes-Marun CS, Pilarski LM, Backhouse CJ. An integrated microfluidic chip for chromosome enumeration using fluorescence in situ hybridization. *Lab Chip.* 2008,8: 2151–6.
231. Meagher RJ, Hatcxh AV, Renzi RF, Singh AK. An integrated microfluidic platform for sensitive and rapid detection of biological toxins. *Lab Chip.* 2008;8:2046–53.
232. He JH, Reboud J, Ji H, Zhang L, Long Y, Lee C. Biomicrofluidic lab-on-chip device for cancer cell detection. *Appl Phys Lett.* 2008;93:1–3.
233. Rosenbluth MJ, Lam WA, Fletcher DA. Analyzing cell mechanics in hematologic diseases with microfluidic biophysical flow cytometry. *Lab Chip.* 2008;8:1062–70.
234. Bao N, Zhan Y, Lu C. Microfluidic electroporative flow cytometry for studying single-cell biomechanics. *Anal Chem.* 2008;80:7714–9.

235. Do J, Lee S, Han J, Kai J, Hong CC, Gao C, et al. Development of functional lab-on-a-chip on polymer for point-of-care testing of metabolic parameters. *Lab Chip*. 2008;8:2113–120.

236. Do J, Ahn CH. A polymer lab-on-a-chip for magnetic immunoassay with on-chip sampling and detection capabilities. *Lab Chip*. 2008;8:542–9.

237. Sista R, Hua Z, Thwar P, Sudarsan A, Srinivasan C, Eckhardt A, et al. Development of a digital microfluidic platform for point of care testing. *Lab Chip*. 2008;8:2091–104.

238. Tang D, Xia B. Electrochemical immunosensor and biochemical analysis for carcinoembryonic antigen in clinical diagnosis. *Microchim Acta*. 2008;163:41–8.

239. Beyor N, Seo TS, Liu P, Mathies RA. Immunomagnetic bead-based cell concentration microdevice for dilute pathogen detection. *Biomed Microdevices*. 2008;10:909–17.

240. Kubán P, Hauser PC. Evaluation of microchip capillary electrophoresis with external contactless conductivity detection for the determination of major inorganic ions and lithium in serum and urine samples. *Lab Chip*. 2008;8:1829–36.

241. Hattersley SM, Duer CE, Greenman J, Haswell SJ. Development of a microfluidic device for the maintenance and interrogation of viable tissue biopsies. *Lab Chip*. 2008;8:1842–6.

242. Kalatzis FG, Exarxhos TP, Giannakeas N, Markoula S, Hatzi E, Rizos P, et al. Point-of-care monitoring and diagnostics for autoimmune diseases. *8th IEEE International Conference on Bioinformatics & Bioengineering*; 2008. pp. 1–8.

243. Marchand G, Broyer P, Ianet V, Delattre C, Foucault F, Menou L, et al. Opto-electronic DNA chip-based integrated card for clinical diagnostics. *Biomed Microdevices*. 2008;10:35–45.

244. Darain F, Yager P, Gan KL, Tjin SC. On-chip detection of myoglobin based on fluorescence. *Biosens Bioelectron*. 2009;24:1744–50.

245. Ohashi T, Mawatari K, Sato K, Tokeshi M, Kitamori T. A micro-ELISA system for the rapid and sensitive measurement of total and specific immunoglobulin E and clinical application to allergy diagnosis. *Lab Chip*. 2009;9:991–5.

246. Brandenburg A, Curdt F, Sulz G, Ebling F, Nestler J, Wunderlich K, et al. Biochip readout system for point-of-care applications. *Sens Actuators B Chem*. 2009;139:245–51.

247. Zhang Y, Bailey V, Puleo CM, Easwaran H, Griffiths E, Herman JG, et al. DNA methylation analysis on a droplet-in-oil PCR array. *Lab Chip*. 2009;9:1059–64.

248. Javanmard M, Talasaz AH, Nemat-Gprgani M, Pease F, Ronaghi M, Davis RW. Electrical detection of protein biomarkers using bioactivated microfluidic channels. *Lab Chip*. 2009;9:1429–34.

249. Lee SH, Kim SW, Kang JY, Ahn CH. A polymer lab-on-a-chip for transverse transcription (RT)-PCR based point-of-care clinical diagnostics. *Lab Chip*. 2008;8:2121–7.

250. Colyer CL, Tang T, Cheim N, Harrison DJ. Clinical potential of microchip capillary electrophoresis systems. *Electrophoresis*. 1997;18:1733–41.

251. Tüdõs AJ, Besselink GAJ, Schasfoort RBM. Trends in miniaturized total analysis systems for point-of-care testing in clinical chemistry. *Lab Chip*. 2001;1:83–95.

252. Schulte TH, Bardell RL, Weigl BH. Microfluidic technologies in clinical diagnostics. *Clin Chim Acta*. 2002;321:1–10.

253. Weigl BH, Bardell RL, Cabrera CR. Lab-on-a-chip for drug development. *Adv Drug Deliv Rev*. 2003;55:349–77.

254. Pilarski LM, Adamia S, Pilarki PM, Prakash R, Lauzon J, Backhouse CJ. Improved diagnosis and monitoring of cancer using portable microfluidics platforms. *IEEE Proceedings of the 2004 International Conference on MEMS, NANO and Smart Systems*; 2004. pp. 340–3.

255. Gardeniers H, Van Den Berg A. Micro- and nanofluidic devices for environmental and biomedical applications. *Int J Environ Anal Chem*. 2004;84(11):809–19.

256. Toner M, Irima D. Blood-on-a-chip. *Annu Rev Biomed Eng*. 2005;7:77–103.

257. Ziober BL, Mauk MG, Falls Em, Chen Z, Ziober AF, Bau HHB. Lab-on-a-chip for oral cancer screening and diagnosis. A22; 2008. pp. 111–21.

258. Sato K, Mawatari K, Kitamori T. Microchip-based cell analysis and clinical diagnosis system. *Lab Chip*. 2008;8:1992–8.

259. Myers FB, Lee LP. Innovations in optical microfluidic technologies for point-of-care diagnostics. *Lab Chip*. 2008;8:2015–31.

12

Microparticle and Nanoparticle Manipulation

Rong Bai and John T.W. Yeow

CONTENTS

12.1 Introduction

Microfluidics and nanofluidics demonstrate increasing and stimulating opportunities of foundational research on processes and systems that control micrometer- and nanometer-scale operations in science and technology. Besides, microfluidics and nanofluidics provide possibilities of a promising platform for developing new systems or chips for ever-growing applications in chemical and biological science and engineering [1].

A number of methods have been developed to manipulate these particles on the basis of particle size, density, electrical charge, and specific immunological surface markers, for example, electrophoretic, dielectrophoretic, magnetophoretic, and optical methods. Because the length of this chapter is limited, all methods and applications cannot be discussed completely. Therefore, this chapter focuses on micrometer- and nanometer-scale particle manipulation on the basis of dielectrophoresis (DEP) and magnetophoresis (MAP).

12.1.1 DEP for Particle Manipulation

The phenomena of DEP that affect micrometer-and nanometer-scale particles are levitation, rotation, and motion, and these attract researchers' attention because they cause higher separation resolution, low device cost, and rapid separation. The phenomenon of DEP was first defined by Pohl [2,3] as the motion of neutral but polarizable particles subjected to nonuniform electric fields. DEP provides an increased measurement precision and sensitivity in the separation and detection of cells with different dielectric properties. It also has low-volume requirement for samples when compared with traditional methods of cell analysis, such as flow cytometry [4], potentially providing an efficient alternative for clinical use.

According to the spatial and temporal features of a nonuniform field, a particle experiences conveyance, namely, translational motion or rotation or both. A nonuniform electrical field induces a DEP force that causes particle conveyance, which moves toward either an electric field maxima (positive DEP [pDEP]) or a minima (negative DEP [nDEP]), depending on polarization status of the particle. A phase-varying nonuniform electric field causes particle rotation (electrorotation [ROT]) and particle transportation (traveling wave DEP [TWD]). When such a field is implemented in a rotating configuration, it causes the particle to rotate. A linear arrangement, also called a traveling wave electric field, causes linear conveyance and particle rotation in the direction of movement. The three electrokinetic effects, DEP, ROT, and TWD, have been extensively studied, both from a theoretical and an experimental perspective, and have been applied usefully in a variety of biotechnological and clinical applications requiring cell manipulation [5–9].

DEP provides many advantages that make it a suitable separation technology for lab-on-a-chip (LOC) systems [10]. First, conventional techniques developed for the semiconductor industry can also be used to fabricate microelectrodes and the remainder of LOC systems. Second, DEP can also be used to discriminate between cells and similar objects on the basis of the complete dielectric makeup of the objects. Finally, the magnitude and the direction of DEP force are related to the properties of particles; therefore, DEP is ideal for particle manipulation systems.

12.1.2 MAP for Particle Manipulation

Similarly, magnetizable micrometer- and nanometer-scale particles experience a force in a nonuniform magnetic field. The phenomena, called MAP, that have an effect on micrometer- and nanometer-scale particles are levitation, rotation, and motion. Therefore, much of the analysis and interpretation of DEP is carried over to the phenomenology of magnetic particles in a nonuniform field. However, most magnetic particles are ferromagnetic and exhibit strong nonlinearity, limiting somehow the efficiency of the DEP ↔ MAP likeness. In addition, conductive particles, magnetic or not, in a nonuniform alternating current

(AC) magnetic field illustrate a diamagnetic-phenomenon response because of the induced eddy current. As widely known, eddy currents tend to exclude the time-varying field from the particles, causing the particle to be repelled from a region of high magnetic field intensity. The DEP $\leftrightarrow$ MAP likeness breaks down here as well because the magnetic field inside the conductive particle is not curl free [11].

Like DEP, MAP also has attracted researchers' attention in the manipulation of magnetic microparticles and nanoparticles. Today, MAP force is combined with microfluidics and nanofluidics in a variety of approaches. MAP force cannot only be exploited to manipulate magnetic objects such as magnetic particles, magnetically labeled cells, or plugs of ferrofluids inside a microchannel, but can also be used to manipulate diamagnetic objects. Usually, magnetic fields are generated by external permanent magnets outside the microchannel, sometimes by microfabricated electromagnets. MAP force is usually used for transportation [12,13], trapping [14,15], separation [13,16–19], and sorting of magnetic/nonmagnetic objects [20,21]. More recently, LOC techniques on the basis of MAP phenomenon have been investigated vastly. Nicole Pamme [22] gave a detailed review of research papers before 2005. Furthermore, Jones built effective moments of nonspherical particles for precise analysis in 2007. Chen et al. [23] developed a three-dimensional model of a portable medical device for magnetic separation. Jegatha et al. [24] designed and fabricated a new device that combined DEP and MAP phenomena to separate particles.

Because the modern inception of LOC and microfluidic/nanofluidic technologies occurred around 1990, portable LOC devices are now used in point-of-care systems, remote settings, and global health [25]. Although these devices are not yet appropriate for use in the extreme resource-poor settings of developing countries, the advances of LOC are in a prime position to tackle the profound issue of global health.

12.1.3 Outline of the Chapter

This chapter reviews recent research and presents a brief outline of the development of DEP and MAP separation on the basis of an LOC platform. In the section "Introduction," a short introduction of micrometer- and nanometer-scale particle manipulation is given. The general theory related to DEP and MAP is presented in the section "Theory." In the section "Applications of DEP," distinct characteristics of LOC based on DEP are discussed first, in which different shapes and structures of electrodes used in DEP are introduced, and some successful strategies and recognized conclusions are also included. Then, new technologies in MAP-based LOC are discussed in the section "Applications of MAP." A brief summary of MAP is given in this section. In the section "Conclusions," conclusions are made, and the challenges and prospects of the DEP and MAP-based LOC technology are discussed as well.

12.2 Theory

It has been known from ancient times that charged objects (such as a rubbed material picking up small items) can induce a force, either attractive or repulsive. In the last few years, researchers have used this force to separate and sequence strands of DNA and hence diagnose diseases. To understand completely how an electric field works on micro- and nanoscale particles, we must first understand the fundamentals of DEP.

12.2.1 Dielectrophoresis

Under uniform electric field conditions, charged particles, such as the negatively charged DNA, will move toward the electrode with the opposite charge. This is the basis for DNA gel separation. Strong and weakly charged DNA strands will move at different speeds toward the positively charged electrode. A pool of different DNA strands could be separated and then analyzed. Neutral particles, however, behave in a different way. Under a constant electric field, induced charges gather and distribute themselves at the surface of the neutral particle. By definition, a neutral particle has the same number of positive and negative charges. As we can see in Figure 12.1, the neutral particle will remain stationary because the forces induced toward the positive and negative electrodes are identical. Nevertheless, when the electric field is nonuniform, the neutral particle will move toward the region of high electric field density if the particle is more permeable than the medium, and the neutral particle will move toward the region of low electric field density if it is less permeable than the medium.

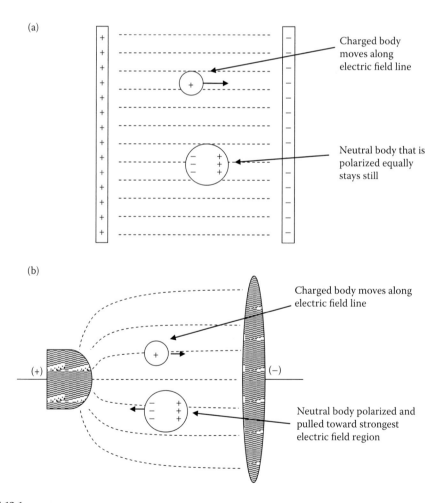

FIGURE 12.1
Charged and neutral bodies under electric field. (a) Uniform electric field applied to neutral and charged bodies. (b) Nonuniform electric field applied to neutral and charged bodies.

The DEP force does not depend on the polarity of the applied electric field and is observed with AC as well as direct current (DC) excitation. On the surface, this fact might seem surprising. Our common sense tells us that reversing the direction of the electric field will reverse the direction of the charged particle. If the electric field alternates in direction, there will be no net electrophoretic movement. However, we must remember that for a neutral body, the charges are induced. When the electric field reverses in direction, the induced dipole in the neutral body will reverse in direction as well. Thus, with DEP, one can avoid problems such as electrode polarization effects and electrolysis at electrodes (for details, see Kang and Li [26]).

12.2.1.1 DEP Levitation

When subjected to a nonuniform electric field E_0, a neutral but polarizable particle, suspended in a medium, experiences a time-averaged DEP force $\vec{F}_{DEP}$, given as [27],

$$\vec{F}_{DEP} = 2\pi R^3 \varepsilon_m \text{Re}[K_e] \nabla E_{0(\text{mrs})}^2, \tag{12.1}$$

where R and ε_m are the radius of the particle and permittivity of the medium, respectively. The polarization (Clausius–Mossotti) factor K_e is defined as

$$K_e(\omega) = \frac{\varepsilon_p^* - \varepsilon_m^*}{\varepsilon_p^* + 2\varepsilon_m^*}, \tag{12.2}$$

where ε^* is the complex permittivity, ω is angular velocity of the electric field, and ∇ is the gradient operator. ε_p^* and ε_m^* are defined as

$$\varepsilon_p^* = \varepsilon_p - j\frac{\sigma_p}{\omega},$$

$$\varepsilon_m^* = \varepsilon_m - j\frac{\sigma_m}{\omega},$$

where σ_m and σ_p are the conductivity of the particle and the medium, respectively.

Equation 12.1 shows that the magnitude of $\vec{F}_{DEP}$ is determined by the following factors: the size of the particle, the conductivity of the medium, the conductivity of the particle, and the magnitude of the electric fields [28]. However, the direction of $\vec{F}_{DEP}$ is determined by only one parameter, $K_e(\omega)$, the real part of the complex Clausius–Mossotti factor. It is the only parameter in Equation 12.1 that can become negative, and the sign of $K_e(\omega)$ will be the indicator of the direction of the $\vec{F}_{DEP}$. When $K_e(\omega) > 0$, $\vec{F}_{DEP}$ is positive, and it directs particles toward regions of high electric field strengths. This attraction effect is called pDEP. When $K_e(\omega) < 0$, an opposite effect, nDEP, dominates, and the particles seek out regions of low electric field strengths. Further analysis of the complex Clausius–Mossotti factor in Equation 12.2 reveals a frequency dependency, which is not apparent at the first glance. Because $K_e(\omega)$ is a complex number, frequency plays an important part in determining the sign of $K_e(\omega)$. By adjusting the frequency of the signals and maintaining all the other variables, the target particles can display opposite behaviors. Frequency analysis of $K_e(\omega)$ in Figure 12.2 reveals distinct regions of pDEP and nDEP. At the crossover frequency, $\vec{F}_{DEP}$ reaches a minimum and the particles display neither pDEP nor nDEP behaviors.

(a)

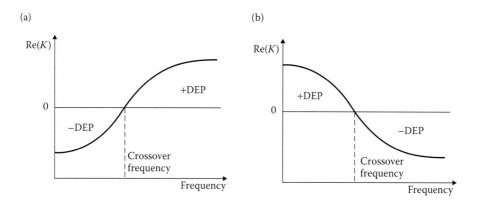

(b)

FIGURE 12.2
Frequency response of $K_e(\omega)$ in different conditions. (a) Frequency response of $K_e(\omega)$ when $\sigma_p < \sigma_m$ and $\varepsilon_p > \varepsilon_m$. (b) Frequency response of $K_e(\omega)$ when $\sigma_p > \sigma_m$ and $\varepsilon_p < \varepsilon_m$. (From Lin, J.T.-Y., Cell manipulations with dielectrophoresis. Master's thesis, University of Waterloo, ON, Canada, 2007.)

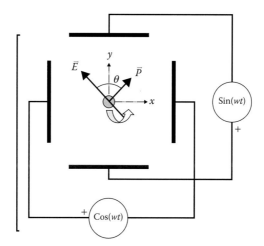

FIGURE 12.3
A neutral particle undergoing ROT when subjected to a rotating electric field. (Cen, E.G., Dalton, C., Li, Y., Adamia, S., Pilarski, L.M., and Kaler, K.V.I.S. A combined dielectrophoresis, traveling wave dielectrophoresis and electrorotation microchip for the manipulation and characterization of human malignant cells. *Journal of Microbiological Methods*, 58:387–401, 2004. With permission.)

12.2.1.2 Electrorotation

In 1982, Arnold and Zimmermann [29] discovered that a rotating electric field can be generated by means of constructing a phase-varying nonuniform electric field. Figure 12.3 shows a simple arrangement of electrodes that, when excited by polyphase AC voltages, creates such a field. When subjected to the electric field, $\vec{E}$, the induced polarization, $\vec{P}$, tries to align with the electric field intensity. A neutral particle placed in the rotating field will rotate either in a cofield direction or in an antifield direction. Cofield rotation occurs when $\vec{P}$ lags $\vec{E}$ by angles ranging from 0° to 180°, whereas antifield rotation occurs when $\vec{P}$ leads $\vec{E}$ by 0° to 180°.

The particle shown in Figure 12.3 undergoes a rotational torque, given by,

$$\vec{\Gamma}_{\text{ROT}} = -4\pi R^3 \varepsilon_m \, \text{Im}[K_e] \nabla E^2_{0(\text{mrs})} \cdot \hat{r}_0,$$

(12.3)

where $\hat{r}_0$ is the unit vector.

12.2.1.3 Traveling Wave DEP

A traveling wave electric field will be produced when a 90° phase-shifted signal sequence is applied to a parallel electrode array. Levitated above a certain height, a particle will undergo smooth translational motion. Close to the electrode surface, a particle may undergo translation simultaneously with a rolling movement. A particle subjected to a traveling wave electric field will experience a force introduced by the field, which is presented as [31],

$$\vec{F}_{\text{TWD}} = -\frac{4\pi R^3 \varepsilon_m \, \text{Im}[K_e] E^2_{0(\text{mrs})} \cdot \hat{r}_0}{\lambda},$$

where λ is the wavelength of the traveling electric field. The value equals the repetitive distance between electrodes of the same phase.

12.2.1.4 Summary of DEP Forces

As discussed above, the differences among the three kinds of DEP forces are summarized as follows:

- The pDEP and the nDEP are static electric fields; ROT and TWD are rotational and translational DEP forces, respectively.
- The target particle under ROT or TWD is in constant motion, whereas under pDEP and nDEP the particle remains stationary once it reaches the local electric field maxima or minima.
- Unlike pDEP and nDEP, the rotational and translational forces are influenced by the imaginary part of the complex Clausius–Mossotti factor, $\text{Im}[K_e]$. ROT is activated with a set of quad-pole sinusoidal voltage waves [29]. Particles under influence of ROT begin to rotate around their central axis. The direction and the speed of the rotation are useful characterization tools to determine the conductivity and the permeability of the particle. ROT experiments are often performed before TWD experiments to determine the optimal frequency and medium conductivity. Unlike the rotational force provided by ROT effect, TWD generates a translational force on the particle [30]. Particles under TWD move across the activated parallel electrodes, and the distance traveled is determined by the total number of the electrodes in the array.
- TWD depends on both $\text{Re}[K_e]$ and $\text{Im}[K_e]$, the real and imaginary part of the complex Clausius–Mossotti factor, respectively. To make TWD to occur, $\text{Re}[K_e]$ is negative to provide a lifting force on the particles by means of nDEP. This is achieved by adjusting the conductivity of the medium to be higher than that of the particle. Further, $\text{Im}[K_e]$ needs to be large enough to provide a translational force. To generate significant $\text{Im}[K_e]$, traveling waves composed of sinusoidal waves of the same frequency with 90° consecutive phase differences are used. In Figure 12.4, the frequency responses of $\text{Im}[K_e]$ and $\text{Re}[K_e]$ are overlapped to produce the shaded TWD zone where $\text{Re}[K_e]$ is negative and $\text{Im}[K_e]$ is sufficiently large.

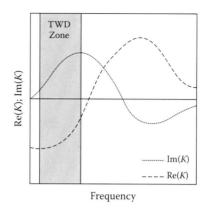

FIGURE 12.4

Typical graph of the Clausius–Mossotti factor versus frequency. (From Lin, J.T.-Y., Cell manipulations with dielectrophoresis. Master's thesis, University of Waterloo, ON, Canada, 2007.)

12.2.2 Magnetophoresis

Like dielectric particles, which experience a force in a nonuniform electric field, magnetizable particles also encounter a force, MAP, in a nonuniform magnetic field. This phenomenon has been exploited in industries, chemistry, biology, and clinics for particle manipulation.

12.2.2.1 Magnetically Linear Particle

Assuming a homogeneous and magnetically linear sphere with radius R and permeability μ_p, the MAP force of this sphere in a nonuniform magnetic field can be written as [11],

$$\bar{F}_{MAP} = 2\pi\mu_m R^3 K(\mu_p, \mu_m)\nabla H_0^2 = \frac{3}{2}V_p\mu_m K(\mu_p, \mu_m)\nabla H_0^2, \tag{12.4}$$

$$K(\mu_p, \mu_m) = \frac{\mu_p - \mu_m}{\mu_p + 2\mu_m}, \tag{12.5}$$

where μ_m and ∇H_0 are permeability of a magnetically linear fluid and the gradient of a nonuniform magnetic field, respectively. V_p is the volume of the sphere.

Note that the MAP force expression has a form similar to Equation 12.1, and the Clausius–Mossotti function remains in its original form (see Equation 12.2) with permeabilities μ_p and μ_m replacing permittivities ε_p and ε_m. Table 12.1 is the comparison between DEP and MAP.

12.2.2.2 Summary of MAP Forces

On the basis of the comparisons between Equations 12.1 and 12.4 and the data in Table 12.1, the fundamental phenomenology of MAP can be summarized as follows:

1. MAP force is proportional to particle volume.
2. MAP force is proportional to the permeability of the suspension medium μ_m.

TABLE 12.1

Summary of Analogous Factors for Dielectric, DC Conductive, and Magnetostatic Problems, All Having Linear Constitutive Laws

Quantity	Dielectric Case	Conductive Case	Magnetostatic Case
Field intensity	$\bar{E}[V/m]$	$\bar{E}[V/m]$	$\bar{H}[A/m]$
Flux density	$\bar{D}[C/m^2]$	$\bar{J}[A/m^2]$	$\bar{B}[T]$
Polarization	$\bar{P}[C/m^2]$	$\bar{P}[C/m^2]$	$\bar{M}[A/m]$
Scalar potential	$\Phi[V]$	$\Phi[V]$	$\Psi[A]$
Linear law coefficient	$\varepsilon[F/m]$	$\sigma[S/m]$	$\mu[H/m]$
Clausius–Mossotti function	$\dfrac{\varepsilon_p - \varepsilon_m}{\varepsilon_p + 2\varepsilon_m}$	$\dfrac{\sigma_p - \sigma_m}{\sigma_p + 2\sigma_m}$	$\dfrac{\mu_p - \mu_m}{\mu_p + 2\mu_m}$

Source: Jones, T. B., *Electromechanics of Particles*. Cambridge University Press, New York, 1995. With permission.

3. MAP force is directed along the gradient of the magnetic field intensity ∇H_0^2.

4. MAP force depends on the magnitude and sign of the Clausius–Mossotti function K, with permeabilities μ_p and μ_m replacing permittivities ε_p *and* ε_m.

Statements 3 and 4 indicate that areas of higher magnetic field intensity may attract or repel magnetizable particles depending on the permeabilities μ_p and μ_m. Similar to the definition of positive and nDEP, Jones [11] defined the following:

- When $K > 0$ (i.e., $\mu_p > \mu_m$), MAP is the positive MAP. Particles are attracted to the magnetic field intensity maxima and repelled from the minima.

- When $K < 0$ (i.e., $\mu_p < \mu_m$), MAP is the negative MAP. Particles are attracted to the magnetic field intensity minima and repelled from the maxima.

12.3 Applications of DEP

DEP has many applications. One of the most important applications of DEP is particle manipulation. As discussed in the section "Theory," the principle is based on the fact that each particle has unique frequency-dependent dielectric properties, which are different from that of other particles. The magnitude and the direction of the DEP force affecting a given particle depend on the conductivity and permittivity of the suspending medium and the frequency and magnitude of the applied field. Therefore, differences in the dielectric properties of particles manifest themselves as variations in the DEP force magnitude or direction, resulting in separation of particles.

As discussed in the section "Theory," DEP force largely depends on the electrode geometries. According to the electrode geometries, current DEP devices exploited for micro- or nanoscale particle manipulation can be categorized as polynomial electrodes, electrode arrays, electrodeless devices, and three-dimensional electrode devices. Each has distinct advantages and drawbacks in particle manipulation.

12.3.1 Polynomial Electrodes

Common polynomial electrodes are used to generate nonuniform electric field. As shown in Figure 12.5, four electrodes are used to produce a quadrupole geometry in which the electrodes are offset by 90°. Voltage is applied by wiring the electrodes diagonally opposite one another identically. Once voltage is applied after sample injection, particles experiencing pDEP move to the electrode edges, and those experiencing nDEP move to the center [9].

The device based on this design was used to separate tobacco mosaic virus (TMV) and herpes simplex virus (HSV) [9]. The result was shown in Figure 12.6. As observed, the HSV was trapped under nDEP force at the field minimum in the center of the electrode array, whereas TMV experienced pDEP and was collected at the high-field

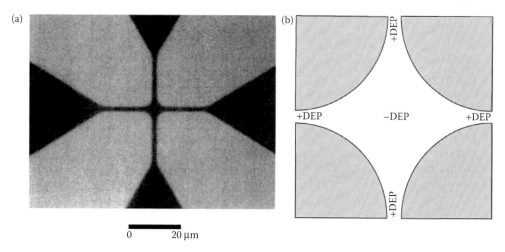

FIGURE 12.5
DEP device on the basis of polynomial electrodes. (a) Photograph of the structure of polynomial electrodes. (b) Schematic graph of generated pDEP and nDEP.

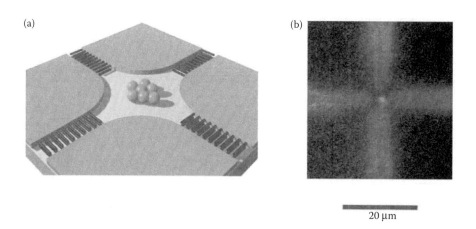

FIGURE 12.6
A photograph and a diagram of the separation of TMV and HSV using polynomial electrodes. The schematic is illustrated in panel a, and the photograph appears in panel b. The TMV (labeled with rhodamine B) can be seen as a red glow in the arms of the electrodes, and the green/yellow HSV (labeled with NBDdihexadecylamine) is visible in the center of the electrode. Both viruses were suspended in an electrolyte of conductivity 10 mS m−1, and the applied potential was 5 V peak-to-peak at a frequency of 6 MHz. (From Morgan, H., Hughes, M.P., and Green, N.G., Separation of submicron bioparticles by dielectrophoresis, *Biophys. J.*, 77, 516–525, 1999. With permission.)

regions, that is, the electrode edges. This results in the physical separation of two particle types.

The main drawback of this method is that particles cannot be easily collected for further analysis.

12.3.2 Electrode Arrays

A parallel track electrode array belongs to the category of electrode arrays. It is a popular choice for generating DEP effects because of its simplicity in construction. A parallel track electrode array is capable of separating, transporting, and trapping particles [27,30]. The electrodes of the parallel track array are rectangular in shape and identical in width. In most cases, the gap between the electrodes is identical in size to the width of an electrode. Because the electrodes are rectangular and parallel to one another, the electric fields in the gap spaces remain constant along the length of the electrodes. Lin and Yeow [8] proposed a novel electrode geometry that enhanced the DEP in 2007. In their work, triangular electrodes were used in the array instead of the usual rectangular electrodes (see Figure 12.7a). The goal of the triangular-shaped electrodes was to create bands of distinct electric field strengths (see Figure 12.7b). Because DEP is a phenomenon of nonuniform electric fields, changes in the pattern of electric fields caused by the difference of the electrode geometry will have significant effects on the behavior of the particles.

According to their results, different particles lined up at different locations, and this could be used for cell separation. Figure 12.8 shows the result of two kinds of particles lined at the different locations and application of potential for separating different bioparticles.

Lin and Yeow [8] also examined the TWD effect. The result, shown in Figure 12.9, demonstrates that the TWD effect moved the particles across the electrodes without breaking the formed lines, if a system was designed without flowing fluids. To move the line of microbeads into a smaller channel, a simple task of extending the underlying electrodes needs to be performed. In addition, DEP signals can be easily switched to TWD signals

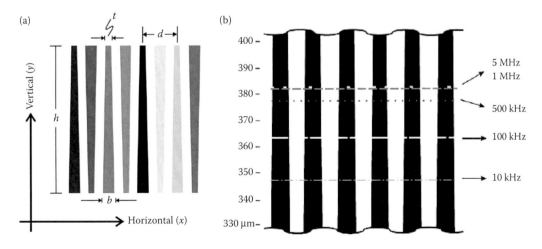

FIGURE 12.7
Relative positions of the microbead bands in a KCl medium. (a) Configuration of a triangular-electrode array. (b) Conceptual drawing of processing particles in multiple lanes. (From Mason, E.A. and Malinauskas, A.P., *Gas Transport in Porous Media: The Dusty Gas Model*, pp. 33, Copyright 1983, Elsevier. With permission.)

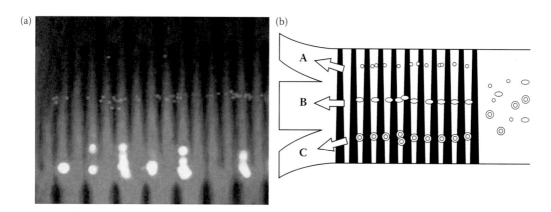

FIGURE 12.8
Microbead experiment results and conceptual diagram of separation. (a) Large red-fluorescent microbeads and small green-fluorescent particles occupying distinct lanes. (b) Conceptual diagram of processing particles in multiple lanes. (From Mason, E.A., and Malinauskas, A.P., *Gas Transport in Porous Media: The Dusty Gas Model*, pp. 33, Copyright 1983, Elsevier. With permission.)

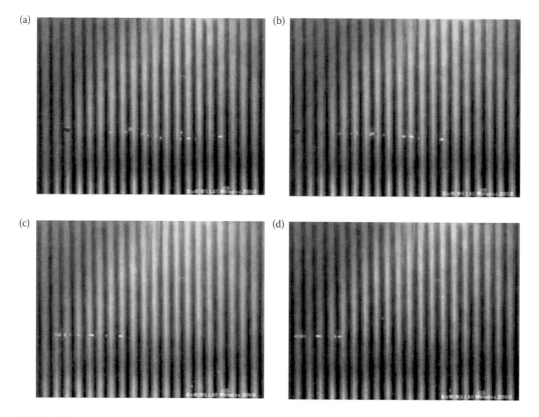

FIGURE 12.9
Microbeads moving to the left when TWD signals are applied. (a–d) Progress of the horizontal movement of microbeads. (From Lin, J.T.-Y., Cell manipulations with dielectrophoresis. Master's thesis, University of Waterloo, ON, Canada, 2007.)

by adjusting the phase differences between the sinusoidal waves without incurring additional complexities.

12.3.3 Electrodeless Devices

In electrodeless devices, insulating posts, such as glass, compress passing electric fields to generate pDEP zones. Because such devices lack metallic electrodes, they are suitable for metal-sensitive biosamples [5,6].

In 2007, Chang et al. [5] designed an electrodeless DEP (EDEP) chip to separate the microparticles into different sizes, and the chip also had the capability of separating the real samples [6]. The DEP force was generated and controlled by providing a nonuniform electric field, which was constricted by the insulator-patterned chip in combination with an AC electric field at frequency 10 kHz and 500 V peak-to-peak set on the two sides of the channel inlet (see Figure 12.10). The EDEP chip was designed to separate the *Escherichia coli* and red blood cells from human whole blood sample via well-controlled DEP force.

Results in Figure 12.11 show that the bacteria and red blood cells could be separated into the higher and lower electric field regions of the EDEP chip in a few seconds. Therefore, a rapid, useful diagnosis tool based on the EDEP method could be applied to the various fields of the bioindustry technology and to the detection and the identification of clinical infections.

12.3.4 Three-Dimensional Electrodes

Cetin et al. [7] designed a novel and simple LOC device for continuous separation of particles by their sizes on the basis of AC-DEP in 2009. The nonuniform electrical field was generated by means of embedded, three-dimensional, asymmetric electrodes inside the LOC device. Particles with different sizes from inlet reservoirs, A and B, were collected at the different exit reservoirs, C and D. Main flow was induced by pressure difference between the inlet and the exit reservoirs. The device could successfully separate the 5- and 10-μm latex particles, yeast cells, and white blood cells as well (see Figures 12.12 and 12.13).

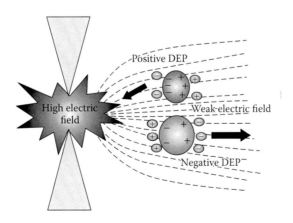

FIGURE 12.10
Electric field lines are focused at the constriction in the electrodeless DEP chip. (From Chang, H.C., Chen, C.H., Cheng, I.F., and Lin, C.C., Manipulation of bioparticles on electrodeless dielectrophoretic chip based on AC electrokinetic control. In 2007 2nd IEEE International Conference on Nano/Micro Engineered and Molecular Systems, pp. 4–10, Bangkok, Thailand, 2007. With permission.)

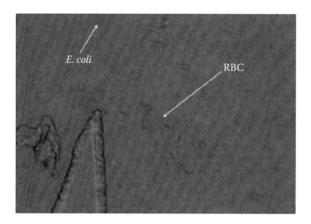

FIGURE 12.11
The *E. coli* and blood cell separated from mixture samples. The applied voltage was 500 V peak-to-peak/mm at frequency 10 kHz. The *E. coli* was trapped at the tip by pDEP, and the blood cells were repelled away from the gap by nDEP. (From Chang, H.C., Chen, C.H., Cheng, I.F., and Lin, C.C., Manipulation of bioparticles on electrodeless dielectrophoretic chip based on AC electrokinetic control. In 2007 2nd IEEE International Conference on Nano/Micro Engineered and Molecular Systems., pp. 4–10, Bangkok, Thailand, 2007. With permission.)

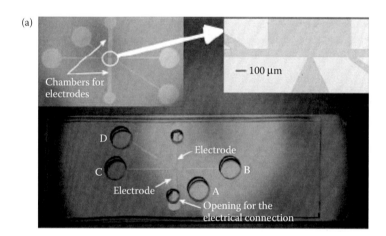

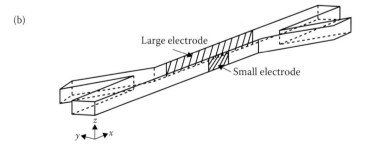

FIGURE 12.12
The design of the AC-DEP chip. (a) AC-DEP chip. (b) Computational domain. (From Cetin, B., Kang, Y.J., Wu, Z.M., and Li, D.Q., Continuous particle separation by size via AC-dielectrophoresis using a lab-on-a-chip device with 3-D electrodes. *Electrophoresis*, 30(5), 766–772, 2009. With permission.)

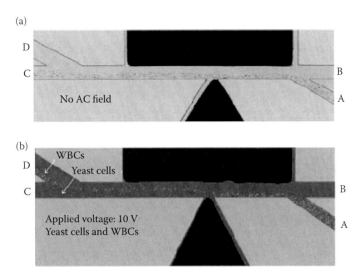

FIGURE 12.13
The comparison of the motion of cells with and without applied electric field. (a) Particle trajectories without electrical field. (b) Separation of WBCs and yeast cells at 10 V. The small dots are the yeast cells, and the larger dots that are very clean are the WBCs. (From Cetin, B., Kang, Y.J., Wu, Z.M., and Li, D.Q., Continuous particle separation by size via AC-dielectrophoresis using a lab-on-a-chip device with 3-D electrodes, *Electrophoresis*, 30(5), 766–772, 2009. With permission.)

12.3.5 Summary of DEP

In this section, four popular and distinct shapes and structures of electrodes generating DEP are presented. The LOC devices based on DEP are demonstrated as well. Each of them has distinct advantages and drawbacks in particle manipulation.

For example, polynomial electrodes excel at generating wells of pDEP and nDEP for trapping cells [9]. In electrodeless devices, insulating posts compress passing electric fields to generate pDEP zones. Because such devices lack metallic electrodes, they are suitable for metal-sensitive biosamples [5,6]. The parallel track electrode array, from the category of electrode arrays, is a popular choice for generating DEP effects because of its simplicity in construction [8,28]. The three-dimensional electrode is a novel and simple LOC device for continuous separation of particles by their sizes on the basis of AC-DEP. Some successful strategies and recognized conclusions are also described in this section.

12.4 Applications of MAP

As mentioned in Section 12.1, MAP has been applied vastly in micro- and nanoparticle manipulation. However, there are two pivotal factors that may affect the MAP application. The first challenge is how to obtain a desiderated magnetic field pattern, strength, and gradient over the confined area of a microfluidic and nanofluidic channel, especially how fast the flux density declines with the distance from the magnet's surface. For particle manipulation in microchannels, usually there are two approaches to generate the MAP force. One is that conventional permanent magnets or electromagnets are allocated outside the microdevices, and another is that permanent magnets or electromagnets (i.e., microcoils) are embedded into the microdevices. The former approach, obviously, has the advantages of easy fabrication and low cost. The latter allows us to control the strength and the pattern of the magnetic field in a confined area and also can make the magnets lie closer to the microchannels than the former does. The second challenge is how to choose manipulated materials.

12.4.1 Selection of Magnets

Permanent magnets used in micro- and nanofluidic applications are usually small neodymium iron boron magnets. Their magnetic flux densities are up to 500 mT at the pole surface. This allows the manipulation of magnetic particles or cells inside a microchannel, even when the magnets are positioned several millimeters far from the microchannel [31].

12.4.1.1 External Magnets

Because of the simplicity and low cost of fabrication and assembly, external magnets still are a common approach exploited for particle manipulation. External magnets may be conventional permanent magnets or electromagnets. Sizes of the external magnets can be selected according to the application.

For example, Wang et al. [19] reported that a permanent magnet contributed to a new, efficient in-droplet magnetic particle concentration and separation method, in which magnetic particles are concentrated and separated into a split droplet. For designing a portable blood-borne magnetic sphere separator, an array of a biocompatible capillary tubing and magnetizable wires is placed in an external magnetic field that is generated by two permanent magnets [23].

Jegatha et al. [24] presented a combined method of DEP and MAP using a small but strong permanent neodymium iron boron (NdFeB) to generate magnetic field within the

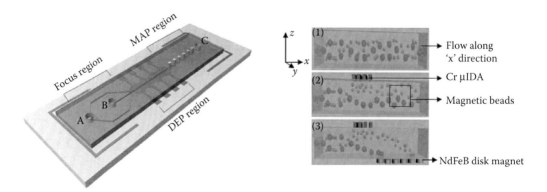

FIGURE 12.14
(a) Integrated microfluidic multiplex detection system comprising three distinct regions namely focus region, DEP region, and MAP region. "A" refers to the buffer solution inlet, "B" refers to the bead solution inlet, and "C" refers to the outlet. (1) without the µIDA and the magnet, sample solution containing two different-sized beads flow homogenously, (2) with only µIDA, beads experiencing pDEP flow near the electrode, and those experiencing nDEP flow away from the electrode, and (3) with both µIDA and the magnet, the sorted beads get trapped at two different magnets/locations within the same microchannel. (From Jegatha, N.K., Choong, K., Hyun, J.P., Ji, Y.K., Tae, S.K., and Sang, K.K., Rapid microfluidic separation of magnetic beads through dielectrophoresis and magnetophoresis, *Electrophoresis*, 30(9), 1457–1463, 2009. With permission.)

microfluidic channel [24]. These magnets were cylindrical with a diameter of 1 mm and a length of 1 mm (see Figure 12.14a). The results are shown in Figure 12.14b.

12.4.1.2 Embedded Permanent Magnet

To precisely control the magnetic field and to reduce the demands for magnetic field strength, researchers started embedding permanent magnets into microfluidic and nano-fluidic devices. For example, a soft magnetic film (CoNiP) is deposited on a substrate to embed the magnets into a microchip [32]. Hiroyoshi et al. designed a device that consists of a reaction chamber channel and two magnet-handling channels for the manipulation of microdroplets containing magnetic beads. Two magnet-handling channels contain embedded permanent magnets (see Figure 12.15). The results can be observed in Figure 12.16. Figure 12.16a shows the schematic of capture amplified DNA on magnetic beads in the droplet. Figure 12.16b shows the effect of the amount of streptavidin-conjugated magnetic beads for the capture of amplified DNA. After PCR was performed using the biotinylated primer, the streptavidin-conjugated magnetic bead mixture was manipulated into the PCR mixture for collection of amplified DNA. The adequate volume of magnetic beads was determined from the tube experiment, and DNA concentration was quantified using PicoGreen. Figure 12.16c shows the results of capturing DNA using 50 µg of streptavidin-conjugated magnetic beads [18].

12.4.1.3 Embedded Electromagnets

For achieving a built-in and time-varying magnetic field, microcoils are fabricated into the microdevice. The common geometry patterns of microcoils are single- or multiloop coils of different shape [17], and the arrays integrated by wires or single- or multiloop coils [13,33–35].

Ramadan and his group have done much research on the fabrication of embedded coils for manipulating magnetic particles and have published the results [13, 34–36]. Now, Ramadan's group can customize the design and fabrication of microcoils to obtain a large magnetic field gradient (up to 20 T/mm) and large magnetic forces (a range of 10^{-8} Newton

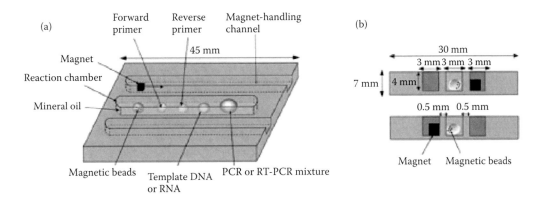

FIGURE 12.15
Schematic of the embedded permanent magnet within micro devices. Perspective view (a) and cross-sectional view (b). The device consists of a reaction chamber and two magnet-handling channels. (From Tsuchiya, H., Okochi, M., Nagao, N., Shikida, M., and Honda, H., On-chip polymerase chain reaction microdevice employing a magnetic droplet-manipulation system, *Sensors and Actuators B-Chemical*, 130(2), 583–588, 2008.)

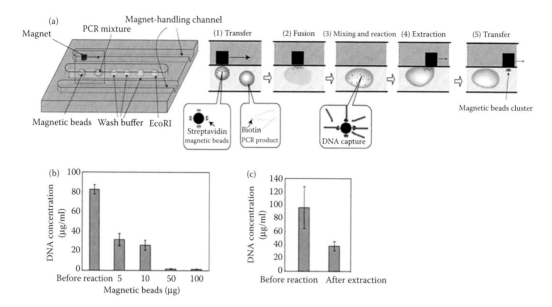

FIGURE 12.16
Capture and purification of amplified DNA using magnetic beads. (a) Schematic showing the capture of amplified DNA on magnetic beads in the droplet. (b) The effect of the amount of streptavidin conjugated magnetic beads for the capture of amplified DNA. After PCR was performed using the biotinylated primer, the streptavidin-conjugated magnetic bead mixture was manipulated into the PCR mixture for collection of amplified DNA. The adequate volume of magnetic beads was determined from the tube experiment, and DNA concentration was quantified using PicoGreen. (c) The results of capturing DNA using 50 μg of streptavidin-conjugated magnetic beads in the experiment. (From Tsuchiya, H., Okochi, M., Nagao, N., Shikida, M., and Honda, H., On-chip polymerase chain reaction microdevice employing a magnetic droplet-manipulation system, *Sensors and Actuators B-Chemical*, 130(2), 583–588, 2008.)

on magnetic particles with a diameter of 1 μm). For a summary of the work of ramadan, see Reference [37] (see Figure 12.17).

12.4.1.4 Hybrid Systems

All of the examples discussed above have a single functionality. To obtain a rapid and precise particle manipulation device, researchers have integrated different functionalities onto a single device.

Lee et al. [37] and Liu [38] presented an integrated circuit (IC)/microfluidic hybrid system for magnetic manipulation of biological cells. The hybrid system consists of an IC and a microfluidic system fabricated on top of it. The IC contains a microcoil array circuit that produces spatially patterned microscopic magnetic fields (Figure 12.18). This allows an efficient simultaneous manipulation of multiple individual bead-bound cells with precise position control (Figure 12.19).

12.4.2 Manipulated Materials

12.4.2.1 Magnetic Beads

Generally, magnetic beads are used in micro- and nanofluidic devices. The size of these beads ranges from a few nanometers to many micrometers. Magnetic beads are a valuable

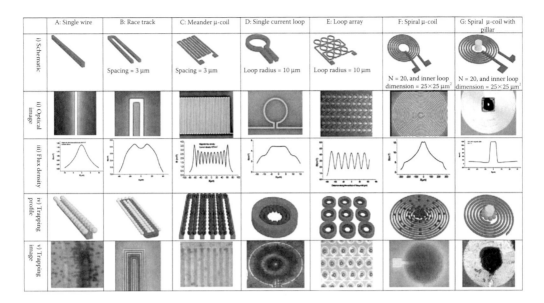

FIGURE 12.17
Summary of the microfabricated coils used for magnetic-particles assembly experiments. All the microcoils are made of a conductor with a cross-sectional area of 10^{-10} m^2 (width = 3 μm, depth = 10 μm). Row 1: a schematic view of the microcoil; row 2: optical image of the microcoil; row 3: the simulated magnetic flux density profile generated by the microcoil; row 4: schematic of the magnetic bead trapping profile because of the magnetic force generated by the microcoil and; row 5: optical image of the magnetic bead trapping by the microcoil at a current intensity of 100 mA, taken after 10 ~ 30s after supplying current to the microcoil. (From Ramadan, Q., Yu, C., and Poenar, D.P., Customized trapping of magnetic particles, *Microfluidics and Nanofluidics*, 6(1), 53–62, 2009.)

tool that is now being applied to various analytical applications, including immunoassays [12,39,40], genetic analysis [41,42], imaging, and drug delivery [43,44]. The majority of these particles are superparamagnetic and are available with carboxyl groups or amino groups on their surface [31].

12.4.2.2 Diamagnetic Objects

As observed in Equation 12.4, it is also possible to manipulate diamagnetic objects with magnetic fields, such that objects experience a force toward magnetic field minima, which makes levitation [14] and trapping [15] possible.

In the study of Frenea-Robin et al. [15], a new method of trapping living cells with contactless diamagnetism was presented. Using a CoPt micromagnet array, regularly spaced magnetic traps were produced, in which cells were confined without any contact, under the effect of negative MAP. Diamagnetic yeast cells were used in an aqueous solution enriched with paramagnetic ions to obtain the result.

Chetouani et al. [14] demonstrated that diamagnetic bodies can be contactless, guided along magnetic grooves, or trapped in magnetic wells both in air and in liquids using diamagnetic levitation.

12.4.3 Summary of MAP

In this section, generation of MAP force by external magnets, embedded magnets, embedded electromagnets, and a hybrid system is presented. The LOC devices based on these

(a)
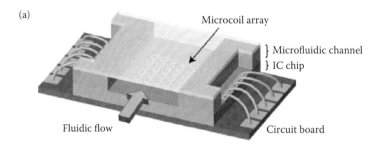

(b)
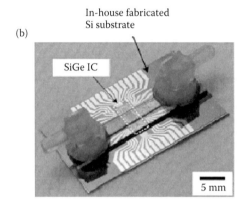

FIGURE 12.18
The structure of IC/microfluidic hybrid system. (a) Conceptual illustration of the IC/microfluidic hybrid system. (b) Image of the first hybrid prototype. (From Lee, H., Liu, Y., Westervelt, R.M., and Ham, D., IC/microfluidic hybrid system for magnetic manipulation of biological cells, *IEEE Journal of Solid-State Circuits*, 41(6), 1471–1480, June 2006. With permission.)

MAP forces are demonstrated as well. Each of them has distinct advantages and drawbacks in particle manipulations. For example, external magnets always have the advantages of ease of fabrication and low cost but lack precise control. Embedded magnets and electromagnets have more accurate control on MAP, but fabrication usually costs more. However, with the development of new IC and microelectromechanical systems (MEMS) fabrication equipment and the needs of LOC in point-of-care applications and in poor countries, hybrid systems will be promising. In addition, manipulated materials are briefly introduced as well.

12.5 Conclusion

DEP and MAP are promising technologies for LOC devices. These can be used to separate, transport, trap, and sort particles. The devices can be easily fabricated using the existing microelectrode or microcoil photolithography techniques. Particle manipulation is achieved by controlling the frequency, voltage, or current applied to microelectrodes or microcoils. The design of the electrodes and coils, the choice of the suspending medium, and the applied peak voltage or current can be predetermined to optimize the operation of the devices.

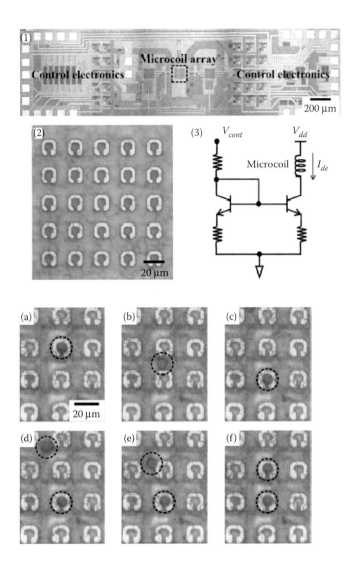

FIGURE 12.19
Hybrid system and its experiment results. (a) (1) Micrograph of the SiGe IC (4mm × 1mm). (2) Microcoil array close-up. The outer and inner diameters of each top-metal coil are 14 and 8 μm, respectively. The center-to-center distance between two adjacent coils is 25 μm. (3) Control electronics for each microcoil. (b) Manipulation of individual beads with the first prototype. (From Lee, H., Liu, Y., Westervelt, R.M., and Ham, D., IC/microfluidic hybrid system for magnetic manipulation of biological cells, *IEEE Journal of Solid-State Circuits*, 41(6), 1471–1480, June 2006. With permission.)

Many methods of particle manipulation based on DEP and MAP have been presented in the sections "Applications of DEP" and "Applications of MAP," respectively. However, most of them have potential drawbacks, which may limit their effectiveness unless conditions are controlled. The main drawback of DEP separation is that the high electric field strength causes electrohydrodynamic instabilities in the movement of the suspending liquid. Undoubtedly, if the strength of the electric field is too high, the electrode gaps are too small, and if the medium is too conductive, permanent damage to electrode devices as well as the formation of bubbles due to electrolysis at the electrode edges may occur. It can be imagined that the effect of Joule heating of the medium can denature the cells [10].

Furthermore, because LOC devices belong to a branch of MEMS, the lack of standardization and the existence of uncertainty in MEMS fabrication are also the obstacles to the commercialization of LOC devices.

However, despite these effects that act to counter the effectiveness of DEP or MAP manipulating techniques, DEP and MAP have demonstrated to be effective for a wide range of particle manipulation by a number of different methods, as discussed in the sections "Applications of DEP" and "Applications of MAP." This fact indicates that the intrinsic advantages of DEP and MAP preponderate the potential drawbacks described above. Those effects do not invalidate DEP- or MAP-based particle manipulation, but those must be considered and controlled well during any experimental design.

With the increase of healthcare cost in the world, new health technologies for point of care systems and for use in developing countries are highly demanded. The LOC based on DEP or MAP is well positioned to contribute to this challenge.

The next phase of the research in DEP and MAP would most likely be focused on the integration of these individual manipulation techniques to form a complete lab-on-a-chip, in which DEP and MAP can be used to transport and separate particles. It can be expected that more and more devices based on DEP and MAP will be designed in the future.

References

1. Wang X, Yang J, Huang Y, Vykoukal J, Gascoyne PRC. Cell separation by dielectrophoretic field-flow-fractionation. *Anal Chem*. 2000;72:832–9.
2. Pohl HA. Some effects of nonuniform [electric] fields on dielectrics. *J Appl Phys*. 1958;29:1182–8.
3. Pohl HA. *Dielectrophoresis*. Cambridge: Cambridge University Press; 1978.
4. Hughes MP. Micro- and nanoelectrokinetics in medicine. *IEEE Eng Med Biol Mag*. 2003;22(6):32–40.
5. Chang HC, Chen CH, Cheng IF, Lin CC. Manipulation of bioparticles on electrodeless dielectrophoretic chip based on AC electrokinetic control. In 2007 2nd IEEE International Conference on Nano/Micro Engineered and Molecular Systems, Bangkok, Thailand; 2007. pp. 4–10.
6. Chou CF, Zenhausem F. Electrodeless dielectrophoresis for micro total analysis systems. *IEEE Eng Med Biol Mag*. 2003;22:62–7.
7. Cetin B, Kang YJ, Wu ZM, Li DQ. Continuous particle separation by size via AC-dielectrophoresis using a lab-on-a-chip device with 3-D electrodes. *Electrophoresis*. 2009;30(5):766–72.
8. Lin JTY, Yeow JTW. Enhancing dielectrophoresis effect through novel electrode geometry. *Biomed Microdevices*. 2007;9(6):823–31.
9. Morgan H, Hughes MP, Green NG. Separation of submicron bioparticles by dielectrophoresis. *Biophys J*. 1999;77:516–25.
10. Hughes MP. Strategies for dielectrophoretic separation in laboratory-on-a-chip systems. *Electrophoresis*. 2002;23(16):2569–82.
11. Jones TB. *Electromechanics of Particles*. New York: Cambridge University Press; 1995.
12. Aytur T, Foley J, Anwar M, Boser B, Harris E, Beatty PR. A novel magnetic beads bioassay platform using a microchip-based sensor for infectious disease diagnosis. *J Immunol Methods*. 2006;314:21–9.
13. Ramadan Q, Yu C, Samper V, Poenar DP. Microcoils for transport of magnetic beads. *Appl Phys Lett*. 2006;88(3):32501.
14. Chetouani H, Jeandey C, Haguet V, Rostaing H, Dieppedale C, Reyne G. Diamagnetic levitation with permanent magnets for contactless guiding and trapping of microdroplets and particles in air and liquids. *IEEE Trans Magn*. 2006;42(10):3557–9.

15. Frenea-Robin M, Chetouani H, Haddour N, Rostaing H, LaforetJ, Reyne G. Contactless diamagnetic trapping of living cells onto a micromagnet array. Annual International Conference of the IEEE Engineering in Medicine and Biology Society; 2008. pp. 3360–3.
16. Pamme N, Eijkel JCT, Manz A. On-chip free-flow magnetophoresis: separation and detection of mixtures of magnetic particles in continuous flow. *J Magn Magn Mater*. 2006 Dec;307(2):237–44.
17. Ramadan Q, Samper V, Poenar D, Yu C. On-chip micro-electromagnets for magnetic-based biomolecules separation. *J Magn Magn Mater*. 2004;281(2–3):150–72.
18. Tsuchiya H, Okochi M, Nagao N, Shikida M, Honda H. On-chip polymerase chain reaction microdevice employing. A magnetic droplet-manipulation system. *Sens Actuators B Chem*. 2008;130(2):583–8.
19. Wang YZ, Zhao YJ, Cho SK. Efficient in-droplet separation of magnetic particles for digital microfluidics. *J Micromech Microeng*. 2007;17(10):2148–56.
20. Gascoyne PRC, Vykoukal JV. Dielectrophoresis-based sample handling in general-purpose programmable diagnostic instruments. *Proc IEEE*. 2004;92(1):22–42.
21. Pamme N, Wilhelm C. Continuous sorting of magnetic cells via on-chip free-flow magnetophoresis. *Lab Chip*. 2006 Aug;6(8):974–80.
22. Nicole Pamme. Magnetism and microfluidics. *Lab Chip*. 2006;6(1):24–38.
23. Chen HT, Bockenfeld D, Rempfer D, Kaminski MD, Rosengart AJ. Three-dimensional modeling of a portable medical device for magnetic separation of particles from biological fluids. *Phys Med Biol*. 2007;52(17):5205–18.
24. Jegatha NK, Choong K, Hyun JP, Ji YK, Tae SK, Sang KK. Rapid microfluidic separation of magnetic beads through dielectrophoresis and magnetophoresis. *Electrophoresis*. 2009;30(9):1457–63.
25. Chin CD, Linder V, Sia SK. Lab-on-a-chip devices for global health: past studies and future opportunities. *Lab Chip*. 2007;7:41–57.
26. Kang YJ, Li DQ. Electrokinetic motion of particles and cells in microchannels. *Microfluid Nanofluidics*. 2009;6(4):431–60.
27. Cen EG, Dalton C, Li Y, Adamia S, Pilarski LM, Kaler KVIS. A combined dielectrophoresis, traveling wave dielectrophoresis and electrorotation microchip for the manipulation and characterization of human malignant cells. *J Microbiol Methods*. 2004;58:387–401.
28. Lin, JT-Y. Cell manipulations with dielectrophoresis [master's thesis]. Ontario (Canada): University of Waterloo; 2007.
29. Arnold WM, Zimmermann U. Rotating-field-induced rotation and measurement of the membrane capacitance of single mesophyll cells of *Avena sativa*. *Z Naturforsch C*. 1982;37:908–15.
30. Cui L, Morgan H. Design and fabrication of travelling wave dielectrophoresis structures. *J Micromech Microeng*. 2000;10:72–9.
31. Nicole Pamme. Continuous flow separations in microfluidic devices. *Lab Chip*. 2007 Dec;7(12):1644–59.
32. Ramadan Q, Uk YS, Vaidyanathan K. Large scale microcomponents assembly using an external magnetic array. *Appl Phys Lett*. 2007;90(17):172502.
33. Ramadan Q, Samper V, Poenar D, Yu C. Magnetic-based microfluidic platform for biomolecular separation. *Biomed Microdevices*. 2006;8(2):151–8.
34. Ramadan Q, Samper V, Poenar DP, Yu C. An integrated microfluidic platform for magnetic microbeads separation and confinement. *Biosens Bioelectron*. 2006;21(9):1693–702.
35. Ramadan Q, Yu C, Samper VD, Puiu DP. Fabrication of three-dimensional magnetic microdevices with embedded microcoils for magnetic potential concentration. *J Microelectromech Syst*. 2006;15(3):624–38.
36. Ramadan Q, Yu C, Poenar DP. Customized trapping of magnetic particles. *Microfluid Nanofluidics*. 2009;6(1):53–62.
37. Lee H, Liu Y, Westervelt RM, Ham D. IC/microfluidic hybrid system for magnetic manipulation of biological cells. *IEEE J Solid-State Circuits*. 2006 Jun;41(6):1471–80.

38. Liu Y. *CMOS Magnetic Cell Manipulator and CMOS NMR Biomolecular Sensor*. Ann Arbor, MI: ProQuest. Vol. 69; 2008.
39. Bronzeau S, Pamme N. Simultaneous bioassays in a microfluidic channel on plugs of different magnetic particles. *Anal Chim Acta*. 2008;609(1):105–12.
40. Nam JM, Thaxton CS, Mirkin CA. Nanoparticle-based bio-barcodes for the ultrasensitive detection of proteins. *Science*. 2003;301:1884–6.
41. Akutsu J, Tojo Y, Segawa O, Obata K, Okochi M, Tajima H, et al. Development of an integrated automation system with a magnetic beadmediatednucleic acid purification device for genetic analysis and gene manipulation. *Biotechnol Bioeng*. 2004;86:667–71.
42. Traverso G, Dressman D, Yan H, Kinzler KW, Vogelstein B. Transforming single DNA molecules into fluorescent magnetic particles for detection and enumeration of genetic variations. *Proc Natl Acad Sci U S A*. 2003;100:8817–22.
43. Lee JH, Huh YM, Jun Y, Seo J, Jang J, Song HT, et al. Artificially engineered magnetic nanoparticles for ultrasensitive molecular imaging. *Nat Med*. 2007;13:95–9.
44. Lewin M, Carlesso N, Tung CH, Tang XW, Cory D, Scadden DT, et al. Tat peptide-derivatized magnetic nanoparticles allow *in vivo* tracking and recovery of progenitor cells. *Nat Biotechnol*. 2000;18:410–4.

13

Optoelectric Particle Manipulation

Aloke Kumar, Stuart J. Williams, Nicolas G. Green, and Steven T. Wereley

CONTENTS

13.1 Introduction

The last four decades have seen vigorous activity aimed at developing schemes for noninvasive manipulation of micro- and/or nanoparticles. Some of these techniques are optical, for example, optical tweezing [1], whereas some rely on alternating current (AC) electrokinetics or thermal gradients, for example, dielectrophoresis (DEP) [2] and thermophoresis [3–5]. The high spatial resolution and the dynamic nature of optical tweezers have been a key to their popularity, and the optical tweezing technique has been used in capturing particles ranging from atoms to micron-sized cells [1]. On the other hand, the ease of microfabrication today has made the incorporation of AC electrokinetic techniques in lab-on-a-chip (LOC) systems attractive. The versatility and the robustness of both these techniques have made them powerful and prevailing tools for a variety of applications requiring noninvasive particle manipulation. Sustained research activity in these fields has seen these techniques develop into entire research areas in their own right.

Innovations in the last decade [6–9] have sought to combine the advantages of optical and electrical manipulation, resulting in the creation of various optoelectric techniques. As the name suggests, optoelectric techniques use both illumination and electric fields for particle and/or fluid manipulation at the micro- and nanoscale. These techniques typically use light to activate and control electrokinetic particle manipulation mechanisms. The simultaneous presence of light and electric fields can result in the appearance of a physical phenomenon unique to these optoelectric systems, such as laser-induced electrothermal flows [10] and light-actuated AC electro-osmosis [11], when compared with traditional AC electrokinetic systems. Moreover, although the use of optical means introduces additional experimental parameters, it also bestows a higher degree of freedom and a dynamic nature upon these optoelectric techniques. Thus, optoelectric techniques define a new class of manipulation methodology, and there is a need to understand the principles of these techniques.

Figure 13.1 categorizes noninvasive manipulation tools into different classes, one such class being optoelectric techniques. Optoelectric techniques refer to a set of different methodologies that have been proposed over the last decade by different research groups.

13.1.1 Various Optoelectric Techniques

One of the first innovations in the field was led by Hayward et al. [8] in 2000, who electrokinetically assembled optically tunable colloidal crystals. They used a simple parallel plate electrode setup, in which a brass electrode served as a cathode and an indium tin oxide (ITO) film served as the anode. Ultraviolet light irradiated the surface of the ITO film after passing through a patterned mask. The microchannel was filled with a fluid containing a dilute suspension of polystyrene (PS) microspheres. The researchers observed that over

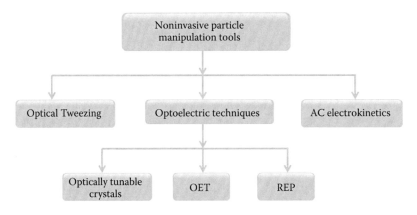

FIGURE 13.1
A chart classifying the newly developed optoelectric techniques.

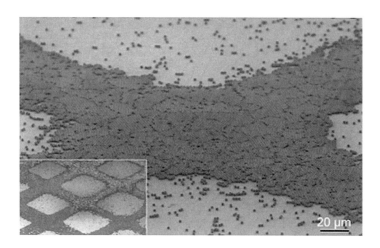

FIGURE 13.2
The assembly of PS beads on an ITO surface using the method outlined by Hayward et al. [8]. (Reprinted with permission from Macmillan Publishers Ltd: *Nature*, Hayward, R.C., Saville, D.A., and Aksay, I.A., Electrophoretic assembly of colloidal crystals with optically tunable micropatterns, *Nature* 404(6773), 56–59, copyright 2000.)

time, the PS beads aggregated on the ITO surface, under the influence of an applied DC field, and the aggregation resulted in dense packed assemblies in patterns similar to the pattern of ultraviolet irradiation (Figure 13.2). It was hypothesized that the illuminated regions on the ITO surface had a higher current density, and research has shown that such a higher current density can lead to spatially selective aggregation [8].

Following this work, Gong and Marr [7] showed that laser-scanned patterns could also be used to create crystals with desired shapes. They showed the existence of poly-crystallinity in the assembled structures, and they showed that not only can nucleation be directed but also crystal annealing and melting can be controlled. Gong and Marr [7] biased the electrodes with an AC electric field (unlike Hayward et al. [8]), and they were able to achieve significantly lower crystal formation times than their predecessors. Before this work, the same research group had established the utility of a simple parallel electrode-based setup in the achievement of tunable crystalline assemblies [12,13].

Another optoelectric technique is the "optoelectronic tweezer" (OET), which was introduced by Chiou et al. [14] in 2003. Figure 13.3 illustrates a typical OET setup. Two substrates are biased with an AC electric field, and one of the substrates in OET is coated with hydrogenated amorphous silicon (a:Si-H), which is a strong photoconductor in the visible spectrum. The a:Si-H-coated substrate is illuminated with light patterns; the illuminated regions can be regarded as "virtual electrodes." Particles can be subsequently manipulated in this platform using DEP. OET has also demonstrated the ability to manipulate particles using other AC electrokinetic forces [15], such as light-actuated AC electro-osmosis and electrothermal flows. Many different applications of OET have been demonstrated including rapid and selective capturing [16] and patterning of nanoparticles [15].

The newest optoelectric manipulation technique is termed rapid electrokinetic patterning (REP). Williams et al. [9] introduced REP in 2008 as a real-time dynamic particle manipulation technique. This technique used an ITO-coated parallel-electrode setup together with an AC electric field and an optical landscape. They showed that in a low AC-frequency

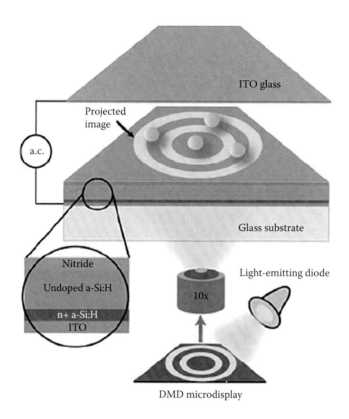

FIGURE 13.3
A typical OET setup. (Reprinted with permission from Macmillan Publishers Ltd: *Nature*, Chiou, P.Y., Ohta, A.T., and Wu, M.C., Massively parallel manipulation of single cells and microparticles using optical images, *Nature* 436(7049), 370–372, copyright 2005.)

region (<200 kHz), particles could be trapped and dynamically configured into various shapes on the electrode surface by using various optical landscapes produced from a laser operating at a wavelength of 1064 nm. The particle sizes that were successfully manipulated ranged from 300 nm to 3.0 µm (later extended to 50 nm PS beads [17]). REP is primarily an electrokinetic technique, in which illumination in the presence of an electric field induces an electrothermal microvortex. This microvortex transports particles to the illuminated locations on the electrode surface, resulting in rapid particle aggregation (Figures 13.4 and 13.5). Aggregations in REP are local to the site of illumination, and thus a changing optical landscape lends a dynamic nature to REP.

13.1.2 Overview

We have seen different optoelectric techniques, which differ from each other in not just implementation but also in the underlying physics. In the following text, we will discuss the various aspects of REP. We will first introduce the fundamental principles of a typical REP setup. In the section "Characteristics of REP," we will provide an overview of some of the characteristics of REP, and in the section "Mechanisms of REP," we will see how a variety of mechanisms work in concert to enable this particle manipulation methodology. A few applications on the basis of REP are reviewed for enhancing the readers' understanding.

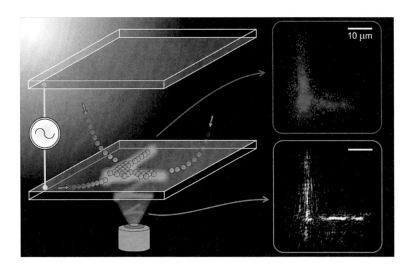

FIGURE 13.4
(See color insert.) An example of the REP manipulation technique. An intense laser hologram is shone on an ITO surface, and the 690-nm red fluorescent PS beads take the shape of the hologram. (From Williams, S.J., Kumar, A., and Wereley, S.T., Electrokinetic patterning of colloidal particles with optical landscapes, *Lab on a Chip* 8(11), 1879–1882, 2008. Reproduced by permission of The Royal Society of Chemistry.)

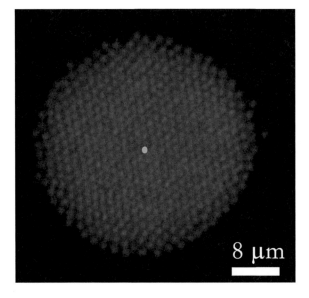

FIGURE 13.5
(See color insert.) Hundreds of red fluorescent 1-μm particles form an REP aggregation on an electrode surface. The green dot in the middle shows the approximate location and size of the focused laser spot, which creates the REP aggregation.

13.2 Materials and Methods

In this section, we outline a simple fabrication procedure for achieving REP in a micro-fluidic device. It should be noted that only the simplest implementation is described, and other implementations are also possible.

13.2.1 Microfluidic Apparatus

The simplest implementation for an REP chip consists of parallel plate electrodes, which are separated by a spacer of microscopic dimensions. Typically, parallel plate electrodes can be easily manufactured by uniformly depositing an appropriate material such as ITO or gold on another substrate such as glass. The choice of the electrode substrate is dictated by the wavelength of optical illumination (see the section "Optical Assembly"). The electrodes are separated by an insulating spacer with microfluidic features (height typically 50–100 µm). The length of the channel can be in the order of few millimeters to allow for fluidic ports. For a substrate such as glass, fluidic ports can be created by drilling with diamond drill bits. An AC field is applied between the two electrodes. The AC frequency, f, is typically less than 200 kHz, and the AC voltage, V, for the above-described configuration would be typically less than 20 V_{pp} (volts peak-to-peak).

13.2.2 Optical Assembly

The purpose of this assembly is to provide the required illumination for REP to occur and also for normal viewing and/or imaging. For the microfluidic chip described in the section "Microfluidic Apparatus," the channel base and lid serve as electrodes. In the REP implementation described by Williams et al. [17], the electrode surfaces were illuminated with tightly focused infrared (1064 nm) laser holograms. A single microscope objective lens can serve for both focusing the laser and viewing of the sample. For the infrared wavelength, the electrodes can be made of ITO. The choice of the electrode material is dictated by the fact that ITO absorbs strongly in the infrared [18], and thus illumination by a highly focused 1064-nm laser would heat the electrode substrate [19]. Heating of the electrode substrate with a highly focused laser illumination leads to large temperature gradients, and these temperature gradients drive electrothermal flows in the presence of an electric field. Hence, if gold is used, then a green laser (532 nm) can be used instead to induce temperature gradients. Note that one of the electrodes necessarily needs to be fabricated from a transparent substrate such as ITO for optical access. The optical intensities typically used in REP are similar to those used in optical trapping. However, an important point of distinction between optical patterns used in REP and optical tweezing is that a high numerical aperture lens is not a necessity. At Purdue's Microfluidics Laboratory, REP has been successfully demonstrated using lenses ranging from a 20× lens to a 100× oil lens. Kumar et al. [10,20,21] and Williams et al. [9,10,17,20–23] have typically used a laser with a net power of approximately 20 mW (w.r.t. the back focal plane of the objective lens) for initiating electrothermal flows and REP aggregation.

13.2.3 Types of Particle

REP is initiated when a particle-laden liquid sample is introduced into the above-described microfluidic channel, and an AC bias is applied between the electrodes with the simultaneous illumination of the electrode surface with an appropriate optical landscape. REP is a versatile technique, and aggregation of particles as diverse as gold particles (200–250 nm), silica beads, and PS particles (50 nm–3.0 μm) has been successfully attempted at Purdue University's Microfluidics Laboratory. Attempts to extend REP for biological particles and other particle types such as quantum dots are ongoing.

These particles are usually suspended in an aqueous solution with electrical conductivity, σ, typically less than 0.1 S/m. Williams et al. [9] used an aqueous solution of very low conductivity ($\sigma \sim 1$ mS/m) in their first demonstration of REP.

13.3 Characteristics of REP

In REP, particle aggregations are manipulated on the surface of an electrode, and hence the technique is inherently two-dimensional in nature. Figure 13.6 shows a series of experimental images that were acquired when (a) both laser illumination and electric field were activated, (b) only the laser was deactivated, and (c) only the electric field was deactivated. Thus, we can see that both the electric field and the laser illumination are required for creating an REP aggregation. Particle aggregations are localized to the site of illumination, and hence although REP produces a large aggregation of particles, it has a "local" nature. The shape of the applied illumination determines the electrothermal microvortex flow, and thus the "local" REP aggregation resembles the shape of the applied illumination landscape. Figure 13.7 shows an "S"-shaped REP pattern, resulting from the accumulation of 690 nm PS beads. The "S" pattern results because of a slowly rotating line illumination. Thus, illumination in REP bestows a dynamic nature on it. Other features (such as number of particles, interparticle distance) of a "local" REP aggregation also depend on a number of different parameters such as AC voltage, AC frequency, illumination geometry, illumination intensity, illumination wavelength, optical absorptivity of the substrate, dielectric properties of the particle, and dielectric properties of the liquid medium.

Recently, Williams et al. [22] performed a characterization of rate of aggregation of 1.0-μm PS particles suspended in a low-conductivity aqueous solution ($\sigma \sim 2$ mS/m) as a function of applied AC voltage, frequency, and optical intensity. They measured the rate of aggregation of the PS particles and found that initially the rate of particle collection increases linearly (<20 s), and then it asymptotically approaches an equilibrium value (Figure 13.8a and b).

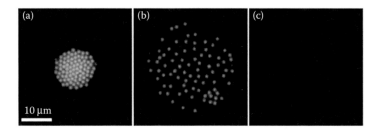

FIGURE 13.6

(See color insert.) (a) An REP aggregation of 1.0-μm PS beads in a low-conductivity electrolyte solution, created by activating infrared illumination and electric field; (b) the laser is subsequently deactivated. The particles are scattered parallel to the electrode surfaces, indicating that (c) the electric field is subsequently deactivated. (From Williams, S.J., Kumar, A., Green, N.G., and Wereley, S.T., Optically induced electrokinetic concentration and sorting of colloids, *Journal of Micromechanics Microengineering* 20, 2010. Reproduced with permission from IOP Publishing.)

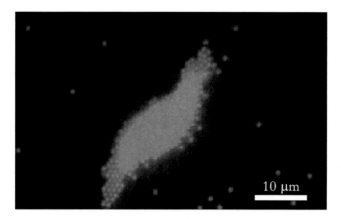

FIGURE 13.7
(See color insert.) An "S"-shaped pattern is formed by 690-nm red fluorescent PS beads. The "S" pattern resulted from a line optical illumination, which was being slowly rotated.

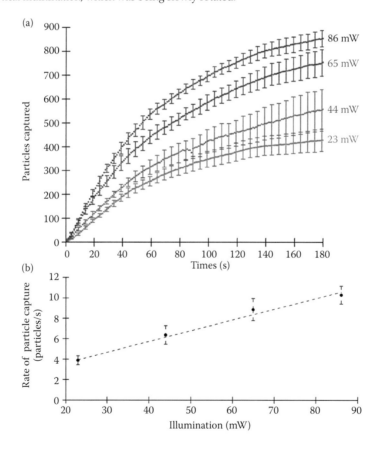

FIGURE 13.8
(a) The number of particles captured in an REP trap as a function of time, at different illumination intensities. The number of particles captured initially increases linearly and finally reaches an equilibrium value. The 1.0-μm PS beads were used. (b) The initial rate of particle capture shows a linear dependence on illumination intensity. (From Williams, S.J., Kumar, A., Green, N.G., and Wereley, S.T., Optically induced electrokinetic concentration and sorting of colloids, *Journal of Micromechanics Microengineering* 20, 2010. Reproduced with permission from IOP Publishing.)

The initial rate of particle capture increases with the applied AC voltage, and this dependence on voltage can be characterized by a second-order polynomial fit. This increase with voltage is expected as the electrothermal microvortex increases in strength with voltage (section "Optically Induced Electrothermal Hydrodynamics"), thus transporting more particles to the REP site. The dependence on AC frequency is more intricate as the behavior of different electrokinetic phenomena changes with AC frequency. Williams et al. [22] also characterized the change in the rate of concentration of particles with AC frequency (Figure 13.9). They investigated the aggregation formation kinetics from 25 to 110 kHz in steps of 15–25 kHz. They observed that above a certain AC frequency, no more REP aggregation could be observed. Such a thresholding phenomenon is caused by the three-dimensional nature of the electrothermal microvortex. At the threshold/ critical frequency, the particle–electrode interaction forces are overcome by the drag from the electrothermal microvortex, and the particles are carried away into the bulk by the microvortex [19].

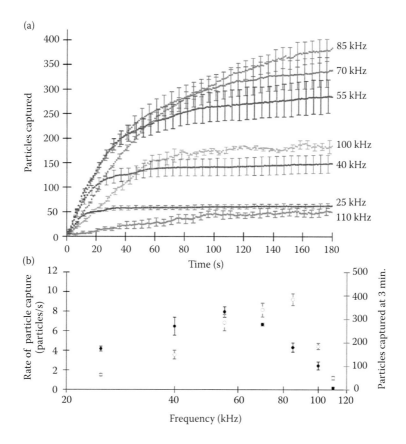

FIGURE 13.9
The number of particles captured in an REP trap as a function of time, at different AC frequencies. As the AC frequency is increased, a threshold or critical frequency is reached above which particle capturing is not observed; (b) as the threshold frequency is approached, the number of particles captured in an REP "trap" also decreases. (From Williams, S.J., Kumar, A., Green, N.G., and Wereley, S.T., Optically induced electrokinetic concentration and sorting of colloids, *Journal of Micromechanics Microengineering* 20, 2010. Reproduced with permission from IOP Publishing.)

13.4 Mechanisms of REP

Colloidal aggregation in REP is a consequence of many different forces acting in concert. Any behavior in REP needs to be understood in terms of three components: (i) colloidal forces that are independent of the applied illumination or electric field, (ii) purely electro-kinetic forces that are independent of laser illumination, and (iii) optically induced electro-thermal flows. Colloidal forces result from an interplay between the van der Waals forces and the electrostatic double-layer interaction, and these forces are ever present in any col-loidal system independent of any applied electric field. These forces can be understood in terms of the classical theory of Derjaguin, Landau, Verwey, and Overbeek [24,25]. Purely electrokinetic forces independent of any laser illumination can result from phenomena such as induced dipole interactions, induced charge electro-osmosis [26], nonequilibrium surface phenomenon [27], and electrohydrodynamic (EHD) forces [28–30]. When the laser is turned off and the particles scatter (Figure 13.1), only the first two types of forces dictate the behavior of the system. However, it is primarily the electrothermal vortex fluid flow that is responsible for the rapid transport of particles from the bulk to the illumination sites [22]. We will first describe optically modulated electrothermal forces and later discuss the nature of the electrokinetic forces.

13.4.1 Optically Induced Electrothermal Hydrodynamics

Fluid flow has been observed at frequencies of approximately 1 MHz and higher [31], where electrode polarization and AC electro-osmosis are negligible. This induced fluid flow is electrothermal in nature [32]. Thermal gradients within the fluid generate localized changes in fluid permittivity and conductivity. The electric field will act upon these dielec-tric gradients, resulting in fluid motion. Electrothermal fluid pumping can occur through Joule heating [31] or optically induced heating [33,34]. For the specific case of REP, a highly focused laser heats the electrode surface, generating sharp thermal gradients necessary to drive electrothermal hydrodynamics.

Williams et al. [35] investigated the heating effects of a highly focused 1064-nm laser on ITO electrode substrates. It was found that heating of the ITO electrode at 1064 nm was the primary source of temperature gradients leading to electrothermal flow. However, opti-cally induced heating is not restricted to this combination of laser wavelength and mate-rial. For example, silver readily absorbs 300 nm of light, gold 400–500 nm, silicon 600–1000 nm, or water itself absorbs in the infrared [36]. Therefore, one is not restricted to use the combination of ITO and 1064 nm to generate sharp thermal gradients.

Changes in temperature will modify the dielectric properties (permittivity and conductiv-ity) of the fluid. An applied electric field will act upon these gradients, resulting in electrother-mal fluid motion. For electrothermal flows, the time-averaged body force is given by [31]

$$\langle f_e \rangle = \frac{1}{2} \mathrm{Re}\left(\frac{\sigma \varepsilon (\alpha - \beta)}{\sigma + i\omega\varepsilon}(\nabla T \cdot E)E^* - \frac{1}{2}\varepsilon\alpha|E|^2 \nabla T \right), \tag{13.1}$$

where Re(...) denotes the real part of the expression, E is the electric field vector , E^* is its complex conjugate, T is the temperature, ω is the angular frequency of the applied electric field, σ is the fluid conductivity, ε is the fluid permittivity, and α and β are the fractional changes of ε and σ with temperature given by $1/\varepsilon$ (dε/dT) and $1/\sigma$ (dσ/dT), respectively.

The first term in Equation 13.1 is the Coulomb force, and the second term is the dielectric force; the former will dominate as the applied AC frequency decreases. Each expression is dependent on local temperature gradients (∇T). In REP, the heating is highly nonuniform because of the highly focused light patterns that vary radically in intensity over a few micrometers. The induced fluid velocity depends on the illumination intensity (thermal gradient), the dielectric properties of the fluid, and the applied AC signal (voltage and frequency). In REP, we assume that optically induced heating is the primary source of nonuniform temperature fields, and we neglect Joule heating. Thus, we see that the fluid velocity (v) scales as the square of the electric field ($|E|^2$).

In a parallel-electrode setup, as that used for REP, laser heating creates a three-dimensional microvortex. This electrothermal microvortex is responsible for the transport of particles from the bulk to the illuminated region. The three-dimensional nature of the electrothermal microvortex also plays a role in the appearance of a threshold or critical frequency. Thus, in REP, the induced electrothermal vortex plays an important role, and hence it is necessary to understand and characterize the nature of the optically induced electrothermal fluid motion.

13.4.1.1 Visualization of the Microvortex

Kumar et al. [10] observed an electrothermal microfluidic toroidal vortex when a uniform electric field is applied parallel to the optical axis (Figure 13.10). It was observed that the flow near an electrode surface resembled a "sink-type" flow (Figure 13.11). Figure 13.12 depicts particle path lines when 200 images were overlayed. Two zones can be noticed— a sink-type region external to the dashed circle and an interior region where path lines terminate. The "sink-type" region displays a predominantly planar flow, which decreases radially away from the center of the vortex. The region interior to the dashed circle is the region where the particles experience significant out-of-plane velocities, and thus particle path lines appear to terminate in this region's periphery. In their work, Kumar et al. [10] established the three-dimensional nature of the vortex and performed two-dimensional flow visualization using

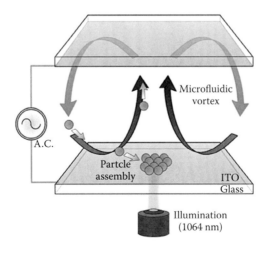

FIGURE 13.10
The electrothermal microvortex causes rapid particle transport to the illuminated site. (From Williams, S.J., Kumar, A., and Wereley, S.T., Electrokinetic patterning of colloidal particles with optical landscapes, *Lab on a Chip* 8(11), 1879–1882, 2008. Reproduced by permission of The Royal Society of Chemistry.)

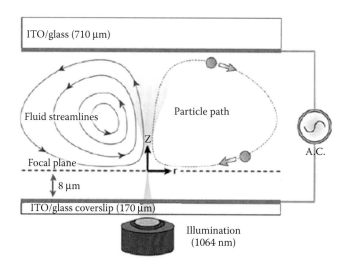

FIGURE 13.11
The setup used by Kumar et al. [10] to explore the nature of the electrothermal flow. (With kind permission from Springer Science + Business Media: *Microfluidics and Nanofluidics*, Experiments on opto-electrically generated microfluidic vortices, 6(5), 2009, 637–646, Kumar, A., Williams, S.J., and Wereley, S.T., Figure 4.)

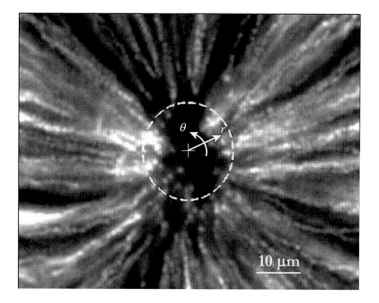

FIGURE 13.12
Particle path lines obtained in the plane close to the bottom electrode surface. The inward radial flow resembles a sink-type flow exterior to the dashed circle. Interior to the dashed circle, out-of-plane velocities cause the path lines to abruptly terminate.

micron-resolution particle image velocimetry in the region external to the dashed circle. The flow itself is highly three-dimensional in nature, and the presence of large velocity gradients presents exceptional challenges to the prevalent three-dimensional flow visualization technique. The three-dimensional structure of the flow is easily visualized with the help of larger particles (~3 µm). Particles suspended in the medium will follow the fluid

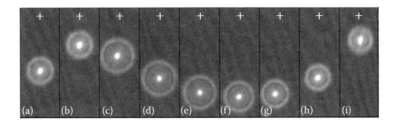

FIGURE 13.13
A 3-μm particle caught in a closed orbit. The "+" represents the laser focus. The diffraction ring patterns can give insight into the three-dimensional nature of the electrothermal microvortex.

streamlines of the vortex. These particles will periodically translate in and out of focus; as they move away from the focal plane, a ring pattern is produced, resulting from diffraction and spherical lens aberration (Figure 13.13). If a diffraction-based extension of the standard micron-resolution particle image velocimetry system is used [37], then the diffraction ring pattern can yield insight into a particle's distance from the focal plane.

The nature of the electrothermal microvortex is of singular importance in REP. The sink-type flow is responsible for rapid particle transport to the illuminated site, and optical modulation of the vortex gives REP a dynamic nature. We will see later that the three-dimensional nature of the vortex also plays an important role in selective particle capture by REP.

13.4.2 Other Mechanisms of REP

The optically modulated electrothermal microvortex plays a constructive role in REP. However, other AC electrokinetic forces such as particle–particle and particle–electrode forces also play an important role (Figure 13.14). Although these forces have been known to exist for quite some time, a comprehensive understanding of these forces is still lacking. We will now undertake a short review of these forces.

13.4.2.1 Electrokinetic Forces Transverse to the Electric Field

Figure 13.14 illustrates a cross-sectional view of the REP aggregation. The electrothermal drag provides drag forces on the particles, both parallel (i.e., transverse to the direction of the applied electric field) and normal to the electrode surface. It has been observed by several researchers that colloidal particles near the electrode surfaces can exhibit motion parallel to the electrode surface, by the simple application of only an applied electric field (both AC and DC) [12,13,28–30,38–46]. Particle clustering leading to densely packed assemblies and other forms of two-dimensional colloidal phases [13] can result from such transverse motion. Because electrostatic interactions between identical particles are expected to be repulsive, such particle clustering was unexpected. Various theories have been proposed to explain such a clustering effect [29,30,45]. Solomentsev et al. [45] argued that particle clustering arises primarily because of localized electro-osmotic flow caused by charges on the particles. However, that study focused on DC electric fields. Trau et al. [46] proposed an EHD model involving bulk-charge densities as the driving mechanism of fluid flow. The model was useful in theoretically explaining the findings by other researchers [28,44]. Ristenpart et al. [28] predicted that these EHD forces scale with the square of the electric field strength and the inverse of AC frequency. A somewhat different mechanism was proposed by Yeh et al. [30]. Their work suggested an

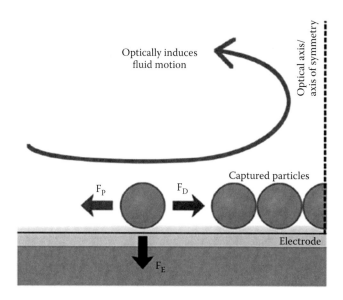

FIGURE 13.14

Drag from the electrothermal microvortex assists in REP. However, interparticle and particle–electrode interactions are also important. (From Williams, S.J., Kumar, A., Green, N.G., and Wereley, S.T., Optically induced electrokinetic concentration and sorting of colloids, *Journal of Micromechanics Microengineering* 20, 2010. Reproduced with permission from IOP Publishing.)

EHD model based on particle-induced distortion of the local electric field. They argued that such particle-induced distortion of the electrical double layer (EDL) would lead to hydrodynamic flow that would promote particle aggregation. This model was supported experimentally by Nadal et al. [47], demonstrating a frequency-dependent nature of aggregation and particle–particle distance. It was later recognized that at very low AC frequencies (<500 kHz), Faradaic currents can play an important role in such particle clustering, and such phenomenon was later incorporated [48–51]. Moreover, the situation is complicated by the fact that an applied electric field can act on its own diffuse charge, causing fluid flow [26]. Such effects can exhibit behavior that is highly nonlinear. Thus, despite much effort, the exact mechanisms behind particle clustering are not fully understood. Generally, it is accepted that clustering is brought about by a complex interplay of a variety of mechanisms. The term EHD flow will be applied collectively to the various theories involving a fluid flow process aiding such an aggregation. The relative contribution of drag from EHD flows to drag from the electrothermal microvortex in REP will change with the various experimental parameters, for example, solution conductivity and AC frequency. However, for low-solution conductivities (<2 mS/m) and AC frequencies greater than 10 kHz, fluid drag from the electrothermal microvortex dominates over other forces. This is evident from the fact that deactivating the laser stops electrothermal motion, leading to particle–particle repulsion and scatter (see video [52]).

13.4.2.2 Electrokinetic Forces Normal to the Electrode Surface

In Figure 13.14, we saw that the electrothermal drag has both transverse and normal components. The normal component tends to "pull" a particle away from the electrode surface. For an REP aggregation to remain stable, the normal drag force must be overcome by a particle–electrode attractive force. In the past, researchers have explored the AC electrokinetic forces acting on a colloidal particle normal to the electrode surface. Fagan et al. [48–50] used total internal reflection microscopy to study the vertical motion of a single PS particle near an electrode surface. They found that the particle exhibited an oscillatory motion that was out of phase with the AC electric field. These vertical oscillations were not only a function of AC frequency but also of the electrolyte. They studied various force-producing mechanisms that could explain the observed behavior and found that fluid flow around the particle could account for a temporally varying vertical force on the particle. However, the variation of the vertical motion with AC frequency could be accounted for only by a combination of various fluid flow mechanisms.

Apart from the vertical or normal forces exerted by the fluid, other mechanisms can also contribute to vertical forces on colloidal particles. Image charges can play a role in the normal particle–electrode interaction forces [53,54]. It has also been shown that bringing two EDLs in close proximity can create attractive or repulsive forces solely because of an interaction of the double layers [55–57]. Because of such complexities, the exact nature of the total normal force exerted on the particles and its dependence on AC frequency are not fully understood. However, for stable REP aggregations, it can be easily concluded that the sum of all these AC electrokinetic forces produces an attractive force to counterbalance the normal drag force from the electrothermal vortex.

13.4.2.3 Dipole–Dipole Forces in REP

We have seen that REP produces an aggregation of closely packed colloidal particles. Colloidal particles suspended in a dielectric medium polarize on the application of an electric field. The polarization of the dielectric particle can be effected through a variety of mechanisms such as electronic, atomic, interfacial, and EDL polarization. The Maxwell–Wagner interfacial polarization mechanism can be an important mechanism to an induced dipole moment, and in fact, this mechanism is a central concept in understanding dielectrophoretic motion of particles. However, in REP the low AC frequencies used (<200 kHz) can also lead to the polarization of the EDL, contributing to particle permittivity and induced dipole moment (Figure 13.15). Irrespective of the mechanism involved in producing an induced dipole moment, the dipole moment results in a repulsive force between particles, as the dipoles (i.e., the particles) align with the direction of the applied field and the proximity

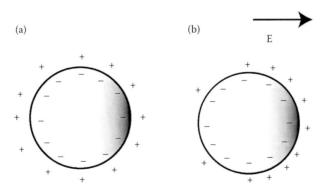

FIGURE 13.15
(a) An equilibrium electrical double layer. (b) On the application of an electric field, the double layer can polarize. The polarization of the EDL around a particle can also contribute to the dipole moment in particles aggregated with REP.

of like charges of neighboring particles causes them to repel from each other. This repulsive force can cause a particle group to scatter when the laser is turned off (see video [52]).

13.5 Applications

REP can be used in a variety of colloidal concentration applications, including creating artificial architectures such as photonic crystals. The frequency-dependent polarization of particles and cells within a prepared sample can be explored without the interference of nonuniform electrokinetic mechanisms such as DEP or AC electro-osmosis. REP simultaneously induces fluid motion to expedite sample concentration, thereby improving the efficiency of LOC sensors. REP is a versatile tool that can be applied to a variety of existing LOC techniques. In this section, we present some applications toward which the technique can be applied.

13.5.1 Trapping of Nanoparticles

Williams et al. [17] expanded REP to enable concentration and patterning of 49- and 100-nm particles. Red fluorescent PS particles 49 and 100 nm in diameter (Duke Scientific, CA) were used. Particles were suspended in an aqueous KCl solution with an electrical conductivity of 0.6 mS/m. The small size of the nanoparticles and the limitations of the optics used did not show individual particles; instead, the concentrated particles resembled fluorescent spots on the electrode surface. A plain ITO/Au chip with a 100× objective lens was used. Although individual particles could not be visualized, micrometer-sized fluorescent spots were observed that increased in intensity with particle concentration; therefore, the overall fluorescence of the experimental image was used to characterize REP as a function of voltage and frequency.

As the particles were concentrated with REP, an increase of fluorescence was observed at the center of the capturing region (Figure 13.16a). As with previous microparticle investigations, nanoparticles could be patterned with focused illumination landscapes using REP (Figure 13.16b).

The rate of particle (100 nm) accumulation on the surface of the gold electrode was investigated with a focused illumination spot having a laser power of 20 mW and an AC signal of 10 kHz and 12.0 V_{pp}. The amplitude of fluorescent intensity for the concentrated particle group is plotted for a period of 25 s (Figure 13.17). The fit is an exponential decay of the form $y = Ae^{-x/t} + y_o$, with $A < 0$ and $y_o > 0$. This result is similar to trends with micrometer-sized particles [22].

The frequency-dependent nature of REP is rather complex as electrothermal pumping, particle–electrode interactions, and particle–particle interactions are each dependent on the applied AC frequency. However, general experimental trends have been previously demonstrated [9,10,17,22]. Concentration at high frequencies (>200 kHz) was negligible as the polarization of the ionic double layer relaxed. The electrothermal microvortex, however, would continue at even higher frequencies (>5 MHz), when Brownian motion dominated particle translation [10]. Particle accumulation improved with decreasing AC frequencies (<200 kHz), reaching a peak before decreasing at even lower frequencies (typically <50 kHz). At lower frequencies, the microfluidic vortex was unable to translate or pattern nanoparticles along the electrode surface; they remained unaffected by this hydrodynamic drag and preferred areas on the surface with scratches that would generate electric

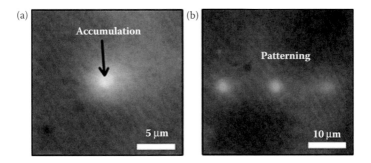

FIGURE 13.16

Experimental images showing the concentration and patterning of 100-nm fluorescent PS particles on the surface of a gold electrode with 1064-nm optical landscapes. (a) A highly focused beam of light and (b) three focused spots were used. (From Williams, S.J., Kumar, A., Green, N.G., and Wereley, S.T., A simple, optically induced electrokinetic method to concentrate and pattern nanoparticles, *Nanoscale* 1, 133–137, 2009. Reproduced by permission of The Royal Society of Chemistry.)

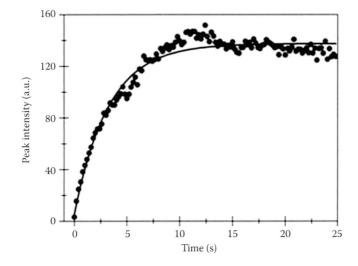

FIGURE 13.17

The rate of 100-nm PS particle concentration on a gold surface with a laser power of 20 mW and an AC signal of 10 kHz and 12.0 V_{pp}. (From Williams, S.J., Kumar, A., Green, N.G., and Wereley, S.T., A simple, optically induced electrokinetic method to concentrate and pattern nanoparticles, *Nanoscale* 1, 133–137, 2009. Reproduced by permission of The Royal Society of Chemistry.)

field nonuniformities. Local AC electro-osmotic and dielectrophoretic forces dominated translation at lower frequencies and inhibited particle capture.

13.5.2 Particle Sorting

13.5.2.1 Polarization of the Particles' Electric Double Layer

We saw earlier that similar to electro-osmotic flow, the ions within the double layer of a particle will migrate when subjected to an electric field. The ions within the double layer will migrate and align themselves in the direction of the electric field, polarizing the particle and thereby contributing to the induced dipole moment. Lyklema et al. [58] derived a model for the polarization of the double layer and included charge effects from the solid surface, the Stern layer, and diffuse layer.

Polarization involves the movement of ions, which take a finite amount of time to translate and form the induced dipole. Therefore, each polarization mechanism exhibits its own characteristic frequency response. At low AC frequencies, these ions have sufficient time to translate and form the most efficient induced dipole. As AC frequency increases, ions do not have adequate time to translate, and polarization will not occur. This decrease in polarizability is referred to as dielectric relaxation. Each polarization mechanism relaxes at different frequencies. Previous dielectric measurements indicate that there was a second dispersion at frequencies lower than that in the Maxwell–Wagner polarization [59–62]. This relaxation was attributed to the polarization of the ionic double layer surrounding the particle. Relaxation of the electric double layer corresponds to the time it takes for the ions to translate around the particle, therefore making its dispersion to occur at lower frequencies than the Maxwell–Wagner relaxation. This dispersion is referred to as α-dispersion in biophysics or low-frequency dielectric dispersion in colloidal and physical chemistry.

Many theories for electric double-layer polarization were developed; however, the relaxation times of the double-layer polarization are typically of the form of the Schwarz's model [62]:

$$\tau_\alpha = a^2/2D, \qquad (13.2)$$

where D is the diffusivity of the ions. Relaxation time is proportional to the surface area of the particle (a^2); therefore, α-dispersion occurs at lower frequencies as the particle diameter increases. Schwarz's model, though, does not account for all the possible mechanisms. Lyklema et al. [63] obtained a similar expression compared with Schwarz, but with an additional factor that accounts for the electro-osmotic contribution of the ion flux of the double layer. DeLacey and White [64] proposed a different relaxation that takes into account the time for ions to traverse the double layer. O'Brien [65] developed a relaxation model at high frequencies.

Regardless of the model used, the frequency-dependent α-relaxation mechanism can be used to characterize particles on the basis of their size and behavior of their ionic double layer. REP takes advantage of the frequency-dependent electric double-layer polarization to enable particle sorting. Williams et al. [22] illustrate such sorting capabilities in a ternary mixture containing 0.5-, 1.0-, and 2.0-μm PS beads.

13.5.2.2 Frequency-Dependent Capture of Particles

The frequency-dependent polarization of the particles' ionic double layer has been used to separate particles [22]. The relative theory states that the relaxation frequency is inversely proportional to particle surface area (a^2)—this trend was confirmed by Williams et al. [22] who demonstrated the frequency-dependent trapping of similar carboxylate-modified PS particles of various diameters (0.5, 1.0, and 2.0 μm). All particle sizes were initially captured at a frequency of 38 kHz (Figure 13.18). At 80 kHz, the 2.0-μm particles were carried away by the vortex, whereas the 0.5- and 1.0-μm beads remained (Figure 13.18b). At 106 kHz, the 1.0-μm particles escaped, whereas the 0.5-μm particles remained (Figure 13.18c).

The maximum frequency at which particles could be captured with REP was measured for each particle diameter (Figure 13.18d). The fit is a second-order polynomial consistent with the theory that particle relaxation time is proportional to the square of its radius (a^2). This means that REP can be used to trap and sort particles on the basis of their frequency-dependent double-layer polarization behaviors.

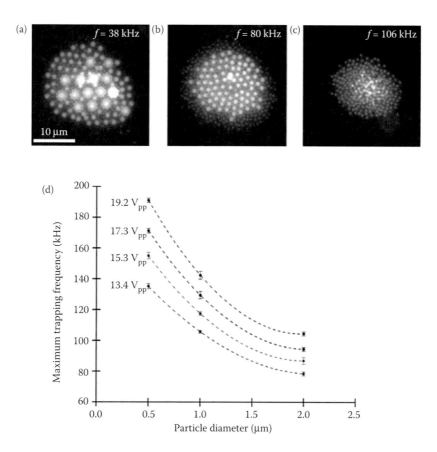

FIGURE 13.18
(a) The 2.0-, 1.0-, and 0.5-μm PS particles were all initially captured at 38 kHz. (b) The 2.0-μm particles were carried away at 80 kHz. (c) Only 0.5-μm particles remained at 106 kHz. (d) The maximum trapping frequency for particles tested. The fit is a second-order polynomial, which is consistent with theory. Error bars represent one standard deviation. (From Williams, S.J., Kumar, A., Green, N.G., and Wereley, S.T., Optically induced electrokinetic concentration and sorting of colloids, *Journal of Micromechanics Microengineering* 20, 2010. Reproduced with permission from IOP Publishing.)

Because particle trapping is a function of ionic double-layer polarizability, particles of similar diameter but different surface chemistries can be separated. Williams et al. [22] demonstrated such separation with carboxylate-modified silica (nonfluorescent) particles and PS (fluorescent) particles each 1.0 μm in diameter. Bulk pressure-driven fluid flow was applied simultaneously with REP. No particles were captured at 200 kHz (14 V_{pp}). At 120 kHz, only PS particles were trapped (Figure 13.19a), whereas both silica and PS particles were captured at 80 kHz (Figure 13.19b). Therefore, the relaxation of silica particles occurred at a lower frequency compared with PS.

13.5.3 Particle Concentration

REP continuously carries suspended particles within the microchannel and then captures them on the surface of the electrode. Suspended colloidal particles 0.49–3.0 μm in diameter have been captured with REP. The diameter range is typical of biologically relevant particles (cells, bacteria, etc.). Subparticles (100 nm) are more pertinent to nanotechnology, in

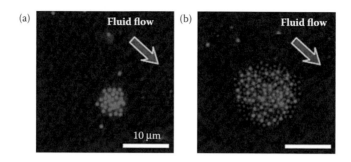

FIGURE 13.19
(a) REP was used to capture 1.0-µm PS (fluorescent) particles only at an applied frequency of 120 kHz, whereas (b) both PS and silica (nonfluorescent) 1.0-µm particles were captured at 80 kHz. Refer to supplementary documentation by Williams et al. [22] for a video demonstration.

which concentration difficulties arise from Brownian motion. REP can be used for dynamic and selective concentration of particles near an electrode substrate and can thus prove to be an important concentration process. The ability to concentrate particles with consistent placement on the substrate surface will enhance many LOC sensing processes or MEMS sensors. Stabilization is essential for accurate diagnostic analysis. The added ability to translate massive particle groups will enhance bead-based assays, bringing biologically modified particles into contact with the surfaces of interest.

13.5.4 Creation of Artificial Architectures

Particles (>0.5 µm) can be irreversibly concentrated in a compact, crystalline aggregation with REP. This arrangement can benefit the creation of precise artificial architectures, such as colloidal crystals. Permanently adhered particles captured on the surface of the electrode with an applied DC voltage and REP have been demonstrated [9], and so it follows that creation and characterization of multilayered colloidal structures with REP will be a future focus.

13.5.5 Microfluidic Mixing and/or Pumping

The optically induced electrothermal microfluidic motion is a technique that can be used to mix or pump fluids within an LOC device. This process was demonstrated previously [10,34,66]. Holographically controlled landscapes or alternative, advanced illumination systems can enable advanced electrothermal pumping methodologies.

13.5.6 Characterization of Electrokinetic Behavior of Particles

Particle capture and sorting with REP is a function of the electrokinetic properties. REP can be used to characterize particles on the basis of electrokinetic behavior. This insight can be extracted from (i) the experimental conditions for successful particle capture or by (ii) investigating particle–particle interactions. The electrothermal fluid drag force exerted on trapped particles can be used as an investigative tool for electrokinetic mechanisms. Nonuniform electrokinetic forces from DEP or ACEO can be minimized with a properly fabricated planar electrode surface.

Kumar and Wereley [67] used REP to initially create a cluster and then the illumination was deactivated. In the absence of the electrothermal microvortex, the dipole–dipole interactions cause the particles to scatter. Using high-speed video microscopy and Delaunay tessellations, Kumar and Wereley [67] showed that the rate of scattering increases with decreasing AC frequency and thus established the presence of a relaxation mechanism in the induced dipole moment generation.

13.6 Future Directions

The last decade has seen the emergence of various optoelectric manipulation techniques. One prime advantage of these optoelectric techniques over their electric counterparts lies in their dynamic nature. Moreover, most of these optoelectric manipulation chips have relatively simple fabrication requirements. The multiple optoelectric techniques that exist today have already been shown to cater to diverse and novel applications. Examples of such applications include a dynamic nanopen [15], a force measurement technique for an ensemble of particles, a method to discriminate normal oocytes [68], and directed crystallization [7]. Although these diverse optoelectric techniques promise to abundantly benefit the microfluidic and biochemical societies, there exists a need for sustained fundamental and applied research in this domain. From a fundamental point of view, an exact description of the multitude of forces in optoelectric techniques is necessary. Application-wise, these techniques need to demonstrate the capability to address various biomedical diagnostic problems in a simple and affordable fashion. As these techniques are constantly growing and reaching out to more applications, we can anticipate increasing and flourishing development in the near future.

Acknowledgments

The authors are grateful for inputs from Prof. Ming Wu at University of California, Berkeley, Prof. P. Y. Chiou at University of California, Los Angeles, Prof. I. A. Aksay at Princeton University, and Prof. David Marr at Colorado School of Mines. Aloke Kumar acknowledges support from the Josephine De Kármán Fellowship and the Bilsland Dissertation Fellowship.

References

1. Ashkin A. Optical trapping and manipulation of neutral particles using lasers. *Proc Natl Acad Sci U S A*. 1997;94(10):4853–60.
2. Pohl HA. *Dielectrophoresis: The Behavior of Neutral Matter in Nonuniform Electric Fields.* Cambridge Monographs on Physics. Cambridge: Cambridge University Press; 1978. p. xii, 579.
3. Geelhoed P, Westerweel J, Kjelstrup S, Bedeaux D. Thermophoresis. In: Li D, editor. *Encyclopedia of Microfluidics and Nanofluidics*. New York: Springer; 2008.
4. Duhr S, Braun D. Two-dimensional colloidal crystals formed by thermophoresis and convection. *Appl Phys Lett*. 2005;86(13):131921.1–3.

5. Duhr S, Braun D. Why molecules move along a temperature gradient. *Proc Natl Acad Sci U S A*. 2006;103(52):19678–82.

6. Chiou PY, Ohta AT, Wu MC. Massively parallel manipulation of single cells and microparticles using optical images. *Nature*. 2005;436(7049):370–2.

7. Gong TY, Marr DWM. Photon-directed colloidal crystallization. *Appl Phys Lett*. 2004;85(17):3760–2.

8. Hayward RC, Saville DA, Aksay IA. Electrophoretic assembly of colloidal crystals with optically tunable micropatterns. *Nature*. 2000;404(6773):56–9.

9. Williams SJ, Kumar A, Wereley ST. Electrokinetic patterning of colloidal particles with optical landscapes. *Lab Chip*. 2008;8(11):1879–82.

10. Kumar A, Williams SJ, Wereley ST. Experiments on opto-electrically generated microfluidic vortices . *Microfluid Nanofluidics*. 2009;6(5):637–46.

11. Chiou PY, Ohta AT, Jamshidi A, Hsu HY, Wu MC. Light-actuated ac electroosmosis for nanoparticle manipulation. *J Microelectromech Syst*. 2008;17(3):525–31.

12. Gong T, Marr DWM. Electrically switchable colloidal ordering in confined geometries. *Langmuir*. 2001;17(8):2301–4.

13. Gong TY, Wu DT, Marr DWM. Two-dimensional electrohydrodynamically induced colloidal phases. *Langmuir*. 2002;18(26):10064–7.

14. Chiou PY, Chang ZH, Wu MC. A novel optoelectronic tweezer using light induced dielectrophoresis. 2003 IEEE/Leos International Conference on Optical MEMS; 2003. pp. 8–9.

15. Jamshidi A, Neale SL, Yu K, Pauzauskie PJ, Schuck PJ, Valley JK, et al. NanoPen: dynamic, low-power, and light-actuated patterning of nanoparticles. *Nano Lett*. 2009;9(8):2921–5.

16. Hwang H, Park JK. Rapid and selective concentration of microparticles in an optoelectrofluidic platform. *Lab Chip*. 2009;9(2):199–206.

17. Williams SJ, Kumar A, Green NG, Wereley ST. A simple, optically induced electrokinetic method to concentrate and pattern nanoparticles. *Nanoscale*. 2009;1:133–7.

18. Yavas O, Ochiai C, Takai M. Substrate-assisted laser patterning of indium tin oxide thin films. *Appl Phys A Mater Sci Process*. 1999;69:S875–8.

19. Kumar A, Kwon J-S, Williams SJ, Green NG, Yip NK, Wereley ST. Optically modulated electrokinetic manipulation and concentration of colloidal particles near an electrode surface. *Langmuir*. 2010;26(7):5262–72.

20. Kumar A, Ewing AH, Wereley ST. Optical tweezers for manipulating cells and particles. In: Li D, editor. *Encyclopedia of Microfluidics and Nanofluidics*. New York: Springer; 2008.

21. Kumar A, Kwon J-S, Williams SJ, Wereley ST. A novel optically driven electrokinetic technique for manipulating nanoparticles. *Proc SPIE*. 2009;7400:74000V.

22. Williams SJ, Kumar A, Green NG, Wereley ST. Optically induced electrokinetic concentration and sorting of colloids. *J Micromech Microeng*. 2010;20:015022.

23. Williams SJ, Kumar A, Wereley ST. Optically induced electrokinetic patterning and manipulation of particles. *Phys Fluids*. 2009;21:091104.

24. Derjaguin BV, Landau L. Theory of the stability of strongly charged lyophobic sols and the adhesion of strongly charged particles in solutions of electrolytes. *Acta Phys Chim U S S R*. 1941;14:633.

25. Verwey EJ, Overbeek JTG. *Theory of Stability of Lyophobic Colloids*. Amsterdam: Elsevier; 1948.

26. Bazant MZ, Squires TM. Induced-charge electrokinetic phenomena: theory and microfluidic applications. *Phys Rev Lett*. 2004;92(6):066101.1–4.

27. Dukhin SS. Nonequilibrium electric surface phenomena. *Adv Colloid Interface Sci*. 1993;44:1–134.

28. Ristenpart WD, Aksay IA, Saville DA. Assembly of colloidal aggregates by electrohydrodynamic flow: kinetic experiments and scaling analysis. *Phys Rev E*. 2004;69(2):021405.

29. Trau M, Saville DA, Aksay IA. Field-induced layering of colloidal crystals. *Science*. 1996;272(5262):706–9.

30. Yeh SR, Seul M, Shraiman BI. Assembly of ordered colloidal aggregates by electric-field-induced fluid flow. *Nature*. 1997;386(6620):57–9.

31. Green NG, Ramos A, Gonzalez A, Castellanos A, Morgan H. Electrothermally induced fluid flow on microelectrodes. *J Electrostat.* 2001;53(2):71–87.

32. Ramos A, Morgan H, Green NG, Castellanos A. AC electrokinetics: a review of forces in micro-electrode structures. *J Phys D Appl Phys.* 1998;31(18):2338–53.

33. Green NG, Ramos A, Gonzalez A, Castellanos A, Morgan H. Electric field induced fluid flow on microelectrodes: the effect of illumination. *J Phys D Appl Phys.* 2000;33(2):L13–7.

34. Nakano M, Katsura S, Touchard GG, Takashima K, Mizuno A. Development of an optoelectrostatic micropump using a focused laser beam in a high-frequency electric field. *IEEE Trans Ind Appl.* 2007;43(1):232–7.

35. Williams SJ. *Optically Induced, AC Electrokinetic Manipulation of Colloids.* Doctoral dissertation in Mechanical Engineering. West Lafayette: Purdue University; 2009.

36. Weber MJ. *Handbook of Optical Materials.* Boca Raton: CRC Press; 2003.

37. Peterson SD, Chuang HS, Wereley ST. Three-dimensional particle tracking using micro-particle image velocimetry hardware. *Meas Sci Technol.* 2008;19(11):115406.

38. Bohmer M. In situ observation of 2-dimensional clustering during electrophoretic deposition. *Langmuir.* 1996;12(24):5747–50.

39. Brisson V, Tilton RD. Self-assembly and two-dimensional patterning of cell arrays by electrophoretic deposition. *Biotechnol Bioeng.* 2002;77(3):290–5.

40. Giersig M, Mulvaney P. Formation of ordered 2-dimensional gold colloid lattices by electrophoretic deposition. *J Phys Chem.* 1993;97(24):6334–6.

41. Giner V, Sancho M, Lee RS, Martinez G, Pethig R. Transverse dipolar chaining in binary suspensions induced by RF fields. *J Phys D Appl Phys.* 1999;32(10):1182–6.

42. Grzybowski BA, Whitesides GM. Dynamic aggregation of chiral spinners. *Science.* 2002;296(5568):718–21.

43. Richetti P, Prost J, Barois P. Two-dimensional aggregation and crystallization of a colloidal suspension of latex spheres. *Europhys Lett.* 1984;45(23):1137–43.

44. Ristenpart WD, Aksay IA, Saville DA. Electrically guided assembly of planar superlattices in binary colloidal suspensions. *Phys Rev Lett.* 2003;90(12):128303.1–4.

45. Solomentsev Y, Bohmer M, Anderson JL. Particle clustering and pattern formation during electrophoretic deposition: a hydrodynamic model. *Langmuir.* 1997;13(23):6058–68.

46. Trau M, Saville DA, Aksay IA. Assembly of colloidal crystals at electrode interfaces. *Langmuir.* 1997;13(24):6375–81.

47. Nadal F, Argoul F, Hanusse P, Pouligny B, Ajdari A. Electrically induced interactions between colloidal particles in the vicinity of a conducting plane. *Phys Rev E.* 2002;65(6):061409.

48. Fagan JA, Sides PJ, Prieve DC. Vertical motion of a charged colloidal particle near an AC polarized electrode with a nonuniform potential distribution: theory and experimental evidence. *Langmuir.* 2004;20(12):4823–34.

49. Fagan JA, Sides PJ, Prieve DC. Evidence of multiple electrohydrodynamic forces acting on a colloidal particle near an electrode due to an alternating current electric field. *Langmuir.* 2005;21(5):1784–94.

50. Fagan JA, Sides PJ, Prieve PC. Vertical oscillatory motion of a single colloidal particle adjacent to an electrode in an ac electric field. *Langmuir.* 2002;18(21):7810–20.

51. Hoggard JD, Sides PJ, Prieve DC. Electrolyte-dependent multiparticle motion near electrodes in oscillating electric fields. *Langmuir.* 2008;24(7):2977–82.

52. Williams SJ, Kumar A, Wereley ST. *Optically Induced Electrokinetic Patterning and Manipulation of Particles.* Available from: http://hdl.handle.net/1813/11399.

53. Hatlo MM, Lue L. The role of image charges in the interactions between colloidal particles. *Soft Matter.* 2008;4(8):1582–96.

54. Lyklema J, Leeuwen HPV, Vliet MV, Cazabat AM. *Fundamentals of Interface and Colloid Science.* San Diego: Academic Press; 1991.

55. Chakraborty S. Electrical double layers interaction. In: Li D, editor. *Encyclopedia of Microfluidics and Nanofluidics.* Berlin: Springer; 2008.

56. Chakraborty S, Srivastava AK. Generalized model for time periodic electroosmotic flows with overlapping electrical double layers. *Langmuir*. 2007;23:12421–8.
57. Das PK, Bhattacharjee S. Electrostatic double layer force between a sphere and a planar substrate in the presence of previously deposited spherical particles. *Langmuir*. 2005;21(10):4755–64.
58. Lyklema J. Fundamentals of interface and colloid science. Vol. 2. London: Academic Press; 1995.
59. Green NG, Morgan H. Dielectrophoresis of submicrometer latex spheres. I. Experimental results. *J Phys Chem B*. 1999;103(1):41–50.
60. Schwan HP. Determination of biological impedances. In: Nastuk WL, editor. *Physical Techniques in Biological Research. Vol. VI. Electrophysiological Methods. Part B*; 1963. pp. 323–407.
61. Schwan HP, Schwarz G, Maczuk J, Pauly H. On low-frequency dielectric dispersion of colloidal particles in electrolyte solution. *J Phys Chem*. 1962;66(12):2626.
62. Schwarz G. A theory of low-frequency dielectric dispersion of colloidal particles in electrolyte solution. *J Phys Chem*. 1962;66(12):2636.
63. Lyklema J, Dukhin SS, Shilov VN. The relaxation of the double-layer around colloidal particles and the low-frequency dielectric-dispersion. I. Theoretical considerations. *J Electroanal Chem*. 1983;143:1–21.
64. DeLacey EHB, White LR. Dielectric response and conductivity of dilute suspension of colloidal particles. *J Chem Soc Faraday Trans*. 1981;77:2007–39.
65. O'Brien RW, Ward DN. Electrophoresis of a spheroid with a thin double layer. *J Colloid Interface Sci*. 1988;121:402–13.
66. Mizuno A, Nishioka M, Ohno Y, Dascalescu LD. Liquid microvortex generated around a laser focal point in an intense high-frequency electric-field. *IEEE Trans Ind Appl*. 1995;31(3):464–8.
67. Kumar A, Wereley ST. Optically induced rapid electrokinetic patterning: a study of the operational regimes and dominant forces. Proceedings of ASME-IMECE 2009, Florida, USA; 2009.
68. Hwang H, Lee DH, Choi WJ, Park JK. Enhanced discrimination of normal oocytes using optically induced pulling-up dielectrophoretic force. *Biomicrofluidics*. 2009;3(1):014103.

14

Microfluidic Particle Counting Sensors

Chan Hee Chon, Hongpeng Zhang, Xinxiang Pan, and Dongqing Li

CONTENTS

14.1 Introduction

The particle counting sensor is an important and widely used device in various areas from environmental (Aalto et al., 2005) to biological applications (Smolen et al., 1983; Amann et al., 1990; Yarnell et al., 1990). For examples, counting dust particles is required in a clean room facility (Wu et al., 1989); counting debris particles is needed for studies of lubricating systems (Miller and Kitaljevich, 2000); counting contaminant particles is key in water purification systems (Bundschuh et al., 2001; Judd and Hillis, 2001); and counting white blood cells is essential for many biomedical diagnostic purposes such as detecting HIV infection (Yarnell et al., 1991; Kannel et al., 1992; Burnett et al., 1999; Vozarova et al., 2002). Particle counting sensors have been developed further to detect a single DNA molecule (Akeson et al., 1999; Kasianowicz et al., 2006) and to analyze DNA contents (Sohn et al., 2000). However, conventional particle counters rely on costly, bulky, and complex instruments and require a large number of samples and reagents. These are the barriers to many particle counting sensors.

Microfluidic devices have been studied extensively in the past decade and have shown enormous potential for portable and low-cost applications especially in medical diagnostics (Akeson et al., 1999; Burnett et al., 1999; Kannel et al., 1992; Sohn et al., 2000; Vozarova et al., 2002; Kasianowicz et al., 2006). The lithographic fabrication technique makes it possible to build inexpensive and small devices integrated with electrodes and sensors, and microfluidic control technologies such as electrokinetics (Li, 2004) are able to control particles and liquid flow in micro- and nanochannels. In addition, microfluidic devices are particularly useful for applications where a very small quantity of samples is available or desired. Microfluidic-based particle counting sensors have great advantages over conventional devices and allow the development of accurate, cheap, and portable particle counting devices.

Particle counting techniques have very diverse characteristics and can be categorized differently depending on counting particle size, particle focusing, flow handling, application, and so forth (Ateya et al., 2008). The size involved in particle counting can range from nanoscale DNA molecules to 100-μm wear debris. Particle focusing includes the use of hydrodynamic fluid, nozzle, and dielectrophoresis. Flow handling schemes include using external syringe pumps and electroosmotic flow. However, in this chapter, we focus on reviewing major advancement and applications of microfluidic particle counting sensors: microfluidic resistive repulse sensors, nanopore resistive sensors (NRS), capacitance particle sensors, light-scattering and light-blocking particle sensors, fluorescence-based particle sensors, and microparticle image velocimetry (micro-PIV) particle sensors. For each type of particle sensor, some applications and the advantages and disadvantages are discussed.

14.2 Coulter Counter

The Coulter counter perhaps is the most popular and typical conventional particle counting device. Its working principle of measuring electric resistive pulse is also used in some microfluidic particle counting sensors. The Coulter counter was invented by Wallace H. Coulter during World War II and patented in 1953 (Coulter). When Coulter worked for the U.S. Navy, he used this technique to count the number of plankton particles that always caused large echoes on sonar. In a Coulter counter, a small aperture on the wall is immersed into a container that has particles suspended in a low-concentration electrolyte solution. Two electrodes are placed before and after the aperture, and a current path is provided by the electrolyte when an electric field is applied (Figure 14.1) and the aperture creates a "sensing zone." As a particle passes through the aperture, a volume of electrolyte equivalent to the immersed volume of the particle is displaced from the sensing zone. This causes a short-term change in the impedance across the aperture. This change can be measured as a voltage pulse or as a current pulse. The pulse height is proportional to the volume of the sensed particle. If a constant particle density is assumed, the pulse height is also proportional to the particle mass. This technology is also called aperture technology.

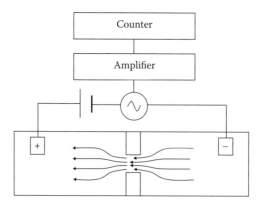

FIGURE 14.1
Schematic of a Coulter counter.

Because the Coulter counter is simple, highly sensitive, and reliable, it is widely applied in many areas including medical instruments used for counting and analyzing blood cells (Horne et al., 2005), proteins (Kulp et al., 2004), and viruses (Wahl-Jensen et al., 2007) besides detecting fine particles and pollen (DeBlois and Bean, 1970). Before the Coulter counter was invented, complete blood counts were carried out manually with a microscope and static samples, which was inaccurate and time consuming. With the Coulter counter, complete blood counts for a large amount of samples can be done in a very short period of time. When Coulter first demonstrated the Coulter counter, it could count red blood cells at a high rate of 6000 particles per second, which revolutionized the science of hematology. Within a decade, literally every hospital laboratory in the United States had Coulter counters, and today every modern hematology analyzer adopts the Coulter principle in some way. For example, Coulter counters may measure the change of impedance (Carbonaro and Sohn, 2005; Jagtiani et al., 2006; Zhe et al., 2007; Wu et al., 2008), conductance (Sohn et al., 2000; Murali et al., 2009), and reflected radiofrequency power (Wood et al., 2005) when a particle passes through the aperture.

After many years of development, the modern Coulter counter is versatile and accurate in particle sizing and counting. The Multisizer™ 4 Coulter Counter® (Beckman Coulter, Fullerton, CA), for example, provides size distribution in number, volume, and surface area in the measurement of particles that range from 0.4 to 1600 μm in diameter. Its aperture dynamic range can reach to 1:40 by diameter, and reproducibility is approximately 1% in accuracy. Although the Coulter counter detects and analyzes particles accurately and reliably, it has still many drawbacks: bulky size, heavy weight, complexity, high power consumption, high cost, and no portability. Especially with urgent public diseases such as severe acute respiratory syndrome and influenza outbreak, there are many needs for simple, low-power, low-cost, and portable particle counting devices.

14.3 Microfluidic Resistive Pulse Sensors

The microfluidic resistive pulse technique applies the basic working principle of the Coulter counter to microchannels for counting micro- and submicron particles. The resistive pulse sensor (RPS) was applied by DeBlois and Bean in 1970 to detect submicron polystyrene beads of 90 nm inside a 0.4- to 0.5-mm-diameter polycarbonate pore. In 1977, DeBlois and Wesley used this technique to sucessfully detect viruses. After the technical advance of microfabrication (Rogers and Nuzzo, 2005), the RPS has been applied to count particles moving in microchannels. Many microfluidic RPS applications involve manipulating and transporting particles by electrokinetics flow in microchannels (Jagtiani et al., 2006), whereas some still use traditional flow control by hydraulic pressure.

The key advantages of the microfluidic RPS include label-free particle detection and simplicity without other peripheral complex instruments, except a simple electric circuit and a micro- or nanoscale-sized channel. Therefore, it is mostly applicable for portable lab-on-a-chip devices to detect biopolymers such as DNA, protein, and blood cells. However, the flow rate of the microfluidic RPS is small, and the sensitivity of the microfluidic RPS is limited by its aperture size, resulting in poor throughput and sensitivity. To overcome these shortcomings, recent researches for the microfluidic RPS focus on two main issues: improvement of sensitivity and enhancement of throughput. The RPS throughput is evaluated by particle flow rate at the aperture or the number of the counted particles at a given time. Its sensitivity is determined directly by the volume ratio of the detected particles and the aperture and can be adjusted by controlling the amplification gain of the electronic circuit or instrument and the noise reduction from the fluidic network and the electronic sensing system.

DeBlois and Bean (1970) were able to detect 90-nm polystyrene spheres, which was equivalent to a minimum volume ratio of 0.06% with submicron pores etched in an irradiated plastic sheet. To improve the sensitivity, Xu et al. (2007) and Sridhar et al. (2008) used a relatively wide polydimethylsiloxane (PDMS) channel of 16 µm with a metal oxide semiconductor field-effect transistor. They detected particles by monitoring the metal oxide semiconductor field-effect transistor drain current modulation instead of the modulation in the ionic current through the sensing channel and achieved a minimum volume ratio of 0.006%, 10 times smaller than DeBlois and Bean's. Recently, Wu et al. (2008) developed a microfluidic RPS method using a mirror-symmetric channel structure and a two-stage differential amplifier (Figure 14.2). They could significantly reduce noise and achieve a much better signal-to-noise ratio. This sensing scheme detected 520-nm-diameter polystyrene particles with a sensing gate size of 20 µm and improved the minimum volume ratio to as low as 0.0004%, which was about 10 times more sensitive than the current commercial Coulter counter with the minimum volume ratio of 0.0037% (Beckman Coulter Multisizer 4). The principle of the symmetric dual channel design was to make noise levels for the output signals (V_{D1} and V_{D2} as indicated in Figure 14.2) from both gate branches identical, and hence the noises could be canceled by a subtraction electronic circuit.

(a)

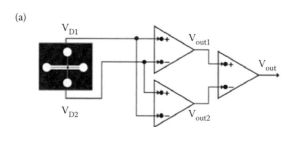

(b)

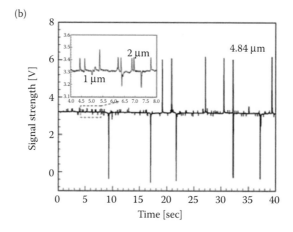

FIGURE 14.2
(a) The schematic of dual channel microfluidic differential RPS and (b) the repulsive pulse signals of 1-, 2-, and 4.84-µm particles in 7.5 mM of sodium borate buffer. The magnified inset shows the signal strength of 1- and 2-µm particles.

However, ideal noise subtraction was not possible because of the realistic limitation to fabricate identical dual channels. Furthermore, this dual channel method would not be able to detect particles when two particles pass the two apertures at the same time because the two signals with similar amplitude would be subtracted by each other and cancelled at the second stage of the differential amplifier.

To improve this microfluidic RPS, Wu et al. (2008) solved the signal cancelation problem as mentioned above by using a single sensing gate and two detecting arm channels next to the sensing gate at both ends (Figure 14.3). They coupled the RPS with a laser fiber-optic fluorescence technique to demonstrate a flow cytometer lab-on-a-chip that was

(a)

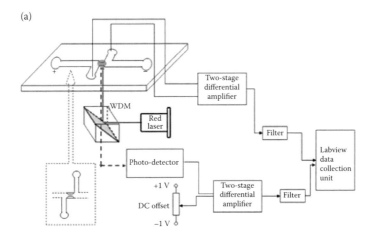

(b)

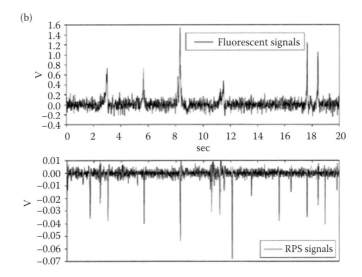

FIGURE 14.3
(a) The schematic of single-channel and two detecting arm channels microfluidic differential RPS and (b) the RPS counter and fluorescent signals for 0.9-μm Nile blue particles mixed with 0.99-μm dragon green particles. The greater RPS peaks are the signals of the 0.99-μm nonfluorescent particles.

able to detect fluorescent and nonfluorescent particles simultaneously, and the RPS signal-to-noise ratio was improved significantly. Two-stage differential amplification was also used to further increase the signal-to-noise ratio for fluorescent signals to detect 0.9-μm fluorescent particles. This flow cytometer chip showed comparable sensitivity for detecting fluorescent and nonfluorescent particles to commercial flow cytometers with a simple, cheap, and compact system on a micro glass slide. Drawbacks of the method are the baseline drifting by multiple-stage differential amplification and the low throughput of single-channel detection.

The throughput of a single-channel Coulter counter is proportional to the square of the diameter of the detecting aperture. When submicron or nanometer-size particles are to be counted, the size of the aperture has to be scaled down to submicron or nanometer levels to maintain the sensitivity. Otherwise, the signal-to-noise ratio will be very low. Although the sensitivity of the single gate microfluidic RPS is much higher, it requires a longer detecting time because of low flow rate in a microchannel and the use of diluted samples to avoid multiple particles flowing together. To overcome this low-throughput issue, multicounting techniques are developed. Carbonaro and Sohn (2005) first demonstrated the simultaneous immunoassays of two different human antigens by integrating multiple artificial pores and the RPS technique on a single chip. Coulter and Hogg (1976) patented the particle analyzing apparatus and method with multiple sensing apertures. However, it was difficult to integrate the detection circuit and the independent power supplies of their systems on one chip. Jagtiani et al. (2006) proposed a multiaperture Coulter counter, which consists of four peripheral reservoirs and a central reservoir (Figure 14.4a). Each peripheral reservoir was connected to the central reservoir through a miniature channel on each polymer membrane. Their result showed that the sensor could detect and count particles through its four sensing apertures simultaneously. However, the four apertures were the maximum number that could be built in a single chip because the chip was occupied by several reservoir blocks and polymer membranes, and no other parts for additional apertures could fit in it as shown in Figure 14.4a. Furthermore, Zhe et al. (2007) proposed a different high-throughput single chip counter using multiple channels operating in parallel with a single common sample reservoir and a power source (Figure 14.4b). A common electrode located at the inlet reservoir worked as a common ground, and four electrodes in each channels functioned as sampling resistors. When a particle passed through a channel, it caused a change of the resistance of the channel and resulted in a voltage pulse across the sampling resistor for that channel. The voltage pulses across

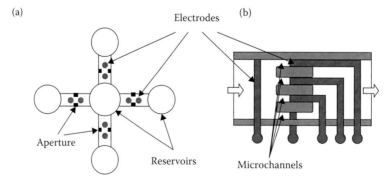

FIGURE 14.4
High-throughput Coulter counters: (a) A multiaperture Coulter counter and (b) a multichannel Coulter counter for microparticle detection.

each sampling resistor could be recorded simultaneously. This counter was capable of differentiating and counting polymethacrylate particles and juniper pollen about three times faster than a single-channel counter. This concept could be extended to multichannel microfluidic chips to improve the counting efficiency.

14.4 Nanopore Resistive Sensor

A nanopore here is referred to as a small pore of nanometer size in an electrically insulating membrane, and an NRS is a nanometer-scale Coulter counter, which uses the nanopore as an aperture to detect a single molecule. The NRS, the Coulter counter, and the RPS share the same principle of particle detection; however, the NRS has attracted many researchers since 1990s because its volumetric ratio can be high enough to detect much smaller particles, such as single molecules by the nanometer-sized pores (Bezrukov et al., 1996; Kasianowicz et al., 1996; Akeson et al., 1999). There are two types of nanopores depending on the type of material, synthetic and natural. A typical natural nanopore is a biological protein channel in a lipid bilayer.

Bezrukov et al. (1996) demonstrated the counting of polymer molecules passing through a single alamethicin pore of 5 and 2 nm in length and diameter, respectively. Kasianowicz et al. (1996) showed that they were able to sense single-stranded RNA and DNA through a 2.6-nm-diameter ion channel in a lipid bilayer membrane. This promising result sparked many studies of DNA translocation and dynamics in biological nanopores. Akeson et al. (1999) detected single DNA and RNA molecules in an α-hemolysin channel driven by an applied electric field (Figure 14.5). Because the small-sized pore can hold only one strand

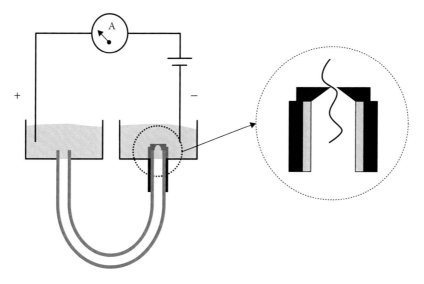

FIGURE 14.5
Horizontal bilayer apparatus: A U-tube connects two baths, and all are filled with KCl buffer. The baths are connected to an Axopatch 200B amplifier by Ag-AgCl electrodes. One end of the tube has a conical tip of a 25-μm aperture. Diphytanoyl phosphatidylcholine/hexadecane bilayers are formed across the aperture, and one or more α-hemolysin channels are inserted into the bilayer to detect single DNA or RNA molecules.

of DNA or RNA at a time, nucleotides within the polynucleotide must pass through the channel or pore in a sequential and single-file order during the translocation. In this process, not only counting but also discriminating between pyrimidine and purine segments along a DNA or RNA molecule could be accomplished. A very important application of this research was the direct sequencing of individual DNA and RNA molecules with a nanopore. In addition, NRS was used to measure small particles such as metal ions, nucleic acids, and other types of polymers in a less than 10-nm channel (Biance et al., 2006).

The geometrical and chemical properties of biological nanopores can be reproducibly controlled by genetic engineering. However, the biological nanopores are not very robust and not size-tunable. Even in a laboratory environment, these can last for several hours only. Therefore, efforts have been made to fabricate artificial, solid-state nanopores to overcome this limitation. The nanopores based on synthetic material are generally made in silicon compound membranes, such as silicon nitride. Manufacturing techniques are either focus ion-beam sculpting or electron beam sculpting. Siwy et al. (2005), Harrell et al. (2006), Wharton et al. (2007), and Sexton et al. (2007) did a series of research on synthetic conical nanopores for biosensing applications. They made a conical pore with a 1.5-mm base diameter and a 40-nm tip diameter and sensed a single-stranded phage DNA of 7250 bp and a double-stranded plasmid DNA of 6600 bp. They also detected protein in the same way. Alternatively, Ito et al. (2004) used a 132-nm multiwall carbon nanotube to detect 28- to 90-nm nanoparticles. Synthetic nanopores are chemically and structurally stable; however, it is still hard to control the size of these nanochannels. Reproducing synthetic nanopores of the same size is hardly possible because of the inherent anisotropic etching process (Wharton et al., 2007). This is because the etching and the etch-stop solutions are mixing and neutralizing each other in the nascent tip, making it difficult to control the etch rate in this critically important region of the nanopore. Additionally, the high manufacturing cost is another obstacle that needs to be overcome for further development and applications of synthetic nanopore particle sensors.

14.5 Microfluidic Capacitance Particle Sensor

The capacitance particle sensor uses a similar principle as that of the Coulter counter. It measures the AC capacitance instead of the DC resistance when a micron or submicron particle passes a sensing gate (aperture). The capacitance sensor is particularly useful for detecting particles in liquids of low electrical conductance because the resistance change due to the passage of a particle is difficult to measure in a poorly conducting liquid (Murali et al., 2009). In the past, the capacitance measurements have generally been used to identify bulk materials and to investigate ensembles of biological cells. However, Sohn et al. (2000) used this technique to detect and to quantify the polarization responses of DNA in the nucleus of single eukaryotic cells. They built an integrated microfluidic device (Figure 14.6) and used a syringe pump to deliver the liquid to the device. A capacitance bridge at a frequency of 1 kHz across the device was used to detect the capacitance change to determine the DNA content of single eukaryotic cells. Additionally, they demonstrated the relationship between the capacitance and the DNA content of a cell. Recently, Murali et al. (2009) adopted the capacitance counter to monitor the wear debris in lubrication oil to avoid the catastrophic system failure of machines in real time. Unlike bulk measurement methods, their method could scan each individual particle and determine the size of particles without being affected by the change of oil properties. Particles from 10 to 25 μm were successfully detected.

(a)

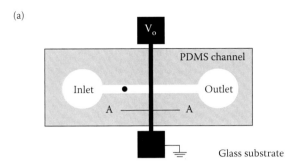

(b)

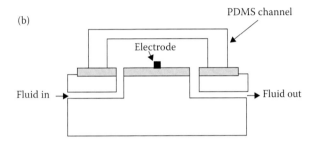

FIGURE 14.6
Schematic of the integrated microfluidic capacitance particle counting sensor: (a) top view and (b) side view along the line A-A the capacitance particle counting sensor.

Similar to a microfluidic RPS, the capacitance particle sensor has advantages in terms of simple sample preparation, cost, size, and robustness. In addition, the capacitance sensor is sensitive enough to probe the polarization response of a wide range of materials, both organic and inorganic, to an external electric field. Furthermore, it can monitor changes in DNA contents and cell-cycle kinetics because DNA is a highly charged molecule, and it will produce a change in capacitance. Eventually, it may serve as a medical diagnostic device to identify the presence of malignancy in very small quantities of tissue, such as tumor cells, and to monitor in real time the effects of pharmacological agents on cell cycle and cell death (Sohn et al., 2000). The capacitance sensor, however, has complications resulting from the charge-screening effects at the electrode–conductive liquid interface in an electronic measurement, which prevent the interpretation of the absolute capacitance value. Besides, because of the AC voltage and the frequency required to detect the capacitance, a frequency modulation controller is necessary, and an external support such as a syringe pump is required to deliver the liquid in the channel. These restrictions in flow control and frequency modulation are some of the obstacles for developing compact microfluidic capacitance particle sensors.

14.6 Light-Scattering and Light-Blocking Particle Sensors

A light-scattering particle sensor and a light-blocking particle sensor are two similar types of light-based particle sensors. They both use laser light sources such as laser diodes to illuminate individual particles that pass through the laser beam. The difference between these two sensors is in how the interaction between the light and the particle is measured for counting particles. When light strikes an object, generally it will be divided into three components. Some light will pass through the object, some will be reflected, and the rest will be absorbed by the object. The portion of these three components is determined by the optical properties of the material composition of the particle.

The light-scattering particle sensor uses the reflected light of the particle. As shown in Figure 14.7a, when the source light hits the particle, some of the light is reflected, and the reflected light can be detected by a photodetector positioned at an spot at a certain angle from the light path. In general, the detector is placed at an angle of 20° to 40° from the light path to the particle. The strength of the detected light signal corresponds to particle size, and the number of pulses of the detected light signal is proportional to the number of particles. On the other hand, the light-blocking particle sensor (Figure 14.7b) measures the light absorbed or reflected away from the detector by the particle. In this arrangement, the light is focused directly onto the detector; when the particle passes between them, the photodetector senses the sudden change in the light intensity caused by a blocking particle. It is obvious that the larger particle will block more light. In principle, the light-scattering particle sensor detects light, whereas the light-blocking sensor detects darkness. The light-scattering method is more sensitive than the light-blocking method. This has the same principle as detecting light in a darkroom that is easier than detecting a dark spot in a bright room because of the diffuse reflectance caused by solid surfaces and particles in space. However, to focus and detect the scattered light at a specific angle, the light-scattering sensing system normally contains more optical elements and complex electronic circuitry that makes the light-scattering sensor more expensive than the light-blocking sensor. The light-blocking sensor is widely used in detecting particles in water (e.g., water treatment application), whereas the light-scattering sensor is mostly used in detection of atmospheric particles. The key disadvantage of light-scattering and light-blocking particle counting sensors is the low sensitivity in comparison to the Coulter counter because the sensitivity of light-based sensors is determined by the particle's surface area, whereas the sensitivity of the Coulter counter is decided by the volume of the particle. In addition, the light-blocking technique has a severe light diffraction problem when particles are not close to the sensor (Nieuwenhuis et al., 2003). According to Nieuwenhuis et al. (2003), the

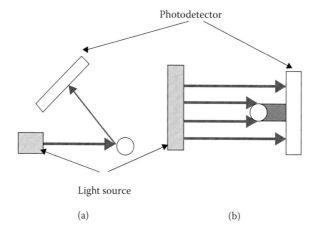

FIGURE 14.7
Illustration of two types of light-based particle sensors: (a) a light-scattering particle sensor and (b) a light-blocking particle sensor.

distance between particles and the light-blocking sensor should be within two times of the particle diameter.

Pamme et al. (2003) researched the counting and sizing of particles and particle agglomerates of C-reactive protein by the laser light-scattering method. They detected scattering light at two different angles of 15° and 45°. The experiment was carried out on a polymethyl methacrylate (PMMA) microchip, which consists of three inlet channels of two buffers and one sample channel and one outlet channel. The fluid was driven by a syringe pump generating a negative pressure with a flow rate of 1 µL/s to cancel the hydrostatic pressure in the inlet reservoirs. A beam splitter and an objective lens were applied to confine the He–Ne laser inside the channel to avoid scattering on the wall. The scattering light signals were sensed by two optical fibers and amplified by a photomultiplier tube. By this method, the authors realized particle detection of 2–9 µm. In addition, size discrimination of particles with a diameter ratio of 1:2 was achieved. On the other hand, the standard deviation from the average scattering light intensity for a given particle population was high, up to 30%, and a wide laser beam led to overall lower scattering light intensity and higher background scattering from the channel walls. These caused problems for measurement of smaller particles because of the lower scattering intensity and the lower signal-to-noise ratio. Xiang et al. (2005) used the light-blocking technique and developed a multifunctional particle detection system with embedded optical fibers in a PDMS chip to detect moving microparticles in a microchannel. They developed a PDMS glass microfluidic chip with two pairs of embedded optical fibers and removed the glass cladding layer of the input and receiving fibers. By filling the gap between the fiber and the fiber channel with PDMS, light leakage from the fiber core and light scattering from the fiber tips were minimized. By using the two-fiber detection method, particle velocity determination as well as particle counting and size identification were achieved. Xiang et al. achieved counting particles of 10, 20, and 25 µm. To get a better signal, both input and receiving fibers must be perfectly aligned. Xiang et al. used a larger size of the receiving fiber to alleviate this problem. However, in real applications, careful handling of the fiber-embedded chip and highly controlled fiber alignment were required.

Schafer et al. (2009) created an all-silicon and glass microfluidic device using femtosecond laser ablation and an anodic bonding technique and applied it for cell counting. This microchannel fabrication needed just a one-step process and shorter fabrication time by femtosecond pulse laser. The optical fibers for light-scattering signal detection were made to directly contact the liquid, causing light to focus on the optical detector by removing the optical free space between the detecting fiber and the channel. The incident light was delivered into the fiber with a 20×, 0.4 NA objective, and the light was delivered through the fiber at the normal direction and collected through the fiber on the opposite side at 14° from the normal and then collimated by an objective and focused onto a photodiode. HeLa cells derived from cervical cancer cells (Van Valen and Maiorana., 1991) were detected by scattered light with the reported lowest power. Conclusively, the system achieved similar particle detecting quality with a lower laser power of 2 mW and a cheaper photodiode. However, this device had difficulty in controlling even the depth and the alignment of the grooves used to guide the optic fibers and had poorer reproductivity than PDMS chips.

Kummrow et al. (2009) developed a microfluidic flow cytometer combined light-scattering detection and fluorescence detection system in a PMMA chip with integrated optical fibers, mirrors, and electrodes for flow cytometric analysis of blood cells. They used ultraprecision milling technique to fabricate different flow cells featuring single-stage and two-stage cascaded hydrodynamic focusing of particles in horizontal and vertical directions by a sheath flow. As shown in Figure 14.8, the first stage decreased the diameter of the sample flow to approximately 30 mm, and the second stage allowed reducing the

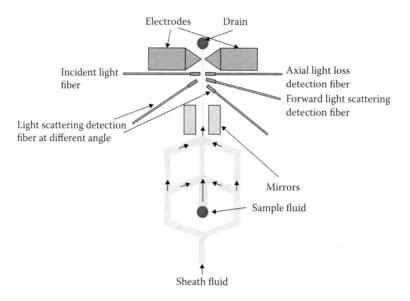

FIGURE 14.8
Schematic of a microfluidic system with two-stage cascaded hydrodynamic focusing, integrated mirrors, optical fibers, and fluidic connection. Multimode fibers serve to detect the axial light loss or scattered light when particles pass the interaction region. Orthogonal light scatter and fluorescence are achieved perpendicular to the joining plane by a microscope objective.

diameter to 5 mm while maintaining stable operation for sample flow rates of up to 20 μL/min. Inserted optic fibers were used to excite fluorescence of stained cells and to detect the axial light loss and the light scatter. Integrated mirrors were used to image the sample flow in the vertical direction, thus proving the efficiency of hydrodynamic focusing in two dimensions. After passing the optical sensor, the particles entered the interaction zone for impedance measurements. Three multimode optic fibers with the angles of 12°, 42°, and 145° from the incident light were used to detect the light scatter. The multimode optic fiber opposite to the incident light served to measure the axial light loss. An argon ion laser was used for forward light at 488 nm and for exciting the fluorescence of stained cells. T-helper lymphocytes labeled by monoclonal antibodies were identified by measuring side scatter and fluorescence. Using this cytometer, monodisperse polystyrene spheres with diameters ranging from 2 to 22 μm were detected.

14.7 Fluorescence-Based Particle Sensor

Fluorescence is an optical phenomenon in which the molecular absorption of a photon triggers the emission of a photon with a longer and less energetic wavelength. The energy difference between the absorbed and the emitted photons is reflected as molecular rotations, vibrations, or heat. This fluorescence characteristic has been applied to count particles. Lin and Lee (2003) proposed a method of fluorescence detection in a microflow cytometer without on-chip fibers. In this system, a PDMS microchip was bonded to a 150-μm-thick glass substrate. Laser was passed through a filter cube and an optical fiber to excite particles inside the microchannel. The excited fluorescence was also detected by the same fiber and translated to electrical signals in the photodetector. The lock-in amplifier amplified the electrical signal and transmitted it to a computer. Embedding the optic fiber into the chip added difficulty and cost to manufacturing the chip. However, the fiber was separated by the glass substrate, and the microchip was disposable. In the experiments, the authors successfully detected and counted 5- to 20-μm fluorescent particles, white blood cells, and yeast cells. The major advantage of this method was that there was no embedded optic fiber. The glass substrate worked as the interface between the detection fiber and the fluorescent particle. However, the detected fluorescent signals were not constant, depending on the distance between the optic fiber detector and the particles (the particles moving closer to the detector produce stronger signals). In addition, the precise alignment of the fiber and the channel was still difficult to realize.

Chen and Wang (2009) reported an on-chip fluorescence detection and counting system in a PDMS microchip. The fluorescence and the size information of the particle were characterized by combining forward-scattering signal and backward fluorescence signal. In the experiments, the particles of four different sizes with diameters ranging from 3.2 to 10.2 μm were distinguished and counted. The relative percentage of the fluorescence-labeled particles could be analyzed by the ratio of the events of fluorescence signals to forward scattered signals. Similar to that developed by Lin and Lee (2003), this system also adopted the externally installed fiber system for disposable chips. However, without a bulk lock-in amplifier, they could detect 3.2–10.2-μm-diameter particles by using an avalanche photodetector and a simple electronic filter circuit. In addition, the chip fabrication process was simpler and cheaper by use of only PDMS soft lithography. However, the same problems of inconsistent fluorescent signals caused by the distance between the photodetector and the particles and the difficulty in fiber alignment still existed.

In general, fluorescence is the one of the most sensitive methods for particle detection; however, for employing this method, particles must have fluorescent characteristics, and nonfluorescent particles have to be labeled with proper fluorophores. Therefore, this technique is widely used to distinguish specific target particles from other particles and often used in combination with other particle-counting techniques (Chen and Wang, 2009; Kummrow et al., 2009; Murali et al., 2009).

14.8 Micro-PIV Particle Sensor

PIV is an optical method used to obtain instantaneous velocity measurements and related properties in fluids. When small seeding particles flow inside a flowing liquid, their motion is traced and computed to show the fluid flow by taking sequential images of the positions of the tracing particles. A typical PIV apparatus consists of a digital camera, a high-power laser, and an optical arrangement to convert the laser output light to a light sheet. A fiber-optic cable often connects the laser to the cylindrical lens setup. The laser acts as a photographic flash for the digital camera, and the particles in the fluid scatter the light. The scattered light is detected by the camera. In the early twentieth century, the German scientist Ludwig Prandtl firstly used particles to study fluids in a systematic manner (Goldstein, 1996), and then the rapid development of laser and camera technology provided the possibility for flow visualization and later on quantified whole flow field measurement. Modern PIV software continues to improve the performance of PIV systems and their applicability to difficult flow measurements. Nowadays, PIV has become one of the most popular instruments for flow measurements in numerous applications.

Yabusaki et al. (1999, 2000) developed image cytometry centrifugation to count leukocytes and later extended their research to theoretically and experimentally study a microfluidic image cytometry by using micro-PIV flow visualization technique (Hirono et al., 2008). In their research, the number and size of particles flowing through a microchannel were measured simultaneously by image sequence analysis. During the experiments, a dilution series of 2-μm polystyrene particle suspensions was measured and compared with the results obtained with conventional Burker–Turk hemocytometry for validation of the particle counting. For particle diameter measurements, the diameters of 2-, 5-, 10-, and 20-μm particles were measured, and the results agreed well with reference values. This method may be used for the quantitative study of platelet aggregation in blood flow, and it would become a powerful diagnostic tool in the future. The image processing procedure demands complicated computational work and microscopic instrument, resulting in high cost, additional postprocessing, and a large system.

14.9 Conclusion

For counting biological or nonbiological particles, a traditional Coulter counter is still the most popular device because of its high sensitivity and throughput. However, because of the increasing need for portability and low cost in wide biological applications such as a handheld flow cytometer and a single DNA analyzer, many recent researches have focused on micro- or nanoscale particle counting sensors using advanced microfluidic technologies to realize portability, low cost, and simplicity.

In this chapter, among many particle sensing techniques, we reviewed only those applicable to microfluidic chips. Microfluidic RPS has been studied and demonstrated to have significantly higher sensitivity than the commercial Coulter counter, at a low cost. Furthermore, the RPS technique extends to nanosize by using natural or synthetic nanopores to detect single DNA/RNA. For applications in liquids of low electrical conductance, microfluidic capacitance measurement has been used instead of resistance measurement for particle counting. Light-scattering and light-blocking methods can also be applied in microfluidic chips, and these have a little lower sensitivity than the RPS technique. However, they are more useful in applications involving higher-density particles. Fluorescent detection has an advantage in identifying specific particles among similar-sized particles because the fluorescence method has higher specificity than any other electrical measurement. Micro-PIV technique can perform particle counting and analysis of the flow field simultaneously.

In addition to the disadvantages of the individual detecting techniques described in this chapter, low throughput is a major hindrance for the application of microfluidic particle counting sensors. A possible way to overcome this obstacle is to use multiple parallel channels; however, the number of parallel sensing microchannels that can be built in a single detecting chip is limited practically. Therefore, the enhancement of counting speed as well as stable and sensitive detection over a longer period of time is among the challenging issues. Furthermore, to meet the demand of the complex analytical detection in real applications, the integration of multiple particle counting methods in a single microfluidic system should be investigated.

Acknowledgment

The research grants from the Canada Research Chairs program (Li) and the Canada Foundation for Innovation (Li) and from China 111 Project (Zhang and Pan) are greatly appreciated.

References

Aalto P, Hameri K, Paatero P, Kulmala M, Bellander T, et al. (2005). Aerosol particle number concentration measurements in five European cities using TSI-3022 condensation particle counter over a three-year period during health effects of air pollution on susceptible subpopulations. *J Air Waste Manage Assoc*. 55:1064–76.

Akeson M, Branton D, Kasianowicz JJ, Brandin E, Deamer DW. (1999). Microsecond time0scale discrimination among polycytidylic acid, polyadenylic acid, and polyuridylic acid as homopolymers or as segments within single RNA molecules. *Biophys J*. 77:3227–33.

Amann RI, Binder BJ, Olson RJ, Chisholm SW, Devereux R, Stahl DA. (1990). Combination of 16S rRNA-targeted oligonucleotide probes with flow cytometry for analyzing mixed microbial populations. *Appl Environ Microbiol*. 56:1919–25.

Ateya DA, Erickson JS, Howel PB Jr, Hilliard LR, Golden JP, Ligler FS. (2008). The good, the bad, and the tiny: a review of microflow cytometry. *Anal Bioanal Chem*. 391:1485–98.

Bezrukov SM, Vodyanoy I, Brutyan RA, Kasianowicz JJ. (1996). Dynamics and free energy of polymers partitioning into a nanoscale pore. *Macromolecules*. 29:8517–22.

Biance A-L, Gierak J, Bourhis B, Madouri A, Lafosse X, Patriarche G, et al. (2006). Focused ion beam sculpted membranes for nanoscience tooling. *Microelectron Eng*. 83:1474–7.

Bundschuh T, Knopp R, Winzenbacher R, Kim JI, Köster R. (2001). Quantification of aquatic nano particles after different steps of Bodensee water purification with laser-induced breakdown detection (LIBD). *Acta Hydrochim Hydrobiol*. 21:7–15.

Burnett AK, Grimwade D, Solomon E, Wheatley K, Goldstone AH. (1999). Presenting white blood cell count and kinetics of molecular remission predict prognosis in acute promyelocytic leukemia treated with all-trans retinoic acid: result of the randomized MRC trial. *Blood*. 93:4131–43.

Carbonaro A, Sohn LL. (2005). A resistive-pulse sensor chip for multianalyte immunoassays. *Lab Chip*. 5:1155–60.

Chen HT, Wang YN. (2009). Optical microflow cytometer for particle counting, sizing and fluorescence detection. *Microfluid Nanofluids*. 6:529–37.

Coulter WH, inventor. (1953). Means for counting particles suspended in a fluid. United States patent US 2656508.

Coulter WH, Hogg WR, inventors. (1976). Methods and apparatuses for correcting coincidence count inaccuracies in a Coulter type of particle analyzer. United States patent US 3949198.

DeBlois RW, Bean CP. (1970). Counting and sizing of submicron particles by the resistive pulse technique. *Rev Sci Instrum*. 41:909.

DeBlois RW, Wesley RKA. (1977). Sizes and concentrations of several type C oncornaviruses and bacteriophase T2 by resistive-pulse technique. *J Virol*. 23:227.

Goldstein RJ. (1996). *Fluid Mechanics Measurements*. 2nd ed. Philadelphia: Taylor & Francis.

Harrell CC, Choi Y, Horne LP, Baker LA, Siwy ZS, Martin CR. (2006). Resistive-pulse DNA detection with a conical nanopore sensor. *Langmuir*. 22:10837–43.

Hirono T, Arimoto H, Okawa S, Yamada Y. (2008). Microfluidic image cytometry for measuring number and sizes of biological cells flowing through a microchannel using the micro-PIV technique. *Meas Sci Technol*. 19:025401.

Horne BD, Anderson JL, John JM, Weaver A, Bair TL, Jensen KR, et al. (2005). Which white blood cell subtypes predict increased cardiovascular risk? *J Am Coll Cardiol.* 45:1638–43.

Ito T, Sun L, Henriquez RR, Crooks RM. (2004). A carbon nanotube-based Coulter nanoparticles counter. *Acc Chem Res.* 37:937–45.

Jagtiani AV, Sawant R, Zhe J. (2006). A label-free high throughput resistive-pulse sensor for simultaneous differentiation and measurement of multiple particle laden analytes. *J Micromech Microeng.* 16:1530–9.

Jagtiani AV, Zhe J, Hu J, Carletta J. (2006). Detection and counting of micro-scale particles and pollen using a multi-aperture Coulter counter. *Meas Sci Technol.* 17:1706–14.

Judd SJ, Hillis P. (2001). Coagulation pretreatment for crossflow microfiltration of upland waters. *Water Res.* 35:2895–904.

Kannel WB, Anderson K, Wilson PWF. (1992). White blood cell count and cardiovascular disease. *Am Med Assoc.* 267:1253–6.

Kasianowicz JJ, Brandin E, Branton D, Deamer DW. (1996). Characterization of individual polynucleotide molecules using a membrane channel. *Proc Natl Acad Sci.* 93:13770–3.

Kulp SK, Yang Y, Hung C, Chen K, Lai J, Tseng P, Fowble JW, et al. (2004). 3-Phosphoinositide-dependent protein kinase-1/Akt signaling represents a major cyclooxygenase-2-independent target for Celecoxib in prostate cancer cells. *Cancer Res.* 64:1444–51.

Kummrow A, Theisen J, Frankowski M, Tuchscheerer A, Yildirim H, Brattke K, et al. (2009). Microfluidic structures for flow cytometric analysis of hydrodynamically focused blood cells fabricated by ultraprecision micromachining. *Lab Chip.* 9:972–81.

Li D. (2004). *Electrokinetics in Microfluidics.* Boston: Elsevier.

Lin C, Lee G. (2003). Micromachined flow cytometers with embedded etched optic fibers for optical detection. *J Micromech Microeng.* 13:447–53.

Miller JL, Kitaljevich D. (2000). In-line oil debris monitor for aircraft engine condition assessment. IEEE Aerospace Conference, Big Sky, Montana, USA.

Murali S, Jagtiani AV, Xia X, Carletta J, Zhe J. (2009). A microfluidic Coulter counting device for metal wear detection in lubrication oil. *Rev Sci Instrum.* 80:016105.

Nieuwenhuis JH, Bastemeijer J, Bossche A, Vellekoop MJ. (2003). Near-field optical sensors for particle shape measurements. *IEEE Sens J.* 3:646–51.

Pamme N, Koyama R, Manz A. (2003). Counting and sizing of particles and particle agglomerates in a microfluidic device using laser light scattering: application to a particle-enhanced immunoassay. *Lab Chip.* 3:187–92.

Rogers JA, Nuzzo RG. (2005). Recent progress in soft lithography. *Materials Today.* 8:50–6.

Schafer D, Gibson EA, Salim EA, Palmer AE, Jimenez R, Squier J. (2009). Microfluidic cell counter with embedded optical fibers fabricated by femtosecond laser ablation and anodic bonding. *Opt Express.* 17:6068–73.

Sexton LT, Horne LP, Sherrill SA, Bishop GW, Baker LA, Martin CR. (2007). Resistive-pulse studies of proteins and protein/antibody complexes using a conical nanotube sensor. *J Am Chem Soc.* 129:13144–52.

Siwy Z, Trofin L, Kohli P, Baker LA, Trautmann C, Martin CR. (2005). Protein biosensors based on biofunctionalized conical gold nanotubes. *J Am Chem Soc.* 127:5000–1.

Smolen AJ, Wright LL, Cunningham TJ. (1983). Neuron numbers in the superior cervical sympathetic ganglion of the rat: a critical comparison of methods for cell counting. *J Neurocytol.* 12:739–50.

Sohn LL, Saleh OA, Facer GR, Beavis AJ, Allan RS, Notterman DA. (2000). Capacitance cytometry: measuring biological cells one by one. *Proc Natl Acad Sci.* 97:10687–90.

Sridhar M, Xu D, Kang Y, Hmelo AB, Feldman LC, Li D, Li D. (2008). Experimental characterization of a metal–oxide–semiconductor field-effect transistor-based Coulter counter. *J App Phys.* 103:104701.

Van Valen LM, Maiorana VC. (1991). HeLa, a new microbial species. *Evol Theory Rev.* 10:71–4.

Vozarova B, Weyer C, Lindsay RS, Pratley RE, Bogardus C, Tataranni PA. (2002). High alanine aminotransferase is associated with decreased hepatic insulin sensitivity and predicts the development of type 2 diabetes. *Diabetes.* 51:455–61.

Wahl-Jensen V, Chapman J, Asher L, Fisher R, Zimmerman M, Larsen T, et al. (2007). Temporal analysis of Andes virus and Sin Nombre virus infections of Syrian hamsters. *J Virol*. 81:7449–62.

Wharton JE, Jin P, Sexton LT, Horne LP, Sherrill SA, Mino WK, et al. (2007). A method for reproducibly preparing synthetic nanopores for resistive-pulse biosensors. *Small*. 3:1424–30.

Wood DK, Oh S-H, Lee S-H, Soh HT, Cleland AN. (2005). High-bandwidth ratio frequency Coulter counter. *App Phys Lett*. 87:184106.

Wu JJ, Miller RJ, Cooper DW, Flynn JF, Delson DJ, Teagle RJ. (1989). Deposition of submicron aerosol particles during integrated circuit manufacturing: experiments. *J Environ Sci*. 32:43–5.

Wu X, Chon CH, Wang Y, Kang Y, Li D. (2008). Simultaneous particle counting and detecting on a chip. *Lab Chip*. 8:1943–9.

Wu X, Kang Y, Wang Y, Xu D, Li D, Li D. (2008). Microfluidic differential resistive pulse sensors. *Electrophoresis*. 29:2754–9.

Xiang Q, Xuan X, Xu B, Li D. (2005). Multi-functional particle detection with embedded optical fibers in a poly(dimethylsiloxane) chip. *Instrum Sci Technol*. 33:597–607.

Xu D, Kang Y, Sridhar M, Hmelo AB, Feldman LC, Li D, et al. (2007). Wide-spectrum, ultrasensitive fluidic sensors with amplification from both fluidic circuits and metal oxide semiconductor field effect transistors. *Appl Phys Lett*. 91:013901.

Yabusaki K, Hirono T, Matsui H. (1999). XII International Symposium on Technological Innovations in Laboratory Hematology, Workshop on New Technology in Laboratory Hematology, Kobe, Japan.

Yabusaki K, Saitoh N, Hirono T, Matshu H, Okada H. (2000). High-sensitive method for counting low concentrations of particles in liquids using centrifugation and image analysis. *Jpn J Appl Phys*. 39:3641–4.

Yarnell JW, Baker IA, Sweetnam PM, Bainton D, O'Brien JR, Whitehead PJ, et al. (1991). Fibrinogen, viscosity, and white blood cell count are major risk factors for ischemic heart disease. The Caerphilly and Speedwell collaborative heart disease studies. *Circulation*. 83:836–44.

Zhe J, Jagtiani AV, Dutta P, Hu J, Carletta J. (2007). A micromachined high throughput Coulter counter for bioparticle detection and counting. *J Micromech Microeng*. 17:304–13.

15

Magnetic-Particle-Based Microfluidics

Ranjan Ganguly, Ashok Sinha, and Ishwar K. Puri

CONTENTS

15.1 Introduction

Progress in microfluidics (Dittrich et al., 2006) is being driven by a growing understanding of the behavior of materials at the microscale and the need for miniaturization that is driven by emerging applications, for example, the demands of fluid and analyte handling in micro–total analytical systems (μ-TAS). Controlled transport of fluids and fluid-borne solids in microfluidic environments is enabled through diverse effects, such as inertial, viscous, surface tension, electrostatic, magnetic, chemical, or molecular interactions. Magnetic-particle-based microfluidics offers an enabling technology to overcome major challenges related to the design of lab-on-a-chip devices. This chapter provides insight into the microfluidic domain that involves magnetic particles, in particular, the transport of ferrofluids and magnetic microspheres.

Magnetic-particle-based microfluidics has numerous advantages, for instance:

(i) Action at a distance: Particles can be manipulated in microchannels by externally imposed magnetic field gradients, without direct physical contact, to overcome surface tension, viscous force, and van der Waals interactions. The ability to manipulate particles from a distance frees space on microfluidic platforms to include additional components or to enable multiplexed batch processing.

(ii) Insensitivity to the biochemical environment: The magnetic force on the particles is relatively insensitive to the biochemical environment of the particles, for example, pH and ionic concentration. This enables the decoupling of the physical phenomena from the biochemical environment, which is not possible using electrophoretic forces alone.

(iii) Insensitivity to dominating physical forces at the microscale: A static magnetic field in a low-velocity flow system has almost no interference from other microscale forces, such as electrostatic, surface tension, Brownian, or van der Waals forces. Thus, a carefully designed microfluidic system can include both a magnetic force and these other forces.

(iv) Magnetic contrast: Most biological entities, with the exception of magnetotactic bacteria and red blood cells, and commonly used chip materials, such as Si, PDMS, PMMA, and μ-TAS working fluids, offer stark magnetic contrast with magnetic particles, offering medium-independent magnetic actuation.

(v) Tailored particles: Perhaps the most compelling reason to use magnetic particles is the range of physical and chemical properties that they may be tailored to. These particles are available in a wide range of sizes from tens of nanometers to a few microns, offering a large and tunable surface to volume ratio.

(vi) Tunable particle surface: Besides being physically maneuverable, these particles can also be functionalized with chemical substances and bioconjugates to accomplish specific tasks. The functional groups coated on the particle surface interact with complementary species in the immediate fluid environment. Biochemical reactions can thus be executed with considerable ease on the mobile substrates. The combination of functionalization and "action at a distance" is a major advantage of magnetic particles in microfluidic bioanalytical devices.

(vii) Magnetic detection: Because magnetic particles offer a significant magnetic contrast with respect to the environment, they can also be used for magnetic detection, for example, using a magnetic relaxation immunoassay (MARIA), a giant magnetoresistance (GMR), or a Hall sensor. Unlike other bioanalytical labels, for example, radioisotope or fluorophores, the magnetic properties of the particles do not deteriorate over time or because of biochemical reactions. Implementing these magnetic diagnostic systems on a biosensor chip requires only marginal addition to the magnetic system required for achieving particle manipulations, thereby eliminating the need for a separate and elaborate diagnostic system, for example, electrochemical or fluorescence detection.

The most common magnetic particles used for microfluidic applications are single domain iron oxide nanoparticles of 5–15 nm diameter. At this size, the particles are superparamagnetic so that that the particle magnetization curves do not show any hysteresis, although their magnetization is comparable with ferri- or ferromagnetic particles. When these particles are coated with an adsorbed surfactant layer and stably suspended in a nonmagnetic liquid carrier, the resulting suspension is called a ferrofluid (Rosensweig, 1985). Although ferrofluids are nonmagnetic in the absence of a magnetic field, the thermally disoriented magnetic moments of the particles are readily aligned with an externally imposed magnetic field, making the fluid magnetically responsive. The superparamagnetic iron oxide nanoparticles (SPIONs) can also be embedded within biodegradable polymers, such as dextran, in the form of 50- to 500-nm aggregates (Kantor et al., 1998) or encapsulated inside 500-nm to 10-µm-sized polystyrene, latex, or silica microspheres that are often termed "magnetic beads" (Rida and Gijs, 2004). The SPION aggregates and magnetic beads possess larger magnetic moments than an isolated nanoparticle but still retain their superparamagnetic nature. The transport of ferrofluids, SPION aggregates, and magnetic beads can be controlled using a suitably tailored magnetic field. Similar to isolated nanoparticles, ferrofluids, SPIONs, and magnetic beads can also be surface functionalized with chemical and biological ligands to accomplish specific bioanalytical tasks.

The magnetic field and its gradient necessary for manipulation of magnetic particles in a microfluidic environment can be provided either by an active device, for example, a miniaturized permanent magnet or electromagnet, or by a passive device, for example, a macroscale biasing magnet in conjunction with microscale magnetizable elements for creating local field gradients. State-of-the-art microfabrication technology offers several possibilities to generate large magnetic field gradients up to 20 T/mm. Several designs are available in the literature, including (1) microelectromechanical systems (MEMS) scale electromagnetic coil with a soft magnetic core (Ramadan et al., 2009), (2) chip-embedded microfabricated current-carrying wires (Deng et al., 2001) and meshes (Lee et al., 2004), or (3) a combination of a permanent magnet and magnetizable inserts (Rida and Gijs, 2004; Furlani, 2006; Bu et al., 2008) placed within short distances to manipulate these particles. Permanent magnets and electromagnets have their own advantages. Permanent magnets can create stronger magnetic force fields that function without an active power source. However, electromagnets generally offer better control because (1) they are switchable, (2) the field strength can be altered by controlling the current, and (3) excellent localization of the force field is possible as the electromagnets can be microfabricated at smaller length scales than the permanent magnets. The force field can be localized using a Halbach array (Häfeli et al., 2007) or a matrix of permanent magnets placed with alternate polarity (Bu et al., 2008). The choice of the magnetic system for a particular application depends on the relevant task of the device. A high-throughput magnetic separator, for instance, would require a stronger magnetic force than can be induced by a permanent magnet array (Bu et al., 2008), whereas a magnetic manipulator that requires an alternating magnetic field would require an electromagnet (Suzuki et al., 2004). Sometimes, a combination of permanent magnets and electromagnets is also used. In such designs, the permanent magnet magnetizes the magnetic particles, whereas the electromagnet provides the necessary guiding force (Beyzavi and Nguyen, 2009). Most designs use soft magnetic materials, typically permalloy or nickel, in the form of chip-embedded posts and linear arrays to augment the local field and its gradients.

Although the process of synthesizing and modifying magnetic particles is not new, their extensive use in microfluidic applications is relatively recent. This chapter provides an insight into the magnetic behavior and the transport of ferrofluids and magnetic microspheres and an overview of their state-of-the-art applications in microfluidic devices. A background of magnetism and magnetic materials that is necessary to explain the behavior of magnetic particles is provided in the section "Magnetism and Magnetic Materials." The section "Dynamics of Magnetic Particles in Microfluidic Environments" discusses the salient attributes of microfluidic magnetic-particle transport. Microfluidic applications of ferrofluids and magnetic microspheres are discussed in the sections "Microfluidic Application of Ferrofluids" and "Microfluidic Application of Magnetic Microspheres," respectively. The promises and the challenges of magnetic-particle-based microfluidics are summarized in the section "Concluding Remarks."

15.2 Magnetism and Magnetic Materials

15.2.1 Fundamentals of Magnetism

Magnetic fields are produced because of electric currents, either free or bound (these two terms are explained soon). For a solenoid of length l and having n turns that carries a current I through the conductor, the magnetic field inside the coil (having n/l turns per unit length and air or vacuum in its core) is represented by $\mathbf{B} = \mu_0 nI/l$ along the axis of

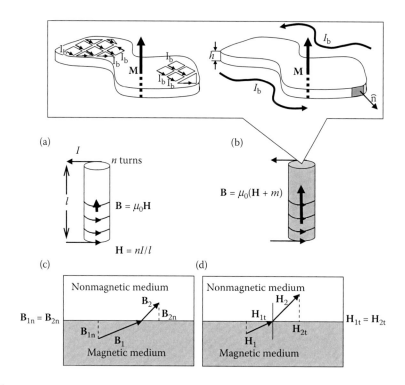

FIGURE 15.1
Magnetic field inside a solenoid with (a) air/vacuum core and (b) iron core. Magnetization of a magnetic material can be attributed to the bound circulating current I_b. For a uniformly magnetized slab of thickness h, the bound current $\mathbf{K}_b$ ($|\mathbf{K}_b| = I_b/h$) is related to the magnetization $\mathbf{M}$ as $\mathbf{K}_b = \mathbf{M} \times \hat{n}$ (inset to panel b). Boundary conditions at the interface of a ferrofluid and a nonmagnetic fluid for the (c) $\mathbf{B}$ and (d) $\mathbf{H}$ vector fields.

the coil (see Figure 15.1a), where μ_0 (= $4\pi \times 10^{-7}$ N/A²) denotes the permeability of vacuum. The quantity nI/l is purely a function of the coil geometry and the current and is denoted by the variable **H** (A/m).* The introduction of a magnetizable material inside the solenoid enhances the magnetic field. This happens because the **H** produced by the current-carrying coil induces a magnetic dipole moment **M** per unit volume in the material, and it adds to **H** to produce the magnetic field **B** = μ_0(**M** + **H**) (see Figure 15.1b).

For a more generalized case of magnetostatics, Ampere's law relates the free current (the current produced by the free charges, e.g., by connecting a voltage source across a circuit, which can be measured by an ammeter) density $\mathbf{J}_f$ with the distribution of resulting **H** as follows:

$$\nabla \times \mathbf{H} = \mathbf{J}_f. \tag{15.1}$$

Magnetization is a property of the material and also a function of the **H** imposed. In electrodynamics, **M** is assumed to be produced by a collection of circulating bound current loops uniformly distributed throughout the material (refer to the top-left inset of Figure 15.1b)—a hypothesis that relates the macroscopic magnetism to the origin of molecular magnetism. Within atoms (or molecules), the orbital and spin motions of electrons and spinning of protons within the nucleus produce magnetic fields. However, the latter contribution is usually neglected because the induced magnetic field of nucleons is typically two thousand times smaller than that due to electrons. For the uniformly magnetized material shown in Figure 15.1b, the adjacent circulating current loops cancel each other, leaving a single ribbon of net current of the bound charges flowing along the edge of the material element (see the right-top inset of Figure 15.1b). It is interesting to note that unlike the free current, the bound current can neither be measured using an ammeter, nor can it cause Joule heating. The magnetization and the circulating current density are related by the cross product $\mathbf{K}_b = \mathbf{M} \times \hat{n}$, where $|\mathbf{K}_b| = I_b/h$. For a more generalized case, in which the magnetization varies spatially (both in magnitude and direction), the relationship is represented as follows (Griffiths, 2004):

$$\nabla \times \mathbf{M} = \mathbf{J}_b. \tag{15.2}$$

Depending upon the magnetic response (i.e., the value of **M**), materials are classified as diamagnetic, paramagnetic, ferromagnetic, ferrimagnetic, antiferromagnetic, and canted antiferromagnetic (Spaldin, 2006). The materials used for magnetic-particle-based microfluidics generally fall under the class of ferro- and ferrimagnetic materials, which are characterized by large magnetization and hysteresis.

The magnetic field **B** is an outgrowth of both the free and the bound current and is related to the total current density **J** (where **J** = $\mathbf{J}_f$ + $\mathbf{J}_b$) as follows:

$$\nabla \times \mathbf{B} = \frac{1}{\mu_0}\mathbf{J}. \tag{15.3}$$

Because magnetic fields are produced by circulating currents (as opposed to the traditional belief of magnetic "charge" or monopoles), the **B** field does not have any "source"

---

* We follow the nomenclature proposed by Griffiths (2004) and quote a line from p. 271 to explain why "Many authors call **H**, not **B**, the 'magnetic field'. Then they have to invent a new word for **B**: the 'flux density', or magnetic 'induction' (an absurd choice, since the term has at least two other meanings in electrodynamics). Anyway, **B** is indisputably the fundamental quantity, so I shall continue to call it the 'magnetic field', as everyone does in spoken language. **H** has no sensible name: just call it '**H**.'" Also the footnote on the same page of the book quotes from A. Sommerfield's *Electrodynamics* (Academic Press, New York, 1952, p. 45): "the unhappy term 'magnetic field' for **H** should be avoided as far as possible. It seems to us that *this term has led into error none less than Maxwell himself ...*"

or "sink," and the magnetic flux lines are always closed in nature. Thus, the **B** field has a solenoidal property:

$$\nabla \times \mathbf{B} = 0. \tag{15.4}$$

Equations 15.1 and 15.4 together describe the governing equations for static magnetic fields in a medium and are also known as Maxwell's equations. At the interface of a magnetic and nonmagnetic media, where a discontinuity of magnetization exists, Equations 15.1 and 15.4 warrant that the tangential components of the **H** vector and the normal components of the **B** vector on either sides of the interface are consistent (refer to Figures 15.1c and 15.1d), that is,

$$\mathbf{H}_{1t} = \mathbf{H}_{2t} \text{ and} \tag{15.5}$$

$$\mathbf{B}_{1n} = \mathbf{B}_{2n}, \tag{15.6}$$

where the suffixes t and n indicate the tangential and normal components at the interface, respectively.

When a magnetic material is exposed to a magnetic field, it experiences a force that occurs because of the magnetization (the bound current) **M**. The volumetric body force, also known as the Kelvin body force (KBF), is expressed as (Zahn, 1979)

$$f = \mu_0(\mathbf{M} \times \nabla)\mathbf{H} + \nabla((1/2)\,\mu_0\mathbf{M} \times \mathbf{M}). \tag{15.7}$$

Physically, this indicates that the bulk of a magnetic material is attracted toward the region of higher magnetic field (see Figure 15.2a). Using tensor notation, the expression of KBF reduces to

$$f_i = \mu_0[M_j(\partial H_i/\partial x_j) + (1/2)\,\partial(M_j\,M_j)/\partial x_i] = M_j(\partial B_j/\partial x_i). \tag{15.8}$$

The total force on a magnetizable body is evaluated using a volumetric integral of the KBF, that is,

$$\mathbf{F} = \iiint \left[\mu_0\left(\mathbf{M} \times \nabla\right)\mathbf{H} + \nabla\left(\frac{1}{2}\mu_0\mathbf{M} \times \mathbf{M}\right) \right] d\vartheta, \tag{15.9a}$$

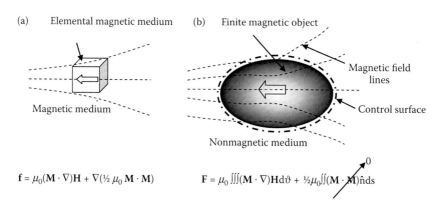

FIGURE 15.2
Magnetic body forces in a nonuniform **B** field on (a) an elemental magnetic medium immersed in the same magnetic medium and (b) a finite magnetic object in a nonmagnetic medium.

where $d\vartheta$ denotes the differential volume. Using the corollary of Gauss divergence theorem, the second term on the right-hand side of Equation 15.9a can be modified to a surface integral, so that

$$\mathbf{F} = \mu_0 \iiint (\mathbf{M} \times \nabla) \mathbf{H} \, d\vartheta + \frac{1}{2} \mu_0 \oiint (\mathbf{M} \times \mathbf{M}) \hat{n} ds, \tag{15.9b}$$

where $\hat{n}ds$ denotes a differential area vector on the control surface. For evaluation of force on a magnetic body of finite size immersed in a nonmagnetic medium, the last term in Equation 15.9b vanishes because the control surface encloses the magnetizable body completely (thus passing through a region outside the magnetic body where $\mathbf{M} = 0$; see Figure 15.2b). Thus, the expression of magnetic force on a magnetic object in a nonmagnetic medium is as follows:

$$\mathbf{F} = \mu_0 \iiint (\mathbf{M} \times \nabla) \mathbf{H} \, d\vartheta. \tag{15.9c}$$

15.2.2 Magnetic Nanoparticles

Although the commonly used magnetic nanoparticles contain ferro- or ferrimagnetic materials, the magnetic behavior of finely divided particles is quite different from what their bulk counterparts exhibit. Mössbauer spectroscopy of ferro- and ferrimagnetite particles (McNab et al., 1968) of diameter ranging from 10 to 16 nm has revealed that although the particles have the same saturation magnetization characteristic of the bulk material, the remnant magnetism is zero. At this size range, the nanoparticles often contain a single domain and therefore have a permanent magnetic moment proportional to their volume. At normal temperatures (~300 K), the thermal Brownian energy of individual particle magnetic domains is much larger than the magnetocrystalline anisotropy energy, and the particles exhibit superparamagnetic behavior (Berkovsky, 1978). The individual particle dipoles are randomly oriented because of thermal agitation, and hence a collection of these nanoparticles does not respond magnetically unless an external magnetic field (that tends to align the individual dipoles of the nanoparticles) is imposed. Depending upon the size of a magnetic nanoparticle and the viscosity of the medium where it is suspended, the magnetic relaxation (ability to follow the direction of the imposed magnetic field) can be Brownian, in which the entire particle rotates physically to align with the imposed field, or Néelian, which means that the magnetic moment inside the particle rotates without any bodily rotation of the particle (refer to Figure 15.3a). The effective relaxation of the magnetic nanoparticle would be dominated by the faster one of the two mechanisms (Ganguly, 2005).

Because the superparamagnetic nanoparticles suspended in ferrofluids (liquid suspension) or embedded inside polystyrene or silica microspheres (solid suspension) behave as collections of individual noninteracting dipoles, the magnetization of the suspension obeys Langevin equation (Rosensweig, 1985), namely,

$$\langle |\mathbf{M}| \rangle = M_{sat} \phi_v \left(\coth(\alpha) - \frac{1}{\alpha} \right), \tag{15.10}$$

where M_{sat} denotes the bulk saturation magnetization of the particle material, ϕ is the volume fraction of magnetic particles in the suspension, and α is the Langevin parameter, which is given by

$$\alpha = \frac{4\pi a_{np}^3}{3} \frac{\mu_0 M_{sat} |\mathbf{H}|}{k_B T}. \tag{15.11}$$

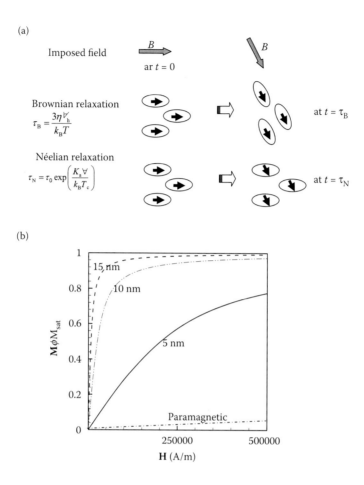

FIGURE 15.3
(a) Two different modes of magnetic relaxation of nanoparticles: (top) Brownian and (bottom) Néelian. τ_0 = Larmor time constant ~10^{-9} s. The effective relaxation time due to the combined action is $\tau_{eff} = \tau_B \tau_N / (\tau_B + \tau_N)$. (b) Magnetization curves for SPIONs. A qualitative comparison of the magnetization of a paramagnetic substance is shown on the same scale.

Here, a_{np} denotes the radius of the nanoparticles, k_B is the Boltzmann constant (=1.3807×10^{-23} J/K), and T is the absolute temperature. Figure 15.3b shows the representative magnetization curves (normalized with respect to the saturation magnetization) for superparamagnetic ferrite nanoparticles following the Langevin equation at 300 K. As evident from Figure 15.3b, an almost linear magnetization curve is observed for particles with a diameter smaller than 5 nm. Thus, the particle magnetization at very small diameters approaches the characteristic magnetization of a paramagnetic substance, but only with a much larger magnetic moment. The magnetic susceptibility χ of the nanoparticles suspension is defined as follows:

$$\chi = \left(\frac{\partial \mathbf{M}}{\partial \mathbf{H}} \right)_T.$$

(15.12)

The Langevin function has low-field ($\alpha \ll 1$) and high-field ($\alpha \gg 1$) asymptotes

$$\lim_{\alpha \to 0} L(\alpha) = \frac{M}{\phi M_{sat}} = \frac{\alpha}{3} = \frac{4\pi a_{np}^3}{9} \frac{\mu_0 M_{sat} |\mathbf{H}|}{k_B T} \quad \text{and} \tag{15.13a}$$

$$\lim_{\alpha \to \infty} L(\alpha) = \frac{M}{\phi M_{sat}} = \left(1 - \frac{1}{\alpha}\right) = \left(1 - \frac{3k_B T}{4\pi a_{np}^3 \mu_0 M_{sat} |\mathbf{H}|}\right), \tag{15.13b}$$

respectively. As can be seen from Equation 15.13a and Figure 15.3b, the low-field **M–H** curve for the nanoparticles suspension is linear, and the susceptibility can be approximated as $\chi \approx \phi - (4\pi a_{np}^3/9)(\mu_0 M_{sat}^2/k_B T)$. Currently available small electromagnets produce fields with strengths much lower than the saturating field (e.g., for magnetite, $\mu_0 M_{sat}$ is 0.56 T). Hence, many theoretical analyses assume a constant magnetic susceptibility for the magnetic nanoparticles suspensions (Smistrup et al., 2005a).

15.2.2.1 Ferrofluids

Ferrofluids are colloidal suspensions of single-domain magnetic nanoparticles, typically of 10-nm diameter, containing Ni, Co, Mg, or Zn compositions of ferrite (Fe_2O_4), magnetite (Fe_3O_4), and maghemite (γ-Fe_2O_3) in a nonmagnetic liquid carrier (Rosensweig, 1985; Odenbach, 2004). Because the suspended particles are superparamagnetic, ferrofluids do not exhibit permanent magnetization until an external magnetic field is imposed. The nanoparticles are often coated with an adsorbed layer of surfactant that prevents particle agglomeration because of the attractive van der Waals force and magnetic dipolar interactions between the particles (see Figure 15.4a). Magnetite nanoparticles can be prepared by chemical coprecipitation of ferrous salts, for example, iron (II) and iron (III) chloride salts in the presence of ammonium hydroxide at pH 9–10. Subsequently, the nanoparticles are stabilized in oleic acid that prevents particle agglomeration by steric (entropic) repulsion (Harris, 2002). Depending on the nature of the particle material, the surfactant, and the carrier liquid, ferrofluids are commercially available (Ferrotec; www.ferrotec.com) with a wide range of physical and magnetic properties. The choice of the liquid carrier, which can be water, mineral oil, organic solvent, ester, and so forth, depends on the application. As mentioned before, the nanoparticles of a ferrofluid can be functionalized with suitable inorganic or organic molecules to serve specific chemical or biological tasks (Kouassi et al., 2005).

In a microfluidic medium, a colloidally stable ferrofluid can behave as a continuum (Cowley and Rosensweig, 1967; Ganguly et al., 2004a,b) or a self-assembled aggregate (Ganguly et al., 2005), or it can even be dispersed as a suspension of finely divided droplets in a background liquid. Such a ferrofluid suspension can be treated using a continuum approach, as the size of the dispersed nanoparticles is comparable with the molecular mean free path of the carrier liquid. The nanoparticle concentration in the ferrofluid is generally small enough ($\phi_v < 0.03$) to neglect the dipole–dipole interactions, so that the Langevin equation describes the fluid magnetization well (Lange, 2002; van Ewijk et al., 2002).

Neglecting the magnetodissipation (Müller and Engel, 1999) and the magnetoviscous effect (Shliomis and Stepanov, 1993), the hydrodynamic equations describing the

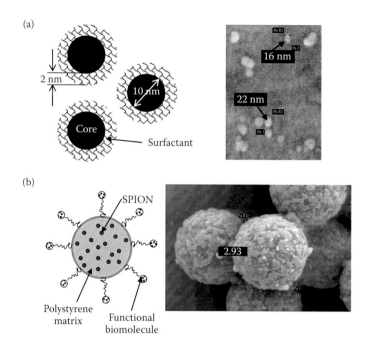

FIGURE 15.4
(Left) Schematic and (right) scanning electron microscope images of (a) ferrofluid and (b) magnetic microspheres.

mass, momentum, and species equations (for cases involving two or more fluids) can be written as

$$\partial\rho/\partial t + \partial(\rho v_j)/\partial x_j = 0, \tag{15.14}$$

$$\partial(\rho v_i)/\partial t + \partial(\rho v_j v_i)/\partial x_j = -\partial p/\partial x_i + \partial\tau_{ij}/\partial x_j + f_i, \quad \text{and} \tag{15.15}$$

$$\partial(\rho Y_f)/\partial t + \partial[\rho v_j Y_f - \rho D \cdot \partial Y_f/\partial x_j + 1/(6\pi\eta a_{np}) \cdot \rho Y_f f_j]/\partial x_j = 0, \tag{15.16}$$

respectively. Here ρ denotes fluid density, v_i is the velocity, p is the pressure, D is the ferrofluid diffusivity, and Y_f is the ferrofluid mass fraction. The viscous stress tensor τ_{ij} that appears in the momentum equation is expressed in terms of the fluid viscosity η and velocity gradient, that is,

$$\tau_{ij} = \eta(\partial v_i/\partial x_j + \partial v_j/\partial x_i) - (2/3)\,\eta\delta_{ij}(\partial v_k/\partial x_k), \tag{15.17}$$

where δ_{ij} denotes the Kronecker delta. The last term in the momentum equation contains the KBF expressed per unit volume $f_i = M_j(\partial B_j/\partial x_i)$ as explained in Equation 15.8. As evident from the expression of KBF, this magnetic body force is experienced only under a nonuniform magnetic field (i.e., in a field having a nonzero gradient). The mass fraction of nanoparticles in the host fluid, Y_f, is linked with the particle volume fraction ϕ through the relation

$$\phi = \rho_l Y_f/[\rho_p - (\rho_p - \rho_l)Y_f], \tag{15.18}$$

where ρ_p and ρ_l denote the densities of the particles and the host liquid, respectively, and the local fluid density ρ is related to the local volumetric fraction of the nanoparticles through the expression

$$\rho = (\rho_p - \rho_l)\phi + \rho_l. \tag{15.19}$$

For a dilute ferrofluid, the diffusivity D of the magnetic nanoparticles (of radius a_n) can be obtained using the Stokes–Einstein equation (Probestein, 1994), that is,

$$D = k_B T/6\pi\eta a_{np}. \tag{15.20}$$

The last term in the species equation (Equation 16) represents magnetophoretic motion, which arises because of the Stokesian migration of superparamagnetic nanoparticles under the magnetic force $f_j = \mu_0 m_k^P \partial H_j/\partial x_k$ (Equation 15.9c). Here, m_k^P denotes the component of the particle moment vector along x_k-axis.

15.2.2.2 Magnetic Microspheres

Ferrous nanoparticles embedded in micron-size polystyrene microspheres are commercially available as magnetic beads (Micromod: www.micromod.de; Kisker: www.kisker-biotech.com; Dynabeads®: www.invitrogen.com; Sphero™: www.spherotech.com). Functionalized magnetic beads are prepared, for example, by emulsion polymerization or dispersion polymerization in a nearly monodispersed suspension (Horák et al., 2007). The small size of the beads can offer a large specific surface for chemical binding. Figure 15.4b shows a schematic and scanning electron microscope micrograph of magnetic microspheres. Functionalizing the bead surface with ligands (antibodies, peptides, or lectins) ensures specific binding with selective biomolecules upon contact. The complex formed from this binding is easily manipulated by a magnetic field. These magnetic particles may attach to (or be engulfed by) cells, thus indicating a technique to manipulate magnetically unresponsive material (Ugelstad et al., 1993; Radbruch et al., 1994; Tibbe et al., 1999).

The magnetic force on an isolated spherical magnetic bead of radius a can be evaluated from Equation 15.9c as follows:

$$\mathbf{F} = \mu_0 \iiint (\mathbf{M} \times \nabla)\mathbf{H} \, d\vartheta = \mu_0 \left(\frac{4}{3}\pi a^3 \right)(\mathbf{M} \times \nabla)\mathbf{H}. \tag{15.21a}$$

The underlying assumption of the Equation 15.21a is that the particle size is small enough so that the magnetic field and its gradient do not vary much within the volume of a particle. For all practical purposes, the force is evaluated not in terms of actual $\mathbf{M}$ and $\mathbf{H}$ within the particle but in terms of $\mathbf{H}_0$ ($= \mathbf{B}_0/\mu_0$), that is,

$$\mathbf{F}_m = \mu_0 \left(\frac{4}{3}\pi a^3 \right)\chi_{eff} \frac{1}{2}\nabla(\mathbf{H}_0 \cdot \mathbf{H}_0). \tag{15.21b}$$

Here $\mathbf{B}_0$ denotes the imposed magnetic field at the location of the particle center measured (or computed) in absence of the particle, and χ_{eff} is the effective magnetic susceptibility of the particle. There is obviously a difference between $\mathbf{H}$ and $\mathbf{H}_0$, which arises because the particles themselves are magnetized in a magnetic field and produce a demagnetizing effect. The effective magnetic susceptibility is different from the intrinsic magnetic susceptibility χ (where $\mathbf{M} = \chi\mathbf{H}_0$) of the microspheres because it takes into account the distortion of the magnetic flux lines around a magnetized material placed in a magnetic field. For a uniformly magnetized spherical particle (see Figure 15.5a), the magnetic field inside the sphere is $\mathbf{B} = \mu_0(\mathbf{H}_0 + 2/3\,\mathbf{M})$ (Griffiths, 2004), and the effective magnetic susceptibility (Smistrup et al., 2005a) is linked to the intrinsic susceptibility of the nanoparticles through the relationship $\chi_{eff} = \chi/(1 + (1/3)\,\chi)$.

Equation 15.21b indicates that there is no net force on an isolated magnetic microsphere in a homogeneous magnetic field. However, in a dense system of particles, there is a magnetic interaction between the particles because a magnetic dipole moment $\mathbf{m} = (4/3)\pi a^3 \chi_{eff}\mathbf{H}_0$ is induced in each microsphere. The local $\mathbf{H}_{0i}$ field for a multiparticle system can be obtained from a linear superposition of the $\mathbf{H}$ fields because of the individual point dipoles and the imposed external $\mathbf{H}_{ex}$, so that

$$\mathbf{H}_{0i} = \mathbf{H}_{ex} + \sum_{j=1}^{n} \mathbf{H}_{ij}, \qquad (15.22)$$

where $\mathbf{H}_{ij}$ represents the vector at the location of the ith particle because of the magnetic dipole moment of the jth particle. The second term on the RHS of Equation (15.22) represents the cumulative contributions of all the n particles that act as induced point dipoles, that is,

$$\mathbf{H}_{ij} = -\frac{m_j}{r_{ij}^3} + \frac{3(m_j \times r_{ij})r_{ij}}{r_{ij}^5}; i \to [1,n]; j \to [0,n]; i\,j, \qquad (15.23)$$

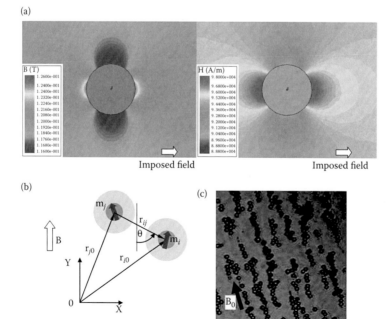

FIGURE 15.5
Demagnetization effects due to the presence of an isolated spherical particle of $\chi = 0.1$ in an externally imposed homogeneous horizontal magnetic field: (left) $\mathbf{B}$ field and (right) $\mathbf{H}$ field (simulation performed using Maxwell® v12.1). When a homogeneous magnetic field is imposed on a dense suspension of magnetic beads, (b) the dipole–dipole interaction leads to (c) the formation of long chains.

where r_{ij} denotes the interparticle distance as shown in Figure 15.5b. For any two magnetic microspheres of identical dipole strength m that are separated by a distance r, the magnetic interaction potential of potential $U_{mag} = -(m \cdot B)$ is (Landau et al., 1984; Biswal and Gast, 2004a) (refer to Figure 15.5b)

$$U_{mag} = \frac{\mu_0 \mu_r}{4\pi} |m|^2 \frac{(1 - 3\cos^2 \delta)}{r^3}, \tag{15.24}$$

where μ_r represents the relative permeability of the liquid in which the beads are suspended, and δ is the angle between the magnetic field vector and the radius vector connecting the two particles. The relation shows that the dipole–dipole interaction force $F (\sim\nabla U_{mag} = \nabla(m \cdot B) \approx (m \cdot \nabla)B)$ is attractive and scales as $1/r^4$, implying that the force becomes much stronger as two microspheres more closely approach each other (Zborowski et al., 1999; Gijs, 2007). Unlike the nanoparticles in a ferrofluid, the dipole–dipole interaction in a suspension of magnetic microspheres cannot be effectively overcome by the thermal Brownian action, and the microspheres form chain-like structures under a homogeneous magnetic field (see Figure 15.5c).

15.3 Dynamics of Magnetic Particles in Microfluidic Environments

An isolated magnetic microsphere suspended in a fluid under an imposed magnetic field gradient experiences following forces: the magnetic force $\mathbf{F}_m$ (see Equation 15.21b), the drag force $\mathbf{F}_d$ exerted by the fluid (because the particle tends to move with a finite velocity relative to the fluid), the gravitational force $\mathbf{F}_g$, and the Brownian force $\mathbf{F}_B$. The motion of a particle (in terms of its absolute velocity V_p) under these forces is described by applying Newton's second law of motion for a particle, that is,

$$\left(\frac{4}{3}\pi a^3 \rho_p\right)\frac{dV_p}{dt} = [\mathbf{F}_m + \mathbf{F}_d + \mathbf{F}_g + \mathbf{F}_B]. \tag{15.25}$$

The gravity force on the particle takes care of buoyancy, such that

$$\mathbf{F}_g = \left(\frac{4}{3}\pi a^3\right)(\rho_p - \rho_l)g, \tag{15.26}$$

where ρ_p and ρ_l denote the densities of the particle and the carrier fluid, and g is the acceleration due to gravity.

The hydrodynamic interactions during particle transport in microchannels arise because of the drag (both skin friction and form) as the particles move through the background fluid with a finite slip velocity. Besides, the drag force is influenced by the presence of walls of the microchannel. This effect is incorporated through a wall drag multiplier K_{wall} such that

$$\mathbf{F}_d = K_{wall} 6\pi a \eta (V - V_p), \tag{15.27}$$

where V_p and V_l denote the absolute velocities of the particle and the carrier fluid, respectively. The values of K_{wall} differ while computing the components of drag force parallel and perpendicular to the wall. The corresponding $K_{wall}^{\parallel}$ and $K_{wall}^{\perp}$ are calculated as suggested by Clift et al. (1978):

$$K_{wall}^{\parallel} = \left[1 - \frac{9}{16}\xi\right]^{-1} \quad \text{and} \quad K_{wall}^{\perp} = \left[1 - \frac{9}{8}\xi\right]^{-1}. \tag{15.28}$$

For the particles sufficiently away from the walls, the value of ξ (the ratio of particle diameter to its distance from the wall) approaches zero, so that the values of $K_{wall}^{\parallel}$ and $K_{wall}^{\perp}$ approach unity.

The random Brownian displacements cause particle trajectories to fluctuate about the deterministic path lines in the carrier-fluid flow field. Although the magnitude of the Brownian force on a spherical particle is not readily available, the random particle displacements measured with respect to the carrier fluid can be described by the following three-dimensional Gaussian probability density function

$$p(x,y,z) = \frac{\exp\left[-3\pi a\eta\left(x^2 + y^2 + z^2\right)/2kT\Delta t\right]}{((2/3)2kT\Delta t)/3a\eta}. \tag{15.29}$$

The extent of the Brownian fluctuation on the overall particle transport depends on the particle size, the carrier fluid viscosity, the temperature, and the sampling time with which the instantaneous positions of a particle are determined. For a 1-μm-diameter particle in a carrier fluid of 0.001 Pa·s viscosity at 300 K, the velocity movements due to thermal Brownian fluctuations are plotted in Figure 15.6 for 100-μs and 1-ms sampling times. For the 1-μm-diameter particle, a velocity of O (100 μm/s) can be calculated without significant error if the particle transport equation (Equation (15. 25)) is solved by neglecting Brownian

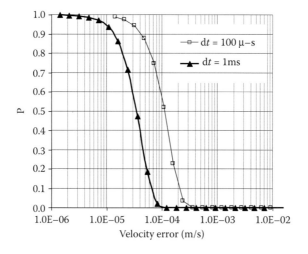

FIGURE 15.6
Effect of the sampling interval (considered during the Lagrangian tracking of a magnetic microsphere) on the probability P of a specified velocity error because of the Brownian fluctuations for a 1-μm-diameter particle in water ($\eta = 0.001$ Ps-s) at 300 K.

motion at 1-ms intervals. Thus, the particle thermal fluctuations for this case are negligible compared with their movements induced by the flow and magnetophoresis (Smistrup et al., 2005a; Furlani and Sahoo, 2006; Sinha et al., 2009).

For a microsphere of 1-μm radius (ρ_p= 1800 kg/m^3) in a fluid having η = 0.001 Pa·s and ρ = 1000 kg/m^3, a slip velocity of 1 mm/s produces a Stokesian drag that is nearly 500 times stronger than the gravity force (Sinha et al., 2007). Moreover, the particle mass being very small (~7.5 × 10^{-15} kg), the inertial effects on the particle trajectories can also be neglected unless the particle acceleration exceeds a very high value (~100 m^2/s, which is unusual for microfluidic applications). Therefore, analyses of magnetophoretic microsystems have typically neglected gravity and inertial effects (Smistrup et al., 2005a; Sinha et al., 2007; Nandy et al., 2008). Consequently, the equation of motion for a particle under the combined influence of magnetic and drag forces is

$$V_p = V + \frac{\mathbf{F}_m}{6\pi a \eta K_{\text{wall}}}. \tag{15.30}$$

Particle trajectories can be solved by numerical integration of Equation 15.30. For a dilute suspension (isolated particles), the time interval for integration is limited to a value small enough such that the carrier fluid and particle velocity do not change appreciably during that interval. In a multiparticle system, particle–particle interactions lead to physical contact between the particles. The unphysical overlapping of particles, which can occur because the particles (that have finite radii) are assumed as point entities for the purpose of Lagrangian tracking, is generally prevented through a velocity-based collision barrier (Dance et al., 2004). The technique imparts the particles a repulsive velocity V^r once the centers of two particles approach close to one particle-diameter distance from each other. If V_{ref} denotes the maximum overlap velocity and R_{ref} the critical distance parameters, the repulsive velocity when $r_{ij} < R_{\text{ref}}$,

$$\mathbf{v}^r_{ij} = \frac{-V_{ref}}{2a}\left(\frac{R_{ref}^2 - r_{ij}^2}{R_{ref}^2 - 4a^2}\right)^2 \mathbf{r}_{ij}. \tag{15.31}$$

Using the notation $\bar{V} = |-v^r_{ij}|/V_{\text{ref}}$, $\bar{R} = R_{\text{ref}}/2a$, $\bar{r} = r_{ij}/2a$, Equation 15.31 assumes the form

$$\bar{V} = \left(\frac{\bar{R}^2 - \bar{r}^2}{\bar{R}^2 - 1}\right)^2 \bar{r}; \ \bar{r}:(1, \infty); \ \bar{R}:(R_{\text{ref}}/2a, \infty). \tag{15.32}$$

The behavior of this collision barrier is described in Figure 15.7. Approaching particles continue undisturbed until they are separated by a distance R_{ref}. Once, $\bar{r} < (r_{ij} < R_{\text{ref}})$, $\bar{V}$ increases with decreasing $\bar{r}$. The selection of $\bar{R}$ determines whether the colliding particles behave as hard or soft spheres. In general, R_{ref} can be selected arbitrarily under the constraint that any further reduction does not significantly alter the dynamics predicted by the simulations, and so herein we use $\bar{R}$ = 1.02. At $\bar{r}$ = 1, $\bar{V}$ = 1 for all $\bar{R}$, which corresponds to a "kissing" configuration. Some particle pairs never attain this "kissing" state because all particles are assigned a V_{ref} on the basis of the fastest approaching particle pair in the domain. The time step for Lagrangian tracking of the particles in a dense suspension is primarily limited by this collision criterion, where a travel distance

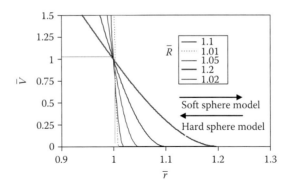

FIGURE 15.7
Velocity-based collision barrier used for preventing particle overlap while simulating a multiparticle system.

per time step is kept below the width of the attraction/repulsion zone of Figure 15.7, or a fraction F of the particle size, whichever is less, that is, $|u_p| \times \delta t < \min[\{R_{ref}(R_a - R'_c)/10\}; (aF)]$.

15.3.1 Single-Particle Dynamics

The transport of an isolated magnetic particle and particles in a very dilute suspension has been investigated by Sinha et al. (2007). Figure 15.8a shows the geometrical and flow arrangement used for the experimental investigation. Sinha et al. used a needle-shaped electromagnet to generate the necessary magnetic field gradient for influencing the particle transport. A magnetic dipole of finite size was represented by a point dipole to produce a similar magnetic field (see Figure 15.8b) and gradient in the region of interest, which is the shaded region in Figure 15.8a. To assess the behavior of a dilute particle suspension, individual particle trajectories were tracked (see Figure 15.8c). Simulations for a combination of the dipole strength $m = 2.28 \times 10^{-7}$ A/m^2 and $r_{mag} = 175$ μm were superposed on the experimentally observed trajectories in the region of interest. Because the primary forces on the particles are the magnetic force and the fluid drag, the particle trajectories are influenced by the particle size and its magnetic susceptibility, the fluid flow velocity, and the fluid viscosity. A length-scale dependence is also expected, which is related to the magnetic field gradient.

Parametric investigations revealed that the important physical parameters influencing the particle trajectories and the capture efficiency follow a power law relationship

$$\Phi \propto f(\chi_{eff}, m, a, U, \eta, r_{mag}), \tag{15.33}$$

where the symbols in the parentheses in the right-hand side denote the magnetic susceptibility, the dipole strength, the particle size, the fluid velocity, the fluid viscosity, and the relevant magnetic length scale, respectively. Physically, Φ indicates the percentage of particle influx that is retained in the domain because of the attractive magnetic force. A dimensionless number is composed of the ratio of the magnetic force to the drag force $\Pi_1 = (\mathbf{F}_m/\mathbf{F}_d) \propto (a^2 \mu_0 m^2 \chi_{eff}/\eta U r_{mag}^7)$. Investigation of the dependence of Φ on Π_1 illustrates the existence of two distinct regimes

$$\Phi = 0.25\,\Pi_1^{0.87} \text{ for } \Pi_1 < 1650 \quad \text{and} \quad \Phi = 0.55\,\Pi_1^{0.48} \text{ for } \Pi_1 > 1650. \tag{15.34}$$

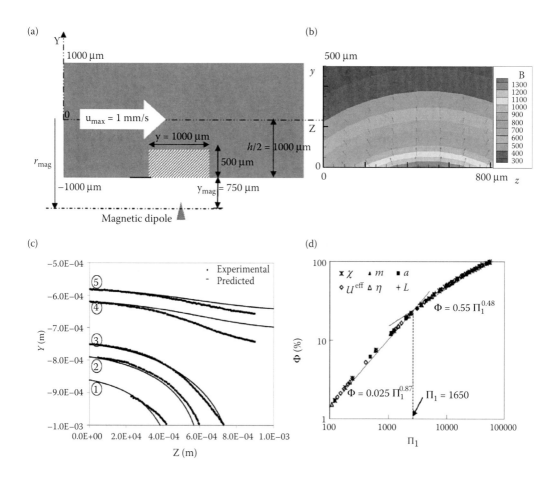

FIGURE 15.8
Single-particle dynamics in a microfluidic channel. (a) The microchannel geometry and the flow arrangement. (b) The modeled magnetic field (in units of Gauss) in the region of interest (the shaded region in panel a). (c) The experimentally observed and computed particle trajectories. (d) Particle capture efficiency Φ as a function of the variable $\Pi_1 = (\mathbf{F}_m/\mathbf{F}_d) \propto (a^2\mu_o m^2\chi/\eta UL^7)$. (Panels a, c, and d are reprinted with permission from Sinha, A., Ganguly, R., De, A.K., Puri, I.K., Single magnetic particle dynamics in a microchannel. *Phys Fluids.* 19: 117102, 2007, Copyright 2007, American Institute of Physics.)

The values of Φ are piecewise linear for each of the two regimes, divided by the point $\Pi_1 = 1650$. A larger Π_1 value represents a stronger magnetic force that results in a high particle capture. Equation (15.34) enables the selection of Π_1 for a desired Φ in microfluidic applications. An analytical study conducted by Nandy et al. (2008) and a numerical study by Modak et al. (2009) showed that the capture efficiency for a purely two-dimensional magnetic field configuration (e.g., that produced by a line dipole, which can be created by a pair of chip-embedded conductors carrying currents in opposite direction) depended on a similar nondimensional variable, $\Pi = (\mathbf{F}_m/\mathbf{F}_d) \propto (a^2\mu_o m^2\chi_{eff}/\eta Uh^5)$. The difference of the exponents, with which the length scale appears in the definition of the nondimensional parameters for the point (r_{mag}^7) and line (h^5) dipole configurations, complies with the fact that the magnetic field diminishes more quickly away from a point dipole, that is, a linear dipole configuration has a longer zone of influence than a point dipole.

15.3.2 Multiparticle Dynamics

We discussed in the section "Magnetic Microspheres" that the particle–particle dipolar interactions are important for a dense suspension under the influence of a magnetic field. If the interparticle distance r_{ij} (see Figure 15.5b) is small, the particle motion has a strong influence on neighboring particles. Figure 15.9 describes the motion of two spherical

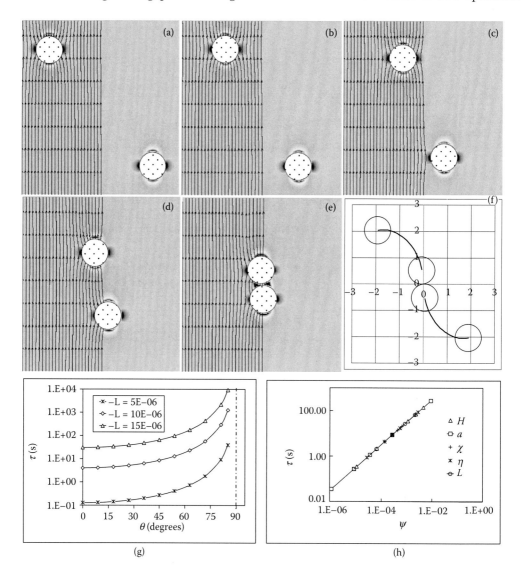

FIGURE 15.9
Dipole–dipole interaction between magnetic particles in a homogeneous magnetic field because of local distortion of magnetic field lines (shown only in half of the domain for clarity). (a–e) Time stamps at $t = 0.0666, 0.3200, 0.4610, 0.4990$, and 0.5030 s, respectively. (f) Trajectory of the particle pair. Simulation carried for $a = 0.5$ μm, $H = 10^5$ A/m, $\eta = 0.001$ Pa·s, $\chi = 0.1$, and initial particle separation $L = 8\sqrt{2}a$. (g) Variation of collision time τ with θ for three different initial separations (L, in meters) between the particles. All other parameters in ψ are kept fixed. (h) Parametric dependence of aggregation time t as functions of $\eta, L, B, \chi_{\text{eff}}$, and a. The simulations are performed for $\theta = 45°$. All the plots collapse on a single curve when the aggregation time is plotted against the group variable $\psi = (\eta L^5)/(\chi^3 B^2 a^5)$.

particles that are initially separated by a distance $L = |r_{ij}|_{t=0} = 8\sqrt{2}a$ with an angle θ (made by the imposed $\mathbf{H}_0$ with r_{ij}) $= \pi/4$. Figure 15.9a through e denotes the instantaneous positions of the particles at t = 0.0666, 0.3200, 0.4610, 0.4990, and 0.5030 s, respectively, whereas the frame (f) describes the particle trajectories as they eventually aggregate. For a magnetic particle pair, the interaction depends strongly on both the relative distance $|r_{ij}|$ between the particles and their angular disposition θ with respect to the imposed magnetic field. Figure 15.9g shows the variation of the time to aggregate (τ) for a particle pair in a quiescent fluid for three different initial separations. For a given initial separation, the time τ is minimum when $\theta = 0$ at $t = 0$. The particle orientation becomes less favorable for aggregation at larger θ when τ increases. At $\theta = \pi/2$, $\tau \rightarrow \infty$; physically, the particles never collide and the interaction is repulsive. This behavior is consistent with macroscopic systems where two bar magnets are most attractive when placed N to S and are repulsive when placed N to N (or S to S). Other physical variables critical to the particle dynamics are η, H_0, a, and χ_{eff}. A parametric investigation of the particle aggregation time for a binary particle system at a given θ (=45°) under different L, η, B, a, and χ_{eff} reveals a linear relationship (see Figure 15.9h) between τ and a nondimensional variable ψ ($=(\eta L^5)/(\chi B^2 a^5)$), which denotes the ratio of viscous force to the magnetic force. The relationship would differ from the linear nature if the particle size is small enough such that the Brownian fluctuation would no longer be negligible.

The scenario is more complicated in a dense suspension, where the neighboring particles may assemble to form a chain or become part of an already existing chain. The aggregation time and the tendency of chain formation in a multiparticle suspension are also functions of concentration. Figure 15.10 shows the snapshots of particle aggregation in a quiescent fluid for two different particle concentrations under a homogeneous field. Clearly, the aggregation time is much smaller, and chain growth rate is much quicker in a denser suspension.

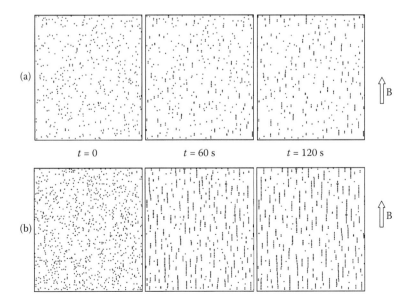

FIGURE 15.10
Particle aggregation and formation of chains aligned in the direction of the imposed homogeneous magnetic field (B = 0.003 T) in a quiescent fluid (η = 0.001 Pa·s) for two different initial concentrations: (a) 10^{16} m^{-3} and (b) 2×10^{16} m^{-3}. Particles have a = 0.5 μm and χ = 0.1.

When a dense suspension of particles flowing through a microchannel is subject to a magnetic field gradient, the particle–particle dipolar interaction becomes responsible for formation of particle aggregates. Experiments and simulations (Sinha et al., 2009) in conditions analogous to those described in Figure 15.11a indicate the growth of an immobilized aggregate of particles on the channel wall near the dipole. Close to the lower channel wall, the fluid velocity is low, whereas the field gradient is a maximum. The magnetic force on a particle in this region thus dominates over the hydrodynamic force. Such a situation is favorable for particle aggregation. As the aggregate size increases (see Figure 15.11b), its periphery moves away from the wall and the dipole. Therefore, hydrodynamic shear gradually become comparable with the magnetic effects, and the aggregate growth rate decreases. Simulations by Sinha et al. (2009) show that the aggregate

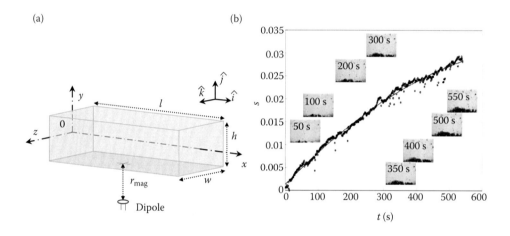

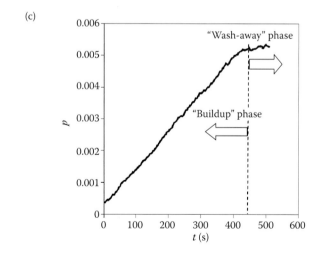

FIGURE 15.11
Particle transport and aggregation in a dense suspension. (a) The flow and the magnetic configurations. (b) Experimentally observed temporal growth rate of the particle aggregates for $a = 2\ \mu m$, $I = 0.5\ A$, $U = 1\ mm/s$, and $c = 10^{12}\ m^{-3}$, inside a 200×1000-μm cross-section microchannel. (c) Simulated growth rate of the aggregate volume fraction, showing the "buildup" and the "wash-away" phases for a field due to a point dipole of strength $m = 10^{-11}\ A/m^2$ and $a = 500\ \mu m$, with $r_{mag} = 125\ \mu m$ and $c = 3 \times 10^{15}\ m^{-3}$.

growth ceases after a time when the rate of magnetic aggregation and hydrodynamic wash away of the particles balance each other, producing a nearly stable aggregate size. They identified (see Figure 15.11c) an aggregate "buildup" phase (when the aggregate grows almost linearly with time) and a "wash-away" phase (characterized by a nearly steady aggregate size).

It is intuitive that, like the single-particle dynamics discussed in the section "Single-Particle Dynamics," the transport of the magnetic particles and the aggregate structure in a dense suspension would also depend on the ratio of the magnetic and viscous forces. Besides, the particle concentration also plays a role because it influences the mean inter-particle distance and the particle flux rate. The stable aggregate size was characterized in terms of the parameter P (the volume fraction of the computational domain occupied by the particle aggregate) that depended on the particle concentration (c), dipole strength (m), and the distance (r_{mag}) of the magnet, flow velocity (U_0), fluid viscosity (η), particle radius (a), and susceptibility (χ_{eff}) as

$$P \propto \Pi = \frac{c^{0.1}a^2}{U_0^{2/3}} \sqrt{\frac{\mu_0 \chi_{eff}}{\eta r_{mag}^7}} = \frac{c^{0.1}a}{U_0^{1/6}} \sqrt{\frac{\mu_0 \chi_{eff} m^2 a^2}{U_0 \eta r_{mag}^7}}. \tag{15.35}$$

The dimensionless ratio Π is related to the $\Pi_1 = (\mathbf{F}_m/\mathbf{F}_d)$ discussed in the section "Single-Particle Dynamics" as

$$\Pi = \frac{ac^{0.1}}{U_0^{1/6}} \sqrt{\Pi_1}. \tag{15.36}$$

15.3.3 Dynamics of Rotating Chains of Magnetic Beads

The previous section discussed the formation of magnetic particle chains in a dense suspension. The dynamic behavior of these self-assembled chains is interesting for microfluidic applications. When a quiescent suspension of magnetic microspheres is subjected to a homogeneous magnetic field, the particles form chains that are aligned with the imposed magnetic field (see Figures 15.5c and 15.10). When the magnetic field is rotated, the particle chain also rotates to follow the rotating field (Figure 15.12a). How closely a chain follows the rotating field depends on the magnitude and rotational speed of the imposed magnetic field, the size and the magnetic property of the particles, and the viscosity of the host fluid.

A chain containing N spherical particles experiences a magnetic torque M^m and an opposing viscous drag M^v (as illustrated in Figure 15.12b) such that (Biswal and Gast, 2004a,b)

$$M^m = \frac{\mu_0 \mu_r}{4\pi} \frac{3|\bar{\mathbf{m}}|^2 N^2}{2(2a)^3} \sin(2\theta) \quad \text{and} \quad M^v = \frac{4}{3} N \pi a^3 \frac{2N^2}{\ln(N/2)} \eta \omega. \tag{15.37}$$

Here, θ denotes the angle between the rotating field and the magnetization vector $\bar{\mathbf{m}}$ of a particle, N is the number of particles in a chain, μ_r is the relative permeability of the medium, η is the fluid viscosity, and ω is the angular velocity of the chains. Two other forces influence the chain integrity. The tensile force F^t acting on the chain arises due to the centrifugal force on the rotating particles (which is negligibly small as compared with the magnetic and viscous forces in our experiments) and promotes chain disintegration. The compressive force F^c occurs due to the magnetic interaction between the particles and tends to hold the chain together.

The response of a chain to the magnetic force is characterized using the Mason number (Biswal and Gast, 2004a,b)

$$\text{Ma} = \frac{\ln(N/2)}{N} \frac{M^v}{M^m} \sin(2\theta) = \frac{\mu_0 32 \omega \eta}{\mu_r \chi_{\text{eff}}^2 |\mathbf{B}_0|^2} \tag{15.38}$$

that compares the magnitudes of the viscous and magnetic forces. When $\mathbf{B}_0 = 0.123\ T$, $\omega = 5$ rpm, $\chi_{\text{eff}} = 0.019$, and $\eta = 0.0009$ in water (i.e., $\mu_r \approx 1$), the particle chains in Figure 15.12a correspond to Ma = 0.003. For these conditions, the microspheres form long unbroken chains that are approximately 10–15 μm (i.e., five to seven particles) long, which rotate in synchronism with the imposed magnetic field. The value of θ for these chains is observed to be very small because the chains closely follow the orientation of the imposed rotating magnetic field, as shown schematically in Figure 15.12a.

The literature (Melle et al., 2002) reports that self-assembled bead chains have the capacity to reduce the sizes of their structures to decrease the viscous drag that they

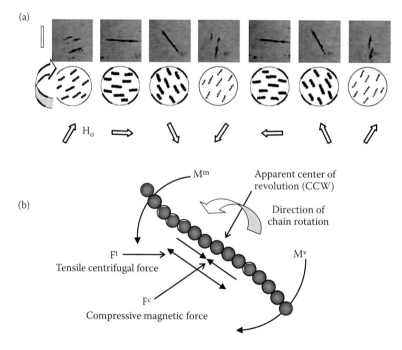

FIGURE 15.12
(a) Rotation of the magnetoresponsive chains as the imposed magnetic field rotates at $\omega = 5$ RPM. Here, the chains are used as localized rotors or stirrers. (The size of the scale on the left is 10 μm.) (b) Schematic showing the forces and moments acting on a particle chain, where M^m denotes the driving magnetic moment, and the reaction of the resistive moment M^v favors the stirring action within the droplet. (Reprinted with permission from Roy, T., Chakraborty, S., Sinha, A., Ganguly, R., Puri, I.K., Magnetic microsphere-based mixers for micro-droplets. *Phys Fluids.* 21: 027101, 2009, Copyright 2009, American Institute of Physics.)

experience while rotating synchronously with a rotating field. Vuppu et al. (2003) experimentally observed that the chains fragment at higher rotational speeds because of the shear forces acting on them. At very low rotational frequencies, chain growth occurs because of the rotation, which provides an opportunity for distant chains to approach each other. At slightly higher and at intermediate frequencies, the chain lengths decrease rapidly because of the increasing viscous forces, whereas the chain lengths decrease slowly at still higher frequencies. They found that the equilibrium chain length follows a power law relation with the rotational frequency of the field. Kang et al. (2007) performed direct numerical simulation of magnetic beads in rotating magnetic fields under a wide range of Ma. At a low Mason number, the chain rotates synchronously like a rigid body, following the field with a small angular lag. In the intermediate Mason number regime $0.03 \leq$ Ma ≤ 0.3 (evaluated as per the definition of Equation 15.41), chain breakup is observed. The number of detached chains increases with increasing Mason number. In the regime of Ma ~ 0.06, the two split chains rejoin together, forming a single chain again, and this splitting and reconnecting takes place in an alternating manner (Kang et al., 2007).

15.4 Microfluidic Application of Ferrofluids

The KBF in ferrofluids is responsible for a wide range of interesting hydrodynamic actuations and field-induced self-assembly behaviors, which can be harnessed for a range of novel microfluidic applications. Table 15.1 outlines some salient microfluidic applications of ferrofluids or colloids of magnetic nanoparticles. The merits of ferrofluid-based microfluidic systems lie primarily in their good magnetic response (actuation), ease of handling in microfluidic systems, and the "action-at-a-distance" feature of the ferrofluids.

TABLE 15.1

Microfluidic Applications of Ferrofluids

	Attributes of Ferrofluids	Applications
1.	Dynamic plugs of ferrofluids in microchannels	Micropump for LOC applications: Hatch et al. (2001), Andò et al. (2009), and Yamahata et al. (2005)
		Microvalve: Hartshorne et al. (2004) and Kim et al. (2006)
		Microplunger for inducing chaotic mixing: Oh et al. (2007)
		Microplunger for high-throughput polymerase chain reaction application: Sun et al. (2008)
		Ferrofluid tunable optical fiber filter: Liao et al. (2005)
2.	Field-induced ferrofluid aggregates	Soft mask for photolithography: Yellen et al. (2004)
		Diffusion mask: Halverson et al. (2005)
		Ferrofluidic nanoprinting: Sen et al. (2000) and Sen (2004)
3.	Ferrofluid-induced mixing	Tsai et al. (2009), Koser et al. (2007)
4	Ferrofluid droplet	Magnetic manipulation of ferrofluid droplets in an immiscible medium or on a flat substrate: Nguyen et al. (2007), Sterr et al. (2008), Guo et al. (2006), and Bormashenko et al. (2008)
5.	Magnetic relaxation of nanoparticles	MARIA: Grossman et al. (2004), Enpuku et al. (2009), and Ku et al. (2008)

15.4.1 Dynamic Plugs of Magnetic Fluids

The combined action of magnetic and surface tension forces can produce stable plugs of ferrofluids in microchannels filled with another immiscible fluid. Using a sequentially switchable electromagnet array, these stable plugs of ferrofluid (see Figure 15.13a) may be externally actuated as reciprocating plungers. When subjected to an axial gradient of a

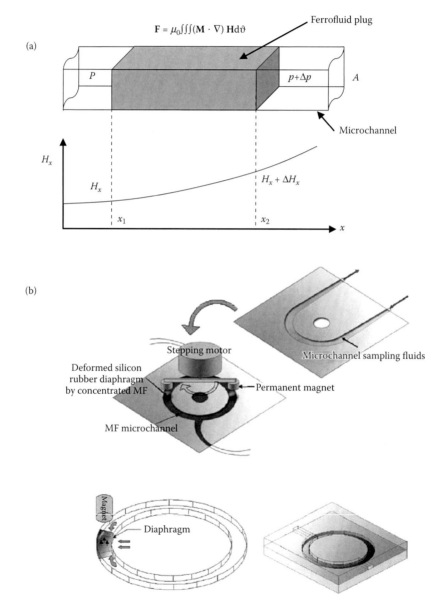

FIGURE 15.13

(a) A small plug of ferrofluid inside a microchannel can act as a reciprocating piston actuator under an applied **H** gradient. KBF on the plug develops a magnetostatic pressure gradient in the direction of positive **H** gradient. (b) Schematic of a ferrofluid-based peristaltic micropump. (Reprinted from *Sensors Actuators A* 128, Kim, E.-G. et al., A study on the development of a continuous peristaltic micropump using magnetic fluids, pp. 43–51, Copyright 2006, with permission from Elsevier.)

magnetic field, a small plug of uniform cross section A, as described in Figure 15.13a, can withstand a pressure difference of

$$\Delta p = \frac{1}{2}\mu_0 \int_{x_1}^{x_2} \mathbf{M}\,d\mathbf{H}. \tag{15.39}$$

For a ferrofluid of average magnetic susceptibility of 0.01 and a $\Delta(H^2) \sim 10^{10}\ A^2/m^2$ imposed across the ferrofluid plug, the estimated pressure difference Δp developed across the plug due to the KBF is of the order of 1 kPa.

The key advantages of using such a ferrofluid plunger in microfluidic architecture are that (1) it offers low-friction reciprocating motion, that is, when exposed to a gradient field, the whole plug responds as a homogeneous magnetic liquid, moving to the region of the highest field; (2) because of its fluid-like nature, there is ease of delivery to the point of requirement inside the microfluidic architecture, and hence, unlike a solid plunger, a ferrofluid does not require device integration at the microfabrication phase; and (3) there can be contactless actuation without any moving components. The primary challenge, however, is to maintain a stable ferrofluid plug that remains integrated at a reasonably large contact angle with the channel wall so that the plugs are not "torn away" into smaller fragments (because of the differential magnetic force on them), and they do not "creep" through the microchannel.

The proper choice of ferrofluids and the suitable surface modification of microchannel walls are essential for successful implementation of ferrofluid plungers in a microfluidic device. Yamahata et al. (2005) reported the use of a water-based ferrofluidic plug in a Y-shaped channel with two passive check valves and a mechanically moving NdFeB permanent magnet to demonstrate a micropumping application that produced 30 µL/min and a back pressure of 2.5 kPa. Three-dimensional ferrofluid aggregates have also been used in microfluidic channels as both pumps and valves (Hartshorne et al., 2004). Kim et al. (2006) used a magnetically actuated ferrofluid plug to deform a silicone rubber diaphragm that created peristaltic pumping action (see Figure 15.13b). Besides continuous pumping, the arrangement could also act as a flow control valve. Sun et al. (2008) used permanent magnet-actuated ferrofluid plugs in closed-loop parallel microchannels to drive DNA samples through the different temperature ranges in a high-throughput polymerase chain reaction (PCR) device. Oh et al. (2007) used dynamic plugs of ferrofluids in a micromixer that consisted of a T-shaped main mixing channel with two parallel subchannels intersecting the main channel. Oscillation of two ferrofluid slugs in the subchannels, induced by external permanent magnet actuation, generates chaotic advection in the main channel flow. Liao et al. (2005) used a ferrofluid plug inside 300-µm- to 1-mm-diameter capillary tube to form a ferrofluid tunable optical fiber filter based on a long-period grating.

15.4.2 Magnetically Assembled Ferrofluid Aggregates

Magnetic-field-induced self-aggregation of ferrofluids occurs through the interplay of magnetic, surface tension, gravity, and hydrodynamic shear forces. When a ferrofluid and a nonmagnetic liquid share a common interface, the application of a normal magnetic field produces a destabilizing effect on the planar free interface, leading to formation of freestanding spikes that have a static hexagonal pattern. This is known as Rosensweig instability (Cowley and Resensweig, 1967; Rosensweig, 1985). The theory behind the formation of ferrofluid structures in idealized geometries has been addressed by Boudouvis et al. (1993), who solved the magnetically augmented Young–Laplace's equation to explain this

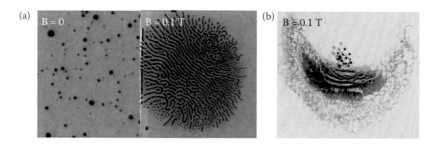

FIGURE 15.14
(a) Labyrinthine structures formed by a ferrofluid in a Hele–Shaw cell in contact with an immiscible liquid.
(b) Magnetically induced patterns exhibited by an oil-based ferrofluid on water surface. The magnetic field acts
perpendicular to the plane of the paper.

behavior. Several other magnetically induced patterns of ferrofluids have been reported in
the literature. Inside an immiscible liquid medium in a Hele–Shaw cell, ferrofluids exhibit
labyrinthine patterns (Sinha et al., 2004) under an imposed magnetic field (see Figure
15.14a). Elborai et al. (2005) demonstrated that controllable, self-forming, quasi-two-dimensional patterns of a ferrofluid can be achieved in an immiscible fluid in Hele–Shaw cells
using in-plane rotating and dc-axial uniform magnetic fields of varying magnitude. When
a sessile drop of ferrofluid placed on a flat substrate is subjected to a spatially varying
magnetic field, the interactions between gravity, surface tension, and magnetic force create
different shapes (Boudouvis and Scriven, 1993; Goldstein et al., 1993; Jang et al., 1999). Field-
induced self-assembly of an oil-based ferrofluid on water surface under a magnetic field is
observed to create intriguing structures (Figure 15.14b). If suitable means are identified for
depositing these structures on a surface and curing those to permanent features, a viable
technique for the bottom–up assembly of MEMS devices would emerge. Yellen et al. (2004)
used a thin ferromagnetic film and a perpendicular biasing field in isolated islands to
direct the assembly of superparamagnetic colloidal particles into two-dimensional arrays.
The principle is proposed as a promising "soft masking" technology that used programmable alignment marks built into the substrate and an opaque ferrofluid to protect or
deprotect selected areas of the magnetically patterned substrate according to a programmable sequence of masking patterns of ferrofluids. Dynamic magnetic aggregation of ferrofluids is also used to mask specific locations on a surface for combinatorial chemistry
or to restrict diffusion of molecules in selected areas in a microfluidic device (Halverson
et al., 2005). Sen et al. (2000) presented theoretical and computational prediction of the possibility to eject single ferrofluid grains from a ferrofluid pool subjected to a strong, homogeneous magnetic field directed perpendicular to the surface by using modest mechanical
impulses. Although their study (Sen, 2004) has shown the theoretical feasibility of achieving nozzle-free, ultrafast, ink-jet printing of single grain (~20 nm) resolution, a matching
experimental demonstration is still awaited in the literature.

15.4.3 Microfluidic Mixing Using Ferrofluids

Magnetic body forces on ferrofluids can be harnessed in a forced flow configuration to
augur cross-stream advection. Ganguly et al. (2004) showed the possibility of generating
thermomagnetic convection in miniaturized channels to enhance the wall heat transfer in
a forced flow. A combined influence of spatially nonuniform magnetic susceptibility (generated in this case due to temperature nonuniformity) of the ferrofluid and an imposed

magnetic field gradient (created by a line dipole) is found to produce strong advective rolls in the flow. The same principle can be extended to auger mixing.

Figure 15.15a shows a proof-of-concept demonstration of KBF-induced mixing in a Y-shaped microchannel. The streams of a water-based ferrofluid (EMG 705) and water, introduced through the two inlet channels, pass through a microchannel in a coflow. In a very low Re flow, as is the case here, the cross-stream mixing is diffusion-dominated.

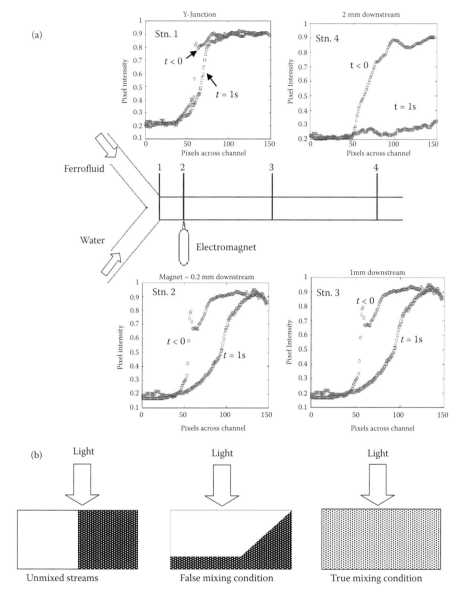

FIGURE 15.15

(a) Magnetically induced mixing using ferrofluid. The pixel intensity distribution across the channel at four axial locations (stations 1 through 4) is reported in the graphs at $t < 0$ and $t = 1$ s. The electromagnet is turned on at $t = 0$. Pixel intensity data indicate low values (dark) across the entire channel width, indicating enhanced mixing of the opaque ferrofluid. (b) A false mixing condition can be indicated if the ferrofluid is stratified near the upper or lower wall due to asymmetric placement of the dipole, thereby blocking the path of the light.

Because, for an average particle diameter of ~10 nm, the ferrofluid diffusivity in distilled water is of the order of 10^{-11} m^2/s (Blūms et al., 1983), the streams are nearly unmixed. The inset plots of Figure 15.15a show the pixel intensity distribution at the different sections of the channel. The round, blue symbols in the pixel intensity plots of Figure 15.15a denote the initial unmixed state. When the magnetic field is applied using an electromagnet with a needle-shaped core tip close to the distilled water stream, the higher susceptibility ferrofluid tries to move closer to the electromagnet, and a cross-stream advection leads to rapid mixing of the two streams. The pixel intensity (denoted by the red, round symbols in the pixel intensity plots of Figure 15.15a) homogenizes across the channel, denoting a well-mixed flow in the downstream section of the electromagnet.

Similar mixing strategies have also been demonstrated by other groups. For example, Tsai et al. (2009) demonstrated that mixing efficiencies can reach more than 90% within three characteristic lengths of the channels using a pair of permanent magnet slabs (one below and one above the channel). However, this arrangement could not rule out the possibility of reporting erroneously high mixing index because the apparent homogenization of the opacity of the microchannel could actually be due to the stratified ferrofluid layers along the top and bottom walls (see Figure 15.15b). Mao and Koser (2007) also proposed a ferrohydrodynamics-induced mixing strategy by introducing a two-phase travelling magnetic wave in the coflow of a ferrofluid and a fluorescein buffer.

15.4.4 Droplet Handling in Microfluidic Platforms

In droplet-based microfluidic applications, samples and reagents are enclosed in microdroplets on flat substrates or in immiscible fluids. Droplet-based microfluidics enjoys several advantages over the traditional flow-through microfluidics. Unlike continuous-flow microfluidics, droplet-based microfluidics eliminates the need for elaborate flow systems, valves, and so forth. Isolated handling of droplets minimizes reagent consumption and contamination. Through independent micromanipulation of discrete droplets, complex procedures can be executed on a microfluidic platform, closely mimicking traditional benchtop protocols. Because each droplet can be independently controlled, highly parallelized, integrated, scalable, and flexible processes can be implemented in microfluidic architectures.

Ferrofluids can be used either as picoliter droplets in an emulsion or in the form of nanoliter to microliter droplets in an immiscible medium or on a substrate. Two unique features of ferrofluids make them an attractive choice for droplet-based microfluidics—the fluid behavior of the ferrofluid allows the use of all conventional microdroplet-generation strategies (Zheng et al., 2004) in microfluidic architecture, whereas the precise maneuverability of droplets using external magnetic fields offers excellent control over the microfluidic manipulation of the droplets.

Nguyen et al. (2007) investigated the kinematics and deformation of ferrofluid droplets driven by planar coils. They demonstrated an effective strategy of controlling the motion of 1.2- to 2.4-mm-diameter ferrofluid droplets immersed in an immiscible medium by changing the sign of the field gradient or the driving current. A similar investigation on controlled transport of cobalt-based ferrofluidic droplets on the surface of an immiscible liquid is reported by Sterr et al. (2008). Using a pair of magnetic fields, one rotating in a vertically oriented plane, and the other alternating along a horizontal axis, they could, in principle, maneuver the drops to any arbitrary position on the whole two-dimensional liquid layer surface. Guo et al. (2006) reported the suspension and motion of ferrofluid droplets on superhydrophobic surfaces in the presence of strong external magnetic fields. Their study also provided a quantitative evaluation of friction between the ferrofluid droplets and the substrate surface. Bormashenko et al. (2008) reported a similar study on the motion of 20–200 µL of ferrofluidic marbles (formed from nanopowders of polyvinyl fluorides and γFe_2O_3) on flat superhydrophobic polymer substrates using an external magnetic field. They were able to impart as high as 25 cm/s velocity to 20 µL droplets while they found that the threshold magnetic force necessary for the drop displacement depends linearly on the drop radius. In contrast to many other strategies (e.g., electrowetting-on-dielectric; Ren et al., 2004), magnetic manipulations do not require prepatterned surfaces or electrical contacts; the droplets can be actuated much faster than is possible in techniques relying on optical switching of surface wetting properties.

15.4.5 Biosensors

The ferrous nanoparticles of a ferrofluid can provide the sites for bioassays through suitable functionalization. The change in particle relaxation behavior can be harnessed to achieve biosensing applications. For example, bioconjugates from the surface epitopes of cells, large DNA molecules, or the complementary surface of a sandwich immunoassay can alter the relaxation behavior of magnetic nanoparticles after attaching to them. The Brownian relaxation time constant of magnetic nanoparticles under a rapid transient magnetic field increases significantly when the particles get attached to entities much larger than their diameter, and the magnetic moments relax via Néel mechanism (Ganguly, 2005). The difference in the relaxation time of a "bound" and "unbound" nanoparticle can be detected using a superconducting quantum interference device. The technique is called MARIA. One salient advantage of MARIA is that quantification of binding sites is possible even in the presence of the residual unbound molecules, thereby eliminating the need for rinsing the background fluid. The technique is also applicable for turbid media (e.g., blood sample), where optical methods (e.g., immunofluorescence) would fail. Fast detection of biological targets using 25–100 nm Fe_3O_4 nanoparticles and a superconducting quantum interference device has been successfully demonstrated in the literature (Grossman et al., 2004; Ku et al., 2008; Enpuku et al., 2009) for small analyte sample volumes and concentration, although a complete microfluidic implementation has not yet been reported.

15.5 Microfluidic Application of Magnetic Microspheres

Several advantageous features of magnetic microspheres have been harnessed to meet different microfluidic challenges in µ-TAS applications. A major class of applications uses biochemical binding of target biomolecules on the surface of functionalized magnetic microspheres and subsequent magnetophoretic separation of the target species in a microfluidic environment. For example, immunomagnetic separation (IMS) via mobile suspended magnetic microbeads provides the advantage of relatively rapid antigen capture as compared with enzyme-linked immunosorbent assay. IMS is a popular analytical technique in which magnetic beads with antibodies immobilized on their surface can bind to target cells, toxins, or other molecules of a test sample.

Figure 15.16 describes the basic steps involved in IMS in a microfluidic environment. The immobilization process requires mixing of the bead and the analyte samples within a finite residence time. Subsequently, a magnetic field gradient concentrates the immobilized

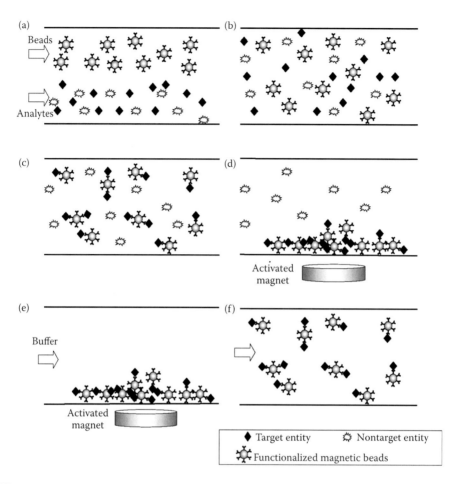

FIGURE 15.16
Basic steps of IMS in a microfluidic configuration. (a) Introduction of the beads and the analyte mixture, (b) mixing of the beads and analytes, (c) binding of the target analyte in the bead surfaces, (d) magnetic separation of the bead and target analyte conjugates, (e) washing of the nontarget entities and background fluid by buffer, and (f) collection of the target analyte–bead conjugates.

sample. Sample debris, nontarget organisms, and molecules are replaced by a buffer. The particles are resuspended in the buffer by deactivating the magnetic field. The resulting suspension contains concentrated molecules or cells of interest that have bound to the antibody attached to the magnetic particles. This concentrated sample is subsequently used for detection. The method enjoys several advantages, for example, it eliminates the complexities of filtration or centrifugation concentration methods common to other techniques and the associated loss of sample integrity or content. Additionally, antibody-coated microspheres offer a large surface area that enhances the sensitivity and kinetics of the reaction. Traditional benchtop IMS techniques have been used to easily and effectively confine and concentrate bacteria and protozoa from a variety of complex biological media, including foods and waste (Luk and Lindberg, 1991; Fratamico et al., 1992; Wright et al., 1994) and environmental water samples (Morgan et al., 1991; Yu, 1998; Kuczynska et al., 2003; Garcia-Aljaro et al., 2005). A microfluidic version of the IMS technique has shown much promise for different μ-TAS applications, including biosensors and cell separators.

The fact that the magnetic microspheres can chaperone biochemical agents, cells, or other entities attached to them to target locations has led to the development of ideas for new devices. This property can be used to transport biochemical agents either to a more active reaction zone or to withdraw the agents from a reacting medium, thus promoting or inhibiting a specific biochemical reaction. It is also possible to magnetically target the tagged molecules to a specific zone in a microfluidic device to achieve a higher analyte concentration. The separation of tagged molecules or biological entities (e.g., bacteria, DNA, protein, or living cells) can be subsequently performed by controlling the formation and transport of these beads in microfluidic devices. This can enable detection processes to work above a threshold concentration.

The transport of magnetic microspheres in a microfluidic channel occurs in a strongly viscosity-dominated regime because the particle Reynolds number (Re = $2\rho_l V_{slip} a/\eta$) is extremely low for these particles. The viscous force exerted by the fluid on the particles poses an equal and opposite reaction on the fluid. For a dense-enough particle concentration, and under strong magnetic field, the two-way particle–fluid interaction can alter the flow field (Modak et al., 2009). For several microfluidic applications, the momentum exchange between the magnetically transported microspheres and the host fluid is harnessed to induce active mixing on a microfluidic platform. Figure 15.17 shows the evidence of magnetic-bead-induced mixing between a stream of fluorescently labeled oligonucleotide sample and a coflowing buffer stream in a simple microfluidic configuration. Initially,

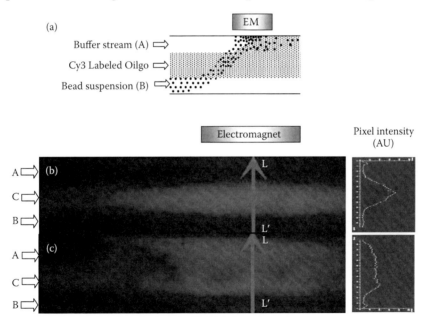

FIGURE 15.17
Magnetic-bead-assisted mixing of a central stream (stream C) of Cy3 fluorophore-labeled biological sample (22 base pair oligonucleotides of diffusivity ~10–11 m²/s) with a coflowing buffer stream (stream A) in a microchannel (width = 600 μm) at a flow Re ~ 0.1. A homogeneous suspension of 2 μm diamagnetic beads (loading = 1.3 mg/mL) is introduced through the stream B. Upon excitation of the electromagnet (EM), the beads move transversely toward the electromagnet, passing over the central stream. Particle–fluid momentum exchange creates advective motion of the central stream, leading to cross-stream mixing of streams C and A. (a) The schematic of the arrangement. Epifluorescence images of the flow: (b) before and (c) after switching on the electromagnet. Transverse distribution of fluorescence intensity plots at the location of the red, vertical arrow LL′ in the two snapshots is shown in the plots on the right.

when the three streams of the buffer (A), the biological sample (C), and the bead suspension (B) are passed through the channel in the absence of any magnetic influence, they remain almost unmixed (Figure 15.17b). Upon activation of the electromagnet, the magnetic beads undergo cross-stream magnetophoretic motion, and the particle–fluid momentum interaction induces a transverse advective component of flow, leading to rapid mixing of the streams (Figure 15.17c).

15.5.1 The Choice of Particles and Functionalizing Agent

The choice of magnetic particles (e.g., the size, density, magnetic particle content) and the functionalizing agent can vary quite largely depending on the application. An extensive review of the physical properties and the task-specific functionalization of commercially available magnetic beads can be found in Häfeli et al. (2005). A wide variety of bioconjugates are commercially available, which when coated on the microspheres enable them to be used as selective tags. Besides, such tags can also have biosynthetic use. For example, streptavidin–biotin conjugation or base pairing between oligonucleotide residues can be used for immobilizing DNA targets on magnetic beads for DNA hybridization (Fan et al., 1999). Streptavidin-coated magnetic microbeads combine readily with biotinylated oligonucleotides through a strong noncovalent biotin–streptavidin bond. In a homogeneous solution, the forward (association) and reverse (dissociation) rate constants are of the order of 10^7 M^{-1} s^{-1} and 10^{-8} s^{-1}, respectively, corresponding to an equilibrium constant K_a = 10^{15} M^{-1} (Green, 1990). However, in the case of a heterogeneous immunoassay, where at least one agent (e.g., the streptavidin) is coated on the surface of solid particles, the binding kinetics is influenced by the size of the analyte molecules and the particles. For example, if the streptavidin is bound on a microsphere ~1 µm and biotin is attached to ~100 base pair DNA molecules in a heterogeneous immunoassay, the association rate constant is found to reduce to ~10^6 M^{-1} s^{-1} (Huang et al., 1996). Albeit the dissociation rate is still negligible in its comparison, the noncovalent streptavidin–biotin bonds are not easily broken even by eluting. A commercially available 2-µm-diameter magnetic bead can have an average of 10^5 streptavidin sites on its surface (Micromod; www.micromod.de), whereas each streptavidin molecule on the microsphere can bind to approximately 0.45–0.78 molecules of the biotinylated agent (Huang et al., 1996). Therefore, each bead can effectively act as an active biochemical reaction site. All these bioanalytical features of magnetic microspheres are realizable at the macro- as well as microscale. Because the biochemical binding forces are essentially short-range forces, selective tagging by the magnetic beads is more effective in MEMS-scale analyzers.

For cell separation applications, an important factor is the degree to which a cell binds to the magnetic particles. McCloskey et al. (2003) provided a mathematical relationship to correlate magnetic labeling to the magnetophoretic mobility of an immunomagnetically labeled cell. They found that the magnetophoretic mobility of an immunochemically labeled cell depends primarily on (1) the antibody binding capacity (the number of antibody binding sites, which depends on the expression level of the targeted antigen molecules on the cell surface), (2) the secondary antibody amplification factor (applicable in a two-step antibody labeling protocol, where a magnetic-particle-bound secondary antibody targets a variety of antigens on the primary antibody on the cell surface), (3) the dipole moment of the particles (which is proportional to the particle volume and its magnetic susceptibility), and (4) the cell size (e.g., a 15-µm-diameter cell can bind magnetic particles 10 times more than a 7-µm-diameter cell does). Depending on these factors, the number of magnetic particles that can bind to a cell ranges to 10^6.

15.5.2 Specific Applications

Availability in a wide range of sizes, tunable magnetic and physical properties, and surface functionalization of the magnetic microspheres have recently led to a wide variety of uses in microfluidic applications. Table 15.2 gives a brief overview of such salient applications in the literature. Although the classifications in Table 15.2 are based on the main functional objectives of the applications (e.g., specific binding and separation, biosensing,

TABLE 15.2

Microfluidic Applications of Magnetic Microspheres

Principle of Application	Specific Applications
1. Specific adsorption to target analyte through surface functionalization and subsequent magnetophoretic separation of the bead–analyte conjugate	*Task based* Cell separation: Mccloskey et al. (2003), Rong et al. (2006), Ramadan et al. (2006), Inglis et al. (2004), Furudi and Harrison (2004), Pamme and Wilhem (2006), and Tsai et al. (2006) Nucleic acid isolation: Liu et al. (2008) and Jiang and Harrison (2000) DNA hybridization/PCR: Fan et al. (1999), Ohashi et al. (2007), and Zaytseva et al. (2005) Immunoassay: Edelstein et al. (2000), Kim and Park (2005), and Mujika et al. (2009) *Design based* Magnetic trap: Smistrup et al. (2005a,b), Choi et al. (2001), Ramadan et al. (2006), Furlani et al. (2006), Modak et al. (2010), Sinha et al. (2007, 2009), Nandy et al. (2008), Drogoff et al. (2008), Jiang and Harrison (2000), Liu et al. (2008), Furdui and Harrison (2004), Bronzeau and Pamme (2008), Dubus et al. (2006), Bu et al. (2008), and Mujika et al. (2009) Flow sorter: Rong et al. (2006), Pekas et al. (2005), Inglis et al. (2004), and Kim and Park (2005). Magnetophoretic FFF and SPLITT: Latham et al. (2005), Tsai et al. (2006), Pamme et al. (2006), and Pamme and Wilhelm (2006)
2. Physical or chemical detection of analytes attached to the magnetic microspheres	Fluorescence-based detection: Kwakye and Baeumner (2003) and Zaytseva et al. (2005) Electrochemical detection: Goral et al. (2006) *Magnetic detection* GMR and spin valve sensors: Graham et al. (2004), Rife et al. (2003), Wirix-Speetjens et al. (2006), Janssen et al. (2008), Mujika et al. (2009) Hall probe sensor: Mihajlovic´ et al. (2007) and Ejsing et al. (2004)
3. Targeted assembly of magnetic beads	DNA separation using columnar sieve of beads: Doyle et al. (2002) Programmable templating of beads on substrates: Yellen and Friedman (2003, 2004), Lyles et al (2004), and Pregibon et al. (2006)
4. Active mixing using controlled movement of magnetic beads or their aggregates	*Enhanced mixing through particle–fluid momentum exchange:* Using rotating chains of magnetic beads: Grumann et al. (2005), Biswal and Gast (2004b), Calhoun et al. (2006), Kang et al. (2007), Lee et al. (2009), and Roy et al. (2009) Using chaotic trajectories of magnetic beads: Wang et al. (2008) Enhancement in bead–analyte contact through controlled manipulation of the magnetic particles: Suzuki et al. (2004), Lehmann et al. (2006), Zolgharni et al. (2007), and Ganguly et al. (2009) Enhanced bead–analyte mixing by using dynamic plug of magnetic beads: Hayes et al. (2001), Rida and Gijs (2004), Lund-Olesen et al. (2008), Moser et al. (2009), and Lacharme et al. (2008, 2009)
5. On-chip manipulation of droplets containing a suspension of magnetic beads	Shikida et al. (2004, 2006), Ohashi et al. (2007), Long et al. (2008), Ohashi et al. (2008), Lehmann et al. (2006, 2007), Wang et al. (2007), and Sista et al. (2008a,b)

droplet manipulation, mixing, etc.), some of the specific applications simultaneously harness more than one functional attribute of the beads. For example, the biosensing applications proposed by Mujika et al. (2009) harness one of the separation techniques in their operation, whereas both the droplet manipulation and the mixing strategies are realized simultaneously in the application proposed by Lehmann et al. (2006).

15.5.2.1 Magnetophoretic Separation

Microfluidic implementation of IMS has attracted immense interest in the recent past, leading to a large variety of applications. Immunomagnetic separation is achieved via different modes (see Figure 15.18), for example, magnetic trap, flow diverter, or split flow thin (SPLITT) fractionation. In the magnetic trap design, a homogeneous suspension of the target analyte (bound to the magnetic bead) enters the channel across which a transverse magnetic field gradient is imposed. The background fluid may also contain some nontarget entities. The magnetic field gradient and particle–particle interaction lead to capture of the magnetic beads (along with the bound target analyte) within the channel bed (see Figure 15.18a). Several designs are proposed in the literature involving electromagnets with soft magnetic cores, permanent magnets, or a combination of both (Choi et al., 2001; Smistrup et al., 2005a,b; Ramadan et al., 2006)

Analyses of magnetophoretic transport in microchannel (Furlani et al., 2006; Sinha et al., 2007; Drogoff et al., 2008; Nandy et al., 2008; Modak et al., 2009, 2010) indicate that the separation efficiency of these traps depends primarily on the ratio of the magnetic and

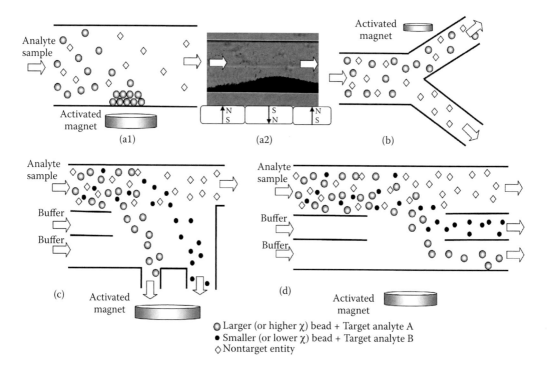

FIGURE 15.18
Salient magnetophoretic separation techniques of magnetic beads in microfluidic channels: (a) magnetic trap (a1, schematic; a2, image of the trapped bead, 6-μm diameter, collected in a 600-μm-wide microchannel using a permanent magnet trap), (b) flow sorter, (c) magnetophoretic FFF, and (d) magnetophoretic SPLITT fractionation.

viscous forces on the magnetic bead–analyte complexes. Magnetic traps are ideal for batch-type operation, where the target analyte immobilized on the magnetic beads is first separated from the background flow as the magnetic particles collect in the region of large magnetic field gradients within the channel (see Figures 15.11b and 15.18a2). The main flow is stopped after the separated aggregate grows beyond a critical size (Sinha et al., 2009). These traps offer the possibility of further *in situ* treatment of the bead-bound analytes such as washing, eluting, and redispersing. Such a trap design is also used for other magnetic-bead-based bioanalytical applications, for example, RNA extraction and reverse transcription (Jiang and Harrison, 2000; Liu et al., 2008), T-cell capture from blood for PCR analysis (Furdui and Harrison, 2004), simultaneous bioassay (Bronzeau and Pamme, 2008), PCR-free DNA detection (Dubus et al., 2006), and biosensor (see the section "Biosensors"). Recently, Bu et al. (2008) extended the magnetic trap design (comprising an array of small NdFeB permanent magnets arranged in a checkerboard pattern with alternating magnetization directions and an array of integrated permalloy elements encapsulated in the bottom of a microfabricated separation chamber) for high-throughput applications.

Flow sorter or diverter designs use a suitably placed magnetic field near the bifurcation of a microchannel. A homogeneous suspension of bead-bound analyte is split such that one outflow stream is richer in the bead–analyte concentration (Figure 15.18b). This design is an effective technique for online purification and enrichment of magnetic-bead-bound analytes (Inglis et al., 2004; Kim and Park, 2005; Pekas et al., 2005; Rong et al., 2006). Modak et al. (2009) characterized the separation efficiency in a T-channel sorter and argued that a cascade of such separators can be used to have a multiplicative enrichment of the sample.

A more selective sorting is possible by magnetic field-flow fractionation (FFF) (Latham et al., 2005) and SPLITT fractionation (Tsai et al., 2006). In both designs, two or more different target analytes are first conjugated to magnetic beads of different magnetophoretic mobility. They are then passed through a wide microchannel across a transverse magnetic field gradient. Because of the difference in mobility, particles develop different transverse velocities and hence segregate into different regions of the flow. In magnetophoretic FFF (see Figure 15.18c), bead–analyte conjugates with different magnetophoretic mobility are segregated along the axial length of the channel (particles with the largest magnetophoretic mobility collecting at the first upstream outlet). In SPLITT (see Figure 15.18d), magnetic particles bound to different target analytes segregate along different transverse streams at the outlet plane (particles with largest magnetophoretic mobility collecting at the stream closest to the magnet). Pamme et al. (2006) report the continuous sorting of magnetic microspheres in a microfluidic magnetic separation device. Cells were passed through a wide but shallow microfluidic chamber deflected from the flow direction by a transverse magnetic field gradient to achieve a free flow fractionation according to the mobility of the magnetic beads. They extended the same technique for separation of magnetic nanoparticles-loaded cells (Pamme and Wilhelm, 2006).

15.5.2.2 Biosensors

There have been a number of attempts toward developing magnetic-bead-based devices that perform the key individual steps of any generic pathogen detection process. For example, Lien et al. (2007) have developed a microfluidic chip that performs purification of Dengue virus serotype 2 through a sequence of incubation. Bead collection, washing, and enrichment on a two-module microfluidic chip achieved almost a fivefold improvement in time (sample preparation was achieved in 10 min and 35 s as opposed to 48 min in a traditional method). Simpler versions of magnetic-bead-based sensors have been proposed by some other researchers who performed bead-based separation and/or detection for specific pathogens (Choi et al., 2000; Choi et al., 2002; Kwakye and Baeumner, 2003; Zaytseva et al., 2005; Goral et al., 2006). Besides providing the sites for bioassays and the ease of transporting, washing, and enriching, the magnetic beads also offer a scope for magnetic detection. According to Mulvaney et al. (2007), it is far simpler to detect a lower number of microbeads with routine optical microscopy (Lee et al., 2000) and magnetic detection (e.g., GMR technique and Hall sensor) than a high concentration of molecular fluorophores, chromophores, or nanoparticles (the other commonly used "tags" for detection). Hence, the use of magnetic microspheres in a microfluidic immunomagnetic assay can offer a viable option for MEMS-based biosensors.

Solid-state magnetic magnetoresistive and Hall sensors are designed to detect the stray fields from magnetic beads. A basic GMR or spin-valve device consists of layers of magnetic films separated by a nonmagnetic conducting layer (Hartmann, 1999). For example, a typical sensor structure consists of Ta (5 nm)/$Ir_{0.8}$ $Mn_{0.2}$ (10 nm)/$Co_{0.9}$ $Fe_{0.1}$ (2.5 nm)/Cu (3.3 nm)/$Co_{0.9}$ $Fe_{0.1}$ (1 nm)/$Ni_{0.82}$ $Fe_{0.12}$ (2 nm)/Ta (5 nm) layers patterned into rectangular shapes of different sizes using photolithography and ion milling (Srinivasan et al., 2009). When the GMR sensor is exposed to even a small component of in-plane magnetic field, the resistance of the intermediate conducting layer changes by a large amount because of a change in spin-dependent electron scattering at the interfaces within the device. Therefore, the stray field components produced by a magnetic bead sitting on a GMR sensor in a homogeneous magnetic field perpendicular to the plane of the sensor are detectable.

The principle of a GMR (or spin valve) sensor-based sandwich immunoassay is described in Figure 15.19a. In the mixer section of the biosensor, the target analyte in the sample, if present, binds to the functionalized beads (step A and B of Figure 15.19a). Subsequently, the target analyte–microsphere conjugates are separated (step C) from the background sample (that is being tested) by suitably placing electromagnetic traps on the channel walls (as discussed in the section "Magnetophoretic Separation"). The wall of the separator is coated with complementary antibody that forms immunochemical bonds with the target antigen of the pathogen (step D). Subsequently, the electromagnetic traps are deactivated, and the background liquid is flushed with a buffer (step E). The presence of a target antigen (e.g., a waterborne pathogen) would cause some magnetic beads to stick to the surface because of the antigen–antibody binding. On the other hand, the absence of the target pathogen in the sample rules out the possibility of immunochemical binding of the functionalized beads to the surface, and the beads are washed out with the buffer. After washing, a homogeneous vertical biasing magnetic field is applied in the separator section (e.g., by placing the biosensor chip in a slot within a pair of permanent magnets). If microspheres attached to the separator wall are still present after the washing, a horizontal component of fringe magnetic field would develop because of particle magnetization. The GMR sensor chip detects this fringe component (step F of Figure 15.19a), which can be correlated to the presence of pathogen in water. In absence of the target pathogen, no magnetic bead would remain attached to the surface, and hence, no signal will be detected at the GMR sensor upon application of a biasing field.

A Hall sensor works on the principle of Lorentz force (Griffiths, 2004) on a charge as it moves through a field perpendicular to it. The Hall probe can detect the stray magnetic field component produced by a magnetic bead under a biasing field in terms of a cross-electrode voltage (Ejsing et al., 2004). For magnetic microbeads and nanoparticles detection, Ejsing et al. (2004) found that the signal from a Hall sensor is 5–10 times weaker than

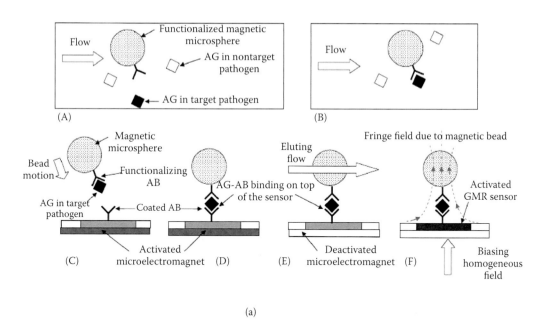

(a)

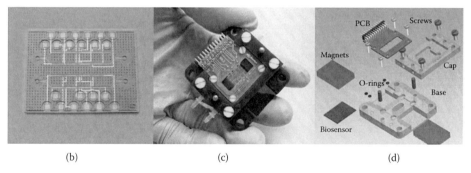

(b) (c) (d)

FIGURE 15.19

(a) Basic steps of a sandwich immunomagnetic assay using GMR sensor and on-chip microelectromagnets: (A) mixing of functionalized beads and analyte, (B) incubation and formation of antibody–antigen bond between bead and target pathogen, (C) magnetic separation of functionalized beads with attached pathogen (nontarget antigens do not form bonds and are washed away with flow), (D) targeted magnetic binding on coated antibody surface leading to antigen–antibody binding, (E) deactivation of electromagnet and washing to remove beads in the absence of immunochemical binding (i.e., when the pathogen is not present), (F) detection of the horizontal component of the fringe field by a GMR sensor under a vertical biasing field if the beads are still attached to the surface after washing (occurs only when the target antigen binds the bead to the surface). Legend: AB = antibody; AG = antigen. (b) A MEMS scale GMR sensor element for the detection of *Escherichia coli* O157:H7, including the microfluidic networks with two inlets/outlets. (c) The packaged biosensor ready to be tested. (d) Three-dimensional schematic of biosensor packaging design. (Panels b, c, and d are reprinted from *Biosens. Bioelectron.* 24, Mujika M. et al., Magnetoresistive immunosensor for the detection of *Escherichia coli* O157:H7 including a microfluidic network, pp. 1253–1258, Copyright 2009, with permission of Elsevier.)

that from a GMR sensor of comparable sensing area; however, the noise from a Hall probe could be as much as 20 times lower.

With solid-state magnetic sensors, a detection sensitivity up to the level of a single particle has been achieved. The sensitivity of the magnetoresistive sensor increases with decreasing surface area of the sensor, but the chemical sensitivity, or the number of analyte molecules that can hybridize to the surface, increases with increasing surface area. When the sensor size is reduced to the size of the bead, it is possible to detect beads with a radius down to 100 nm or smaller (Gijs, 2004). Successful single-particle detections have been reported for magnetic beads having diameters of 2.8 μm with a GMR (Rife et al., 2003), 2 μm with a spin valve sensor (Wirix-Speetjens et al., 2006), and 1.2 μm with a semiconductor Hall probe sensor (Mihajlović et al., 2007). An integrated magnetic manipulation-cum-GMR detection device by Janssen et al. (2008) demonstrated a single-particle resolution for even 1 μm particles traveling across the sensor surface. The principle of solid-state magnetic detection has been extensively used in microfluidic biosensors; for example, detection of waterborne pathogens (Mujika et al., 2009; see Figures b through d), surveillance of biological threats (Edelstein et al., 2000), point-of-care diagnosis (Schotter et al., 2009), or cell counting in chip cytometer (Roh et al., 2009).

15.5.2.3 Targeted Assembly

Highly localized magnetic fields can be used to direct, confine, and pattern magnetic particles in a controlled fashion. BioMEMS applications of functionalized magnetic beads can use patterned substrates to provide active support for biochemical reactions in biosensor arrays and for protein coupling. Doyle et al. (2002) used self-assembled magnetic bead chains that were immobilized across the flow in a microchannel to separate large, uplex DNA.

Dynamic self-assembly in confined microgeometries offers the possibility of reconfigurable multifunctional microdevices and may also suggest new protocols for fabricating three-dimensional microsystems and nanosystems. Yellen et al. (2003) and Yellen and Friedman (2004) demonstrated a programmable self-assembly method for the placement of two or more different types of superparamagnetic colloidal beads onto lithographically defined micromagnets and a combination of micromagnets and microwell templates. Lyles et al. (2004) used an electrostatic self-assembly method to pattern individual magnetic particles and long chains of magnetic microbeads on a flat substrate. Long, diagonal strips of magnetic bead chains can be embedded in a microchannel wall to form magnetic speed bumps in flow cells, similar to the arrangement proposed by Inglis et al. (2004), to allow cross-flow separation of magnetic beads on the basis of size and magnetic susceptibility. Pregibon et al. (2006) presented an approach to the direct patterning of biologically and magnetically active microbeads using the photopolymerization of the bead containing hydrogel precursors in nonbiofouling polymer scaffolds. They were able to achieve dispersed and packed patterns of exposed, covered, or protein-decorated magnetic beads on polyethylene glycol surfaces using a homogeneous perpendicular magnetic field or a combination of the magnetic field and shallow channels on the substrate. They also used the polyethylene-glycol-encapsulated magnetic patterns for filtering bead-bound B cells from T cells and for direct capture of B cells on exposed bead patterns.

15.5.2.4 Active Mixing

Several versions of magnetic-bead-based active mixers have been proposed in the literature. Some of them rely on the added advection force caused by particle–fluid momentum exchange, whereas others offer controlled manipulation of the magnetic beads in such a manner that the beads are transported into the analyte stream and mixed well with the same. Grumann et al. (2005) developed a centrifugal microfluidic platform on which they demonstrated accelerated mixing in batch-mode stopped flow. They had a microstrucured disk functioning as a mixing chamber mounted on a rotating macroscopic driving unit.

Magnetic beads that were prefilled into the mixing chamber were periodically deflected radially inward or outward because of stationary magnetic mounts as the mixing chamber rotated. Advection induced by the relative motion of the beads and the host fluid promoted mixing.

Biswal and Gast (2004b) proposed a novel mixing strategy using rotating chains of magnetic microspheres for flow-through systems. Viscous interactions between the particle chains (as they rotate under a rotating magnetic field) and the surrounding fluid lead to advective motion in the microchannel. This advection was found to enhance the otherwise diffusion-limited mixing. They also observed that such mixing was effective only within a frequency range. Outside this range, either the advection was not accomplished or the chains were unstable. The field-induced dynamic chain formation of magnetic microspheres and the mixing induced by the chain rotation under a rotating magnetic field have been computationally investigated through direct numerical simulation (Kang et al., 2007) and lattice Boltzmann methods (Calhoun et al., 2006).

Roy et al. (2009) experimentally demonstrated this strategy to be effective in augmenting mixing in micron-size droplets (see Figure 15.20a). They quantified the mixing of an optically opaque dye inside a sessile droplet through a mixing index parameter (Lu et al., 2002),

$$C' = \sqrt{(1/N) \sum_{1}^{N} ((P - \bar{P})^2 / \bar{P}^2)} \ , \quad \text{for } 0 \le P \le 256, \tag{15.40}$$

where the pixel value $\bar{P} = (1/N) \sum_{i}^{N} P$ was assumed to be proportional to the local concentration of the dye, and N denoted the total number of pixels of the recorded images. Initially, when the dye was relatively unmixed, the variance of the pixel density was large, and C' had a correspondingly large value (C'_0). As mixing progressed, the hue–intensity distribution became more uniform so that for perfect mixing the value of C approached a steady low value C'_∞ (see Figure 15.20b). When the mixing index parameter was normalized as

$$C = \frac{(C' - C'_\infty)}{(C'_0 - C'_\infty)}, \tag{15.41}$$

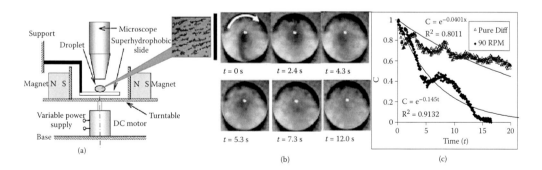

FIGURE 15.20
The droplet-based micromixer using rotating chains of magnetic microspheres. (a) The schematic (inset: magnetic beads forming chains). (b) Mixing of a blue dye due to the clockwise rotating bead chains (see Figure 15.12a for a zoomed-in view of the rotating chains) under a rotating magnetic field. (c) Time evolution of the mixing index, which shows improvement in mixing under a rotating field.

the C versus t curve varied from unity (unmixed state) to zero (fully mixed state), exhibiting a typical exponential time decay $C(t) \sim \exp(-t/\tau)$, where τ denotes the characteristic mixing time that is calculated from the hue–intensity histograms (see Figure 15.20c). Their results predicted that the extent of mixing depends on a dimensionless Mason number Ma (Equation 15.41), which represents the ratio of the viscous torque to the magnetic torque.

The study of Roy et al. showed that the mixing half-life decreased with increasing Ma to assume a minimum near Ma = 0.1 and then increased for larger Ma values. Similar results have been reported by Calhoun et al. (2006), who determined a critical Ma = 0.12 through LB simulations, and Kang et al. (2007), who observed an optimum Ma = 0.002 (equivalent to Ma = 0.06 as per Equation 15.41). Such a mixing behavior is correlated with the occurrence of rotating and corotating flows caused by alternating chain breakup and reformations in the regime of the optimum Ma, when the fastest mixing is observed. Below this optimum Ma value, mixing is limited by the extent of momentum transfer between the chains and the fluid, whereas above the optimum Ma, it is limited by chain deformation and breakdown. Lee et al. (2009) used rotating chains of 4-μm-diameter magnetic beads inside a microliter-volume mixing chamber and claimed a high mixing index (~96%) in a very short distance (~800 micron).

A different mixing strategy was proposed by Suzuki et al. (2004), who used a magnetic-bead-based chaotic mixer to enhance the transverse spreading of magnetic beads in a microchannel (and hence improve the bead–analyte contact) at a very low Reynolds number (~10^{-2}). They used embedded microconductors in a serpentine microchannel to create a time-dependent magnetic field. The magnetic force on the particles is adjusted by modulating the input current sequence through the microconductors in such a way that the magnetic particles experience a periodic push and pull as they negotiate the bends of the serpentine channel. A supporting numerical study showed that this arrangement is able to create a stretching and folding of the material lines, which is a manifestation of chaos. The resulting particle distribution at the downstream of the device shows effective mixing within 10 mixing blocks (i.e., combination of a pair of electrodes and a serpentine bend). Zolgharni et al. (2007) carried out a numerical investigation in a two-dimensional microchannel configuration, in which microcoils were placed alternately on either side of the channel. Passing a periodically timed current through the coils resulted in magnetic particle streak lines undergoing repeated stretching and folding, leading to chaotic mixing. Another numerical study by Wang et al. (2008) showed that magnetic particles steered in a microchannel using a carefully modulated alternating magnetic field enhance mixing in a microchannel via coupled particle–fluid momentum exchange. They also showed that the magnetic field actuation at a certain frequency results in maximal mixing. This optimum switching frequency depends on the channel's lateral dimension and the applied magnetic force. The maximum efficiency is obtained at a relatively high operating frequency for large magnetic actuation forces and narrow microchannels. If the magnetic particles are actuated with a much higher or lower frequency than the optimum switching frequency, they add only limited agitation to the fluid flow and do not significantly enhance mixing.

Another technique proposed for achieving effective bead–analyte contact in microchannels uses dynamic plugs of magnetic beads that are strategically held in a constriction of a microchannel by localized magnetic field gradients. Hayes et al. (2001) demonstrated the operation of a small-volume heterogeneous immunoassay, where the analyte sample was perfused through a magnetically immobilized plug of magnetic bead to promote on-bead binding. Rida and Gijs (2004), Lund-Olesen et al. (2008), and Moser et al. (2009) have further improved this concept by using an AC magnetic-field-modulated dynamic plug of beads to enhance mixing and analyte binding. A similar approach by Lacharme et al. (2008, 2009) used self-assembled chains of magnetic beads placed across the flow in a periodically structured microchannel to improve bead–analyte contact.

15.5.2.5 On-Chip Droplet Manipulation

The on-chip manipulation of droplets consisting of a suspension of magnetic microspheres can merge, mix, and separate droplets on a microfluidic platform. This principle is schematically described in Figures 15.21a and 15.21b. When a suspension of magnetic particles

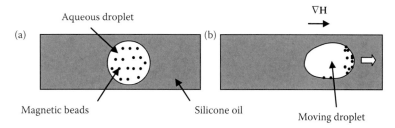

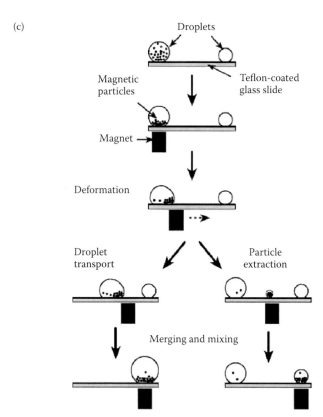

FIGURE 15.21

The principle of manipulation of submerged droplets containing magnetic beads. (a) The droplet in the absence of magnetic field gradient. (b) Particles migrate to one end of the droplet under an imposed magnetic field gradient, transmit the force through the interface, and drag the droplet in the direction of the field gradient. (c) Droplet manipulation on a hydrophobic surface using a permanent magnet. (From Long, Z., Shetty, A.M., Solomon, M.J. and Larson, R.G., Fundamentals of magnet-actuated droplet manipulation on an open hydrophobic surface, *Lab Chip* 9: 1567–1575, 2009. Reproduced by permission of The Royal Society of Chemistry.)

inside a liquid droplet is subjected to a magnetic field gradient, the particles experience a magnetic force. However, because the interfacial energy barrier at the droplet interface is often large, the particles remain confined within the droplet. The magnetic force on the particles is thus transmitted through the interface to the entire droplet, thereby moving the droplet. Shikida et al. (2004, 2006) presented a system using the advantages of manipulation of magnetic droplets in a microfluidic structure, with a sliding external magnet for actuation of droplets and confining barriers for localizing them.

Ohashi et al. (2007) manipulated droplets containing magnetic beads and PCR materials over a linear thermal cycler using reciprocating permanent magnets to achieve a droplet-based PCR. Long et al. (2009) investigated droplet movement, coalescence, and splitting (extraction) that was actuated by magnetic beads internalized in an oil-coated aqueous droplet on an open hydrophobic surface using an external magnet (Figure 15.21c). They explored the range of conditions under which magnetic force actuation can be used for microdroplet manipulation and identified the conditions under which one might easily switch between bead extraction and droplet transport. They also correlated the influence of particle type, droplet size surface tension, viscosity, and effect of oil coating on droplets on kinematics of droplets. Ohashi et al. (2008) applied magnetic field gradients to sample micro- or nanoliter volume droplets from electrostatically anchored mother droplets.

Lehmann et al. (2006, 2007) developed a system for the two-dimensional magnetic manipulation of aqueous droplets suspended in silicone oil as a platform for on-chip bioanalysis. Superparamagnetic microparticles of 250-nm to 6-μm diameter suspended inside the aqueous droplets were subjected to a combination of homogeneous field (created by a soft magnetic sheet under the substrate with permanent magnets at its ends) superposed by a two-dimensional field gradient (created by passing current through a planar microcoil on a four-layer printed circuit board that forms the substrate). The resulting magnetic force was sufficient for displacing, merging, mixing, and separating the droplets on the chip without the use of external moving magnets or parts. Additionally, they (Lehmann et al., 2006, 2007) used hydrophilic/hydrophobic spots on the substrates for immobilizing aqueous reagents, with which microdroplets containing the magnetic beads are merged, mixed, and separated to follow the required bioanalytical steps for an enzyme-based detection procedure. A method combining electrowetting-on-dielectric and magnetic manipulation of droplets has been proposed in the literature for in-droplet magnetic particle separation and concentration (Wang et al., 2007), heterogeneous sandwich immunoassays (Sista et al., 2008a), and point-of-care testing (Sista et al., 2008b).

15.6 Concluding Remarks

The use of magnetic particles, in the size range of a few nanometers to a few tens of microns, is rapidly gaining interest for microfluidic applications because of their unique and advantageous physicochemical properties. The wide range of available size and magnetic property, the opportunity of tunable surface functionalization, and the ease and precision of magnetic manipulation even in the presence of other competing physicochemical forces favor the use of magnetic particles in novel microfluidic applications. Traditional microfluidic challenges, for example, rapid mixing, precise and selective separation, and targeted transport, can be effectively addressed using magnetic particles. Shrinking the magnetic-particle-based immunoassay to flow-through as well as digital microfluidic environments have been successfully demonstrated in the literature. Magnetic particles and particle suspensions have been used effectively to perform the major bioanalytical steps, for example, analyte handling, incubation, reaction, separation, and washing on the same microfluidic platform. A relatively mature technology of integrating solid-state magnetic sensors into microfluidic chips has offered the possibility of developing a vast range of biosensors that can be operated in conjunction with existing microelectronic devices. Therefore, the importance of magnetic-particle-based microfluidics in future μ-TAS devices cannot be overemphasized.

However, the promises of the magnetic particles are not present without challenges. One problem of a static magnetic field is its strong inverse relationship with distance, meaning that the magnetic actuation becomes weak within a very short distance, particularly for fields produced by chip-based electromagnets. For example, the field produced by an electromagnetic trap may fail to ensure 100% capture of particles across its entire cross section. Such a problem can be overcome by traveling wave fields, which are created by passing multiphase currents through chip-embedded conductors (Liu et al., 2007). Conventional magnetophoretic separation techniques, a translating periodic potential energy landscape, can simultaneously separate different types of superparamagnetic beads on the basis of only marginal differences in their size, geometric properties, and/or the type of biological microorganisms attached to the bead (Yellen et al., 2007).

Another practical difficulty of magnetic-particle-based system is that a large particle concentration often leads to undesirable particle clustering, which not only reduces the available site on the particles for the bioanalytical reactions but also alters the magnetophoretic transport characteristics. Particle aggregation can be controlled by suitably altering the solution pH. Lai et al. (2009) took this feature in stride to devise a strategy that exploits the reversible change in size (due to aggregation and disaggregation under a low and a high pH, respectively) and magnetophoretic mobility of smart magnetic nanoparticles and perform controlled bioseparation in a continuous flow operation.

Settling of the particles or even clogging of the microfluidic channels of a device is another challenge that needs to be taken care of in any magnetophoretic microsystem, particularly when the particle size is large. This can be overcome by embedding a piezoelectric vibrator in the storage wells of the beads on the microfluidic chip. However, for multiplexed systems, where beads with different surface functionalizations are used, one would require multiple vibrators (for storage of each kind of beads), making the device extremely complicated. An alternate strategy could be to use a single storage of streptavidin-coated beads, which can be flown through several parallel channels or in-line incubation units, in which they immunochemically bind in situ (Ganguly et al., 2010; Modak et al., 2010) with several different specific functional molecules (target or probe analytes), before taking part in the intended bioanalytical processes of the multiplexed system.

Despite the numerous advantageous features, the use of magnetic nano- and microparticles in the LOC context is still largely at the proof-of-concept level, primarily because of these practical challenges. The task of the ongoing and future research in the area of magnetic-particle-based microfluidics will continue to be a better characterization of the particle transport in microfluidic environment. The success will lay not only in avoiding features such as particle agglomeration and settling in the system but also in harnessing them intelligently, for example, by building mesoscopic architecture through field-assisted self-assembly (Ganguly and Puri, 2007). Advances in the materials research toward improving magnetic property, surface functionalization, and response to biochemical environment are also expected to tailor the particles better for their widespread microfluidic applications.

Acknowledgments

The first author acknowledges the research funding from the BRNS, Department of Atomic Energy (BRNS-YSR Award No. 2005/20/36/7-BRNS) and the Department of Science and Technology (SERC Fast Track Project for Young Scientists Scheme, Grant No. SR/FTP/ETA-07/2006), Government of India, for the relevant research by his group, which is presented in the chapter. Research support from Dr. Steffen Hardt (Center of Smart Interfaces, TU Darmstadt), who hosted the first author during a postdoctoral fellowship from the Alexander von Humboldt Stiftung, Germany, is also gratefully acknowledged.

References

Andò B, Ascia A, Baglio S, Beninato A. (2009). The "one drop" ferrofluidic pump with analog control. *Sensors Actuators A*. 156:251–256.

Berkovsky BM. (1978). Some aspects of theoretical modeling of thermomechanics of magnetic fluids. In: Berkovsky BM, editor. *Proceedings of the International Advanced Course and Workshop on Thermomechanics of Magnetic Fluids*. Washington (DC): Hemisphere; pp. 149–57.

Beyzavi A, Nguyen TN. (2009). One-dimensional actuation of a ferrofluid droplet by planar microcoils. *J Phys D Appl Phys*. 42:015004.

Biswal SL, Gast AP. (2004a). Rotational dynamics of semiflexible paramagnetic particle chains. *Phys Rev E*. 69:041406.

Biswal SL, Gast AP. (2004b). Micromixing with linked chains of paramagnetic particles. *Anal Chem*. 76:6448–55.

Blūms E, Pļaviņš J, Chukhrov A. (1983). High-gradient magnetic separation of magnetic colloids and suspensions. *J Magn Magn Mater*. 39:147–51.

Bormashenko E, Pogreb R, Bormashenko Y, Musin A, Stein T. (2008). New investigations on ferrofluidics: ferrofluidic marbles and magnetic-field-driven drops on superhydrophobic surfaces. *Langmuir*. 24:12119–22.

Boudouvis AG, Scriven LE. (1993). Sensitivity analysis of hysteresis in deformation of ferrofluid drops. *J Magn Magn Mater*. 122:254–8.

Boudouvis AG, Puchalla JL, Scriven LE. (1993). Interaction of capillary wetting and fringing magnetic field in ferrofluid systems. *J Colloid Interface Sci*. 124:677–87.

Bronzeau S, Pamme N. (2008). Simultaneous bioassays in a micro-fluidic channel on plugs of different magnetic particles. *Anal Chim Acta*. 609(1):105–12.

Bu M, Christensen TB, Smistrup K, Wolff A, Hansen MF. (2008). Characterization of a microfluidic magnetic bead separator for high-throughput applications. *Sens Actuators A*. 145–6:430–6.

Calhoun R, Yadav A, Phelan P, Vuppu A, Garcia A, Hayes M. (2006). Paramagnetic particles and mixing in micro-scale flows. *Lab Chip*. 6:247–57.

Choi J-H, Oh KW, Thomas JH, Heineman WR, Halsall HB, Nevin JH, et al. (2002). An integrated microfluidic biochemical detection system for protein analysis with magnetic bead-based sampling capabilities. *Lab Chip*. 2:27–30.

Choi J-W, Ahn CH, Bhansali S, Henderson T. (2000). A new magnetic bead-based, filterless bio-separator with planar electromagnet surfaces for integrated bio-detection systems. *Sens Actuators B*. 68:34–9.

Choi J-W, Liakopoulos TM, Ahn CH. (2001). An on-chip magnetic bead separator using spiral electromagnet with semi-encapsulated permalloy. *Biosens Bioelectron*. 16:409–26.

Clift R, Grace JR, Weber ME. (1978). *Bubbles, Drops, and Particles*. New York: Academic Press.

Cowley MD, Rosensweig RE. (1967). The interfacial stability of a ferromagnetic fluid. *J Fluid Mech.* 30(4):671–88.

Dance SL, Climent E, Maxey MR. (2004). Collision barrier effects on the bulk flow in a random suspension. *Phys Fluids.* 16:828–31.

Deng T, Whitesides GM, Radhakrishnan M, Zabow G, Prentiss M. (2001). Manipulation of magnetic microbeads in suspension using micromagnetic systems fabricated with soft lithography. *Appl Phys Lett.* 2001;78:1775–7.

Dittrich PS, Tachikawa K, Manz A. (2006). Micro total analysis systems. Latest advancements and trends. *Anal Chem.* 78:3887–907.

Doyle PS, Bibette J, Bancaud A, Viory J-L. (2002). Self-assembled magnetic matrices for DNA separation chips. *Science.* 295:2237.

Drogoff BL, Clime L, Veres T. (2008). The influence of magnetic carrier size on the performance of microfluidic integrated micro-electromagnetic traps. *Microfluid Nanofluidics.* 5:373–81.

Dubus S, Gravel JF, Le Drogoff B, Nobert P, Veres T, Boudreau D. (2006). PCR-free DNA detection using a magnetic bead supported polymeric transducer and microelectromagnetic traps. *Anal Chem.* 78:4457–64.

Edelstein RL, Tamanaha CR, Sheehan PE, Miller MM, Baselt DR, Whitman LJ, et al. (2000). The BARC biosensor applied to the detection of biological warfare agents. *Biosens Bioelectron.* 14:805–13.

Ejsing L, Hansen MF, Menon AK, Ferreira HA, Graham DL, Freitas PP. (2004). Planar Hall effect sensor for magnetic micro- and nanobead detection. *App. Phys. Lett.* 84:4729–31.

Elborai S, Kim DK, He X, Lee SH, Rhodes S, Zahn M. (2005). Self-forming, quasi-two-dimensional, magnetic-fluid patterns with applied in-plane-rotating and dc-axial magnetic fields. *J Appl Phys.* 97:10Q303.

Enpuku K, Tokumitsu H, Sugimoto Y, Kuma H, Hamasaki N, Tsukamoto A, et al. (2009). Fast detection of biological targets with magnetic marker and SQUID. *IEEE Trans Appl Supercond.* 19:844–7.

Fan ZH, Mangru S, Granzow R, Heaney P, Ho W, Dong QP, et al. (1999). Dynamic DNA hybridization on a chip using paramagnetic beads. *Anal Chem.* 71:4851–9.

Fratamico PM, Schultz PJ, Buchanan RL. (1992). Rapid isolation of *Escherichia coli* O157:H7 from enrichment cultures of foods using an immunomagnetic separation method. *Food Microbiol.* 9:105–13.

Furdui VI, Harrison DJ. (2004). Immunomagnetic T cell capture from blood for PCR analysis using microfluidic systems. *Lab Chip.* 4:614–8.

Furlani EP. (2006). Analysis of particle transport in a magnetophoretic microsystem. *J Appl Phys.* 99:024912.

Furlani EP, Sahoo Y. (2006). Analytical model for the magnetic field and force in a magnetophoretic microsystem. *J Phys D Appl Phys.* 39:1724–32.

Ganguly R. (2005). Ferrofluid transport analyses for thermal, biomedical and MEMS applications. PhD dissertation. Chicago: University of Illinois.

Ganguly R, Gaind AP, Puri IK. (2005). A strategy for the assembly of 3-d mesoscopic structures using a ferrofluid. *Phys Fluids.* 17:057103.

Ganguly R, Hahn T, Hardt S. (2010). Magnetophoretic mixing for in-situ immunochemical binding on magnetic beads in a microfluidic channel. *Microfluid Nanofluidics.* 8:739–53.

Ganguly R, Puri IK. (2007). Field-assisted self-assembly of superparamagnetic nanoparticles for biomedical, MEMS and bioMEMS applications. *Adv Appl Mech.* 41:293–335.

Ganguly R, Sen S, Puri IK. (2004a). Heat transfer augmentation in a channel with a magnetic fluid under the influence of a line-dipole. *J Magn Magn Mater.* 271:63–73.

Ganguly R, Sen S, Puri IK. (2004b). Thermomagnetic convection in a square enclosure using a line dipole. *Phys Fluids.* 16:2228–36.

Garcia-Aljaro C, Bonjoch X, Blanch AR. (2005). Combined use of an immunomagnetic separation method and immunoblotting for the enumeration and isolation of *Escherichia coli* O157 in wastewaters. *J Appl Microbiol.* 98:589–97.

Gijs MAM. (2004). Magnetic bead handling on-chip: new opportunities for analytical applications. *Microfluid Nanofluidics.* 1:22–40.

Gijs MAM. (2007). Magnetic beads in microfluidic systems: towards new analytical applications. In: Hardt S, Schönfeld F, editors. *Microfluidic Technologies for Miniaturized Analysis Systems*. New York: Springer; pp. 244–6.

Goldstein RE, Jackson DP, Langer SA. (1993). Dynamics of pattern-formation in magnetic fluids. *J Magn Magn Mater*. 122:267–70.

Goral VN, Zaytseva NV, Baeumner AJ. (2006). Electrochemical microfluidic biosensor for the detection of nucleic acid sequences. *Lab Chip*. 6:414–21.

Graham DL, Ferreira HA, Freitas PP. (2004). Magnetoresistive-based biosensors and biochips. *Trends Biotechnol*. 22:455–62.

Green NM. (1990). Avidin and streptavidin. *Methods Enzymol*. 184:51–67.

Griffiths DJ. (2004). *Introduction to Electrodynamics*. New Delhi: Prentice Hall.

Grossman HL, Myers WR, Vreeland VJ, Bruehl R, Alper MD, Bertozzi CR, et al. (2004). Detection of bacteria in suspension by using a superconducting quantum interference device. *Proc Natl Acad Sci U S A*. 101:129–34.

Grumann M, Geipel A, Riegger L, Zengerle R, Ducrée J. (2005). Batch-mode mixing on centrifugal microfluidic platforms. *Lab Chip*. 5:560–5.

Guo Z-G, Zhou F, Hao JC, Liang YM, Liu WM. (2006). "Stick and slide" ferrofluidic droplets on superrhydrophobic surfaces. *Appl Phys Lett*. 89:081911.

Häfeli UO, Gilmour K, Zhou A, Lee S, Hayden ME. (2007). Modeling of magnetic bandages for drug targeting: Button vs. Halbach arrays. *J Magn Magn Mater*. 311:323–9.

Häfeli UO, Lobedann MA, Steingroewer J, Moore LR, Riffle J. (2005). Optical method for measurement of magnetophoretic mobility of individual magnetic microspheres in defined magnetic field. *J Magn Magn Mater*. 293:224–39.

Halverson D, Yellen B, Kalghatgi S, Friedman G. (2005). Use of ferrofluid patterns as diffusion masks. *Mater Res Soc Symp Proc*. 877:183–8.

Harris L. (2002). Polymer stabilized magnetite nanoparticles and poly(propylene oxide) modified styrene-dimethacrylate networks. PhD dissertation. Blacksburg: Virginia Polytechnic Institute and State University.

Hartmann U. (1999). *Magnetic Multilayers and Giant Magnetoresistance*. New York: Springer.

Hartshorne H, Backhouse CJ, Lee WE. (2004). Ferrofluid-based microchip pump and valve. *Sens Actuators B*. 99:592–600.

Hatch A, Kamholz AE, Holman G, Yager P, Böhringer KF. (2001). A ferrofluidic magnetic micropump. *J. Microelectromech Syst*. 2:215–221.

Hayes MA, Polson NA, Phayre AN, Garcia AA. (2001). Flow-based microimmunoassay. *Anal Chem*. 73:5896–902.

Horák D, Babič M, Macková H, Beneš MJ. (2007). Preparation and properties of magnetic nano- and microsized particles for biological and environmental separations. *J Sep Sci*. 30:1751–72.

Huang S-C, Stump MD, Weiss R, Caldwell KD. (1996). Binding of biotinylated DNA to streptavidin-coated polystyrene latex: effect of chain length and particle size. *Anal Biochem*. 237:115–22.

Inglis DW, Riehn R, Austin RH, Sturm JC. (2004). Continuous microfluidic immunomagnetic cell separation. *Appl Phys Lett*. 85:5093–5.

Jang IJ, Horng HE, Chiou YC, Hong C-Y, Yu JM, Yang HC. (1999). Pattern formation in microdrops of magnetic fluids. *J Magn Magn Mater*. 201:317–20.

Janssena XJA, van IJzendoorn LJ, Prins MWJ. (2008). On-chip manipulation and detection of magnetic particles for functional biosensors. *Biosens Bioelectron*. 23:833–8.

Jiang G, Harrison DJ. (2000). mRNA isolation in a microfluidic device for eventual integration of cDNA library construction. *Analyst*. 125:2176–9.

Kang TG, Hulsen MA, den Toonder JMJ, Anderson PD, Meijer HEH. (2007). Chaotic mixing induced by a magnetic chain in a rotating magnetic field. *Phys Rev E*. 76:066303.

Kantor AB, Gibbons I, Miltenyi S, Schmitz J. (1998). Magnetic cell sorting with colloidal superparamagnetic particles, in *Cell Separation Methods and Applications*, ed. D. Recktenwald, and A. Radbruch, 153. New York: Marcel Dekker.

Kim E-G, Oh JG, Choi B. (2006). A study on the development of a continuous peristaltic micropump using magnetic fluids. *Sens Actuators A*. 128:43–51.

Kim KS, Park J-K. (2005). Magnetic force-based multiplexed immunoassay using superparamagnetic nanoparticles in microfluidic channel. *Lab Chip*. 5:657–64.

Kouassi GK, Irudayaraj J, McCarty G. (2005). Activity of glucose oxidase functionalized onto magnetic nanoparticles. *Biomagn Res Technol*. 3:1–10.

Ku BY, Chan ML, Ma Z, Horsley DA. (2008). Frequency-domain birefringence measurement of biological binding to magnetic nanoparticles. *J Magn Magn Mater*. 320:2279–83.

Kuczynska E, Boyer DG, Shelton DR. (2003). Comparison of immunofluorescence assay and immunomagnetic electrochemiluminescence in detection of *Cryptosporidium parvum* oocysts in karst water samples. *J Microbiol Methods*. 53:17–26.

Kwakye S, Baeumner A. (2003). A microfluidic biosensor based on nucleic acid sequence recognition. *Anal Bioanal Chem*. 376:1062–8.

Lacharme F, Vandevyver C, Gijs MAM. (2008). Full on-chip nanoliter immunoassay by geometrical magnetic trapping of nanoparticle chains. *Anal Chem*. 80:2905–10.

Lacharme F, Vandevyver C, Gijs MAM. (2009). Magnetic beads retention device for sandwich immunoassay: comparison of off-chip and on-chip antibody incubation. *Microfluid Nanofluidics*. In press. DOI: 10.1007/s10404–009-0424–7.

Lai JJ, Nelson KE, Nash MA, Hoffman AS, Yager P, Stayton PS. (2009). Dynamic bioprocessing and microfluidic transport control with smart magnetic nanoparticles in laminar-flow devices. *Lab Chip*. 9:1997–2002.

Landau L, Lifshitz E, Pitaevskii L. (1984). *Electrodynamics of Continuous Media*. Oxford: Pergamon.

Lange A. (2002). Kelvin force in a layer of magnetic fluid. *J Magn Magn Mater*. 241:327–9.

Latham AH, Freitas RS, Schiffer P, Williams ME. (2005). Capillary magnetic field flow fractionation and analysis of magnetic nanoparticles. *Anal Chem*. 77:5055–62.

Lee GU, Metzger S, Natesam M, Yanavich C, Dufrene YF. (2000). Implementation of force differentiation in the immunoassay. *Anal Biochem*. 287:261–71.

Lee H, Purdon AM, Chu V, Westervelt RM. (2004). Controlled assembly of magnetic nanoparticles from magnetotactic bacteria using microelectromagnets arrays. *Nano Lett*. 4:995–8.

Lee SH, van Noort D, Lee JY, Zhangb BT, Park TH. (2009). Effective mixing in a microfluidic chip using magnetic particles. *Lab Chip*. 9:479–82.

Lehmann U, de Courten D, Vandevyver C, Parashar VK, Rida A, Gijs MAM. (2006). Two-dimensional magnetic manipulation of microdroplets on a chip as a platform for bioanalytical applications. *Sens Actuators B*. 117:457–63.

Lehmann U, Hadjidj S, Parashar VK, Vandevyver C, Gijs MAM. (2007). On-chip antibody and colorimetric detection in a magnetic droplet manipulation system. *Microelectron Eng*. 84:1669–72.

Liao W, Chen X, Chen Y, Pu S, Xia Y, Li Q. (2005). Tunable optical fiber filters with magnetic fluids. *Appl Phys Lett*. 87:151122.

Lien K-Y, Lin JL, Liu CY, Lei HY, Lee GB. (2007). Purification and enrichment of virus samples utilizing magnetic beads on a microfluidic system. *Lab Chip*. 7:868–75.

Liu C, Lagae L, Borghs G. (2007). On-chip separation of magnetic particles with different magnetophoretic mobilities. *J Appl Phys*. 101:024913.

Liu C-J, Lien KY, Weng CY, Shin JW, Chang TY, Lee GB. (2008). Magnetic-bead-based microfluidic system for ribonucleic acid extraction and reverse transcription processes. *Biomed Microdevices*. 11:339–50.

Long Z, Shetty AM, Solomon MJ, Larson RG. (2009). Fundamentals of magnet-actuated droplet manipulation on an open hydrophobic surface. *Lab Chip*. 9:1567–75.

Lu L-H, Ryu KS, Liu C. (2002). A magnetic microstirrer and array for microfluidic mixing. *J Microelectromech Syst*. 11:462–9.

Luk JC, Lindberg AA. (1991). Rapid and sensitive detection of *Salmonella* (O:6,7) by immunomagnetic monoclonal antibody-based assays. *J Immunol Methods*. 137:1–8.

Lund-Olesen T, Buus BB, Howalt JG, Hansen MF. (2008). Magnetic bead micromixer: influence of magnetic element geometry and field amplitude. *J Appl Phys*. 103:07E902.

Lyles BF, Terrot MS, Hammond PT, Gast AP. (2004). Directed patterned adsorption of magnetic beads on polyelectrolyte multilayers on glass. *Langmuir*. 20:3028–31.

Manciu FS, Manciu M, Sen S. (2000). Possibility of controlled ejection of ferrofuid grains from a magnetically ordered ferrofuid using high frequency non-linear acoustic pulses—a particle dynamical study. *J Magn Magn Mater*. 220:285–92.

Mao L, Koser H. (2007). Overcoming the diffusion barrier: ultra-fast micro-scale mixing via ferrofluids. Transducers Eurosensors '07—4th International Conference on Solid-State Sensors, Actuators and Microsystems. Article No.: 4300511; pp. 1829–32.

McCloskey KE, Chalmers JJ, Zborowski M. (2003). Magnetic cell separation: characterization of magnetophoretic mobility. *Anal Chem*. 75:6668–874.

McNab TK, Fox RA, Boyle AJF. (1968). Some magnetic properties of magnetite (Fe3O4) microcrystals. *J Appl Phys*. 39:5703–11.

Melle S, Calderón OG, Fuller GG, Rubio MA. (2002). Polarizable particle aggregation under rotating magnetic fields using scattering dichroism. *J Colloid Interface Sci*. 247:200–9.

Mihajlović G, Aledealat K, Xiong P, Von Molnár S, Field M, Sullivan GJ. (2007). Magnetic characterization of a single superparamagnetic bead by phase-sensitive micro-Hall magnetometry. *Appl Phys Lett*. 91:172518.

Modak N, Datta A, Ganguly R. (2009). Cell separation in a microfluidic channel using magnetic microspheres. *Microfluid Nanofluidics*. 6:647–60.

Modak N, Kejriwal D, Nandy K, Datta A, Ganguly R. (2010). Experimental and numerical characterization of magnetophoretic separation for MEMS-based biosensor applications. *Biomed Microdevices*. 12:23–4.

Modak N, Datta A, Ganguly R. (2010). Numerical analysis of transport and binding of a target analyte and functionalized magnetic microspheres in a microfluidic immunoassay, *J. Physics D: Appl. Phys*. 43:485002;1–12.

Morgan JAW, Winstanley C, Pickup RW, Saunders JR. (1991). Rapid immunocapture of *Pseudomonas putida* cells from lake water by using bacterial flagella. *Appl Environ. Microbiol*. 57:503–9.

Moser Y, Lehnert T, Gijs MAM. (2009). Quadrupolar magnetic actuation of superparamagnetic particles for enhanced microfluidic perfusion. *Appl Phys Lett*. 94:022505.

Mujika M, Arana S, Castaño E, Tijero M, Vilares R, Ruano-López JM, et al. (2009). Magnetoresistive immunosensor for the detection of *Escherichia coli* O157:H7 including a microfluidic network. *Biosens Bioelectron*. 24:1253–8.

Müller HW, Engel A. (1999). Dissipation in ferrofluids: mesoscopic versus hydrodynamic theory. *Phys Rev E*. 60(6):7001–9.

Mulvaney SP, Cole CL, Kniller MD, Malita M, Tamanaha CR, Rife JC, et al. (2007). Rapid, femtomolar bioassays in complex matrices combining microfluidics and magnetoelectronics. *Biosens Bioelectron*. 23:191–200.

Nandy K, Chaudhuri S, Ganguly R, Puri IK. (2008). Analytical model for the magnetophoretic capture of magnetic microspheres in microfluidic devices. *J Magn Magn Mater*. 320:1398–405.

Nguyen N-T, Beyzavi A, Ng KM, Huang X. (2007). Kinematics and deformation of ferrofluid droplets under magnetic actuation. *Microfluid Nanofluidics*. 3:571–9.

Odenbach S. (2004). Recent progress in magnetic fluid research. *J Phys Condens Matter*. 16:R1135–50.

Oh D-W, Jin JS, Choi JH, Kim HY. (2007). A microfluidic chaotic mixer using ferrofluid. *J Micromech Microeng*. 17:2077–83.

Ohashi T, Kuyama H, Hanafusa N, Togawa Y. (2007). A simple device using magnetic transportation for droplet-based PCR. *Biomed Microdevices*. 9:695–702.

Ohashi T, Kuyama H, Suzuki K, Nakamura S. (2008). Control of aqueous droplets using magnetic and electrostatic forces. *Anal Chim Acta*. 612:218–25.

Pamme N, Wilhem C. (2006). Continuous sorting of magnetic cells via on-chip free-flow magnetophoresis. *Lab Chip*. 6:974–80.

Pamme N, Eijkel JCT, Manz A. (2006). On-chip free-flow magnetophoresis: separation and detection of mixtures of magnetic particles in continuous flow. *J Magn Magn Mater*. 307:237–44.

Pekas N, Gragner M, Tondra M, Popple A, Porter MD. (2005). Magnetic particle diverter in an integrated microfluidic format. *J Magn Magn Mater*. 293:584–8.

Pregibon DC, Toner M, Doyle PS. (2006). Magnetically and biologically active bead-patterned hydrogels. *Langmuir*. 22:5122–8.

Probestein RF. (1994). *Physicochemical Hydrodynamics*. New York: John Wiley and Sons, Inc.

Radbruch A, Mechtold B, Thiel A, Miltenyi S, Pfluger E. (1994). High-gradient magnetic cell sorting. *Methods Cell Biol·* 42:387–403.

Ramadan Q, Poenar DP, Yu C. (2009). Customized trapping of magnetic particles. *Microfluid Nanofluidics*. 6:53–62.

Ramadan Q, Samper V, Poenar D, Yu C. (2006). Magnetic-based microfluidic platform for biomolecular separation. *Biomed Microdevices*. 8:151–8.

Ren H, Fair RB, Pollack MG. (2004). Automated on-chip droplet dispensing with volume control by electro-wetting actuation and capacitance metering. *Sens Actuators B*. 98:319–27.

Rida A, Gijs MAM. (2004). Manipulation of self-assembled structures of magnetic beads for microfluidic mixing and assaying. *Anal Chem*. 76:6239–46.

Rife JC, Miller MM, Sheehan PE, Tamanaha CR, Tondra M, Whitman LJ. (2003). Design and performance of GMR sensors for the detection of magnetic microbeads in biosensors. *Sens Actuators A*. 107:209–18.

Roh JW, Son OT, Lee YT, Lee KI, Jung HI, Lee W. (2009). Highly sensitive spin-valve devices for chip-cytometers. *Physica Status Solidi A Appl Res*. 206:1636–40.

Rong R, Choi JW, Ahn CH. (2006). An on-chip magnetic bead separator for biocell sorting. *J Micromech Microeng*. 16:2783–90.

Rosensweig RE. (1985). *Ferrohydrodynamics*. Cambridge: Cambridge University Press.

Roy T, Chakraborty S, Sinha A, Ganguly R, Puri IK. (2009). Magnetic microsphere-based mixers for micro-droplets. *Phys Fluids*. 21:027101.

Schotter J, Shoshia A, Brueckl H. (2009). Development of a magnetic lab-on-a-chip for point-of-care sepsis diagnosis. *J Magn Magn Mater*. 10:1671–5.

Sen S. (2004). Nanoprinting with nanoparticles: concept of a novel inkjet printer with possible applications in invisible tagging of objects. *J Dispersion Sci Technol*. 25:523–8.

Shikida M, Inouchi K, Honda H, Sato K. (2004). Magnetic handling of droplet in micro chemical analysis system utilizing surface tension and wettability. Techical digest. Maastricht (Netherlands): IEEE MEMS; pp. 359–62.

Shikida M, Takayanagi K, Inouchi K, Honda H, Sato K. (2006). Using wettability and interfacial tension to handle droplets of magnetic beads in a micro-chemical-analysis system. *Sens Actuators B*. 113:563–9.

Shliomis MI, Stepanov V. (1993). Rotational viscosity of magnetic fluids—contribution of the Brownian and Néel relaxational processes. *J Magn Magn Mater*. 122:196–9.

Sinha A, Ganguly R, De AK, Puri IK. (2007). Single magnetic particle dynamics in a microchannel. *Phys Fluids*. 19:117102.

Sinha A, Ganguly R, Puri IK. (2004). Magnetically assembled 3-D mesoscopic patterns using a suspension of superparamagnetic nanoparticles. Proceedings of the 3rd ASME Integrated Nanosystems—Design, Synthesis and Applications, Pasadena, CA. Nano2004–46091.

Sinha A, Ganguly R, Puri IK. (2009). Magnetic separation from superparamagnetic particle suspensions. *J Magn Magn Mater*. 321:2251–6.

Sista RS, Eckhardt AE, Srinivasan V, Pollack MG, Palanki S, Pamula VK. (2008a). Heterogeneous immunoassays using magnetic beads on a digital microfluidic platform. *Lab Chip*. 8:2188–96.

Sista R, Hua Z, Thwar P, Sudarsan A, Srinivasan V, Eckhardt A, et al. (2008b). Development of a digital microfluidic platform for point of care testing. *Lab Chip*. 8:2091–104.

Smistrup K, Hansen O, Bruus H, Hansen MF. (2005a). Magnetic separation in microfluidic systems using microfabricated electromagnets-experiments and simulations. *J Magn Magn Mater*. 293:597–604.

Smistrup K, Kjeldsen BG, Reimers JL, Dufva M, Petersen J, Hansen MF. (2005b). On-chip magnetic bead microarray using hydrodynamic focusing in a passive magnetic separator. *Lab Chip.* 5:1315–9.

Spaldin NA. (2006). *Magnetic Materials: Fundamentals and Device Applications.* Cambridge: Cambridge University Press.

Srinivasan B, Li Y, Jing Y, Xu Y, Yao X, Xing C, et al. (2009). A detection system based on giant magnetoresistive sensors and high-moment magnetic nanoparticles demonstrates zeptomole sensitivity: potential for personalized medicine. *Angew Chem Int Ed Engl.* 48:2764–7.

Sterr V, Krauß R, Morozov KI, Rehberg I, Engel A, Richter R. (2008). Rolling ferrofluid drop on the surface of a liquid. *New J Phys.* 10:063029.

Sun Y, Nguyen NT, Kwok YC. (2008). High-throughput polymerase chain reaction in parallel circular loops using magnetic actuation. *Anal Chem.* 80:6127–30.

Suzuki H, Ho CM, Kasagi N. (2004). A chaotic mixer for magnetic bead-based micro cell sorter. *J Microelectromech Syst.* 13:779–90.

Tibbe AG, de Grooth BG, Greve J, Liberti PA, Dolan GJ, Terstappen LWMM. (1999). Optical tracking and detection of immunomagnetically selected and aligned cells. *Nat Biotechnol.* 17:1210–3.

Tsai H, Fang YS, Fuh CB. (2006). Analytical and preparative applications of magnetic split-flow thin fractionation on several ion-labeled red blood cells. *Biomagn Res Technol.* 4:6.

Tsai T-H, Liou DS, Kuo LS, Chen PH. (2009). Rapid mixing between ferro-nanofluid and water in a semi-active Y-type micromixer. *Sens Actuators A.* 153:267–73.

Ugelstad J, Stenstad P, Kilaas L, Prestvik WS, Herje R, Berge A, et al. (1993). Monodisperse magnetic polymer particles—new biochemical and biomedical. *Blood Purif.* 11:349–69.

van Ewijk GA, Vroege GJ, Phillipse AP. (2002). Susceptibility measurements on a fractionated aggregate-free ferrofluid. *J Phys Condens Matter.* 14:4915–25.

Vuppu AK, Garcia AK, Hayes MA. (2003). Video microscopy of dynamically aggregated paramagnetic particle chains in an applied rotating magnetic field. *Langmuir.* 19:8646–53.

Wang Y, Zhao Y, Cho SK. (2007). Efficient in-droplet separation of magnetic particles for digital microfluidics. *J Micromech Microeng.* 17:2148–56.

Wang Y, Zhe J, Chung BTF, Dutta P. (2008). A rapid magnetic particle driven microstirrer. *Microfluid Nanofluidics.* 4:375–89.

Weitschies W, Kotitz R, Bunte T, Trahms L. (1997). Determination of relaxing or remanent nanoparticle magnetization provides a novel binding-specific technique for the evaluation of immunoassays. *Pharm Pharmacol Lett.* 7:5–8.

Wirix-Speetjens R, Fyen W, de Boeck J, Borghs G. (2006). Single magnetic particle detection: experimental verification of simulated behavior. *J Appl Phys.* 99:103903.

Wright DJ, Chapman PA, Siddons CA. (1994). Immunomagnetic separation as a sensitive method for isolating *Escherichia coli* O157 from food samples. *Epidemiol. Infect.* 113:31–9.

Yamahata C, Chastellain M, Parashar VK, Petri A, Hofmann H, Gijs MAM. (2005). Plastic micropump with ferrofluidic actuation. *J Microelectromech Syst.* 14:96–102.

Yellen B, Friedman G. (2004). Programmable assembly of colloidal particles using magnetic microwell templates. *Langmuir.* 20:2553–9.

Yellen B, Friedman G, Feinerman A. (2003). Printing superparamagnetic colloidal particle arrays on patterned magnetic film. *J Appl Phys.* 93:7331–3.

Yellen BB, Fridman G, Friedman G. (2004). Ferrofluid lithography. *Nanotechnology.* 15:S562–5.

Yellen BB, Friedman G, Barbee KA. (2004). Programmable self-aligning ferrofluid masks for lithographic applications. *IEEE Trans Magn.* 40:2994–6.

Yellen BB, Erb RM, Son HS, Hewlin R Jr, Shangb H, Lee GU. (2007). Traveling wave magnetophoresis for high resolution chip based separations. *Lab Chip.* 7:1681–8.

Yu H. (1998). Use of an immunomagnetic separation–fluorescent immunoassay (IMS–FIA) for rapid and high throughput analysis of environmental water samples. *Anal Chim Acta.* 376:77–81.

Yu H, Raymonda JW, McMahon TM, Campagnari AA. (2000). Detection of biological threat agents by immunomagnetic microsphere-based solid phase fluorogenic- and electro-chemiluminescence. *Biosens Bioelectron.* 14:829–40.

Zahn M. (1979). *Introduction to Electromagnetic Field Theory: A Problem Solving Approach.* New York: John Wiley and Sons, Inc. Chapter 5, p. 369, Eq. (9). for zero free current (i.e., $J_f = 0$) implies that $f_i = \mu_0 M_j \cdot (\partial H_i / \partial x_j) + \frac{1}{2} \mu_0 \cdot \partial / \partial x_i (M_j M_j)$. Using Maxwell's equation for $J_f = 0$, i.e., $\partial H_i / \partial x_j = \partial H_j / \partial x_i$, one gets, $f_i = M_i (\mu_0 \partial H_j / \partial x_i + \mu_0 \partial M_j / \partial x_i) = M_j \partial B_j / \partial x_i$.

Zaytseva NV, Goral VN, Montagnab RA, Baeumner AJ. (2005). Development of a microfluidic biosensor module for pathogen detection. *Lab Chip.* 5:805–11.

Zborowski M, Sun L, Moore LR, Williams PS, Chalmers JJ. (1999). Continuous cell separation using novel magnetic quadrupole flow sorter. *J Magn Magn Mater.* 194:224–30.

Zheng B, Tice JD, Ismagilov RF. (2004). Formation of droplets of alternating composition in microfluidic channels and applications to indexing of concentrations in droplet-based assays. *Anal Chem.* 76:4977–82.

Zolgharni M, Azimi SM, Bahmanyar MR, Balachandran W. (2007). A numerical design study of chaotic mixing of magnetic particles in a microfluidic bio-separator. *Microfluid Nanofluidics.* 3:677–87.

16

The Influence of Microfluidic Channel Wettability on PEM Carbon Paper Fuel Cell

S. AlShakhshir, X. Li, and P. Chen

CONTENTS

16.1 Introduction

Microfluidics deals with fluid flow in submillimeter-sized systems under external influences. Over the past fifteen years, enormous advances in microfluidics have led to the development of numerous microfluidic devices, such as valves, pumps, and multiplexers. The rapid development in microfluidic technology led to significant advancements in many industrial applications. In particular, microfluidics plays an essential role in fuel cell performance as related to the transport of fluids, water management, and temperature distribution.

The fuel cell is an electrochemical device that converts the chemical energy of reactants directly into electrical energy. The polymer electrolyte membrane (PEM) fuel cell is one of the most common types of fuel cells (Larminie and Dicks, 2003). It has smaller volume and lighter weight compared with other fuel cell types. It operates within a relatively low temperature range between the freezing point and the boiling point of water. This contributes to its quick start-up and shut down. In addition, the electrolyte is a solid material, which makes the technology attractive for portable and automotive applications (Wallmark et al., 2002). All of these characteristics justify the reality that approximately 90% of fuel cell research and development work involves the PEM fuel cell (Cropper et al., 2004). Given the practicality of PEM fuel cells, most of the research and development activities are crossing the academic boundaries to the automotive and electronics industries, such as cars and buses, as well as for a very wide range of portable applications and also for combined heat and power systems (Larminie and Dicks, 2003).

Currently, the major technical problem with the development of PEM fuel cells is water management (Nguyen and White, 2003). Water blockage in the gas channels results in lowering of the cell performance. Removing this blockage requires higher gas stream velocity to force the liquid water out of the cell. This requires significant power consumption for more air compression. In a serpentine flow channel design, which became the industry standard in PEM fuel cells, the power needed to purge the flow blockage reaches up to 35% of the fuel cell stack output. Some have used pumps to increase the power, but the root of the problem lies in the surface dynamics of the cathode region. Therefore, facilitation of liquid water removal from the flow channel surfaces can have a significant impact on enhancing the PEM fuel cell performance and achieving cost reduction (Li et al., 2007).

16.2 PEM Fuel Cell

A schematic of a PEM fuel cell is shown in Figure 16.1, illustrating its operational principles. A PEM fuel cell is composed of a number of major components, namely, a bipolar plate, a gas diffusion layer (GDL), a catalyst layer (CL), and a membrane; each component has its own specific role in completing the PEM fuel cell operating cycle.

Pure fully humidified hydrogen enters the anode channel and diffuses through the porous electrode toward the CL. At the anode CL, electro-oxidation of H_2 takes place with the production of protons (H^+) and electrons (e^-) due to the effect of platinum that exists in the CL. The reaction takes place according to hydrogen oxidation reaction (HOR) as follows:

$$H_2 \rightarrow 2H^+ + 2e^- \tag{16.1}$$

The unique property of the membrane electrolyte allows only protons to be transported to the cathode side and prohibits electron transport. Hence, shorting of the cell is avoided. Electrons are forced to travel through the external circuit, delivering electrical energy to the external load on their way to the cathode.

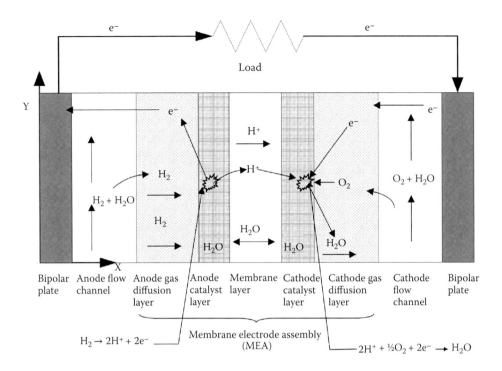

FIGURE 16.1
Schematic and operational principles of PEM fuel cell (unpublished work).

At the cathode side, the transferred protons and the energy-depleted electrons combine with oxygen in the cathode CL to produce water according to the following oxygen reduction reaction (ORR):

$$\tfrac{1}{2} O_2 + 2H^+ + 2e^- \rightarrow H_2O. \tag{16.2}$$

The HOR reaction is slightly endothermic, and the ORR reaction is strongly exothermic. Consequently, heat is generated. Combining the anode reaction and the cathode reaction together gives the overall reaction in the PEM fuel cell:

$$H_2 + \tfrac{1}{2}O_2 \rightarrow H_2O + \text{heat} + \text{electric energy}. \tag{16.3}$$

Perfluorosulfonic acid is the most commonly used membrane material for PEM fuel cell (Larmine and Dicks, 2003). When the membrane gets hydrated, the hydrogen ions (H^+) become mobile by bonding to the water molecules to form hydronium ions; these ions move between the sulfonic acid sites. The water content of the polymer electrolyte is essential for proton conduction; if the membrane becomes dehydrated, it will no longer be protonically conductive. Nafion 112, 115, and 117 from DuPont are commonly used membranes in PEM fuel cells. The CL is a thin layer (several microns to several tens of microns thick) on either side of the membrane. It usually consists of microscale carbon particles. Each particle can support nanoscale platinum (Pt) catalyst particles, which are loosely embedded in a matrix of an ionomer. HOR occurs in the anode CL, and ORR occurs in the cathode CL. The electrochemical reaction is not evenly distributed over the CL; therefore, the Pt particles must be properly distributed in the CL to maximize the reaction efficiency and minimize the cost (Mehta et al., 2003).

The GDL typically consists of randomly aligned carbon fibers (carbon paper) or woven spun yarns (carbon cloth). The GDL is a highly porous layer with typical thickness between 200 and 300 μm. It transports the reactants toward the reaction sites and provides structural support for the CL. Furthermore, it provides a path for electron transfer. The GDL plays significant role in water and heat removal from the reaction sites of the PEM fuel cell. In particular, considerable research is directed toward water removal from the cathode GDL (Moreira et al., 2003). The water is then discharged from the cell through the flow field channels. Therefore, the wettability of the channel surface is important for liquid water removal.

16.2.1 PEM Fuel Cell Performance

The performance of the fuel cell is shown in the form of current density, J, versus cell potential plots. This plot, known as the polarization curve, is shown in Figure 16.2. This curve was generated by measuring the performance of a 10×10-cm PEM fuel cell at the University of Waterloo PEM fuel cell testing laboratory. The variation of individual PEM fuel cell voltage versus current density is found from the maximum cell voltage and the various voltage losses. The sources of these losses, which are also called polarization, irreversibility, or overvoltage, originate from (a) activation polarization, (b) ohmic polarization, and (c) concentration (mass transport) polarization. The summation of these overpotentials is known as the cell overpotential, η_{cell}.

The maximum cell voltage, or reversible voltage, V_{rev}, represents an ideal cell performance. It is independent of the quantity of current drawn from the cell. However, irreversible voltage losses occur in real fuel cells as illustrated in Figure 16.2.

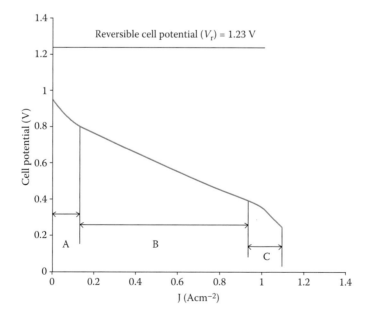

FIGURE 16.2
The 10×10-cm PEM fuel cell polarization curve showing three regions: (A) activation polarization, (B) ohmic polarization, and (C) concentration polarization (unpublished work).

Figure 16.2 demonstrates that the PEM fuel cell polarization curve can be divided into three regions: A, B, and C. In region A, the reaction rate loss occurs. This region is called the activation polarization region, which dominates at low current densities. It is present when the rate of the electrochemical reaction at an electrode surface is controlled by sluggish electrode kinetics. Activation losses increase as current drawn from the cell is increased (Barbir, 2005). In region B, ohmic polarization dominates because of the resistances of (1) the PEM to the ion transfer and (2) of the rest of the cell assembly to the electron transfer. Hence, the cell voltage drops linearly as current drawn is increased (Springer et al., 1993). Region C is known as the concentration polarization region because the reactant concentration at the reaction sites decreases when current drawn from the cell is increased because of the limited rate of mass transfer. This loss becomes significant at high current densities (Springer et al., 1996).

It comes as no surprise that the performance of the cathode is one of the key factors affecting the performance of the fuel cell. Hence, mass transport limitation is imposed by liquid water, especially at high current densities. It is often difficult to remove the product water from the cathode side of the fuel cell, which leads to the compromised transfer of oxygen to the reaction sites through the GDL (Natarajan and Nguyen, 2006). The liquid water formed on the cathode CL transports through the GDL to reach the flow channels. It is then removed from the gas channels by the cathode gas stream. Because of the high surface tension effect, liquid water clogs the flow channels and fills the pores of the GDL. The imbalance between water production rate at the reaction sites and water removal rate from the flow channels leads to water flooding in the flow channels and GDL. Thus, water management is critical for PEM fuel cell.

Changing the flow channel surface wettability properties reduces the required stream velocity to drive the flow out of the channel. Thus, the cell performance increases and the cost of the fuel cell system decreases. Further, polytetrafluoroethylene (PTFE) was used extensively for increasing the GDL hydrophobicity (Wang et al., 2008). Because PTFE is a nonconductive polymer, it reduces the GDL conductivity and porosity, limiting the amount of gases transported through the GDL to the reaction sites. Thus, another coating material is required to modify the GDL surface wettability.

16.3 Advantages and Limitations

In two-phase flow in mini-sized channels, the capillary force is mostly negligible compared with the inertia and viscous forces. However, as the cross-sectional area of the flow channel gets smaller, which is 1×1 mm in PEM fuel cell, the capillary effect starts to play an important role in determining the behavior of two-phase flow patterns. In this case, the interfacial tensions between solid–liquid (γ_{SL}) and solid–gas (γ_{SV}) should be taken into account in addition to the surface tension between liquid and gas (γ_{LV}). In other words, surface properties of the channel walls and the GDL surface as well as combinations of the gas and the liquid are additional important factors to be considered in determining the flow behavior in the gas flow channel.

16.3.1 Wettability of Solid Surfaces

Surface properties have a significant effect on PEM fuel cell performance. Flooding of GDL and flow field channels as well as membrane dehydration are the major problems causing performance losses. These problems can be handled by adjusting the surface wettability of the GDL and the flow field channels. An introduction to wettability is presented here followed by its application in PEM fuel cells in the succeeding sections.

The wettability and the water repellency of the solid surface are important material properties. They strongly depend on both surface composition and surface roughness (Miller et al., 1950). Although surface wettability indicates the hydrophilic characteristic of the surface, water repellency indicates hydrophobic characteristics. Hydrophilicity is important in many diagnosis and medical device applications. Many industrial products require superhydrophilic, antifogging, and self-cleaning surfaces; others need to be hydrophilic. On the other hand, various industrial products have to be superhydrophobic or hydrophobic. Chemical reactions or bond formation through water on a superhydrophobic surface is limited. Accordingly, other phenomena such as adherence of snow, oxidation, and current conduction are expected to be inhibited on such a surface (Bikerman, 1996).

The wettability of the solid surface may be evaluated by the contact angle given by Young's equation:

$$\cos\theta = \frac{\gamma_{SV} - \gamma_{SL}}{\gamma_{LV}}, \tag{16.4}$$

where γ_{SL}, γ_{SV}, and γ_{LV} are interfacial free energies per unit area of the solid–liquid, solid–gas, and liquid–gas interfaces, respectively. The maximum contact angle can be obtained on a flat surface merely by lowering the surface energy (Girifalco and Good, 2008). The lowest recorded surface energy is 6.7 mJ $\cdot$ m^{-2}. It characterizes a surface with regularly aligned closest-hexagonal-packed trifluoromethyl (CF_3) groups (Nishino et al., 1999). The calculated contact angle for this surface is 120° (Nakajima et al., 1999). This angle is relatively small compared with the superhydrophobic contact angle 150° (Onda et al., 1996). Other techniques should be used with surface coating to increase the contact angle to a superhydrophobic contact angle. Surface roughening is one of these techniques used to increase the contact angle; therefore, the hydrophobicity of the surface will increase (Adamson and Gast, 1997).

Young's equation (Equation 16.4) is applicable only on a flat surface. Modifications are required to account for rough surfaces. Wenzel proposed a model describing the contact angle θ' at a rough surface. He modified Young's equation as follows (Wenzel, 1949):

$$\cos\theta' = \frac{r(\gamma_{SV} - \gamma_{SL})}{\gamma_{LV}} = r\cos\theta, \tag{16.5}$$

where r is the roughness factor, defined as the ratio of the actual area of a rough surface to the geometric projected area. Because the r value is greater than unity, the surface roughness enhances the hydrophobicity of the hydrophobic surfaces.

A water droplet on a solid surface can be made to move along a solid surface by using a gradient of some type (Adamson and Gast, 1997). On the thermal gradient, the drop will move from the warm side to the cool side. This move takes place because liquid–gas surface tension is affected by temperature. As temperature increases, surface tension decreases, and *vice versa*. On each area element at the liquid–gas interface, there are two forces pulling in opposite directions, trying to reduce the surface area of the drop. Because surface tension decreases with increasing temperature, the droplet is driven toward higher tension value (the coldest one); in other words, tension pulling in the cold direction is stronger than the one pulling in the warm direction. On a wettability gradient, a drop of a hydrophilic substance will move from the hydrophobic end to the hydrophilic end. This is attributed to the fact that the total energy of the system is at a minimum when the drop is at the hydrophilic end of the gradient (Subramanian et al., 2005).

16.3.2 PEM Fuel Cell Gas Channel Wettability

Extensive research has been done on bipolar-plate surface modification; most of this work focused on improving corrosion resistance and electrical conductivity (Akundy and Iroh, 2001; Borub and Vanderborgh, 1995; Davies et al., 2000; Hermann et al., 2005; Hung et al., 2006; Mehta and Cooper, 2003; Wang and Turner, 2004; Wang et al., 2003; Wind et al., 2002). However, some researchers gave more attention to the surface wettability of the bipolar plates.

Li et al. (2004) began the coating process with chemical etching of a 1.5-mm-thick 316 stainless steel bipolar plate to form flow channels. Then the surface was coated using the hollow cathode discharge ion plating method with TiN as the coating material. They measured the water contact angle of the TiN-coated 316 stainless steel. It was close to the value of the graphite contact angle (90°) (Schrader, 1975), whereas the water contact angle of the noncoated 316 stainless steel was 60°. This result indicates that the 316 stainless steel has higher surface energy and that it more readily floods the cathode side than graphite and the TiN-coated 316 stainless steel bipolar plate.

Lee et al. (2005) used the electrochemical theory for 316 stainless steel surface treatments. The work specimen is the anode, and it is immersed in the electrolyte. When it is connected with the cathode, metallic ions are released from the work specimen to form a passive film. This film increases the capability of corrosion resistance. The surface morphology becomes smoother and shiny. The surface roughness is gently improved and is exhibited as a hydrophobic property. This may improve the flow of gas and water in the gas channel of the bipolar plate.

Tanigucshi and Yasuda (2005) used plasma polymerization for surface coating of titanium and stainless steel plates. The substrate was treated using a combined process of plasma polymerization and sand-blast pretreatment. The water contact angle due to these combined processes was higher than that due to plasma polymerization only. Sand blasting offered significant improvement in water repellency of the coated surface. This is attributed to the increase in surface roughness of the sand-blasted bipolar plate metals. This result was in accordance with that of Nakajima et al. (2001). Furthermore, the coated channels of a PEM fuel cell with sand blasting followed by plasma polymerization showed an improvement in the peak power. This improvement led to effective flow in the coated channel at a low oxygen flow rate.

Low oxygen flow rate is important for improvement of the efficiency of the fuel cell system. Hence, high flow rate results in low oxidant utilization and larger power consumption for driving an air compressor or a blower to supply air as an oxidant to the fuel cell. Moreover, blocking the gas flow channel with condensed liquid water results in serious degradation as electrode area, reactant utilization, and humidifying temperature increase (Borroni-Bird, 1996).

16.3.3 PEM Fuel Cell GDL Wettability

Wettability of the GDL is one of the properties that have a dominant role in controlling the transported water through the GDL. This property is controlled by adding a hydrophobic agent such as PTFE to increase its hydrophobicity and to enhance water removal.

The hydrophobic pore surface distorts the molecular force balance at the line of contact, which results in forcing the liquid water to move toward an unstable state. This leads to a higher capillary pressure within the pore. Unlike the hydrophilic treatment, water is preferentially adsorbed by the fiber surface of the hydrophilic pores (Bevers et al., 1996).

Bevers et al. (1996) coated a 9×9-cm carbon paper sample by PTFE. To coat the paper with PTFE, the sample was slowly lowered into a PTFE suspension, never faster so that the suspension could adsorb the paper. The paper was left standing in the suspension for 5 min and then removed. To guarantee a uniform distribution of PTFE, the paper was laid out flat on a square arrangement of 13 needles (pointed ends up) to dry and then sintered in a sintering oven at a temperature less than 200°C. Bevers et al. concluded that PTFE contents correlate negatively with the conductivity and the diffusion rate, although the sintering temperature correlates positively with the diffusion rate and negatively with the conductivity. This result was in agreement with the results by Paganin et al. (1996). They prepared many GDL samples with different PTFE loadings. The 15 wt.% sample showed the best cell performance.

Giorgi et al. (1998) coated the GDL with PTFE. A homogeneous suspension was prepared by mixing and stirring carbon with an appropriate amount of PTFE dispersion in an ultrasonic bath at room temperature for 25 min. They obtained the best performance at the lowest PTFE loading. On the other hand, it was not possible to reduce the PTFE content to zero to avoid the electrode flooding for lack of hydrophobicity. Further, they pointed out that minimum amount of PTFE in the GDL is necessary to bond the carbon particles together.

Lim and Wang (2004) treated a GDL with fluorinated ethylene propylene (FEP) ranging from 10 to 40 wt.%. The GDL carbon paper was slowly dipped into an FEP suspension diluted to 20 wt.% with deionized water. To obtain a uniform FEP distribution inside the GDL, the sample was heat-treated and sintered. The contact angle measurements indicated a similar level of hydrophobicity among GDLs impregnated with different amounts of FEP ranging from 10 to 40 wt.%. However, the contact angle was found to be a strong function of temperature, with the value close to 80°C water temperature. Furthermore, the 10 wt.% FEP GDL cell gave the highest cell power density. This referred to the fact that an excessive FEP impregnation results in significant blockage of surface pores by thin FEP films and hence a highly restricted surface for reactant transport and product removal.

Wang et al. (2008) studied the effect of PTFE content on the contact angle. In addition to that, they studied the influence of carbonization treatment on the contact angle for the same PTFE contact. In their work, the cell with 10 wt.% PTFE carbonized GDL showed the best performance, and this attributed to the highest contact angle 137° ± 1° for this sample. The contact angle value for hydrophobically treated GDLs does not depend only on the PTFE content but also on the uniformity of the PTFE distribution. For the noncarbonized carbon paper with low PTFE loading, PTFE was mainly accumulated on the cross positions of carbon fibers. This results in a low contact angle value. However, the carbonization process results in coating both the stem and the cross positions of carbon fibers. This leads to higher contact angle values. In their work, the sample was carbonized by dipping the GDL sample into 20 wt.% aqueous sucrose solution for 6 h and then sintered in a tube furnace at 400°C under argon ambience to prevent oxidation. To carbonize all the sucrose, the process was repeated several times. After that the carbonized and noncarbonized samples were dipped into PTFE emulsion with different concentrations to obtain different PTFE loadings.

Pai et al. (2006) used CF_4 plasma treatment to improve the hydrophobicity of the active carbon fiber (ACF) mats. After CF_4 plasma treatment, the ACF mats were dip-coated in 10 wt.% PTFE solutions. Their results showed that the CF_4-treated samples had the best performance compared with the untreated ones. This attributed to the fact that the surface gas diffusion pores of the CF_4 plasma-treated ACFs were apparently less sealed or blocked by excessive hydrophobic material residuals. In addition to that, the water contact angles of the CF_4 plasma-treated ACFs and of the nontreated ones were measured to be $132.8° \pm 0.2°$ and $128.4° \pm 0.2°$, respectively.

16.4 Visualization of Two-Phase Flow in PEM Fuel Cell Channels

The results of electro-osmotic drag of water from the humidified H_2 gas stream at the anode side through the membrane and the electrochemical water formation at the cathode GDL are the net accumulation of excess water in the cathode side of the MEA and the dehydration of the anode side. In addition, the back diffusion from the cathode to the anode due to the water concentration gradient is inadequate to keep the anode side hydrated at high current densities (Nguyen and White, 1993). Furthermore, the water droplets are generated at the channel inlet and on the surface of the channel because of the condensation of the cathode fully humidified air stream. This attributed to the heat loss at the connection pipeline and behind the flow channel surface in the PEM fuel cell stack. Also, the water content increases at the MEA cathode side to high levels at high current density. Thus, the cathode GDL floods and liquid water accumulation can occur in the cathode channels.

Advanced and expensive experimental techniques including neutron radiography, nuclear magnetic resonance, and gas chromatographic measurements have been used to investigate water transportation and distribution inside an operating PEM fuel cell. These technologies have the ability to test a real closed cell without any modifications in the cell design. However, it is hard to investigate the real-time liquid water distribution and removal in a PEM fuel cell. An optical diagnostics technique was applied on the fuel cell for visualizing water droplet buildup in, and removal from, the channels and investigating the liquid water dynamics in a flooded cell with much higher special and temporal resolution as shown in Figure 16.3. However, optically accessible PEM fuel cell technology is limited by inherent nature, including the change of the channel surface conditions, the fogging of the windows due to higher temperature and an almost fully saturated gas stream in the flow channels, and different electrical and thermal conductivities due to the window material (Yang et al., 2004).

Tüber et al. (2004) conducted an experiment with a PEM fuel cell having simple bipolar plates with two gas channels. They observed that if the gas flow rates were not sufficient to keep droplets out of the channel either by evaporation or forced convection, a blockage occurred, causing 25% drop in the current density.

Yang et al. (2004) built an optical PEM fuel cell using two clear polycarbonate plates placed outside the current collector plates to constrain the gas flow and two stainless steel end plates that compressed the entire optical cell. They showed a sequence of photographs, looking through the top of the transparent PEM fuel cell cathode gas channel onto the GDL surface. Between 0 and 180 s, two discrete water droplets formed in the channel grew continuously on the GDL. By 480 s, the droplets grew to the point where their surfaces contacted, causing them to merge and then coalesce with more hydrophilic channel walls. Between 480 and 540 s, the drop on the side wall is expelled to an annular flow regime, and new droplets begin to emerge from a close location to the first two.

They observed that water droplets forming in the gas channels may bridge the walls of the channels under fixed operating conditions. This leads to a partial or complete gas flow

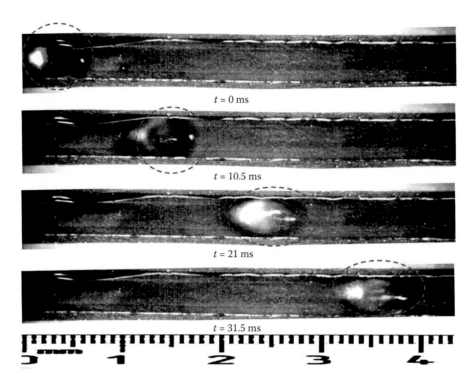

FIGURE 16.3
Real time of water droplet removal from 4 × 4-mm hydrophobically coated graphite channel, 6× magnification (JL = 0.0008 m/s, JV= 3.63 m/s) (unpublished).

blockage. They photographed a complete gas flow channel blockage in their study. This blockage can hinder the reactant supply to the membrane; therefore, the performance will be degraded significantly (Liu et al., 2006).

Kim et al. (2005) designed a transparent PEM fuel cell with 25 cm² active area to allow for the visualization of the cathode channel from the top to study fuel cell performance characteristics. Two-phase flow due to the electrochemical reaction of the fuel cell was experimentally investigated. The images photographed by a charge-coupled device camera at various cell temperatures (30–50°C) and different inlet humidification levels were presented in this study. Results indicated that the flooding on the cathode side first occurs very close to the exit of the cathode flow channel. As a result, the operating temperature of the fuel cell increased, and it was found that water droplets evaporated easily because of increased saturation vapor pressure and that it might have an influence on lowering the flooding level.

Liu et al. (2006) used three transparent PEM fuel cells to investigate the liquid water and water flooding inside the PEM fuel cell. Plexiglas was used as a transparent material at the cathode side. The three transparent cells have different flow field channel designs: parallel, interdigitated, and cascade flow fields. The effects of flow field, cell temperature, cathode gas flow rate, and operation time on water buildup and cell performance were studied, respectively.

Their results indicated that the liquid water columns accumulated in the cathode flow channels could reduce the effective electrochemical reaction area; this led to mass transfer limitation, resulting in low performance of the cell. The water in flow channels at high temperature was much less than that at low temperature. When water flooding occurred, the increasing cathode flow rate removed excess water and led to good cell performance. The water and gas transfer were enhanced, and water removal was easier in the interdigitated channels and cascade channels than in the parallel channels. The cell performances of the fuel cells that were designed with an interdigitated flow field or a cascade flow field were better than that designed with a parallel flow field. The images of liquid water in the cathode channels were recorded at different operating times. The images of liquid water removed from the channels were also recorded by a high-speed video camera.

Ma et al. (2006) designed a transparent PEM fuel cell with a single straight channel to study the liquid water transport in the cathode channel. Through this study, they monitored water buildup and removal in the channel online directly. The real time taken for water buildup was determined. Furthermore, the water removal velocity was determined. The pressure drop between the inlet and the outlet of the channel (ΔP) was monitored during the fuel cell operation, and ΔP was recorded. ΔP increased with the increase of water content in the channel, and a ΔP sharp decline corresponds to the release of water blockage.

Kumbur et al. (2006) achived simultaneous visualization of both sides and top views of a water droplet inside a 5×4-mm channel to determine the droplet behavior. They developed an empirical correlation of the surface tension of a droplet on surface diffusion media as a function of PTFE content on the basis of experimental data. Furthermore, they observed that the removal of relatively taller droplets is easier than that of relatively spread-out droplets and water films because of the squared dependence of the drag force acting on the droplet on height and the linear dependence of the surface adhesion force on droplet chord length.

Theodorakakos et al. (2006) investigated the detachment of water droplets from a carbon porous material surface under the influence of an air stream, which flows around them inside a 2.7×7-mm channel. They indicated that the shape of the droplet changes dynamically from its static position, until the droplet finally loses contact with the wall surface and is swept away by air.

Bazylak et al. (2008) used an experimental apparatus that consists of the gas flow channel apparatus on the fluorescence microscope stage. The GDL is placed between a Plexiglas base and a polydimethylsiloxane channel structure of dimensions 3.7×4.5 mm. The relatively large channel dimensions were chosen such that droplet emergence and transport could be studied in the absence of sidewall effects. A silicone rubber gasket was placed between the GDL and the Plexiglas base to prevent leakage. Air was delivered to the gas channel and controlled with a rotameter. Liquid water was injected through the bottom surface of the GDL using a syringe pump connected to the Plexiglas base with Teflon tubing. Liquid water was introduced from one side of the GDL from a single localized source, and the images were captured using an upright fluorescence microscope through-plane evolution of liquid water transport. To facilitate fluorescence imaging, a fluorescence dye was used to tag the liquid phase.

They observed that individual droplets emerge, grow, and detach from the GDL. However, it was commonly observed that over time these droplets leave residual liquid water particles on the GDL, which provide pinning sites for other droplets. Droplets become pinned to the GDL because of their high surface roughness and high contact angle hysteresis. Furthermore, a droplet may detach more easily and roll away due to surface hydrophobicity. Moreover, droplets sitting on this highly rough surface experience fewer tendencies for detachment because of longer contact lines between the droplet and fibers and the presence of contact angle hysteresis. In addition, the emergence and the detachment of individual droplets were followed by slug formation and channel flooding.

16.5 Liquid Water Removal from the Cathode Side

Water management is crucial to the effective operation of a PEM fuel cell. Water is a product of a fuel cell reaction and is carried out of the cell during the operation of the cell. Both the fuel and air entering the fuel cell must be humidified; this additional water keeps the PEM hydrated. The humidity of the gases must be carefully controlled, as too little water prevents the membrane from conducting the H⁺ ions well and thereby the cell current drops.

If the airflow passing the cathode is too slow, the air will not be able to carry all the produced water at the cathode out of the cell, thus causing the cathode to flood. In that case, cell performance is not maximized as not enough oxygen is able to penetrate the excess liquid water to reach the cathode catalyst sites.

Interfacial and transport phenomena play a vital role in water management in a fuel cell. The following physical dynamics and surface properties involved in the management and distribution of water within a PEM fuel cell are critical in understanding fuel cell design.

- Structure and transport in porous GDL
- Water formation in the GDL by capillary condensation
- Water removal from the GDL by capillary interaction and forces acting on the water droplet
- Channel design
- The role of wetting and contact angle in the optimization of the GDL surface to enable easier removal of water and to reduce operation cost

16.6 Structure and Transport in Porous GDL

The GDL typically consists of randomly aligned carbon fibers or woven spun yarns, which have high porosity. In addition to the carbon GDL, metallic GDL or metal mesh diffusers have also been proposed as GDL (Fushinobu et al., 2006). For the purpose of liquid water removal, a GDL can be characterized by two different parameters: microstructure and pore surface wettability. Carbon-fiber-based porous materials are categorized into two types: nonwoven carbon paper and woven carbon cloth.

Although the carbon cloth-type GDL may present higher power performance than that of paper-type GDL, there are considerable academic and industrial interests in using paper-type GDL. This interest is driven by cost advantage with the use of nonwoven substrates. Furthermore, it is more convenient to fabricate a microporous layer (MPL) or CL directly onto them (Antolini et al., 2002). A scanning electron microscope micrograph of a carbon paper is shown in Figure 16.4.

In spite of the crucial role of the GDL in a PEM fuel cell, less attention has been paid to the GDL. The GDL covers a wide range of functions within an MEA area (Antolini et al., 2002). These functions, listed below, represent the backbone of the PEM fuel cell operation principles necessary to generate electricity:

- Providing mechanical support to the CL and membrane
- Providing a passage for reactant access from the flow field channel to the CL

Carbon fiber stems

Cross position of carbon fibers

10 µm

FIGURE 16.4
Raw GDL consisting of randomly aligned carbon fibers and carbon stem 2000× magnification (unpublished work).

- Providing a passage for product water removal from the CL to the flow field channel
- Providing a conductive media for electrons between the CL and the flow field rib
- Transferring the heat generated in the electrochemical reaction in the CL to the flow field ribs

Performing these functions in a good manner requires some modifications to the morphological, mechanical, and wetting properties of the GDL.

The GDL can be coated with an additional layer. The commonly used layer is an MPL, which consists of carbon or graphite particles mixed with a polymeric binder such as PTFE. The GDL surface is irregular without an MPL, and the carbon particles are very close to each other, which are attributed to the pore size. Generally speaking, the pore size of the MPL (100–500 nm) is smaller than that of the GDL (10–100 µm). Therefore, the electric contact resistance with the adjacent CL can be reduced. Furthermore, a small pore size has higher capillarity, causing the water to flow more easily. In the mean time, small pores reduce the gas permeability. On the other end, large pore size provides better gas transport and lower mass transport resistance.

In terms of hydrophobicity, the MPL is exclusively hydrophobic in nature. Given the hydrophobic nature of MPLs, it provides effective wicking of liquid water from the CL, resulting in much smaller water droplets. These droplets are less likely to clog and flood inside the CL. In addition, the MPL keeps the membrane hydrated by pushing water away from the GDL (Lin and Nguyen, 2005). Adding the MPL on the cathode GDL may be more important than adding it on the anode GDL because more severe flooding takes place at the cathode. At high current densities, the electrochemical reaction rate is faster than the amount of reactants supplied, specifically the oxidizer. Therefore, the reaction rate is limited by the transport rate of the oxidizer to the reaction sites, which is sandwiched between the GDL and the membrane. Water blockage in some pores of the GDL further limits the reactant transport. As such, the GDL is commonly coated with PTFE to provide a highly hydrophobic surface for easy removal of liquid water, as shown in Figure 16.5.

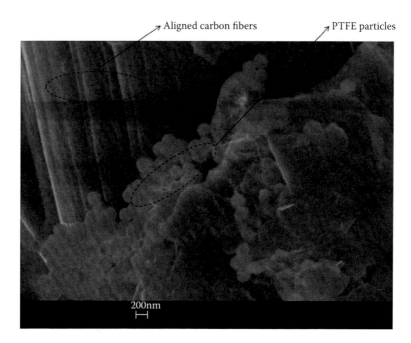

FIGURE 16.5
PTFE-coated GDL showing the PTFE particles on the carbon fibers at 32× magnification.

16.7 Flow Regimes in PEM Fuel Cell Diffusion Medium

Lenormand (1988, 1990) classified the drainage that takes place in porous media into three regimes depending on the viscosity ratio (M) and the capillary number (Ca). Drainage is the displacement of a wetting fluid by the injection of a nonwetting fluid. When drainage takes place in a porous media, three types of fluid flow regimes occur depending on M and Ca values:

$$M = \frac{\mu_{nw}}{\mu_w},\tag{16.6}$$

$$Ca = \frac{v\mu_{nw}}{\sigma},\tag{16.7}$$

where μ_w and μ_{nw} are the wetting and nonwetting fluid viscosities, v is the fluid velocity, and σ is the surface tension.

These regimes are viscous fingering, capillary fingering, and stable displacement flow. If the viscosity of the injected fluid is lower than that of the displaced fluid, the injected fluid permeates irregularly through the porous medium. This results in the formation of multiple conduits or fingers of approximately the same size. This regime is known as viscous fingering flow. On the other hand, when the injected fluid has higher viscosity than the displaced fluid, two flow regimes can be predicted depending on the injection flow rate. If the injection flow rate is relatively slow, the injected fluid again generates irregular conduits within the porous media. The resulting flow distribution is characterized by the formation of a few fingers of different sizes and is referred to as capillary fingering flow. However, if the injection flow rate is relatively high, the injected fluid permeates evenly through the porous media without finger formation. The resulting flow distribution is known as stable displacement flow.

Litster et al. (2006) investigated the water distribution inside the diffusion medium (DM) of the PEM fuel cell using fluorescence microscopy. In their experiment, distilled water with a fluorescence dye was injected into a DM from the bottom surface, obtaining a "through-plane" percolation. The fingering type of pattern distribution was predicted in the DM. Evidence of nonuniform water distribution was found using magnetic resonance imaging by Tsushima et al. (2005), neutron imaging by Satija et al. (2004), and numerically by Sinha et al. (2007). They made a simple calculation for the water generated in a fuel cell under normal operating conditions using the drainage phase diagram. Their findings suggested that the main regime expected in a fuel cell is capillary fingering. In this calculation, the effect of different pore sizes along the different layers was not taken into account. Kimball et al. (2008) suggested that water will flow along the DM only through the largest pores.

Medici and Allen (2009) explored the Ca–M phase diagram of fuel cell DM using a pseudo-Hele–Shaw experimental setup as shown in Figure 16.6. This diagram will provide a fundamental resource for understanding the dynamics of the diffusion process and transport characteristics taking place inside the DM as well as a characterization method for DMs in a PEM fuel cell.

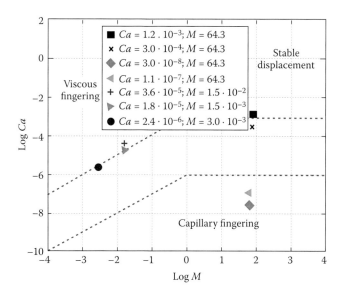

FIGURE 16.6
Points in the drainage phase diagram explored experimentally. Dashed lines represent the hypothetical limits for each flow regime. (From *J. Power Sources*, 191, Medici E.F., Allen J.S., Existence of the phase drainage diagram in proton exchange membrane fuel cell fibrous diffusion media, 417–427, Copyright 2009, with permission from Elsevier.)

16.8 Water Formation by Capillary Condensation in the CL

Capillary condensation is one of the best known examples of fluid phase transition in confined spaces. It is defined as a phenomenon whereby a gas at chemical potential $\phi < \phi_{sat}$ condenses to a more dense, liquid-like phase that fills the pore. In other words, this phenomenon takes place inside the confined spaces when the chemical potential difference $(\Delta\phi > 0)$

$$\Delta\phi = \phi_{sat} - \phi = \frac{2\gamma_{LV} \times \cos\theta}{L(\rho_L - \rho_v)}, \tag{16.8}$$

where γ_{LV} is the surface tension of liquid and gas, θ is the contact angle for a single substrate, ρ_L is the liquid density, ρ_v is the gas density, and L is the distance separating the two parallel substrates of infinite surface area (slit).

In the CL, when water vapor pressure exceeds the saturation level, condensation starts, forming a tree-like liquid water percolation as depicted in Figure 16.7.

The Young's equation for wall fluid tension is given by Evans (1990) as follows:

$$\gamma_{WV} = \gamma_{WL} + \gamma_{LW} \times \cos\theta, \tag{16.9}$$

where W is the capillary wall in the GDL.

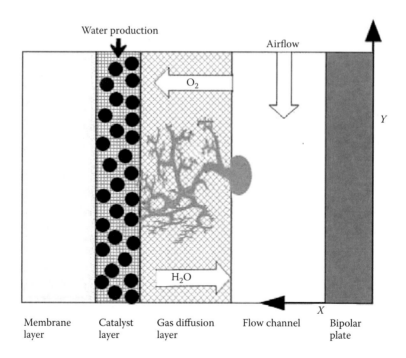

FIGURE 16.7
Transport process in hydrophobic cathode GDL of PEM fuel cell. (From *Int. J. Heat and Mass Transfer*, 46, Nam J.H., Kaviany M., Effective diffusivity and water-saturation distribution in single- and two-layer PEMFC diffusion medium, 4595–4611, Copyright 2003, with permission from Elsevier.)

It is important to note that under ideal gas conditions, Equation 16.9 reduces to the well-known Kelvin equation for condensation

$$\Delta\phi = RT \ln(P_{sat} - P), \tag{16.10}$$

where P_{sat} is the saturation pressure.

16.9 Formation of Water Droplets in the GDL

There are two possible modes for liquid removal from the GDL surface. These modes depend on the air velocity in the gas channel. The first is by drag force exerted from the gas flow when the droplet size is small and gas velocity is high. The second is by capillary interactions with the channel sidewalls when the gas velocity is low and the droplets grow to a size comparable with the flow channel.

16.9.1 Liquid Removal by Drag Force

To develop a definitive relationship between the droplet detachment diameter and the air velocity in the channel, it is instructive to consider forces acting on a single droplet at the GDL surface. The ratio of gravitational force to surface tension is given by the bond number, Bo, defined as

$$Bo = \frac{\Delta\rho g d_d^2}{\sigma}, \tag{16.11}$$

where $\Delta\rho$ is the density difference between the liquid and the gas, g is the gravitational acceleration, d_d is the droplet diameter, and σ is the interfacial surface tension. The maximum droplet diameter before the droplet touches and interacts with the channel bottom wall is given by the geometric relation, $2h/(1 - \cos\theta)$, where h is the channel depth and θ is the GDL contact angle.

The surface adhesion force, holding the drops on the GDL surface, is governed by the surface tension, contact angle, and diameter of the droplet. At mechanical equilibrium of a sessile droplet, the force balance on it is given by Young's equation 16.8.

The forces acting on the water droplet in a shear flow are shown in Figure 16.8. The adhesion force (F_s), which is a result of surface tension, is given by (Kumbur, 2006)

$$F_s = \sigma_{LV} d_d \sin\theta [\cos(\theta_r) - \cos(\theta_a)]. \tag{16.12}$$

Let us assume that the advancing and the receding contact angles are equal, and then the advancing angle is

$$\theta_a = \theta + \Delta\theta \tag{16.13}$$

and the receding angle is

$$\theta_r = \theta - \Delta\theta, \tag{16.14}$$

where $\Delta\theta$ is the contact angle hysteresis.

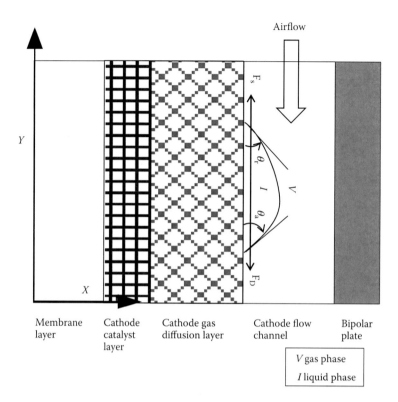

FIGURE 16.8
Schematic of force balance on a water droplet in a shear flow in a PEM fuel cell channel.

By substituting Equations 16.13 and 16.14 in Equation 16.12, the total adhesion force becomes

$$F_s = 2\sigma_{LV}d_d \sin^2(\theta)\sin(\Delta\theta). \tag{16.15}$$

The drag force (F_D) is described as

$$F_D = \frac{1}{2}C_D\rho V^2 A_P, \tag{16.16}$$

where C_D is the drag coefficient on the droplet, V is the channel air velocity, and A_P is the projected area of the droplet normal to the flow direction

$$A_P = \frac{d_d^2}{4}\left(\theta - \frac{1}{2}\sin(2\theta)\right). \tag{16.17}$$

Clift et al. (1978) implemented the following empirical correlation to estimate C_D.

$$C_D\frac{24}{Re}(1 + 0.1925\,Re^{0.63}) = \frac{4.62}{Re^{0.37}}(1 + 5.2\,Re^{-0.63}). \tag{16.18}$$

It is worth mentioning that the first term in Equation 16.18 describes the Stoke's flow approximation for creeping flow. Re is known as the Reynolds number, and it is defined in the following equation:

$$Re = \frac{V d_d}{v_g} \tag{16.19}$$

where v_g is the gas kinetic viscosity.

It should be pointed out that the C_D equation cannot be entirely applied because of two reasons: first, although the droplet is spherical in shape, it is not a full sphere. Second, unlike Stoke's flow, where a uniform approaching velocity is present, the approaching velocity in this study has a parabolic profile starting from zero at the droplet base. Therefore, a correction factor, K, has to be imposed to account for the aforementioned two effects. Hence, the drag force on the droplet becomes

$$F_D = \frac{1}{2} K C_D \rho V^2 A_P. \tag{16.20}$$

For a droplet to detach from the GDL surface, the drag force has to overcome the adhesion force. When the drag force on the droplet equals the adhesion force, the diameter at this point becomes the detachment droplet diameter. When the droplet diameter is bigger than the detachment droplet diameter, the drag force is larger than the adhesion force, and the droplet detaches from the surface and starts to flow along with the gas stream.

Reducing the adhesion force will reduce the required drag force to detach the droplet from the GDL surface. Changing the surface properties from hydrophilic to hydrophobic is not the only technique for reducing the adhesion force. Adding surfactant is an emerging technique to reduce the adhesion force. However, this is a relatively new technique with hardly any published work (Paolo et al., 2009).

The detachment droplet diameter is an important design factor for gas channel dimension. If the channel size is smaller than the detachment droplet diameter, the droplet will touch the channel walls before detaching from the GDL surface. In this case, the droplet will form an additional solid–liquid contact, with a contact angle equal to that of the liquid water-channel surface. Consequently, the drag force has to be bigger to overcome the effects of the sidewalls.

For design purposes, the detachment droplet diameter (d_d) can be calculated by following the expression of adhesion and drag forces, Equations 16.15 and 16.20, respectively,

$$d_d = K' c_D Re^2, \tag{16.21}$$

where K' is an experimental coefficient. It is evaluated experimentally by combining surface wetting properties into a single coefficient (Yamali and Merte, 2002).

16.9.2 Liquid Removal by Capillary Action

Inside the GDL, liquid water is driven by capillary wicking action. This capillary action is a result of capillary pressure distribution, P_c, which is defined as the difference between gas and liquid-phase pressures, P_g and P_l, respectively,

$$P_c = P_g - P_l. \tag{16.22}$$

In hydrophobic GDL, the capillary pressure is negative; hence, the liquid pressure is larger than the gas-phase pressure, whereas in hydrophilic media, the gas-phase pressure is higher than the liquid-phase pressure. Further, the liquid pressure increases with the fraction of void spaces occupied by liquid water. Therefore, a liquid pressure gradient is formed from higher to lower liquid saturation regions. This pressure gradient becomes the driving force for liquid water flow inside the GDL, as schematically demonstrated in Figure 16.9.

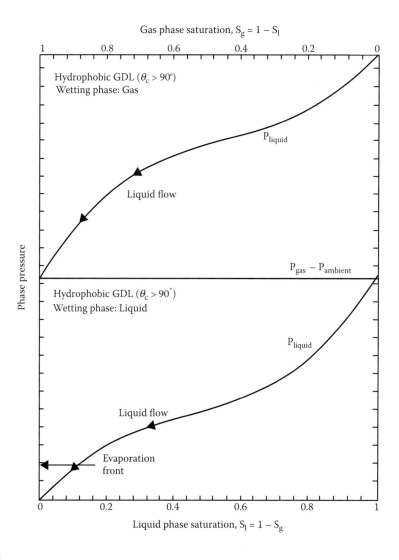

FIGURE 16.9
The schematic of liquid and gas-phase pressure profiles in hydrophilic and hydrophobic porous media. (From Pasaogullari U. and Wang C.Y., Liquid water transport in gas diffusion layer of polymer electrolyte fuel cells, *Int. J. Electrochem. Soc.*, 151(3), 2004, A399–A406. Reproduced by permission of the Electrochemical Society.)

In PEM fuel cell, the liquid saturation is higher at the CL than at the GDL channel interface. This is attributed to water generation as a result of the electrochemical reaction in the CL and the electro-osmotic drag. As a result, the liquid pressure gradient formed in the GDL drives liquid water from the action sites toward the open channel.

It is noticed in Figure 16.9 that although the relative magnitude of liquid pressure to gas pressure is different in hydrophobic GDL than in the hydrophilic one, both media provide capillary action to drive liquid water from the inside to its surface. However, the slope of the capillary pressure is different, depending on whether water is the wetting or nonwetting phase. The higher slope of the capillary pressure near the evaporation front in hydrophobic GDL is indicative of more effectiveness of this type of medium for water removal.

16.10 Channel Design

Many parameters should be considered in designing the flow field channels. These parameters can be determined using a design procedure based on fluid dynamic principles and the surface tension of the material used. These parameters include channel cross section, channel dimensions, and channel wettability.

16.10.1 Channel Cross-Sectional Shape

A flow channel cross-sectional shape is determined by the manufacturability of its shape and the cost of manufacturing. Many choices of the cross-sectional shape are proposed, from the simple rectangular or square shape to the triangular, trapezoidal, semicircular, or any other shape that might be applied to the flow channel (Li, 2006). However, graphite is the typical material used in the conventional bipolar plate, which is hard and brittle; hence, it is hard to machine the flow channel on it. Consequently, fabricating the flow channels on the bipolar plate is a time-consuming and expensive process, which contributes significantly to the total cost of a PEM fuel cell stack (Li, 2005).

To reduce the cost, the channel fabrication process should be simplified. The geometrical shape of the cross section has to be categorized as simple geometry. Rectangle and square cross-sectional shapes have traditionally been chosen for the flow channel design and fabrication because of their geometrical simplicity.

16.10.2 Channel Dimensions

The channel dimension can be categorized into small dimensions and a large dimension. The small dimensions are channel width (a), channel depth (b), and the land area width (w). The length of the channel (l) is the only large dimension as illustrated in Figure 16.10.

16.10.2.1 Channel—Small Dimension Calculations

The channel width, a, is often chosen on the basis of the need for distributing the reactant gas over the active cell surface. The distance between the channels, w, is decided on the basis of the need for current collection. The channel depth, b, is determined on the basis of the consideration of the flow regime. Therefore, the design or selection of the flow channel dimensions should consider the cell operating conditions as well as the cell structural parameters. Typically, the GDL and the bipolar plate materials are highly conductive electronically, whereas the reactant gas transport is relatively lower; hence, the ratio of the land width to the channel width is typically in the range of 0.8–1.0 (Li, 2007). The typical values for Reynolds numbers in PEM fuel cells are within the laminar flow regime (Maharudrayya et al., 2004).

For laminar flow, the accepted correlation for the hydrodynamic entrance length is

$$\frac{L_e}{D_h} \approx 0.056\,Re, \tag{16.23}$$

where L_e is the hydrodynamic entrance length and D_h is the hydraulic diameter.

The maximum entrance length typically encountered in a PEM fuel cell can be estimated. The channel cross section is illustrated in Figure 16.10. The maximum depth (b) and width (a) are 1.1 mm.

16.10.2.2 Channel Length Calculations

The pressure drop in a pipe can be calculated by the relation

$$\Delta p = \rho g h_f, \tag{16.24}$$

where

$$h_f = f\,\frac{lV^2}{2D_h g}. \tag{16.25}$$

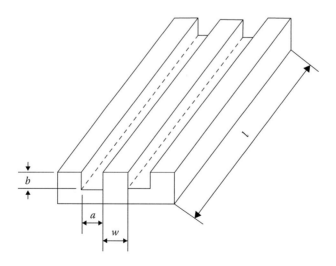

FIGURE 16.10
Schematic showing the channel dimensions in details. (From *J. Power Sources*, 163, Li X., Sabir I., Park J., A flow channel design procedure for PEM fuel cells with effective water removal, 933–942, Copyright 2007, with permission from Elsevier.)

The laminar friction factor (f) for square cross section is given as (White, 1986)

$$f = \frac{56.91}{\mathrm{Re}_{D_h}}. \tag{16.26}$$

Therefore, after substituting Equations 16.26 and 16.25 into Equation 16.24, the pressure drop expression for square cross section ($a = b$) becomes

$$\Delta p = 28.455 \left(\frac{\mu \dot{m}}{\rho} \right) \left(\frac{l}{a^4} \right). \tag{16.27}$$

Then the flow channel length for square cross section can be determined using the following equation:

$$l = \frac{\Delta p a^4 \rho}{28.455 \mu \dot{m}}. \tag{16.28}$$

Similarly, a rectangular cross-sectional channel can be calculated by

$$l = \frac{\Delta p \rho (ab)^3}{C \mu \dot{m} (a+b)^2}, \tag{16.29}$$

where μ is the viscosity and $\dot{m}$ is the mass flow rate. $C = f \mathrm{Re}_{D_h}$ is a function of the aspect ratio (b/a) for rectangular flow channels, and it is evaluated by White (1986). When the aspect ratio is equal to 1, which is the typical case in the channel cross-sectional design, $C = 56.91$; thus, the channel length can be calculated directly from the following equation:

$$l = 0.1405 \frac{\Delta p \rho (ab)^3}{\mu \dot{m} (a+b)^2}. \tag{16.30}$$

Channel length can be calculated from above equations if the required pressure drop is known.

16.10.3 Channel Wettability

Surface tension and wettability play a particularly important role in water removal from fuel cell minichannels as illustrated in Figure 16.11. The graphite channel is in contact with the GDL at the rip areas; thus, three of the channel walls are graphite surface (hydrophilic), and the fourth one is the GDL surface (hydrophobic). A hydrophobic GDL surface will favor the removal of water from the cell more than a hydrophilic one. This is because on a hydrophobic surface, the contact angle between the liquid droplet and the surface is large (>90°). Therefore, little wetting occurs, which means that the water droplet formed on the GDL surface does not spread on the surface of the GDL, hence enabling easier detachment of the water droplet from the surface. Furthermore, having the GDL surface as hydrophobic is beneficial because it means that very little drag force is needed to over-come the force of adhesion. However, there is a limit to the amount of air pumped into the cell because excess humidified air can cause flooding of the cathode and also can increase operation cost. Over the years, researchers and scientists have sought ways to optimize the surface of the GDL to minimize the amount of air pumped into the cell.

The appropriate design of flow channels built on the bipolar plates is critical to the tack-ling of water management. Serpentine graphite flow channel layout (Quan et al., 2005) is the most widely used one, often regarded as "industry standard," because under the same operating and design conditions, PEM fuel cells with serpentine flow channels tend to have the best performance and durability. Recent numerical and experimental investiga-tion (Zhu et al., 2008) indicated that for serpentine flow channels, the reactant gas experi-ences a pressure drop and concentration change along the channel. This is attributed to the hydrophilic nature of the graphite channel as shown in Figure 16.12.

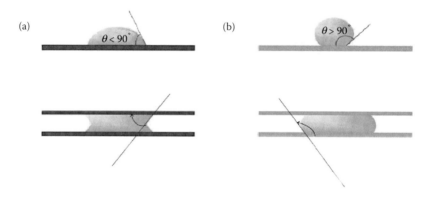

FIGURE 16.11
Contact angle for liquid–solid interface for a single droplet on a flat solid surface and for a liquid plug inside the channel: (a) hydrophilic and (b) hydrophobic. (With kind permission from Springer Science + Business Media: *Micro-Mini Fuel Cells-Fundamentals and Applications Minor,* Water transport dynamics in fuel cell micro-channels, 2008, pp. 153–170, G.X. Zhu, P. Oshkai, P.C. Sui, N. Djilali, (Eds.: S. Kakac, L. Vasiliev, A. Pramuanjaroenkij).

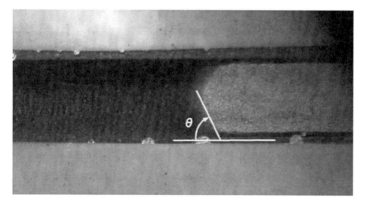

FIGURE 16.12
Contact angle for water–raw graphite interface in 4×4-mm cross-sectional graphite channel, $\theta = 65°$.

Reducing the surface tension in the graphite channel will reduce the pressure drop across the channel and will consequently enhance the cell performance.

Because the channel length is much larger than its cross section (by 100 times), the pressure drop at the corresponding location between the adjacent channels becomes substantial. A significant pressure gradient is thus set up across the porous electrode (much larger than the pressure gradient along the channel direction), resulting in considerable cross leakage flow between the adjacent channels.

The significant cross leakage flow through the porous electrode induces a strong convection in the electrode, bringing the reactant gas to the CL. As a result, an electrochemical reaction occurs, and the removal of water from the reaction sites and electrodes ensues. This driving force across the curved surface is expressed by the following equation:

$$\Delta P = \frac{2\gamma_{LV} \cos \theta}{r},$$ (16.31)

where r is the channel radius.

16.10.4 Nature of Capillary Material

Explaining the role of hydrophilic channels in water removal from fuel cell field flow channels requires an understanding of capillary actions. On the hydrophilic surface (where the contact angle is small), the water wets the capillary walls causing a rise in the capillary tube due to the positive change in pressure across the curved surface, as illustrated in Figure 16.13a. On the contrary, the hydrophobic surface has a high contact angle, with little or no wetting of the walls. This repellent nature causes a capillary depression as illustrated in Figure 16.13b.

It is clear that both phenomena may be used to design the capillary walls to channel water constructively. Water flow direction must be optimized, such that water flows out of the cell and not back into the electrodes. In that case, the airflow from Equation 16.31 must be considered to deduce a relationship with the directionality of water flow, s,

$$s^2 = \frac{2\gamma}{\Delta \rho g} = rh.$$ (16.32)

The quantity of s defined in Equation 16.32 represents the capillary constant or capillary length; $\Delta\rho$ is the density difference between the liquid and the gas, g is the gravitational

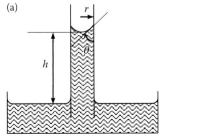
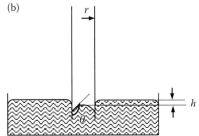

FIGURE 16.13
Rise of water in capillary tubes: (a) hydrophilic walls and (b) hydrophobic walls.

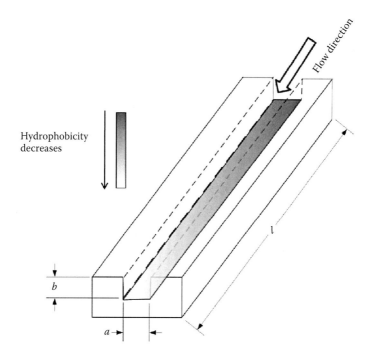

FIGURE 16.14
Channel design with different surface wettability gradients, from higher to lower hydrophobicity.

acceleration, r is the channel radius, and h is the extent of the change of the surface level of the liquid from the original liquid level. Because the water flows from lower surface tension to higher surface tension, increasing s indicates the favorable water direction.

Combining the effect of the channel wettability and the nature of the capillary of the graphite channel requires an optimized channel surface design. The designed surface should be hydrophobic to reduce pressure drop across the channel by reducing the surface tension, and the surface hydrophobicity gradient should be reduced across the channel to direct the water out of the cell (from lower to higher surface tension) as illustrated in Figure 16.14. This design might result in higher water removal from the PEM fuel cell cathode side, and consequently enhance the cell performance.

16.11 Experimental Diagnostics of the Effect of Different Surface Wettabilties on PEM Fuel Cell Performance

PEM fuel cell performance measurement is one of the diagnostic tools to understand the effect of different surface wettabilities of the fuel cell components on the nature of water removal in PEM fuel cell. Performance tests were conducted on two different 10 × 10-cm active areas. PEM fuel cells with different cathode GDLs were assembled in-house as shown in Figure 16.15. The first cell has a 0 wt.% PTFE carbon paper (raw GDL), and the second cell has 25 wt.% PTFE carbon paper coated on one side and not coated on the other side.

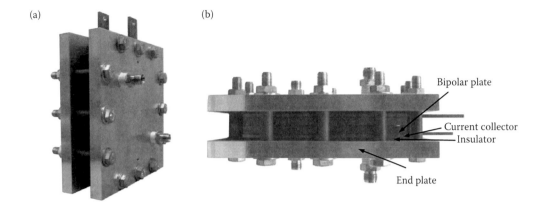

FIGURE 16.15
The 10 × 10-cm active area PEM fuel cell manufactured and assembled in-house: (a) vertical position and (b) horizontal position showing some components of the cell.

The carbon paper is coated using PTFE 60 wt% solution at University of Waterloo in the Nano-Bio Interfacial Engineering Group Labs. The solution is diluted to 15 wt% with deionized water and then brushed on one side of the noncoated carbon paper. The above sample was cured at ambient condition for more than 2 h. Finally, the coated GDL was washed by methanol and water, and its weight was measured before and after coating. Further, the contact angle on the coated surface is 140° and the one on the noncoated surface is 90°.

The assembly procedures and the material components for both cells are the same except for the cathode GDL as mentioned above. Moreover, in the second cell, the PTFE-coated surface of the cathode GDL is facing the MEA, and the noncoated one is facing the flow field channels.

As a part of the ongoing experimental diagnostics for different surface wettabilities of the cathode GDL and the flow field channels, the performance test is conducted for both cells under the same running conditions: $T = 70°C$, relative humidity (100%), pressure drop is 235 kPa across the anode and cathode side, and the stoichiometric ratio is 2 for O_2 and 1.2 for H_2. The performance results are presented in Figure 16.16.

Figure 16.16 shows that at high current density, the PTFE-coated GDL has better performance than the noncoated one. Both GDLs have the same performance at low current densities. Although the improved influence of GDL coating has been demonstrated earlier, the current work uses a one-sided coated GDL to investigate the effect of different surface wettabilities of the GDL and/or the flow field channel on water removal from the cathode side. Moreover, the highest current density for the noncoated carbon paper reaches 1.1 A cm^{-2} and the coated one reaches 1.3 A cm^{-2}. In other words, for the whole cell, it reaches 110 A for the noncoated GDL and 130 A for the coated GDL. This is a considerable improvement for a 10 × 10-cm PEM fuel cell despite the fact that the PTFE-coated GDL allows for higher contact resistance and lower porosity.

At high current density, the liquid water generated by the electrochemical reaction is significantly increased. This could significantly reduce the effective porosity of the GDL to hinder oxygen transport to the CL. The removal rate of liquid water due to capillarity-induced motion can be enhanced by an increase in hydrophobicity. Furthermore, the wettability gradient in the coated GDL is from the PTFE-coated side (hydrophobic) to the

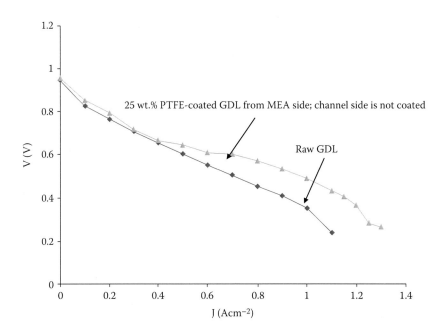

FIGURE 16.16
PEM fuel cell performance comparison with the coated and noncoated cathode GDL.

noncoated side (hydrophilic), and because the hydrophobic side is facing the MEA, the wettability gradient enhances water removal from the reaction sites. This provides an on-time water removal and enhances the cell performance.

16.12 Conclusions

The PEM fuel cell is an electrochemical device used to facilitate the movement of electrons, producing useful energy and liberating water as an exhaust. It operates at relatively low temperature, which makes it eligible for practical uses such as in the transportation industry and for computer devices. Water is the only emission product of a PEM fuel cell, which is of prime commercial importance due to growing environmental concerns.

As the proton conductivity of a PEM is directly proportional to its water content, it must have sufficient water. It is evident that a level of water must remain in the fuel cell in order for ion conductivity to occur. However, too much water can short-circuit the fuel cell because of electrode flooding and can cause degradation of the catalyst membrane. Excess water causes flooding of the electrolyte, and in the process, blocks the pores in the electrode. Therefore, it is essential to maintain the water balance.

To tackle the problem of water management, one has to consider the formation and removal of water at the various interfaces. The techniques for calculating water formation and removal rates were discussed throughout this chapter. Several relationships displaying the formation of water by capillary condensation, which depends on the water-saturation level, have been presented. It is affected by the surface interactions with water at the GDL interface. Water can be removed from these layers (as a whole) by capillary action, which is caused by the pressure difference across the curved surface that water forms at the interface of the GDL. Microsized droplets are removed due to forces acting on the individual droplet as it relates to the air velocity. Therefore, one can calculate the rate at which water leaves the fuel cell. The proper rate to design the air velocity as well as the coating materials of the capillary walls can then be determined.

The presented experimental diagnostics for different surface wettabilities of the cathode GDL showed that the one-sided coating of GDL in the cathode side provides an effective water removal at high current density. Therefore, the cell performance is enhanced.

Acknowledgments

The authors thank Yongxin Wang and Ehab Abu Ramadan for their assistance, and they are grateful for all the support provided by the members of our research groups.

The research of our work in this chapter is financially supported by the Natural Sciences and Engineering Research Council of Canada (NSERC), the Canadian Foundation of Innovation (CFI), and the Canada Research Chairs program.

Nomenclature

Variables

V_r	Reversible cell potential	(V)
J	Current density	(A/cm^2)
a	Channel width	(mm)
b	Channel depth	(mm)
w	Land width	(mm)
P	Pressure	(kPa)
T	Temperature	(°C)
R	Universal gas constant (8.3143)	(J/mol K)
l	Channel length	(mm)
L	Hydrodynamic entrance length	(mm)
D_h	Hydraulic diameter	(m)
$\dot{m}$	Mass flow rate	(kg/s)
Δp	Pressure drop in channels	(kPa)
s	Directionality	
A_p	Projected area of the droplet	(m^2)
Bo	Bond number	
C_D	Drag coefficient	
d_d	Detachment diameter of the droplet	(m)
F_S	Work of adhesion	(N)
F_D	Drag force	(N)
g	Gravitational acceleration	(m/s^2)
K	Correction factor	

Re	Reynolds number	
v	Channel air velocity	(m/s)
ν_g	Gas kinetic viscosity	
P_l	Pressure in the liquid phase	(kPa)
P_g	Pressure in the gas phase	(kPa)
P_c	Capillary pressure	(kPa)
P_{sat}	Vapor pressure at saturation	(kPa)
r	Channel radius	(mm)
L	Distance between the two parallel substrates of infinite surface area	(mm)
h	The extent of the change of the surface level of the liquid from the original liquid surface level	(mm)
J	Superficial velocity	

Abbreviations

DM	Diffusion media
PEM	Polymer electrolyte
MEA	Membrane electrode assembly
PTFE	Polytetrafluoroethylene
Pt	Platinum
Ca	Capillary number
M	Viscosity ratio
W	Wall
GDL	Gas diffusion layer

Subscripts

a	Advancing
r	Receding
w	Wetting
nw	Nonwetting
g	Gas
O_2	Oxygen
H_2	Hydrogen
L	Liquid phase
V	Gas phase

Greek Letters

μ	Viscosity	(kg/m s^{-1})
ρ	Density	(kg/m^3)
η	Overpotential	(V)
ϕ	Chemical potential	
γ	Surface tension	(N/m)
σ	Interfacial surface tension	(N/m)
θ	Contact angle	(°)

References

Adamson AW and Gast AP. (1997). *Physical Chemistry of Surfaces*. 6th ed. New York: John Wiley & Sons; p. 353.

Akundy GS, Iroh JO. (2001). Polypyrrole coatings on aluminium: synthesis and characterization. *Polymer*. 42:9665–9.

Barbir F. (2005). *PEM Fuel Cells: Theory and Practice*. Burlington, MA: Academic Press; pp. 40–4.

Bazylak A, Sinton D, Djilali N. (2008). Dynamic water transport and droplet emergence in PEMFC gas diffusion layers. *J Colloid Interface Sci*. 176:240–6.

Bevers D, Rogers R, Bradke MV. (1996). Examination of the influence of PTFE coating on the properties of carbon paper in PEFCs. *J Power Sources*. 63:193–201.

Bikerman JJ. (1996). Sliding of drops from surfaces of different roughness. *Polym Eng Sci*. 36:1849–55.

Billings RE, Sanchez M. (1995). Solid polymer fuel cells: an alternative to batteries in electric vehicles—an overview. *Int J Hydrogen Energy*. 20(1):521–9.

Borroni-Bird CE. (1996). Fuel cell commercialization issues for light-duty vehicle applications. *J Power Sources*. 61:33–48.

Borub RL, Vanderborgh NE. (1995). Design and testing criteria for bipolar plate materials for PEM fuel cell applications. *Mater Res Soc Symp Proc*. 393:151–5.

Cropper MAJ, Geiger S, Jollie DM. (2004). Fuel cells: a survey of current developments. *J Power Sources*. 131:57–61.

Davies DP, Adcock PL, Turpin M, Rowen SJ. (2000). Bipolar plate materials for solid polymer fuel cells. *J Appl Electrochem*. 30:101–5.

Evans R. (1990). Fluids adsorbed in narrow pores: phase equilibria and structure. *J Phys Condens Matter*. 2:8989–9007.

Giorgi L, Antolini E, Pozio A, Passalacqua E. (1998). Influence of the PTFE content in the diffusion layer of low-Pt loading electrodes for polymer electrolyte fuel cells. *Electrochim Acta*. 43(24):3675–80.

Girifalco LA, Good RJ. (2008). A theory for the estimation of surface and interfacial energies. I. Derivation and application to interfacial tension. *J Phys Chem*. 61(07):904–8.

Hermann A, Chaudhuri A, Spagnol P. (2005). Bipolar plates for PEM fuel cells: a review. *Int J Hydrogen Energy*. 30:1297–302.

Hung Y, El-Khatib KM, Tawfik H. (2006). Testing and evaluation of aluminium coated bipolar plates of PEM fuel cells operating at 70°C. *J Power Sources*. 163(1):509–13.

Kim H-S, Ha T-H, Park S-J, Min K, Kim M. (2005). Visualization study of cathode flooding with different operating conditions in a PEM unit fuel cell. Proceedings of FUELCELL2005, The Third International Conference on Fuel Cell Science, Engineering and Technology, Ypsilanti, Michigan; May 23–25.

Kimball E, Whitaker T, Kevrekidis YG, Benzinger JB. (2008). Drops, slugs, and flooding in polymer electrolyte membrane fuel cells. *AIChE J*. 54(5):1313–32.

Kumbur EC, Sharo KV, Mench MM. (2006). Liquid droplet behaviour and instability in a polymer electrolyte fuel cell flow channel. *J Power Sources*. 161:333–45.

Larminie J, Dick A. (2003). *Fuel Cell System Explained*. 2nd ed. Chichester, West Sussex: John Wiley & Sons; pp. 69–73.

Lenormand R, Touboul E, Zarcone C. (1988). Numerical models and experiments on immiscible displacements in porous media. *J Fluid Mech*. 189:165–87.

Lenormand R. (1990). Liquids in porous media. *J Phys Condens Matter*. 2:SA79–88.

Lee SJ, Huang CH, Lai JJ, Chen YJ. (2005). Corrosion-resistant component for PEM fuel cells. *J Power Sources*. 131:162–8.

Li M, Luo S, Zeng C, Shen J, Lin H, Cao C. (2004). Corrosion behaviour of TiN coated type 316 stainless steel in simulated PEMFC environments. *Corrosion Sci*. 46:1369–80.

Li X. (2006). *Principles of Fuel Cells*. New York: Taylor & Francis.

Li X, Sabir I. (2005). Review of bipolar plates in PEMFC: flow field designs. *Int J Hydrogen Energy*. 30:359–71.

Li X, Sabir I, Park J. (2007). A flow channel design procedure for PEM fuel cells with effective water removal. *J Power Sources*. 163:933–42.

Lim C, Wang CY. (2004). Effects of hydrophobic polymer content in GDL on power performance of a PEM fuel cell. *Electrochim Acta*. 49:4149–56.

Litster S, Sinton D, Djilali N. (2006). Ex situ visualization of liquid water transport in PEM fuel cell gas diffusion layers. *J Power Sources*. 154(1):95–105.

Liu X, Guoa H, Ma C. (2006). Water flooding and two-phase flow in cathode channels of proton exchange membrane fuel cells. *J Power Sources*. 156:267–80.

Ma HP, Zhang HM, Hua J, Cai YH, Yi BL. (2006). Diagnostic tool to detect liquid water removal in the cathode channels of proton exchange membrane fuel cells. *J Power Sources*. 162:469–73.

Maharudrayya JS, Deshpande AP. (2004). Pressure loss in laminar flow through serpentine channels in fuel cell stacks. *J Power Sources*. 138:1–13.

Medici EF, Allen JS. (2009). Existence of the phase drainage diagram in proton exchange membrane fuel cell fibrous diffusion media. *J Power Sources*. 191:417–27.

Mehta V, Cooper JS. (2003). Review and analysis of PEM fuel cell designed manufacturing. *J Power Sources*. 144:32–53.

Miller JD, Veeramasunent S, Drelich J, Yalamanchili MR. (1950). Effect of roughness as determined by atomic force microscopy on the wetting properties of PTFE thin films. *J Colloid Sci*. 5(4):349–59.

Minor G, Zhu X, Oshkai P, Sui PC, Djilali N. (2008). Water transport dynamics in fuel cell micro-channels. In: Kakac S, Vasiliev L, Pramuanjaroenkij A. editors. *Micro-Mini Fuel Cells—Fundamentals and Applications*. Netherlands: Springer; pp. 153–70.

Moreira J, Ocampo AL, Sebastian PJ, Smit MA, Salazar MD, del Angel P, et al. (2003). Influence of hydrophobic material content in the gas diffusion electrodes on the performance of a PEM fuel cell. *Int J Hydrogen Energy*. 28:625–7.

Nakajima A, Hashimoto K, Watanabe T. (1999). Transparent super-hydrophobic coating films with photo catalytic activity. *Kokagaku*. 30(3):199–206.

Natarajan D, Nguyen TV. (2006). Three-dimensional effects of liquid water flooding in the cathode of a PEM fuel cell. *J Power Sources*. 115:66–80.

Nam JH, Kaviany M. (2003). Effective diffusivity and water-saturation distribution in single- and two-layer PEMFC diffusion medium. *Int J Heat Mass Transf*. 46:4595–611.

Nguyen T, White R. (1993). A water and heat management model for proton exchange membrane fuel cells. *J Electrochem Soc*. 140:2178–86.

Nishino T, Meguro M, Nakamae K, Matsushita M, Ueda Y. (1999). The lowest surface free energy based on -CF3 alignment. *Langmuir*. 15:4321–3.

Onda T, Shibuichi S, Satoh N, Tsujii K. (1996). Super-water-repellent fractal surfaces. *Langmuir*. 12(9):2125–7.

Paganin VA, Ticianelli EA, Gonzalez ER. (1996). Development and electrochemical studies of gas diffusion electrodes for polymer electrolyte fuel cells. *J Appl Electrochem*. 26:297–304.

Pai YH, Ke JH, Huang HF, Lee CM, Zen JM, Shieu FS. (2006). CF4 plasma treatment for preparing gas diffusion layers in membrane electrode assemblies. *J Power Sources*. 161:275–81.

Paola GS, Cinzia C, Giovanni D, Luca O, Luca Z, Renato P, et al. (2009). Effect of different substrates, inks composition and rheology on coating deposition of microporous layer (MPL) for PEM-FCs. *Catal Today*. 147:S30–5.

Pasaogullari U, Wang CY. (2004). Liquid water transport in gas diffusion layer of polymer electrolyte fuel cells. *J Electrochem Soc*. 151(3):A399–406.

Satija R, Jacobson DL, Arif M, Werner SA. (2004). In situ neutron imaging technique for evaluation of water management systems in operating PEM fuel cells. *J Power Sources*. 129(2):238–45.

Schrader ME. (1975). Ultrahigh vacuum techniques in the measurement of contact angles. IV. Water on graphite (0001)12. *J Phys Chem*. 79(23):2508–15.

Sinha PK, Mukherjee PP, Wang CY. (2007). Impact of GDL structure and wettability on water management in polymer electrolyte fuel cells. *J Mater Chem*. 17(30):3089–103.

Springer TE, Wilson MS, Gottesfeld S. (1993). Modeling and experimental diagnostics in PEM fuel cells. *J Electrochem Soc*. 140(12):3513–26.

Springer TE, Zawodzinski TA, Wilson MS, Gottesfeld S. (1996). Characterization of PEM fuel cells using AC impedance spectroscopy. *J Electrochem Soc*. 143(2):587–99.

Subramanian RS, Moumen N, McLaughlin JB. (2005). Motion of a drop on a solid surface due to a wettability gradient. *Langmuir*. 21:11844–9.

Taniguchi A, Yasuda K. (2005). Highly water-proof coating of gas flow channels by plasma polymerization for PEM fuel cells. *J Power Sources*. 141:8–12.

Theodorakakos A, Ous T, Gavaises M, Nouri JM, Nikolopoulos N, Yanagihara H. (2006). Dynamics of water droplets detached from porous surfaces of relevance to PEM fuel cells. *J Colloid Interface Sci*. 300:673–87.

Tsushima S, Teranishi K, Nishida K, Hirai S. (2005). Water content distribution in a polymer electrolyte membrane for advanced fuel cell system with liquid water supply. *Magn Reson Imaging*. 23(2):255–8.

Tüber K, Pocza D, Hebling C. (2004). Visualization of water build-up in the cathode of a transparent PEM fuel cell. *J Power Sources*. 124:403–14.

Wallmark C, Alvfors P. (2002). Design of stationary PEFC system configurations to meet heat and power demands. *J Power Sources*. 106:83–92.

Wang ED, Shi PF, Du CY. (2008). Treatment and characterization of gas diffusion layers by source carbonization for PEMFC applications. *Electrochem Commun*. 10:555–8.

Wang H, Sweikart MA, Turner JA. (2003). Stainless steel as bipolar plate material for polymer electrolyte membrane fuel cells. *J Power Sources*. 115:243–51.

Wang H, Turner JA. (2004). Ferritic stainless steels as bipolar plate material for polymer electrolyte membrane fuel cells. *J Power Sources*. 128:193–200.

Wenzel RN. (1949). Surface roughness and contact angle. *J Phys Chem*. 53(9):1466–7.

White FM. (1986). *Fluid Mechanics*. 2nd ed. New York: McGraw Hill.

Wind J, Spah R, Kaiser K, Bohm G. (2002). Metallic bipolar plates for PEM fuel cells. *J Power Sources*. 105:256–60.

Yamali C, Merte CH Jr. (2002). Theory of dropwise condensation at large subcooling including the effect of the sweeping. *Heat Mass Transf*. 28:191–202.

Yang XG, Zhang FY, Lubawy AL, Wang CY. (2004). Visualization of liquid water transport in a PEFC. *Electrochem Solid-State Lett*. 7(11):A408–11.

17

Biologically Inspired Adhesives

Animangsu Ghatak

CONTENTS

17.1 Introduction

The history of adhesion technology dates back to the beginning of human civilization when Stone Age man used pitch [1] (prepared by pyrolysis of wood or bark) to glue blades to the shaft of his hunting tool. Ancient Egyptians used a kind of plaster adhesive to preserve mummies. Today, adhesives are ubiquitous in applications ranging from a spaceship to a postal stamp or from a therapeutic patch for transdermal drug delivery to the manufacture of computer chips. Everywhere some kind of an adhesive is used for joining two surfaces or objects. The natural world around us does not lack examples of adhesives. In fact, the feet of most animals such as arthropods, lizards, and amphibians use a variety of mechanisms for adhesion, which not only allow these animals to stick to different surfaces (rough and smooth, hydrophilic and hydrophobic, or dry and wet) but also allow them to accomplish other activities, for example, running up and down on the smooth surface of a leaf while carrying food, darting rapidly on a wall for catching a prey, or in the world of fantasy, like Spider-Man, jumping from one skyscraper to another, at ease, instantly constructing his own web.

What causes adhesion between two objects? Obviously, the thermodynamic work of adhesion, W, which is a measure of how two surfaces interact, is an important parameter. It is defined as the amount of energy released when two surfaces having surface energies γ_{sa} and γ_{sb} adhere together to form an interfacial energy γ_{ab}: $W = \gamma_{sa} + \gamma_{sb} - \gamma_{ab}$. This is the thermodynamic amount of energy involved in the adhesion process in a nondissipative system. Usually, this energy varies between 40 and 200 mJ/m^2, when the predominant force across the interface is dispersion, polar, hydrogen bonding, or acid–base interaction. However, most applications demand the energy of separation of the two surfaces, commonly defined as the fracture energy, to far exceed the thermodynamic work of adhesion by the order of 1–2 J/m^2. In conventional adhesives, enhancement in fracture energy is achieved by rendering the adhesive viscoelastic, so that during separation of the adherents, the viscous flow at the bulk of the adhesive results in dissipation of energy. Besides, yielding of the adhesive material, plastic deformation, and interfacial friction also contributes to energy dissipation and enhancement of adhesion at the interface. Natural adhesives, however, use a variety of mechanisms other than viscoelasticity to remain effective. For example, animals such as geckos and house lizards have developed the mechanism of dry adhesion, which interacts with the substrate via the van der Waals force [2,3]. Insects such as the beetle *Chrysolina polita* hold on to a surface via frictional resistance against shear in a manner very similar to mechanical hooks [4,5].

Different species of ants, for example, the Asian weaver ants (*Oecophylla smaragdina*) [6], and many species of frogs, for example, the Trinidadian tree frogs from the genus *Hyla*, are known to adhere to a surface by the capillary action of a thin layer of liquid [7–9] that is secreted from glands underneath their adhesive pads. Spiders are known to secrete a glue that wets the tiny fibers of their web [10,11], and as a flying prey hits the web, it gets stuck to it. There are sea animals such as barnacles and zebra mussels that achieve adhesion under water by secreting a protein that wets a submerged surface displacing water [12]. The protein hardens over time, resulting in strong fracture toughness of the interface [13]. The large variety of these mechanisms allows living organisms not only to adhere strongly onto a surface but also to detach easily from the surface for fast locomotion. These adhesives undergo many thousand cycles of adhesion and debonding in their lifetime without failure. Even dirt and humidity do not appear to affect their performance to any significant degree [14]. In contrast, most conventional adhesives lack these qualities. They can hardly be used more than once and can create strong adhesion initially, but the adhesion strength decreases over time and eventually fails. Most adhesives succumb to adverse environmental conditions of heat, moisture, and particulate matter. Adhesion may be strong against normal load but fails when subjected to shear and *vice versa*. It is because of these practical limitations of the conventional adhesive that extensive research is being carried out to learn about various adhesion mechanisms that are prevalent in nature and how they can be implemented to design artificial adhesives useful for clean and instant adhesion, repeatable application, and resistance against extreme conditions.

In what follows, we have presented a brief review of different patterned adhesives that have been developed based on these naturally occurring adhesion mechanisms. The principle of splitting the adhesive surface as observed in nature has been discussed in detail, followed by demonstrations using model adhesive layers to enhance adhesion via crack arrest and initiation mechanisms. The issues associated with optimal fabrication of fibrillar surfaces have been discussed in the context of mimicking of the dry adhesion mechanism of geckos. The challenges and the possible methodology of mimicking the structural features of the terminal cap of the gecko setae have been presented in detail. Following this overview on dry adhesion, the wet adhesion mechanism used in the feet of many insects has been discussed in the light of frictional resistance against shear fracture. Finally, the possible role of subsurface microstructures, for example, fluidic vessels present in the feet of many living creatures, has been explored by illustrating model adhesives embedded with fluid-filled microchannels. Similarly, contribution of internal self-adhesion of these subsurface structures to overall enhancement of adhesion has been demonstrated by the description of adhesive layers embedded with stacked microchannels with internal surface patterns. In essence, we describe recent advancements made in adhesion science that have led to the design and fabrication of many different types of smart adhesives, essentially inspired by the mechanisms prevalent in biology.

17.2 Classical Adhesion Experiments

Before discussing in detail the different physical mechanisms of adhesion that have so far been explored and those that have inspired the design of a variety of textured adhesives, it is appropriate to briefly describe the classical adhesion tests that have been developed for characterizing the effectiveness of adhesive surfaces. In Figure 17.1, we present the schematic of some of these experiments.

17.2.1 Peel Experiment

This is one of the simplest and most widely used experiments for measuring the adhesion strength of an interface, as the geometry of this experiment closely matches many practical situations. Here, the adhesive layer remains strongly bonded to a plastic backing, which is brought in contact with the surface of a rigid or flexible adherent and is then peeled off at a desired peel angle θ. The peel experiment is carried out at different peeling rates, and the peel force F is measured, which is used for calculating the

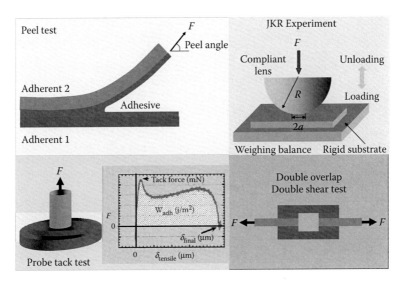

FIGURE 17.1
Schematic of different classical adhesion tests.

adhesion strength as $G = F(1 - \cos\theta)$. Various peel experiments have been developed, depending on the applications, for example, T-peel, cleavage peel, climbing drum, and floating roller techniques. Approximate analysis of the peel geometry suggests that the debonding stress at the interface of the two adherents is maximum when close to the contact line but varies sinusoidally with exponentially decreasing amplitude when it is away from the contact line [15,16]. In other words, during peeling, not the whole portion of the adhesive but the region at the vicinity of the contact line is maximally stressed.

17.2.2 Probe Tack Test

Although in peel geometry the adhesive is not uniformly loaded everywhere, many applications demand that the adhesive performance be tested by straining it uniformly in the whole contact area. Here, a rigid flat plate indenter is brought in contact with a layer of adhesive bonded to a rigid substrate. The indenter is then pulled off the adhesive at a constant rate. The adhesion strength of the adhesive is obtained by calculating the area under the resultant force versus the displacement plot [17–19]. In addition, one obtains other interesting parameters useful for designing an adhesive: the maximum nominal stress and the maximum nominal strain at which the debonding occurs.

17.2.3 The Johnson, Kendall, and Roberts Contact Experiment [20]

Although the above tests are useful for measuring the effective adhesion strength of an adhesive interface, these tests are not rigorous enough for accurately estimating the actual thermodynamic work of adhesion at the interface. For example, in the peel experiment, the estimated adhesion strength often includes the plastic deformation of the backing [16]. Similarly, in the probe tack test, debonding ensues by the appearance of cavities within the contact area, which renders theoretical analysis difficult. To study the effect of the interface, a variant of the probe tack test has been proposed with a rigorous theory. Here a soft hemispherical [20,21] or hemicylindrical indenter [22] is brought in contact with a soft or rigid adherent and pressed against it. In the other version of the experiment, a rigid indenter is brought in contact with a semi-infinite block of the adhesive. The contact load is measured both during loading and unloading of the contactors, along with the measurement of the contact area. A rigorous theoretical analysis then yields the adhesion strength of the adhesive as well as the effective modulus of the adherents. The theory has also been developed for finite thickness of the soft adhesive layer [23]. This test has the advantage that a very small area of contact appears between the two adherents; as a result, adhesion measurement is minimally influenced by surface heterogeneity. Accurate measurement of interfacial interactions [24], especially the hysteresis [25] of the adhesive surface can be carried out using this test.

17.2.4 Shear Test

Many applications demand that the adherents be shear loaded as opposed to the earlier geometries in which the adhesive has to resist failure against the normal load. To estimate this particular feature of the adhesive, single- and double-overlap shear tests and other variants of the test have been proposed. Here again, the shear load is measured at different shearing rates from which the adhesion strength of the adhesive is obtained by measuring the area under the load versus displacement plot.

17.3 Splitting of Contact Area between Adhesive and Substrate

Natural surfaces as well as most engineering surfaces can hardly be smooth always. In fact, roughness of these surfaces ranges over several different length scales spanning over several orders of magnitude [14]. On the other hand, most interactions between adherents die out within a very short separation distance, within a few nanometers [26]. Therefore, for adhesion to occur, it is imperative that the surfaces come into atomic contact, which necessitates that at least one of the surfaces be soft enough, in fact be liquid-like, so that it can follow the roughness of the other. The second possibility is that the adhesive may be made of a hard material but is fragmented hierarchically from macroscopic dimensions to fine hairlike structures, which are flexible enough to follow the surface roughness. Although most conventional adhesives are of the former kind, natural adhesives have perfected the art of using either or both of these mechanisms through millions of years of optimization and natural selection. For example, Figure 17.2 shows the contact organs of various insects that differ substantially in their shapes and forms [27]. Although the adhesive pads of some insects appear hemispherical (Figure 17.2a) or ellipsoidal (Figures 17.2c and 17.2f), many of which are made of soft and pliable tissue, these have evolved to form hair-like structures, yet for others the pad consists of flat, smooth, polygonal outgrowths that remain separated by deep trenches from which a liquid is secreted to keep the contact area wet. Figure 17.2a depicts a hemispherical adhesive pad called *pulvillus* found in insects such as *P. apterus* and *Coreus marginatus* of the family Heteroptera. The pulvillus is smooth but extremely soft so

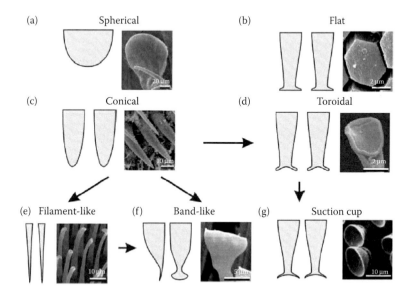

FIGURE 17.2
Shapes of attachment devices in nature and their hypothetical evolution paths (shown as arrows) [27]. (a) Bug *Pyrrhocoris apterus*, smooth pulvillus. (b) Grasshopper *Tettigonia viridissima*, surface of the attachment pad. (c) Fly *Myathropa florea*, unspecialized hairs on the leg. (d) Fly *Calliphora vicina*, seta of the pulvilli. (e) Beetle *Harmonia axyridis*, seta of the second tarsal segment. (f) Beetle *Chrysolina fastuosa*, seta of the second tarsal segment. (g) Male beetle *Dytiscus marginatus*, suction cups on the vertical side of the foreleg tarsi. (From Spolenak, R., Gorb, S. N., Gao, H., Arzt, E. Effects of contact shape on the scaling of biological attachments, *Proc. R. Soc. London Ser. A.* 461, 2005, 305–319. With permission.)

that it can generate strong adhesion on the smooth surface of stems and leaves of plants [6]. Questions arise as to how much force is required to separate the pulvillus from a smooth substrate and how many such attachment devices would generate strong enough adhesion to allow an insect of ~100 mg to hang upside down from these surfaces [28]. Importantly, these quantities are to be estimated for a finite total area of contact because every species has only a limited surface area available for adhesion by virtue of its particular shape and size. The critical force F_c that is required to separate the hemispherical pulvillus from a smooth substrate can then be estimated under the assumption that the spherical adhesive loads normal to the surface without any shearing or peeling action. From the theory of contact mechanics [29–31], this force is estimated as $F_c = (3/2)\pi RW$, where W is the work of adhesion that depends on the specific intermolecular interaction at the interface and R is the radius of contact of the two adherents. Assuming that the adherents here interact via the van der Waals force of attraction, a representative value of work of adhesion [32] is $W = 50$ mJ/m^2. Furthermore, noting that the average radius of contact area $R = 100$ μm, the critical load is calculated as $F_c = 2.5$ mg, which is significantly smaller than the weight of an insect ~100 mg. Therefore, to hang from a ceiling, the insect would require a large number of attachment pads. In fact, the insect has to account for the eventual failure of some contacts, possible damage of some seta, and occasional loss of contact due to the presence of dust speckles [33–35]. An estimation of the required number of such attachment pads can be made by considering that the area of contact is split into n equal areas, which implies that the radius of contact of each such small segment is $R' = R/\sqrt{n}$. The critical force required to separate such a segmented surface is then estimated as $F_c' = n(3\pi/2)(R/\sqrt{n})W = (3\sqrt{n}/2)\pi RW$. This simple analysis presents us an interesting result that the contact load for separation of the adherents increases if the contact area is split into smaller segments although the total contact area remains unaltered [32]. Similar results can be deduced for different other shapes of adhesive pads. For example, Figure 17.2b shows the smooth, flat adhesive surface of grasshopper *Tettigonia viridissima* [35]. In the language of adhesion science, such a surface is called *flat punch* for which the separation force can be estimated as [26] $F_c' = n^{1/4}\sqrt{8\pi ER^3 W}$. Among other parameters, this relation predicts that the pull-off force F_c' increases with the effective modulus E of the adhesive and the number of segments n, although its dependence on n is less pronounced than that for the hemispherical pad. It then implies that to achieve the same separation force, the contact area is required to be divided into more smaller segments as is indeed observed for the insect *T. viridissima*; its hexagonal segments are at least 40 times smaller than that for *P. apterus*, which has the hemispherical pulvillus [32,33]. It has been observed that the unit size of the adhesive segment consistently decreases, that is, the number density N of such units increases as the volume of the living creature increases. In fact, the areal density of the terminal attachment device is found to increase with the body mass m of the living creatures as $N(\text{m}^{-2}) \sim m(\text{kg})^{2/3}$ over several orders of magnitude of m and for species of different lineages, for example, for fruit flies of mass 10 mg to insects of ~1 g to the gecko of mass ~100 g. Although for the gecko the setal density is ~10^{13} per m^2, for beetles it decreases to ~10^{10} per m^2. In other words, setal density is directly linked to the adhesion requirement of the organism. This mechanism is known in the literature as "contact splitting," which rationalizes patterning; however, the hexagonal outgrowths of *T. viridissima* are not made of one of the hardest materials available in nature, for example, keratin, which has a modulus of 1 GPa. In fact, it is made of a very soft material of elastic modulus ~30 kPa, which does not quite ensure enhanced pull-off force as suggested by the above expression for F_c'. Nevertheless, a soft material is necessary for the insect as it renders its adhesive pad pliable enough to comply with the roughness of surface. It is worth mentioning here that unlike the assumptions made for the "contact-splitting" mechanism, for most natural adhesives, neither the adhesive is pulled off normal to the surface nor does it make smooth contact with the substrate. Furthermore, many adhesive pads are either shaped like suction cups, as shown in Figure 17.2g for beetle *Dytiscus marginatus*, or have a cup-like structure at the end of their hairs (Figures 17.2d and 17.2e), for which the dependence of total pull-off force on the segment number n vanishes as can be depicted by the equation [27] $F_c' = \Delta P \pi R^2$. Here, ΔP is the vacuum pressure inside the suction cup. Thus, while the "contact-splitting" mechanism brings out the fundamental necessity for patterning an adhesive, it possibly misses out many other subtleties related to the mechanics of adhesion. In other words, it is not enough to merely split an adhesive surface; the shape of the adhesive segments, their deformability, and the spatial length scale of the substrate are also equally important design parameters. The following experiment brings out these issues quite elaborately.

17.4 Effect of Surface Pattern on Adhesion Mechanism

The influence of surface structures on adhesion has been investigated by a variety of contact-mechanics experiments, for example, indentation of a model adhesive using a rigid or a soft indenter, peeling off a flexible adherent, interfacial shearing under constant load, and so forth [36–39]. Figure 17.3 summarizes the results obtained from experiments in which an adhesive layer, a thin elastomeric film of polydimethylsiloxane (PDMS), bonded to a rigid substrate is patterned with incisions made using a sharp razor blade [36,37]. The incisions are created as a one-dimensional discontinuity spaced at regular intervals of 0.5–5.0 mm along the length or width of the film or as a two-dimensional mesh. The incisions are made either throughout the thickness of the adhesive layer or only to a desired depth; the incisions are made normal to the surface of the film or are inclined to a desired degree. Thus, several different types of geometric patterns are generated on the adhesive using the incisions. Although the simple incision does not mimic the hierarchical fibrillar structure or even the polygonal outgrowths of the naturally occurring adhesives, it conforms to the basic principle of these structures to alter the mechanics of adhesion. In a typical set of experiments, elastomeric layers with controlled thickness ($h = 40$–200 µm) and shear modulus ($\mu = 0.2$–3.1 MPa) remain strongly bonded to a rigid substrate, for example, a microscope slide, whereas a flexible plate, for example, a microscope coverslip of flexural rigidity $D = 0.2$–1.2 Nm, is brought in complete contact with it. It is worth mentioning here that flexural rigidity [40] is a measure of deformability of thin plates and is defined as $D = \mu_p t_p^3 / 6(1 - v_p)$, where μ_p, t_p, and v_p are, respectively, shear modulus, thickness, and Poisson's ratio of plate. The flexible plate is then lifted from its hanging end at a displacement-controlled experiment using a motion control stage while the applied load is measured using a weighing balance interfaced with a computer. The plate is flexible enough that a reasonably small load is enough to peel it off the adhesive but rigid enough that it undergoes small bending during peeling. Small bending ensures applicability of a simple theory and thus allows precise estimation of the interfacial adhesion strength directly from the force versus displacement measurements. In a similar experiment, the adhesive film bonded to the flexible backing of rigidity D_1 can be peeled off a second flexible plate of rigidity D_2. It is easy to show that this experiment is no different from the former with the flexible backing replaced by one of infinite rigidity and the flexible adherent replaced by one of rigidity [41,42] $D_{eq} = (D_1 D_2)/(D_1 + D_2)$.

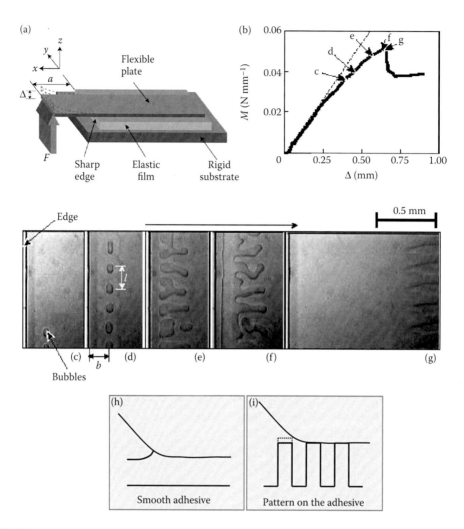

FIGURE 17.3
(a) Schematic of the peel experiment in which a flexible plate is lifted at a controlled rate off an elastic adhesive film, which remains bonded to a rigid substrate. Peeling is carried out from the sharp edge of the film. (b) The applied torque, $M = Fa$, plotted against the displacement Δ of the hanging end of the plate. (c–g) Video micrographs depicting sequence of morphology changes in the contact zone as the peeling force is increased. The images correspond to a film of thickness $h = 40\ \mu m$ and shear modulus $\mu = 1.0$ MPa and a plate of rigidity $D = 0.02$ Nm. (h and i) The schematic of the contact between the adherent plate and the adhesive depicts the deformation for smooth adhesive layers and the ones with patterns. (Adapted from Ghatak, A., Mahadevan, L., Chung, J., Chaudhury M. K. and Shenoy, V., Peeling from a biomimetically patterned thin elastic film. *Proc. R. Soc. London Ser. A.* 460, 2004, 2725–2735. Reproduced with permission from The Royal Society.)

On a smooth layer of adhesive, used as control, the peeling front propagates continuously with continuous lifting of the contacting plate that corroborates with our everyday experience with most adhesion applications, for example, the ubiquitous adhesive tape. However, on a patterned adhesive, the peeling does not progress continuously but intermittently with complete arrests at the vicinity of the incisions followed by catastrophic initiations beyond a critical peel-off load. Figure 17.3b shows the corresponding torque versus displacement plot, which shows that the lifting torque increases with the increase in the displacement of plate while the crack front remains arrested. For adhesive layers with multiple parallel incisions oriented in a direction perpendicular to that of propagation of the peeling front, multiple peaks appear corresponding to crack arrest and initiation from each incision. In contrast, for a smooth and unstructured adhesive, the first peak corresponds to crack initiation from the edge of the adhesive, beyond which the torque required to drive the crack decreases to a smaller value than that required to initiate it. It is interesting to note that during peeling from the sharp edge of the incision, stress concentration occurs at a distance away from the edge of the film. As a result, a crack initiates by the appearance of cavities that grow with continued lifting of the plate and eventually coalesces, leading to the appearance of the cusp-shaped crack (Figures 17.3c through 17.3g). Following linear elasticity theory for an incompressible elastic material, this distance has been shown to increase with a characteristic length scale [41] $q^{-1} = (Dh^3/12\mu)^{1/6}$. The length scale q^{-1} obtained as the ratio of deformability of the plate and that of the elastic film has a significance: it is a measure of the spacing between the incisions below which the effect of individual incision is not felt. Typical values of q^{-1} can be obtained by putting representative values for various parameters: $D = 0.02$ Nm, $\mu = 1.0$ MPa, and $h = 100$ µm; q^{-1} is calculated as 350 µm. This fact is demonstrated in Figure 17.4a in which the adhesive surface is "chocolate" patterned, with the interincision spacing maintained at 200 µm and the depth of incision at 3 µm. Peeling the flexible plate off this layer does not result in the characteristic stick-slip behavior. In fact, the crack propagates smoothly with a constant torque applied on the plate, but its value remains significantly larger than what is observed on a smooth adhesive, implying a stronger adhesion at the interface.

The adhesion strength for these experiments is estimated by measuring the area under the corresponding force versus displacement plots. Figure 17.4b shows the G values calculated for adhesive layers of increasing thickness $h = 40–1000$ µm and maximum depth of incision 100 µm. These values are considerably larger than those calculated for smooth

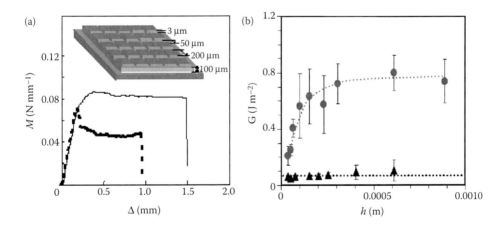

FIGURE 17.4
(a) The solid line represents M versus Δ plot for peeling off a pattern adhesive as shown in the inset. The dashed line represents that for a smooth layer of adhesive. (Adapted from Ghatak, A., Mahadevan, L., Chung, J., Chaudhury M. K. and Shenoy, V., Peeling from a biomimetically patterned thin elastic film. *Proc. R. Soc. London Ser. A.* 460, 2004, 2725–2735. Reproduced with permission from The Royal Society of Chemistry.) (b) Adhesion strength calculated from corresponding force versus displacement plots for adhesives with increasing thickness. The filled circles and triangles represent experiments with patterned and smooth adhesives, respectively. (Adapted from Chung, J. Y. and Chaudhury, M. K., Roles of discontinuities in bio-inspired adhesive pads. *J. R. Soc. Interface* 2, 2005, 55–61. Reproduced with permission from The Royal Society.)

layers represented by the dashed line. In fact, *G* increases with *h*, eventually attaining a plateau value that is ~10 times of that obtained for smooth adhesives.

A question arises, what leads to this enhanced adhesion strength of the patterned adhesive? In other words, how does the energy get dissipated in an elastic system in the absence of any viscous effect? This has to do with the incisions or the discontinuities in the adhesive that act as a barrier for elastic energy to get transferred from one portion of the layer to the next. For a crack to propagate, the tensile stress at the crack tip is always required to be maintained at a critical level. For smooth propagation on a continuous layer, elastic energy at the crack tip gets always transferred ahead of it (Figure 17.3h), resulting in smooth propagation of the crack. However, on an incision-patterned film, the elastic energy does not get transferred. As a result, while the portion of the film ahead of the crack is sufficiently stressed to drive the crack, the portion behind relaxes under zero load (Figure 17.3i). Thus, each segment of the adhesive film has to be separately stretched in order to drive the crack on it; similarly, each portion relaxes back without transferring its elastic energy, leading to its dissipation. This situation is very similar to the classical Lake–Thomas effect [43] in the context of rupture of polymeric chains: although all the bonds in a polymer chain are strained, when a bond breaks, all the others relax under zero load, leading to dissipation of energy.

17.5 Dry Adhesion at Animal Feet

Although the "contact-splitting" mechanism directly shows that for many different shapes of adhesive pads, splitting or patterning the adhesive surface should indeed enhance adhesion, the "crack initiation and crack arrest mechanism" shows more fundamentally how surface patterns actually help in arresting the peel front from propagating. It is then appropriate to examine to what extent these mechanisms may be applicable for biological adhesives. However, instead of describing the biological adhesives in general, we focus here on the pads of the most amazing climbers in nature, the geckos, whose adhesive is very minutely split to the length scale of a few nanometers.

The adhesive pads of geckos have been a subject of much interest because of their versatility and effectiveness [44,45]. The gecko has four legs, each having five toes that are nimble enough to allow it to create both normal loading and peeling action on a substrate (Figure 17.5). The adhesive pad on each toe consists of rows of overlapping lamellae that remain supported by blood vessels. The lamellae contain millions of hair-like structures called *setae*, which are made of an elastic protein called *keratin*. Similar to the adhesive pads of many insects, the hairs are hierarchically structured with an approximate diameter of 20 µm and length 130 µm. Each of these hairs branches out into thousands of smaller fibers that flatten out in the end with spatula-like structures of approximate width 200 nm and thickness 5 nm. Although the agility of the gecko's toes allows it to bend around a centimeter-scale curvature, the blood vessels beneath the adhesive pads can orient the lamellae, suitably allowing them to follow closely millimeter-scale roughness. The setae can negotiate milli- to microscale, and the thin and flexible spatulae can perfectly conform with nanoscale roughness. Experiments show that the gecko's hair can adhere well to both highly hydrophobic and hydrophilic surfaces, signifying that the dominant adhesion mechanism here is the van der Waals force of interaction [46–49]. Each of these hairs creates approximately 10^{-7} N force; millions of hairs together create 10 N/cm², which is enough to allow the gecko to walk upside down on the ceiling. The van der Waals interaction also ensures that for

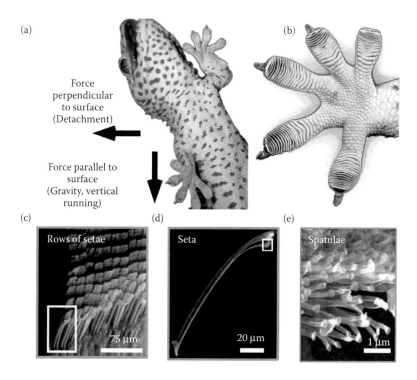

FIGURE 17.5
(a–e) The hierarchical structure of the adhesive pad at the gecko feet. (From Autumn, K. and Peattie. A. M., Mechanisms of adhesion in geckos, *Integr. Comp. Biol.* 42, 2002, 1081–1090, with permission of Oxford University Press.)

gecko hair, not the surface chemistry but the geometrical features such as size and shape of the adhering element play an important role. In addition, the material properties of the adhesive are also important. Because the gecko's foot is elastic, as it is made up of an elastic protein called keratin, it allows strong adhesion and quick debonding from a surface, which is essential for rapid locomotion on vertical or horizontal walls. A question arises, how does the gecko keep its adhesive surface clean? Experiments with actual gecko feet and spherical dust particles show that this property of the gecko also has to do with the nanoscopic structure of the setae because of which individual dust particles attach to fewer setae than that required to overcome the adhesion between the particles and the substrate [50]. Besides, in the dynamic situation during locomotion, rolling and sliding of the particles and relative movement of hairs help keeping the contact area of adhesion low [51]. To summarize, seemingly conflicting demands are met because of the hierarchical structure of the gecko hairs that interact with the substrate via van der Waals force.

17.6 Gecko-Inspired Artificial Mesoscopic Polymeric Hairs

Mimicking gecko hairs is an attractive proposition in the context of bioinspired adhesive; nevertheless, its design and fabrication are not trivial. Despite apparent understanding of the fundamental mechanism of adhesion, difficulty arises because of the conflicting characters that epitomize, the bioadhesives. For example, although it is imperative that the surface coverage of the hairs be large enough for enhancement of adhesion at the interface, it should be small enough that the hairs do not collapse on each other, thereby rendering them ineffective for repeated use; similarly, although it is essential that these hairs be flexible enough and soft so that they all attach together to an uneven surface, in fact, all real surfaces are rough or uneven, yet they should be hard enough that they do not undergo permanent deformation, yielding, and eventual breakage during cyclic adhesion and debonding. Noting that the flexibility of fibers increases with their length, the threshold length h_f of fibers beyond which they are expected to self-adhere has been worked out [52–54] by balancing the bending energy of the fibers and the energy of adhesion of a pair of them: $h_f < 2.133(\mu a^{5/2} w^{3/2}/W_A)^{1/3}$, where $2a$ is the diameter of the circular cross-section of fibers that remain separated by a distance $2w$. The dependence on modulus μ and the adhesion energy W_A suggests that the materials of construction of the gecko-inspired hairs also demand contrasting requirements, for example, the hairs should cause strong adhesion with a substrate yet should not get contaminated by dust and particulate matter. Fabricating submicroscopic hairs on a large area is also an issue. In fact, although there has been some success in generating a small area pattern with regular cylinders, generating hairs with a large length-to-diameter ratio and generating patterns with nonuniform and nonregular cross-sectional hairs with hierarchical architecture has not yet been possible.

Figure 17.6a shows an example of gecko hairs prepared by using a replica-molding technique [55]. Here, a suitable template, for example, an AFM tip, was used to generate a negative pattern on a suitable material, for example, wax. By cross-linking or solidifying a liquid polymer on this mold, an array of hairs could be generated. Similar polyimide hairs were also prepared by different lithographic routes. AFM measurement of the normal load of separation from each hair was found to be ~200 nN, which shows that for a typical spatial density of three to four hairs per square micrometer, the adhesion strength should exceed 500 nN/µm². However, in a typical experiment in which the polymeric hairs are generated on a hard substrate, for example, silicon wafer, the adhesion strength is found to be far less, ~10 µN, implying that only a small fraction of hairs adhere to the substrate. Figure 17.6b shows that self-adhesion of hairs indeed acts as a spoiler for adhesion, rendering the adhesive unsuitable for reuse; importantly, the hairs do not all attach to the surface. The adhesion strength, however, improves drastically when the hairs are transferred to a soft backing, as in a tape, which signifies that the number of hairs that participate in the adhesion process increases severalfold because of the compliance of the backing.

(a) 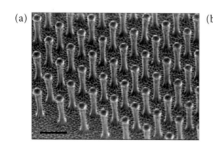   (b)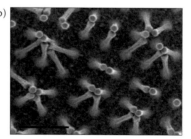

FIGURE 17.6
(a) Scanning electron microscopy of typical artificial gecko hairs prepared using replica-molding technique. (b) Self-adhesion leading to collapse of flexible hairs. Scale bars represent 2 µm. (Reprinted by permission from Macmillan Publishers Ltd. Geim, A. K., Dubonos, S. V., Grigorieva, I. V., Novoselov, K. S., Zhukov, A. A. and Shapoval, S. Yu., Microfabricated adhesive mimicking gecko foot-hair, *Nat. Mater.* 2, 461–463, copyright 2003.)

Nevertheless, the geometry of these artificial pillars does not follow that of their natural counterpart quite adequately. Importantly, in natural adhesives, the spatulae-like terminal flap on the hairs plays a critical role that is missed by the artificial ones [56–59]. In fact, this terminal structure plays multiple roles: it enhances the area of contact at the interface, yet it keeps the neighboring pillars separated enough, so that collapsing of hairs would result in a large elastic energy penalty both for the pillars and the flaps. Thus, self-adhesion of the pillars and resultant loss of usability of the adhesive are prevented. Furthermore, thinness of the flap enhances the flexibility and local compliance, which allows intimate and tenacious contact with the adherent. In other words, for a more compliant system, the peeling load that gets transferred to the crack tip decreases, enhancing the arresting effect on the interfacial crack. In fact, the spatulae may have other functionalities, for example, the self-cleaning effect of the adhesive pad, directional adhesion, and so forth, which have not yet been studied, explored, or mimicked.

Notwithstanding these benevolent characteristics, the fabrication of the flap on the top of each pillar is a difficult task to accomplish. Therefore, instead of creating a flap for each pillar, the above problem has been circumvented by creating a continuous thin elastic film on the hairy surface [60,61]. Figure 17.7 shows the schematic of such a construct of the adhesive surface. Here, first the array of pillars is generated via lithographic routes on a soft base in which the pillars remain arranged in hexagonal packing. The array is then placed on a thin liquid film of PDMS, which is spin-coated on a substrate. By capillary action, the hairs get partially moistened by the liquid, which is then cured thermally, resulting in an array of pillars that remains coated by a thin elastic film. Although such a design does not quite represent the setal system in biology, it performs many of its activities.

The effectiveness of this adhesive is demonstrated by contact-mechanics experiments, in which a spherical indenter is pressed against the adhesive layer at displacement-controlled loading while the contact load is measured using a weighing balance. Typical load versus

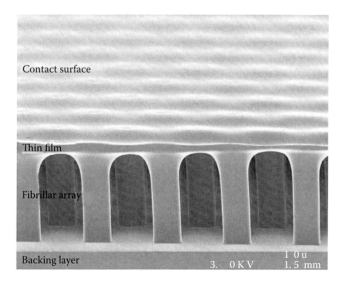

FIGURE 17.7
Synthetic fibrillar adhesion surface, including spatular features. Shown is an array of micropillars with a terminal film. (From Glassmaker, N. J., Jagota, A., Hui, C.Y., Noderer, W. L. and Chaudhury, M. K., Biologically inspired crack trapping for enhanced adhesion, *Proc. Nat. Acad. Sci.* 104, 10786–10791, Copyright 2007, National Academy of Sciences U.S.A.)

displacement plots during cyclic loading and unloading show that at the same indentation depth, the magnitude of the compressive force decreases during unloading. Eventually, the compressive load turns tensile because of adhesion between the indenter and the sample. The area under the loop of the load versus the displacement plot is a measure of the hysteresis. The area of contact viewed through an inverted optical microscope also shows a significant difference from that of a flat control sample. Although for the flat control adhesive, the contact area is expectedly circular with increasing radius of contact with increase in contact load; for the fibrillar surface the contact area appears hexagonal because of the corresponding arrangement of the fibers. Although retracting, for the control sample, the load immediately starts to decrease; however, for the fibrillar surface, the contact line gets pinned. These observations signify an increase in compliance for the fibrillar sample that, in fact, increases with an increase in the spacing between the fibers. Experiments show that it is possible to enhance the adhesion strength by a factor 9 over what is achieved on a smooth adhesive surface.

17.7 Wet Adhesion: Role of Liquid Film on Adhesive Pad

Now, there is ample evidence that for many living organisms, dry adhesion mediated by various interactions, for example, the van der Waals force, is not the only mechanism of sticking to the substrate. Especially, for insects such as cockroaches, grasshoppers, and many others that possess a soft and smooth attachment device, the dry adhesion mechanism is supplemented by a nonvolatile liquid secretion at the contact surface [62–65]. These insects have also developed a sophisticated mechanism by which the liquid is delivered in a controlled manner at the contact surface. For example, in bush cricket *T. viridissima*, behind the adhesive pad called *euplantulae*, there exist epidermal cells that secrete a liquid that is transported to the pad surface through the trenches that exist between their hexagonal outgrowths [64]. In flies, at the base of each pulvillus, there are large cells that secrete the liquid into the spongy cuticle layer. From there, it is delivered to the contact surface by porous channels. For the damselfly, liquid is secreted through its porous setae. Not much is known about the chemical character of the liquid, except that it is an emulsion consisting of both hydrophilic and hydrophobic phases that possibly allow the liquid to spread on surfaces of a wide range of wettability [65]. It is conjectured that insects possibly take advantage of both the surface tension and the viscous effects of the sandwiched liquid film to guard against pull-off and shear loads that act normal and parallel to the surface, respectively. Marine animals such as echinoderms depend on liquid secretion for both adhesion and debonding from a substrate [66]. Although the adhesive secretion is delivered through the cuticle of its attachment pad to the surface of the substrate, the de-adhesion secretion is released within the cuticle. The de-adhesion liquid is an enzyme because of which the outer layer is left behind on the substrate as a fuzzy coat, during separation from the substrate.

In the context of adhesion of terrestrial animals, the exact role of the liquid film is, however, far from clear because the liquid film is expected to influence the static and dynamic friction at the interface with a substrate in a conflicting manner. For example, the presence of a lubricating fluid between the adhesive pad and the substrate should actually decrease the shear resistance of the adherent; in other words, the liquid film has a negative effect on the static friction [5]. However, it is also true that the frictional effect of the liquid film should increase with the rate of shearing, implying that dynamic friction should increase with the sliding velocity. Further complication arises when we consider the amount of liquid required to be present at the contact area for rough and smooth surfaces. For example, for a smooth substrate, both the shearing load and the pull-off load should increase with a decrease in the thickness of the liquid film. However, if the substrate is rough, existence of a thicker film is more desirable.

For an ideal situation of two smooth adherents perfectly wetted by a liquid film, the above set of expectations can be summarized in the form of the following mathematical representation: the frictional resistance exerted by the liquid film of thickness h_f and viscosity η against shear over a distance L can be written by simplifying the two-dimensional Navier–Stokes equation under the lubrication approximation [67], as $\tau \sim \eta V L / h_f^2$, where V is the relative sliding velocity of one plate with respect to the other. Here, it is assumed that the liquid film wets both the adherents. In the language of fluid mechanics, it is termed as the "no-slip" condition, adherence to which, in the context of insect adhesion, depends on the chemical nature of the secreted liquid. Assuming that the liquid completely wets both the surfaces, we can obtain the capillary pressure generated by the liquid film $\Delta p_c \sim 2\gamma_f / h_f$, in which γ_f is the surface tension of the liquid and $2/h_f$ is the curvature of the meniscus of the sandwiched liquid film. In other words, the pressure in the liquid is lower than the atmospheric pressure by Δp_c, implying that the corresponding amount of normal force needs to be applied to pull off the two surfaces. Thus, both the viscous and the capillary effects show that the thickness of the liquid film present between the two adhering surfaces is an important parameter in the context of resistance against failure. In addition, the insect must also guard against the loss of liquid due to drying, which can happen only if the exactly right amount of liquid is present at the interface. As mentioned earlier, the liquid is supplied by subsurface fluidic vessels that terminate into tiny pores open to the atmosphere; it has been suggested that the diameter of these channels and that of the pores that are used for delivering the liquid between the adhering surfaces determine the amount of liquid to be present at the contact surface [68,69]. For example, according to the above definition of capillary pressure, for a channel diameter w smaller than the gap $h_f (w < h_f)$ between the surfaces, the pressure in the liquid inside the channel ($p_{channel} = P_{atm} - 2\gamma/w$) becomes smaller than that between the contacting surfaces ($p_{gap} = P_{atm} - 2\gamma/h_f$), which results in liquid being sucked from the gap between the surfaces to inside the channel; this effect gets reversed for $h_f < w$. These arguments are, however, very simplistic, as the quantity of liquid present in the contact area is a function of many other parameters, for example, roughness of the surfaces, contact pressure or normal preload, and rate of drying of the liquid from the gap between the surfaces. As an example, an increase in compressive preload on the adhesive pad leads to an increase in secretion of liquid, which adversely affects the resistance against normal and shear load; however, normal load should increase the frictional resistance at the interface, which prevents shear failure.

Let us examine how these conflicting situations actually affect insect adhesion. Controlled experiments on insects, for example, *O. smaragdina*, show that the dynamic frictional force increases linearly with the velocity of sliding, corroborating with the expression of liquid viscous stress τ as presented earlier [70]. In fact, ants can modulate the resistance to shear by varying the contact area with the substrate and can generate a shear force as large as 500 times their own body weight [5]. The resistance against pull-off load is, however, smaller than the shear resistance, which possibly occurs because the capillary effect is less pronounced. However, experiments also indicate the existence of a large static friction, almost 200 times their own body weight, which cannot be accounted for by the liquid viscous stress. On the contrary, the presence of liquid should actually decrease the static friction. Preliminary experiments show that the static friction does not change with temperature, although the dynamic frictional resistance decreases as the temperature is increased. These observations imply that the resistance against shear failure possibly has less to do with the liquid film present at the interface of the adhesive pad of the insect and the substrate but more to do with the liquid-less dry friction at the interface. In fact, experiments with rubber-like deformable material, indeed, show the existence of a static friction and frictional resistance increasing nonlinearly with the sliding velocity [71–73]. At any sliding velocity, friction increases with decrease in temperature. In a biological context, the cuticle of the adhesive pad is indeed a deformable rubber-like material that slides on a rigid substrate. Thus, the "rubber friction" model for the insect adhesive pad admits doubt as to what extent "wet adhesion" may be playing any role for many insects. Clearly, more definitive experiments are needed to resolve these issues.

17.8 Frictional Resistance against Shear Failure

Nevertheless, dry or wet frictional resistance against shear failure is important for most biological adhesives as the terrestrial animals such as insects, geckos, and the frogs have to negotiate with a variety of surfaces including vertical walls. Friction, in fact, is implemented via several length scales of the microstructure of the adhesive pads: although in the macroscopic length scale it is the claws of the feet that cling onto rough surfaces like a hook, in the microscopic length scale the spatulae at the end of the setae become effective via adhesion with the surface. In other words, the origin of friction is same as that of rubber friction, that is, intermolecular interactions, for example, the van der Waals interaction. How effective are the synthetic surface-patterned and fibrillar adhesive surfaces in imitating this aspect of biological adhesives? As described earlier, polymeric pillars with a high aspect ratio are rather soft and susceptible to self-adhesion and collapse, which render them unsuitable for generating strong shear resistance in repeated applications. In contrast, the hairs of gecko are made of keratin having a modulus in the range of 1–2 GPa. The stiffness of these biomaterials has motivated the use of carbon nanotubes as the material for generating hairy surfaces [74–79] (Figure 17.8). The carpet of multiwalled, vertically aligned, carbon nanotubes (diameter 10–20 nm, length ~65 μm) grown on rigid and soft backing

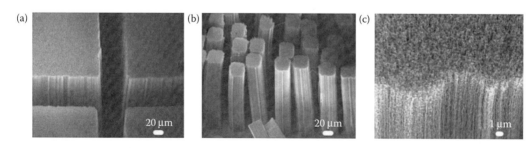

FIGURE 17.8
Scanning electron microscopy images of densely packed carbon nanotube bundles grown on a micropatterned surface. The images are taken at different spatial resolutions. (From Sethi, S., Ge, L., Ci, L., Ajayan, P., M. and Dhinojwala, A., Gecko-inspired carbon nanotube-based self-cleaning adhesives, *Nano Lett.* 8, 2008, 822–825. With permission.)

has shown significantly large adhesion in peeling and shear experiments on a variety of surfaces including Teflon: $\sim 1.6 \pm 0.5 \times 10^{-2}$ nN/m^2, which is about four times that generated by gecko setae. The genesis of this large adhesion strength has been linked not only to the van der Waals interaction but also to the energy of dissipation during elongation of the nanotubes. This implies that the adhesion strength here depends on the length and diameter of the carbon nanotubes, the packing of the nanotube bundle, the size and shape of the bundle, and the stiffness of the backing tape. In addition to adhesion, the roughness of the nanotube carpet is also important for imparting superhydrophobicity [75,76] to the adhesive surface. Thus, the nanotube carpet serves the dual purpose of generating large adhesion strength and superhydrophobicity, which renders the adhesive less susceptible to particulate contamination, easily washable with water, and thus suitable for repeated use.

It should be noted that the nanotubes interact with the adhering surface at its top surface, that is, via tip contact, whereas for the purpose of shear resistance, it may be more desirable if the hairs are allowed to interact at their lateral surface, that is, via side contact. Thus, the true contact area for side contact is larger than that for the tip contact, and the contact area can increase with the prenormal load on adhesive because of the increased bending of the hairs [77–80]. This effect has been demonstrated by fabrication of a micropillar array made of polypropylene, which shows significant enhancement of shear adhesion. The shear adhesion, in fact, increases with shear load on the adhesive and also with repeated use because of the increase in the contact area.

17.9 Role of Subsurface Microstructure of Adhesive

So far, our discussions have focused on the surface patterns of the adhesive, shape and size of setae, number density of attachment hairs, and so forth. We have also discussed the role of wet adhesion as mediated by the secreted liquid at the interface. We will now look into what role the microstructures, particularly internal structures of the adhesive pads, play on adhesion. For example, shock-freezing experiments of the adhesive pad of insect *T. viridissima* [81,82] have shown (Figure 17.9) that its hexagonal pads are of diameter 5 µm, each of which remains separated from its neighbor by a trench of width ~1 µm. The hexagonal pads, however, do not sit directly onto the feet of the insect but via a block of dendritic, hierarchically branched rod-like structures in which main rods of diameter 1.12 µm finally split into branches of diameter ~0.08 µm. The rods of the dendritic mesh remain oriented at an angle of approximately 70° to the surface and remain immersed in a liquid called *hemolymph*. It is easy to think that such a structure provides flexibility and degrees of freedom to the pad, allowing it to negotiate with surfaces of different orientations and microroughness features. Along with these dendritic structures, these adhesive pads also have subsurface air pockets and liquid-filled vessels as shown in Figure 17.10. Not much is known about the role of these subsurface fluid chambers in adhesion, except that they supposedly store the liquid to be secreted at the contact surface through pores located at the trenches.

The direct role of subsurface liquid-filled vessels on interfacial adhesion is more evident for insects such as *Rhodnius prolixus* [4], the "kissing bug," which is known to suck blood. It has been known for long that the adult insects have an orange-colored fleshy pad at the lower end of the tibia of their first two pair of legs, in which they store blood. It has been observed that the adult can climb the smooth, shiny surface of a glass jar, whereas the nymphs without such a fleshy sticking organ fail to do so. Although this organ is known

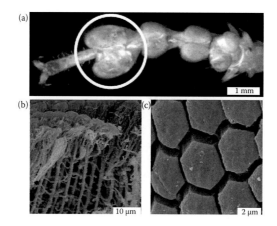

FIGURE 17.9
Microstructure of adhesive pad of bush cricket *Tettigonia viridissima*. (a) Complete leg section containing the adhesive pads. (b) A shock-frozen pad is fractured to show the dendritic mesh that supports the adhesive pad. (c) Top view of the pad surface. (From Scherge, M. and Gorb, S. N., Using biological principles to design MEMS, *J. Micromech. Microeng.* 10, 2000, 359–364. With permission.)

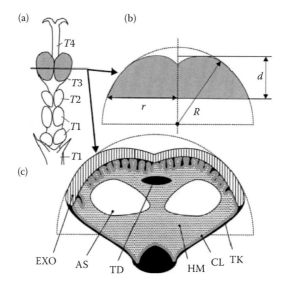

FIGURE 17.10
(a and b) Schematic of a typical adhesive pad of insect *Tettigonia viridissima*. (c) Cross section of the pad shows, among other things, the presence of air sacs (AS) and regions containing hemolymph (HM). (With kind permission from Springer Science + Business Media: *J. Comp. Physiol. A.*, Ultrastructural architecture and mechanical properties of attachment pads of *Tettigonia viridissima* (Orthoptera Tettigoniidae), 186, 2000, 821–831, Gorb, S., Yuckan, J. and Scherge, M.)

to be an elastic sac distended with fluid, it is not clear how it helps the insect to cling to smooth surfaces. Similarly, the spider tarantula is known to have silk-producing glands at its feet [11]. These glands generate fine hairs that the spider uses to adhere to rough vertical surfaces. Although the role of these surface hairs on adhesion is very similar to that in other insects, the role of the subsurface silk-producing glands on adhesion has not yet been recognized.

17.10 Adhesion on Microchannel Embedded Adhesives

Nevertheless, the internal fluid-filled structures in the adhesive layers may be significant, for they not only help in manipulating the contact pressure but also modulate the effective material properties, for example, elastic modulus as shown by the experiments presented in Figure 17.11. Here, microchannels of different diameters are embedded inside an adhesive layer by using suitable filaments and cylindrical rods of a variety of diameter and cross-sectional shapes as templates [83,84]. The vertical locations of the channels within the layers are controlled by placing the templates using suitable spacers. The adhesive layers with the embedded microstructures remain bonded to a rigid substrate and are subjected to a peeling experiment similar to that presented in Figure 17.3a. For air-filled microchannels, the peeling front does not propagate smoothly but rather intermittently, signifying that the channels act as a barrier for propagation of the peel front similar to the surface incisions on films. In other words, the elastic energy in the adhesive behind this front does not get fully transferred in the portion ahead of it because of the presence of the channel. Figure 17.11b shows the peeling torque $M = F \cdot a$ plotted against displacement

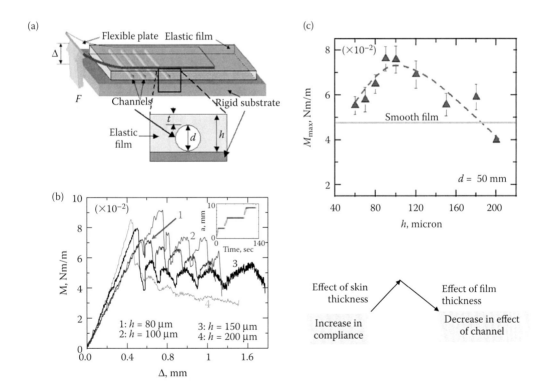

FIGURE 17.11
(a) Schematic of the peeling experiment using microfluidic adhesive. (b) Typical plot of peeling torque $M = F \cdot a$ against Δ of the flexible plate. The shear modulus of the film is $\mu = 1.0$ MPa, and flexural rigidity of plate is $D = 0.02$ Nm. (c) Variation of the M_{max} with h. Triangles represent the maximum torque at which the crack initiates, and error bars represent standard deviation. (From Majumder, A., Ghatak, A. and Sharma, A., Microfluidic adhesion induced by subsurface microstructures, *Science* 318, 2007, 258–261. Reprinted with permission of AAAS.)

Δ of the flexible plate, where a is the distance of the peeling front from the line of application of the load F. Similar to the incision-patterned films, this plot also shows that at the vicinity of the channels, the peeling front gets arrested so that M increases with Δ until a critical torque is reached, beyond which the crack initiates again with catastrophic decrease in the torque. As a result, several peaks appear in this plot. The first one corresponds to the formation of the cusp-shaped crack at the edge of the film and the subsequent ones appear because of the arresting effect of the buried channels. The adhesive strength for this layer is estimated by calculating the area under the corresponding force versus the displacement plot. Similar experiments can be carried out with adhesive layers of different thicknesses, with embedded channels of different diameters, different vertical locations, interchannel spacing, and so forth. A question arises as to what extent these geometric features affect adhesion. Figure 17.10c shows the average maximum torque M_{max} plotted against film thickness h, in which the channel diameter remains constant at $d = 50$ μm and channels remain located maximally buried within the film and filled with air. For this set of experiments, M_{max} initially increases with h in consistence with that observed for the incision-patterned films, but when the film thickness becomes sufficiently large, the effect of the channel gets vanishingly small at the surface so that the M_{max} value approaches that for the smooth, unpatterned adhesive surface. Besides the thickness of the layers, other geometric parameters in question are the lateral spacing s between the channels and their vertical location within the layer. Experiments on layers with varying lateral spacing between channels show that the effect of individual channels is not felt when the channels remain close enough, for example, $s < 2$–3 mm, so that a constant peeling torque is observed. This distance can be linked to the characteristic length scale q^{-1} defined earlier in the context of the incision-patterned adhesive. Interestingly, for an adhesive layer of given thickness, the vertical location of the channel within it does not seem to influence the adhesion significantly. In fact, experiments with different sets of film thickness and channel diameter suggest that the effective barrier for spatial transfer of elastic energy is solely characterized by the fractional thickness of the adhesive at the location of the channel, that is, $1 - d/h$. Thus, the design parameters that emerge for the adhesive embedded with air-filled channels are layer thickness, h, the net adhesive thickness at the location of the channels, $h - d$, and the characteristic length scale q^{-1}.

It is to be noted that for air-filled channels, the pressure inside the channel remains atmospheric. However, the mechanics of adhesion alter significantly when the channels are filled with liquids such as silicone oil, which wets the surface of the elastomer. Figure 17.12a depicts the top view of a typical adhesive layer in which the oil fills in the channel by capillary action. Oleophilicity of the surface of the channel leads to negative Laplace pressure at the interface of the oil and atmosphere. As a result, the pressure in oil p_0 decreases below the atmospheric pressure p_1 by $\Delta p = p_1 - p_0 = 4\gamma/d$. Expectedly, the pressure inside the channel alters the stress field in the soft deformable wall around it: although, far away from the channel, the pressure in the elastic layer remains atmospheric, that is, p_1, in the thin skin covering the channel, decreases to p_0. The skin then behaves very similar to an elastic, which buckles and bulges out under compressive axial loads, as shown by the cross-sectional images of the adhesive layer in Figure 17.12b; the cross section of the channel also changes from being circular. Figure 17.12c shows the vertical deformation δ of the surface of the skin normalized by film thickness, δ/h, plotted along direction x, which is perpendicular to the direction of orientation of the channels. The deformation flattens out for films with a large thickness, for example, $h = 300$ μm, but sharpens out to form narrow spikes for thinner films, for example, $h = 90$ μm and $d = 50$ μm. It is easy to show that for thicker adhesives, $h > 120$ μm and $d = 50$ μm; the trace of the surface bulge can indeed be

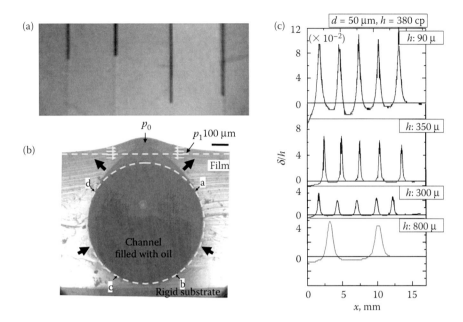

FIGURE 17.12

(a) Top view of a typical adhesive: the images of embedded microchannels being filled with silicone oil by capillary pressure. (b) Magnified image of the liquid meniscus shows that the liquid pressure inside the channel decreases from atmospheric pressure because of the capillary effect. This differential pressure causes the thin film above the channel to deform by the classical Euler buckling instability. (c) Magnified images of the microscopic deformation of the surface of the film are obtained using optical profilometry. The figure shows the dimensionless height δ/h along the spatial direction normal to the orientation of the channels. (From Majumder, A., Ghatak, A. and Sharma, A., Microfluidic adhesion induced by subsurface microstructures, *Science* 318, 2007, 258–261. Reprinted with permission of AAAS.)

fitted to the elliptic equation derived for classical Euler buckling instability of a curved elastic. However, for thinner films, the effect of capillary pressure is so strong that the trace of deformation for these films is not captured well by the elliptic equation. In fact, such sharp deformation of the adhesive surface does not allow the flexible plate to come in complete contact with the film. For the thicker films, however, the flexible plate comes in complete contact.

Peeling of a flexible plate off thick films, $h = 500$–1000 µm, embedded with liquid-filled channels, $d = 450$–810 µm, results in M versus Δ plots (Figure 17.13a), which consist of intermittent peaks very similar to those obtained for air-filled channels. However, here the height of the peaks far exceeds those observed earlier, signifying that the axial compression-induced buckling of the adhesive skin results in considerable enhancement of its local compliance so that the adhesive becomes persistent to remain in contact with the contacting plate. In essence, at the location of the channel, the adhesive arrests the propagation of a crack because of the increase in its compliance; however, ahead of the channel, its compliance decreases, requiring smaller torque for driving the crack. As a result, when the crack initiates at a threshold torque, a portion of the adhesive skin covering the channel relaxes under a significantly smaller load, which leads to dissipation of energy. It is important to emphasize that the liquid inside the channel does not flow during the propagation of the crack over the channel and therefore does not lead to any viscous dissipation. In

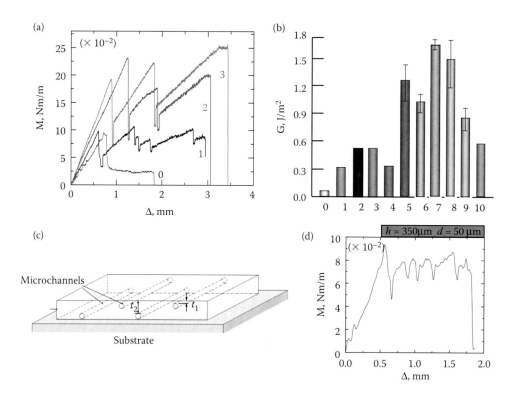

FIGURE 17.13

(a) Typical M versus Δ plots for unstructured adhesive of thickness $h = 300$ μm (curve 0), adhesive layers ($h = 300$ μm) with microchannels ($d = 50$ μm) filled with air (curve 1), and adhesive layers ($h = 570$ and 750 μm) with channels ($d = 530$ and 710 μm) filled with oil of viscosity = 380 cP (curves 2 and 3, respectively). (b) Bar chart depicts adhesion strength G obtained in a variety of experiments. For example, bars 1–4 correspond to microchannels of diameter $d = 50$ μm embedded in a film of thickness $h = 300$ μm filled with air and liquids of different viscosities 5–5000 cP. (c) Schematic of an adhesive film embedded with channels arranged in two different layers. (d) When the peel experiment is done on this adhesive with the channels of the bottom layer filled with oil while those at the top contain air at atmospheric pressure, the M versus D plots are characteristic of crack arrests and initiations at the location of the channels. (From Majumder, A., Ghatak, A. and Sharma, A., Microfluidic adhesion induced by subsurface microstructures, *Science* 318, 2007, 258–261. Reprinted with permission of AAAS.)

fact, experiments with liquids of wide range of viscosity (100–1000 cP) do not show any difference in the peeling torque, nor does the rate of lifting plate (3–100 mm/s) affect it significantly, corroborating the above hypothesis that viscous dissipation does not play any role. The bar chart in Figure 17.13b showing the adhesion strength of different cases brings out this point. Here, bar 1 represents peeling off an unstructured adhesive, whereas bars 2–4 represent those ($d = 50$ μm, $h = 300$ μm) with embedded channels filled with silicone oils of viscosity 120, 1000, and 5000 cP, respectively. Clearly, the adhesion is enhanced for adhesives embedded with oil-filled channels. It is to be noted that because very low viscous oils consist of small polymeric chains, they can diffuse into the cross-linked network altering the rheology of the adhesive and consequently decreasing adhesion strength G. For high-viscosity oils, G also decreases. Here the elastic property of long polymeric molecules is more pronounced as against the effect of Laplace pressure, which results in a decrease in G. Bars 5–10 show the values of G for various sets of h and d at an intermediate viscosity of oil, 380 cP. For all theses cases, the channel remains maximally buried in the adhesive.

The data reveal nonmonotonic dependence of G on these two geometric parameters h and d. For example, for a constant channel diameter, with increase in the thickness of the adhesive, G first increases and then decreases. Similarly, with an increase in the diameter of the channel, the G value decreases. This not-so-simplistic picture signifies that the stress field around the channel varies nonlinearly with the geometry, resulting in a wide range of adhesion strengths at the interface. For example, bar 7 represents $G = 1800$ mJ/m², which is approximately 25 times larger than that for an unstructured adhesive represented by bar 1, measuring only $G = 60$ mJ/m². In other words, without incorporating any viscous effect, the adhesion strength is enhanced by orders of magnitude. In fact, complexity of the stress field allows one to accomplish many other feats. For example, Figure 17.12c and d show that the channels are arranged in two different layers within the adhesive. Channels in either of these two layers or in both can be filled with oil. When the top layer is filled with oil, keeping those at the bottom layer filled with air, the deformation of the film becomes large enough not to allow the adherent to come into contact with the adhesive. However, when the channels in the bottom layer are filled with oil, keeping those at the top layer unfilled, the compliance of the adhesive as well as its ability to remain stuck to the adherent is enhanced locally. Thus, the multilayer arrangement of the channels in the adhesive layer can allow the adhesive layer to be used as both a release coating and a strong adhesive.

17.11 Self-Adhesion of Microstructures

It is interesting to note that controlling fluidic pressure inside the subsurface air pockets and liquid vessels may not be the only way that these structures can alter surface adhesion; it will be shown now that self-adhesion of hairs may be the other mechanism of manipulating adhesion. Although this observation contrasts with several artificial hairy surfaces that suffer from self-adhesion, leading to clustering of hairs and loss of adhesion, biological hairy surfaces do not suffer from such deficiencies. Figure 17.14 depicts the internal microstructure of the adhesive pad of bush cricket *T. viridissima* [81,82], which shows that when subjected to a normal load, its hairs bend and come in to contact with each other, but the branched structure of the hairs does not allow collapse of the fibers. A question arises if this adhesion and detachment phenomenon occurring at the interior of the adhesive can lead to energy dissipation for biological adhesives. The answer is yes, which has been demonstrated by designing multilayer adhesives as shown in Figure 17.15 [85]. Here, thin flexible plates, for example, transparency sheets, glass coverslip, and aluminum foil of varying thicknesses (10–100 µm) and widths (2–8 mm), are used as templates for embedding microchannels in the PDMS elastomeric layers. In addition, these templates are formed into different shapes as shown in Figure 17.13 to generate several different channel geometries. Multiple layers of channels are prepared by stacking several such layers of template with spacers of suitable heights placed in between. It is to be noted that with the elastomer being elastic, there is little hysteresis when two surfaces of the microchannels adhere and detach from each other. However, hysteresis can be induced by creating surface reacting groups by further processing of these samples. For example, these elastic blocks are extracted in a Soxhlet chamber using chloroform as a solvent for approximately 8 h followed by drying them completely in a normal atmospheric condition for a sufficiently long period, for example, 24–48 h. In this process, the sol portion of the elastomer was removed, leaving surface-bound tethered chains. Contact of two such surfaces results in a variety of intermolecular interactions, which finally enhance

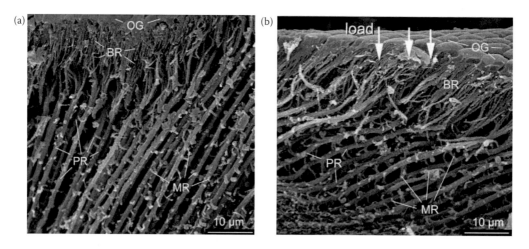

FIGURE 17.14
Hairy internal structure of pads of the insect *Tettigonia viridissima*. Under the load, the hairs come in to close contact with each other, but not in complete contact because of branched structure. (With kind permission from Springer Science+Business Media: *J. Comp. Physiol. A.*, Ultrastructural architecture and mechanical properties of attachment pads of *Tettigonia viridissima* (Orthoptera Tettigoniidae), 186, 2000, 821–831, Gorb, S., Yuckan, J. and Scherge, M.)

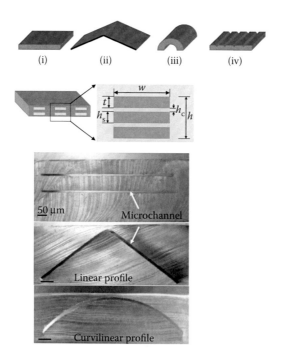

FIGURE 17.15
(i–iv) Templates of different shapes used for generating the stack of microchannels within the adhesive. The stack of microchannels. (From Arul, E. P. and Ghatak, A., Bioinspired design of a hierarchically structured adhesive, *Langmuir* 25, 2009, 611–617.)

the adhesion strength. This mechanism is known as *self-adhesion hysteresis* [86–89], which occurs multiple times at the contacting surface of the walls of the stack of microchannels.

The schematic of the experiment is presented in Figure 17.16, in which the adhesive layer with an embedded stack of microchannels remains bonded to a rigid substrate and a hemispherical indenter is brought in contact with it in a displacement-controlled experiment. The contact area is observed using a microscope, and the applied load is measured using a weighing balance. The adhesion strength is estimated by calculating the force P versus displacement Δ plot at maximum indentation and by dividing it by the maximum area of contact, $\Delta E \sim \oint P d\Delta / A_{max}$.

Figure 17.17 summarizes the effect of self-adhesion hysteresis at the contact of internal walls of microchannels. Curve 1 of Figure 17.17a shows that during cyclic loading and unloading, the P versus Δ data for a layer without any embedded structure superimposes, resulting in negligible hysteresis. However, for one with embedded structures, the load neither increases continuously with indentation depth nor decreases monotonically during retraction. For example, curve 2 of Figure 17.17a shows that in the displacement range A $\rightarrow$ B, the slope of P versus Δ plot remains significantly smaller than that of curve 1 because here indentation occurs against thin films 1 and 2 supported only at the two sides of the microchannel. At point B, the displacement of films 1 and 2 is large enough that it contacts surface 2', following which at B $\rightarrow$ C the adhesive behaves like a single continuous block; therefore, P increases with the same slope as in curve 1. The adhesive behaves similarly in the reverse cycle in regime C $\rightarrow$ D; as a result, the load data superimpose that during loading. In this range, debonding essentially occurs at interface 1. Debonding of interface 1 occurs also in the regime F $\rightarrow$ G during which the unloading data also superimposes the loading data. However, hysteresis occurs during debonding of interface 2–2' in the regime D $\rightarrow$ F, resulting in difference in the load versus displacement plot during loading and unloading. Thus, loading–unloading curves superimpose everywhere except in the region A'BDEF. The net hysteresis of the adhesive is estimated as 277 ± 48 mJ/m^2, which is a significant enhancement over that of the control adhesive. No hysteresis is observed if the interface 2–2' does not separate as in Figure 17.17b in which the skin thickness of the channel is small, for example, $t=440$ µm, or the width of the channel is sufficiently large

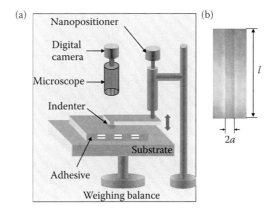

FIGURE 17.16
(a) Schematic of the setup in which an adhesive layer bonded to a rigid substrate is indented in displacement-controlled experiment using a hemicylindrical indenter. The contact load was measured using a weighing balance. (b) Typical image of the area of contact of the indenter and the adhesive. (From Arul, E. P. and Ghatak, A., Bioinspired design of a hierarchically structured adhesive, *Langmuir* 25, 2009, 611–617.)

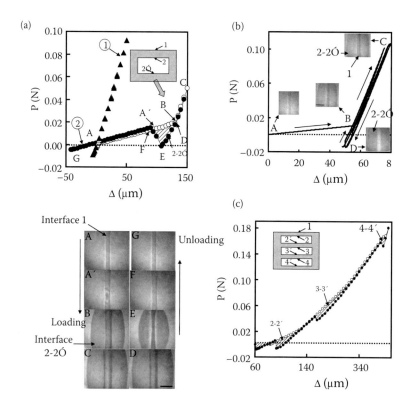

FIGURE 17.17

(a) Curve 1: Typical load versus displacement plot for an unstructured adhesive layer ($h = 3$ mm). Curve 2: Indentation on a layer with an embedded rectangular channel of height $h_c = 100$ μm and width $w = 8$ mm, with skin thickness $t = 800$ μm. The shaded areas A'BDEF represent the hysteresis between loading and unloading curves; the corresponding interfaces are depicted by optical micrographs A–G. (b) During cyclic loading and unloading, the interface 2–2' does not separate when the skin thickness is reduced beyond a critical value. Load versus displacement plot for a layer with a single layer of microchannel with $t = 440$ μm and $w = 8$ mm demonstrates it. (c) Loading and unloading data correspond to an elastic film embedded with multiple layers of microchannels with skin thickness and interchannel spacing $t = h_s = 800$ μm, channel height $h_c = 100$ μm, and channel width $w = 8$ mm. The scale bar in optical micrographs represents 500 μm. The blocks in panels a and d represent the cross-sectional view of the adhesive layer. (From Arul, E. P. and Ghatak, A., Bioinspired design of a hierarchically structured adhesive, *Langmuir* 25, 2009, 611–617.)

so that in repeated loading–unloading experiments, the film with the embedded channels behaves as a single unstructured film. Thus, optimization of the geometrical length scales is important for self-adhesion hysteresis to occur. For a stack of multilayer of channels, the hysteresis occurs at each internal layer, thus enhancing it significantly. For example, for three-layer channels as in Figure 17.17c, the total hysteresis is estimated to be 1556 ± 134 mJ/m². Notice that here the depth of indentation is quite large (~600 μm) to ensure that the internal wall of each channel comes into contact.

Figure 17.18 shows the dependence of hysteresis on indentation depth Δ and the number n of layers of microchannels in the stack. Typically, ΔE increases with both n and Δ until a plateau value ΔE_{max} is reached at Δ_{max} (as shown by the dotted lines); ΔE_{max} increases with both n and h_s. Figure 17.18 shows that for five embedded channels within the adhesive, the plateau value can be as high as $\Delta E_{max} = 4412 \pm 90$ mJ/m². The hysteresis is further increased when the internal surfaces of the channels are micropatterned by using suitable templates.

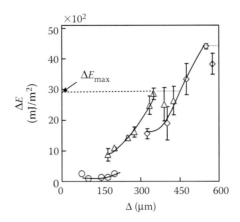

FIGURE 17.18
Adhesion hysteresis as a function of depth of indentation and number of layers. The data represent channel height $h_c = 100$ μm, skin thickness $t = 1000$ μm, and interchannel spacing equal to 1000 μm. The symbols ○, △, and ◇ represent the data for one, three, and five layers of channels, respectively. The solid lines are a guide to the eye. (From Arul, E. P. and Ghatak, A., Bioinspired design of a hierarchically structured adhesive, *Langmuir* 25, 2009, 611–617.)

Thus, the multilayered channel system, in essence, incorporates features of at least three different length scales: the channel walls of thickness 400–1500 μm, the micropatterns of 15–50 μm inscribed on the internal walls of the channels, and finally, the molecular length scale of dangling chains at the internal and external surfaces of the adhesive. In each of these length scales, adhesion hysteresis increases because of a distinct physical phenomenon. Besides, by suitably altering spatial orientation and density of these patterns, it is possible to achieve a spatially varying hysteresis that can be useful for a variety of applications, particularly, ones involving autonomous locomotion.

17.12 Summary

Traditionally, the adhesion strength of an adhesive has been enhanced either by altering its rheology, for example, its viscoelasticity, or by incorporating suitable chemical moiety at the adhesive surface. Although both these processes involve chemical means for controlling the adhesion characteristics, biological adhesives present in nature use a variety of mechanisms that essentially alter their geometrical features. The most important aspect of these adhesives is their hierarchical structures, the length scales of which span the millimeter to nanometer range, that is, over six orders of magnitude. These microstructures, on the one hand, alter the deformability of the adhesive surface, rendering it more pliant to the contacting surface; on the other hand, the microstructures maintain the integrity of the macroscopic features of the adhesive pad under reasonably large normal and shear loading conditions. Another important aspect of biological adhesives is the redundancy in their design, which implies that the actual adhesive area being used by the animal may be an order of magnitude smaller than what is available at its adhesive pad. Redundancy, as mentioned earlier, is needed to account for local damage and defect of some of the hairs and to guard against failure because of dirt and adverse environmental conditions. Importantly, redundancy is present not only in individual setal level but also in regard to the mechanisms of adhesion. For example, insects not only use a dry adhesion mechanism, that is, the van der Waals force of adhesion, but also wet adhesion on the basis of surface tension at the meniscus, interfacial friction between the insect pad and the substrate, and so forth. Another important aspect is multifunctionality, for example, the adhesive pad at the feet of animals not only generates strong adhesion but also helps in locomotion on rough and three-dimensional-oriented terrains, for example, on a vertical wall or ceiling. On such surfaces, the adhesive pad also helps in attaining dynamic stability against occasional shocks [90]. Similarly, the hierarchical structures not only enhance adhesion but also create hydrophobicity, which helps to keep the adhesive clean of particulate contaminations. In this chapter, we have briefly reviewed some of these mechanisms and recent attempts to incorporate them in the design and manufacture of man-made adhesives. Research in this area has led to the development of a new range of bioinspired adhesives [91,92], which will be useful in a variety of applications, including the ones involving repeated application of the adhesive, something difficult to accomplish using viscoelasticity as the dominant mechanism. In addition, this research has led to many new developments, for example, in contrast to soft, easily flowable, viscoelastic materials, the application of materials with a wide range of moduli as adhesive, for example, soft rubber-like purely elastic material, to hard polypropylene fibers, to highly strong carbon nanotubes. In contrast to chemical formulation of the adhesive, the more challenging engineering problems that have emerged today are how to create texture at the surface of the adhesive and its bulk, that is, over a large area, and how to generate hierarchical patterns. The goal is not only to create strong adhesion and easy release using the same adhesive layer but also to impart multifunctionality to the adhesive. Maintaining the integrity of these novel adhesion mechanisms under extreme conditions, for example, under physiological conditions remains another important research goal to be achieved [93,94]. We can confidently say that recent research in this area has led to the understanding of the essential physics and mechanics of adhesion, and also there has been some success in mimicking these mechanisms. As a result, in some cases, the adhesion strength exceeding that of the gecko has been achieved. However, in most cases, the adhesive is not multifunctional, which is in contrast to their natural counterparts. The lifetime of the adhesives has not been ascertained. In most situations, the adhesive has patterns in one or two different length scales, but the true hierarchical structure with a wide range of length scales has not yet been developed. Importantly, the issue of scale-up is completely open; whatever has been achieved in the laboratory is yet to be produced in the commercial field for practical applications. We predict a significant amount of research in these directions in the near future.

Acknowledgments

AG thanks the research initiation grant from the IIT, Kanpur and the financial assistance from the Department of Science and Technology, Government of India.

References

1. Sauter F, Hayek E, Moche W, Jordis U. Identification of betulin in archaeological tar. *Z Naturforsch C*. 1987;42(11):1151–2.
2. Hiller U. Comparative studies on the functional morphology of two gekkonid lizards. *J Bombay Nat Hist Soc*. 1976;73:278–82.
3. Autumn K, Peattie AM. Mechanisms of adhesion in geckos. *Integr Comp Biol*. 2002;42:1081–90.
4. Gillett JD, Wigglesworth VB. The climbing organ of an insect, *Rhodnius prolixus* (Hemiptera Reduviidae). *Proc R Soc Lond B Biol Sci*. 1932;111:364–76.
5. Stork NE. Experimental analysis of adhesion of *Chrysolina polita* (Chrysomelidae: Coleoptera) on a variety of surfaces. *J Exp Biol*. 1980;88:91–107.
6. Federle W, Brainerd EL, McMahon TA, Hölldobler B. Biomechanics of the movable pretarsal adhesive organ in ants and bees. *Proc Natl Acad Sci U S A*. 2001;98(11):6215–20.
7. Federle W, Barnes WJP, Baumgartner W, Drechsler P, Smith JM. Wet but not slippery: boundary friction in tree frog adhesive toe pads. *J R Soc Interface*. 2006;3:689–97.
8. Smith JM, Barnes WJP, Downie JR, Ruxton GD. Structural correlates of increased adhesive efficiency with adult size in the toe pads of hylid tree frogs. *J Comp Physiol A*. 2006;192:1193–204.
9. Barnes WJP, Oines C, Smith JM. Whole animal measurements of shear and adhesive forces in adult tree frogs: insights into underlying mechanisms of adhesion obtained from studying the effects of size and scale. *J Comp Physiol A*. 2006;192:1179–91.
10. Lin LH, Edmonds DT, Vollrath F. Structural-engineering of an orb-spiders web. *Nature*. 1995;373:146–8.
11. Gorb SN, Niederegger S, Hayashi CY, Summers AP, Vötsch W, Walther P. Biomaterials: silk-like secretion from tarantula feet. *Nature*. 2006;443:407.
12. Naldrett MJ. The importance of sulphur cross-links and hydrophobic interactions in the polymerization of barnacle cement. *J Marine Biol Assoc U.K.* 1993;73(3):689–702.
13. Kamino K, Inoue K, Maruyama T. Barnacle cement proteins—importance of disulfide bonds in their insolubility. *J Biol Chem*. 2000;275(35):27360–5.
14. Scherge M, Gorb SN. *Biological Micro- and Nanotribology: Nature's Solutions*. Heidelberg (Germany): Springer; 2001.
15. Kaelble DH. Theory and analysis of peel adhesion: bond stress and distribution. *Trans Soc Rheol*. 1960;4:45–73.
16. Gardon JL. Peel adhesion. II. A theoretical analysis. *J Appl Polym Sci*. 1963;7:643.
17. Kendall K. The adhesion and surface energy of elastic solids. *J Phys D Appl Phys*. 1971;4:1186.
18. Ganghoffer JF, Gent AN. Adhesion of a rigid punch to a thin elastic layer. *J Adhes*. 1995;48:75–84.
19. Creton C, Lakrout H. Micromechanics of flat-probe adhesion tests of soft viscoelastic polymer films. *J Polym Sci B Polym Phys*. 2000;38:965–79.
20. Johnson KL, Kendall K, Roberts AD. Surface energy and the contact of elastic solids. *Proc R Soc Lond A Math Phys Sci*. 1971;324:301–13.
21. Chaudhury MK, Whitesides GM. Direct measurement of interfacial interactions between semi-spherical lenses and flat sheets of poly(dimethylsiloxane) and their chemical derivatives. *Langmuir*. 1991;7(5):1013–25.
22. Chaudhury MK, Weaver T, Hui CY, Kramer EJ. Adhesive contact of cylindrical lens and a flat sheet. *J Appl Phys*. 1996;80(1):30–7.
23. Shull KR, Dongchan A, Mowery CL. Finite-size corrections to the JKR technique for measuring adhesion: soft spherical caps adhering to flat, rigid surfaces. *Langmuir*. 1997;13(6):1799–804.
24. Chaudhury MK, Owen MJ. Correlation between adhesion hysteresis and phase state of monolayer films. *J Phys Chem*. 1993;97(21):5722–6.

25. Chaudhury MK, Whitesides GM. Correlation between surface free energy and surface constitution. *Science*. 1992;255:1230–2.
26. Israelachvili J. *Intermolecular and Surface Forces*. 2nd ed. London: Academic Press; 1992.
27. Spolenak R, Gorb SN, Gao H, Arzt E. Effects of contact shape on the scaling of biological attachments. *Proc R Soc Lond A Math Phys Sci*. 2005;461:305–19.
28. Gladun D, Gorb SN. Insect walking techniques on thin stems. *Arthropod Plant Interact*. 2007;1:77–91.
29. Johnson KL. *Contact Mechanics*. Cambridge: Cambridge University Press; 1985.
30. Shull KR. Contact mechanics and the adhesion of soft solids. *Mater Sci Eng R Rep*. 2002;36:1–45.
31. Chaudhury MK, Whitesides GM. Direct measurement of interfacial interactions between semispherical lenses and flat sheets of poly(dimethylsiloxane) and their chemical derivatives. *Langmuir*. 1991;7:1013–25.
32. Arzt E, Gorb S, Spolenak R. From micro to nano contacts in biological attachment devices. *Proc Natl Acad Sci U S A*. 2003;100:10603–6.
33. Arzt E. Biological and artificial attachment devices: lessons for materials scientists from flies and geckos. *Mater Sci Eng C*. 2006;26:1245–50.
34. Gao H, Yao H. Shape insensitive optimal adhesion of nanoscale fibrillar structures. *Proc Natl Acad Sci U S A*. 2004;101(21):7851–6.
35. Gorb SN, Jiao Y, Scherge MJ. Ultrastructural architecture and mechanical properties of attachment pads in *Tettigonia viridissima* (Orthoptera Tettigoniidae). *J Comp Physiol A*. 2000;186:821–31.
36. Ghatak A, Mahadevan L, Chung J, Chaudhury MK, and Shenoy V. Peeling from a biomimetically patterned thin elastic film. *Proc R Soc Lond A Math Phys Sci*. 2004;460:2725–35.
37. Chung JY, Chaudhury MK. Roles of discontinuities in bio-inspired adhesive pads. *J R Soc Interface*. 2005;2:55–61.
38. Crosby AJ, Hageman M, Duncan A. Controlling polymer adhesion with "pancakes." *Langmuir*. 2005; 21:11738–43.
39. Thomas T, Crosby AJ. Controlling adhesion with surface hole patterns. *J Adhes*. 2006;82:311–29.
40. Landau LD, Lifshitz EM. *Theory of Elasticity*. 3rd revised ed. *Course of Theoretical Physics*. Vol. VII. New York: Pergamon Press; 1986.
41. Ghatak A, Mahadevan L, Chaudhury MK. Measuring the work of adhesion between a soft confined film and a flexible plate. *Langmuir*. 2005;21:1277–81.
42. Ghatak A. Unpublished results.
43. Lake GJ, Thomas AG. The strength of highly elastic materials. *Proc R Soc Lond A Math Phys Sci*. 1967;300:108–19.
44. Ruibal R, Ernst V. The structure of the digital setae of lizards. *J Morphol*. 1965;117:271–94.
45. Russell AP. A contribution to the functional morphology of the foot of the tokay *Gekko gecko* (Reptilia, Gekkonidae). *J Zool (Lond)*. 1975;176:437–76.
46. Autumn K, Liang YA, Hsieh ST, Zesch W, Chan WP, Kenny TW, et al. Adhesive force of a single gecko foot-hair. *Nature*. 2000;405:681–5.
47. Autumn K, Sitti M, Peattie A, Hansen W, Sponberg S, Liang YA, et al. Evidence for van der Waals adhesion in gecko setae. *Proc Natl Acad Sci U S A*. 2002;99:12252–6.
48. Autumn K. Gecko adhesion: structure, function, and applications. *MRS Bull*. 2007;32:473–8.
49. Autumn K, Gravish N. Gecko adhesion: evolutionary nanotechnology. *Philos Trans R Soc Lond A*. 2008;366:1575–90.
50. Hansen WR, Autumn K. Evidence for self-cleaning in gecko setae. *Proc Natl Acad Sci U S A*. 2005;102:385–9.
51. Kohli R. Particles on surfaces 7: detection, adhesion, and removal. In: Mittal KL, editor. *Proceedings of the International Symposium on Particles on Surfaces: Detection, Adhesion and Removal*. Vol. 7. Utrecht (The Netherlands): VSP; 2002. pp. 113–49.
52. Jagota A, Bennison SJ. Mechanics of adhesion through a fibrillar microstructure. *Integr Comp Biol*. 2002;42:1140–5.

53. Hui CY, Glassmaker NJ, Tang T, Jagota A. Design of biomimetic fibrillar interfaces: II. Mechanics of enhanced adhesion. *J R Soc Interface*. 2004;1:35–48.

54. Glassmaker NJ, Jagota A, Hui CY, Kim J. Design of biomimetic fibrillar interfaces: I. Making contact. *J R Soc Interface*. 2004;1:23–33.

55. Geim AK, Dubonos SV, Grigorieva IV, Novoselov KS, Zhukov AA, Shapoval SY. Microfabricated adhesive mimicking gecko foot-hair. *Nat Mater*. 2003;2:461–3.

56. Greiner C, Campo A, Arzt E. Adhesion of bioinspired micropatterned surfaces: effects of pillar radius, aspect ratio, and preload. *Langmuir*. 2007;23:3495–502.

57. Rizzo NW, Gardner KH, Walls DJ, Keiper-Hrynko NM, Ganzke TS, Hallahan DL. Characterization of the structure and composition of gecko adhesive setae. *J R Soc Interface*. 2006;3:441–51.

58. Huber G, Gorb SN, Spolenak R, Arzt E. Resolving the nanoscale adhesion of individual gecko spatulae by atomic force microscopy. *Biol Lett*. 2005;1:2–4.

59. Huber G, Mantz H, Spolenak R, Mecke K, Jacobs K, Gorb SN, et al. Evidence for capillarity contributions to gecko adhesion from single spatula nanomechanical measurements. *Proc Natl Acad Sci U S A*. 2005;102:16293–6.

60. Glassmaker NJ, Jagota A, Hui CY, Noderer WL, Chaudhury MK. Biologically inspired crack trapping for enhanced adhesion. *Proc Natl Acad Sci U S A*. 2007;104:10786–91.

61. Noderer WL, Shen L, Vajpayee S, Glassmaker NJ, Jagota A, Hui CY. Enhanced adhesion and compliance of film-terminated fibrillar surfaces. *Proc R Soc Lond A Math Phys Sci*. 2007;463:2631–54.

62. Walker G, Yue AB, Ratcliffe J. The adhesive organ of the blowfly, *Calliphora vomitoria*: a functional approach (Diptera: Calliphoridae). *J Zool (Lond)*. 1985;205:297–307.

63. Attygalle AB, Aneshansley DJ, Meinwald J, Eisner T. Defence by foot adhesion in a chrysomelid beetle (*Hemisphaerota cyanea*): characterization of the adhesive oil, *Zoology*. 2000;103:1–6.

64. Jiao Y, Gorb S, Scherge M. Adhesion measured on the attachment pads of *Tettigonia viridissima* (Orthoptera, Insecta). *J Exp Biol*. 2000;203:1887–95.

65. Vötsch W, Nicholson G, Müller R, Stierhof YD, Gorb S, Schwarz U. Chemical composition of the attachment pad secretion of the locust *Locusta migratoria*. *Insect Biochem Mol Biol*. 2002;32:1605–13.

66. Santos R, Gorb S, Jamar V, Flammang P. Adhesion of echinoderm tube feet to rough surfaces. *J Exp Biol*. 2005;208:2555–67.

67. Batchelor GK. *Introduction to Fluid Dynamics*. Cambridge: Cambridge University Press; 1967.

68. Barnes J. Tree frogs and tire technology. *Tire Technology International*. 1999:42–7.

69. Persson BNJ. Wet adhesion with application to tree frog adhesive toe pads and tires. *J Phys Condens Matter*. 2007;19:376110.

70. Persson BNJ. Biological adhesion for locomotion: basic principles. Personal communication.

71. Federle W, Baumgartner W, Hölldobler B. Biomechanics of ant adhesive pads: frictional forces are rate- and temperature dependent. *J Exp Biol*. 2004;206:67–74.

72. Greenwood JA, Tabor D. The friction of hard sliders on lubricated rubber: the importance of deformation losses. *Proc Phys Soc Lond*. 1958;71:989–1001.

73. Schallamach A. A theory of dynamic rubber friction. *Wear*. 1963;6:375–82.

74. Vorvolakos K, Chaudhury MK. The effects of molecular weight and temperature on the kinetic friction of silicone rubbers. *Langmuir*. 2003;19:6778–87.

75. Yurdumakan B, Raravikar NR, Ajayan PM, Dhinojwala A. Synthetic gecko foot-hairs from multiwalled carbon nanotubes A. *Chem Commun*. 2005;30:3799–801.

76. Ge L, Sethi S, Ci L, Ajayan PM, Dhinojwala A. Carbon nanotube-based synthetic gecko tapes. *Proc Natl Acad Sci U S A*. 2007;104(26):10792–5.

77. Sethi S, Ge L, Ci L, Ajayan PM, Dhinojwala A. Gecko-inspired carbon nanotube-based self-cleaning adhesives. *Nano Lett*. 2008;8(3):822–5.

78. Qu L, Dai L, Stone M, Xia Z, Wang ZL. Carbon nanotube arrays with strong shear binding-on and easy normal lifting-off. *Science*. 2008;322:238–42.

79. Majidi C, Groff RE, Maeno Y, Schubert B, Baek S, Bush B, et al. High friction from a stiff polymer using microfiber arrays. *Phys Rev Lett*. 2006;97:076103.

80. Lee J, Majidi C, Schubert B, Fearing RS. Sliding-induced adhesion of stiff polymer microfibre arrays. I. Macroscale behaviour. *J R Soc Interface*. 2008;5:835–44.

81. Schubert B, Lee J, Majidi C, Fearing RS. Sliding-induced adhesion of stiff polymer microfibre arrays. II. Microscale behaviour. *J R Soc Interface*. 2008; 5:845–53.

82. Gorb S, Yuckan J, Scherge M. Ultrastructural architecture and mechanical properties of attachment pads of *Tettigonia viridissima* (Orthoptera Tettigoniidae). *J Comp Physiol A*. 2000;186:821–31.

83. Scherge M, Gorb SN. Using biological principles to design MEMS. *J Micromech Microeng*. 2000;10:359–64.

84. Verma MKS, Majumder A, Ghatak A. Embedded template-assisted fabrication of complex microchannels in PDMS and design of a microfluidic adhesive. *Langmuir*. 2006;22(24):10291–5.

85. Majumder A, Ghatak A, Sharma A. Microfluidic adhesion induced by subsurface microstructures. *Science*. 2007;318:258–61.

86. Arul EP, Ghatak A. Bioinspired design of a hierarchically structured adhesive. *Langmuir*. 2009;25(1):611–7.

87. Silberzan P, Perutz S, Kramer EJ, Chaudhury MK. Study of the self-adhesion hysteresis of a siloxane elastomer using the JKR method. *Langmuir*. 1994;10:2466–70.

88. Choi GY, Kim S, Ulman A. Adhesion hysteresis studies of extracted poly(dimethylsiloxane) using contact mechanics. *Langmuir*. 1997;13:6333–8.

89. Mason R, Koberstein J. Adhesion of PDMS elastomers to functional substrates. *J Adhes*. 2005;81:765–89.

90. Mason R, Emerson J, Koberstein J. Self-adhesion hysteresis in polydimethylsiloxane elastomers. *J Adhes*. 2004;80:119–43.

91. Ghatak A, Majumder A, Kumar R. Hysteresis of soft joints embedded with fluid-filled microchannels. *J R Soc Interface*. 2009;6:203–8.

92. Creton C, Gorb SN. Sticky feet: from animals to materials. *MRS Bull*. 2007;32:466–8.

93. Northen MT, Greiner C, Arzt E, Turner KL. A Gecko-inspired reversible adhesive. *Adv Mater*. 2008;20:1–5.

94. Lee H, Lee BP, Messersmith PB. A reversible wet/dry adhesive inspired by mussels and geckos. *Nature*. 2007;448:338–41.

95. Mahdavi A, Ferreira L, Sundback C, Nichol JW, Chan EP, Carter DJD. et al. A biodegradable and biocompatible gecko-inspired tissue adhesive, *Proc Natl Acad Sci U S A*. 2008;105:2307–12.

18

Microfluidics for Aerospace Applications

Surya Raghu

CONTENTS

18.1 Introduction

Development of microfluidic devices for aerospace applications has been mainly focused on three areas relevant to the aerospace industry: aerodynamic flow control in aerospace vehicles, micropower generators (fuel cells and microturbines), and micropropulsion devices. For these applications, dimensions up to 1 mm are considered "microfluidic" scales, although some of them at the millimeter scale are sometimes referred to as "mesoscale" or miniature devices. There is also considerable work on microfluidic devices for extraterrestrial applications such as advanced environmental control and life-support systems. In this chapter, we present an overview of these applications.

18.2 Aerodynamic Flow Control

The aerodynamic design of future aerospace vehicles will be greatly influenced by active flow control technologies available for separation and drag reduction, jet engine inlet and exhaust systems, thrust vectoring, cavity flow/acoustics, jet noise reduction, and propulsion devices such as jet engines and rockets. Before the introduction of microelectromechanical systems (MEMS)-based devices, other devices such as vibrating ribbons, acoustic and piezoelectric actuators, and suction and blowing systems have been extensively used for demonstrating flow control schemes. Although these devices have demonstrated significant levels of flow control in laboratory-scale experiments, integration of such systems into the actual hardware of aerospace vehicles has remained a real challenge.

MEMS-based flow control systems comprising sensors, actuators, and integrated electronic systems offer great promise for flow control and are being implemented in a variety of flow situations to modify free- and wall-bounded shear layers (Ho and Tai, 1998), control mixing (Rathnasingam and Breuer, 1997; Wiltze and Glezer, 1998), and for separation control (Parekh et al., 2002; Smith et al., 1998). Microfluidic jets have been used for thrust vectoring (Raman et al., 2001), suppression of jet impingement noise (Alvi et al., 2002), and control resonant cavity oscillations (Raman et al., 1999; Rizetta and Visbal, 2002; Stanek et al., 2002a, 2002b). Some examples of microfluidic actuators and their applications to active flow control are discussed in the following section.

18.2.1 Synthetic Jets

A commonly used technique is the zero-net-mass-flux "synthetic jet" actuator (Smith and Glezer, 1998). A schematic of a synthetic jet is shown in Figure 18.1. During the downward movement of the diaphragm, the ambient fluid is drawn in from all directions, and in the upward movement of the diaphragm, the fluid is ejected in the axial direction with a higher momentum. The diaphragm is driven using a piezoelectric material or other means. Velocities of up to 30 m/s have been achieved using these synthetic jets. The MEMS fabrication techniques permit production of arrays of actuators and easy integration of electronics for phase-controlled actuation.

Smith and Glezer (1997) have demonstrated that synthetic jet actuators placed at the edge of a large nozzle could produce thrust vectoring, thus replacing large movable deflectors currently used for thrust vectoring applications. Smith et al. (1998) and Parekh et al. (2002) have shown that synthetic jets can be used to modify the airfoil characteristics by creating a "virtual aerodynamic shaping" of the airfoil. Parekh et al. (2002) have shown that by using a pair of two-dimensional synthetic jets at the leading edge of a thick airfoil, significant improvement in the lift/drag characteristics (length–depth ratio of 3.63

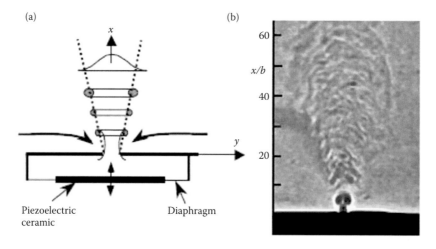

FIGURE 18.1
(a) Schematic and (b) photo of a synthetic jet. (From Smith, B.L., Glezer, A., The formation and evolution of synthetic jets. *Phys. Fluids* 31, 1998, 2281–2297. With permission.)

compared with 0.86 in unforced case) can be obtained. A detailed review of synthetic jets can be found in Glezer and Amitay (200x).

Although synthetic jets have shown significant promise in low-speed flows, they do not provide the momentum coefficients needed for flow control at higher flight speeds. The amplitude and frequency of operation is dependent on the diaphragm characteristics. MEMS-based devices do not have very high amplitudes even at their resonant frequencies, and hence it is difficult to obtain higher momentum coefficients. Further, the need for the open cavity to be exposed to the external flow field makes this device susceptible to clogging by dust or debris. The deposit of dust on the vibrating diaphragm also changes its vibration characteristics. The mechanically vibrating diaphragm in the synthetic jet is also subjected to a variety of external environmental conditions such as large changes in temperature, which are likely to affect the performance of the actuators.

18.2.2 Supersonic Microjets

Supersonic microjets of 400 μm size placed circumferentially at the edge of a round jet have been used for control of jet impingement noise produced by the resonant oscillations of an impinging jet (Alvi et al., 2000). Figure 18.2 shows the images of the shadowgraphs of the structure of the impinging jet with and without flow control. One can clearly see the suppression of the large structures with the microjets on. Noise reduction of up to 10 dB was produced by this technique. Control of flow separation in curved ducts (Kumar and Alvi, 2003a) and diffusers (Kumar and Alvi, 2003b) has also been demonstrated with significant success using similar microjets.

18.2.3 Fluidic Oscillators

One method of producing high-frequency perturbations in a flow is by the use of fluidic oscillators. These are nonmoving-parts devices that depend on the bistable nature of fluidic jets to produce oscillations (Viets, 1975). Submillimeter-scale fluidic oscillators have been developed to produce high-frequency (up to 10 kHz) oscillating jets with very low mass flow rates (~10–3 kg/s) (Raghu, 2001; Stouffer, 1985). Such fluidic jets can be used to

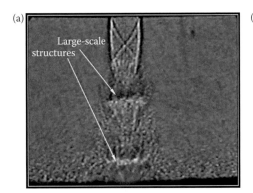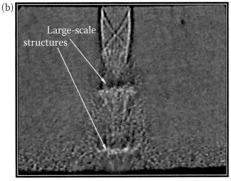

FIGURE 18.2
Phase-averaged shadowgraph images of an impinging jet. NPR = 3.7 (ideally expanded, M = 1.5), h/d = 4.5. Sixteen 400-μm microjets operating at ~100 psi. (a) No control; (b) With microjet control. (Courtesy of Prof. Farrukh Alvi, Florida State University; Alvi, F. S., Elavarasan, R. and Shih, C. Garg. G. and Krothapalli, A., Active control of supersonic impinging jets using microjets, *AIAA Journal*, 41, 2003, 1347–1355.)

produce high-frequency forcing of shear layers and have been used to demonstrate flow control in a variety of situations as described below.

Exploratory experiments have been conducted using miniature (~1 mm) fluidic devices to suppress jet–cavity interaction tones (Raman et al., 1999). Figure 18.3 shows our test bed for evaluation of the fluidic excitation technique. The test case was the flow-induced resonance produced by an $M = 0.69$ jet flowing over a cavity with a length–depth ratio of 6. When located at the upstream end of the cavity floor, the miniature fluidic device was successful in suppressing cavity tones by as much as 10 dB with mass injection rates of the order of only 0.12% (~10^{-3} kg/s) of the main jet flow as shown in Figure 18.4. Steady upstream mass flow addition at the same levels as those for fluidic excitation affected cavity tones only marginally (1-dB reduction). The results not only provide an example of the effectiveness of fluidic excitation but also offer grounds for believing that vast possibilities exist for the use of fluidic excitation in aeroacoustic control.

Figures 18.5a and 18.5b show the jet thrust vectoring obtained by Raman et al. (2001a) using miniature high-frequency fluidic actuators with very low mass flow (~10^{-3} kg/s). Up to 15° of thrust vectoring was obtained by this technique. This technique has the advantage of no large, moving parts such as deflectors or paddles, which have high static and dynamic loads that can be potential causes of structural damage.

Work conducted at the University of Arizona using fluidic actuators indicates successful control of flow separation on a 25% trailing edge flap on an NACA0015 airfoil (Lucas et al., 2008). An array of fluidic actuators was placed on the main part of the airfoil such that there was tangential injection of airflow at the end of the main wing section. A substantially large increase in performance was obtained at $\delta f = 30°$ (Figure 18.6), where even

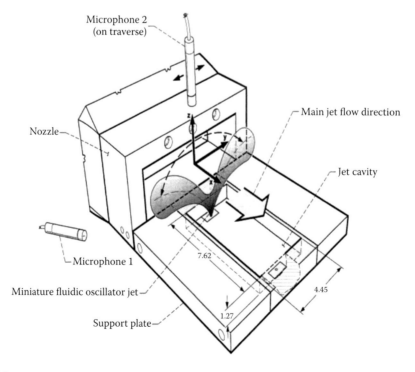

FIGURE 18.3
Configuration for cavity noise suppression. Note: All dimensions are in cm.

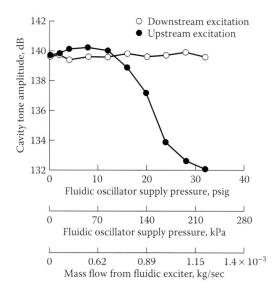

FIGURE 18.4
Cavity noise suppression as a function of nozzle flow rate.

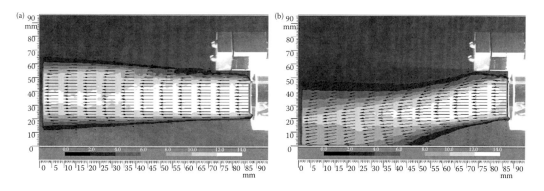

FIGURE 18.5
Time-averaged flow field using particle image velocimetry. The fluidic actuator is on the top of the jet. (a) The fluidic actuator switched off and (b) the actuator switched on.

a $C\mu = 0.42\%$ caused a significant increase in lift and reduction in drag. However, the smaller momentum coefficient was not able to entirely prevent the flow separation over the flap at larger α. At $C\mu = 1.69\%$, flow separation was prevented over the entire range of useful values of α without increasing α_{stall}. For example, at $\alpha = -3°$, the lift coefficient was increased by 0.5, and for $C_l = 1$, the drag coefficient was reduced by 0.07.

Scale-down designs of these fluidic actuators up to nozzle sizes of 125×250 µm have been fabricated and tested. Although the fluidic actuators have shown great promise in laboratory-scale experiments, the effectiveness of such actuators in full-scale applications has not been evaluated. Scaling issues and energy requirements are yet to be resolved for such applications.

Culley et al. (2003) have shown that pulsed injection using fluidic diverters in the trailing edge of a stator vane can be used to reduce the separation in a low-speed axial compressor

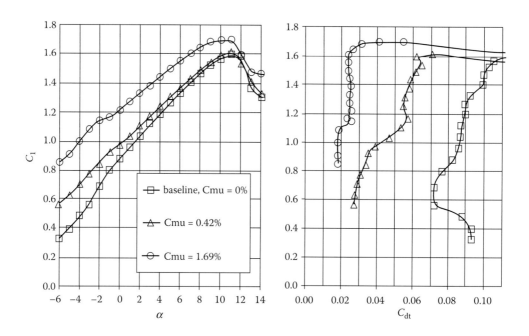

FIGURE 18.6
Lift and total drag polar at $F064_{flap} = 30°$. (From Lucas, N., Taubert, L., Woszidlo, R., Wygnanski, I., and McVeigh, M., Oscillating discrete jets as tools for separation control, AIAA-2008–3868, AIAA Flow Control Conference, Seattle, Washington, June 23–26, 2008. With permission.)

configuration. They used both an externally driven pulsed injection system as well as a fluidic diverter specially designed for that purpose. Recently, low-frequency pulsed injection in the form of microjets and slots for film cooling has been demonstrated to be very effective in reducing the mass flow requirements needed for the purpose (Ou and Rivir, 2006).

Along the lines of synthetic jets and fluidic oscillators, combustion-based flow actuators have been recently developed by Crittenden et al. (2001) to obtain high-momentum jets for flow control. A small amount of premixed fuel and oxidizer is injected into a small combustion chamber where it is ignited with a spark. The resulting combustion generates a high-momentum pulsed jet of the hot gases. It has been proposed by Crittenden et al. that arrays of such jets can be used for flow control. A recent advancement is also the combination of a microcombustor with a microfluidic oscillator (Crittenden and Raghu, 2009) to take advantage of both the high-momentum and the high-frequency oscillator.

18.2.4 Turbulent Boundary Layer Control

To control the turbulent boundary layer, one needs to sense the near-wall structures such as the low-speed streaks, the low-speed ejections, and the high-speed sweeps (Rao et al., 1971; Robinson, 2001) and take appropriate control action such as momentum addition to accelerate the fluid to prevent a burst from occurring. An excellent analysis of the number density of sensors and actuators for such control action for a typical flight Reynolds number is provided by Gad el Hak (1999). It is estimated that that one would need a sensor/actuator density of approximately 1.5 million/m^2 (with a 260-μm span-wise spacing and a 2000-μm longitudinal spacing and a frequency response of up to 18 kHz). Sensors can be local-pressure or shear-stress measurement devices, whereas actuators can be blowing, suction, vibrating, and heating elements. MEMS-based technology offers the opportunity to achieve such a high-density system of sensors and actuators, although we are quite far from achieving it. As an example, a control scheme using shear stress sensors and synthetic jets as actuators has been developed for turbulent boundary layer control (Rathnasingam and Breuer, 1997).

18.3 Micropropulsion

In the area of space applications, the growing trend has been the miniaturization of satellites: microsatellites (up to 100 kg), nanosatellites (up to 10 kg), picosatellites (up to 1 kg), and femtosatellites (less than 100 g). The role of these small satellites is to perform specific functions (e.g., global positioning systems) rather than to serve as generic platforms, which is typical of larger satellites. Distributed space systems consisting of formation flights of micro- or femto-sized satellites are being developed (Leitner et al., 2002) to enhance certain measurement and imaging capabilities while reducing launch and maintenance costs. MEMS-based cold gas, plasma, and combustion-based thrusters are being developed for attitude control of such satellites (Bayt and Breuer, 2001; Lewis et al., 2000). One of the main criteria in developing such microthrusters is the impulse-bit size—defined as the smallest impulse one can obtain using a microthruster. Impulse bits of 10^{-4}–10^{-6} N·s are desirable for satellites of the order of 1 kg mass (Lewis et al., 2000). Such "digital propulsion" devices can deliver precise impulse bits for microsatellite applications such as insertion, station keeping, attitude control, and disposal.

A schematic of a microthruster is shown in Figure 18.7. The length of the entire thruster is of the order of a few millimeters, and the nozzle exit diameter is of the order of 250–400 μm. The propellant is placed in the propellant chamber and is initially sealed with a diaphragm. A small electric heating element is provided at the base of the propellant chamber. When the heater is switched on, the propellant ignites and the increased pressure ruptures the diaphragm and then produces an impulse by the flow of the gas through the nozzle. It is easy to conceive of a large array of such thrusters in a 1 × 1-cm chip using microfabrication techniques as shown in Figure 18.8.

Depending on the needs, solid, liquid, or gaseous propellant can be used in this system. The increased pressure due to ignition will rupture the diaphragm, and the nozzle will deliver the impulse to the microsatellite. Specific impulse in the range of 200–250 s has been achieved (Lewis et al., 2000), and the aim is to achieve specific impulse of 1000 s.

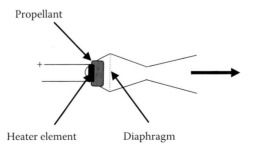

FIGURE 18.7
Schematic of a microthruster.

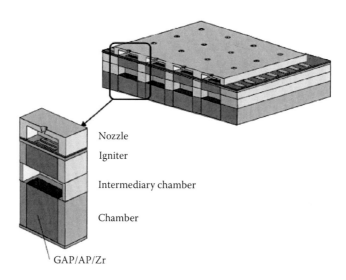

Nozzle

Igniter

Intermediary chamber

Chamber

GAP/AP/Zr

FIGURE 18.8
Schematic view of an array of nozzles. (From Rossi, C., Micropyros, IST–1999–29047, Final Report, 2003. With permission.)

The "digital" design has operational advantages such as nonmoving parts and multiple propellant options and variable plenum and throat dimensions for programmable thrust and impulse delivery. Microthrusters also have applications in the propulsion and trajectory control systems of microaerial vehicles. An excellent review of micropropulsion for space applications is provided by Micci and Ketsdever (2000).

18.4 Micropower Generation

There is an increased demand for sustainable power sources in the range of 1–15 W for use in both terrestrial and space applications. Some common examples are miniature cameras, communication equipment such as cell phones, laptop computers, and remotely operating telemetry-capable sensors. The battery technology has reached its saturation, and fuel cells offer a great promise in this direction. Of all the different designs available for fuel cells, the direct methanol fuel cells (DMFCs) seem to have the greatest promise because of the high-energy density of methanol (5960 W-h/kg). A schematic of a DMFC is shown in Figure 18.9.

The operation of the DMFC is based on the chemical reactions as shown below:

$$CH_3OH + H_2O \rightarrow CO_2 + 6\ H^+ + 6\ e^-\quad \text{(at the Pt-Ru anode)}$$

$$3/2\ O_2 + 6\ H^+ + 6\ e^- \rightarrow 3\ H_2O\quad \text{(at the Pt cathode)}$$

$$CH_3OH + 3/2\ O_2 \rightarrow CO_2 + 2\ H_2O\quad \text{(cell reaction)}$$

One molecule of methanol (CH_3OH) and one molecule of water (H_2O) together store six atoms of hydrogen. When fed as a mixture into a direct methanol air fuel cell (DMFC), they react to generate one molecule of CO_2, 6 protons (H^+), and 6 electrons to generate

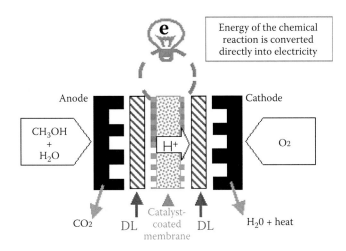

FIGURE 18.9
A PEM-based DMFC. (www.mtimicrofuelcells.com).

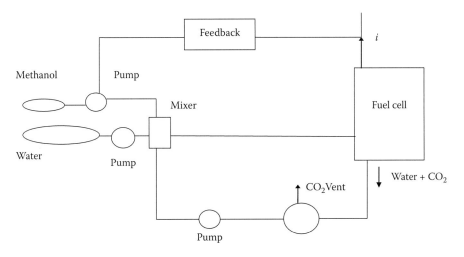

FIGURE 18.10
Schematic of a power plant based on DMFC.

a flow of electric current. The protons and the electrons generated by methanol and water react with oxygen to generate water. The methanol–water mixture provides an easy means of storing and transporting hydrogen, which is much better than storing liquid or gaseous hydrogen in storage tanks. Unlike hydrogen, methanol and water are liquids at room temperature and can be easily stored in thin-walled plastic containers. A typical power plant on the basis of DMFC is shown in Figure 18.10.

All the components shown in the schematic above have to be fabricated in a single unit using microfabrication techniques, and work is under progress in this direction. Some of the requirements for such micropower units are minimum parasitic power consumption, minimum pressure losses in the fluid paths, minimum space requirements, easy access for electronic interfacing, and ease of fabrication of the part.

Fluid handling and thermal management are the biggest challenges faced in designing such a system. Pumping and mixing of concentrated methanol and water, handling two-phase (liquid + bubbles) flow in the microchannels without gas-accumulating effects, separation of the gas phase, and feedback control of the methanol concentration in the system need to be addressed in the design.

18.5 Microfluidics for Life Support and Environmental Control in Space Applications

The miniature microfluidic systems or "lab-on-a-chip" technology renders itself very useful for many extraterrestrial applications such as life-support systems, environmental monitoring, and astrobiology (de Mello, 2004). Examples of applications include temperature and humidity control in spacecrafts, trace contaminant monitoring and removal, water quality monitoring, and *in vitro* physiological monitoring and drug delivery devices (Alexander and Ostrach, 2003; Roediger, 2006). Microfluidic systems have a significant advantage over conventional systems in terms of space, power, and weight requirements, which are a premium in all extraterrestrial applications. The Mars Organic Analyzer built to probe Mars dust for life-based amino acids is an excellent example of the application of microfluidic technology for extraterrestrial probes (Skelly et al., 2005).

18.6 Conclusions

Microfluidics technology will have a large impact on the future developments in the aerospace industry. In the area of aerodynamics, microfluidics will have applications in the design of aerodynamic surfaces such as wings, control surfaces, jet noise control, and flow in turbomachinery. Micropropulsion devices will be used for satellite attitude corrections and formation flights of micro to femto-sized satellites. High-energy-density micropower generation devices such as fuel cells will consist of microfluidic components and thermal management devices. Extraterrestrial missions will contain many microfluidic systems for both life support and analysis of compounds.

References

Alexander JID, Ostrach S. (October 2003). Physicochemical transport processes in biological systems and advanced human support technologies. NASA National Center for Microgravity Research.

Alvi FS, Elavarasan R, Shih C, Garg G, Krothapalli A. (2003). Active control of supersonic impinging jets using microjets. *AIAA J.* 41(7):1347–55.

Bayt R, Breuer K. (2001). System design and performance of hot and cold supersonic microjets. 39th AIAA Aerospace Sciences Meeting and Exhibit, 2001 January 8–11; Reno, Nevada. AIAA Paper No.: 2001-0721.

Coe DJ, Allen MG, Trautman MA, Glezer A. (1994). Micromachined jets for manipulation of macro-flows. Technical digest. Proceedings of the Solid State Sensor and Actuator Workshop, Hilton Head Islands, SC. Ohio: Transducer Research Foundation, Cleveland Heights; pp. 243–7.

Crittenden TM, Raghu S. (2009). Combustion powered actuator with integrated high frequency oscillator. *Int J Flow Control*. 1(1):87–97.

Crittenden TM, Glezer A, Funk R, Parekh D. (2001). Combustion-driven jet actuators for flow control. AIAA Paper No.: 2001-2768.

Culley DE, Bright MM, Prahst PS, Strazisar AJ. (2003). Active flow separation control of a stator vane using surface injection in a multistage compressor experiment. NASA/TM-2003-212356, GT2003-38863ASME, IGTI Turbo Expo 2003; 2003 June 16–19; Atlanta, Georgia.

de Mello AJ, Stone BM. (2004). Unit Operations & Application of Microfluidic Systems for Remote Analysis Workshop on Mars Astrobiology Science and Technology; 2004 September 8–10; Carnegie Institution of Washington.

Elwenspoek M, Jansen H. (1998). *Silicon Micromachining*. Cambridge: Cambridge University Press.

Gad-el-Hak M. (1999). The fluid mechanics of microdevices—the Freeman Scholar Lecture. *J Fluids Eng*. March;121:5.

Ho CM. (2001). Fluidics—the link between micro and nano sciences and technologies. IEEE-0-7803-5998-4/01.

Ho C-M, Tai Y-C. (1998). Micro-electro-mechanical-systems (MEMS) and fluid flows. *Annu Rev Fluid Mech*. 30:579–612.

Janson SW, Helvajian H, Breuer K. MEMS, Microengineering and Aerospace Systems Center for Microtechnology. AIAA Paper No.: 99-3802.

Kumar V, Alvi F. (2003a). Active control of flow separation using supersonic microjets. ASME-FED2003-45129. Proceedings of 4th ASME-JSME Joint Fluids Engineering Conference; 2003 July 6–11; Honolulu, Hawaii.

Kumar V, Alvi F. (2003b). Use of supersonic microjets for active separation control in diffusers. 33rd AIAA Fluid Dynamics Conference, Orlando, Florida. AIAA Paper No.: 2003-4160.

Leitner J, Frank B, Folta D, Carpenter R, Moreau M, How J. (February 2002). Formation Flight in Space. GPS World.

Lewis DH Jr, Janson SW, Cohen RB, Antonsson EK. (2000). Digital micropropulsion. *Sens Actuators A Phys*. 80(2):143–54.

Lucas N, Taubert L, Woszidlo R, Wygnanski I, McVeigh M. (2008). Oscillating discrete jets as tools for separation control. AIAA Flow Control Conference; 2008 Jun 23–26; Seattle, Washington. AIAA Paper No.: 2008-3868.

McNeely MR, Spute MK, Tusneem NA, Oliphant AR. (2000). Hydrophobic microfluidics. Proceedings of SPIE, Hilton Head.

Micci MM, Ketsdever AD. (2000). Micropropulsion for small spacecraft. In: *Progress in Astronautics and Aeronautics Series*, Vol. 187. Reston (VA): AIAA.

Ou S, River RB. (2006). Shaped-hole film cooling with pulsed secondary flow. GT2006–90272.

Parekh D, Palaniswamy S, Goldberg U. (2002). Numerical simulation of separation control via synthetic jets. AIAA Flow Control Conference, Colorado. AIAA Paper No.: 2002-3167.

Raghu S. (2001). Feedback-free fluidic oscillator. United States patent US 6253782.

Raghu S, Raman G. (1999). Miniature fluidic devices for flow control. ASME FEDSM 99-7256.

Raman G. (1997). Using controlled unsteady fluid mass addition to enhance jet mixing. *AIAA J*. 35:647–56.

Raman G, Cornelius D. (1995). Jet mixing control using excitation from miniature oscillating jets. *AIAA J*. 33:365–8.

Raman G, Cornelius D. (October 1996). Multiple fluidic devices provide flow-mixing control. NASA Technical Briefs, Vol. 20. pp. 91–2.

Raman G, Raghu S. (2000). Miniature fluidic devices for flow and noise control. Fluids 2000; 2000 Jun 19–22, Denver, CO. AIAA Paper No.: 2000-2554.

Raman G, Raghu S, Bencic TJ. (1999). Cavity resonance suppression using miniature fluidic oscillators. 5th AIAA/CEAS Aeroacoustics Conference; 1999 May 10–12; Seattle, WA. AIAA Paper No.: 99-1900.

Raman G, Rice EJ, Cornelius D. (1994). Evaluation of flip-flop jet nozzles for use as practical excita-
    tion devices. *J Fluids Eng.* 116:508–15.
Raman G, Packiarajan S, Papadopoulos G, Weissman C, Raghu S. (2001a). Jet thrust vectoring
    using miniature fluidic actuator. FEDSM2001-18057. ASME-FED 2001 Meeting; New Orleans,
    Louisiana.
Raman G, Mills A, Othman S, Kibens V. (2001b). Development of powered resonance tube actua-
    tors for active flow control. Proceedings of ASME-FEDSM, 2001 Fluids Engineering Division
    Summer Meeting; 2001 May 29–Jun 1; New Orleans, Louisiana.
Rao KN, Narasimha R, Badri Narayanan MA. (1971). The 'bursting' phenomenon in a turbulent
    boundary layer. *J Fluid Mech.* 48:339–52.
Rathnasingham R, Breuer KS. (1997). System identification and control of a turbulent boundary layer.
    Phys Fluids. 9(7):1867–9.
Rizzetta DP, Visbal MR. (2002). Large-Eddy simulation of supersonic cavity flowfields including flow
    control. 32nd AIAA Fluid Dynamics Conference; 2002 Jun 24–26; St. Louis, MO. AIAA Paper
    No.: 2002-2853.
Roediger T. (2006). Microfluidic technology for human space flight. Alumni Conference, International
    Space University; 2006 August 3–4.
Rossi C. (2003). Micropyros. IST–1999-29047. Final Report.
Rossi C, Orieux S, Larangot B, Do Conto T, Estève D. (2002). Design, fabrication and modeling of solid
    propellant microrockets—application to micropropulsion. *Sens Actuators A.* 99(1–2):125–33.
Rossi C, Chaalane A, Larangot B, Conédéra V, Briand D, Pham P-Q, et al. (2003). Pyrotechnical
    microthrusters for space application. Proceedings of the International Workshop on Micro and
    Nanotechnology for Power Generation and Energy Conversion Applications—PowerMEMS
    2003; 2003 December 4–5; Tohoku, Japan.
Sakaue H, Gregory JW, Sullivan JP, Raghu S. (2002). Porous pressure-sensitive paint for character-
    izing unsteady flow fields. *AIAA J.* June;40(6):1094–8.
Santiago JG, Wereley ST, Meinhart CD, Beebe DJ, Adrian RJ. (1998). A particle image velocimetry
    system for microfluids. *Exp Fluids.* 25:316–9.
Skelley AM, Scherer JR, Aubrey AD, Grover WH, Ivester RHC, Ehrenfreund P, et al. (2005).
    Development and evaluation of a microdevice for amino acid biomarker detection and analysis
    on Mars. *Proc Natl Acad Sci U S A.* January;102(4):1041–6.
Smith BL, Glezer A. (1997). Vectoring and small-scale motions effected in free shear flows using syn-
    thetic jet actuators. 35th Aerospace Sciences Meeting and Exhibit; 1997 January 6–10; Reno, NV.
    AIAA Paper No.: 97-0213.
Smith BL, Glezer A. (1998). The formation and evolution of synthetic jets. *Phys Fluids.* 31:2281–97.
Smith D, Amitay M, Kibens V, Parekh D, Glezer A. (1998). Modification of lifting body aerodynamics
    using synthetic jet actuators. AIAA Paper No.: 98-0209.
Srinivasan R, Saffarian HM, Raghu S. (2000). A lightweight all-polymer fuel cell power source. United
    States patent US 6872486 (03/29/05).
Stanek MJ, Sinha N, Seiner JM, Pearce B, Jones MI. (2002a). High frequency flow control—suppres-
    sion of aero-optics in tactical directed energy beam propagation & the birth of a new model
    (part I). 33rd AIAA Plasmadynamics & Lasers Conference; 2002 May 20–23; Maui, Hawaii.
    AIAA Paper No.: 2002-2272.
Stanek MJ, Raman G, Ross JA, Odedra J, Peto J, Alvi F, et al. (2002b). High frequency acoustic sup-
    pression—the role of mass flow, the notion of superposition, and the role of inviscid instabil-
    ity— a new model (part II). 8th AIAA/CEAS Aeroacoustics Conference & Exhibit; 2002 June
    17–19; Breckenridge, Colorado. AIAA Paper No.: 2002-2404.
Stouffer RD. (1985). Liquid oscillator device. United States patent US 4508267.
Viets H. (1975). Flip-flop jet nozzle. *AIAA J.* 13:1375–9.
Wilson JR. (2003). Major new thrust for MEMS engines. *Aerosp Am.* February:34–8.
Wiltze JM, Glezer A. (1988). Direct excitation of small scale in free shear flows. *Phys Fluids.*
    10(8):1988.

19

Chemically Reacting Flows at the Microscale

Achintya Mukhopadhyay

CONTENTS

19.1 Introduction

Microscale combustors are being extensively investigated owing to their potential use as power sources in portable electronic devices such as laptop computers and mobile phones (Fernadez-Pello, 2002). The major advantage offered by microscale combustors is the high specific energy of fuels (energy per unit weight), particularly liquid fuels (~45 MJ/kg), compared with that of batteries (~0.6–1.2 MJ/kg). This feature makes microscale combustion systems attractive as high-specific-energy microelectromechanical system (MEMS) power systems, owing to reduced weight and increased operational life (Fernandez-Pello, 2002). Although microfluidic devices, even at the MEMS scale, have characteristic lengths large enough to be in the continuum regime, the extremely small length scales involved alter the characteristics of transport phenomena significantly from those of large-scale devices.

A major issue in the design of microscale combustors is to sustain the flame in these small devices. This is due to the high surface to volume ratio, leading to high heat loss from the combustion region and low residence time available for chemical reactions due to small volume of the combustor. Innovative designs that have sought to address both these issues have been reported. The present chapter addresses the thermal and chemical issues in microscale reacting flows. The organization of the chapter is as follows. First, the fundamentals of propagation and extinction of flames are discussed to highlight the challenges involved at the microscale. This is followed by a discussion on thermal approach (heat recirculation) of stabilizing microscale flames followed by a chemical approach (catalytic combustion). Finally, some interesting novel applications of microscale combustors are presented. However, this chapter is not intended to be an exhaustive review on combustion at microscales. Rather, the objective of this chapter is to highlight the fundamental issues in microscale combustion, particularly, flame stabilization. Consequently, only such works that help to elucidate these issues have been cited from the literature.

19.2 Propagation of Laminar Premixed Flames

Most combustion phenomena are characterized by a localized region of intense chemical reaction known as flame. Flames are primarily of two types: premixed and nonpremixed (Law, 2006; Turns, 2000). In a premixed flame, the reactants (fuel and oxidizer) are mixed at the molecular level before any significant chemical reaction takes place. On the other hand, in nonpremixed flames, the reactants are initially separated and have to be transported by convection and diffusion to a common region where the chemical reaction takes place. In addition, combustion may also take place in a flameless mode, where the chemical reaction occurs in a distributed manner over a large region (Turns, 2000). Microcombustors mostly use premixed flames, although in certain cases of weak reactions, the combustion may approach the flameless mode also. Owing to the small length scales involved, these flames are mostly laminar.

Premixed flames are characterized by propagation of the flame into the unburned mixture at a definite rate. Thus, propagation of the flame is cardinal to the behavior of premixed flames. For measuring the rate of propagation of the flame, an appropriate coordinate system has to be fixed to the propagating flame. The speed of the unburned mixture relative to the flame is known as *flame speed*, S_u (Turns, 2000). For a flame propagating freely through a quiescent mixture, the flame speed is the rate of propagation of the flame through the mixture. On the other hand, for a flat flame stabilized on a burner, this is equal to the rate at which the unburned mixture reaches the flame. Flame speed is influenced by several factors such as flame curvature, flow nonuniformities, heat loss, and ambient pressure and most importantly by the chemical properties of the fuel and the air–fuel ratio (stoichiometry) of the reacting mixture. To isolate the effects of fuel chemistry and stoichiometry, flame speeds are often computed for freely propagating planar adiabatic flames. Flame speeds for such flames are often referred to as laminar burning velocity (S_u^0).

The flame structure consists of two distinct regions, the preheat zone and the reaction zone (Law, 2006). In the preheat zone, both convection and diffusion of energy and species are dominant and balance each other. In the reaction zone, chemical reaction and diffusion of energy and species are the major phenomena and balance each other. Because the reaction zone is much thinner than the preheat zone, the flame thickness can be approximated as the thickness of the preheat zone only. The mass flux through the flame and the flame thickness scales as follows (Law, 2006):

$$(f^0)^2 \sim \frac{(\lambda/C_p)\,\omega_b^0}{Ze} \tag{19.1}$$

$$(l_D^0)^2 \sim \frac{(\lambda/C_p)Ze}{\omega_b^0} \sim \frac{\rho_u \alpha Ze}{\omega_b^0}. \tag{19.2}$$

Because $f^0 = \rho_u S_u^0$, the laminar burning velocity scales as

$$(S_u^0)^2 \sim \frac{\alpha \omega_b^0}{Ze \rho_u}. \tag{19.3}$$

A more detailed integral analysis (Turns, 2000) yields the following expressions for the above quantities:

$$S_u^0 = \left[-2\alpha(v+1) \frac{\bar{\omega}^0}{\rho_u} \right]^{1/2} \tag{19.4}$$

$$l_D^0 = \left[\frac{-2\rho_u \alpha}{(v+1)\bar{\omega}^0} \right]^{1/2}. \tag{19.5}$$

Typical values of S_u^0 and l_D^0 for hydrocarbon–air flames at atmospheric pressures are $O(1-100)\,\mathrm{cm/s}$ and $O(0.1-1)\,\mathrm{mm}$, respectively. Because $\omega_b^0 \sim p^n \exp(-T_a/T_b^0)$ and $\rho_u \sim p/T_u$, where n is the order of the chemical reaction, the effects of pressure and temperature on burning velocity and flame thickness can be expressed as

$$S_u^0 \sim \left[\frac{\alpha p^{n-1} T_u \exp(-T_a/T_b^0)}{Ze} \right]^{1/2} \tag{19.6}$$

$$l_D^0 \sim \left[\frac{T_u \alpha Ze}{p^{n-1} \exp(-T_a/T_b^0)} \right]^{1/2}. \tag{19.7}$$

The above relation shows that for the order of the reaction $n > 1$, flame thickness decreases and burning velocity increases with an increase in pressure or a decrease in unburned gas temperature. The exponential function implies a strong nonlinear dependence on the temperature of burned gases. Consequently, the rate of reaction increases sharply with an increase in the temperature. This has important implications for extinction of the flame because of heat loss, which is discussed in the subsequent sections.

19.3 Extinction of Ducted Premixed Flames

As discussed in the previous section, the exponential dependence of reaction rate on temperature in the reaction zone implies that a small reduction in the temperature of the reaction zone, which is of the order of the temperature, T_b^0, of the burned gas, can lead to a major reduction in the reaction rate. Asymptotic analysis by Law (2006) shows that although adiabatic flames can propagate at a finite speed even with a very small concentration of reactants, a reduction in flame temperature of the order $O(T_b^2/T_a)$ leads to a reduction in burning flux of the order $O(1)$. The burning mass flux is related to the heat loss from the flame as follows (Law, 2006):

$$\tilde{f}^2 \ln \tilde{f}^2 = -\tilde{L}_v. \tag{19.8}$$

In the above equation,

$$\tilde{f} = \frac{f}{f_0} \quad \text{and} \quad \tilde{L}_v = \frac{\lambda/C_P}{(f^0)^2 \, (T_{ad}^2/T_a) q_C Y_u} L_v.$$

The corresponding reduction in the flame temperature is given by

$$\tilde{T}_b = \tilde{T}_{ad} - \varepsilon^0 \theta. \tag{19.9}$$

In the above equation, $\varepsilon^0 = T_{ad}^2/T_a$ and $\theta = \tilde{L}_v/\tilde{f}^2$.

For the adiabatic limit ($\tilde{L}_v = 0$), $\tilde{f} = 1$. With increase in heat loss to nonzero values, the mass flux decreases. However, no solution is possible beyond a critical value of heat loss $\tilde{L}_{v,ext} = 1/2e \approx 0.18$. This heat flux is the extinction limit of heat loss. The mass flux at extinction is given by $\tilde{f}_{ext} = e^{-1/2} \approx 0.607$. Equation 19.8 has two solutions, which approach each other as $\tilde{L}_v$ approaches $\tilde{L}_{v,ext}$. However, the second solution $\tilde{f} < \tilde{f}_{ext}$ represents a physically unrealizable unstable solution.

For ducts and channels, considering heat transfer across the duct or channel as the only mechanism for heat loss, Turns (2000) showed that there is a minimum width of the channel below which flames cannot be sustained. This minimum dimension, known as *quenching distance*, is given by $d_{ext} = \sqrt{b} l_d^0$, where $b \gg 2$ is a factor that accounts for steep temperature gradients near the wall due to convective heat transfer. Daou and Matalon (2002) examined the extinction of laminar premixed flames in narrow and wide channels. They expressed heat loss in the form

$$\frac{\partial \tilde{T}}{\partial y} - \frac{k}{\beta} \tilde{T} = 0. \tag{19.10}$$

Here k/β represents the strength of heat loss by conduction in the transverse direction (y-direction). They characterized channels as narrow or wide based on the ratio of flame thickness to channel width ε. Asymptotic analysis reveals that for narrow channels, $\varepsilon \gg 1$, to the leading order, in a frame of reference moving with the average flow velocity, mean temperature and concentration profiles and the extinction conditions are identical to those obtained from the solution of a one-dimensional problem with volumetric heat loss of intensity $\tilde{L}_v = k\varepsilon^2$. The results show that there exists a critical width of the channel (ε^*) below which complete extinction is possible due to increased heat loss. For a quiescent mixture, complete extinction occurs for $\varepsilon > \varepsilon^* = 0.135$. This corresponds to a value of approximately 7.5 times the value of l_D^0 as defined in Equation 19.5. Because the flame thickness is of the order of 0.1–1 mm, typical combustor dimensions are vulnerable to complete extinction due to heat loss. The dimensionless mass flux at extinction is approximately 0.6 (very close to the asymptotic value obtained for one-dimensional flame). The extinction limit of heat loss increases with an increase in channel width. For moderately wide channels, the flame is only partially extinguished even for $k \to \infty$. For adiabatic walls, the flame spans the entire channel width. As k increases, the flame extinguishes near the walls and withdraws toward the center. The dead space near the wall, where the flame is extinguished, can extend up to several times the flame thickness. The presence of flow directed toward the unburned ($u_0 < 0$) or burned ($u_0 > 0$) gases alters the extinction characteristics.

It is found that the results are more sensitive to heat loss when the flow is directed toward the burned gas ($u_0 > 0$) than when it is directed toward the unburned gas ($u_0 < 0$). Figure 19.1 shows the burning rate for different values of channel width, flow velocity, and heat loss. In the figure, $k = 0$ corresponds to adiabatic wall whereas $k \to \infty$ corresponds to walls maintained at the inlet temperature of the reactants.

Cui et al. (2004) investigated the effect of Lewis number (ratio of thermal to mass diffusivity) of the reactant mixture on flame propagation in wide and narrow channels with adiabatic walls. For narrow channels, where the flame thickness was comparable to or greater than the channel width, transverse diffusion was important. Hence, Lewis number had a minimal role in flame propagation. However, for moderate to large channels, the burning rate is strongly affected by the Lewis number of the reactant mixture. In addition, flame tip opening or dead space near the walls was observed for Lewis number much less than unity. Chakraborty et al. (2008) showed that for moderately wide channels, at Lewis numbers significantly less than unity and at high heat loss, local extinction is observed in

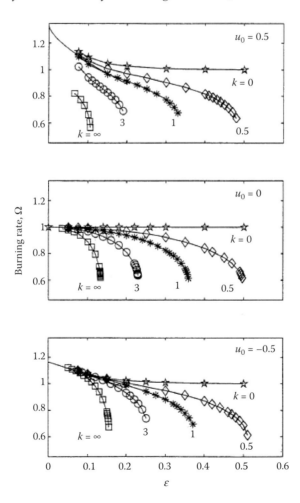

FIGURE 19.1
Burning rate Ω plotted against ε for selected values of k and for $u_0 = -0.5, 0,$ and 0.5. (Reprinted from *Combustion and Flame*, 128, Daou, J. and Matalon, M., Influence of conductive heat losses on the propagation of premixed flames in channels, pp. 321–339, Copyright 2002, with permission from Elsevier.)

flames both near the wall and the channel axis, at leading to simultaneous tip opening and dead space near the walls. This led to the formation of multicellular flame with "funnel-like" shape. For Lewis numbers greater than unity, the effect of Lewis number on reaction rate gets reversed as heat loss increases.

Gutkowski et al. (2008) investigated the shapes of propane flames under different conditions in narrow channels. Their experiments showed that the flame propagation velocity in narrow channels first increases sharply with an increase in channel width, reaches a maximum, and then falls off as the channel width is further increased. At near-quenching conditions, the flame propagation speed was found to be very sensitive to small changes in mixture composition. The propagation velocity of the flame at near-quenching condition was found to be significantly less than that for adiabatic flames. However, for rich propane–air mixtures (Le < 1), propagation velocity increased because of curvature to values close to adiabatic flame speeds.

19.4 Effect of Heat Recirculation

The heat transfer from the flame zone plays an important role in the propagation and stabilization of the flame within a duct. Apart from the role of heat loss in flame extinction, as discussed in the previous section, heat transferred from the flame follows a complicated path through the structure from the postflame region to the preflame region. This preheats the reactants and helps in the combustion process. The early thermal model of flame propagation by Mallard and Le Chatelier (1883) states that the laminar flame speed, S_u, is proportional to the square root of the product of thermal diffusivity of unburned reactants and the chemical reaction rate, ω, as follows:

$$S_u = \sqrt{\alpha \frac{T_b - T_i}{T_i - T_u} \omega} \tag{19.11}$$

This theory applies to freely propagating flame. Leach et al. (2006) showed that the heat transfer path can be expressed as a thermal resistance network as shown in Figure 19.2a. In the figure, R_r, R_s, and R_{bl} represent the thermal resistances to conduction heat transfer in the gas and the structure and convective heat transfer from the gas to the structure, respectively. The resistances are given by

$$R_r = \frac{l_R}{k_r A_r}, \quad R_s = \frac{l_R}{k_s A_s}, \quad \text{and} \quad R_{bl} = \frac{2H}{k_r \mathrm{Nu} A_{conv}}. \tag{19.12}$$

Extending Mallard and Le Chatelier's theory to the above resistance network, Leach et al. (2006) derived an expression for the reaction zone thickness, assuming $l_R = S_u/\omega$,

$$l_R = \sqrt{\beta \alpha \frac{T_b - T_i}{T_i - T_u} \frac{1}{\omega}}. \tag{19.13}$$

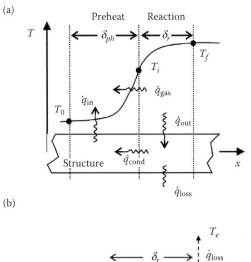

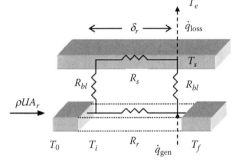

FIGURE 19.2
(a) Schematic of thermal exchange of a flame in close contact with a structure. q_{gas} and q_{cond} represent the heat conducted from the reaction zone to the preheat zone. (b) Thermal resistor network for visualizing heat exchange between the gas and the structure and heat loss to the environment. (From Leach, T.T., Cadou, C.P., and Jackson, G.S., Effect of structural conduction and heat loss on combustion in micro-channels, *Combustion Theory Model*, 10, 85–103, 2006. With permission.)

In the above equation, β is given by

$$\beta = \frac{1 + \zeta\lambda\left(1 + \dfrac{2H^2}{l_R^2}\dfrac{1}{\text{Nu}}\right)}{1 + 2\zeta\lambda\dfrac{H^2}{l_R^2}\dfrac{1}{\text{Nu}}}. \tag{19.14}$$

For no heat transfer (Nu = 0) or infinite channel height ($H \rightarrow \infty$), $\beta \rightarrow 1$ and the classical result of Mallard and Le Chatelier is retrieved. On the other hand, for large Nusselt numbers and small channel height, $\beta \rightarrow 1 + \zeta\lambda$, which is always greater than 1. This indicates that for small channel width and large heat transfer, the reaction zone becomes broader and is determined by the ratio of thermal conductivity of the gas and the structure and relative areas for axial heat conduction. The presence of the conductive path leads to additional heat transport from the flame zone to the unburned mixture upstream. This preheating of the reactants causes intensification of the burning rate. Consequently, these burners can be termed "excess enthalpy" or "superadiabatic" burners. Leach et al. (2006) extended their model to include the heat loss in determining the reaction zone thickness as

$$l_R = \frac{\dot{q}}{\rho C_p (T_i - T_u) \omega}. \tag{19.15}$$

In Equation 19.15, the reaction rate is adjusted to account for the "effective" flame temperature, T_b, as

$$\omega = \omega_{ad} \exp\left(-\frac{E_a}{RT_b} + \frac{E_a}{RT_{ad}}\right). \tag{19.16}$$

The rate of heat transferred from the reaction zone to the preheat zone is given as the difference between the heat generated by the reaction and the heat lost to the ambience. The analysis of Leach et al. (2006) shows that at large channel heights, the flame thickness approaches that for freely propagating flames and is primarily determined by the thermal conductivity of the gas. As the channel height decreases, the role of thermal conductivity of the solid wall starts becoming important. At about a channel height equal to approximately 10 times the thickness of the reaction zone (~5 mm), the transition from gas-to structure-dominated heat transfer occurs. This clearly establishes the importance of considering the heat recirculation through the wall in the analysis of microscale combustors.

Norton and Vlachos (2003, 2004) numerically simulated premixed combustion of methane (CH_4) and propane (C_3H_8) in microscale combustors. They showed that disappearance of stable flame in microchannels occurs either through blowout or extinction. Both are caused by lack of upstream heat transfer to the incoming reactants. In blowout, the reaction zone shifts significantly downstream, although the reaction rate is only slightly reduced. On the other hand, during extinction, the maximum reaction rate reduces and the reaction zone broadens, although the reaction zone shifts only slightly downstream. In blowout, the major heat loss occurs with the hot exhaust gases leaving the combustor, whereas in extinction, excess heat loss occurs through the wall to the surroundings. These works identified thermal conductivity of the combustor structure (wall), dimensions of the channel and the wall, heat loss to the ambience, and flow velocity as the key parameters in determining the flame characteristics. Thicker walls allow more upstream heat propagation and thus allow higher velocities and less conductive materials. However, increasing the channel gap width enhances the stability for highly conductive materials but decreases the stability with respect to blowout. These results indicate the need for microscale combustors with thick walls. The wall conductivity plays a dual role. It reduces the thermal resistance to both axial heat conduction from the reaction zone to the unburned gases upstream and transverse heat conduction across the wall to the ambience. Thus, ideally, the walls should have anisotropic thermal conductivity that allows axial heat conduction but inhibits transverse heat conduction. This would allow preheating of the reactants but minimize the heat loss to the ambience. The flow velocity also has a dual role. Increase in flow velocity decreases heat transfer upstream and promotes blowout. At the same time, decrease in flow velocity increases the conduction heat transfer and broadens the reaction zone causing excessive heat loss and extinction. Thus, flames can be stabilized within an optimum band of velocities, depending on the wall's thermal conductivity.

Barrios et al. (2008) investigated the role of the duct thickness on the quenching of premixed flames propagating in narrow channels. They identified a heat loss parameter,

$\beta = \rho^* C^* h (1 + h/2a)$, for characterizing the flame behavior, where ρ^* and C^* denote the ratio of densities and specific heats of the wall material and the gas, and h and a denote the dimensionless wall thickness and duct radius (both scaled with the flame thickness), respectively. Because of the large value of density ratio, this parameter has a very large value for most practical conditions. For such conditions, the wall temperature at the flame location is close to the temperature of the unburned gas. Heat loss to the wall becomes very important, and quenching in this regime is mainly due to thermal effects.

Ju and Xu (2005) investigated flame propagation and extinction in mesoscale channels theoretically and experimentally. They found that for zero flow rate of the gas, the variation of propagation velocity with heat loss is similar to that of freely propagating flames discussed earlier. The flame is not sustained when the heat loss is above a critical parameter. Below that critical value of the heat loss parameter, for each value of the heat loss parameter, two solutions exist. One solution, higher than the propagation speed at critical heat loss, is stable, whereas another solution, lower than the propagation velocity at critical heat loss, is unstable. Since only stable solutions are actually found in practice, the propagation velocity is maximum at adiabatic conditions and decreases monotonically as the heat loss is increased till the critical value is reached. However, in the presence of gas flows, the dependence is different from that at zero flow rate as shown in Figure 19.3. At very low flame speeds, a new flame branch exists. This new flame branch is caused by the recirculation of heat from burned to unburned gas through the solid wall. For high values of wall thermal capacity, this branch exists only for low values of heat loss, whereas at lower values of the wall thermal capacity, this lower branch exists for high values of heat loss (or small-diameter tubes). This implies that at small scales, for low values of wall thermal capacity, flames can exist for tubes with diameter much smaller than the extinction diameter or with heat loss much larger than the critical heat loss obtained for freely propagating flames. The existence of this slow flame regime was experimentally demonstrated by Ju and Xu (2005, 2006) and Maruta et al. (2005).

Maruta et al. (2005) also observed a new dynamic flame regime in narrow channels with temperature gradients in the wall, in which the flames were repeatedly extinguished and reignited. They termed this regime flames with repetitive extinction and ignition (FREI). This regime was observed for mixture velocities between the conventional flame and the slow flame regimes. This experimental observation was successfully predicted by the theoretical work of Minaev et al. (2007) and the numerical work of Kessler and Short (2008). The work of Kessler and Short (2008) shows that as the heat from the exit of the channel propagates down the microburner walls, a reaction wave is driven rapidly down the channel toward the inlet via a sequence of oscillatory ignition and quenching transients. They also demonstrated a wide range of flame dynamics like FREI, oscillatory combustions, and combinations of the two in the presence of initial temperature gradients in the wall. All these features have been observed experimentally.

In a novel design of microcombustors, Kumar et al. (2007a, 2007b, 2008) and Fan et al. (2009) considered flame propagation in reacting mixtures in radial microchannels formed by two heated parallel circular disks. The design was motivated by a proposed design of microcombustors for microscale gas turbines (Wu et al., 2005). The reacting mixture is admitted near the axis and it flows radially outward. An advantage of this arrangement is that as the mixture flows outward, heat is transferred from the wall. The increased temperature implies an increase in flame speed as the mixture moves radially outward. On the other hand, the flow velocity decreases as the mixture moves radially outward. This generates a mechanism for stabilizing the flame at a given radial location. However, for channel gaps less than the quenching distance, a variety of steady and unsteady flame

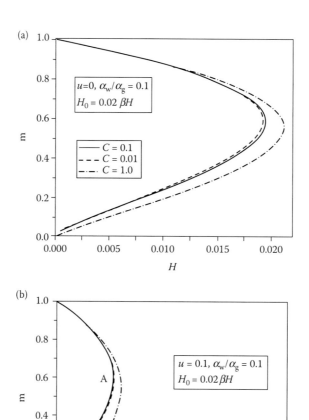

FIGURE 19.3
Dependence of flame speed on normalized heat transfer rate at (a) $U = 0$ and (b) $U = 0.1$ for various wall heat capacities. (Reprinted from *Proceedings of the Combustion Institute*, 30, Ju, Y. and Xu, B., Theoretical and experimental studies on mesoscale flame propagation and extinction, pp. 2445–2453, Copyright 2005, with permission from Elsevier.)

regimes were observed under different mixture equivalence ratios and mixture velocities as shown in Figures 19.4 and 19.5.

Heat recirculation generates regions of locally high temperature and thus facilitates combustion outside the range of conventional flammability limits. Such combustion is referred to as superadiabatic combustion. Ronney (2003) developed a simplified model of heat-recirculating combustors. In this work, he considered two configurations: a once-through conducting tube (in which the reactants enter from one end and leave from the other) and a counterflow configuration where the unburned and burned gases flow through parallel passages in opposite directions, separated by a conducting wall. In both the configurations, the reaction took place at one end of the tube. He simplified the analysis by assuming a well-stirred reactor model for the reacting zone. Extinction

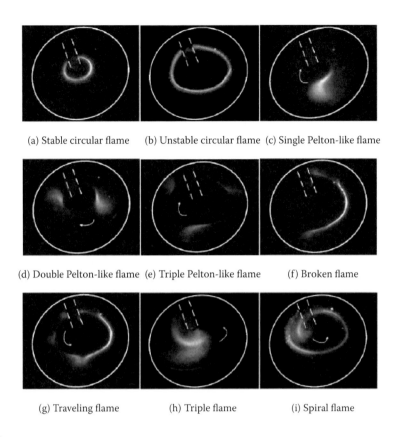

(a) Stable circular flame   (b) Unstable circular flame (c) Single Pelton-like flame

(d) Double Pelton-like flame (e) Triple Pelton-like flame   (f) Broken flame

(g) Traveling flame   (h) Triple flame   (i) Spiral flame

FIGURE 19.4
Photos of various flame patterns observed in radial microchannels: (a) stable circular flame, (b) unstable circular flame, (c) single Pelton-like flame, (d) double Pelton-like flame, (e) triple Pelton-like flame, (f) broken flame, (g) traveling flames, (h) triple flames, and (i) spiral flame. Solid and dashed white lines indicate the position of the top plate and mixture delivery tube, respectively. Arrows indicate the rotating direction of the rotating flame patterns. (Reproduced from *Proceedings of the Combustion Institute*, 32, Fan, A., Minaev, S., Sereshchenko, E., Fursenko, R., Kumar, S., Liu, W. and Maruta, K., Experimental and numerical investigations of flame pattern formations in a radial microchannel, pp. 3059–3066, Copyright 2009, with permission from Elsevier.)

limits were predicted at both high and low mass flow rates. At large mass flow rates, the well-known blow-off limit of well-stirred reactors was attained, whereas at low mass flow rates, the flame was extinguished due to heat losses, which are unimportant at higher flow rates. The extinction at low mass flow rates was obtained only if wall conduction was considered. With wall conduction, some heat is transferred from the reaction zone to the gas and from there it is ultimately lost to the ambience, leading to the low mass flow rate limit. The counterflow configuration was found to be a superior configuration of heat-circulating burner as lower fuel concentrations are needed to sustain combustion at a given mass flow rate. Ju and Choi (2003) investigated two flames propagating in opposite directions in parallel channels separated by a thin wall and obtained closed-form solutions for flame speed, flame temperature, and distance between two flames. The heat recirculation through the coupling between the two flames drastically increased the flame speed and extended the flammability limit. Flames adjusted their position depending on heat recirculation, heat loss to ambience, and fuel concentration. Two flame modes, a fast mode and a slow one, were observed. With an increase in the separation

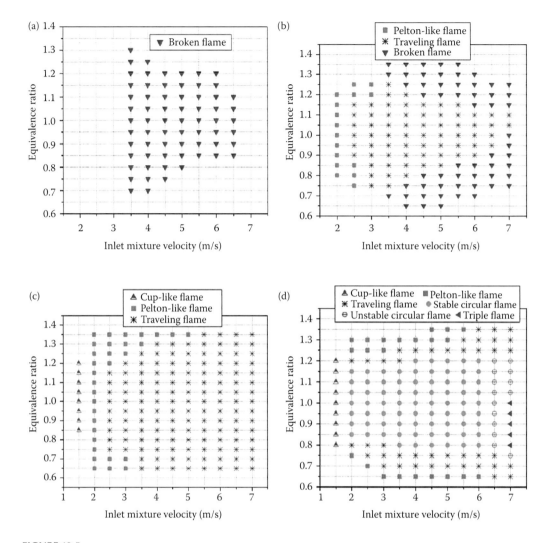

FIGURE 19.5
Regime diagram for flame patterns of different spacings between plates (a) 0.5 mm, (b) 1 mm, (c) 1.5 mm, and (d) 2 mm. (Reproduced from *Proceedings of the Combustion Institute*, 32, Fan, A., Minaev, S., Sereshchenko, E., Fursenko, R., Kumar, S., Liu, W. and Maruta, K., Experimental and numerical investigations of flame pattern formations in a radial microchannel, pp. 3059–3066, Copyright 2009, with permission from Elsevier.)

between the flames, the speed of the fast mode decreased whereas that of the slow mode increased. A maximum flame separation was observed beyond which flames were not obtained for equivalence ratios below the fundamental limit. Schoegl and Ellzey (2007) studied superadiabatic combustion in combustors of finite length consisting of two parallel channels separated by a conducting wall. They considered both coflow and counterflow configurations. Their study showed that the dominant heat transfer mechanisms for coflow and counterflow configurations are axial conduction and heat transfer across the wall, respectively. Consequently, the flame stability is most sensitive to axial conduction and interfacial heat transfer in the former case and to the length of the combustor and interfacial heat transfer in the latter case. The results established the superiority of the counterflow configuration. The counterflow burners could operate well beyond the conventional flammability limits and at velocities much higher than the laminar flame speed. In addition, the heat-recirculating mechanism was found to anchor the flames in a self-stabilizing process over a large range of flow velocities at a given equivalence ratio. This allowed a large turn-down ratio of the burner.

An efficient design of microscale combustors that have been widely investigated is the "Swiss-Roll" combustor. A Swiss-Roll combustor is a counterflow combustor in which the combustor volume is at the center, and a pair of long spiral channels is provided for heat recirculation from burned gases to unburned gases. Richecoeur and Kyritsis (2005) investigated flame propagation in a curved duct, which was motivated by the Swiss-Roll geometry. They found that the flame behavior in mesoscale curved ducts is significantly different from that of large-scale straight tubes. They investigated tube diameters of the order of flame thickness and Reynolds number of the range of 10–100. A stable oscillation, including frequent extinction and reignition, similar to FREI observed by Maruta et al. (2005), was observed at intermediate Reynolds number for which the flame neither moved upstream nor remained in a stable position. The precise range of Reynolds number for which the oscillatory regime could be observed was a strong function of the tube curvature. Figure 19.6 shows the various combustion regimes for a straight tube and a curved tube. It is observed that the stable flame regime is almost doubled.

Kim et al. (2005) investigated flame stability in small Swiss-Roll combustors that can be used as potential heaters. They investigated different designs of combustors, each of them having an outer diameter of 64 mm. As the size of the combustor decreased, its efficiency increased and NO_x emission decreased while the emission of CO increased. However, with addition of a catalytic reactor at the exit, the emission of CO could be eliminated. The study also investigated the effect of gas velocity on the stability of the flame. Figure 19.7a shows the minimum velocity required for flame stabilization at different equivalence ratios for different combustion designs. Stable flames were obtained for conditions above each curve. "S," "W," and "D" represent standard, wide, and deep combustors, respectively. However, blowout was not observed in any of these cases. Kim et al. (2007) extended the earlier work to investigate the scale effects on the performance of Swiss-Roll combustors.

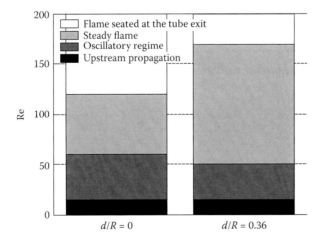

FIGURE 19.6
Stabilization regimes in a 4-mm ID tube as a function of Re for $d/R = 0.36$ and for a straight duct (d and R represent the tube diameter and tube radius of curvature, respectively). (Reproduced from *Proceedings of the Combustion Institute*, 30, Richecoeur and Kyritsis, Copyright 2005, with permission from Elsevier.)

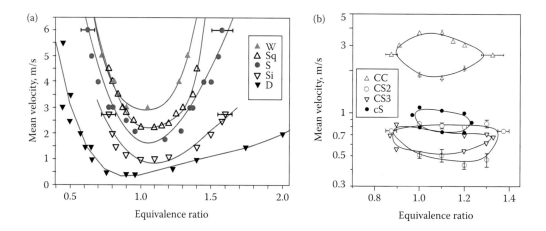

FIGURE 19.7

(a) Flame stabilization conditions for various combustors and different heat transfer conditions. (Reproduced from *Combustion and Flame*, 141, Kim, N.I., Kato, S., Kataoka, T., Yokomori, T., Maruyama, S., Fujimori, T., et al. Flame stabilization and emission of small Swiss-roll combustors as heaters, pp. 229–240, Copyright 2005, with permission from Elsevier.) (b) Experimental results of flame stabilization of coin-size combustors. (Reproduced from *Proceedings of the Combustion Institute*, 31, Kim, N.I., Aizumi, S., Yokomori, T., Kato, S., Fujimori, T. and Maruta, K., Development and scale effects of Swiss-roll combustors, pp. 3243–3250, Copyright 2007, with permission from Elsevier.)

They reduced the combustors to coin-size combustors, having outer diameters of 26 mm (CS2, CS3, and CC combustors) and 20 mm (cS combustor). Figure 19.7b shows the stability region for these combustors. Contrary to the cases shown in Figure 19.7a, in these combustors, an upper limit of flow velocity above which blowout occurs exists.

Chen and Buckmaster (2004) modeled heat transfer and combustion in Swiss-Roll combustors. They unwrapped the spiral channels into straight sections and assumed Poiseuille flow in the channels to eliminate the need for solving momentum equations. Their model, consequently, did not capture the effect of curvature of the channels. They found that with an increase of Reynolds number or equivalence ratio, the reaction rate increases and the flame moves away from the center toward the inlet. Increasing wall thermal conductivity (or equivalently wall thickness) causes the extinction limit to increase, although the effect becomes less important at higher Reynolds number. Kim et al. (2007) developed a quasi-one-dimensional model for Swiss-Roll burner, in which they considered the effect of spiral geometry. They used the model to design miniaturized combustors of different size and explained their experimental observations. Kuo and Ronney (2007) developed a two-dimensional model of Swiss-Roll combustor using commercial CFD software FLUENT. The model predicts flame structures that are distinctly different from that of conventional premixed flames. The model predicted extinction limits that were in reasonable quantitative agreement with experiments. For higher Reynolds number (Re > 500), inclusion of turbulent flow and transport was necessary for this agreement. Heat conduction along the heat exchanger wall and thermal radiation had major influence on the extinction limits. In addition to the weak-burning extinction limits, strong-burning limits in which the reaction zone moved out of the combustor center toward the inlet were also predicted by the numerical model, in agreement with experiments. Several factors such as turbulence, radiation, and wall conduction played important roles in the predicted extinction limits, but their relative importance depends on the flow's Reynolds number.

19.5 Catalytic Effects

Apart from heat recirculation, use of catalysts also plays a major role in flame stabilization in ducts and channels whose dimensions are smaller than the quenching distance. Raimondeau et al. (2002) showed that both heat loss and radical quenching at the walls play crucial roles in flame propagation in microchannels. Kim et al. (2006) carried out experiments to assess the roles of heat loss and radical quenching on flame extinction in microchannels. Their results indicate that at low wall temperatures (100°C–350°C), quenching is mainly determined by heat loss, with wall conditions playing an insignificant role. At temperatures above 400°C, flame quenching depends on the heterogeneous reaction of the radicals at the surface, and the quenching distance increases with the surface temperature. However, for wall temperatures above 600°C, the homogeneous reactions start to dominate the radical quenching at the walls and flames can resist quenching. In catalytic reactors, the presence of catalysts on the combustor wall sustains combustion over a longer range and also prevents radical quenching. The high surface to volume ratio of the microscale reactors increases the flux of the reactants to the catalytic surface. Maruta et al. (2002) investigated the extinction limits of combustion in a microtube in the presence of catalysts such as platinum. They observed that the extinction curve showed U-shaped characteristics in which the extinction limit increased/decreased with Reynolds number in the small/large Reynolds number range. The former was caused by heat loss, whereas the latter was due to insufficient residence time in the reaction zone (blowout). The heat loss and the blowout limits were characterized by small/large amount of Pt(s) on the wall. From the study, the authors concluded that for catalytic microburners, stoichiometric mixtures with exhaust gas recirculation were more effective in lowering the flame temperatures compared with the use of lean mixtures. Chen et al. (2007) numerically simulated hydrogen–air combustion in a catalytic microtube. They identified competing effects of catalytic reactions on homogeneous combustion in a microtube. On the one hand, heterogeneous reactions consume part of the fuel, thereby weakening the homogeneous reaction. On the other hand, high temperature and radicals produced on the wall strengthen the homogeneous reaction and help sustaining the heat loss. They found that at low velocities and relatively large diameters, the homogeneous (gas phase) reaction dominates the heterogeneous (surface) reaction, although the former is affected by the latter. As velocity is increased and tube diameter is decreased, heterogeneous reaction starts to dominate the homogeneous reactions to the extent that homogeneous combustion cannot be sustained within the tube. However, the gas phase reactions still affect the intermediate-species concentrations. Further increase in velocity and reduction in tube diameter make the homogeneous (gas phase) reactions insignificant. They observed that in a catalytic combustor, the homogeneous reactions are initiated and sustained by the heat from the heterogeneous reactions. Consequently, wall thermal conductivity has a less important role than in noncatalytic reactors.

19.6 Applications of Microscale Combustors

Research interest in microscale combustors has been fuelled by a number of potential MEMS applications. The major areas of these applications are to produce power or to be used as fuel reformers in microfuel cells. A MEMS-based gas turbine power generator has been developed at Massachusetts Institute of Technology (Spadaccni et al., 2003; Wu et al., 2005), which produced 10–50 W of power in a volume of less than 1 cm^3 and consumed 7 g of fuel per hour. Such a device represents a 10-fold increase in power density over conventional batteries. At the University of California, Berkeley, a mini Wankel engine has been developed (Fu et al., 2001a, 2001b) with sizes that are well below those that are commercially available. Microrockets for applications such as microsatellites have been developed using both solid (Rossi et al., 2002) and liquid (Yetter et al., 2001) propellants. Weinberg et al. (2002) developed thermoelectric devices coupled with counterflow heat-recirculating combustors for power generation in microscale devices, with improved thermal efficiency. A microthermophotovoltaic system has been developed as a novel power MEMS concept (Yang et al., 2003a, 2003b). The device consists of a micro-SiC combustor, where hydrogen is burnt and emitters (radiators) and photovoltaic cells in which 3–10 W of electric power is developed in a package of volume less than 1 cm^3.

19.7 Conclusions

Combustion in microscale burners promises to be a major energy source for MEMS devices, particularly because of the extremely high specific energy in fuels. However, owing to the small residence time of the reactants in the reaction zone and the high surface to volume ratio, it is difficult to sustain the flame in microscale devices. Flame stabilization in devices with dimensions smaller than the quenching distance has been achieved primarily through two mechanisms, thermal and chemical. In the thermal approach, heat recirculation from the burned gas to the unburned gas through the solid structure helps stabilizing the flame. In the chemical approach, the combustor walls are treated with special chemical agents, which, on one hand, prevent quenching of active radicals at the wall while, on the other hand, promote heterogeneous chemical reactions at the wall, generating heat and radicals that help sustaining combustion in the gas phase. Microscale combustors have been used to develop a number of novel applications such as microscale gas turbines, microscale internal combustion engines, and microscale thermophotovoltaic and thermoelectric devices.

Nomenclature

A_{conv} Area for convection between gas and wall in Equation 19.12 (m^2)

A_r Cross-sectional area for conduction in gas in Equation 19.12 (m^2)

A_s Cross-sectional area for conduction in solid structure in Equation 19.12 (m^2)

C_p Specific heat at constant pressure for gas (J/kg K)

f Mass flux through flame (kg/m^2 s^{-1})

f^0 Mass flux through freely propagating planar adiabatic flame (kg/m^2 s^{-1})

H Channel height in Equation 19.12 (m)

k_r Thermal conductivity of gas in Equation 19.12 (W/mK)

k_s Thermal conductivity of solid structure in Equation 19.12 (W/mK)

l_D^0 Thickness of freely propagating planar adiabatic flame (m)

l_R Reaction zone thickness in flame (m)
L_V Rate of volumetric heat loss from flame (W/m³)
Nu   Nusselt number (dimensionless)
p   Pressure (Pa)
q_c Heating value of fuel (J/kg)
S_u Flame speed (m/s)
S_u^0 Flame speed of freely propagating planar adiabatic flame (m/s)
T_a Activation temperature (K)
T_{ad} Adiabatic flame temperature (K)
$\tilde{T}_{ad}$ Dimensionless adiabatic flame temperature ($= C_p T_{ad}/q_C Y_u$)
T_b Temperature of burned gas (K)
$\tilde{T}_b$ Dimensionless burned-gas temperature ($= C_p T_b/q_C Y_u$)
T_b^0 Temperature of burned gas for freely propagating planar adiabatic flame (K)
T_i Ignition temperature (K)
T_u Temperature of unburned gas (K)
u_0 Velocity of unburned mixture (m/s)
Y_u Fuel mass fraction in unburned mixture (dimensionless)
Ze   Zeldovich number (dimensionless)
α Thermal diffusivity of gas (m²/s)
λ Thermal conductivity of gas (W/mK)
ν Kinematic viscosity of gas (m²/s)
ρ_u Density of unburned gas (kg/m³)
ω Volumetric fuel consumption at the flame (kg/m³ s⁻¹)
ω_{ad} Volumetric fuel consumption rate for adiabatic flame (kg/m³ s⁻¹)
ω_b^0 Volumetric fuel consumption rate at burned-gas temperature for freely propagating adiabatic flame (kg/m³ s⁻¹)
$\bar{\omega}^0$ Volumetric fuel consumption rate at average gas temperature for freely propagating adiabatic flame (kg/m³ s⁻¹)

References

Barrios E, Prince JC, Trevino C. (2008). The role of duct thickness on the quenching process of pre-mixed flame propagation. *Combust Theory Model*. 12:115–33.

Chakraborty S, Mukhopadhyay A, Sen S. (2008). Interaction of Lewis number and heat loss effects for a laminar premixed flame propagating in a channel. *Int J Therm Sci*. 47:84–92.

Chen G-B, Chen C-P, Wu C-Y, Chao Y-C. (2007). Effect of catalytic walls on hydrogen/air combustion in a microtube. *Appl Catal A Gen*. 332:89–97.

Chen M, Buckmaster J. (2004). Modelling of combustion and heat transfer in 'Swiss roll' micro-scale combustors. *Combust Theory Model*. 8:701–20.

Cui C, Matalon M, Daou J, Dold J. (2004). Effects of differential diffusion on thin and thick flames propagating in channels. *Combust Theory Model*. 8:41–64.

Daou J, Matalon M. (2002). Influence of conductive heat losses on the propagation of premixed flames in channels. *Combust Flame*. 128:321–39.

Fan A, Minaev S, Sereshchenko E, Fursenko R, Kumar S, Liu W, et al. (2009). Experimental and numerical investigations of flame pattern formations in a radial microchannel. *Proc Combust Inst*. 32:3059–66.

Fernandez-Pello AC. (2002). Micropower generation using combustion: issues and approaches. *Proc Combust Inst.* 29:883–99.

Fu K, Knobloch AJ, Martinez FC, Walther DC, Fernadez-Pello C, Pisano AP, et al. (2001a). Design and fabrication of a silicon-based MEMS rotary engine. ASME International Mechanical Engineering Congress and Exposition, New York.

Fu K, Knobloch A, Martinez F, Walther D, Fernandez-Pello AC, Pisano A, et al. (2001b). Design and experimental results of small-scale rotary engines. Proceedings of the 2001 International Mechanical Engineering Congress and Exposition (IMECE); 2001 November 11–16; New York: American Society of Mechanical Engineers.

Gutkowski A, Tecce L, Jarosinski J. (2008). Some features of propane-air flames under quenching conditions in narrow channels. *Combust Sci Tech.* 180:1772–87.

Ju Y, Choi CW. (2003). An analysis of sub-limit flame dynamics using opposite propagating flames in mesoscale channels. *Combust Flame.* 133:483–93.

Ju Y, Xu B. (2005). Theoretical and experimental studies on mesoscale flame propagation and extinction. *Proc Combust Inst.* 30:2445–53.

Ju Y, Xu B. (2006). Effects of channel width and lewis numbers on multiple flame regimes and propagation limits in mesoscale. *Combust Sci Tech.* 178:1723–53.

Kessler DA, Short M. (2008). Ignition and transient dynamics of premixed flames in microchannels. *Combust Theory Model.* 12:809–29.

Kim KT, Lee DH, Kyon S. (2006). Effects of thermal and chemical surface-flame interaction on flame quenching. *Combust Flame.* 146:19–28.

Kim NI, Kato S, Kataoka T, Yokomori T, Maruyama S, Fujimori T, et al. (2005). Flame stabilization and emission of small Swiss-roll combustors as heaters. *Combust Flame.* 141:229–40.

Kim NI, Aizumi S, Yokomori T, Kato S, Fujimori T, Maruta K. (2007). Development and scale effects of Swiss-roll combustors. *Proc Combust Inst.* 31:3243–50.

Kuo CH, Ronney PD. (2007). Numerical modeling of non- adiabatic heat-recirculating combustors. *Proc Combust Inst.* 31:3277–84.

Kumar S, Maruta K, Minaev S. (2007a). Pattern formation of flames in radial microchannels with lean methane-air mixtures. *Phys Rev E.* 75:016208.

Kumar S, Maruta K, Minaev S. (2007b). On the formation of multiple rotating Pelton-like flame structures in radial microchannels with lean methane-air mixtures. *Proc Combust Inst.* 31:3261–8.

Kumar S, Maruta K, Minaev S, Fursenko R. (2008). Appearnce of target patterns and spiral flames in radial microchannels with CH_4–air mixtures. *Phys Fluids.* 20:024101–1–024101–11.

Law CK. (2006). *Combustion Physics.* New York: Cambridge University Press.

Leach TT, Cadou CP, Jackson GS. (2006). Effect of structural conduction and heat loss on combustion in micro-channels. *Combust Theory Model.* 10:85–103.

Mallard E, Le Chatelier HL. (1883). *Ann Mines.* 4:379.

Maruta K, Kataoka T, Kim NI, Minaev S, Fursenko R. (2005). Characteristics of combustion in a narrow channel with a temperature gradient. *Proc Combust Inst.* 30:2429–36.

Maruta K, Takeda K, Ahn J, Borer K, Sitzki L, Ronney PD, Deutschmann O. (2002). Extinction limit of catalytic combustion in microchannels. *Proc Combust Inst.* 29:957–63.

Minaev S, Maruta K, Fursenko R. (2007). Nonlinear dynamics of flame in a narrow channel with a temperature gradient. *Combust Theory Model.* 11(2):187.

Norton DG, Vlachos DG. (2003). Combustion characteristics and flame stability at the microscale: a CFD study of premixed methane/air mixtures. *Chem Eng Sci.* 58:4871–82.

Norton DG, Vlachos DG. (2004). Combustion characteristics and flame stability at the microscale: a CFD study of propane/air microflame stability. *Combust Flame.* 138:97–107.

Raimondeau S, Norton D, Vlachos DG, Masel RI. (2002). Modeling of high temperature microburners. *Proc Combust Inst.* 29:901–7.

Richecoeur F, Kyritsis DC. (2005). Experimental study of flame stabilization in low Reynolds and Dean number flows in curved mesoscale ducts. *Proc. Combust. Inst.* 30:2419–2427.

Ronney PD. (2003). Analysis of non-adiabatic heat-recirculating combustors. *Combust Flame.* 135:421–39.

Rossi C, Orieux S, Larangot B, Do Conto T, Estève D. (2002). Design fabrication and modeling of solid propellant microrocket—application to micropropulsion. *Sens Actuators A Phys*. 99:125–33.

Schoegl I, Ellzey JL. (2007). Superadiabatic combustion in conducting tubes and heat exchangers of finite length. *Combust Flame*. 151:142–59.

Spadaccini CM, Mehra A, Lee J, Zhang X, Lukachko S, Waitz IA. (2003). High power density silicon combustion systems for micro gas turbine engines. *J Eng Gas Turbines Power*. 125:709–19.

Turns SR. (2000). *An Introduction to Combustion: Concepts and Applications*. Singapore: McGraw Hill International Edition.

Weinberg FJ, Rowe DM, Min G, Ronney PD. (2002). On thermoelectric power conversion from heat recirculating combustion systems. *Proc Combust Inst*. 29:941–7.

Wu M, Hua J, Kumar K. (2005). An improved micro-combustor design for micro gas turbine engine and numerical analysis. *J Micromech Microeng*. 15:1817–23.

Yang WM, Chou SK, Shu C, Xue H, Li ZW, Li DT, et al. (2003a). Microscale combustion research for application to micro thermophotovoltaic systems. *Energy Convers Manag*. 44:2625–34.

Yang WM, Chou SK, Shu C, Li ZW, Xue H. (2003b). Research on micro thermophotovoltaic power generators. *Solar Energy Mater Solar Cells*. 80:95–104.

Yetter RA, Yang V, Milius DL, Aksay IA, Dryer FL. (2001). Development of a Liquid Propellant Microthruster for Small Spacecraft. Eastern States Section Meeting of the Combustion Institute; 2001 December 2–5; Hilton Head Island, SC.

20

Methane Solubility Enhancement in Water Confined to Nanoscale Pores

Mery Diaz Campos, I. Yucel Akkutlu, and Richard F. Sigal

CONTENTS

20.1 Introduction

It is well known that phase equilibrium conditions in porous media depend on the pore size. For example, the freezing point of water is lowered in small pores, so that measurements of the electrical conductivity of subsurface rocks in permafrost regions show shale zones that are dominated by micro- to nanometer size pores to be conductive but adjacent sands with much larger size pores to be very resistive. In a porous medium characterized by the presence of small pores and occurrences of formation water (e.g., free and inherent moisture), such as coal and shale, gas can be stored not only in its gaseous phase but also as dissolved gas in the pore water. For the gas dissolved in water in large pores, it is known that the dissolved gas is only a small fraction of the amount of gas the pore could hold if it was primarily gas that was not water-saturated and was at a pressure equal to the hydrostatic pressure. The amount of dissolved gas in water depends on its solubility, which should increase in smaller pores. The amount of dissolved gas in water then may represent a significant portion of the total gas reserve. The latter argument suggests that we consider gas solubility in water confined to small pores.

The solubility of methane in water, which is given in the standard tables, has been established through measurements in macroscopic pressure cells. In this chapter, the solubility of gas in water confined to a small pore is estimated by two different methods. The first is an analytical method where simple but reasonable approximations are used to show that in very small pores CH_4 solubility should be somewhat larger than the macroscopic measurement values. The analytical calculation is based on the assumption that a minimum number of gas molecules is required for an aggregate to behave like a classical fluid with a surface tension and, from the definition of solubility, when a solution is minimally super-saturated, that number in the aggregate must be only a small percentage of the number of gas molecules left in solution. An implicate assumption in the calculation is that water wets the pore wall, that is, it is hydrophilic. Depending on the values of parameters in this calculation, the mass of gas per unit volume of water that is contained in a very small water-filled pore could approach the density of gas at high hydrostatic pressure or even that of liquid methane. This result is analogous to the large amount of methane stored in a methane hydrate where the water cages act to hold the gas in place.

The second solubility estimate for methane in a small pore is made using equilibrium molecular dynamics (MD) simulations. There is a widespread opinion in the literature that breakthroughs in solution chemistry tend to come from computer simulations and particularly from free-energy calculations. It is now possible to evaluate with a reasonable accuracy the free-energy change accompanying the dissolution of a solute such as methane in a liquid solvent such as water. One of the techniques that have been used widely for this purpose is the test-particle method originally developed by Widom (1963). A so-called test particle is introduced in a random location in a solvent, and the difference in potential energy before and after the insertion is computed. For pairwise interactions, this would be the interaction of potential energy between the randomly placed test particle and the N particles that the solvent system is composed of minus the potential energy of the solvent itself. Although the Widom test-particle method involves finding by a random process a configuration of the water system in which a methane molecule could be accommodated, the process of finding a cavity can be significantly improved using the excluded volume map sampling (EVMS) method. In our case, the pore walls comprise graphene, that is, which consists of layers of carbon atoms, which is not strongly hydrophilic. (The contact angle of the water droplet on the graphite surface is estimated to be 86° using molecular simulation; Werder et al., 2003.) Also, a significant methane solubility enhancement is predicted for nanometer size pores using the test-particle method. It is perhaps remarkable that enhanced solubility is observed for a wide range of pore-wall wettability.

20.2 Methane Solubility in Water Confined to Small Pores

The solubility at a given temperature and pressure of methane in water is defined as the maximum saturation of CH_4 that the water can hold. At this saturation, any small decrease in pressure or addition of more gas molecules will lead to a drop (bubble) of gas forming in the solution. When the CH_4 solution is confined to a small container, a limited amount of gas is available to form the CH_4 drop. This combined with the conditions that the gas drop has a minimum size and the initial drop formed must contain only a small fraction of the total amount of gas leads to the increase in solubility of methane. The minimum size follows from the need for the gas drop to contain enough molecules to behave as a separate phase with a well-defined surface.

For a spherical drop of gas in a brine solution, the gas pressure P_g is related to the water pressure P_w by Equation 20.1

$$P_g = P_w + \frac{2\gamma}{r_b},$$

(20.1)

where γ is the gas–water surface tension and r_b is the radius of the gas drop. The second term is referred to as the capillary pressure. Using a common equation of state for gas, the gas pressure in the drop is given by

$$P_g = z\left(\frac{n_b}{V_b}\right)RT, \tag{20.2}$$

where z is the compressibility factor, n_b is the number of moles of CH_4 contained in the drop, V_b the volume of the spherical drop, R the universal gas constant, and T is the absolute temperature (McCain, 1990). The compressibility factor z is a function of pressure and temperature but can be evaluated numerically using commercially available computer codes. For this work, the NIST program *Supertrap* was used to evaluate z. Combining Equations 20.1 and 20.2 gives

$$n_b = \left(\frac{(4/3)\pi r_b^3}{zRT}\right)\left(P_w + \frac{2\gamma}{r_b}\right). \tag{20.3}$$

The moles of gas in the methane drop are sourced from the CH_4 dissolved in the water. Before the formation of the first gas drop, the brine had a saturation of ρ_{si} moles of methane per unit volume. After the formation of the first gas drop, its saturation is ρ_s, where ρ_s is the solubility of methane in brine at P_w and T. The supersaturated value ρ_{si} and the saturated value ρ_s can be related by Equation 20.4,

$$\rho_{si} = (1 + \alpha)\rho_s. \tag{20.4}$$

In Equation 20.4, α is positive and, in the classical limit, small because after the formation of the gas drop, the methane in the drop is in equilibrium with the methane in solution, which forces the saturation to be ρ_s, and before the formation of the drop, the solution could be arranged to be minimally supersaturated.

Equation 20.4 implies that for a container of volume V_p

$$n_b = \alpha \rho_s V_p. \tag{20.5}$$

This leads to Equation 20.6, which provides a relationship to estimate the solubility of CH_4 in brine in a small pore,

$$\rho_s = \left(\frac{(4/3)\pi r_b^3}{zRT\,\alpha V_p}\right)\left(P_w + \frac{2\gamma}{r_b}\right). \tag{20.6}$$

Examination of Equation 20.6 shows that ρ_s increases with increasing r_b and decreasing α. The value of r_b must be large enough for the gas in the drop to behave like a single-phase fluid with a well-defined surface tension.

The critical point for methane is −82.6°C, 666 psi (4.59 mPa). At this point, liquid methane has a density of 0.029 mol/cm³ (0.464 g/cm³). Methane gas has a density of 0.023 mol/cm³ (0.368 g/cm³) at 20,000 psi (138 mPa) and room temperature. The liquid density provides a reasonable upper limit to how closely methane molecules can be packed in the gas drop. At this density, a 0.5-nm radius sphere contains 9 molecules, a 1.0-nm radius sphere contains 73 molecules, and a 2.0-nm radius sphere contains 584 molecules of CH_4. Nine molecules are almost certainly too few to form an independent gas phase, and 584 should be more than adequate. This implies that the minimum drop size is in the order of 1 nm.

The more difficult parameter to estimate is α. There is not an obviously easy way to provide a good bound on it. By the definition of solubility, $\rho_{si} - \rho_s$ should be approximated by a differential in the limit of large water and gas volumes. For an undersaturated system, on an average, CH_4 molecules will be uniformly distributed. However, there will be statistical fluctuations in CH_4 density. As the saturation approaches the supersaturated state, these density fluctuations provide the nuclei for the formation of gas drops. The larger the value of α, the larger the percentage of the CH_4 molecules that have to be contained in a single density fluctuation. For α equal to 0.1, 10% of the available CH_4 would have to be contained in a single density fluctuation.

Equation 20.6 can be used to examine gas solubility for shale gas reservoir conditions. For the calculations that follow, the minimum drop size has been taken as 1 nm. The most prolific shale gas reservoir is currently the Barnett shale. The major productive areas of the Barnett are located near Dallas, Texas. The Barnett reservoir has a temperature of approximately 80°C and a water pressure of 4000 psi (27.6 mPa). The water in the Barnett, as with most reservoirs, is saline. For this pressure and temperature, a reasonable value for γ is 25 dyn/cm (Schowalter, 1979).

In the reservoir using this value, $2\gamma/r_b$ for a 1-nm spherical drop has a value of 7250 psi (50 mPa), so the total gas pressure in a 1-nm drop is 11,250 psi (862 mPa). Then from Equation 20.3, the number of molecules in a 1-nm radius methane drop is 43. A value of z of 1.52 has been used in this calculation. At 4000 psi (27.6 mPa) and 80°C, solubility of methane in pure water as determined by measurement in a large cell is 18 SCF/bbl. At 50,000 ppm salinity, this is reduced by a factor of 0.8 (McCain, 1990). In standard units, this is 1.76×10^{-3} g/cm³. Also, methane gas at 4000 psi (27.6 mPa) and 80°C has a density of 1.49×10^{-1} g/cm³.

Figure 20.1 shows the estimated solubility of methane as a function of α for various pore sizes at these reservoir conditions. The results clearly indicate that solubility rapidly increases with decreasing α and decreasing pore size. Table 20.1 presents the methane solubility for a range of pore volumes for $r_b = 1.0$ nm and $\alpha = .01$. Note that for a container size of 50 nm, we estimate solubility less than the bulk value. This suggests that an α value of 0.01 is not unreasonable to use for the pore volumes in Table 20.1. With that assumption, the methane stored as dissolved gas in brine for a 1000-nm³ pore volume is comparable with the amount of methane that would be stored in the pore if it was just filled with gas at the same reservoir conditions.

20.3 Molecular Simulation Approach

Molecular simulations have made enormous strides in recent years and are gradually becoming a ubiquitous tool in science and engineering. This is due to the development of new methods for description of complex fluids and materials and due to the availability of high-speed and large-capacity computers to an increasing number of researchers.

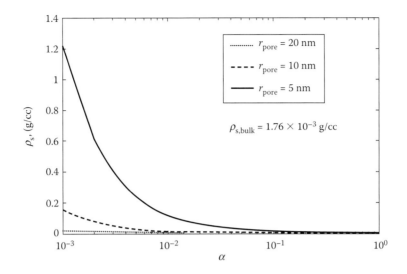

FIGURE 20.1
Methane solubility as a function of α for various spherical pore sizes at 4000 psi pressure and 80°C temperature.

TABLE 20.1

Methane Solubility for a Range of Pore Volumes at 4000 psi Pressure and 80°C Temperature ($r_b = 1.0$ nm and $\alpha = .01$)

Container Size (nm)	V_p (nm³)	ρ_s (g/cm³)
10	1000	117×10^{-3}
25	15625	6.40×10^{-3}
40	64000	1.76×10^{-3}
50	125000	0.928×10^{-3}

Today, molecular models of, for example, multicomponent mixtures are common (Nath et al., 1998), or predictions of structural and thermodynamic properties for large molecules are considered to be reliable (Karayiannis et al., 2002). Furthermore, molecular simulations are being used more frequently to construct virtual experiments in cases where controlled laboratory experiments are difficult and too costly, if not impossible, to perform. Molecular investigations of the phase transformation of materials under extreme pressures are good examples of the latter application (Lacks, 2000).

The main objective in this study is to predict accurately the solubility of methane in bulk water and in water confined to micropore using the test-particle method combined with the excluded volume map sampling (EVMS) method. The output obtained from molecular simulation is then used, with appropriate time correlation functions, to calculate certain macroscopic quantities of interest such as simple thermodynamic averages (e.g., kinetic, potential and internal energies, temperature, and pressure) and transport coefficients related to the dynamics of fluid (e.g., shear and bulk viscosities and diffusive heat and mass transfer coefficients). In our case, this macroscopic quantity is the Ostwald coefficient, which can be related to Henry's law constant of solubility. A detailed description of the numerical method can be found in classical books by Allen and Tildesley (2007) and Frenkel and Smit (2002).

20.3.1 Fluid Model and Potential Function

Simulations of fluid in the canonical (NVT) ensemble are carried out at 75.5°C using the Leapfrog algorithm and Nosé-Hoover Thermostat. Typically, the bulk fluid system is composed of 891 water molecules in a three-dimensional parallelepiped computation cell of 3.85 × 3.46 × 2.08 nm furnished with periodic boundary conditions, resulting in 0.961 g/cm³ water density at 75.5°C.

There are a number of atomistic models for water (Chaplin, 2004). The so-called extended simple-point charge water model (SPC/E) characterized by three point masses with an O–H bond of 0.1 nm and H-O-H angle equal to 109.47° was used in this study. The centered electron charges on the oxygen and hydrogen are –0.8476e and +0.4238e, respectively. The O–H bond is kept rigid during the simulations using SHAKE algorithm. A time step of 0.2 fs was assigned to maintain good energy conservation.

The Lennard–Jones (LJ) interaction is used to represent the van der Waals forces between the oxygen–oxygen atoms of the two SPC/E water:

$$U_{\mathrm{LJ}}(r_{ij}) = 4\varepsilon \left[\left(\frac{\sigma}{r} \right)^{12} - \left(\frac{\sigma}{r} \right)^{6} \right], \tag{20.7}$$

where $\varepsilon = 0.6504$ kJ/mol and $\sigma = 3.1656$ Å are the depth of the potential well and the distance between the two water molecules, respectively (when its potential energy is at a minimum), and r is the interparticle distance at real time during the simulation. Calculations of long-range water interactions were performed using the Ewald sum technique. The total interaction energy between the water molecules consists of the LJ potential and the Coulombic potential on the basis of classical electrostatics: $U_{\mathrm{TOTAL}} = U_{\mathrm{LJ}} + U_{\mathrm{C}}$, where U_{C} between two molecules i and j is represented as the sum of Coulomb interactions acting among the charged points in the following way:

$$U_{\mathrm{C}} = \sum_{O,H} \frac{q_I q_J}{r_{ij}^{IJ}}, \tag{20.8}$$

where r_{ij}^{IJ} is the distance between site I of the molecule i and site J of the molecule j.

A united-atom carbon-centered LJ potential (based on the optimized potentials for liquid simulations, OPLS-UA) force field, $\varepsilon = 1.2309$ kJ/mol and $\sigma = 3.73$ Å, has been used in the simulations of methane. Methane was considered as a spherical molecule, and the water–methane interaction was described by the optimized Konrad-Lankau (2005) potential (see Table 20.2). Methane–methane interaction is not required because methane is the test particle. Fluid–fluid and fluid–solid interactions are also of the LJ type and act between

TABLE 20.2

LJ Potential Parameters

Interacting Molecules	ε (kJ/mol)	σ (Å)
SPC/E water–SPC/E water	0.6504	3.1656
OPLS-UA methane–SPC/E water	1.0131	3.56
Graphite–graphite	0.2353	3.4
Graphite–OPLS-UA methane	0.8947	3.4478
Graphite–SPC/E water	0.3912	3.2828

methane–water, water–water, graphite–water, and graphite–methane, respectively. The LJ interactions were cut off at 0.9 nm.

20.3.2 Small Pore Model

Graphite is modeled as a slit-like pore, using the 10–4-3 conventional potential where each plane of the graphitic surface, that is, graphene, consists of stacked planes of carbon atoms separated by a 3.4-Å distance. The carbon–carbon separation on the plane along a hexagon side is 0.142 nm. The distance between two graphene layers that confine the fluids is considered as the width of the slit-pore system. The solubility study is numerically performed at four separate confined systems with pore width values of 6.9, 11.6, 16.2, and 20.8 Å. As the pore width is increased, the number N of water molecules varies from 297 until 891, which is presented in Table 20.3. Standard Lorentz–Berthelot combination rule is used to obtain the corresponding LJ parameters and describe the interactions of graphite–methane and graphite–water:

$$\sigma_{ij} = \frac{1}{2}(\sigma_{ii} + \sigma_{jj}) \quad \text{and} \quad \varepsilon_{ij} = \sqrt{\varepsilon_{ii}\varepsilon_{jj}}, \tag{20.9}$$

where $\sigma_{ii}, \varepsilon_{ii}$ and $\sigma_{jj}, \varepsilon_{jj}$ correspond to the distance and energy parameters of the pure chemical species case. The distance and energy parameters for describing graphite–methane interaction are modified by a constant of 0.97 and 1.66, respectively. The total length of each MD run was 400 picoseconds.

20.3.3 Infinite Methane Dilution and the Widom Test Particle with EVMS

Usually, the solubility of a solute (2) in a liquid solvent (1) is measured by the Ostwald coefficient, $L = \rho_2^l / \rho_2^g$, where ρ_2^l and ρ_2^g are the number densities of the solute in the liquid and gas phases in equilibrium, respectively. The condition of equilibrium provides another expression for L, which is

$$L = e^{-\left(\mu_2^{*l} - \mu_2^{*g}\right)/k_B T}, \tag{20.10}$$

where μ_2^{*l} and μ_2^{*g} are the excess chemical potentials of the solute in the liquid and gas phases, respectively. Hence, in general, the evaluation of the solubility requires knowledge of the excess chemical potential of the solute in both phases.

TABLE 20.3

Calculated Values of the Solubility Parameters γ_l, L, and Henry's Law Constant k_H at 75.5°C

Pore Width (Å)	6.9	11.6	16.2	20.8	Bulk (Numerical)	Bulk (Empirical)*
$\gamma_l \sim L \ (\times 10^3)$	357 ± 54	198 ± 34	135 ± 14	62 ± 13	26 ± 4	
k_H (kb)	4.3	7.8	11.5	25.1	60.2	63
ρ (g/cm³)	0.960	0.960	0.960	0.961	0.961	0.961
N	297	495	693	891	891	

Note: N is the number of water particles per each slit pore and bulk system.

When the system density in the gas phase is negligible, the Ostwald coefficient becomes identical to the liquid phase activity coefficient γ_l, and the only required quantity for its evaluation is the excess chemical potential of the solute in the liquid phase alone (Guillot and Guissani, 1993):

$$L \cong \gamma_l = e^{-\mu_2^{*l}/k_B T}. \tag{20.11}$$

In statistical mechanics, the excess chemical potential μ^* of one molecule of solute in a fluid solvent, that is, the infinite dilution limit, can be given in terms of the Helmholtz free-energy difference between the system composed of (N + 1) molecules and the one composed of N solvent molecules:

$$\mu^* = \Delta F = -Tk_B \ln\left(\frac{Z_{T,N+1}}{V Z_{T,N}}\right). \tag{20.12}$$

Here, $Z_{T,N}$ is the configurational partition function at temperature T and volume V. Using test-particle method described by Widom (1963), Equation 20.10 can be rewritten as

$$\mu^* = -k_B T \ln\left\langle e^{-\phi/k_B T}\right\rangle_N = -k_B T \ln \gamma, \tag{20.13}$$

where ϕ is the potential energy difference between a system composed of (N + 1) molecules and that containing only N solvent molecules. Consequently, it is possible in practice to introduce methane, as a test particle, at a fixed vacant location into the solvent and evaluate the Boltzman factor $\exp(-\phi/k_B T)$, while moving the whole system using the MD simulation.

In addition to the Ostwald coefficient, the solubility of gases in liquids can be expressed in terms of Henry's law constant, k_H. In the liquid phase, k_H is simply related to the activity coefficient γ_l with the expression:

$$k_H = \frac{\rho RT}{\gamma_l}, \tag{20.14}$$

where ρ is the number density of the pure solvent.

Before calculating the potential energy of the interaction between the test particle that has to be inserted and the solvent, a map is made of the previous solvent system in equilibrium; this map is done by dividing the system into small boxes that can contain at most one molecule of solvent. Each box is labeled depending on whether it contains a particle or is empty. This map is used to check whether there is an empty cavity or space for the particle to be inserted. In our case, we divided each configuration into a series of cubic subcells with a length of 3.1656 Å and inserted the methane molecule in an empty cavity.

Before the insertion of methane, the system was equilibrated at 75.5°C using Berendsen thermostat algorithm. After the equilibrium is reached, the Nosé–Hoover thermostat algorithm is used to control the system temperature and carry out the real canonical ensemble

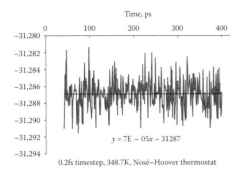

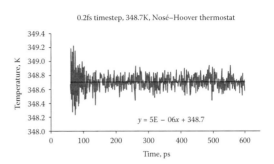

FIGURE 20.2

(Left) Total energy and (right) temperature evaluation over timer for a pure water system reaching thermodynamic equilibrium in a nanopore at 75.5°C.

on the N and $N + 1$ particle fluid systems. The equilibrium is assumed to be achieved if no drift was observed in the time-independent quantity, such as the total energy of the system. Figure 20.2 shows the total energy of the water system under equilibrium conditions in a nanoslit pore at 75.5°C. At frequent intervals during the simulation, we generate a coordinate with $N + 1$ particles by using the excluded map sampling method. We then compute the Boltzman factor $\exp(-\phi/k_B T)$, sampling the last 360 canonical configurations for each confined system. The interval for the insertion was every 50 ps. In the case of the bulk model, a trajectory of 120 ps was sampled for analyzing the Boltzman factor.

20.3.4 Pore Walls with Controlled Wettability

When a small amount of two immiscible fluids are put in to contact with a flat solid surface, three phases are delimited by a certain *triple contact line*. Because each interface has its own free-energy per unit area, the fluid–solid interaction energy will determine whether the solid surface is not wet, partially wet, or totally wet by the correspondent fluid. The partial or total fluid–solid wetting can be defined as surface hydrophilicity/hydrophobicity depending on whether the fluid is water or methane. For low-energy surfaces such as graphite, hydrophobicity can be driven by a strong methane–wall interaction as well as by a weak water–wall interaction, promoting an easier sliding for the water particles across the solid (Roselman, 1976).

In principle, a given surface is wetted completely if the solid–fluid attraction (i.e., van der Waals interaction near the surface) is much higher than the fluid–fluid attraction. Using MD simulation, Cao et al. (2006) have earlier shown that a small slit-like pore structure can have an enhanced surface hydrophobicity because of changes in the liquid–solid interaction. In this study, we varied the LJ energy parameter for the potential interaction between water and graphite as follows:

$$\varepsilon_{f-s} = c\,\varepsilon_{water\text{-}graphite'} \tag{20.15}$$

where the coefficient c takes the values of 0.4, 0.6, and 1.0 to see the hydrophobicity (when $c = 0.4$ and 0.6) of the wall over the solubility.

20.4 Numerical Results and Discussion

For each system modeled, values of the solubility parameters, γ_l and L, and the Henry's law constant, k_H, of the methane activity in water are estimated at 75.5°C and reported in Table 20.3. The results also include the estimated water density values at the same temperature. At 75.5°C, the density of the water vapor coexisting with the liquid phase is considered negligible (Guillot and Guissani, 1993); therefore, methane activity coefficient in the water vapor phase, γ_g, is assumed to be ~1 and the Ostwald coefficient $L \sim \gamma_l$. The corresponding experimental data for the bulk (i.e., without pore) are from Economou et al. (1998). It is clear that the numerically obtained density and k_H values of this study compare quite well with the experimental measurement. If the authors of the experimental work were to calculate the liquid phase activity coefficient γ_l, they would have predicted 25×10^3, which is also a value quite close to that obtained in this study using MD simulation. Because of the lack of experimental and simulation results of methane solubility in water under confinement and because the methodology we carried out for each confined system follows the same process than the bulk one, we consider our results valid, although they may carry out large statistical errors. Evaluation of Henry's law constant for methane in water by Economou et al. (1998) illustrates that inserting a molecule in an energetically favorable position by using the Widom process causes a larger statistical error, 10%–22%.

Figure 20.3 shows the estimated methane solubility ($L \sim \gamma_l$) in water for varying slit pore widths and compares with the bulk case. Notice that there is enhancement in solubility when the fluid system is confined to smaller pores. When the pore length is in 1- to 2-nm range, the estimated solubility is approximately two to eight times more. The solubility in a slit pore of 0.7 nm, a size comparable with macromolecular openings in coal and kerogen, is enhanced with a factor of 13.7. Comparing our simulation results for the solubility enhancement with the MD study of the temperature dependence of methane solubility in water, developed by Guillot and Guissani (1993) in a bulk system, the Ostwald coefficient that we have estimated for a slit pore of 0.7 nm at 75.5°C corresponds to the one that Guillot and Guissani (1993) have found at 336.8°C.

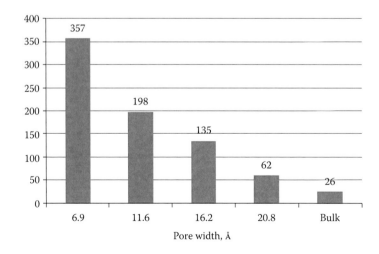

FIGURE 20.3
Estimated Ostwald coefficient, L ($\times 10^3$), of methane in water confined to slit pores compared with the solubility in bulk water.

Table 20.4 and Figure 20.4 show the effect of varying pore-wall wettability on methane solubility in water. Here, the wall wettability is controlled by changing the coefficient c in Equation 20.15. When c takes values smaller than 1.0, the system represents a hydrophobic wall because the water–wall interaction has been minimized. In the same way, when c is larger than unity, the interactions represent a hydrophilic wall. de Gennes (1985) explains that most liquids achieve complete wetting with higher surface–fluid than fluid–fluid interactions near the surface ($U_{SF} > U_{FF}$). Hence, for complete wetness, the surface–fluid interaction must be higher than the fluid–fluid interaction. Graphite–water interaction in our case corresponds to 0.3912 kJ mol^{-1}, whereas water–water interaction corresponds to 0.6504 kJ mol^{-1}. An increment in c yields a higher graphite–water interaction, and therefore the wall is more water-wet. On the other hand, the water wettability of the wall is decreased by decreasing c so that it is more hydrophobic or CH$_4$-wet.

On the basis of the results, we observe significant pore-wall wettability effect on solubility. When the coefficient is 0.4, for example, a factor of 1764/135 = 13.1 increase occurs with respect to the base value case (with c = 1.0) and a factor of 1764/26 = 67.9 increase occurs with respect to the solubility in bulk water; on the other side of the wettability scale, when c increases up to 1.5, then methane solubility increases by a factor of 24.3 with respect to

TABLE 20.4

Effect on the Solubility Parameters γ_l and L at 75.5°C by Changing the Wall–Water Interaction (Pore Width = 1.62 nm)

c	0.4	0.6	1.0	1.2	1.5
$\gamma_l \sim L$, ($\times 10^3$)	1764	304	135	2025	3283
k_H (bars)	876	5090	11,479	763	471

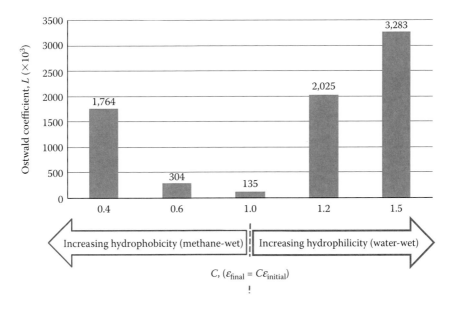

FIGURE 20.4
Estimated Ostwald coefficient, L ($\times 10^3$), of methane in water confined to pore with varying pore-wall wettability.

the case when c is equal to unity. It is clear that the pore-wall wettability has a significant influence on the enhancement of methane solubility regardless of the surface's hydrophilicity. The enhancement is more dramatic when the wall is more water-wet, however. This is because methane–methane interaction is much bigger than water–water interaction.

20.5 Conclusions

It is found that the level of confinement in pores affects the thermodynamics of fluids significantly. Consequently, as the pore size reaches nanoscale, the dissolved amount of methane in water becomes considerably larger. The solubility is extremely sensitive to the changes in the small pore-wall wettability.

References

Allen MP, Tildesley DJ. (2007). *Computer Simulation of Liquids*. London: Oxford University Press.

Cao BY, Chen M, Guo ZY. (2006). *Phys Rev E Stat Nonlin Soft Matter Phys*. 74:66311.

Chaplin M. (2004). Available from: http://www.lsbu.ac.uk/water.

Economou IG, Boulougouris GC, Errington JR, Panagiotopoulos AZ, dTheodorou DN. (1998). *J Phys Chem B*. 102:8865.

de Gennes PG. (1985). *Rev Mod Phys*. 57:827.

Frenkel D, Smit B. (2002). Understanding molecular simulation—from algorithms to applications. In: *Computational Science Series*. San Diego: Academic Press.

Guillot B, Guissani Y. (1993). A computer simulation study of the temperature dependence of the hydrophobic hydration. *J Chem Phys*. 99(10):8075–94.

Karayiannis NC, Giannousaki AE, Mavrantzas VG, Theodorou DN. (2002). Atomistic Monte Carlo simulation of strictly monodisperse long polyethylene melts through a generalized chain bridging algorithm. *J Chem Phys*. 117:5465–72.

Konrad O, Lankau T. (2005). Solubility of methane in water: the significance of the methane–water interaction potential. *J Phys Chem B*. 109:23596–604.

Lacks DJ. (2000). First-order amorphous–amorphous transformation in silica. *Phys Rev E*. 84(20):4629–32.

McCain WD. (1990). *The Properties of Petroleum Fluids*. 2nd ed. Tulsa: PennWell Books.

Nath SK, Escobedo FA, de Pablo JJ. (1998). On the simulation of vapor–liquid equilibria for alkanes. *J Chem Phys*. 108(23):9905–11.

Roselman IC, Tabor D. (1976). *J Phys D Appl Phys*. 9:2517.

Schowalter T. (1979). *Am Assoc Per Geol Bull*. 63(5).

Werder T, Walther JH, Jaffe RL, Halicioglu T, Koumoutsakos P. (2003). On the water–carbon interaction for use in molecular dynamics simulations of graphite and carbon nanotubes. *J Phys Chem B*. 107:1345–52.

Widom B. (1963). Some topics in the theory of fluids. *J Chem Phys*. 39(11):2808–12.

Zwanzig RW. (1954). High-temperature equation of state by a perturbation method: I. Nonpolar gases. *J Chem Phys*. 22(8):1420–6.

Index

A

Printed and bound by CPI Group (UK) Ltd, Croydon, CR0 4YY

18/10/2024

01776236-0020